Birkhäuser

Birkhäuser Advanced Texts Basler Lehrbücher

This series presents, at an advanced level, introductions to some of the fields of current interest in mathematics. Starting with basic concepts, fundamental results and techniques are covered, and important applications and new developments discussed. The textbooks are suitable as an introduction for students and non–specialists, and they can also be used as background material for advanced courses and seminars.

Tanja Eisner • Bálint Farkas

A Journey Through Ergodic Theorems

Birkhäuser

Tanja Eisner
Mathematisches Institut
Universität Leipzig
Leipzig, Germany

Bálint Farkas
Fakultät für Mathematik und
Naturwissenschaften
Bergische Universität Wuppertal
Wuppertal, Germany

ISSN 1019-6242 ISSN 2296-4894 (electronic)
Birkhäuser Advanced Texts Basler Lehrbücher
ISBN 978-3-031-96505-0 ISBN 978-3-031-96506-7 (eBook)
https://doi.org/10.1007/978-3-031-96506-7

Mathematics Subject Classification: 28D05, 37A05, 37A25, 37A44, 37B05, 11B25, 37A30, 47A35, 37B20

This book is published under the imprint Birkhäuser, www.birkhauser-science.com by the registered company Springer Nature Switzerland AG
The registered company address is: Gewerbestrasse 11, 6330 Cham, Switzerland

Dedicated to our teacher, Rainer Nagel, a great mathematician and a wonderful friend

Preface

This book is devoted to ergodic theorems, both modern and classical, which form a core topic of ergodic theory and even gave rise to this mathematical field. During the past decades, one has been witnessing a spectacular development, and today ergodic theorems are subjects of very active and manifold research. Before we describe the content of this book, let us explain how it all started and begin with a historical account on how ergodic theorems emerged.

Around 1880, the Austrian physicist and philosopher L. Boltzmann (1844–1906) formulated his famous *ergodic hypothesis* [169], or *Ergodenhypothese.*[1] Roughly speaking, he hypothesized that the phase space motion of the ideal gas goes through each physically feasible state. This postulate, called by Maxwell and his followers the *principle of continuity of paths*, was used by Boltzmann to deduce that, after restricting to the physically feasible states and independently of the initial state, the phase space motion will visit, in the average, each fixed region of the phase space proportionally often (time mean) to the volume of the region (phase space mean). The ergodic hypothesis was shortened to the slogan "time mean equals space mean." Even though the original "ergodic hypothesis" was criticized, it was indispensable for many calculations in statistical mechanics at that time, and a number of researchers were involved in the quest for a mathematical justification.

Among the critics, however, were Lord Kelvin [639] and Poincaré [854] who formulated serious doubts concerning the validity of the ergodic hypothesis. Then around 1913 Plancherel [850] and Rosenthal [882] proved independently that the ergodic hypothesis fails in its original form. Already in 1894 Poincaré indicated a possible rectification of this postulate by requiring that any state comes arbitrarily close, even if not precisely, to any other state. (We will mathematically return to this idea in Chap. 3.) This suggestion became P. and T. Ehrenfest's [334, Page 3] *quasi-*

[1] Boltzmann used the word "Ergode" without explanation of its meaning. P. and T. Ehrenfest (in the footnote page 30 of [334]) provide the following etymology: ergon $\sim$ "work" and odos $\sim$ "path," "way." This and other interpretations are discussed, e.g., by Mathieu in [775] and by Gallavotti in [459].

ergodic hypothesis. However, it is believed and supported by some evidence that this property was already suggested by Boltzmann himself, see, e.g., the historical article [1029] by von Plato.

Even though the original ergodic hypothesis appeared to lose its general significance, the ingenious ideas of Boltzmann did not get lost and had enormous impact outside of physics as well. The first mathematical confirmation of the hypothesis "time mean equals space mean" for *ergodic* systems is due to von Neumann [1027] in what is known today as the *mean ergodic theorem*. His proof used the operator theoretic approach to dynamics suggested by Koopman [658]. This was quickly followed by the "complete" proof of the quasi-ergodic hypothesis for ergodic systems by Birkhoff [156] (his result, the *pointwise ergodic theorem*, was actually published before von Neumann's). See Birkhoff and Koopman [157], Zund [1076], Bergelson [108], and Moore [804] for some more historical details on these fascinating developments. Note that Carleman [225] developed an operator theoretic approach independently of Koopman and came up in [226] with a proof of the mean ergodic theorem for differentiable flows several months after von Neumann (with an attempt before von Neumann in [225]), see von Neumann [1028] for a discussion.[2] Also A. Weil was a part of the story and discovered the idea of this linearization,[3] but did not pursue the quest for the ergodic theorem any further.[4] Going back to the ergodic hypothesis, we remark that the crucial property of ergodicity of the ideal gas is still a complex topic with no unique answers, see, e.g., Lebowitz and Penrose [712], Patrascioiu [845], Szász [967], Simányi and Szász [947], Simányi [946], Çömez [229].

Based on these landmark results ergodic theory was established as a mathematical discipline treating both the time continuous and the time discrete settings. Today it is a powerful and beautiful area with connections to and interactions with many (partly unexpected) areas of mathematics such as probability theory, functional analysis, number theory, combinatorics, group theory, topology, algebra, or graph theory. Hereby, ergodic theorems play a fundamental role and contributed greatly to the development of ergodic theory and its visibility, importance, and recognition among mathematicians. For interdisciplinary aspects and impact of ergodic theory outside of mathematics see, e.g., Moore [804].

For instance, Bourgain [178] discovered a deep connection between *subsequential ergodic theorems* and harmonic analysis. Furstenberg [445] built a bridge between additive combinatorics and *multiple ergodic theorems*, presenting a new and much more flexible proof of Szemerédi's celebrated theorem on the existence of arithmetic progressions in large sets of integers. His approach led very quickly

[2] As pointed out by Karlsson [630, Introduction], while earlier publications on ergodic theorems give credits to Carleman's work in the context of the mean ergodic theorem (see, e.g., Wiener [1045], Riesz [872] and Hopf [542] in addition to von Neumann [1028]), it is unfortunately hardly mentioned anywhere in modern works on the subject.

[3] See his collected works [1038, pp. 522–523].

[4] Some historical account on the controversy of priority can be found in the thesis of Küster, see [700, Präludium].

to various generalizations of Szemerédi's result as well as to several other proofs from different areas of mathematics connecting them all to ergodic theory. Finally, Green and Tao (2004), building heavily on ideas of Furstenberg and others, proved the existence of arbitrarily long (finite) arithmetic progressions in the set of primes, later extended to other combinatorial patterns, see [485, 486]. Last but not the least, *weighted ergodic theorems* play an important role in modern number theory via the relatively recent Sarnak conjecture (2010) on the random behavior of the Möbius function.

In this book we give a taste of the classical ergodic theorems as well as of the three directions of modern ergodic theorems mentioned above in the discrete time setting. We shall not pursue the physical interpretation of mathematical truths in this book, and present applications of mathematics in mathematics. This is not a monograph on ergodic theorems let alone on ergodic theory. In particular, we do not touch several important topics such as (semi)group actions (in particular (semi)flows), non-commutative and multiplicative ergodic theorems, smooth ergodic theory, random dynamical systems, ergodic theory of Markov processes, ergodic theorems for infinite measure spaces, or non-singular transformations, and even measure-theoretic entropy and rates of convergence are mentioned only tangentially.

Let us now be more precise about what exactly we mean by ergodic theorems, and present a selection of developments and results that are covered in this book.

Classical Ergodic Theorems

We start with the mathematical model originated from the physical motivation[5] leading to the classical ergodic theorems due to von Neumann and Birkhoff.

Let (X, μ) be a probability space, and let $T : X \to X$ be a measurable and **measure-preserving** transformation, the latter meaning that

$$\mu(T^{-1}B) = \mu(B) \text{ for each measurable } B \subset X.$$

Then, of course, for each positive (meaning ≥ 0) integer n also the n times iterated transformation $T^n : X \to X$ has both of these properties.

The following interpretation connects this to the previously mentioned evolution of physical systems. The set X is the **phase space**, $x \in X$ is a possible **state** of the system, the measure $\mu(A)$ of a set A describes the probability that a system is in a state that belongs to A. [6] The time evolution here is discrete: The transformation

[5] For a detailed discussion how this model comes from physical laws (in the time-continuous case), see, e.g., Petersen [848, Sect. I.1.2.A].

[6] In the time continuous case and when the dynamics is described by Hamiltonian differential equations, the existence of an invariant measure is assured by Liouville's theorem [848, Thm. I.2.1].

$T : X \to X$ describes the dynamics, i.e., what happens with a given initial state $x \in X$ within one time step. The sequence $x, Tx, T^2x, \dots$ describes the whole path (phase space motion) that the system goes through as time evolves.

For a measurable function $f : X \to \mathbb{C}$ consider the arithmetic averages

$$A_N f := \frac{1}{N} \sum_{n=0}^{N-1} f \circ T^n, \ N = 1, 2, \dots,$$

also called **Cesàro averages** or **Cesàro means**. (In the above interpretation, f can represent an observation of a physical characteristic of the system and is called **observable**.)

In this interpretation, $f(x)$ is a (physical) measurement of the state x. For example, if $f = \mathbf{1}_A$, the characteristic function of a subset A of the phase space, then $f(x)$ tells whether or not the state x belongs to the region A. For this choice $A_N \mathbf{1}_A(x)$ yields the average number of time steps (until time N) that the phase space motion starting in x spends in the region A. By the ergodic hypothesis this should converge, as N goes to infinity, to the phase space volume $\mu(A)$ of A.

Using the ideas of Koopman [658] (which were developed independently by Carleman [225]), von Neumann [1027] established his result called today the *mean ergodic theorem*. A decisive observation of Koopman [658] and implicitly of Carleman [225] was to consider a linear operator instead of a non-linear transformation T, allowing to exploit the arsenal of functional analysis and operator theory, both of which were blossoming at that time. Indeed, the mapping $f \mapsto f \circ T$ acts as a *linear operator* on the vector space $\mathscr{L}^0(X)$ of measurable functions, and it is easily seen to preserve the p-(semi)norm for every $p \in [1, \infty]$, so it leaves the space $\mathscr{L}^p(X, \mu)$ of p-integrable functions invariant. Furthermore, since T preserves μ, this linear operator carries over to the corresponding Banach spaces $\mathrm{L}^p(X, \mu)$, in particular to the Hilbert space $\mathrm{L}^2(X, \mu)$. On all these Banach spaces it becomes an isometry, called the **Koopman operator** of the transformation T. It is convenient and today quite common to write Tf instead of $f \circ T$, implying that the Koopman operator of the transformation T is also denoted by T. The properties of the Koopman operator will be discussed in Chap. 2 in detail.

Here comes the announced result of J. von Neumann. (We formulate it for the time discrete, not necessarily invertible, case as opposed to von Neumann's original consideration of invertible, time continuous flows, see also Carleman [225, 226] and von Neumann [1028].)

Theorem (von Neumann) *For a probability space (X, μ) consider the Hilbert space $\mathrm{L}^2 := \mathrm{L}^2(X, \mu)$. Let $T : X \to X$ be a measure-preserving transformation, and let P be the orthogonal projection onto the fixed space*

$$F := \{f \in \mathrm{L}^2 : Tf = f\}.$$

Then for each $f \in \mathrm{L}^2$

$$\lim_{N\to\infty} \frac{1}{N} \sum_{n=0}^{N-1} T^n f = Pf \text{ in the } \mathrm{L}^2\text{-norm.}$$

Suppose that the fixed space F consists of constants only (such transformations are precisely the *ergodic* ones, cf. Chap. 3). In this case von Neumann's theorem takes the form

$$\lim_{N\to\infty} \frac{1}{N} \sum_{n=0}^{N-1} T^n f = \int_X f \, \mathrm{d}\mu \cdot \mathbf{1} \text{ in the } \mathrm{L}^2\text{-norm for each } f \in \mathrm{L}^2$$

(here $\mathbf{1}$ is the constant 1 function). This is an instance of "time mean equals space mean," i.e., a variant of the ergodic hypothesis. Specializing $f = \mathbf{1}_A$ and integrating over measurable $B \subset X$ with respect to μ yields

$$\lim_{N\to\infty} \frac{1}{N} \sum_{n=0}^{N-1} \mu(T^{-n} A \cap B) = \mu(A)\mu(B).$$

This implies that if $\mu(A) > 0$ and $\mu(B) > 0$, then there is an infinite sequence $(n_k)_{k\in\mathbb{N}}$ such that $\mu(T^{-n_k} A \cap B) > 0$ for every $k \in \mathbb{N}$, i.e., a positive portion of B visits the region A after n_k iterations. Since no matter how small the measures of A and B are and where these sets are located, this can be considered as a weak variant of the quasi-ergodic hypothesis. It justifies, in a certain sense, Poincaré's ideas mentioned above, see Chap. 3. Note that, without assuming ergodicity, it was known before von Neumann's result that for every set A with positive measure there exists an infinite sequence $(n_k)_{k\in\mathbb{N}}$ with $\mu(T^{-n_k} A \cap A) > 0$ for each $k \in \mathbb{N}$, known as **Poincaré's recurrence principle**,[7] see Chap. 2.

Birkhoff's result [156] provides us more detailed information and assures pointwise convergence.

Theorem (Birkhoff) *If* $T : X \to X$ *is a measure-preserving transformation on a probability space* (X, μ)*, then for each measurable function* $f : X \to \mathbb{C}$ *with* $\int_X |f| \, \mathrm{d}\mu < \infty$ *the limit*

$$\lim_{N\to\infty} \frac{1}{N} \sum_{n=0}^{N-1} (T^n f)(x) \text{ exists for } \mu\text{-almost every } x \in X.$$

[7] Poincaré [853] discussed this principle in 1890 in the physical context and Carathéodory [224] proved its abstract version in 1919.

(As in the case of von Neumann, the original version was formulated for the time-continuous case.) This in combination with von Neumann's result yields for each measurable $A \subset X$ and each ergodic transformation T that

$$\lim_{N\to\infty} \frac{1}{N} \sum_{n=0}^{N-1} \mathbf{1}_{T^{-n}A}(x) = \mu(A) \text{ for almost every } x \in X.$$

Since the sum on the left-hand side is precisely the average number of return times to A (or the average time the system spends in A) up to time N, this makes the quasi-ergodic hypothesis a theorem for ergodic systems and almost all initial states.

These results of von Neumann and Birkhoff (which we will prove in Chap. 7) have generalizations that reach well beyond their original scope. We pick here one quick application of these classical ergodic theorems leading to an important number theoretic property of sequences, called *equidistribution*. Consider the interval $[0, 1]$ equipped with the Lebesgue measure and the transformation $T : [0, 1) \to [0, 1)$, $Tx := \{x + \alpha\}$ for some fixed irrational number α, with $\{\cdot\}$ denoting the fractional part. This transformation preserves the Lebesgue measure, and can be proved to be ergodic. If $f : [0, 1] \to \mathbb{C}$ is continuous with $f(0) = f(1)$ (we call each such f 1-periodic), then so is Tf. In analogy to the mean ergodic theorem, Weyl proved in [1043] that for each continuous 1-periodic function the averages

$$\frac{1}{N} \sum_{n=0}^{N-1} T^n f$$

converge as $N \to \infty$ in the **supremum norm**, in particular, pointwise everywhere (see also Sect. 7.3 below). The limit then must be the same as given by von Neumann's theorem, i.e., $\int_0^1 f(t)\,\mathrm{d}t \cdot \mathbf{1}$. It follows that

$$\frac{1}{N} \sum_{n=0}^{N-1} f(\{n\alpha\}) \to \int_0^1 f(t)\,\mathrm{d}t \text{ as } N \to \infty.$$

This means that the sequence $(\{n\alpha\})_{n\in\mathbb{N}}$ is so well-behaved that sampling an arbitrary continuous periodic function f at these points and building the average of the values of f at these points approximate the integral of f. The sequence $(\{n\alpha\})_{n\in\mathbb{N}}$ is then called **equidistributed** modulo 1. This property of the sequence $(\{n\alpha\})_{n\in\mathbb{N}}$ was discovered by Bohl [168], Sierpiński [939], and Weyl [1042]. Equidistribution has been studied intensely since then and plays an important role in number theory, dynamics, group theory, and combinatorics, and will be discussed in more detail in Chap. 10. More applications of the classical ergodic theorems will be given in Chap. 9.

Multiple Ergodic Theorems

Multiple ergodic theorems are inspired by Szemerédi's celebrated[8] result [970] from 1975 on arithmetic progressions in large sets of integers.

***Theorem** **(Szemerédi)** Let $E \subset \mathbb{N}$ be of positive upper density, i.e., satisfying*

$$\limsup_{N\to\infty} \frac{|E \cap \{1, \ldots, N\}|}{N} > 0.$$

Then E contains arbitrarily long arithmetic progressions.

Two years later, Furstenberg presented his groundbreaking[9] ergodic theoretic proof [445] where he reformulated the above number theoretic statement as a result on the behavior of multiple ergodic averages

$$\frac{1}{N}\sum_{n=0}^{N-1} T^n f \cdot T^{2n} f \cdots T^{kn} f, \tag{0.1}$$

in particular their correlation with the original function f, known as *multiple recurrence*. One year later, Furstenberg and Katznelson [450] gave an alternative ergodic theoretic proof of multiple recurrence using so-called *characteristic factors*, see also Furstenberg et al. [455], providing a multidimensional version of Szemerédi's theorem for $\mathbb{N}^d$. These ergodic theoretic proofs opened a door to results in many directions. More improvements of Szemerédi's result appeared such as the existence of polynomial progressions, and many proofs of Szemerédi's theorem from different areas of mathematics followed, connecting each of them to ergodic theory. For example, Gowers [481] gave a Fourier analytic proof of Szemerédi's theorem using so-called Gowers uniformity norms. Finally, Green and Tao showed in 2004 [485] (see also [486]) that the primes (which have density zero) contain arbitrarily long arithmetic progressions using the ideas of Furstenberg and others as a key ingredient. This was later extended to polynomial patterns in the primes by Tao and Ziegler [991, 995]. Ergodic theoretic methods turned out to be powerful and helped to discover other, purely combinatorial, results such as the density version of the Hales–Jewett theorem by Furstenberg and Katznelson [454] (which will not be discussed in this book). Multiple recurrence and hence the ergodic theoretic approach to Szemerédi's theorem will be treated in Chap. 13.

[8] Szemerédi received the 2012 Abel Prize *"for his fundamental contributions to discrete mathematics and theoretical computer science, and in recognition of the profound and lasting impact of these contributions on additive number theory and ergodic theory"* (2012 Abel Prize citation, https://abelprize.no).

[9] Furstenberg received the 2020 Abel Prize together with G. Margulis *"for pioneering the use of methods from probability and dynamics in group theory, number theory and combinatorics"* (2020 Abel Prize citation, https://abelprize.no).

It took longer to show *multiple convergence*, i.e., L^2-convergence of the averages (0.1) extending von Neumann's ergodic theorem to the multiple setting. To do that, Host and Kra [549] developed their structure theory pushing the idea of Furstenberg and Katznelson of characteristic factors to crystallize out a well-structured relevant factor for multiple averages using the so-called *Gowers–Host–Kra seminorms*. An alternative proof of multiple convergence was given by Ziegler [1065]. Then multiple convergence was generalized by Tao [976] to commuting transformations and by Walsh [1031] to transformations generating a nilpotent group. However, a structure theory for this general case is still missing.

We will prove L^2-norm convergence of triple averages in Chap. 9 in detail, while we discuss some of the necessary tools for the proof of the general mean convergence in Chaps. 14 and 15.

In this book we will discuss[10] four proofs of multiple recurrence (and thereby of Szemerédi's theorem), namely Tao's modification of the Furstenberg–Katznelson proof and Furstenberg's original proof in Chap. 13 as well as two more proofs using (the easy part of) the Host–Kra theory in Chap. 15.

Note that, despite the described progress on norm convergence, pointwise convergence of multiple ergodic averages is not well studied. While Bourgain [184] showed that double ergodic averages converge pointwise, the case of triple ergodic averages remains open.

Subsequential Ergodic Theorems

Subsequential ergodic theorems deal with the behavior of orbits along a given subsequence and answer the following question. Let (k_n) be a subsequence of $\mathbb{N}$ and $p \in [1, \infty)$. Do the averages

$$\frac{1}{N}\sum_{n=0}^{N-1} T^{k_n} f$$

converge as $N \to \infty$ for every (invertible) measure-preserving system (X, μ, T) and every $f \in L^p(X, \mu)$? Here, one again distinguishes between norm and pointwise convergence, leading to generalizations of von Neumann's and Birkhoff's ergodic theorem, respectively.

In a physical interpretation, this could mean that one looks at the state of the system not exactly at each (say) second, but with some irregularities due to physical uncertainties, or just due to convenience. As a mathematical example one could observe the system at (say) each second numbered by a prime number. In both

[10] Furstenberg's original proof is mentioned only partially and very briefly while the other proofs are presented in detail.

situations one desires to determine the convergence of the above averages and, of course, its limit.

There is an easy characterization of *mean good subsequences*, i.e., subsequences for which convergence in norm holds. Using this and the equidistribution theory, it is quite easy to show that polynomial sequences of the form $(P(n))$ for a polynomial $P : \mathbb{N}_0 \to \mathbb{N}_0$, the sequence of primes (p_n) and polynomials of primes of the form $(P(p_n))$ are mean good subsequences. Most random subsequences, made precise by Bourgain's return times theorem [182, 186], are mean good. These results and examples of mean bad subsequences are discussed in Chaps. 12 and 16.

Again, it occurs that pointwise convergence is much harder to show even for the simple polynomial sequence (n^2). Translating the problem into the Fourier analytic language, Bourgain [179, 181, 183] showed in 1988–1989 that polynomial sequences (for certain p) and the sequence of primes are pointwise good, extended by Wierdl [1047] to all $p > 1$ and Wierdl [1048] and Nair [817, 818] to polynomials of primes. (Note that for $p = 1$ pointwise convergence fails dramatically in general as shown later by Buczolich and Mauldin [207] in 2010 and LaVictoire [708] in 2011.) A large class of pointwise good sequences for bounded functions is given by Bourgain's return time theorem [182, 186]. We briefly discuss pointwise good and universally bad sequences in Chap. 12.

Weighted Ergodic Theorems

Weighted ergodic theorems go back to the following result by Wiener and Wintner [1046] being a generalization of Birkhoff's ergodic theorem.

Theorem (Wiener–Wintner [1046]) *Let $T : X \to X$ be a measure-preserving transformation on a probability space (X, μ) and let $f \in \mathscr{L}^1(X, \mu)$ be an integrable function. Then there is a set $X' \subset X$ of full measure such that the averages*

$$\frac{1}{N}\sum_{n=0}^{N-1} \lambda^n (T^n f)(x)$$

converge as $N \to \infty$ for every $x \in X'$ and every $\lambda \in \mathbb{C}$ with $|\lambda| = 1$.

We prove this and its topological counterparts in Chap. 16. For various other proofs and extensions we refer to the book by Assani [44]. The linear weights (λ^n) were generalized to polynomial weights by Lesigne [732, 734] and to the more general class of nilsequences by Host and Kra [552]. These results use the idea of characteristic factors introduced by Furstenberg for multiple ergodic averages.

So it is natural to ask which complex sequences are good weights, i.e., to study sequences $(a_n) \subset \mathbb{C}$ such that the weighted averages

$$\frac{1}{N}\sum_{n=0}^{N-1} a_n T^n f$$

converge as $N \to \infty$ for every measure-preserving system (X, μ, T) and every function $f \in \mathscr{L}^p(X, \mu)$ for a given $p \in [1, \infty)$. We discuss such weights in Chap. 16. The p-norm convergence is again easy to characterize, so one is mostly interested in pointwise convergence. Here, various different classes of good weights have been discovered, some coming from number theory. For example, the von Mangoldt function provides one of rare examples of a good unbounded weight for $p > 1$, and the Möbius function is a good weight for $p = 1$. The last fact is closely related to a recent conjecture by Sarnak [909] stating that the Möbius function is asymptotically orthogonal to every deterministic sequence. This topic is briefly touched upon in Chap. 17.

Other Books

Our book systematically treats both classical and (an update on) all three types of modern ergodic theorems mentioned above, providing an invitation to the beautiful area of ergodic theorems.

Partially, ergodic theorems appear in many books. The classical book about them is [683] of U. Krengel from 1985. Also, classical ergodic theorems are treated in all books on ergodic theory such as Halmos [514], Walters [1033], Cornfeld et al. [260], Petersen [848], Nadkarni [814], Silva [942], Kalikow and McCutcheon [617], Kerr and Li [640] as well as Eisner et al. [342], and many others.

Books that cover multiple recurrence and variants of Furstenberg's original proof of Szemerédi's theorem are, for example, Furstenberg [446], Petersen [848], McCutcheon [783], Einsiedler and Ward [336], Glasner [468], Tao [977, Ch. 2], Kerr and Li [640], while multiple convergence is discussed in detail in the recent book by Host and Kra [553]. Subsequential ergodic theorems are treated in Krengel [683] and Rosenblatt and Wierdl [881], while Assani [44] focuses on the Wiener–Wintner theorem as a special case of weighted ergodic theorems.

Finally, we mention a few books on topics that we do not cover, although this list is by no means exhaustive.

Ergodic theorems for group actions are discussed in Moulin Ollagnier [808], Tempelman [996], Glasner [468], Einsiedler and Ward [336], Gorodnik and Nevo [478], Kerr and Li [640]. Infinite ergodic theory is treated in, e.g., Krengel [683], Aaronson [4], Eigen et al. [335], Kesseböhmer et al. [641], Urbański et al. [1011]. For various versions of the multiplicative ergodic theorem we refer to Arnold [35], Krengel [683], Barreira and Pesin [76], Viana [1020], Barreira [75]. Ergodic theorems for random dynamical systems are discussed in, e.g., Kifer [647], Liu and Qian [754], Arnold [35]. For smooth ergodic theory see, e.g., Bowen [191], Cornfeld et al. [260], Mañé [768], Pollicott [855], Katok and Hasselblatt [634], Barreira and Pesin [77, 78], Qian et al. [858]. Operator ergodic theorems can be found in, e.g., Hille and Phillips [534], Dunford and Schwartz [324], Foguel [393], Krengel [683], Engel and Nagel [369], Eisner et al. [342] and the chapter by Cohen and Lin [252]. For ergodic theorems for stochastic processes, we refer to Doob [314], Neveu [821], Foguel [393], Meyn and Tweedie [790], Shields [937], Norris [827], Kulik [694], Silvestrov [945] among many others and for non-commutative ergodic theorems to Krengel [683]. Finally, entropy is discussed in many books on ergodic theory, e.g., Billingsley [152], Parry [840, 843], Ornstein [836], Walters [1033], Moulin Ollagnier [808], Sinai [949], Viana and Oliveira [1021], Coudène [262], Kerr and Li [640], Hawkins [523], Dajani and Kalle [270].

The recent encyclopedic volume on ergodic theory [943] edited by Danilenko and Silva covers many topics in ergodic theory, few of which are covered briefly also in the present book, such as operator ergodic theorems, Host–Kra structure theory, and Sarnak's conjecture.

How to Use This Book

This textbook is intended for students and researchers interested in the beautiful area of ergodic theorems. It is divided in three parts according to increasing degree of difficulty. In Part I we introduce basic notions of ergodic theory and present some fundamental notions and results from Fourier analysis and operator theory. Part II is devoted to classical ergodic theorems and some of their applications and includes some preparatory material for Part III such as a discussion of equidistribution and topological groups. In the most challenging Part III we present the promised three directions of modern ergodic theorems, namely subsequential, multiple and weighted ergodic theorems, including a discussion of nilsystems and finishing with a recent conjecture of Sarnak from number theory. The chapters about ergodic theorems are augmented with a number of bibliographical remarks and notes for further study, it is advisable to skip these parts at first reading.

The book can be read linearly with several possibilities for shortcuts according to the reader's interests. We begin with a preliminary chapter, with the purpose of indicating the minimal requirements to be able to follow this book. This collection of results is far from being complete, it serves only as a guide, and the knowledgeable reader may skip it without any harm. The reader familiar with the classical ergodic theorems can start reading from Chaps. 9 or 10. The first two parts (up to Chap. 10) can serve as a basis for an introductory course, leading to more advanced topics in Part III. In that part Chaps. 12, 13–15, and 16–17, covering subsequential, multiple, and weighted ergodic theorems, respectively, are independent from each other, and can be used for seminars or more advanced courses. Here are some examples of the paths through the book starting after Chap. 9:

Subsequential ergodic theorems:	Chapters 10, 12
Multiple ergodic theorems:	Chapters 11, 13–15
Weighted ergodic theorems:	Chapters 10, 16, 17

Acknowledgment This book profited from discussions with and input from many people. We are greatly indebted to our teacher and friend Rainer Nagel for introducing the beauty of ergodic theory to us, for the idea to write this book, and for invaluable discussions and advices. We owe Michael Lin a great debt of gratitude for generously sharing his ideas with us and for his numerous valuable comments that shaped the manuscript substantially. Asgar Jamneshan and Henrik Kreidler have read large parts of the manuscript, we heartily thank them for their careful reading and many improvement suggestions. We are indebted to Dustin Anglewitz, Patrizio Bifulco, Simon Collette, Noa Bihlmaier, Nils Dietrich, Leon Duensing, Aurelia Dymek, Nikolai Edeko, Ulrich Groh, Rainer Nagel, Lukas Rönisch, Leo Versteegen, Jürgen Voigt for reading and correcting parts of this book. We are very grateful to Vitaly Bergelson, Zoltán Buczolich, Jochen Glück, Markus Haase, Anders Karlsson, Ben Krause, Sasha Leibman, Vladimir Müller, Felix Pogorzelski, Máté Wierdl, Pavel Zorin-Kranich for helpful discussions, comments, and references, as well as to Michael Karbach and Rainer Verch for interesting discussions about the ergodic hypothesis. The book also profited from our interactions with the participants of the 22nd Internet Seminar "Ergodic Theorems" from the academic year 2018–2019. We thank them all. B.F. acknowledges the support received from the DFG (project number 451698284). We are very grateful to the referees for careful reading and inspiring comments and suggestions which considerably improved the presentation of this book (including simplifications of some proofs) and to senior Birkhäuser editor Dorothy Mazlum for her patience and encouragement. Last but not least, we wish to thank our families for their constant support and boundless understanding.

Leipzig, Germany — Tanja Eisner
Wuppertal, Germany — Bálint Farkas

Contents

Chapter 1
Preliminaries

We assume the reader to be familiar with the basics of functional analysis, topology and measure theory and explain here some notation as well as recall classical facts for later reference. For details we recommend, e.g., the books by Billingsley [153], Rudin [889, 890], Royden and Fitzpatrick [887], Kadets [606], and Haase [508]. These will be also our main sources of reference.

For this book we assume true the basic axioms of set theory including the axiom of choice. We use the notation $\mathbb{N} := \{1, 2, \ldots, \}$ for the set of natural numbers and set $\mathbb{N}_0 := \{0\} \cup \mathbb{N}$, while the meaning of $\mathbb{Z}$, $\mathbb{Q}$, $\mathbb{R}$ or $\mathbb{C}$ is as usual. We write $\mathbb{T} := \{z \in \mathbb{C} : |z| = 1\}$ for the unit circle.

1.1 Some Functional Analysis

Vector spaces, unless otherwise stated, are considered over the complex field $\mathbb{C}$. For given sets X, Y, the set of mappings from X to Y is denoted by Y^X. Thus $\mathbb{C}^Y$ is the vector space of all complex-valued functions on Y (with the usual pointwise operations). The **kernel** and the **range** of a linear mapping T between vector spaces is denoted by $\ker(T)$ and $\operatorname{ran}(T)$, respectively.

Banach Spaces

The usual notation for the norm on a normed space E is $\|\cdot\|_E$, or simply $\|\cdot\|$ if the context allows. The open ball in E centered at $x \in E$ with radius $r \geq 0$ is denoted by $\mathrm{B}_E(x, r)$.

T. Eisner, B. Farkas, *A Journey Through Ergodic Theorems*, Birkhäuser Advanced Texts Basler Lehrbücher, https://doi.org/10.1007/978-3-031-96506-7_1

For $p \in [1, \infty)$ and for a sequence $x = (x_n)_{n\in\mathbb{Z}} \in \mathbb{C}^{\mathbb{Z}}$ of complex numbers one defines

$$\|x\|_p := \Big(\sum_{n\in\mathbb{Z}} |x_n|^p\Big)^{1/p} \text{ and } \|x\|_\infty := \sup_{n\in\mathbb{Z}} |x_n|$$

and for $p \in [1, \infty]$

$$\ell^p(\mathbb{Z}) := \big\{x : x \in \mathbb{C}^{\mathbb{Z}},\ \|x\|_p < \infty\big\}.$$

Then $\ell^p(\mathbb{Z})$ is a Banach space with the norm $\|\cdot\|_p$ and the coordinatewise operations. For $p \in [1, \infty)$ the space $\ell^p(\mathbb{Z})$ is contained in the space $c_0(\mathbb{Z})$ of null-sequences, which in turn is a closed subspace of $\ell^\infty(\mathbb{Z})$. Similarly one defines $\ell^p(\mathbb{N}_0)$ etc. See [889, 5.1–5.4] and [508, Ch. 2,5].

Linear Operators and Functionals

Given normed spaces E, F and a continuous, linear operator $S : E \to F$, the **operator norm** $\|S\|$ is given by

$$\|S\| := \|S\|_{E\to F} := \sup\{\|Sx\|_F : \|x\|_E \leq 1\}.$$

A linear operator $S : E \to F$ is called a **contraction**, or **contractive**, if $\|S\| \leq 1$, and an **isometry**, or **isometric**, if $\|Sx\| = \|x\|$ holds for every $x \in E$. Note that every isometry is a contraction.

Given two Banach spaces E and F, the space of all continuous (bounded), linear operators from E to F is denoted by $\mathscr{L}(E, F)$ and we use the notation $\mathscr{L}(E)$ in case $F = E$. With the operator norm $\|\cdot\|_{E\to F}$ the vector space $\mathscr{L}(E, F)$ becomes a Banach space. The operator norm is submultiplicative, in particular $\mathscr{L}(E)$ is a Banach algebra, with the identity operator I_E as unit.

Given $m \in \ell^\infty(\mathbb{Z})$ and $E \in \{c_0(\mathbb{Z}), \ell^p(\mathbb{Z}) : p \in [1, \infty]\}$ the **multiplication operator**

$$M_m : E \to E, \quad (x_n)_{n\in\mathbb{Z}} \mapsto (m_n x_n)_{n\in\mathbb{Z}}$$

is bounded and linear on E, its operator norm satisfies $\|M_m\| = \|m\|_\infty$.

If $F = \mathbb{C}$, then the Banach space $E' := \mathscr{L}(E, \mathbb{C})$ is called the **dual space** of E, whose elements are called **linear functionals**. Different (extension and separation) versions of the **Hahn–Banach theorem** guarantee existence of linear functionals with various properties. In particular, one has the following separation result, where we use the pairing notation

$$\langle x, x' \rangle := x'(x)$$

for $x \in E$, $x' \in E'$.

Theorem 1.1 (Hahn–Banach) *Let E be a normed space, $F \subset E$ a closed subspace and $x \in E \setminus F$. Then there is $x' \in E'$ such that $F \subset \ker x'$ and $\langle x, x' \rangle \neq 0$.*

In particular, E' **separates points** of E, i.e., for different $x, y \in E$ there is $x' \in E'$ with $x'(x) \neq x'(y)$.

The adjoint of an operator $S \in \mathscr{L}(E, F)$ is the operator $S' \in \mathscr{L}(F', E')$ given by $S'x' = x' \circ S$ for $x' \in F'$. See, e.g., [890, Ch. 4], [889, 5.16–5.21], and [508, Ch. 2, 16].

We will use the following fact quite often.

Proposition 1.2 (Standard Approximation Argument) *Let E, F be Banach spaces and let $D \subset E$ be a dense subset. For a sequence of operators $(S_n)_{n\in\mathbb{N}} \subset \mathscr{L}(E, F)$ the following assertions are equivalent.*

(i) *For each $x \in E$ the sequence $(S_n x)_{n\in\mathbb{N}}$ is convergent in F.*
(ii) *$\sup_{n\in\mathbb{N}} \|S_n\| < \infty$ and for each $x \in D$ the sequence $(S_n x)_{n\in\mathbb{N}}$ converges in F.*

In this case, the limit operator S given by $Sx := \lim_{n\to\infty} S_n x$ belongs to $\mathscr{L}(E, F)$.

For one implication here one needs the **principle of uniform boundedness**: For Banach spaces E, F a subset $\mathcal{T} \subset \mathscr{L}(E, F)$ is uniformly bounded if and only if it is pointwise bounded, see, e.g., [508, §15.2]. The proof relies on the following result.

Theorem 1.3 (Baire) *Let (X, d) be a complete metric space, and for $n \in \mathbb{N}$ let $G_n \subset X$ be dense open sets. Then $\bigcap_{n\in\mathbb{N}} G_n$ is dense in X. If $X = \bigcup_{n\in\mathbb{N}} F_n$ with F_n closed, then there is $m \in \mathbb{N}$ with F_m having non-empty interior.*

See [889, 5.5–5.10], [890, Thm. 2.2], and [508, Ch. 15].

Hilbert Spaces

The reader is assumed to be familiar with the basics of Hilbert space theory, see, e.g., [890, Ch. 12], [889, Ch. 4], [508, Ch. 8, 12, 13] or [606, Ch. 12]. We denote the scalar product on a Hilbert space H by $(\cdot|\cdot)_H$, or simply by $(\cdot|\cdot)$ if no confusion can arise. The space $\ell^2(\mathbb{Z})$ is a Hilbert space with the scalar product

$$(x|y) = \sum_{n\in\mathbb{Z}} x_n \overline{y}_n \quad (x, y \in H).$$

Two vectors $x, y \in H$ in a Hilbert space H are **orthogonal**, denoted by $x \perp y$, if $(x|y) = 0$. For a subset $A \subset H$ we write $x \perp A$ if $x \perp a$ for every $a \in A$. The orthogonal (complement) of a set $A \subset H$ is $A^\perp := \{x \in H : x \perp A\}$. If F is a closed subspace in H, then $F^\perp$ is indeed a complement of F in the sense that $H = F \oplus F^\perp$, i.e., $F \cap F^\perp = \{0\}$ and $F + F^\perp = H$. The **Riesz–Fréchet theorem** states that for every Hilbert space H and for every $\varphi \in H'$ there is a unique $z \in H$ such that $(x|z) = \langle x, \varphi \rangle$ for all $x \in H$. As a consequence, the mapping $H \to H'$, $z \mapsto (\cdot|z)$ is surjective. It is also isometric and conjugate linear, and we may identify H and H' under this mapping. The Hilbert space adjoint of an operator $S \in \mathscr{L}(H, K)$ is denoted by $S^* \in \mathscr{L}(K, H)$, it is the uniquely determined linear operator such that $(Sx|y)_K = (x|S^*y)_H$ for each $x \in H$, $y \in K$. Recall that an operator $U \in \mathscr{L}(H, K)$ is called **unitary** if $U^*U = I_H$ and $UU^* = I_K$, and $T \in \mathscr{L}(H)$ is **self-adjoint** if $T = T^*$.

For a set I and $n \in I$ let $e_n \in H$ be pairwise orthogonal unit vectors. **Bessel's inequality** states that

$$\|x\|^2 \geq \sum_{n\in I} |(x|e_n)|^2 \text{ for every } x \in H.$$

If for every $x \in H$ one has equality here, we call $(e_n)_{n\in I}$ an **orthonormal basis**, and we can write

$$x = \sum_{n\in I} (x|e_n)e_n \text{ for each } x \in H,$$

where the series is unconditionally convergent in H. See [889, Ch. 4] and [508, Ch. 1,8, Sec. 12.2]. We remark that Bessel's inequality is also true for semi-scalar products $(\cdot|\cdot)$.

Projections

Recall that a linear operator $P : E \to E$ on a Banach space E is a **projection** if it is idempotent, i.e., $P^2 = P$. Recall further that for a bounded projection P on a Banach space E one has the decomposition $E = \ker P \oplus \operatorname{fix} P$ into two P-invariant closed subspaces on which P acts as 0 and I, respectively. Moreover, a bounded projection P on a Hilbert space is orthogonal (i.e., $\ker P$ and $\operatorname{fix} P = \operatorname{ran} P$ are orthogonal) if and only if it is contractive and if only if it is self-adjoint.

The closed graph theorem implies the next result.

Proposition 1.4 *Let E be a Banach space and $E_0, E_1 \subset E$ closed subspaces such that $E_0 \cap E_1 = \{0\}$. The mapping*

$$P : E_0 + E_1 \to E_0 + E_1, \quad e_0 + e_1 \mapsto e_0$$

is a projection. The subspace $E_0 + E_1$ is closed if and only if P is bounded.

Spectrum

Let E be a Banach space. The **spectrum** of an operator $S \in \mathscr{L}(E)$ is the set

$$\begin{aligned}\sigma(S) &:= \{\lambda \in \mathbb{C} : \lambda I - S : E \to E \text{ is not bijective}\} \\ &= \{\lambda \in \mathbb{C} : \lambda I - S \text{ is not invertible in } \mathscr{L}(E)\},\end{aligned}$$

where the equality of the two sets follows from the open-mapping theorem, while the **resolvent set** is $\rho(S) := \mathbb{C} \setminus \sigma(S)$. We abbreviate $\lambda I - S$ simply as $\lambda - S$. The set $\sigma(S) \subset \mathbb{C}$ is compact and non-empty (provided $\dim E \neq 0$), and the mapping, called the **resolvent mapping**, $\rho(S) \to \mathscr{L}(E)$, $\lambda \mapsto (\lambda - S)^{-1}$ is continuous (and even analytic on $\rho(S)$). The spectrum of $S \in \mathscr{L}(E)$ satisfies

$$\sigma(S) \subset \overline{\mathrm{B}}(0, r(S)),$$

with

$$r(S) = \lim_{n\to\infty} \|S^n\|^{1/n} = \inf_{n\in\mathbb{N}} \|S^n\|^{1/n},$$

the **spectral radius** of S. One has $r(S) = \inf\{s > 0 : \sigma(S) \subset \mathrm{B}(0, s)\}$.

In particular, the spectrum of every contraction is contained in the closed unit disc. Moreover, for isometries one has the following dichotomy.

Proposition 1.5 (Spectrum of Isometries) *Let E be a Banach space and $S \in \mathscr{L}(E)$ be an isometry. If S is invertible, then $\sigma(S) \subset \mathbb{T}$, otherwise $\sigma(S)$ is the whole closed unit disc.*

The following result relates the spectrum of an operator S to the spectrum of a polynomial[1] of S or a resolvent of S, respectively.

Proposition 1.6 (Spectral Mapping Theorem) *For a bounded, linear operator $S \in \mathscr{L}(E)$ on a Banach space E and a polynomial P with complex coefficients*

$$\sigma(P(S)) = P(\sigma(S))$$

holds. Moreover, for every $a \in \rho(S)$ we have

$$\sigma((a - S)^{-1}) = \{(a - \lambda)^{-1} : \lambda \in \sigma(S)\}.$$

The **point spectrum** of $S \in \mathscr{L}(E)$ is the set

$$P\sigma(S) := \{\lambda \in \mathbb{C} : \lambda I - S : E \to E \text{ is not injective}\},$$

whose elements are called **eigenvalues**. Clearly, $P\sigma(S) \subset \sigma(S)$.

[1] For a polynomial $P \in \mathbb{C}[\cdot]$ with $P(z) = a_0 + a_1 z + \ldots + a_d z^d$ and a bounded linear operator $S \in \mathscr{L}(E)$, the operator $P(S) \in \mathscr{L}(E)$ is defined by $P(S) := a_0 I + a_1 S + \ldots + a_d S^d$.

Compact Operators

Let E, F be Banach spaces. A linear operator $S : E \to F$ is called **compact** if $\overline{SB_E(0,1)} \subset F$ is compact. A compact operator is bounded, and a bounded operator is compact if and only if for each bounded sequence (x_n) in E the sequence (Sx_n) has a Cauchy (or, equivalently, convergent) subsequence. The set $\mathscr{K}(E, F)$ of compact linear operators is a closed subspace of $\mathscr{L}(E, F)$ (with respect to the operator norm). The Banach space $\mathscr{K}(E) := \mathscr{K}(E, E)$ is in fact an ideal in $\mathscr{L}(E)$, i.e., for $S, T \in \mathscr{L}(E)$ we have $ST \in \mathscr{K}(E)$ provided either S or T is in $\mathscr{K}(E)$. The identity operator $I \in \mathscr{L}(E)$ is compact if and only if $\dim E < \infty$. **Schauder's theorem** tells that $S \in \mathscr{L}(E, F)$ is compact if and only if $S' \in \mathscr{L}(F', E')$ is compact.

Theorem 1.7 (Riesz–Schauder) *Let $S \in \mathscr{K}(E)$. Then $\sigma(S) \setminus \{0\} = P\sigma(S) \setminus \{0\}$, this set is at most countable and has 0 as the only possible accumulation point. Moreover, for every $\lambda \in \mathbb{C} \setminus \{0\}$ the space $\ker(\lambda - S)$ is finite dimensional.*

See, e.g., [606, Sec. 11.3].

Hilbert–Schmidt Operators

Let H, K be Hilbert spaces and $T \in \mathscr{L}(H, K)$. Moreover, let $(e_n)_{n\in I}$ and $(f_m)_{m\in J}$ be orthonormal bases in H and K, respectively. We then have

$$\sum_{n\in I} \|Te_n\|^2 = \sum_{n\in I}\sum_{m\in J} |(Te_n|f_m)_K|^2 = \sum_{n\in I}\sum_{m\in J} |(e_n|T^* f_m)_H|^2 = \sum_{m\in J} \|T^* f_m\|^2.$$

This implies that the expressions on the left- and on the right-hand side are independent of the particular choice of the orthonormal bases, and we can set

$$\|T\|_{\mathrm{HS}} := \Big(\sum_{n\in I} \|Te_n\|^2\Big)^{\frac{1}{2}}$$

for any choice of the orthonormal basis $(e_n)_{n\in I}$ in H. One has $\|T\| \le \|T\|_{\mathrm{HS}}$. The operator T is called **Hilbert–Schmidt** if $\|T\|_{\mathrm{HS}} < \infty$. The set $\mathrm{HS}(H, K)$ of Hilbert–Schmidt operators is a $\mathbb{C}$-vector space and a Hilbert space with the scalar product

$$(T|S)_{\mathrm{HS}} := \sum_{n\in I} (Te_n|Se_n)_K,$$

and $\|T\|_{\mathrm{HS}} = (T|T)_{\mathrm{HS}}^{\frac{1}{2}}$, the Hilbert–Schmidt norm. Hilbert–Schmidt operators are compact, and finite rank operators form a dense subspace of $\mathrm{HS}(H, K)$ with respect to the Hilbert–Schmidt norm. An important class of Hilbert–Schmidt operators are formed by integral operators of the following kind. Let (X, μ) be a measure space and $k \in \mathrm{L}^2(X \times X, \mu \times \mu)$. The linear operator T defined for $f \in \mathrm{L}^2(X, \mu)$ by

$$Tf = \int_X k(x, \cdot) f(x)\, \mathrm{d}\mu(x)$$

is Hilbert–Schmidt with $\|T\|_{\mathrm{HS}} = \|k\|_2$. If k satisfies $k(x, y) = \overline{k(y, x)}$ for a.e. $x, y \in X$, then T is self-adjoint. Spectral theory of self-adjoint Hilbert–Schmidt operators will be discussed in some detail in Sect. 5.5. See, e.g., [508, Ch. 11, 12].

Weak Topologies

Let E be a Banach space. A set $U \subset E'$ is called **weak* open** if for each $x' \in U$ there are $N \in \mathbb{N}$, $x_1, \ldots, x_N \in E$ and $\varepsilon > 0$ such that

$$U'_{\varepsilon, x_1, \ldots, x_N} := \{y' : |\langle x_j, x' - y'\rangle| < \varepsilon \text{ for } j = 1, \ldots, N\} \subset U.$$

This way a (Hausdorff) topology is defined on E', called the **weak* topology** and denoted by $\sigma(E', E)$. Note that a sequence $(x'_n)_{n\in\mathbb{N}}$ in E' is weak* convergent to $x' \in E'$ if and only if $\langle x, x'_n\rangle \to \langle x, x'\rangle$ as $n \to \infty$ for each $x \in E$.

The following classical fact shows the importance of the weak* topology, see [606, Sect. 17.1.4, 17.2.4] and [890, Thms. I.3.15, I.3.17].

Theorem 1.8 (Banach–Alaoglu)

(a) *The unit ball* $\overline{\mathrm{B}}_{E'} := \{x' \in E' : \|x'\| \le 1\}$ *is weak* compact.*
(b) *If E is separable, then the weak* topology of* $\overline{\mathrm{B}}_{E'}$ *is metrizable. Consequently, every norm bounded sequence* $(x'_n)_{n\in\mathbb{N}}$ *possesses a subsequence* $(x'_{n_j})_{j\in\mathbb{N}}$ *which is convergent in the weak* topology.*

The **weak topology** $\sigma(E, E')$ on E is defined by specifying the open sets as follows. A set $U \subset E$ is called **weakly open** if for each $x \in U$ there are $N \in \mathbb{N}$, $x'_1, \ldots, x'_N \in E'$ and $\varepsilon > 0$ such that

$$U_{\varepsilon, x'_1, \ldots, x'_N} := \{y : |\langle x - y, x'_j\rangle| < \varepsilon \text{ for } j = 1, \ldots, N\} \subset U.$$

Note that a sequence $(x_n)_{n\in\mathbb{N}}$ in E is weakly convergent to $x \in E$ if and only if $\langle x_n, x'\rangle \to \langle x, x'\rangle$ as $n \to \infty$ for each $x' \in E'$.

Since we can identify a Hilbert space H with its dual H', the weak and weak* topologies coincide for Hilbert spaces. In particular the closed unit ball $\overline{\mathrm{B}}_H$ of a

Hilbert space H is weakly compact. Moreover, each bounded sequence $(x_n)_{n\in\mathbb{N}}$ possesses a weakly convergent subsequence. See [890, Ch. 3–4].

Reflexivity

Given a normed space E and $x \in E$, the mapping $\delta_x : E' \to \mathbb{C}$, $\delta_x(x') := x'(x)$ is a linear functional with $\|\delta_x\| = \|x\|$. It follows that the mapping $\iota_E : E \to E''$, $x \mapsto \delta_x$ in the bidual space is a linear isometry. The normed space E is called **reflexive** if ι_E is surjective. A reflexive normed space is complete, and reflexivity is an isomorphism invariant. The dual, closed subspaces and quotients of reflexive spaces are reflexive. The closed unit ball of a Banach space E is compact for the weak topology $\sigma(E, E')$ if and only if E is reflexive, this is **Kakutani's theorem**. The spaces ℓ^p (and therefore $\ell^p(\mathbb{N}_0)$ and $\ell^p(\mathbb{Z})$), $p \in (1, \infty)$, are reflexive just as Hilbert spaces or the spaces $\mathrm{L}^p(X, \mu)$ for $p \in (1, \infty)$ (cf. below), while the spaces ℓ^∞ and the infinite dimensional $\mathrm{L}^\infty(X, \mu)$ are not. See [890, §4.5] and [606, 17.2].

Convex Sets

Let C be a convex set in a vector space. A point $x \in C$ is called an **extreme point** of C if for each $t \in (0, 1)$, $z, y \in C$ with $x = ty + (1 - t)z$ it follows that $y = z = x$. The set of extreme points of C is denoted by $\mathrm{Ex}(C)$.

Theorem 1.9 (Krein–Milman) *Let X be a locally convex space (e.g., $X = E$ with the norm topology, $X = E$ with $\sigma(E, E')$, or $X = E'$ with $\sigma(E', E)$), and let $C \subset X$ be a compact, convex subset. Then C is the closed, convex hull of* $\mathrm{Ex}(C)$*, i.e., C is the smallest closed, convex set containing* $\mathrm{Ex}(C)$.

See [890, Thm. 3.23] and [606, Sect. 18.1.2].

1.2 Some Topology

We assume familiarity with basic notions of point set topology at least in metric spaces (open and closed sets, boundary, limits of sequences, Cauchy sequences, continuity, etc). At a few places we shall need general topological spaces and the most basic notions and results, such as connectedness, compactness and basic properties of continuous functions on (compact) topological spaces. We stress here that under a **compact space** we understand a topological space for which every open cover has finite subcover *and* it is *Hausdorff* (this terminology is not uniformly used in the literature). In Chap. 11 also locally compact spaces will appear shortly, whose definition also includes the Hausdorff property. The contents of Chapter Two of the book [532] by Hewitt and Stromberg will be more than sufficient for our purposes, but we also recommend Engelking's treatise [370].

Continuous Functions

For a set X and a function $f : X \to \mathbb{C}$ we define $\|f\|_\infty := \sup_{x\in X} |f(x)|$. Given a topological space $(X, \mathcal{O})$ (or a metric space (X, d)) the space of bounded and continuous functions

$$\mathrm{C_b}(X) := \{f : X \to \mathbb{C} : \ f \text{ is continuous with } \|f\|_\infty < \infty\}$$

is a Banach space, and even a unital Banach algebra with the norm $\|\cdot\|_\infty$, the constant one function as unit and the pointwise operations. If K is a compact space, then each continuous function $f : K \to \mathbb{C}$ is bounded (actually, there is $x \in K$ with $|f(x)| = \|f\|_\infty$), and we write $\mathrm{C}(K) := \mathrm{C_b}(K)$.

The next result applied to disjoint closed sets $\{x\}$ and $\{y\}$ yields that $\mathrm{C}(K)$ separates the points of K.

Proposition 1.10 (Urysohn's Lemma) *Let K be a compact space, and let $A, B \subset K$ be disjoint, closed subsets. Then there is $f \in \mathrm{C}(K)$ such that $f(K) \subset [0, 1]$ and $f(a) = 1$, $f(b) = 0$ for every $a \in A$, $b \in B$.*

Point separation is also crucial in the next famous result, see [606, Sect. 18.2.2].

Theorem 1.11 (Stone–Weierstraß) *For a compact space K, every point separating, conjugation invariant, unital subalgebra of $\mathrm{C}(K)$ is dense in $\mathrm{C}(K)$.*

A function f defined on the unit circle $\mathbb{T} = \{z \in \mathbb{C} : |z| = 1\}$ is called a **trigonometric polynomial** if it is a finite linear combination of the functions $\mathbf{z}^n : z \mapsto z^n$, $n \in \mathbb{Z}$. For $d \in \mathbb{N}$ and $j = 1, \ldots, d$ consider the functions $\mathbf{z}_j : \mathbb{T}^d \to \mathbb{C}$, $(z_1, \ldots, z_d) \mapsto z_j$. A function of the form $\mathbf{z}^n := \mathbf{z}_1^{n_1} \cdots \mathbf{z}_d^{n_d}$, $n = (n_1, \ldots, n_d) \in \mathbb{Z}^d$ is called a **trigonometric monomial**, and a finite linear combination of such monomials is a **trigonometric polynomial** on $\mathbb{T}^d$; these functions form the vector space $\mathrm{P}(\mathbb{T}^d)$. The Stone–Weierstraß theorem immediately implies the following.

Proposition 1.12 *The vector space $\mathrm{P}(\mathbb{T}^d)$ of trigonometric polynomials is dense in $\mathrm{C}(\mathbb{T}^d)$.*

For a function $f : X \to \mathbb{R}$ we use the notation

$$[f > \alpha] := f^{-1}((\alpha, \infty)), \quad [f < \alpha] := f^{-1}((-\infty, \alpha)), \quad [f = \alpha] = f^{-1}(\{\alpha\}),$$

etc. The support of a continuous function $f : X \to \mathbb{C}$ on a topological space X is $\mathrm{supp}(f) := \overline{[f \neq 0]}$, where $\overline{A}$ is the closure of A. See [889, 2.3–2.13] and [508, Ch. 3,4].

Locally Compact Spaces

Recall that a topological space X is called **locally compact** if it is *Hausdorff* and every point in X has a compact neighborhood. In this case each point in X has a neighborhood basis consisting of compact sets (which is sometimes taken as the definition of local compactness). A locally compact space is called **σ-compact** if it is a union of countably many of its compact subsets (e.g., $\mathbb{R}^d$),

We denote $\mathrm{C_c}(X)$ the $\mathbb{C}$-vector space of complex-valued continuous functions on X with compact support. Then $\mathrm{C_c}(X)$ is an algebra of functions (with pointwise operations) and is quite rich as Urysohn's lemma tells. Let us denote by $\mathbf{1}_A$ the characteristic function of a set $A \subset X$, i.e.,

$$\mathbf{1}_A(x) = \begin{cases} 1 & \text{if } x \in A, \\ 0 & \text{if } x \notin A. \end{cases}$$

The following version of Urysohn's lemma is valid in locally compact spaces.

Theorem 1.13 (Urysohn) *Let X be a locally compact space and let $K \subset U \subset X$ with K compact and U open. Then there is $f \in \mathrm{C_c}(X)$ with*

$$\mathbf{1}_K \leq f \leq \mathbf{1}_U.$$

In particular, the algebra $\mathrm{C_c}(X)$ separates the points of X. See, e.g., [532, Thm. 6.80].

1.3 Some Measure Theory

We assume familiarity with the elementary theory of positive measures, the Lebesgue integral, and with the most fundamental results: Lebesgue's dominated convergence theorem, Beppo Levi's monotone convergence theorem, Fatou's lemma. The construction of product measures, and the theorems of Fubini and Tonelli are regarded as well-known. See, e.g., [153, 532, 606, 889]. Here we only fix some notation, terminology and recall some fundamental results.

Set Systems

Let $\mathcal{X}$ be a σ-algebra over a set X, the pair $(X, \mathcal{X})$ is then called a **measurable space**. If the σ-algebra $\mathcal{X}$ is understood from the context, then we refer to sets $A \in \mathcal{X}$ simply as **measurable**. The power set of a given set X is denoted by $\mathcal{P}(X)$. A set system $\mathcal{E} \subset \mathcal{P}(X)$ is called **$\cap$-stable** if it is closed under taking finite intersections of its elements.

Theorem 1.14 (Dynkin) *Let X be a set, let $\mathcal{E} \subset \mathcal{P}(X)$ be $\cap$-stable and let $\mathcal{D} \subset \mathcal{P}(X)$ have the following properties:*

(1) $X \in \mathcal{D}$,
(2) *if* $A \in \mathcal{D}$, *then* $A^c := X \setminus A \in \mathcal{D}$,
(3) *if* $A_n \in \mathcal{D}$ *for each* $n \in \mathbb{N}$ *and these sets are pairwise disjoint, then* $\bigcup_{n\in\mathbb{N}} A_n \in \mathcal{D}$,

i.e., a so-called ***Dynkin-system****. Then the σ-algebra $\sigma(\mathcal{E})$ generated by $\mathcal{E}$ is contained in $\mathcal{D}$.*

See, e.g., [153, Thm. 3.2].

Measure and Integral

Let $\mathcal{X}$ be a σ-algebra over a set X and define

$$\mathscr{L}^0(X, \mathcal{X}) := \{f : X \to \mathbb{C} \text{ measurable}\},$$

where $\mathbb{C}$, as everywhere in this book, is equipped with the Borel σ-algebra.

A positive **measure** $\mu : \mathcal{X} \to [0, \infty]$ is a σ-additive function with $\mu(\emptyset) = 0$. (Usually, we only use the term "measure" for positive measures, but sometimes it is also important to stress positivity.) In this case (X, μ) is called a **measure space**. If the σ-algebra needs to be stressed, we write $(X, \mathcal{X}, \mu)$. The measure μ is called **finite** if $\mu(X) < \infty$ and **σ-finite** if X can be covered by at most countably many measurable sets with finite measure. A finite (positive) measure μ satisfying $\mu(X) = 1$ is called a **probability** (measure) and (X, μ) is called a **probability space**.

The integral of $f \in \mathscr{L}^0(X, \mathcal{X})$ with respect to μ (if it exists) is denoted by

$$\int_X f \,\mathrm{d}\mu \text{ or } \int_X f(x) \,\mathrm{d}\mu(x).$$

Given a measure space (X, μ) we define for $p \in (0, \infty)$ and $f \in \mathscr{L}^0(X, \mathcal{X})$

$$\|f\|_p := \Big(\int_X |f|^p \,\mathrm{d}\mu \Big)^{1/p} \text{ and } \|f\|_\infty := \inf\{\alpha : |f| \le \alpha \ \mu\text{-a.e.}\}.$$

We also set for $p \in (0, \infty]$

$$\mathscr{L}^p(X, \mu) := \{f \in \mathscr{L}^0(X, \mathcal{X}) : \|f\|_p < \infty\},$$

which are vector spaces. Moreover, for $p \geq 1$ we have that $\|\cdot\|_p$ is a seminorm on $\mathscr{L}^p(X, \mu)$. Define the equivalence relation $\sim_\mu$ by $f \sim_\mu g$ if $[f \neq g]$ is μ-null set. After identifying $\sim_\mu$-equivalent functions, in the case $p \geq 1$, we obtain the Banach spaces $\mathrm{L}^p(X, \mu)$ with norm $\|\cdot\|_p$, and the vector space $\mathrm{L}^0(X, \mu)$. Depending on the context we may write $\mathrm{L}^p(X)$, $\mathrm{L}^p(\mu)$, or even L^p. If the σ-algebra needs to be stressed, we write $\mathrm{L}^p(X, \mathcal{X}, \mu)$. If (X, μ) is a finite measure space, then for $1 \leq p \leq r \leq \infty$ we have $\mathrm{L}^r(X, \mu) \subset \mathrm{L}^p(X, \mu)$ with continuous embedding. This can be proven, e.g., with the help of Hölder's inequality. The following general version of **Hölder's inequality** will be also useful: For $0 < p_0, p_1, \ldots, p_m \leq \infty$ with $\frac{1}{p_0} = \frac{1}{p_1} + \cdots + \frac{1}{p_m}$ and $f_1, \ldots, f_m \in \mathrm{L}^0(X, \mu)$ we have

$$\|f_1 \cdots f_m\|_{p_0} \leq \|f_1\|_{p_1} \cdots \|f_m\|_{p_m}. \tag{1.1}$$

One of **F. Riesz'** many theorems states that any Cauchy sequence in the Banach space L^p possesses an almost everywhere convergent subsequence (with a little subtlety of choosing representatives from the equivalence classes). See [889, Ch. 1–3,7] and [508, Ch. 7].

For $p \in (1, \infty)$ the dual of the Banach space $\mathrm{L}^p(X, \mu)$ is $\mathrm{L}^q(X, \mu)$ with $\frac{1}{p} + \frac{1}{q} = 1$ under the following isometric isomorphism

$$\begin{aligned} &J_p : \mathrm{L}^q(X, \mu) \to \mathrm{L}^p(X, \mu)', \\ &\quad J_p(g)(f) = \langle f, g \rangle = \int_X fg \,\mathrm{d}\mu \quad (f \in \mathrm{L}^p,\ g \in \mathrm{L}^q). \end{aligned} \tag{1.2}$$

The same is true for $p = 1$ and $q = \infty$, for example in the case when μ is a σ-finite measure.

A sometimes convenient way to compute the L^p-norm is the following: If $p \in (0, \infty)$ and $f \in \mathrm{L}^0(X, \mu)$, then

$$\|f\|_p^p = \int_0^\infty pt^{p-1} \mu([|f| > t]) \,\mathrm{d}t. \tag{1.3}$$

If (f_n) is a sequence of measurable functions on a finite measure space $(X, \mathcal{X}, \mu)$ converging to f almost everywhere, then for every $\varepsilon > 0$ there is $A \subset X$ with $\mu(X \setminus A) < \varepsilon$ such that $f_n \to f$ uniformly on A. This is **Egorov's theorem**, see, e.g., [606, §3.2.4].

Given $m \in \mathrm{L}^\infty(X, \mu)$ and $p \in [1, \infty]$ the **multiplication operator**

$$M_m : \mathrm{L}^p(X, \mu) \to \mathrm{L}^p(X, \mu), \quad f \mapsto mf$$

is bounded and linear on $\mathrm{L}^p(X, \mu)$, its operator norm satisfies $\|M_m\| = \|m\|_\infty$. Of course, this generalizes the notion of multiplication operators on sequence spaces from the above.

Carathéodory–Hahn Theory

An **outer measure** over a given set X is a mapping

$$\varphi : \mathcal{P}(X) \to [0, \infty]$$

on the power set $\mathcal{P}(X)$ of X with the following properties:

(1) $\varphi(\emptyset) = 0$,
(2) φ is monotone,
(3) φ is σ-subadditive.

Given an outer measure $\varphi : \mathcal{P}(X) \to [0, \infty]$ we say that a subset $A \subset X$ is φ-measurable if

$$\varphi(H) = \varphi(H \cap A) + \varphi(H \setminus A) \text{ for all } H \subset X.$$

Theorem 1.15 (Carathéodory) *For an outer measure φ on the set X the set $\mathcal{M}_\varphi$ of all φ-measurable subsets of X is a σ-algebra, φ is σ-additive on $\mathcal{M}_\varphi$ and $(X, \mathcal{M}_\varphi, \varphi|_{\mathcal{M}_\varphi})$ is a (complete) measure space.*

A set system $\mathcal{E} \subset \mathcal{P}(X)$ is called a **semi-ring** if it satisfies the following two conditions:

(1) $\mathcal{E}$ is $\cap$-stable and $\emptyset \in \mathcal{E}$.
(2) If $A, B \in \mathcal{E}$, then $A \setminus B$ is a disjoint union of members of $\mathcal{E}$.

Theorem 1.16 (Carathéodory–Hahn Extension) *Let $\mathcal{E} \subset \mathcal{P}(X)$, and let $\alpha : \mathcal{E} \to [0, \infty]$ satisfy $\emptyset \in \mathcal{E}$ and $\alpha(\emptyset) = 0$. Then $\varphi_\alpha : \mathcal{P}(X) \to [0, \infty]$ defined by*

$$\varphi_\alpha(A) := \inf\Big\{ \sum_{n\in\mathbb{N}} \alpha(E_n) : (E_n)_{n\in\mathbb{N}} \subset \mathcal{E},\ A \subset \bigcup_{n\in\mathbb{N}} E_n \Big\} \qquad (A \in \mathcal{P}(X))$$

is an outer measure. Furthermore, if $\mathcal{E}$ is a semi-ring and $\varphi_\alpha : \mathcal{E} \to [0, \infty]$ is σ-additive on $\mathcal{E}$, then the σ-algebra $\sigma(\mathcal{E})$ generated by $\mathcal{E}$ is contained in the σ-algebra $\mathcal{M}_{\varphi_\alpha}$ of φ_α-measurable sets and φ_α is an extension of α. If X can be covered by at most countably many sets from $\mathcal{E}$ with finite measures, this extension is unique.

See, e.g., [153, Sec. 3]. The combination of the previous results is useful for constructing measures with certain prescribed properties. We record a particular case, also called the **Hahn–Kolmogorov extension theorem**, for later reference.

Corollary 1.17 (Carathéodory Extension Theorem) *Let $\mathcal{E} \subset \mathcal{P}(X)$ be a semi-ring with $X \in \mathcal{E}$ and let $\alpha : \mathcal{E} \to [0, \infty)$ be σ-additive on $\mathcal{E}$. Then there is a unique (finite) measure on the σ-algebra $\sigma(\mathcal{E})$ generated by $\mathcal{E}$ which extends α.*

Measures on Compact Spaces

To be able to apply methods from functional analysis to measures we extend the notion "measure" to complex-valued functions.

Let $(X, \mathcal{X})$ be a measurable space, i.e., $\mathcal{X}$ be σ-algebra over the set X. A **complex measure** is a σ-additive function $\mu : \mathcal{X} \to \mathbb{C}$ with $\mu(\emptyset) = 0$ whereas a **signed measure**[2] μ is a σ-additive function $\mu : \mathcal{X} \to [-\infty, +\infty]$. So a real-valued complex measure is a signed measure, and finite-valued signed measure is a (real-valued) complex measure. In this context we often use the word "positive measure" for complex or signed measures that have values in $[0, \infty)$ or $[0, \infty]$ only, respectively. The **total variation** (or **modulus**) $|\mu|$ of a measure μ is defined by

$$|\mu|(A) := \sup \Big\{ \sum_{n\in\mathbb{N}} |\mu(A_n)| : (A_n)_{n\in\mathbb{N}} \text{ is a measurable partition of } A \Big\}.$$

A positive measure equals its total variation and a signed measure is real-valued if and only if $|\mu|(X) < \infty$.

We denote by $\mathfrak{M}(X, \mathcal{X})$ the set of all complex measures on $(X, \mathcal{X})$. Here are the most important properties.

Proposition 1.18 (a) *For $\mu \in \mathfrak{M}(X, \mathcal{X})$ and $A \in \mathcal{X}$ we have $|\mu(A)| \le |\mu|(A)$.*
(b) *$|\mu| : \mathcal{X} \to [0, \infty)$ is a finite positive measure.*
(c) *The set $\mathfrak{M}(X, \mathcal{X})$ is a $\mathbb{C}$-vector space and a Banach space if endowed with the norm $\|\cdot\| : \mu \mapsto |\mu|(X)$, called the* ***total variation (norm)****.*

Let $\mu \in \mathfrak{M}(X, \mathcal{X})$ and let $f : X \to \mathbb{C}$ be a simple function, i.e., of the form $f = \sum_{j=1}^N a_j \mathbf{1}_{A_j}$ with $A_1, \ldots, A_N \in \mathcal{X}$ pairwise disjoint and $a_1, \ldots, a_N \in \mathbb{C}$. Then we can set

$$\int_X f \, \mathrm{d}\mu := \sum_{j=1}^N a_j \mu(A_j),$$

and quickly prove that this expression is indeed well-defined and that $(f, \mu) \mapsto \int_X f \, \mathrm{d}\mu$ is bilinear. Since

$$\Big| \sum_{j=1}^N a_j \mu(A_j) \Big| \le \sum_{j=1}^N |a_j| |\mu(A_j)| \le \sum_{j=1}^N |a_j| |\mu|(A_j) = \|f\|_{\mathrm{L}^1(|\mu|)},$$

[2] We follow here the terminology as in the book [516] by Halmos, but note that many authors use the term "signed measure" for what we call here real-valued measure.

by density we can extend the integral $\int_X \cdot \, \mathrm{d}\mu$ to $f \in \mathrm{L}^1(|\mu|)$ with linearity in f remaining true. For a bounded measurable function f the mapping $\mu \mapsto \int_X f \, \mathrm{d}\mu$ is linear. See [889, Ch. 6].

An important special case is when $X = K$ is a compact space and $\mathcal{X} = \mathcal{B}(K)$ is the Borel σ-algebra generated by the open sets in K. Every continuous function $f : K \to \mathbb{C}$ is then measurable. A finite positive measure $\mu : \mathcal{B}(K) \to \mathbb{C}$ is said to be **regular** if for each $B \in \mathcal{B}(K)$

$$\mu(B) = \sup\{\mu(F) : F \subset B,\ F \text{ compact}\} = \inf\{\mu(G) : B \subset G,\ G \text{ open}\}.$$

A complex Borel measure μ is said to be regular if $|\mu|$ is regular. The set of regular, complex Borel measures is denoted by $\mathrm{M}(K)$. It can be shown that $\mathrm{M}(K)$ is a Banach space with the total variation norm, and if K is metrizable, then any finite positive measure is regular, i.e., $\mathrm{M}(K) = \mathfrak{M}(K, \mathcal{B}(K))$. As a consequence of regularity and Urysohn's lemma we obtain for signed measures $\mu, \nu \in \mathrm{M}(K)$

$$\mu \le \nu \quad \Longleftrightarrow \quad \text{for every positive } f \in \mathrm{C}(K) : \quad \int_K f \, \mathrm{d}\mu \le \int_K f \, \mathrm{d}\nu,$$

and for general complex measures $\mu, \nu \in \mathrm{M}(K)$

$$\mu = \nu \quad \Longleftrightarrow \quad \text{for every } f \in \mathrm{C}(K) : \quad \int_K f \, \mathrm{d}\mu = \int_K f \, \mathrm{d}\nu.$$

The following fundamental theorem allows the application of functional analytic techniques when studying complex measures on compact spaces.

Theorem 1.19 (Riesz–Markov–Kakutani) *Let K be a compact space. The mapping $J : \mathrm{M}(K) \to \mathrm{C}(K)'$ defined by*

$$(J\mu)f := \int_K f \, \mathrm{d}\mu \quad (\mu \in \mathrm{M}(K),\ f \in \mathrm{C}(K))$$

is an isometric isomorphism of Banach spaces.

By virtue of this result we identify the Banach spaces $\mathrm{M}(K)$ and $\mathrm{C}(K)'$, and write

$$\langle f, \mu \rangle = \int_K f \, \mathrm{d}\mu.$$

See [889, 2.14–2.18 and Thm. 6.19] and [606, Sect. 8.4.2].

The **support** $\mathrm{supp}(\mu)$ of a positive measure $\mu \in \mathrm{M}(K)$ is the closed set

$$\mathrm{supp}(\mu) := \big\{x \in K : \mu(U) > 0 \text{ for each open set } U \text{ with } x \in U\big\}.$$

If $\text{supp}(\mu)$ is a singleton, then μ is a scalar multiple of a **Dirac measure**, i.e., $\mu = c\delta_a$ for some $c \in \mathbb{C}$ and $a \in K$, where $\delta_a(A) = 1$ if $a \in A$ and $\delta_a(A) = 0$ if $a \notin A$.

Proposition 1.20 *Let K be a compact space and let $\mu \in \mathrm{M}(K)$ be positive.*

(a) *We have $\mu(\text{supp}(\mu)^c) = 0$, and if a measurable function $f : K \to \mathbb{C}$ vanishes on $\text{supp}(\mu)$, then*

$$\langle f, \mu \rangle = \int_K f \, \mathrm{d}\mu = 0.$$

(b) *If $L \subset K$ is a closed subset and for every $f \in \mathrm{C}(K)$ with $f|_L = 0$ also $\langle f, \mu \rangle = 0$ holds, then $\text{supp}(\mu) \subset L$.*

The next result is a functional analytic description of the support of a measure.

Proposition 1.21 *Let $K \subset \mathbb{C}$ be a compact set, and let $\mu \in \mathrm{M}(K)$ be positive. Consider the multiplication operator $M_{\mathbf{z}} : \mathrm{L}^2(K, \mu) \to \mathrm{L}^2(K, \mu)$ given by $(M_{\mathbf{z}} f)(z) := zf(z)$. Then $M_{\mathbf{z}}$ is bounded, linear and its spectrum satisfies $\sigma(M_{\mathbf{z}}) = \text{supp}(\mu)$.*

We will use the following important information about the set $\mathrm{M}_1(K)$ of regular Borel probability measures on K.

Proposition 1.22 *Let K be a compact space.*

(a) *The set $\mathrm{M}_1(K)$ is a weak* compact, convex subset of $\mathrm{M}(K)$.*
(b) *The extreme points of $\mathrm{M}_1(K)$ are precisely the Dirac measures.*
(c) *The set $\mathrm{M}_1(K)$ is the closed convex hull of the Dirac measures with respect to the weak* topology.*

Radon Measures on Locally Compact Spaces

A (positive) measure on the Borel algebra $\mathcal{B}(X)$ of a locally compact space X is called a **Radon measure** if for each compact set $K \subset X$ we have $\mu(K) < \infty$, and for each Borel set $B \subset X$

$$\mu(B) = \inf\{\mu(G) : B \subset G, \ G \text{ open}\}$$

and for each set $A \subset X$ which is open or which satisfies $\mu(A) < \infty$

$$\mu(A) = \sup\{\mu(K) : K \subset A, \ K \text{ compact}\}.$$

Such a measure gives rise to a linear functional $\varphi_\mu : \mathrm{C_c}(X) \to \mathbb{C}$ by setting

$$\varphi_\mu(f) := \int_X f \, \mathrm{d}\mu \quad (f \in \mathrm{C_c}(X)).$$

In fact, φ_μ is a positive linear functional, i.e., one that maps positive functions to positive reals. The Riesz–Markov–Kakutani reverses this construction.

Theorem 1.23 (Riesz–Markov–Kakutani) *Let X be a locally compact space and $\varphi : \mathrm{C_c}(X) \to \mathbb{C}$ a positive linear functional. Then there is a unique Radon measure μ on the Borel algebra $\mathcal{B}(X)$ such that for each $f \in \mathrm{C_c}(X)$*

$$\varphi(f) := \int_X f \, \mathrm{d}\mu.$$

See, e.g., [889, Thm. 2.14]. Note that for a σ-compact, locally compact space X each Radon measure (representing a positive linear functional on $\mathrm{C_c}(X)$ as in Theorem 1.23) is σ-finite.

Lebesgue's Decomposition and the Radon–Nikodym Theorem

We begin with recalling the definition of absolute continuity and singularity of complex measures.

Definition 1.24 (a) Let $\mu, \nu \in \mathfrak{M}(X, \mathcal{X})$ be two positive measures. If each μ-null set is at the same time a ν-null set, then ν is said to be **absolutely continuous** with respect to μ. We denote this by writing $\nu \ll \mu$. If additionally $\mu \ll \nu$, then we call the measures μ and ν **equivalent**, denoted by $\mu \simeq \nu$.
(b) Let $\mu, \nu \in \mathfrak{M}(X, \mathcal{X})$ be two complex measures. We say that ν is absolutely continuous with respect to μ if $|\nu| \ll |\mu|$.
(c) A pair of complex measures $\mu, \nu \in \mathfrak{M}(X, \mathcal{X})$ is called **(mutually) singular** if there is a set $A \in \mathcal{X}$ with $|\mu|(A) = |\nu|(A^c) = 0$. We denote this relation by writing $\mu \perp \nu$.

The following properties are easy to check.

Lemma 1.25 *For a σ-algebra $\mathcal{X}$ over a set X and $\lambda, \mu, \nu \in \mathfrak{M}(X, \mathcal{X})$ the following assertions hold.*

(a) *$\mu \ll \nu$ and $\nu \ll \lambda$ implies $\mu \ll \lambda$.*
(b) *$\simeq$ is an equivalence relation on $\mathfrak{M}(X, \mathcal{X})$.*
(c) *If $\mu \perp \nu$ and $\mu \ll \nu$, then $\mu = 0$.*
(d) *If $\mu \ll \nu$ and $\nu \perp \lambda$, then $\mu \perp \lambda$.*
(e) *If $\mu \perp \nu$, then $|\mu + \nu| = |\mu| + |\nu|$.*
(f) *If $\lambda \ll \mu + \nu$, then there are $\mu', \nu' \in \mathfrak{M}(X, \mathcal{X})$ with $\lambda = \mu' + \nu'$ and $\mu' \ll \mu$, $\nu' \ll \nu$.*
(g) *The sets $\{\sigma \in \mathfrak{M}(X, \mathcal{X}) : \sigma \ll \mu\}$ and $\{\sigma \in \mathfrak{M}(X, \mathcal{X}) : \sigma \perp \mu\}$ are norm closed subspaces in $\mathfrak{M}(X, \mathcal{X})$.*

For the next result see, e.g., [889, Thms. 6.7–6.14].

Proposition 1.26 *For a set X, a σ-algebra $\mathcal{X}$ over X, a complex measure $\mu \in \mathfrak{M}(X, \mathcal{X})$, $f \in \mathrm{L}^1(X, |\mu|)$, and $A \in \mathcal{X}$ define*

$$\nu(A) := \int_A f \, \mathrm{d}\mu.$$

Then the following assertions hold.

(a) *ν is a complex measure which is absolutely continuous with respect to μ. We use the notation $f \cdot \mu$ for this measure ν.*
(b) *For every bounded measurable function $g : X \to \mathbb{C}$*

$$\int_X g \, \mathrm{d}\nu = \int_X gf \, \mathrm{d}\mu.$$

The same assertion holds for each $g : X \to \mathbb{C}$ with $fg \in \mathrm{L}^1(X, |\mu|)$.
(c) *There is a Borel measurable function $h : X \to \mathbb{T}$ such that $\mu = h \cdot |\mu|$. This is called the* **polar decomposition** *of μ. Moreover, we have $|\mu| = \overline{h} \cdot \mu$.*
(d) *The total variations satisfy $|\nu| = |f| \cdot |\mu|$. In particular, $\|\nu\| = \|f\|_{\mathrm{L}^1(X,|\mu|)}$.*
(e) *Suppose $\mu \in \mathfrak{M}(X, \mathcal{X})$ is a positive measure. Then the mapping $J_\mu : \mathrm{L}^1(X, \mu) \to \mathfrak{M}(X, \mathcal{X})$, $J_\mu f := f \cdot \mu$ is a linear isometry preserving the modulus, i.e., $J_\mu |f| = |f \cdot \mu|$. Moreover, f is real-valued if and only if $J_\mu f$ is a signed measure, $f \geq 0$ if and only if $f \cdot \mu$ is a positive measure, and $\mathrm{ran}(J_\mu) = \{\nu : \nu \ll \mu\}$.*

Given a signed measure μ on $\mathcal{X}$, there are measurable sets $P, N \in \mathcal{X}$ with $P \cap N = \emptyset$, $X = P \cup N$ and such that for every $A \in \mathcal{X}$ one has $\mu(A \cap P) \in [0, \infty]$ and $\mu(A \cap N) \in [-\infty, 0]$. The sets P, N with this property are unique up to $|\mu|$-null sets, and yield the so-called **Jordan decomposition**. If we set $\mu^+(A) := \mu(A \cap P)$ and $\mu^-(A) := -\mu(A \cap N)$ for each $A \in \mathcal{X}$, then we obtain $\mu = \mu^+ - \mu^-$, the so-called **Hahn decomposition**, where $\mu^+ \perp \mu^-$. We also have $|\mu| = \mu^+ + \mu^-$. See [889, 6.6] and [606, Sect. 7.1.2].

The assertions (c) and (e) above are based on the following classical result, see [889, 6.7–6.14] and [606, Sect. 7.1.6].

Theorem 1.27 (Lebesgue Decomposition and Radon–Nikodym Theorem) *Let $\nu, \mu \in \mathfrak{M}(X, \mathcal{X})$ and let μ be positive.*

(a) *There are unique measures ν_a and ν_s with $\nu_s \perp \mu$, $\nu_a \ll \mu$ and $\nu = \nu_a + \nu_s$.*
(b) *There is $f \in \mathrm{L}^1(\mu)$ such that $\nu_a = f \cdot \mu$, i.e.*

$$\nu_a(A) = \int_A f \, \mathrm{d}\mu \text{ for each } A \in \mathcal{X}.$$

Moreover, if ν is positive (or real-valued), then so are ν_a, ν_s and f.

If $\nu \ll \mu$, then $\nu_s = 0$ in (a), so $\nu = f \cdot \mu$. The function f is called the **Radon–Nikodym derivative** of ν with respect to μ. For a signed measure $\mu \in \mathfrak{M}(X, \mathcal{X})$ we have $\mu^+ = \mathbf{1}_P \cdot |\mu|$ and $\mu^- = \mathbf{1}_N \cdot |\mu|$. If $\mu, \nu \in \mathfrak{M}(X, \mathcal{X})$, μ is positive and $\nu = f \cdot \mu$, then $\nu^+ = f^+ \cdot \mu$ and $\nu^- = f^- \cdot \mu$, where $f^+ \max(f, 0)$, $f^- = \max(-f, 0)$.

Part I
The Fundamentals

Chapter 2
Measure-Preserving Systems

Basic objects of ergodic theory are invariant measures, measure-preserving transformations/systems and the corresponding Koopman operator, which will be introduced in this chapter. As a first elementary but surprising result, we present Poincaré's recurrence theorem. We also discuss various examples showing connections to other areas such as group theory, probability theory, number theory and topology.

2.1 Measure-Preserving Systems and Recurrence

Measure-preserving systems are fundamental objects for ergodic theory.

Definition 2.1

(a) Let (X, μ) be a probability space, and let $T : X \to X$ be a measurable transformation. Then T is called **μ-preserving** (and μ is called **T-invariant**) if

$$\mu(T^{-1}B) = \mu(B)$$

holds for every measurable $B \subset X$. (Here $T^{-1}B$ is the preimage of B under T.) In this case, the triple (X, μ, T) is called a **measure-preserving system**. If the underlying σ-algebra $\mathcal{X}$ needs to be stressed, we write $(X, \mathcal{X}, \mu, T)$.

(b) More generally, a measurable map $T : X \to Y$ between two measure spaces (X, μ) and (Y, ν) is called **measure-preserving** if $\mu(T^{-1}B) = \nu(B)$ for every measurable $B \subset Y$.

(c) Let μ be a positive measure on the σ-algebra $\mathcal{X}$ over X, let $\mathcal{Y}$ be a σ-algebra over Y, and let $T : X \to Y$ be measurable. The **push-forward measure** $T_*\mu$ of μ is defined by $T_*\mu(B) := \mu(T^{-1}B)$ for $B \in \mathcal{Y}$.

T. Eisner, B. Farkas, *A Journey Through Ergodic Theorems*, Birkhäuser Advanced Texts Basler Lehrbücher, https://doi.org/10.1007/978-3-031-96506-7_2

Remark 2.2

(a) A measurable map $T : X \to Y$ between two measure spaces (X, μ) and (Y, ν) is measure-preserving if and only if $T_*\mu = \nu$.
(b) If T is μ-preserving, then the images of sets cannot have smaller measure than the sets themselves:

$$\inf\{\mu(C) : TB \subset C,\ C \text{ measurable}\} \geq \mu(B) \quad \forall \text{ measurable } B \subset X,$$

where TB denotes the image of B under T. This inequality can be strict, see, for instance, Example 2.18. In particular, $\mu(TB) \geq \mu(B)$, whenever both B and TB are measurable.
(c) In this book we will deal only with *probability*-preserving transformations (i.e. with finite invariant measures), so unless stated specifically otherwise, the terms "measure-preserving" and "probability-preserving" are used synonymously. For the study of the more general situation we refer to, e.g., the books [683] by Krengel and [4] by Aaronson.
(d) A measurable map $T : X \to Y$ between two measure spaces (X, μ) and (Y, ν) is called **non-singular**[1] if $\nu(B) = 0$ implies $T_*\mu(B) = 0$ (i.e., if $T_*\mu \ll \nu$, see Definition 1.24). Every measure-preserving transformation is, of course, non-singular. In this book we shall only look at ergodic theorems for measure-preserving systems, and refer the reader to, e.g., Krengel's book [683] for the study of non-singular transformations. In particular, the problem of the existence of an invariant probability measure for a non-singular transformation is treated in, e.g., [683, Sec. 3.4].
(e) If (X, μ, T) is a measure-preserving system and $k \in \mathbb{N}$, then setting $T^k = T \circ \cdots \circ T$ (composition k-times), we obtain a measure-preserving system (X, μ, T^k) called the kth **iterate** of T (or of (X, μ, T)).

The following famous recurrence result is remarkably easy to prove.

Theorem 2.3 (Poincaré's Recurrence Theorem) *For a measure-preserving system (X, μ, T) and $A \subset X$ satisfying $\mu(A) > 0$, there is $n \in \mathbb{N}$ such that $\mu(A \cap T^{-n}A) > 0$.*

Proof Let $A \subset X$ be measurable such that $\mu(A \cap T^{-n}A) = 0$ for every $n \in \mathbb{N}$. Since T is μ-preserving, we have

$$\mu(T^{-k}A \cap T^{-(n+k)}A) = 0 \text{ for every } n \in \mathbb{N},\ k \in \mathbb{N}_0.$$

This means that the sets $A, T^{-1}A, T^{-2}A, \ldots$ are pairwise disjoint up to null sets. But they all have the same measure $\mu(A)$. Since μ is a finite measure, this can only happen if $\mu(A) = 0$. ■

---

[1] Some authors use a stronger definition requiring that $\nu(B) = 0$ if and only if $T_*\mu(B) = 0$ (i.e., if $T_*\mu \simeq \nu$, see Definition 1.24).

We will see strengthenings of Theorem 2.3 later in Theorems 7.6, 9.19 and 13.5, see also Remark 7.7 and Sect. 13.9 for more results and references.

Given a measurable set $A \subset X$, we define the **time of first return** to A by

$$n_A : X \to \mathbb{N} \cup \{\infty\}, \quad n_A(x) := \min\{k \in \mathbb{N} : T^k x \in A\}$$

(with the convention $\min(\emptyset) = \infty$), of course, this describes the time of first *return* of x to A only if $x \in A$. It is easy to show that n_A is a measurable mapping.

Poincaré's result shows in particular that for every $A \subset X$ with positive measure there is at least one $x \in A$ which is **recurrent** to A, i.e., $n_A(x) < \infty$. In fact, *almost every* $x \in A$ is **infinitely recurrent** to A, i.e., comes back to A infinitely often. So, in particular, n_A is almost everywhere finite on A.

Corollary 2.4 (Infinite Recurrence) *Let (X, μ, T) be a measure-preserving system and let $A \subset X$ be measurable. Then almost every $x \in A$ satisfies $T^n x \in A$ for infinitely many $n \in \mathbb{N}$.*

Proof We first show that almost every $x \in A$ is recurrent to A. Consider the measurable set

$$B := A \setminus \bigcup_{k=1}^{\infty} T^{-k}A = A \cap \Big(\bigcap_{k=1}^{\infty} T^{-k}A^c\Big)$$

of all points of A which never return to A, where A^c denotes the complement of the set A. For every $n \in \mathbb{N}$ we have

$$B \cap T^{-n}B \subset A \cap T^{-n}A \cap \Big(\bigcap_{k=1}^{\infty} T^{-k}A^c\Big) = \emptyset.$$

Thus Poincaré's recurrence theorem implies $\mu(B) = 0$.

We now show that almost every $x \in A$ is infinitely recurrent to A. We have

$$A \subset \bigcup_{n=1}^{\infty} T^{-n}A \cup B \tag{2.1}$$

with the above null set B. For each $k \in \mathbb{N}$ it follows that

$$T^{-1}A \subset \bigcup_{n=2}^{\infty} T^{-n}A \cup T^{-1}B, \ \ldots, T^{-(k-1)}A \subset \bigcup_{n=k}^{\infty} T^{-n}A \cup T^{-(k-1)}B.$$

Inserting this into (2.1) leads to

$$A \subset \bigcup_{n=k}^{\infty} T^{-n} A \cup N_k$$

for every k, where each N_k is a null set. Since $\cup_{k=1}^{\infty} N_k$ is again a null set, almost every $x \in A$ is infinitely recurrent to A. ■

The following strengthening of Poincaré's theorem was discovered by M. Kac in 1947, see [599].

Theorem 2.5 (Kac) *Let (X, μ, T) be a measure-preserving system and let $A \subset X$ be measurable. Then*

$$\int_A n_A \, \mathrm{d}\mu = \mu\Big(\bigcup_{k \in \mathbb{N}_0} T^{-k} A \Big).$$

Proof For $k \in \mathbb{N}$ we set $A_k := [n_A = k] \cap A$. Then the sets A_k are pairwise disjoint, $A = \bigcup_{k \in \mathbb{N}} A_k$ (up to null sets) by Corollary 2.4 and $A \cap \bigcap_{k=1}^{n-1} T^{-k} A^c = \bigcup_{k \geq n} A_k$ (up to null sets) for every $n \in \mathbb{N}$. It is easy to see by induction that for each $N \in \mathbb{N}$

$$1 = \sum_{n=1}^{N} \mu\Big(A \cap \bigcap_{k=1}^{n-1} T^{-k} A^c \Big) + \mu\Big(\bigcap_{k=0}^{N-1} T^{-k} A^c \Big).$$

Whence we conclude

$$\begin{aligned}\mu\Big(\bigcup_{k=0}^{\infty} T^{-k} A \Big) &= \sum_{n=1}^{\infty} \mu\Big(\bigcup_{k \geq n} A_k \Big) = \sum_{n=1}^{\infty} \sum_{k=n}^{\infty} \mu(A_k) \\ &= \sum_{k=1}^{\infty} \sum_{n=1}^{k} \mu(A_k) = \sum_{k=1}^{\infty} k \mu(A_k) = \int_A n_A \, \mathrm{d}\mu,\end{aligned}$$

and that was to be proved. ■

The following lemma helps to prove the measure-preserving property of a transformation.

Lemma 2.6 *Let $(X, \mathcal{X}, \mu)$ be a probability space, and let $\mathcal{E}$ be a generator of $\mathcal{X}$ which is closed under $\cap$. Assume that $T : X \to X$ is a measurable transformation satisfying $\mu(T^{-1}B) = \mu(B)$ for every $B \in \mathcal{E}$. Then T is measure-preserving.*

Proof Define $\mathcal{D} := \{B : \ B \in \mathcal{X}, \ \mu(T^{-1}B) = \mu(B)\}$. It is easy to see that $\mathcal{D}$ has the following three properties.

(1) $X \in \mathcal{D}$.
(2) If $A \in \mathcal{D}$, then $A^c \in \mathcal{D}$.
(3) If $A_n \in \mathcal{D}$ for each $n \in \mathbb{N}$ and these sets are pairwise disjoint, then $\bigcup_{n \in \mathbb{N}} A_n \in \mathcal{D}$.

Since by assumption $\mathcal{E} \subset \mathcal{D}$, Dynkin's theorem, see Theorem 1.14, implies that the σ-algebra $\mathcal{X}$ generated by $\mathcal{E}$ is contained in $\mathcal{D}$. ■

Before presenting another useful characterization of the measure-preserving property we make an elementary but important remark.

Remark 2.7 For a map $T : X \to Y$, a subset $A \subset Y$ and its characteristic function $\mathbf{1}_A$ the simple but very useful relation

$$\mathbf{1}_A \circ T = \mathbf{1}_{T^{-1}A} \tag{2.2}$$

holds.

Proposition 2.8 *Let (X, μ), (Y, ν) be probability spaces and let $T : X \to Y$ be measurable.*

(a) *For every bounded (or positive) measurable function $f : Y \to \mathbb{C}$*

$$\int_X f \circ T \, \mathrm{d}\mu = \int_Y f \, \mathrm{d}T_*\mu.$$

(b) *The transformation T is measure-preserving if and only if*

$$\int_X f \circ T \, \mathrm{d}\mu = \int_Y f \, \mathrm{d}\nu \tag{2.3}$$

for every bounded, measurable function $f : Y \to \mathbb{C}$. In this case, (2.3) *holds for every $f \in \mathscr{L}^1(Y, \nu)$.*

Proof (a) By relation (2.2) for characteristic functions and linearity we obtain the statement for simple functions. If $f : Y \to \mathbb{C}$ is a bounded (or positive) measurable function, then there is a sequence $(f_n)_{n \in \mathbb{N}}$ of simple functions converging pointwise to f with $|f_n| \le |f_{n+1}| \le |f|$ for every $n \in \mathbb{N}$. We obtain, by Lebesgue's or Beppo Levi's theorem, that

$$\int_X f \circ T \, \mathrm{d}\mu = \lim_{n \to \infty} \int_X f_n \circ T \, \mathrm{d}\mu = \lim_{n \to \infty} \int_Y f_n \, \mathrm{d}T_*\mu = \int_Y f \, \mathrm{d}T_*\mu.$$

(b) Suppose (2.3) is satisfied for every bounded, measurable function. Let $A \subset Y$ be measurable and consider $f := \mathbf{1}_A \in \mathrm{L}^\infty(Y, \nu)$. Then by the hypothesis

$$\nu(A) = \int_Y \mathbf{1}_A \, \mathrm{d}\nu = \int_X \mathbf{1}_A \circ T \, \mathrm{d}\mu = \int_X \mathbf{1}_{T^{-1}A} \, \mathrm{d}\mu = \mu(T^{-1}A),$$

thus T is measure-preserving.

Next, suppose that T is measure-preserving, i.e., $T_*\mu = \nu$. Each $f \in \mathrm{L}^1(Y, \mu)$ can be decomposed into real and imaginary, and then into positive and negative parts, all of which are positive and belong to $\mathscr{L}^1(Y, \nu)$. By part (a) and linearity we obtain (2.3). ∎

The following definition of invertibility is natural, although it might appear somewhat technical at first glance.

Definition 2.9

(a) Let (X, μ) and (Y, ν) be two probability spaces. A measurable map $\theta : X \to Y$ is called **essentially invertible** (or **point invertible**) if there are measurable sets $X' \subset X$ and $Y' \subset Y$ of full measure such that $\theta : X' \to Y'$ is bijective and $\theta^{-1} : Y' \to X'$ is measurable (with respect to the trace σ-algebras).
(b) A measure-preserving system (X, μ, T) is called **point invertible** if T is essentially invertible.
(c) Two measure-preserving systems (X, μ, T) and (Y, ν, S) are called **point isomorphic** if there is an essentially invertible map $\theta : X \to Y$ such that $S \circ \theta = \theta \circ T$ μ-a.e.

Remark 2.10

(a) A measurable map $\theta : X \to Y$ is essentially invertible if and only if there exists a measurable map $\eta : Y \to X$ with $\eta \circ \theta = \mathrm{id}_X$ a.e. and $\theta \circ \eta = \mathrm{id}_Y$ a.e., see Exercise 2.1. In this case η is called the **essential inverse** of θ. Two essential inverses of θ coincide except on a null set, and we denote them by the same symbol θ^{-1}.
(b) If a measure-preserving system (X, μ, T) is essentially invertible, then the transformation T^{-1} is automatically μ-preserving as well and thus gives rise to a new measure-preserving system (X, μ, T^{-1}) called the (essential) **inverse system**. Intuitively, passing from T to T^{-1} corresponds to reversing the time.

2.2 Examples

We present here some basic examples of measure-preserving systems. This list of examples will be augmented during this book.

Example 2.11 (Finite State Space) Let X be a finite non-empty set and let μ be the normalized counting measure on X, i.e.,

$$\mu(A) = \frac{|A|}{|X|},$$

where $|\cdot|$ stands for the cardinality of the given set. Then a transformation $T : X \to X$ is μ-preserving if and only if T is a bijection, in which case T is essentially invertible.

Example 2.12 (Torus Rotation) Consider the unit circle

$$\mathbb{T} := \{z \in \mathbb{C} : |z| = 1\},$$

which is a compact subset of $\mathbb{C}$. On the Borel σ-algebra $\mathcal{B}(\mathbb{T})$ of $\mathbb{T}$ we consider the normalized arclength-measure m, for which $\mathrm{m}(\mathbb{T}) = 1$. Note that $\mathbb{T}$ with the multiplication of complex numbers is a compact, commutative group (see Example 2.14 for the terminology), and the measure m is invariant under each left rotation

$$\tau_a : \mathbb{T} \to \mathbb{T}, \quad x \mapsto ax.$$

For any $a \in \mathbb{T}$ we therefore obtain the essentially invertible measure-preserving system $(\mathbb{T}, \mathrm{m}, \tau_a)$, abbreviated as $(\mathbb{T}, \mathrm{m}, a)$. Note that m is the only probability measure which is invariant under all rotations τ_a simultaneously. More precisely, if $a \in \mathbb{T}$ is **rational** (i.e., a root of unity) there are many τ_a-invariant measures different from m, while for a **irrational** (i.e., not a root of unity) the only τ_a-invariant probability measure is m. See Exercise 2.2.

The additive analogue of the previous example is the following.

Example 2.13 (Translation Mod 1) Let $\alpha \in [0, 1)$. Consider $[0, 1)$ equipped with the Lebesgue measure λ and $T_\alpha : [0, 1) \to [0, 1)$ given by

$$T_\alpha x = x + \alpha \mod 1,$$

where we use the usual notation $y \bmod 1 := y - \lfloor y \rfloor$. Then $([0, 1), \lambda, T_\alpha)$ is an essentially invertible measure-preserving system, abbreviated as $([0, 1), \lambda, \alpha)$.

This translation system is isomorphic to the rotation system $(\mathbb{T}, \mathrm{m}, a)$ from Example 2.12 via the map $e : [0, 1) \to \mathbb{T}$ given by $e(x) := \mathrm{e}^{2\pi \mathrm{i} x}$ for $x \in [0, 1)$.

These two examples are special instances from the following important class.

Example 2.14 (Group Rotation) Consider a group G which is also equipped with a topology such that the multiplication

$$G \times G \to G, \quad (x, y) \mapsto xy$$

and the inversion

$$G \to G, \quad x \mapsto x^{-1}$$

are continuous. Then G is called a **topological group**. For each $a \in G$, as in Example 2.12, we can consider the left rotation

$$\tau_a : G \to G, \quad x \mapsto ax.$$

Since the group is not assumed to be commutative, it makes sense to define the right rotation as well:

$$\rho_a : G \to G, \quad x \mapsto xa.$$

These are all continuous, hence measurable mappings with respect to the Borel σ-algebra $\mathcal{B}(G)$. If G is a **compact topological group** (or a **compact group** for short), i.e., if G is in addition a compact topological space, such as $\mathbb{T}$, $\mathbb{T}^d$ or their additive analogues, then there is a unique regular probability measure on $\mathcal{B}(G)$ invariant under each left and right rotation, see, e.g, [342, Thm. G.10] or [516, Ch. XI]. This measure is called the **Haar measure** of G and denoted by m_G. Moreover, m_G is invariant under the inverse mapping. In Chap. 11 we shall provide a little bit more details about Haar measures, even on locally compact groups.

Thus, for each $a \in G$ we obtain two essentially invertible measure-preserving systems $(G, \mathrm{m}_G, \tau_a)$ and $(G, \mathrm{m}_G, \rho_a)$. Note that the systems $(G, \mathrm{m}_G, \rho_a)$ and $(G, \mathrm{m}_G, \tau_{a^{-1}})$ are isomorphic via the inverse mapping. The measure-preserving systems corresponding to the left rotations τ_a are abbreviated as (G, m_G, a).

Important special cases of group rotations are the higher dimensional torus rotations $(\mathbb{T}^d, \mathrm{m}^d, a)$, where $\mathrm{m}^d = \mathrm{m} \times \cdots \times \mathrm{m}$ is the product measure on $\mathcal{B}(\mathbb{T}^d)$ and $a \in \mathbb{T}^d$.

Example 2.15 (Skew Rotation and Skew Shift) Consider a torus rotation system $(\mathbb{T}, \mathrm{m}, a)$. Since m is invariant under each (left = right) rotation τ_x, the transformation

$$T_a : \mathbb{T}^2 \to \mathbb{T}^2, \quad (x, y) \mapsto (ax, xy)$$

is m^2-preserving. Indeed, by Fubini's theorem, for every bounded, measurable function $f : \mathbb{T}^2 \to \mathbb{C}$

$$\begin{aligned}\int_{\mathbb{T}^2} f(ax, xy)\, \mathrm{d}x\, \mathrm{d}y &= \int_{\mathbb{T}} \Big(\int_{\mathbb{T}} f(ax, xy)\, \mathrm{d}y \Big)\, \mathrm{d}x = \int_{\mathbb{T}} \Big(\int_{\mathbb{T}} f(ax, y)\, \mathrm{d}y \Big)\, \mathrm{d}x \\ &= \int_{\mathbb{T}} \Big(\int_{\mathbb{T}} f(ax, y)\, \mathrm{d}x \Big)\, \mathrm{d}y = \int_{\mathbb{T}^2} f(x, y)\, \mathrm{d}x\, \mathrm{d}y\end{aligned}$$

holds, so by Proposition 2.8 T_a is measure-preserving. (Integration with respect to the Haar measure is often denoted as above.) We thus obtain the measure-preserving system $(\mathbb{T}^2, \mathrm{m}^2, T_a)$, called a **skew rotation**.

The additive version of it is a **skew shift** on $[0, 1)^2$ with the 2-dimensional Lebesgue (= Haar) measure given by

$$(x, y) \mapsto (x + \alpha \bmod 1, x + y \bmod 1)$$

for a fixed $\alpha \in [0, 1)$.

A fundamentally different class of examples is described next.

Example 2.16 (Shifts) Let (Y, ν) be a probability space and consider the infinite product $X = Y^{\mathbb{Z}}$, $\mathcal{X}$ the product σ-algebra generated by **cylinder sets**, i.e., sets of the form

$$A = \cdots \times Y \times Y \times A_{-N} \times \cdots \times A_0 \times A_1 \times \cdots A_N \times Y \times \cdots \times Y \times \cdots$$

with measurable $A_{-N}, \ldots, A_N \subset Y$, and product measure μ given by

$$\mu(A) = \prod_{j=-N}^{N} \nu(A_j)$$

on such cylinder sets. Recall that μ is uniquely determined by this property, see Corollary 1.17. Define now the left shift

$$T : X \to X, \quad (x_n)_{n\in\mathbb{Z}} \mapsto (x_{n+1})_{n\in\mathbb{Z}}.$$

Then (X, μ, T) is an essentially invertible measure-preserving system, which we call the **two-sided shift with state space** (Y, ν) (use Lemma 2.6). If one replaces in the above the set of integers by the natural numbers, and modifies the definition of cylinder sets accordingly, one obtains the **one-sided shift with state space** (Y, ν). Apart from the trivial cases, one-sided shifts are not essentially invertible, see Exercise 2.4.
Let Y be a separable metric space. Then there exists a bounded metric d inducing its topology. So in this case if $\mathcal{Y} = \mathcal{B}(Y)$, then X is also a metric space with the (bounded) metric $d_X(x, y) := \sum_{n=-\infty}^{\infty} 2^{-|n|} d(x_n, y_n)$, and the product σ-algebra $\mathcal{X}$ is the Borel σ-algebra. We leave the proof as Exercise 2.3.

The particular case when Y is finite has its own name.

Example 2.17 (Bernoulli Shift) If $Y = \{0, 1, \ldots, k-1\}$ for some fixed $k \in \mathbb{N}$ and $\nu(\{j\}) = p_j$ for some probability vector $p \in \mathbb{R}^k$, then the corresponding shift from the previous Example 2.16 is called a **Bernoulli shift** and is denoted by $B(p_0, \ldots, p_{k-1})$ and the corresponding measure is called the **Bernoulli measures** associated with p. Of course, one can consider the analogous example of the shift on $Y^{\mathbb{N}}$. Here, Y is called the **alphabet** with k **letters**, and elements of $X = Y^{\mathbb{Z}}$ (or $X = Y^{\mathbb{N}}$, respectively) are called **infinite words**.

Example 2.18 (Bernoulli Shift and the Doubling Map) Consider the one-sided Bernoulli shift $B(1/2, 1/2)$ from Example 2.17 and the interval $[0, 1]$ with the Lebesgue measure λ. The **doubling map** T on $[0, 1]$ given by $Tx := 2x \bmod 1$ preserves λ. The map $\theta : \{0, 1\}^{\mathbb{N}_0} \to [0, 1]$ given by

$$\theta(t_0, t_1, \ldots) := \sum_{n=0}^{\infty} \frac{t_n}{2^{n+1}}$$

corresponds to the binary representation of numbers and provides an isomorphism between the Bernoulli shift and the measure-preserving system $([0, 1], \lambda, T)$, see Exercise 2.5. (Intuitively it is clear that shifting the binary representation of a number to the left corresponds to the doubling the number modulo 1). Note that the isomorphism θ is not bijective due to the fact that the binary representation is unique for almost all but not all numbers.

Remark 2.19 It was a long standing open question in ergodic theory whether the (two-sided) Bernoulli shifts $B(\frac{1}{n}, \dots, \frac{1}{n})$ and $B(\frac{1}{m}, \dots, \frac{1}{m})$ are non-isomorphic whenever $n \neq m$. This was confirmed by Kolmogorov only in 1958 using the notion of the measure-theoretic entropy (see [656]), now known as the Kolmogorov–Sinai entropy. However, contrary to what the first impression would tell, the Bernoulli shifts $B(\frac{1}{4}, \frac{1}{4}, \frac{1}{4}, \frac{1}{4})$ and $B(\frac{1}{2}, \frac{1}{8}, \frac{1}{8}, \frac{1}{8}, \frac{1}{8})$ are isomorphic. This was proven by Mešalkin in 1959, see [789]. This isomorphism problem for essentially invertible Bernoulli shifts was completely settled by Ornstein in 1970 who showed that two two-sided Bernoulli shifts are isomorphic if and only if they have the same Kolmogorov–Sinai entropy, see [834]. The interested reader can consult, for example, the survey [1039] by Weiss.

Example 2.20 (Markov Shifts) Again, we consider, for $k \in \mathbb{N}$, $X = \{0, \dots, k-1\}^{\mathbb{N}}$ with the product σ-algebra $\mathcal{X}$. Let M be a $k \times k$ **row-stochastic matrix** with entries m_{ij}, $i, j = 0, \dots, k-1$, i.e., its rows are probability vectors. By Perron's theorem (see, e.g., [82, Thm. 5.6] and [149, Thm. 8.3.11]) there is a probability vector $p = (p_j)$ with $M^{\top} p = p$. As in Example 2.16 (and Example 2.17) we can find a measure μ on $\mathcal{X}$ with the following prescribed values on cylinder sets $A = \{j_0\} \times \dots \times \{j_n\} \times \prod_{j=n+1}^{\infty} \{0, \dots, k-1\}$:

$$\mu(A) = p_{j_0} m_{j_0 j_1} \cdots m_{j_{n-1} j_n}.$$

It is Exercise 2.13 to work out the details and to show that the so constructed μ is in fact an invariant measure for the shift $T : X \to X$. It is called the **Markov measure** corresponding to the **transition matrix** M and the probability vector p.

The following shows the connection between measure-preserving shift systems and stationary sequences.

Example 2.21 (Stationary Sequences) Let $(\Omega, \mathbb{P})$ be a probability space and for $j \in \mathbb{Z}$ let $f_j : \Omega \to \mathbb{R}$ be measurable functions, i.e., **random variables**. The sequence $(f_j)_{j \in \mathbb{Z}}$ is called a **stochastic process** and we suppose that it is **stationary**, i.e., that for every $m \in \mathbb{N}$, every $n_1, \dots, n_m \in \mathbb{Z}$, all Borel sets $B_1, \dots, B_m \subset \mathbb{R}$ and every $k \in \mathbb{Z}$ one has

$$\mathbb{P}\Big(\bigcap_{j=1}^{m} f_{n_j}^{-1}(B_j)\Big) = \mathbb{P}\Big(\bigcap_{j=1}^{m} f_{n_j+k}^{-1}(B_j)\Big). \tag{2.4}$$

Such $(f_j)_{j\in\mathbb{Z}}$ is called a **stationary sequence**.[2]

There is a natural way to produce (many) stationary sequences from a point invertible measure-preserving system. Let (X, μ, T) be such a measure-preserving system and let $f : X \to \mathbb{R}$ be measurable. Define $f_j := f \circ T^j$ for $j \in \mathbb{Z}$. Then the sequence $(f_j)_{j\in\mathbb{Z}}$ is a stationary sequence. Indeed, by the measure-preserving property of T

$$\mu\Big(\bigcap_{j=1}^{m} f_{n_j}^{-1}(B_j)\Big) = \mu\Big(\bigcap_{j=1}^{m} T^{-n_j}(f^{-1}(B_j))\Big)$$
$$= \mu\Big(T^{-k}\bigcap_{j=1}^{m} T^{-n_j}(f^{-1}(B_j))\Big) = \mu\Big(\bigcap_{j=1}^{m} f_{n_j+k}^{-1}(B_j)\Big)$$

holds for every $m \in \mathbb{N}$, every $n_1, \ldots, n_m \in \mathbb{Z}$, all Borel sets $B_1, \ldots, B_m \subset \mathbb{R}$ and every $k \in \mathbb{Z}$. Thus a measure-preserving system gives rise to a stationary sequence. It turns out that this construction can be reversed, see e.g., Doob [314, pp. 452–457]. Let $(\Omega, \mathbb{P})$ be a probability space, where $\mathbb{P}$ is a (regular) Borel measure on a compact metric space Ω (such spaces are called **regular probability spaces**), and let $(f_j)_{j\in\mathbb{Z}}$ be a stationary sequence on it. Then there is a (in a certain sense unique) $\mathbb{P}$-preserving transformation T on $(\Omega, \mathbb{P})$ such that $f_j = f_0 \circ T^j$ for $j \in \mathbb{Z}$, i.e., the stationary sequence arises from the measure-preserving system $(\Omega, \mathbb{P}, T)$ in the previously described manner. Note that [314, Sec. IV.1] constructs not a point-transformation but a transformation T on the measure-algebra (i.e., a set-transformation modulo null sets); by the regularity assumption the existence of a point-transformation can be assured; cf. Sect. 8.5 below. Note that both constructions work if we replace $\mathbb{Z}$ by $\mathbb{N}$ or $\mathbb{N}_0$.

Thus we see that (for regular probability spaces) every stationary sequence corresponds to a measure-preserving system with a distinguished measurable function *on the same space*. The following concrete construction provides correspondence between stationary sequences and concrete systems, namely shift systems, where the set, the transformation (and the function) are always the same. This shows that to study stationary sequences is the same as to study shift-invariant probability measures on $\mathbb{R}^{\mathbb{Z}}$. Note that also here one can replace $\mathbb{Z}$ by $\mathbb{N}$ or $\mathbb{N}_0$.

Example 2.22 (Shifts and Stationary Sequences) Let $(\Omega, \mathbb{P})$ be a probability space and let $f_j : \Omega \to \mathbb{R}$, $j \in \mathbb{Z}$ form a stationary sequence. Consider $X := \mathbb{R}^{\mathbb{Z}}$ with the Borel algebra (i.e., the product σ-algebra with $\mathcal{B}(\mathbb{R})$ in each of the components) and define $f : \Omega \to X$ by

$$f(\omega) := (\ldots, f_{-1}(\omega), f_0(\omega), f_1(\omega)\ldots), \text{ i.e., } (f(\omega))_j = f_j(\omega).$$

[2] In Doob's book such sequences are called **strictly stationary**, see [314, page 94].

For a Borel set $A \subset X$ set $\mu(A) := \mathbb{P}(f^{-1}(A))$, i.e., μ is the push-forward probability measure corresponding to f. Consider finally the left shift on $\mathbb{R}^{\mathbb{Z}}$ given by $T(x_j)_{j\in\mathbb{Z}} := (x_{j+1})_{j\in\mathbb{Z}}$. Then (X, μ, T) is a measure-preserving system. To see this take a cylinder set $B \subset \mathbb{R}^{\mathbb{Z}}$ and observe that $f^{-1}(B)$ is of the form $\bigcap_{j=1}^{m} f_{n_j}^{-1}(B_j)$. Thus the stationarity property (2.4) together with the definition of μ implies

$$\mu(B) = \mathbb{P}(f^{-1}(B)) = \mathbb{P}(f^{-1}(T^{-1}B)) = \mu(T^{-1}B),$$

i.e., T is indeed μ-preserving. Moreover, for the projection $\pi_0 : X \to \mathbb{R}$ defined by $\pi_0(x_j)_{j\in\mathbb{Z}} := x_0$ we have

$$\pi_0((T^j f)(\omega)) = (f(\omega))_j = f_j(\omega)$$

for every $j \in \mathbb{R}$.

Conversely, let $(\mathbb{R}^{\mathbb{Z}}, \mathcal{B}(\mathbb{R}^{\mathbb{Z}}), \mu, T)$ be a measure-preserving system, where T denotes the left shift on $\mathbb{R}^{\mathbb{Z}}$. For $j \in \mathbb{Z}$ denote $\pi_j : \mathbb{R}^{\mathbb{Z}} \to \mathbb{R}$ the projection onto the jth coordinate. From Example 2.21 we obtain the stationary sequence $(\pi_j)_{j\in\mathbb{Z}}$ (note that $\pi_j = \pi_0 \circ T^j$ for $j \in \mathbb{Z}$). Finally, $(\mathbb{R}^{\mathbb{Z}}, \mu, T)$ arises from $(\pi_j)_{j\in\mathbb{Z}}$ as in the first part of this example.

It turns out that, via a shift construction, each measure-preserving system can be *extended* to an essentially invertible one. The following covers a particular case of this general statement.

Proposition 2.23 (Invertible Point Extension) *Let (X, μ, T) be a measure-preserving system such that $TX = X$. Then there is a point invertible measure-preserving system (Y, ν, S) and a (surjective) measure-preserving map $\pi : Y \to X$ such that*

$$\pi \circ S = T \circ \pi \textit{ almost everywhere.} \tag{2.5}$$

We leave the proof of this result as Exercise 2.8, but refer also to [336, Exercise 2.1.7] and [260, Sec. 10.4]. The property stated in (2.5) for a measure-preserving map π defines the notion of a point factor (and point extension). In this situation we say that (X, μ, T) is a **point factor** of (Y, ν, S) and in turn (Y, ν, S) is a **point extension** of (X, μ, T). In Sect. 8.5 we shall return to invertible extensions, after having discussed a more general notion of factors and extensions in Chap. 8.

Next we turn to one of the many connections to number theory.

Example 2.24 (Gauss Transformation) Let $X := [0, 1)$ and let $T : X \to X$ denote the **Gauss transformation** defined by $T(0) := 0$ and

$$Tx := \tfrac{1}{x} - \left\lfloor \tfrac{1}{x} \right\rfloor = \tfrac{1}{x} \bmod 1 \text{ for } x \neq 0.$$

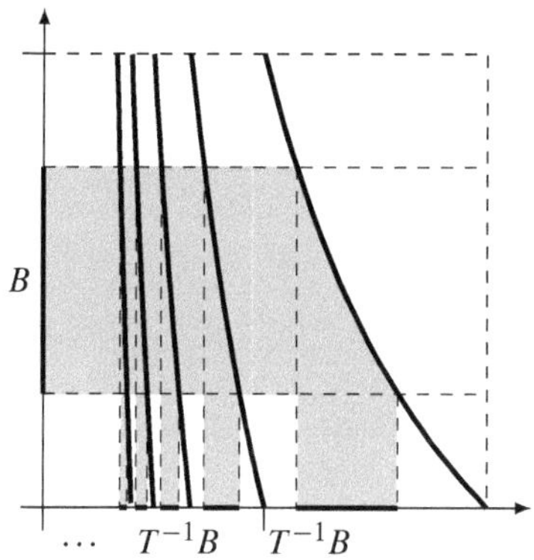

We see that $Tx = \frac{1}{x} - n$ whenever $x \in (\frac{1}{n+1}, \frac{1}{n}]$ and therefore

$$T^{-1}\{y\} = \left\{\tfrac{1}{y+n} : n \in \mathbb{N}\right\} \text{ for } y \in (0, 1). \tag{2.6}$$

We finally define the measure μ by

$$\mu(A) := \frac{1}{\log 2} \int_A \frac{\mathrm{d}x}{x+1}, \qquad A \in \mathcal{B}([0, 1)).$$

It is Exercise 2.6 to show that (X, μ, T) is a measure-preserving system, called the **Gauss system**.

This system is directly connected, by Proposition 2.25 below, to simple continued fractions which we briefly discuss here. A **(simple) continued fraction** is a formal expression of the form

$$[a_0; a_1, a_2, \ldots] := a_0 + \cfrac{1}{a_1 + \cfrac{1}{a_2 + \cdots}}$$

with $a_0 \in \mathbb{Z}$ and (a_n) a finite or infinite sequence in $\mathbb{N}$. Continued fractions lead to a way of representing a real number as a sequence of integers different from the standard expansion using base 10 (or any other base). For a given $x \in \mathbb{R}$ the corresponding continued fraction is obtained by the following recursive procedure.

We set $a_0 := \lfloor x \rfloor$, $r_0 := x - a_0$. Suppose $a_0 \in \mathbb{Z}, a_1, \ldots, a_n \in \mathbb{N}$ and $r_0, \ldots, r_n \in [0, 1)$ have been defined. If $r_n = 0$, we stop and the construction of the continued fraction $[a_0; a_1, \ldots, a_n]$ is finished. Otherwise, define $a_{n+1} := \lfloor \frac{1}{r_n} \rfloor$ and $r_{n+1} := \frac{1}{r_n} - a_{n+1}$, and continue repeating this procedure. In this way we obtain a finite or infinite sequence of numbers a_n with $a_0 \in \mathbb{Z}$ and $a_1, a_2, \ldots \in \mathbb{N}$ and write $x = [a_0; a_1, a_2, \ldots]$. For example, the golden ratio ϕ satisfies $\phi = [1; 1, 1, 1, \ldots]$ (Exercise 2.6). Already the ancient Greeks (notably Euclid, Archimedes of Syracuse and Diophantos of Alexandria) knew about these miraculous objects, but the first treatment that went beyond sheer numerical tricks is due to R. Bombelli [170]. The systematic study was started by P. Cataldi [227], and he is considered by many as *the real discoverer.* A number of giants of mathematics were involved in the development of the theory of continued fractions, such as Euler, Lagrange, Galois, and, of course, Gauss. Here we list some properties of continued fractions to give a flavour of their theory. The interested reader is referred, e.g., to [644] and [208].

(1) A real number x is rational if and only if its continued fraction representation is finite, see Exercise 2.6. Moreover, the continued fraction of x is eventually periodic if and only if x is a quadratic irrational.
(2) Conversely, every infinite sequence $(a_j)_{j\in\mathbb{N}_0}$ with $a_0 \in \mathbb{Z}$ and $a_1, a_2, \ldots \in \mathbb{N}$ leads to a continued fraction of an irrational number. Thus we have a bijection between irrational numbers and infinite continued fractions. Note that the representation of rational numbers is not unique since

$$[a_0; a_1, a_2, \ldots, a_n + 1] = [a_0; a_1, a_2, \ldots, a_n, 1].$$

(3) Let $x = [a_0; a_1, a_2, \ldots]$. Define the nth **convergent** of the continued fraction as the reduced fraction

$$\frac{p_n}{q_n} := a_0 + \cfrac{1}{a_1 + \cfrac{1}{a_2 + \cdots + \cfrac{1}{a_n}}},$$

p_n, q_n being coprime. Then one has

$$\frac{p_2}{q_2} \leq \frac{p_4}{q_4} \leq \cdots \leq x \leq \cdots \leq \frac{p_3}{q_3} \leq \frac{p_1}{q_1}$$

with $\frac{p_{2n}}{q_{2n}} \nearrow x$ and $\frac{p_{2n+1}}{q_{2n+1}} \searrow x$. For every $n \in \mathbb{N}$, the rational number $\frac{p_n}{q_n}$ satisfies

$$\left| x - \frac{p_n}{q_n} \right| \leq \frac{1}{q_n^2}$$

and is the best approximation of x by rationals with denominator less than or equal to q_n. Generally, bounded continued fractions (i.e., bounded sequences $(a_n)_{n\in\mathbb{N}_0}$) correspond to so-called *badly approximable* numbers, the golden ratio representing the worst case.

The next proposition explains the relation of continued fractions to the Gauss transformation, see Exercise 2.6.

Proposition 2.25 *Consider the Gauss system* $([0, 1), \mu, T)$, *and let* $x \in [0, 1) \setminus \mathbb{Q}$ *have the continued fraction representation* $x = [a_0; a_1, a_2, \ldots]$. *Then for each* $n \in \mathbb{N}$

$$a_n = \left\lfloor \frac{1}{T^{n-1}x} \right\rfloor.$$

An analogous statement holds for rational numbers as well.

2.3 The Koopman Operator

We now associate to each measure-preserving system a *linear* operator.

Definition 2.26 Let (X, μ, T) be a measure-preserving system. For a measurable function $f : X \to \mathbb{C}$ define

$$S_T f := f \circ T.$$

Then $S_T f$ is again a measurable function and S_T defines a linear operator on the vector space $\mathscr{L}^0(X)$ of measurable functions on X. The operator S_T acts actually on $\sim_\mu$-equivalence classes of functions, and hence defines a linear operator $S_T : \mathrm{L}^0(X, \mu) \to \mathrm{L}^0(X, \mu)$, called the **Koopman operator** (induced by T). Here and later, $\mathrm{L}^0(X, \mu)$ denotes the set of $\sim_\mu$-equivalence classes of measurable functions on X.

In the same manner, if $T : X \to Y$ is measure-preserving for the probability spaces (X, μ) and (Y, ν), a linear operator $S_T : \mathrm{L}^0(Y, \nu) \to \mathrm{L}^0(X, \mu)$ is defined by $S_T f := f \circ T$. Again S_T can be called the Koopman operator induced by T. By Proposition 2.8 the operator $S_T : \mathrm{L}^p(Y, \nu) \to \mathrm{L}^p(X, \mu)$ is a linear isometry for every $p \in [1, \infty)$. The same holds also for $p = \infty$, and the Koopman operator has the following important algebraic properties, see Exercise 2.10.

Proposition 2.27 (Properties of Koopman Operators) *For probability spaces (X, μ), (Y, ν), a measure-preserving transformation $T : X \to Y$, and functions $f, g \in \mathrm{L}^0(Y, \nu)$ the following assertions hold.*

(a) *$f \geq g$ implies $S_T f \geq S_T g$. In particular, $f \geq 0$ implies $S_T f \geq 0$ (positivity).*
(b) *$S_T(f \cdot g) = S_T f \cdot S_T g$ (multiplicativity).*
(c) *$S_T|f| = |S_T f|$.*
(d) *$S_T \mathbf{1} = \mathbf{1}$. More generally, $S_T \mathbf{1}_A = \mathbf{1}_{T^{-1}A}$ for every measurable $A \subset Y$.*
(e) *If T is essentially invertible, with inverse T^{-1}, then $S_T : \mathrm{L}^p(Y, \nu) \to \mathrm{L}^p(X, \mu)$ is an isometric isomorphism with inverse $S_{T^{-1}}$. In particular, in this case for $p = 2$ the operator S_T is unitary.*

A measure-preserving system (X, μ, T) shall be called **invertible** (or **Markov invertible**) if its Koopman operator T on $\mathrm{L}^p(X, \mu)$ is invertible for some or, equivalently,[3] all $p \in [1, \infty]$. So every essentially invertible system is invertible, while the converse is false in general, see Exercise 2.11. It turns out that this notion

[3] See Exercise 2.11.

of invertibility, referring to functions instead of points, is handier for the study of ergodic theorems. More information about invertibility and its precise connection to essential invertibility will be given in Chap. 8.

Remark 2.28 (Perron–Frobenius Operators) We briefly mention another related operator that can be attached to transformations between measure spaces, and refer to the book [304, Ch. 4] by Ding and Zhou or to Sarig's lecture notes [906] for details and applications. Suppose T is a non-singular transformation on a σ-finite measure space (X, μ) (for example T is a measure-preserving transformation on a probability space, as usually assumed in this book). For $f \in \mathrm{L}^1(X, \mu)$ consider the measure $T_*(f \cdot \mu)$ which is absolutely continuous with respect to μ, and denote the corresponding Radon–Nikodym derivative by Pf. It is easy to see that P is a positive linear operator on $\mathrm{L}^1(X, \mu)$, called the **Perron–Frobenius operator** or **transfer operator** of T. This operator P is thus characterized by the property

$$\int_A Pf \,\mathrm{d}\mu = \int_{T^{-1}A} f \,\mathrm{d}\mu \text{ for each measurable } A \subset X \text{ and } f \in \mathrm{L}^1(X, \mu).$$

The adjoint P' of P on $\mathrm{L}^\infty(X, \mu)$ is the restriction of the Koopman operator S_T onto[4] $\mathrm{L}^\infty(X, \mu)$, i.e., for $f \in \mathrm{L}^1(X, \mu)$, $g \in \mathrm{L}^\infty(X, \mu)$

$$\int_X Pf \cdot g \,\mathrm{d}\mu = \int_X f \cdot S_T g \,\mathrm{d}\mu.$$

More on adjoints of Koopman operators of measure-preserving transformations will follow in Chap. 8.

We close this section with the following important convention.

Notation 2.29 From now on, following many authors, most of the time we will write T for both a measure-preserving transformation and the induced Koopman operator. Thus we use the convention $(Tf)(x) = f(Tx)$.

An exception will be the Koopman operator of the left (or right) rotation by a on a compact topological group from Example 2.14 which we denote by L_a or R_a, respectively.

[4] Note that since T is non-singular, the Koopman operator $S_T : f \mapsto f \circ T$ leaves $\mathrm{L}^\infty(X, \mu)$ invariant.

2.4 Topological Dynamical Systems and Invariant Measures

An important class of measure-preserving systems arises from continuous maps on compact spaces.

Definition 2.30 For a compact[5] space K and a continuous mapping

$$T : K \to K,$$

the pair (K, T) is called a **topological (dynamical) system**. It is called **invertible** if T is bijective (in which case the inverse of T is automatically continuous).

Note that a continuous map $T : K \to K$ is measurable with respect to the Borel σ-algebra on K.

Example 2.31

(1) If $L = \{0, \dots, k-1\}$ and $K = L^{\mathbb{Z}}$ or $K = L^{\mathbb{N}}$, then K with the product topology is a compact space, and the shift T, $(x_n) \mapsto (x_{n+1})$ from Example 2.17 is a continuous mapping. So (K, T) is a topological system.

(2) For $K = \mathbb{T}$ and $T = \tau_a$, $x \mapsto ax$, from Example 2.12, $(\mathbb{T}, \tau_a)$ is topological system, abbreviated as $(\mathbb{T}, a)$.

(3) More generally, given a compact topological group and $a \in G$, we obtain the topological system (G, τ_a), abbreviated as (G, a).

(4) Let (K, T) be a topological system, let L be a compact space, and let

$$\rho : K \times L \to L$$

be continuous. The mapping

$$T_\rho : K \times L \to K \times L, \quad T_\rho(x, y) := (T(x), \rho(x, y))$$

is continuous, hence we obtain a topological system $(K \times L, T_\rho)$, called the ρ**-skew product** of (K, T) with L.

Let us iterate T_ρ. For $x \in K$ and $y \in L$, we write $\rho_x(y) = \rho(x, y)$ and obtain inductively that

$$T_\rho^n(x, y) = (T^n(x), \rho_{T^{n-1}(x)} \circ \cdots \circ \rho_{T(x)} \circ \rho_x(y)).$$

Define the mapping

$$\widehat{\rho} : \mathbb{N}_0 \times K \to L^L, \qquad \widehat{\rho}(n, x) := \rho_{T^{n-1}(x)} \circ \cdots \circ \rho_{T(x)} \circ \rho_x,$$

[5] Recall that we assume compact spaces to be Hausdorff.

called the **cocycle** of the skew product (in this context, in some parts of the literature, the mapping ρ itself is called the cocyle of the skew product). We have

$$T_\rho^n(x, y) = (T^n(x), \widehat{\rho}(n, x)(y)),$$

and the following relation holds

$$\widehat{\rho}(0, x) = \mathrm{id}, \quad \widehat{\rho}(m + n, x) = \widehat{\rho}(m, T^n(x)) \circ \widehat{\rho}(n, x)$$

for every $n, m \in \mathbb{N}_0$, and $x \in K$. To the analogous example in the measure-preserving setting we shall return in Example 8.2 below.

(5) If in the forgoing example $\rho(x, y) = S(y)$ is independent of $x \in K$, i.e., $T_\rho(x, y) = (T(x), S(y))$ then one arrives at the **product** $(K \times L, S \times T)$. The cocycle satisfies $\widehat{\rho}(n, x) = S^n$ for every $n \in \mathbb{N}_0$ and $x \in K$.

(6) If in the previous example (4) we take $L = G$, a compact group and ρ is given by left rotations, i.e., $\rho(x, g) = \varphi(x)g$ for some continuous function $\varphi : K \to G$, then the skew product $(K \times G, T_\rho)$ is called a compact **group extension**. In fact, the mapping $\mathbb{T}^2 \to \mathbb{T}^2$, $(x, y) \mapsto (ax, xy)$ in Example 2.15 is of this form, and the corresponding system $(\mathbb{T}^2, a)$ is called the (topological) **skew rotation**. For a group extension the cocycle has the form

$$\widehat{\rho}(n, x)(g) = \varphi(T^{n-1}(x)) \cdots \varphi(T(x))\varphi(x)g.$$

The additive version of this example is the **skew shift** $([0, 1)^2, \alpha)$ with $\alpha \in \mathbb{R}$.

(7) Given a topological system (K, T) and $k \in \mathbb{N}$, we call the topological system (K, T^k) the kth **iterate** of (K, T).

More examples of topological systems will appear in later chapters.

Similarly to the case of measure-preserving transformations, a continuous mapping $T : K \to L$ between compact spaces induces a linear contraction

$$S_T : \mathrm{C}(L) \to \mathrm{C}(K), \quad S_T f := f \circ T.$$

For a topological system (K, T), S_T is called the **Koopman operator**. It satisfies the algebraic properties (a)–(c) and the first part of (d) from Proposition 2.27.

Recall that the dual space of $\mathrm{C}(K)$ is identified with the space $\mathrm{M}(K)$ of regular Borel measures on K via the mapping $\mu \mapsto \int_K \cdot \, \mathrm{d}\mu$ (see Theorem 1.19).

Let $T : K \to K$ be a continuous transformation. As before we denote for $\mu \in \mathrm{M}(K)$ the **push-forward measure** by $T_*\mu(B) := \mu(T^{-1}B)$ for $B \in \mathcal{B}(K)$ and call μ **T-invariant** if $T_*\mu = \mu$. Proposition 2.8 (a) implies that for each $f \in \mathrm{C}(K)$

$$\int_K f \, \mathrm{d}T_*\mu = \int_K f \circ T \, \mathrm{d}\mu.$$

From this we conclude

$$\langle f, S_T'\mu\rangle = \langle S_T f, \mu\rangle = \int_K S_T f \,\mathrm{d}\mu = \int_K f \circ T \,\mathrm{d}\mu = \int_K f \,\mathrm{d}T_*\mu = \langle f, T_*\mu\rangle$$

for every $f \in \mathrm{C}(K)$, i.e., $T_*\mu = S_T'\mu$. We thus obtain the following important fact.

Proposition 2.32 *Let (K, T) be a topological system. A measure $\mu \in \mathrm{M}(K)$ is T-invariant if and only if $\mu \in \operatorname{fix} S_T' := \ker(I - S_T')$.*

The following result of Krylov and Bogolyubov [687] shows in particular that every topological system gives rise to at least one measure-preserving system. (Recall that T satisfies $S_T\mathbf{1} = \mathbf{1}$.)

Theorem 2.33 (Krylov–Bogolyubov) *Let (K, T) be a topological system and let $f \in \mathrm{C}(K)$ satisfy $f \neq 0$ and $S_T f = f$. Then there exists a T-invariant probability measure μ on K with $\langle f, \mu\rangle \neq 0$.*

Proof Let $x \in K$ such that $f(x) \neq 0$ and consider the probability measures

$$\nu_n := \frac{1}{n}\sum_{j=0}^{n-1} \delta_{T^j x}.$$

Since the set $\mathrm{M}_1(K)$ of all probability measures on K is weak* compact, by Proposition 1.22 the sequence $(\nu_n)_{n\in\mathbb{N}}$ possesses a cluster point μ which is again a probability measure. For $g \in \mathrm{C}(K)$ we have

$$\begin{aligned}\Big|\langle g \circ T, \nu_n\rangle - \langle g, \nu_n\rangle\Big| &= \Big|\frac{1}{n}\sum_{j=0}^{n-1}\Big(\langle g, \delta_{T^{j+1}x}\rangle - \langle g, \delta_{T^j x}\rangle\Big)\Big| \\ &= \frac{1}{n}\Big|\langle g, \delta_{T^n x}\rangle - \langle g, \delta_x\rangle\Big| \leq \frac{2\|g\|_\infty}{n}.\end{aligned}$$

Since μ is a cluster point, we obtain $|\langle g \circ T, \mu\rangle - \langle g, \mu\rangle| = 0$ for each $g \in \mathrm{C}(K)$, implying that μ is T-invariant. Since $S_T f = f$ and $\langle f, \nu_n\rangle = \langle f, \delta_x\rangle \neq 0$, we conclude $\langle f, \mu\rangle = \langle f, \delta_x\rangle \neq 0$, as required. ■

Remark 2.34 The previous result has a more abstract, functional-analytic generalization: Let S be a (linear) contraction on a Banach space. Then $\operatorname{fix}(S') := \ker(S' - I)$ separates $\operatorname{fix}(S). = \ker(S - I)$. The proof of this fact is not hard, but we only shall return to it when we discuss mean ergodicity in Chap. 7, see Proposition 7.15.

For some topological systems (K, T) there is just one T-invariant probability measure on K, such systems are called **uniquely ergodic**, and for others there are uncountably many of them. (Indeed, every convex combination of two T-invariant probability measures on K is again so.) More discussion on this phenomenon comes in Chap. 3. Here we just revisit two of the above examples of topological systems.

Example 2.35 (Examples of Invariant Measures)

(a) As indicated in Example 2.12 and Exercise 2.2, a topological system of the form $(\mathbb{T}, a)$ has a unique invariant measure (that is m) if and only if a is not a root of unity. In Theorem 3.40 and in the discussion after that we shall see a generalization of this fact.
(b) On the other hand, given $L := \{0, \ldots, k-1\}$ and $K = L^{\mathbb{N}}$, consider the shift system (K, T). Then any **Bernoulli measure** from Example 2.17 is an invariant measure. For any row-stochastic matrix M and probability vector p with $M^{\top} p = p$ the corresponding Markov measure from Example 2.20 is an invariant measure for T.
(c) Consider $K := [0, 1]$ and the mapping T defined on K by $T(x) := x^2$. Then (K, T) is a topological system. The Dirac measures δ_0 and δ_1 are both invariant under T (as well as their convex combinations).

In summary, we see that a topological system (K, T) gives rise to, in general, more than one measure-preserving system (K, μ, T). In Sect. 8.5 we shall briefly discuss that it is, in a certain sense, possible to switch back from the measure-preserving situation to the topological one by looking at so-called topological models.

Also for topological systems (K, T) we will write for the Koopman operator S_T on $\mathrm{C}(K)$ only T, cf. Notation 2.29.

Exercises

Exercise 2.1 (Essential Invertibility) Prove the assertion of Remark 2.10(a).

Exercise 2.2 (Invariant Measures for Rotations)

(a) Let $a \in \mathbb{T}$ be irrational, i.e., not a root of unity. Show that the set $\{1, a, a^2, a^3, \ldots\}$ is dense in $\mathbb{T}$.
(b) Let $a \in \mathbb{T}$ be fixed. Describe all τ_a-invariant measures on $\mathbb{T}$, where τ_a is the rotation by a discussed in Example 2.12. (Hint: Distinguish the cases of rational and irrational a. For irrational a use (a).)

What changes if one replaces the rotation from Example 2.12 by the translation from Example 2.13?

Exercise 2.3 (Shifts) Prove the assertion at the end of Example 2.16.

Exercise 2.4 (Shifts and Bernoulli Shifts) Prove that for $k \geq 2$ and a probability vector $p \in \mathbb{R}^k$ with $p_j > 0$ for each $j \in \{0, \ldots, k-1\}$ the one-sided Bernoulli shift is not essentially invertible. More generally, describe all cases when a one-sided shift is essentially invertible.

Exercise 2.5 (Doubling Map) Prove that the doubling map preserves the Lebesgue measure. Show that the Bernoulli shift and the doubling map from Example 2.18 are isomorphic.

Exercise 2.6 (Gauss System and Continued Fractions)

(a) Show that the Gauss system defined in Example 2.24 is measure-preserving. (Hint: Use (2.6).)
(b) Show that a real number is rational if and only if its continued fraction representation is finite.
(c) Show that the golden ratio satisfies $\phi = [1; 1, 1, 1, \ldots]$. Which number has the representation $[1; 2, 2, 2, \ldots]$?
(d) Prove Proposition 2.25 by induction.

Exercise 2.7 (Baker's (Cut and Paste) Transformation) Consider $X := [0, 1]^2$ and the transformation $T : X \to X$ given by

$$T(x, y) := \begin{cases} (2x, y/2), & x \in [0, \frac{1}{2}), \\ (2x - 1, (y + 1)/2), & x \in [\frac{1}{2}, 1]. \end{cases}$$

(a) What does T do geometrically? Why is it called baker's transformation?
(b) Show that T preserves the two-dimensional Lebesgue measure. This provides another example of a measure-preserving system.

Exercise 2.8 (Invertible Point Extensions, Natural Extension) Let (X, μ, T) be a measure-preserving system with $TX = X$. Let

- $Y := \{y \in X^{\mathbb{Z}} : y_{n+1} = Ty_n \text{ for all } n \in \mathbb{Z}\}$,
- S the left shift on Y, i.e., $S(y_n)_{n\in\mathbb{Z}} = (y_{n+1})_{n\in\mathbb{Z}} = (Ty_n)_{n\in\mathbb{Z}}$,
- $\mathcal{Y}$ the smallest left shift invariant σ-algebra over Y such that the projection maps $\pi_k : Y \to X$ given by $\pi_k(y) := y_k$ for $k \leq 0$ are measurable,
- ν the measure defined by

$$\nu(\{y \in Y : y_{-n_1} \in A_1, \ldots, y_{-n_k} \in A_k\}) := \mu(T^{-n_1}A_1 \cap \cdots \cap T^{-n_k}A_k)$$

  for $n_1, \ldots, n_k \in \mathbb{N}_0$ and extended to $\mathcal{Y}$ (by Carathéodory's theorem).

Show that $\mathcal{Y}$ is invariant under the right shift and that (Y, ν, S) is a point invertible measure-preserving system and a point extension[6] of (X, μ, T) via the surjective map $\pi := \pi_0$.

[6] The system (Y, ν, S) is called the **natural extension** of (X, μ, T) in [260, Sec. 10.4].

Exercise 2.9 (Borel–Cantelli Lemma) Let (X, μ) be a probability space and (A_n) be a sequence of measurable sets. Prove the following assertions.

(a) If

$$\sum_{n=1}^{\infty} \mu(A_n) < \infty,$$

then

$$\mu\Big(\bigcap_{n=1}^{\infty} \bigcup_{k=n}^{\infty} A_k\Big) = 0.$$

(b) If

$$\sum_{n=1}^{\infty} \mu(A_n) = \infty,$$

and the sets A_n are **independent**, i.e., satisfy $\mu(\bigcap_{n\in F} A_n) = \prod_{n\in F} \mu(A_n)$ for every finite, non-empty $F \subset \mathbb{N}$, then

$$\mu\Big(\bigcap_{n=1}^{\infty} \bigcup_{k=n}^{\infty} A_k\Big) = 1.$$

Exercise 2.10 (Koopman Operator on L^0) Prove that the Koopman operator of a measure-preserving system is isometric on the space $\mathrm{L}^\infty(X, \mu)$. Prove also the algebraic properties from Proposition 2.27 of Koopman operators on $\mathrm{L}^0(X, \mu)$.

Exercise 2.11 (Koopman Operators and Markov Invertibility) Show that a Koopman operator T of a measure-preserving system (X, μ, T)

(a) satisfies $\|Tf\|_p = \|f\|_p$ for every $f \in \mathrm{L}^0(X, \mu)$ and every $p \in [1, \infty]$, where these p-norms can also be infinity.
(b) is invertible on $\mathrm{L}^p(X, \mu)$ for some $p \in [1, \infty]$ if and only if it is invertible on $\mathrm{L}^p(X, \mu)$ for all $p \in [1, \infty]$.

Show further that Markov invertibility does not imply point invertibility in general. (Hint: Consider a finite system with an appropriate σ-algebra of measurable sets.)

Exercise 2.12 (Koopman Operator on $\mathrm{C}(K)$) Let (K, T) be a topological dynamical system.

(a) Prove the relevant algebraic properties of the Koopman operator on $\mathrm{C}(K)$ from Proposition 2.27. How should one modify property (e)?
(b) When is the Koopman operator an isometry on $\mathrm{C}(K)$?

Exercise 2.13 (Markov Measures) Work out the details of Example 2.35 (b).

Exercise 2.14 (Invariant Measures) Let (K, T) be a topological system and let $\mu \in \mathrm{M}(K)$ be a T-invariant complex measure, i.e., $T'\mu = \mu$. Prove that $|\mu|$ is a T-invariant measure.

Chapter 3
Minimality and Ergodicity

Given a dynamical system, topological or measure-preserving, we now study proper regions of the state space which stay invariant under the dynamics. More precisely, we are interested in the absence of such regions, since their existence would contradict the "ergodic hypothesis" as formulated by Boltzmann. Indeed, in this case the phase space motion is confined in there, hence cannot reach any state outside of that region. Also "time mean equals space mean" would fail since the phase space motion does not spend any time outside of this invariant region.

3.1 Invariant Sets for Topological Systems

We begin with some definitions for a topological system (K, T). For a point $x \in K$ and a subset $A \subset \mathbb{N}_0$ (or $A \subset \mathbb{Z}$ if T is invertible) we define

$$T^A x := \{T^n x : n \in A\},$$

and call the set $\mathrm{orb}_+(x) := T^{\mathbb{N}_0} x$ the (forward) **orbit** of x. A subset $F \subset K$ is called (forward) **invariant**,[1] if $\mathrm{orb}_+(x) \subset F$ for each $x \in F$. If T is invertible, then the **two-sided orbit** of x is $\mathrm{orb}(x) := T^{\mathbb{Z}} x$, and a subset $F \subset K$ is **two-sided invariant** if $\mathrm{orb}(x) \subset F$ for each $x \in F$. A closed, invariant set F yields, by restriction to F, a **subsystem** $(F, T|_F)$, denoted by (F, T). In the same manner two-sided invariant sets of an invertible system yield invertible subsystems. If the system (K, T) is invertible, then any non-empty, closed, invariant set contains a closed, non-empty, two-sided invariant set (see Exercise 3.1). A point $x \in K$ is called (forward) **topologically transitive** if $\overline{\mathrm{orb}_+(x)} = K$, and **topologically two-**

[1] Sometimes such a set is called **absorbing**, and then the term "A is invariant" has the different meaning: $T^{-1}A = A$, i.e., both A and A^c are absorbing.

T. Eisner, B. Farkas, *A Journey Through Ergodic Theorems*, Birkhäuser Advanced Texts Basler Lehrbücher, https://doi.org/10.1007/978-3-031-96506-7_3

sided transitive if $\overline{\mathrm{orb}}(x) = K$. Finally, the system (K, T) is called **minimal** if the only closed, invariant sets are the trivial ones, namely $\emptyset$ and K. Different minimal subsystems of a system (K, T) are disjoint, see Exercise 3.2.

Remark 3.1 The closure of an invariant set is easily seen to be invariant, so in particular $\overline{\mathrm{orb}_+}(x)$ is an invariant set, and the system (K, T) is minimal if and only if every point $x \in K$ is topologically transitive. Analogously, an invertible system (K, T) is minimal if and only if every point $x \in K$ is topologically two-sided transitive (see Exercise 3.2). This gives an informal relation to the *quasi ergodic hypothesis* of the Ehrenfests since in such systems the orbit of every point comes arbitrarily close to any other point.

We introduce the following terminology. Let (K, T) and (L, S) be topological systems. A **homomorphism** between (K, T) and (L, S) is a continuous map $\pi : K \to L$ such that $\pi \circ T = S \circ \pi$. We denote this by writing $\pi : (K, T) \to (L, S)$. Such a homomorphism is called an **isomorphism** if it is also a homeomorphism between K and L. Isomorphisms take orbits to orbits and closed, invariant sets to closed, invariant sets. This can be exploited in the following characterization of minimality for group rotations. An isomorphism of a system onto itself is called an **automorphism**.

Proposition 3.2 *Let G be a compact group and let $a \in G$. The following assertions are equivalent.*

(i) *The group rotation system (G, a) is minimal.*
(ii) *The subsemigroup $\{a^n : n \in \mathbb{N}_0\}$ is dense in G.*
(iii) *The cyclic subgroup $\langle a \rangle := \{a^n : n \in \mathbb{Z}\}$ is dense in G.*

Minimality, therefore, is a strong condition for such systems. For example, it implies that the group G is Abelian, see Exercise 3.5.

Proof The implications (i)⇒(ii)⇒(iii) are clear. To prove that (iii) implies (i) take $x \in G$. Then $\mathrm{orb}(x) = \rho_x \mathrm{orb}(1) = \rho_x \langle a \rangle$, where ρ_x is the multiplication by x from the right and hence an automorphism of (G, a). This implies $\overline{\mathrm{orb}}(x) = G$, i.e., every point in G is topologically two-sided transitive. By Remark 3.1, the proof is complete. ■

Important examples of minimal systems are provided by certain torus rotations.

Proposition 3.3 (Kronecker) *The system $(\mathbb{T}, a)$ is minimal if and only if $a \in \mathbb{T}$ is not a root of unity, i.e., if the cyclic subgroup $\langle a \rangle$ is infinite.*

Proof If a is a root of unity, then $\mathrm{orb}_+(1)$ is finite and invariant, so that $(\mathbb{T}, a)$ cannot be minimal. If a is not a root of unity, then by Exercise 2.2(a) $\mathrm{orb}_+(1)$ is dense in $\mathbb{T}$, and by Proposition 3.2 we obtain the minimality of $(\mathbb{T}, a)$. ■

The following result is a useful characterization of minimality using open sets instead of points.

Proposition 3.4 (Characterization of Minimality) *A topological system (K, T) is minimal if and only if for each non-empty, open set $U \subset K$ there is $N \in \mathbb{N}_0$ such* *that*

$$K = \bigcup_{n=0}^{N} T^{-n} U. \tag{3.1}$$

Proof Suppose (K, T) is minimal, and let $U \subset K$ be non-empty and open. Consider the set

$$F := \bigcap_{n \in \mathbb{N}_0} T^{-n}(K \setminus U)$$

of points which never visit U. This is a closed, invariant set and $F \neq K$ since $U \cap F = \emptyset$. It follows that $F = \emptyset$, i.e., $K = \bigcup_{n \in \mathbb{N}_0} T^{-n} U$, and by compactness we can take a finite subcover. This proves (3.1).

Conversely, let F be a non-trivial, closed, invariant subset of K, then $U := K \setminus F$ is non-empty, open and violates condition (3.1). ∎

It turns out that in a minimal system every point visits each non-empty, open set even infinitely often. Moreover, the sequence (or set) of visiting times has *some structure*. This is the content of the next proposition.

Proposition 3.5 *Let (K, T) be minimal, and let $U \subset K$ be a non-empty, open set. Then for each $x \in K$ the set*

$$R_U(x) := \{n \in \mathbb{N} : T^n x \in U\} \subset \mathbb{N}$$

of return times[2] *to U is* ***syndetic*** *(or* ***relatively dense****), i.e., there is $N \in \mathbb{N}$ such that $[k, k+N] \cap R_U(x) \neq \emptyset$ for every $k \in \mathbb{N}$.*

Proof By Proposition 3.4 there is $N \in \mathbb{N}$ such that $K = \bigcup_{j=0}^{N} T^{-j} U$. This implies that for each $x \in K$ and $k \in \mathbb{N}$ there is $j \in \{0, \ldots, N\}$ such that $T^k x \in T^{-j} U$, meaning $x \in T^{-(k+j)} U$. It follows that $R_U(x) \cap [k, k+N] \neq \emptyset$. ∎

A point $x \in K$ is called **almost periodic** if for each non-empty, open set $U \subset K$ with $x \in U$ the set of return times $R_U(x)$ is syndetic. By the above, every point in a minimal system is almost periodic.

Proposition 3.6 (Characterization of Almost Periodic Points) *For a topological system (K, T) and a point $x \in K$ the following assertions are equivalent.*

(i) *The point x is almost periodic.*
(ii) *The set $\overline{\mathrm{orb}_+(x)}$ is minimal.*
(iii) *The point x is contained in a minimal subsystem.*

---

[2] Of course, if $x \notin U$ this should be called the set of *visiting* times, but the terminology *return times* is more widespread.

Proof The implication (ii)⇒(iii) is trivial and (iii)⇒(i) follows from Proposition 3.5, so we only need to prove (i)⇒(ii). Let U be an open set containing x, and let $F := \overline{\mathrm{orb}_+(x)}$. Since x is almost periodic, there is $N \in \mathbb{N}$ such that $T^k x \in \bigcup_{j=0}^{N} T^{-j}U \subset \bigcup_{j=0}^{N} T^{-j}\overline{U}$ for each $k \in \mathbb{N}$. It follows that $F \subset \bigcup_{j=0}^{N} T^{-j}\overline{U}$. As a consequence, for any $y \in F$ there is $j \in \{0, \dots, N\}$ with $T^j y \in \overline{U}$. This, being true for every open neighbourhood U of x, implies $x \in \overline{\mathrm{orb}_+(y)}$ and $\overline{\mathrm{orb}_+(x)} = \overline{\mathrm{orb}_+(y)}$. Minimality of $\overline{\mathrm{orb}_+(x)}$ follows. ∎

Here is a strong form of recurrence for points in minimal systems.

Proposition 3.7 (Almost Periodic Points in Metrizable Systems are Recurrent) *Let (K, T) be a topological system with metrizable K, and let $x \in K$ be almost periodic. Then x is* ***recurrent****, i.e., there is a subsequence $(n_k)_{k\in\mathbb{N}}$ in $\mathbb{N}$ such that $T^{n_k}x \to x$ as $k \to \infty$.*

We leave the proof as Exercise 3.6.

Proposition 3.8 *Let (K, T) be a topological system.*

(a) *Let (L, S) be another topological system and let $\pi : (K, T) \to (L, S)$ be a homomorphism. If $x \in K$ is almost periodic in (K, T), then so is $\pi(x)$ in (L, S).*
(b) *Suppose (L, T) is a subsystem of (K, T). If $x \in L$ is almost periodic in (K, T), then it is almost periodic in (L, T).*

Again, we leave the proof of these results as Exercise 3.6.

Corollary 3.9 *In a compact group rotation system (G, a) every point is almost periodic.*

Proof The subgroup $H := \overline{\langle a \rangle}$ provides the group rotation subsystem (H, a), which is minimal by Proposition 3.2. This implies that 1 is almost periodic, thus for each $x \in K$ the homomorphic image $x = \rho_x(1)$ is almost periodic. ∎

The following result is known as Birkhoff's Recurrence Theorem [155].

Theorem 3.10 (Birkhoff) *Each topological system (K, T) contains a minimal subsystem and therefore an almost periodic point. In particular, if K is metrizable, then it contains at least one recurrent point.*

Proof Consider the set $\mathcal{F} := \{F : F \subset K \text{ closed, invariant, } F \neq \emptyset\}$, which is partially ordered by inclusion $A \subset B$. This partially ordered set satisfies the conditions of Zorn's lemma, and therefore has a minimal element F, which provides a minimal subsystem (F, T). By Proposition 3.6, any point $x \in F$ is almost periodic. The rest follows from Proposition 3.7. ∎

Example 3.11 The topological system $([0, 1], T)$ with T given by $Tx := x^2$ has two minimal subsystems and two recurrent points.

For a characterization of almost periodicity and recurrence for shifts see Exercise 3.8.

3.2 Invariant Sets for Measure-Preserving Systems

Because of the presence of null sets (which are anyway negligible), the whole business of invariant sets for measure-preserving systems becomes somewhat more subtle. We begin with some terminology.

Let (X, μ) be a probability space and $A, B \subset X$ measurable. We say that $A \subset B$ **up to a null set** if $\mu(A \setminus B) = 0$. Analogously, we say that $A = B$ up to a null set if $\mu(A \triangle B) = 0$, where $A \triangle B = (A \setminus B) \cup (B \setminus A)$ is the symmetric difference of A and B. The simple inequality

$$|\mu(A) - \mu(B)| \leq \mu(A \triangle B) \tag{3.2}$$

will be very useful. Note that being equal up to a null set defines an equivalence relation on measurable sets.

Definition 3.12 (Invariant Sets) Let (X, μ, T) be a measure-preserving system. A measurable set $A \subset X$ is called (T-)**invariant** if $A = T^{-1}A$ up to a null set.

Remark 3.13 By the measure-preserving property (and recall that we only consider probability spaces if not stated otherwise), a measurable set A is invariant if and only if $T^{-1}(A) \subset A$ holds up to a null set, and this holds if and only if $A \subset T^{-1}(A)$ holds up to a null set. This in turn, in terms of the Koopman operator, is equivalent to $T\mathbf{1}_A = \mathbf{1}_A$ almost everywhere, see Exercise 3.9.

The set of invariant sets is denoted by $\mathcal{X}_T$, and it is a sub-σ-algebra of the underlying σ-algebra $\mathcal{X}$ (see Exercise 3.9), called the **invariant sub-σ-algebra**. In particular, we have $A^c \in \mathcal{X}_T$ if $A \in \mathcal{X}_T$. A non-trivial invariant set A thus gives rise to two systems $(X, \mu_A/\mu(A), T)$ and $(X, \mu_{A^c}/\mu(A^c), T)$, where for $B \in \mathcal{X}$ the measure μ_B is defined by

$$\mu_B(C) := \mu(B \cap C) \text{ for each } C \in \mathcal{X}.$$

Definition 3.14 (Ergodicity) The measure preserving system (X, μ, T), and also the transformation T, are called **ergodic** if $\mu(A) \in \{0, 1\}$ holds for each invariant set $A \subset X$.

We now present the first examples of ergodic (and non-ergodic) systems.

Example 3.15 (Finite Systems) Let X be a finite set with the normalized counting measure μ and let $T : X \to X$ be bijective. Then T is ergodic if and only if it is a cyclic permutation.

Example 3.16 (Identity is (Almost) Never Ergodic) Let (X, μ) be a probability space. Then $\mathrm{id} : X \to X$ is ergodic if and only if there is no measurable set $A \subset X$ with $0 < \mu(A) < 1$. This shows that, in contrary to the topological case where every system contains a minimal subsystem, not every measure-theoretic system has an ergodic subsystem. For example, $([0, 1], \lambda, \mathrm{id})$ has no ergodic subsystem.

An important class of ergodic transformations are (Bernoulli) shifts (cf. Example 2.16). We need the following auxiliary result, for a proof see [153, Thm. 11.4 and Cor.].

Lemma 3.17 *Let $(X, \mathcal{X}, \mu)$ be a probability space and let $\mathcal{E}$ be an algebra of sets and a generator of the σ-algebra $\mathcal{X}$. Then for every set $A \in \mathcal{X}$ and for every $\varepsilon > 0$ there is $B \in \mathcal{E}$ such that $\mu(A \triangle B) < \varepsilon$.*

Proposition 3.18 (Shifts are Ergodic) *Let (Y, ν) be a probability space. Both the one-sided and the two-sided shifts with state space (Y, ν) are ergodic.*

Proof We consider only the case of one-sided shifts (X, μ, T), the invertible case being analogous. Let $\mathcal{E}$ be the algebra of finite unions of cylinder sets (which generates the product σ-algebra $\mathcal{X}$). Let $A \in \mathcal{X}$ be an invariant set, let $\varepsilon > 0$ be arbitrary, and let $B \in \mathcal{E}$ be such that $\mu(A \triangle B) < \varepsilon$ (use Lemma 3.17). Then B is of the form

$$B = B_N \times Y \times Y \cdots, \quad \text{where } B_N \subset Y^N.$$

Note that $T^{-n}B = Y^n \times B_N \times Y \times Y \cdots$, so for every $n > N$ we have

$$\mu(B \cap T^{-n}B) = \mu(B) \cdot \mu(T^{-n}B) = \mu(B)^2$$

by the definition of the product measure. Thus we obtain by (3.2) for every $n > N$

$$\begin{aligned}
|\mu(A) - \mu(A)^2| &\le |\mu(A) - \mu(B \cap T^{-n}B)| + |\mu(B \cap T^{-n}B) - \mu(A)^2| \\
&\le \mu(A \triangle (B \cap T^{-n}B)) + |\mu(B)^2 - \mu(A)^2| \\
&\le \mu(A \triangle B) + \mu(A \triangle T^{-n}B) + |\mu(B)^2 - \mu(A)^2| \\
&= \mu(A \triangle B) + \mu(T^{-n}A \triangle T^{-n}B) + |\mu(B)^2 - \mu(A)^2| \\
&\le 2\varepsilon + |\mu(B)^2 - \mu(A)^2| \le 2\varepsilon + 2|\mu(A) - \mu(B)| \le 4\varepsilon.
\end{aligned}$$

This, being true for every $\varepsilon > 0$, implies $\mu(A) = \mu(A)^2$, hence $\mu(A) \in \{0, 1\}$. ■

Proposition 3.19 (Characterization of Ergodicity) *For a measure-preserving system (X, μ, T) the following assertions are equivalent.*

(i) *The system is ergodic.*
(ii) *Every $A \subset X$ with $\mu(A) > 0$ satisfies*

$$\bigcap_{k \in \mathbb{N}_0} \bigcup_{n \ge k} T^{-n}A = X \text{ up to a null set.}$$

(iii) *Every* $A \subset X$ *with* $\mu(A) > 0$ *satisfies*

$$\bigcup_{n\in\mathbb{N}_0} T^{-n}A = X \text{ up to a null set.}$$

(iv) *For each pair of sets* $A, B \subset X$ *with* $\mu(A), \mu(B) > 0$ *there is* $n \in \mathbb{N}$ *with*

$$\mu(T^{-n}A \cap B) > 0.$$

Proof Suppose (i) and $\mu(A) > 0$. For $k \in \mathbb{N}_0$ define $A_k := \bigcup_{n\geq k} T^{-n}A$. Then $T^{-1}(A_k) = A_{k+1} \subset A_k$ and therefore A_k is an invariant set with $\mu(A_k) \geq \mu(T^{-k}A) = \mu(A) > 0$. The assumption implies $\mu(A_k) = 1$, and by intersecting $\mu(\bigcap_{k\in\mathbb{N}} A_k) = 1$, i.e., (ii) follows. The implication (ii) $\Rightarrow$ (iii) is trivial.

Assume (iii) and let $A, B \subset X$ be measurable satisfying $\mu(A) > 0$ and $\mu(T^{-n}A \cap B) = 0$ for each $n \in \mathbb{N}$. For $k \in \{0, 1\}$ let again $A_k := \bigcup_{n\geq k} T^{-n}A$; by assumption $A_0 = X$ up to null sets. Since $A_0 = X = T^{-1}A_0 = A_1 \subset A_0$, it follows that $A_0 = A_1$ up to a null set. We obtain

$$0 = \mu\left(\bigcup_{n\in\mathbb{N}}(B \cap T^{-n}A)\right) = \mu(B \cap A_1) = \mu(B \cap A_0) = \mu(B \cap X) = \mu(B),$$

proving (iv).
To see the implication (iv)$\Rightarrow$(i) take $B := A^c$ in (iv) for an invariant set $A \subset X$, to conclude that $\mu(A) \in \{0, 1\}$. ■

Remark 3.20 Condition (iii) in the above characterization means that almost every $x \in X$ visits every set A with positive measure at least once, and condition (ii) means that almost every $x \in X$ visits every such A infinitely often.

Remark 3.21 (Average Return Time) By Kac's Theorem 2.5 we obtain for ergodic systems (X, μ, T) and $A \subset X$ with $\mu(A) > 0$ that the average (first) return time to A is

$$\frac{1}{\mu(A)}\int_A n_A \, d\mu = \frac{1}{\mu(A)}.$$

Example 3.22 (Recurrence in Random Literature) We now discuss a concrete consequence of Proposition 3.19 which might appear somewhat counterintuitive at the first glance. Imagine someone typing randomly on a typewriter which has 90 different typesetting symbols. The work starts now. This can be modeled by the one-sided Bernoulli shift $B(1/90, \ldots, 1/90)$ which is ergodic by Proposition 3.18. An infinite piece of literature is then described by each of the sequences $x \in \{0, \ldots, 89\}^{\mathbb{N}}$. Some of these sequences correspond to total nonsense, others contain poems of Goethe and Pushkin, novels of Joyce, yet another contain large blocks of digits of π, or a mixture of these, etc. Let $a_1a_2 \cdots a_N$ be the text of your favorite book. Since the cylinder set

$$A := \{x \in \{0, \dots, 89\}^{\mathbb{N}} : x_1 = a_1, \dots, x_N = a_N\}$$

has positive measure (precisely $1/90^N$), we see, by Proposition 3.19, that almost every $x \in \{0, \dots, 89\}^{\mathbb{N}}$ visits A infinitely many times under the shift-dynamics, meaning that the text of your favorite book occurs in almost every infinite word infinitely often. This means that the person will almost surely type your favorite book infinitely often. What changes (and what does it mean for the typist) if we replace here the one-sided Bernoulli shift by a two-sided one?

We close this section with the following famous and very useful structural result about ergodic systems.

Theorem 3.23 (Kakutani–Rokhlin Lemma) *Let (X, μ, T) be an ergodic*[3] *measure-preserving system. Suppose that X contains sets of arbitrarily small positive measure. Then for each $N \in \mathbb{N}$ and for each $\varepsilon > 0$ there is a measurable set $B \subset X$ such that the sets $B, T^{-1}B, \dots, T^{-(N-1)}B$ are pairwise disjoint (such N-tuple is then called a* ***Rokhlin tower*** *of height N) and*

$$\mu\Big(\bigcup_{n=0}^{N-1} T^{-n}B\Big) \geq 1 - \varepsilon.$$

Proof For given $N \in \mathbb{N}$ and $\varepsilon > 0$ fix a subset $A \subset X$ with $\mu(A) < \frac{\varepsilon}{N}$. For $n \in \mathbb{N}_0$ define

$$A_n := \{x \in X : n \text{ is the first time of visit of } x \text{ in } A\}.$$

In particular, $A_0 = A$, $A_1 = T^{-1}A \setminus A$, $A_2 = T^{-2}A \setminus (A_0 \cup A_1)$, etc. By ergodicity we have

$$X = \bigcup_{n\geq 0} T^{-n}A = \bigcup_{n\geq 0} A_n,$$

up to null sets (this will not be mentioned in this proof anymore). The sets A_n, $n \in \mathbb{N}_0$ are pairwise disjoint, and $X \setminus A = \bigcup_{n\geq 1} A_n$. Now for $j \in \{0, \dots, N-1\}$ consider the pairwise disjoint sets

$$\begin{aligned} B_j &:= \{x \in X : \text{the first visiting time of } x \text{ in } A \text{ is at least } N \text{ and } \equiv j \bmod N\} \\ &= \bigcup_{k>1} A_{kN+j}. \end{aligned}$$

[3] One can replace ergodicity by the **aperiodicity** (cf. the paragraph after Definition 12.33), the proof for this case can be found in Halmos' book [514, p. 71], see also [617, Sec. 2.4].

It is easy to see that for each $j \in \{0, \dots, N-1\}$ one has $B_j \subset T^{-j} B_0 \subset B_j \cup (A \cup \dots \cup T^{-(j-1)} A)$. Whence we conclude that $T^{-j} B_0 \cap B_0 = \emptyset$ for $j = 1, \dots, N-1$, i.e., with $B := B_0$ we obtain the Rokhlin tower $B, \dots, T^{-(N-1)} B$ of height N. Finally we have

$$\bigcup_{n=0}^{N-1} T^{-n} B \supset \bigcup_{n=0}^{N-1} B_n = \bigcup_{n \geq N} A_n = X \setminus (A_0 \cup \dots \cup A_{N-1}),$$

and therefore

$$\mu\Big(\bigcup_{n=0}^{N-1} T^{-n} B\Big) \geq 1 - \sum_{n=0}^{N-1} \mu(A_n) \geq 1 - N\mu(A) > 1 - \varepsilon,$$

finishing the proof. ∎

3.3 Koopman Operator and Ergodicity

For a linear operator $S : V \to V$ on a vector space V we define the **fixed space** by

$$\operatorname{fix}(S) := \{f \in V : Sf = f\} = \ker(S - I),$$

and call each $f \in \operatorname{fix}(S)$ **invariant under** S. Recall that for a measure-preserving system, constant functions are always invariant under the Koopman operator. The analogue of the following result in the time continuous case is due to von Neumann, see [1027, p. 78] (see also Carleman [225, p. 5]).

Proposition 3.24 (Ergodicity via Invariant Functions) *Let (X, μ, T) be a measure-preserving system and let $p \in [1, \infty)$. The Koopman operator on $\mathrm{L}^0(X, \mu)$ is also denoted by T. The following assertions are equivalent.*

(i) *The system (X, μ, T) is ergodic.*
(ii) *If $f : X \to \mathbb{C}$ is measurable with $Tf = f$ μ-a.e., then there is $c \in \mathbb{C}$ such that $f = c\mathbf{1}$ μ-a.e.*
(iii) *Every invariant function $f \in \mathrm{L}^p(X, \mu)$ is constant. (In other words, $\operatorname{fix}(T) = \mathbb{C}\mathbf{1}$ in L^p.)*
(iv) *Every invariant function $f \in \mathrm{L}^\infty(X, \mu)$ is constant.*

Recall that we use the convenient notation $[f < 1]$ to denote the set $\{x : f(x) < 1\}$ etc.

Proof (i)⇒(ii): Let $f : X \to \mathbb{C}$ be measurable with $Tf = f$ μ-a.e., and assume without loss of generality that f is real-valued (otherwise we pass to the real and imaginary parts). For $s \in \mathbb{R}$ we set $A_s := [f < s]$, which is easily seen to be a T-invariant set. By the ergodicity assumption in (i) we have that $\mu(A_s) \in \{0, 1\}$. Of course, there is at least one $s \in \mathbb{R}$ with $\mu(A_s) > 0$, i.e., with $\mu(A_s) = 1$. Define

$$c := \inf\{s : \mu(A_s) = 1\},$$

where the set appearing here on the right-hand side is clearly bounded from below. For $s < c$ we have $\mu(A_s) = 0$, i.e., $f \geq c$ almost everywhere, while for $s > c$ we have $\mu(A_s) = 1$, i.e., $f \leq c$ almost everywhere. Altogether we have $f = c$ almost everywhere.
The implications (ii)⇒(iii)⇒(iv) are trivial. So it remains to prove (i) assuming (iv). Let $A \subset X$ be invariant. We have $T\mathbf{1}_A = \mathbf{1}_{T^{-1}A} = \mathbf{1}_A$, i.e., the function $\mathbf{1}_A$ is invariant under T. By (iv), $\mathbf{1}_A$ must be constant, so $\mu(A) \in \{0, 1\}$. ∎

The study of the ergodicity of torus rotations becomes now very easy.

Proposition 3.25 *The torus rotation $(\mathbb{T}^d, \mathrm{m}^d, a)$ with $a = (a_1, \ldots, a_d) \in \mathbb{T}^d$ is ergodic if and only if $a_1, \ldots, a_d$ are* ***rationally independent****, i.e., satisfy*

$$\Big((n_1, \ldots, n_d) \in \mathbb{Z}^d \text{ and } a_1^{n_1} a_2^{n_2} \cdots a_d^{n_d} = 1\Big) \implies n_1 = n_2 = \cdots = n_d = 0.$$

Proof The monomials $\mathbf{z}^n$ with $n = (n_1, \ldots, n_d) \in \mathbb{Z}^d$ given by

$$\mathbf{z}^n(z_1, \ldots, z_d) := z_1^{n_1} \cdots z_d^{n_d}$$

form an orthonormal basis of $\mathrm{L}^2(\mathbb{T}^d, \mathrm{m}^d)$, by Proposition 1.12. Let f be an invariant function of the Koopman operator T on $\mathrm{L}^2(\mathbb{T}^d, \mathrm{m}^d)$. Then expanding f with respect to the orthonormal basis leads to the equality

$$\sum_{n \in \mathbb{Z}^d} b_n \mathbf{z}^n = f = Tf = \sum_{n \in \mathbb{Z}^d} a^n b_n \mathbf{z}^n$$

of the L^2-convergent series for some $(b_n)_{n \in \mathbb{Z}^d} \in \ell^2(\mathbb{Z}^d)$, where $a^n := a_1^{n_1} \cdots a_d^{n_d}$. By comparing the coefficients we obtain $b_n = a^n b_n$ for every $n \in \mathbb{Z}^d$.

If $f \notin \mathbb{C}\mathbf{1} = \mathbb{C}\mathbf{z}^0$, then there is $n \in \mathbb{Z}^d \setminus \{0\}$ with $b_n \neq 0$, implying $1 = a^n = a_1^{n_1} \cdots a_d^{n_d}$. This proves that in case of rational independence, $\mathrm{fix}(T) = \mathbb{C}\mathbf{1}$, and ergodicity follows by Proposition 3.24.

For the converse implication, assume that $a_1, \ldots, a_d$ satisfy $a_1^{n_1} \cdots a_d^{n_d} = 1$ with $n = (n_1, \ldots, n_d) \neq (0, \ldots, 0)$. Then the corresponding monomial $\mathbf{z}^n$ is not constant and satisfies $T\mathbf{z}^n = \mathbf{z}^n$, implying that the system (X, μ, T) is not ergodic by Proposition 3.24. ∎

3.4 Spectral Properties of Koopman Operators

In this section we study the spectrum and the point spectrum of Koopman operators.

Recall for a vector space V and a linear operator $S : V \to V$ the notation

$$P\sigma(S) := \big\{\lambda \in \mathbb{C} : \; Sf = \lambda f \text{ for some } f \in V \setminus \{0\}\big\}$$

for the set of eigenvalues of S, called the **point spectrum** of S.

The following gives information on the point spectrum of the Koopman operator of an ergodic system.

Proposition 3.26 (Point Spectrum of the Koopman Operator for Ergodic Systems) *Let (X, μ, T) be an ergodic measure-preserving system, and denote by T the Koopman operator on $\mathrm{L}^0(X, \mu)$. Then the set $P\sigma(T)\cap\mathbb{T}$ of its unimodular eigenvalues is a subgroup of $\mathbb{T}$, each such eigenvalue has one-dimensional eigenspace, and a corresponding eigenfunction can be taken unimodular. Moreover, eigenfunctions of T form a group.*

Here, and in the following, *unimodular* means having absolute value 1. Since constant functions are always T-invariant, $1 \in P\sigma(T)$ and, in particular, $P\sigma(T) \neq \emptyset$.

Proof Let $\eta \in P\sigma(T)\cap\mathbb{T}$ have an eigenfunction g which, together with f, has non-zero constant absolute value. By the algebraic properties of the Koopman operator (see Proposition 2.27) we obtain that $T(f\overline{g}) = T(f)\overline{T(g)} = (\lambda\overline{\eta})f\overline{g}$. Since $f\overline{g}$ is non-zero, it is an eigenfunction to the eigenvalue $\lambda\overline{\eta}$. If $\eta = \lambda$, ergodicity yields $f\overline{g} = c\mathbf{1}$ for some $c \in \mathbb{T}$, i.e., $f = cg$, showing the one-dimensionality of the eigenspace. Moreover, we see that for $\lambda, \eta \in P\sigma(T) \cap \mathbb{T}$ also $\overline{\eta}$ and $\lambda\eta$ belong to $P\sigma(T) \cap \mathbb{T}$, i.e., this set is a subgroup of $\mathbb{T}$. ■

Remark 3.27 For $p \in [1, \infty)$ and a measure-preserving system (X, μ, T), the Koopman operator is an isometry on $\mathrm{L}^p(X, \mu)$ and therefore its point spectrum is a subset of the unit circle. Hence, if one replaces $\mathrm{L}^0(X, \mu)$ by $\mathrm{L}^p(X, \mu)$ in the above proposition, one can just write $P\sigma(T)$ instead of $P\sigma(T) \cap \mathbb{T}$.

In general (without the ergodicity assumption), one can only show the cyclicity of the spectrum.

Proposition 3.28 (Unimodular Point Spectrum Is Cyclic) *For a measure-preserving system (X, μ, T) denote by T the Koopman operator on $\mathrm{L}^0(X, \mu)$. Then the set $P\sigma(T) \cap \mathbb{T}$ of its unimodular eigenvalues is* ***cyclic****, i.e., for each $\lambda \in P\sigma(T) \cap \mathbb{T}$ and $n \in \mathbb{Z}$ one has $\lambda^n \in P\sigma(T) \cap \mathbb{T}$.*

We leave the proof of this proposition as Exercise 3.11.

The following property of the point spectrum of the Koopman operator on L^p-spaces will be very useful later.

Proposition 3.29 (Eigenfunctions of Koopman Operators for General Systems) *Let (X, μ, T) be a measure-preserving system, $p \in [1, \infty)$ and denote by T the Koopman operator on $\mathrm{L}^p(X, \mu)$.*

(a) *For every $\lambda \in P\sigma(T)$, the set $\ker(\lambda - T) \cap \mathrm{L}^\infty(X, \mu)$ is dense in $\ker(\lambda - T)$, i.e., bounded eigenfunctions are dense in the space of all L^p-eigenfunctions. In particular, for every eigenvalue there is a corresponding bounded eigenfunction.*
(b) *$P\sigma(T) \subset \mathbb{T}$ is a union of subgroups of $\mathbb{T}$ and is independent of p.*

Proof Since T is an isometry on $\mathrm{L}^p(X, \mu)$, each of its eigenvalue belongs to $\mathbb{T}$.

(a) We first observe that for every $f \in \operatorname{fix}(T)$ and every $n \in \mathbb{N}$, one has $\mathbf{1}_{[|f|\leq n]} \in \operatorname{fix}(T)$. Indeed, the T-invariance of f implies

$$(T\mathbf{1}_{[|f|\leq n]})(x) = \mathbf{1}_{[|f|\leq n]}(Tx) = \mathbf{1}_{[|f|\leq n]}(x) \text{ for a.e. } x \in X.$$

Take now $f \in \ker(\lambda - T)$, i.e., $Tf = \lambda f$. By the algebraic property of the Koopman operator from Proposition 2.27 (c) we have $T|f| = |Tf| = |\lambda||f| = |f|$, i.e., $|f| \in \operatorname{fix}(T)$. For $n \in \mathbb{N}$ consider $f_n := f\mathbf{1}_{[|f|\leq n]}$. These functions are (essentially) bounded and converge to f. Moreover, each f_n is an eigenfunction corresponding to λ since by the above observation

$$Tf_n = Tf \cdot T\mathbf{1}_{[|f|\leq n]} = \lambda f \cdot \mathbf{1}_{[|f|\leq n]} = \lambda f_n,$$

whence denseness follows.
The first statement in (b) follows as in the ergodic case, see Proposition 3.26, while the independence from p follows from (a). ∎

We now study the spectrum of Koopman operators and begin with the following simple fact due to Ionescu Tulcea [568].

Lemma 3.30 *Let S be an isometry on a Hilbert space H. If for every $N \in \mathbb{N}$ there exists $f \in H$ with $f \neq 0$ such that the vectors $f, Tf, \dots T^N f$ are pairwise orthogonal, then $\mathbb{T} \subset \sigma(T)$.*

Proof Let $\omega \in \mathbb{T}$ and take $N \in \mathbb{N}$ together with the corresponding $f \in H$ given by the assumption. Define $f_N := \sum_{n=0}^{N} \omega^{-n} T^n f$ and observe $\|f_N\| = \sqrt{N+1}\|f\|$ by orthogonality. Compute further

$$\|\omega f_N - Tf_N\| = \|\omega f - \omega^{-N} T^{N+1} f\| \leq 2\|f\| = \frac{2\|f_N\|}{\sqrt{N+1}}.$$

Thus we see that $\omega I - T$ is not bounded below and therefore $\omega \in \sigma(T)$ for every $\omega \in \mathbb{T}$. ∎

Lemma 3.31 *Let (X, μ, T) be an invertible ergodic measure-preserving system with Koopman operator T on $\mathrm{L}^2(X, \mu)$ such that $T^n \neq I$ for every $n \in \mathbb{N}$. Then X contains measurable sets with arbitrarily small positive measure.*

Proof We claim first that each measurable $A \subset X$ with $\mu(A) > 0$ contains a measurable $B \subset A$ with $0 < \mu(B) < \mu(A)$. To see this we suppose the contrary, i.e., that there is a measurable A such that $\mu(B) \in \{0, \mu(A)\}$ for every measurable subset B of A. Thus for each $n \in \mathbb{N}$ we have $\mu(T^{-n}A \cap A) \in \{0, \mu(A)\}$. If for every $n \in \mathbb{N}$ we had $\mu(T^{-n}A \cap A) = 0$, then for ever $n, m \in \mathbb{N}$ with $m \leq n$ we would have $\mu(T^{-n}A \cap T^{-m}A) = \mu(T^{-(n-m)}A \cap A) - 0$, and that would mean that $T^{-n}A$ and $T^{-m}A$ are disjoint up to null sets. Since all these sets have the same positive measure as A, we would have a contradiction to $\mu(X) = 1$. Thus there must exist an $n \in \mathbb{N}$, and we take immediately the minimal one, with $\mu(T^{-n}A \cap A) = \mu(A)$.

The set

$$A \cup T^{-1}A \cup \cdots \cup T^{-(n-1)}A$$

is T-invariant, and since it has positive measure it must be the whole of X up to null sets. Since T is invertible (and measure-preserving) each of the sets $T^{-n}A$ have the property that whenever $C \subset T^{-n}A$ is measurable, then $\mu(C) \in \{0, \mu(A)\}$. It follows that each $f \in \mathrm{L}^2(X, \mu)$ is constant on each of $A, T^{-1}A, \ldots, T^{-(n-1)}A$, which partition X completely. Finally we conclude $T^n = I$ on $\mathrm{L}^2(X, \mu)$, a contradiction, and our first claim is proved.
We now show that for every $\varepsilon > 0$ there is a measurable subset $A \subset X$ with $\mu(A) < \varepsilon$. Take $A_0 := X$ and suppose that the measurable sets $A_0, \ldots, A_k$ have been constructed with $A_{j+1} \subset A_j$ and $0 < \mu(A_{j+1}) \leq \frac{1}{2}\mu(A_j)$ for each $j = 0, \ldots, k-1$. Let $B \subset A_k$ be a measurable subset with $0 < \mu(B) < \mu(A_k)$ and set $A_{k+1} := B$ if $\mu(B) \leq \frac{1}{2}\mu(A_k)$ and set $A_{k+1} := A_k \setminus B$ if $\mu(B) > \frac{1}{2}\mu(A_k)$. Thus one constructs the infinite sequence $A_0, A_1, \ldots$ recursively, keeping the above properties. Whence we conclude $0 < \mu(A_k) \to 0$ as $k \to \infty$. The proof is complete. ∎

We can now prove the following result due to Ionescu Tulcea [567] and Foiaş [394]. Note that this result remains true also for non-ergodic transformations. This extension is also due to Ionescu Tulcea, see [568] for a proof and further generalizations.

Proposition 3.32 *Let (X, μ, T) be an invertible ergodic measure-preserving system with Koopman operator T on $\mathrm{L}^2(X, \mu)$ such that $T^n \neq I$ for every $n \in \mathbb{N}$. Then $\sigma(T) = \mathbb{T}$.*

Proof By the assumption we are in the position to apply Lemma 3.31 and then the Kakutani–Rokhlin lemma (Theorem 3.23) yielding for every $N \in \mathbb{N}$ a Rokhlin tower $B, T^{-1}B, \ldots, T^{-(N-1)}B$ of height N with $\mu(B) > 0$. Then the function $f := \mathbf{1}_B$ satisfies the condition of Lemma 3.30, so we conclude $\mathbb{T} \subset \sigma(T)$. The reverse inclusion is true, since T is unitary on $\mathrm{L}^2(X, \mu)$. ∎

Using deeper results from functional analysis which we use as black box, we show that one can drop the ergodicity assumption in Proposition 3.32, thus providing a purely operator theoretic proof of Ionescu Tulcea's result [568, Cor. 2]. Note that the following description of the spectrum follows also from the work of Hölz, see [540, Cor. 2.8], he showed namely that a power-bounded lattice homomorphism on a Banach lattice is periodic if and only if its spectrum is a proper subset of $\mathbb{T}$.

Theorem 3.33 (Spectrum of Koopman Operators on L^p) *Let (X, μ, T) be a measure-preserving system and $p \in [1, \infty]$. If its Koopman operator T on $\mathrm{L}^p(X, \mu)$ is*

(a) *not invertible, then $\sigma(T)$ is the whole closed unit disc.*
(b) *invertible and not periodic, then $\sigma(T) = \mathbb{T}$.*
(c) *periodic, then $\sigma(T)$ is a finite union of subgroups of $\mathbb{T}$ of roots of unity.*

Moreover, the above distinction and the spectrum of T are independent of p.

For the proof we will need the following notion. A set $M \subset \mathbb{C}$ is called **cyclic** if for every $z = |z|\mathrm{e}^{\mathrm{i}\varphi} \in M$ one has $|z|\mathrm{e}^{\mathrm{i}n\varphi} \in M$ for every $n \in \mathbb{N}$.

Proof (a) follows from Proposition 1.5.
To show (b) and (c), let T be invertible, in which case $\sigma(T) \subset \mathbb{T}$ again by Proposition 1.5. Moreover, analogously to Proposition 3.28, by Schaefer[4] [915, V.4.4] $\sigma(T)$ is a cyclic subset of $\mathbb{T}$.
We now claim that either $\sigma(T) = \mathbb{T}$ or $\sigma(T)$ consists of finitely many roots of unity. If $\sigma(T)$ contains a number which is not a root of unity, then by Exercise 2.2(a) and the closedness of the spectrum $\sigma(T) = \mathbb{T}$. Moreover, if $\sigma(T)$ contains a sequence of roots of unity which converges to a root of unity, then by the cyclicity property $\sigma(T)$ contains a sequence of roots of unity which converges to 1. This implies again by the cyclicity property and the closedness of the spectrum that $\sigma(T) = \mathbb{T}$. The claim is proved.
Now, by the spectral mapping theorem (Proposition 1.6), $\sigma(T)$ consists of finitely many roots of unity if and only if there exists $k \in \mathbb{N}$ with $\sigma(T^k) = \{1\}$. Since both T^k and T^{-k} are isometries and hence contractions, by Gelfand's $T = I$ theorem (see, e.g., [337, Thm. II.2.12]) $\sigma(T^k) = \{1\}$ holds if and only if $T^k = I$.
Regarding the last assertion, it is clear that periodicity of the Koopman operator T on $\mathrm{L}^p(X, \mu)$ does not depend on p. Moreover, also invertibility of T does not depend on p by Exercise 2.11. The proof is complete. ■

For the spectrum of Koopman operators of topological dynamical systems where the situation is a little more complex (cf. Exercise 3.11), see Scheffold [917].

3.5 More on Invariant Measures for Topological Systems

We now study the relation between topological and measure-preserving systems from the point of view of ergodicity properties of invariant measures.

Let (K, T) be a topological system. Let $\mathrm{M}_1(K, T)$ be the set of T-invariant regular, Borel probability measures. By the Krylov–Bogoljubov Theorem 2.33, $\mathrm{M}_1(K, T)$ is non-empty. An invariant probability measure $\mu \in \mathrm{M}_1(K, T)$ is called **ergodic** if $(K, \mathcal{B}(K), \mu, T)$ is an ergodic measure-preserving system.

Recall the notions of absolute continuity and singularity of measures, e.g., from Sect. 1.3.

Proposition 3.34 (Ergodicity of Invariant Measures) *Let (K, T) be a topological system and let $\mu, \nu \in \mathrm{M}_1(K, T)$.*

(a) *If μ is ergodic and $\nu \ll \mu$, then $\mu = \nu$.*

[4] Schaefer proved that the spectrum $\sigma(S)$ is cyclic for every lattice homomorphism S on a Banach lattice.

(b) *If $\mu \neq \nu$ and both are ergodic, then they are mutually singular.*
(c) *The measure $\mu \in \mathrm{M}_1(K, T)$ is ergodic if and only if μ is an extreme point of the (weak* compact, convex) set $\mathrm{M}_1(K, T)$.*

Proof (a) Since $\nu \ll \mu$ there is $f \in \mathrm{L}^1(\mu)$ with $f \geq 0$ and $\nu = f \cdot \mu$ (Theorem 1.27). We prove that $f = \mathbf{1}$ in $\mathrm{L}^1(\mu)$. We show first that $A := [f < 1]$ is an invariant set for (K, μ, T). To see this we compute

$$\begin{aligned}\int_{A \cap T^{-1}A} f \,\mathrm{d}\mu + \int_{A \setminus T^{-1}A} f \,\mathrm{d}\mu &= \nu(A) = \nu(T^{-1}A)\\ &= \int_{A \cap T^{-1}A} f \,\mathrm{d}\mu + \int_{T^{-1}A \setminus A} f \,\mathrm{d}\mu.\end{aligned}$$

As a consequence we obtain

$$\int_{A \setminus T^{-1}A} f \,\mathrm{d}\mu = \int_{T^{-1}A \setminus A} f \,\mathrm{d}\mu. \tag{3.3}$$

Since $A \setminus T^{-1}A \subset [f < 1]$, $T^{-1}A \setminus A \subset [f \geq 1]$ and

$$\mu(A \setminus T^{-1}A) = \mu(A) - \mu(A \cap T^{-1}A) = \mu(T^{-1}A) - \mu(A \cap T^{-1}A) = \mu(T^{-1}A \setminus A),$$

we conclude from (3.3) that $\mu(A \setminus T^{-1}A) = 0 = \mu(T^{-1}A \setminus A)$, implying $A = T^{-1}A$ up to a μ-null set, and the invariance of A is proven.

By ergodicity of μ we have $\mu(A) \in \{0, 1\}$. If $\mu(A) = 1$ were true, then the inequality $\nu(X) = \int_X f \,\mathrm{d}\mu = \int_A f \,\mathrm{d}\mu < \mu(A) = 1$ would provide a contradiction. Thus $\mu(A) = 0$ must hold, meaning that $f \geq \mathbf{1}$ μ-almost everywhere. Since $\nu = f \cdot \mu$ is a probability measure, we conclude $f = \mathbf{1}$ μ-almost everywhere.
(b) By the Lebesgue decomposition $\nu = \nu_a + \nu_s$ holds with uniquely determined positive measures ν_a, ν_s such that $\nu_a \ll \mu$ and $\nu_s \perp \mu$. Clearly, $T_*\nu_a \ll T_*\mu = \mu$. We now show that $T_*\nu_s \perp T_*\mu = \mu$. Let A be such that $\nu_s(A) = 0 = \mu(A^c)$. Define $B := \bigcap_{n=0}^{\infty} T^{-n}A$. Then $T^{-1}B \supset B$, and $\nu_s(B) = 0 = \mu(B^c)$ since μ is T-invariant. Moreover, we have by $T^{-1}B \setminus B \subset A^c$ and $\nu_a \ll \mu$ that $\nu_a(T^{-1}B) = \nu_a(B)$ and therefore

$$\begin{aligned}\nu_s(B) + \nu_a(B) &= \nu(B) = \nu(T^{-1}B)\\ &= \nu_s(T^{-1}B) + \nu_a(T^{-1}B) = \nu_s(T^{-1}B) + \nu_a(B).\end{aligned}$$

It follows $\nu_s(T^{-1}B) = \nu_s(B) = 0$. Since $\mu(T^{-1}B^c) = 0$, it follows that $T_*\nu_s \perp \mu$.

Since $\nu = T_*\nu = T_*\nu_a + T_*\nu_s$, the uniqueness of the Lebesgue decomposition implies now $T_*\nu_a = \nu_a$. Therefore ν_a is an invariant measure with $\nu_a \ll \mu$. If $\nu_a = 0$, then $\nu = \nu_s \perp \mu$, and we are done. If $\nu_a \neq 0$, the probability measure $\nu_a/\nu_a(X)$ is absolutely continuous with respect to μ, so by part (a) $\nu_a/\nu_a(X) = \mu$.

This implies $\mu \ll \nu_a \leq \nu$. Since also ν is ergodic, this implies again by part (a) $\mu = \nu$.
(c) That $\mathrm{M}_1(K, T)$ is convex and weak* compact is left as Exercise 3.12. Let $\mu \in \mathrm{M}_1(K, T)$ and let A be an invariant set of (K, μ, T). Then also A^c is T-invariant. If $\mu(A) \in (0, 1)$, then both $\mu_A/\mu(A)$ and $\mu_{A^c}/\mu(A^c)$ are T-invariant, probability measures, and μ is a non-trivial convex combination of these, showing $\mu \notin \mathrm{Ex}(\mathrm{M}_1(K, T))$. Conversely, suppose that (K, μ, T) is ergodic and $\mu = (1-t)\mu_1 + t\mu_2$ for some $t \in (0, 1)$ and $\mu_1, \mu_2 \in \mathrm{M}_1(K, T)$. Then $\mu_1 \ll \mu$, so from part (a) we deduce $\mu_1 = \mu$, and hence $\mu_2 = \mu$. ■

Remark 3.35 By the Krein–Milman Theorem 1.9, the set $\mathrm{Ex}(\mathrm{M}_1(K, T))$ of extreme points is non-empty, i.e., there is always an ergodic measure $\mu \in \mathrm{M}_1(K, T)$. Furthermore, we see that $\mathrm{M}_1(K, T)$ is a singleton if and only if there is a unique *ergodic* measure $\mu \in \mathrm{M}_1(K, T)$. Recall that in the latter case we call the topological system (K, T) **uniquely ergodic**.

Next we connect invariant measures and invariant sets for topological systems. Recall, see Sect. 1.3, that the **support** $\mathrm{supp}(\mu)$ of a positive measure $\mu \in \mathrm{M}(K)$ is the closed set

$$\mathrm{supp}(\mu) := \{x \in K : \mu(U) > 0 \text{ for each open set } U \text{ with } x \in U\}.$$

Proposition 3.36 (Invariant Measures vs. Invariant Sets) *Let (K, T) be a topological system.*

(a) *For $\mu \in \mathrm{M}_1(K, T)$ we have $T\mathrm{supp}(\mu) = \mathrm{supp}(\mu)$. In particular, $\mathrm{supp}(\mu)$ is a closed, invariant set in the topological system (K, T).*
(b) *Let $L \subset K$ be a closed, invariant set. Then there is an ergodic measure $\mu \in \mathrm{M}_1(K, T)$ with $\mathrm{supp}(\mu) \subset L$.*
(c) *The system (K, T) is minimal if and only if $\mathrm{supp}(\mu) = K$ for every ergodic $\mu \in \mathrm{M}_1(K, T)$.*

Proof (a) We first prove that $T\mathrm{supp}(\mu) \subset \mathrm{supp}(\mu)$. Let $y \in T\mathrm{supp}(\mu)$ and let U be an open set with $y \in U$. Then there is $x \in \mathrm{supp}(\mu)$ with $Tx = y$, and by continuity there is an open set V with $x \in V$ and $TV \subset U$. Since $V \subset T^{-1}(TV) \subset T^{-1}U$ and by the definition of the support, we obtain $0 < \mu(V) \leq \mu(T^{-1}U) = \mu(U)$. It follows that $y \in \mathrm{supp}(\mu)$.

To see the converse inclusion $\mathrm{supp}(\mu) \subset T\mathrm{supp}(\mu)$ let $f \in \mathrm{C}(K)$ vanish on the compact set $T\mathrm{supp}(\mu)$ but otherwise arbitrary. Then

$$\int_K f \,\mathrm{d}\mu = \int_K f \,\mathrm{d}T_*\mu = \int_K f \circ T \,\mathrm{d}\mu = \int_{\mathrm{supp}(\mu)} f \circ T \,\mathrm{d}\mu = 0.$$

By Proposition 1.20 we obtain that $\mathrm{supp}(\mu) \subset T\mathrm{supp}(\mu)$.

(b) By Remark 3.35 there is an ergodic $\nu \in \mathrm{M}_1(L, T)$. For $B \in \mathcal{B}(K)$ define $\mu(B) := \nu(B \cap L)$. Then $\mu \in \mathrm{M}_1(K, T)$ and $\mathrm{supp}(\mu) \subset L$. To show that μ is ergodic, let $A \subset K$ be an invariant set for (K, μ, T). Then $0 = \mu(A \setminus T^{-1}A) = \nu((A \cap L) \setminus (T^{-1}(A \cap L)))$. The ergodicity of ν implies $\mu(A) = \nu(A \cap L) \in \{0, 1\}$.
(c) To show that under the support condition (K, T) is minimal we can apply part (b). Conversely, suppose that (K, T) is minimal. Then by part (a) $\mathrm{supp}(\mu) = K$ must hold for every invariant, in particular, for every ergodic measure μ. ■

We obtain the following immediate consequence.

Corollary 3.37 (Unique Ergodicity and Minimality) *A uniquely ergodic topological system with strictly positive invariant measure is minimal.*

Remark 3.38 A minimal topological system need not be uniquely ergodic, see Raimi [862]. And, in general, the support of an ergodic measure need not be minimal, see Chou [242].

The following is our first encounter with an ergodic theorem, the general study is postponed to Chap. 7.

Proposition 3.39 *For $a \in \mathbb{T}^d$ consider the torus rotation $(\mathbb{T}^d, a)$, and denote its Koopman operator on $\mathrm{C}(\mathbb{T}^d)$ by L_a. Then for each $f \in \mathrm{C}(\mathbb{T}^d)$ the limit*

$$\lim_{N\to\infty} \frac{1}{N} \sum_{n=0}^{N-1} L_a^n f$$

exists in $\mathrm{C}(\mathbb{T}^d)$. If $a_1, \ldots, a_d$ are rationally independent, then the limit equals $\int_{\mathbb{T}^d} f \,\mathrm{dm}^d \cdot \mathbf{1}$.

Proof For $N \in \mathbb{N}$ let $A_N := \frac{1}{N}\sum_{n=0}^{N-1} L_a^n$. Then $A_N \in \mathscr{L}(\mathrm{C}(\mathbb{T}^d))$ with $\|A_N\| \le 1$ for each $N \in \mathbb{N}$. Therefore, it suffices to prove the convergence of $A_N f$ for f in the dense subspace $\mathrm{P}(\mathbb{T}^d)$ of trigonometric polynomials (see Proposition 1.2). We may also restrict our attention to trigonometric monomials $\mathbf{z}^m$, where $m \in \mathbb{Z}^d$. We then have $L_a\mathbf{z}^m = a_1^{m_1} \cdots a_d^{m_d} \cdot \mathbf{z}^m$. If $b := a_1^{m_1} \cdots a_d^{m_d} = 1$, we readily see that $A_N\mathbf{z}^m = \mathbf{z}^m$ for each $N \in \mathbb{N}$. Otherwise we obtain

$$A_N\mathbf{z}^m = \frac{1}{N}\sum_{n=0}^{N-1} b^n\mathbf{z}^m = \frac{1}{N}\frac{b^N - 1}{b-1}\mathbf{z}^m \to 0 \text{ as } N \to \infty.$$

The last assertion follows by linearity and by the above calculation. Indeed, in the case of rational independence for every $m \in \mathbb{Z}^d \setminus \{0\}$ the limit of $A_N\mathbf{z}^m$ equals $0 = \int_{\mathbb{T}^d} \mathbf{z}^m \,\mathrm{dm}^d \cdot \mathbf{1}$, and the limit of $A_N\mathbf{z}^0 = \mathbf{1}$ equals $\mathbf{1} = \int_{\mathbb{T}^d} \mathbf{z}^0 \,\mathrm{dm}^d \cdot \mathbf{1}$. ■

As a corollary we obtain the promised classical result.

Theorem 3.40 (Kronecker) *The torus rotation* $(\mathbb{T}^d, a)$ *with some given* $a = (a_1, \ldots, a_d) \in \mathbb{T}^d$ *is minimal if and only if* $a_1, \ldots, a_d$ *are rationally independent. In this case* $(\mathbb{T}^d, a)$ *is uniquely ergodic, and the Haar measure is the unique (ergodic) invariant measure.*

Proof If $a_1^{m_1} \cdots a_d^{m_d} = 1$ for some $m = (m_1, \ldots, m_d) \in \mathbb{Z}^d \setminus \{0\}$, then

$$F := \{z \in \mathbb{T}^d : z_1^{m_1} \cdots z_d^{m_d} = 1\}$$

is a non-trivial, closed, invariant set. Therefore $(\mathbb{T}^d, a)$ is not minimal.

Conversely, suppose $a_1^{m_1} \cdots a_d^{m_d} \neq 1$ for each $m \neq 0$. Proposition 3.25 and the hypothesis yield that m^d is an ergodic measure for $(\mathbb{T}^d, a)$. Let $\mu \in \mathrm{M}_1(\mathbb{T}^d, a)$ be any ergodic measure. By the previous proposition and by the dominated convergence theorem $\langle f, \mu\rangle = \langle A_N f, \mu\rangle \to \langle f, \mathrm{m}^d\rangle\langle \mathbf{1}, \mu\rangle = \langle f, \mathrm{m}^d\rangle$ as $N \to \infty$ for every $f \in \mathrm{C}(\mathbb{T}^d)$. This implies $\mu = \mathrm{m}^d$. Since $\mathrm{supp}(\mathrm{m}^d) = \mathbb{T}^d$, Proposition 3.36 yields that $(\mathbb{T}^d, a)$ is minimal. ■

Thus the following characterization of torus rotations is a direct corollary of Theorem 3.40, Propositions 3.2 and 3.25.

Corollary 3.41 *For the torus rotation* $(\mathbb{T}^d, a)$ *with some* $a = (a_1, \ldots, a_d) \in \mathbb{T}^d$ *endowed with the Haar measure, the properties transitivity, minimality, ergodicity, unique ergodicity and rational independence of* $a_1, \ldots, a_d$ *are equivalent.*

The analogous equivalence for general group rotation systems will be given in Theorem 11.26 below, with and yet another generalization to **nilsystems** in Chap. 14.

3.6 Weak and Strong Mixing

Next we discuss two important properties which a measure-preserving system may or may not have. Recall that by Proposition 3.4 for an ergodic system (X, μ, T) and for sets $A, B \subset X$ with $\mu(A) > 0$ and $\mu(B) > 0$ there are infinitely many $n \in \mathbb{N}$ such that $\mu(T^{-n}A \cap B) > 0$. That is to say, for infinitely many times the region A is "mixed" with region B under the dynamics. The following two properties present even stronger requirements.

Definition 3.42 (Mixing and Weak Mixing) Let (X, μ, T) be a measure-preserving system. The system (and the transformation T itself) is called **(strongly) mixing** if for each pair $A, B \subset X$ of measurable sets

$$\lim_{n\to\infty} \mu(T^{-n}A \cap B) = \mu(A)\mu(B),$$

and **weakly mixing** if for each measurable set $A \subset X$ there is a subsequence $(n_k)_{k\in\mathbb{N}}$ of $\mathbb{N}$ such that for each measurable set $B \subset X$

$$\lim_{k\to\infty} \mu(T^{-n_k} A \cap B) = \mu(A)\mu(B).$$

For further notions of mixing such as α-mixing, partial mixing, mild mixing and sequence mixing we refer to Friedman and Ornstein [435, 436], Blum et al. [161], Furstenberg and Weiss [456], Friedman and Thomas [437], Friedman and King [433].

Remark 3.43 Clearly, every strongly mixing system is weakly mixing. Moreover, every weakly mixing system is ergodic. Indeed, if A is an invariant set in a weakly mixing system, then for the corresponding subsequence $(n_k)_{k\in\mathbb{N}}$

$$\mu(A) = \lim_{k\to\infty} \mu(T^{-n_k} A \cap A) = \mu(A)^2.$$

It follows that $\mu(A) \in \{0, 1\}$, implying ergodicity. We will see in Remark 6.30 that every weakly mixing system (X, μ, T) is in fact **totally ergodic**, i.e., (X, μ, T^k) is ergodic for every $k \in \mathbb{N}$.

Thus we have the implications

$$\text{mixing} \implies \text{weak mixing} \implies \text{total ergodicity} \implies \text{ergodicity}.$$

All the converse implications are false. Indeed, a two-points rotation is ergodic, but not totally ergodic since its square is the identity. Moreover, there are many natural totally ergodic systems, such as ergodic rotations on $\mathbb{T}$, that are not weakly mixing, see Example 3.46 and also Example 8.40. To show that weak mixing does not imply mixing is much more difficult. Halmos [513] showed in 1944 that for the interval [0, 1] endowed with the Lebesgue measure (or, more generally, for every non-atomic **regular probability space**, i.e., compact metric space with Borel algebra and a Borel probability measure, see Remarks 8.48, 8.50 below) the set of invertible weakly mixing transformations is large in the sense of Baire category (a dense G_δ set), while Rokhlin [877] showed in 1948 that strongly mixing transformations form a small (meager) set. Here, the considered topology comes from the weak (= strong) operator topology of corresponding Koopman operators and their adjoints and makes the set of all invertible measure-preserving transformations on the given regular probability space into a **complete metric** space. These results imply that a "typical" invertible measure-preserving transformation is weakly mixing but not mixing and initiated the study of "typical" properties of (invertible) measure-preserving transformations. For more such properties see, e.g., the book by Nadkarni [813] as well as Katok [633], del Junco and Lemańczyk [282], Ageev [13], de la Rue and de Sam Lazaro [274], Bashtanov [81].

The first explicit example of non-mixing but weakly mixing transformation was published by Chacon [236] only in 1969 (with a precursor by Kakutani and von Neumann, see Kakutani [613]).

Although the definition of weak mixing given above is natural and simple, it is not the standard one. Also for applications it is not very useful. In Sect. 6.5 we will give an equivalent definition and several characterizations of weak mixing which are much easier to handle, and in Chap. 8 this list of characterizations will be augmented even more. In particular, we will show in Proposition 6.25 that the corresponding sequence (n_k) can be chosen to have density 1 (which is usually included in the definition of weak mixing). Here we just mention the following characterization of *mixing*.

Proposition 3.44 (Characterization of Strong Mixing) *Let (X, μ, T) be a measure-preserving system with Koopman operator T on $\mathrm{L}^0(X, \mu)$. For $p, q \in [1, \infty]$ satisfying $\frac{1}{p} + \frac{1}{q} = 1$ let $D \subset \mathrm{L}^p(X, \mu)$ and $D' \subset \mathrm{L}^q(X, \mu)$ be subsets whose linear hull is dense in the respective spaces. Let finally $\mathcal{E}$ and $\mathcal{E}'$ be $\cap$-stable generators*[5] *of the underlying σ-algebra. Then the following assertions are equivalent.*

(i) *The system (X, μ, T) is strongly mixing.*
(ii) *For every $A \in \mathcal{E}$ and $B \in \mathcal{E}'$*

$$\lim_{n\to\infty} \mu(T^{-n}A \cap B) = \mu(A)\mu(B).$$

(iii) *For every $f \in D$ and every $g \in D'$*

$$\lim_{n\to\infty} \langle T^n f, g\rangle = \langle f, \mathbf{1}\rangle\langle \mathbf{1}, g\rangle. \tag{3.4}$$

(iv) *For every $f \in \mathrm{L}^p(X, \mu)$ and every $g \in \mathrm{L}^q(X, \mu)$*

$$\lim_{n\to\infty} \langle T^n f, g\rangle = \langle f, \mathbf{1}\rangle\langle \mathbf{1}, g\rangle.$$

Proof The implications (i)⇒(ii) and (iv)⇒(iii) are trivial, while the implication (iv)⇒(i) follows by specializing $f = \mathbf{1}_A$ and $g = \mathbf{1}_B$ in (iv).
(iii)⇒(iv): This is a standard approximation argument, see Exercise 3.15.
(ii)⇒(i): Consider the subsets $F := \mathrm{lin}\{\mathbf{1}_A : A \in \mathcal{E}\}$ and $F' := \mathrm{lin}\{\mathbf{1}_B : B \in \mathcal{E}'\}$, both of which are dense in $\mathrm{L}^2(X, \mu)$. Hypothesis (ii) yields the equality (3.4) for $f = \mathbf{1}_A$, $g = \mathbf{1}_B$ with $A \in \mathcal{E}$, $B \in \mathcal{E}'$. By linearity, (3.4) holds for all $f \in F$ and $g \in F'$. By the implication (iii)⇒(iv) we obtain (iv) for $p = q = 2$, and then (i) follows.

[5] Recall that $\mathcal{E}$ is $\cap$-stable if $A, B \in \mathcal{E}$ implies $A \cap B \in \mathcal{E}$, see Sect. 1.3.

(i)⇒(iv): Apply the already proven implication (iii)⇒(iv) to the space $D = D'$ of simple functions, for which (3.4) holds by linearity and hypothesis (i). ∎

An analogous characterization of weakly mixing systems is true (see Proposition 6.26), but for the proof some further techniques are needed that are introduced in the next chapters. An important point to make here is that if for some measurable $A \subset X$ there is a subsequence $(n_k)_{k\in\mathbb{N}}$ of $\mathbb{N}$ such that for each $B \subset X$ measurable set $\lim_{k\to\infty} \mu(T^{-n_k}A \cap B) = \mu(A)\mu(B)$, then there is even a subsequence $(n'_k)_{k\in\mathbb{N}}$ of density 1 along which this convergence holds.[6] We close this chapter with the following two examples.

Example 3.45 (Shifts are Mixing) Each (one- or two-sided) shift with arbitrary state space (Y, ν), in particular, each Bernoulli shift $B(p_0, \ldots, p_{k-1})$ is mixing and hence weakly mixing. A proof is based on the equivalence (i)⇔(ii) in the previous characterization, see Exercise 3.16. Moreover, Markov shifts from Example 2.20 are mixing if and only if there exists $n \in \mathbb{N}$ such that M^n is strictly positive, where M denotes the corresponding transition matrix, see, e.g., Petersen [848, Thm. 2.5.6] or [342, Thm. 9.10]. In fact, every mixing Markov shift is isomorphic to a Bernoulli shift by Friedman and Ornstein [434].

As mentioned above, simple examples of not weakly mixing systems are torus rotations.

Example 3.46 (Torus Rotations are not Weakly Mixing) A rotation system $(\mathbb{T}, \mathrm{m}, a)$ is not weakly mixing. We leave the proof of this as Exercise 3.17.

As a matter of fact, non-trivial group rotations are not weakly mixing. We shall return to this fact in Example 8.40, after we discussed the *Kronecker factor* of measure-preserving systems in Sect. 8.3.

Remark 3.47 (Exact Transformations) One calls a measure-preserving system $(X, \mathcal{X}, \mu, T)$, or the transformation T **exact** if the σ-algebra

$$\bigcap_{n\in\mathbb{N}} T^{-n}\mathcal{X}$$

is trivial, i.e., it contains sets of measure 0 or 1 only. An exact transformation is mixing,[7] see, e.g., Walters [1033, §4.12], but the converse is not true. No non-trivial, essentially invertible system is exact, but some of such systems are mixing such as a two-sided Bernoulli shift. We leave the details as Exercise 3.18.

---

[6] This amazing fact will be shown in Chap. 6. That $(n'_k)_{k\in\mathbb{N}}$ has density 1 means $\lim_{N\to\infty} \frac{|\{k: n'_k \le N\}|}{N} = 1$. Density, its properties, and related notions will be discussed in Sect. 4.3.

[7] For a stronger property of strong convergence of (T'^n) for the adjoint T' of the Koopman operator T see Horowitz [545] with an abstract operator theoretic proof due to Derriennic [294].

Exercises

Exercise 3.1 (Invariant Sets) Let (K, T) be an invertible topological system and let $A \subset K$ be a non-empty, closed, invariant set. Show that there exists a non-empty, closed, two-sided invariant set $B \subset A$.

Exercise 3.2 (Minimality) Prove the characterization of minimality via the orbits as explained in Remark 3.1. Prove also that different minimal subsystems are disjoint.

Exercise 3.3 (Subgroups of $\mathbb{T}$) Prove that a subgroup G of $\mathbb{T}$ is either finite or it is dense in $\mathbb{T}$.

Exercise 3.4 (Minimality and the Koopman Operator) Prove that if a topological system (K, T) is topologically transitive (in particular, if it is minimal), then for the Koopman operator T on $\mathrm{C}(K)$ one has $\mathrm{fix} T = \ker(I - T) = \mathbb{C}\mathbf{1}$. Prove the analogous statement for topologically two-sided transitive invertible systems.

Exercise 3.5 (Group Rotations) Prove that if a group rotation (G, a) is minimal, then G is Abelian. Is the converse true?

Exercise 3.6 (Almost Periodic Points) Prove Propositions 3.7 and 3.8.

Exercise 3.7 (Periodic Points) Let (K, T) be a topological system. A point $x \in K$ is called **periodic** in (K, T) if there is $p \in \mathbb{N}$ such that $T^p x = x$. Show that every periodic point is almost periodic. Is the converse true? Present an example of a topological system without any periodic point.

Exercise 3.8 (Almost Periodic and Recurrent Points for Shifts) Consider the compact space $K = \{0, 1\}^{\mathbb{N}}$ and the one-sided, shift system (K, T) from Example 2.31.

(a) Prove that a point $x \in K$ is almost periodic if and only if for each finite subword y of x there is $\ell \in \mathbb{N}$ such that the gap between each two subsequent occurrences of y is of length at most ℓ.
(b) Give an example of an almost periodic but not periodic point $x \in K$ (cf. Exercise 3.7).
(c) Give an example of a point $x \in K$ which is recurrent but not almost periodic (cf. Proposition 3.7).

What changes if one considers an alphabet with more letters and/or a two-sided shift?

Exercise 3.9 (Invariant Sets) Let (X, μ, T) be a measure-preserving system. Prove that a set $A \in \mathcal{X}$ is invariant if and only if $A \subset T^{-1}A$ up to a null set, if and only if $T^{-1}A \subset A$ up to a null set, and if and only if $T\mathbf{1}_A = \mathbf{1}_A$ a.e. for the Koopman operator T. Show moreover that the T-invariant sets form a σ-algebra.

Exercise 3.10 (Ergodicity) Let (X, μ, T) be a measure-preserving system. Prove that it is ergodic if and only if each measurable set $A \subset X$ with $T^{-1}A = A$ satisfies $\mu(A) \in \{0, 1\}$.

Exercise 3.11 (Spectrum of Koopman Operators)

(a) Prove Proposition 3.28.
(b) Give an example of a measure-preserving system such that the spectrum of the Koopman operator on $\mathrm{L}^0(X, \mu)$ is not a group.
(c) Compute the spectrum of the Koopman operator of the topological dynamical system (K, T), where $K = \{1, 2, 3\}$ and φ is given by $1 \mapsto 2 \mapsto 3 \mapsto 2$. Modify this example to obtain a topological system for which the spectrum of the Koopman operator is an arbitrary set of the form $M \cup \{0\}$, where M is a union of finitely many subgroups of $\mathbb{T}$ of roots of unity.
(d) Present an example of a topological dynamical system for which the spectrum of the corresponding Koopman operator equals $\mathbb{T} \cup \{0\}$.

Exercise 3.12 (Invariant Measures) Let (K, T) be a topological system. Show that $\mathrm{M}_1(K, T)$ of all T-invariant regular, Borel, probability measures on K is a convex and weak* compact subset of $\mathrm{M}(K)$.

Exercise 3.13 (Rotations on the Unit Disc) Consider the closed unit disc $\overline{\mathbb{D}} := \{z \in \mathbb{C} : |z| \leq 1\}$, and let $a \in \mathbb{T}$ be not a root of unity. The rotation $\tau_a : \overline{\mathbb{D}} \to \overline{\mathbb{D}}, z \mapsto az$ is continuous, and gives rise to the invertible topological system $(\overline{\mathbb{D}}, \tau_a)$. Describe all closed, τ_a-invariant sets, all invariant probability measures, and determine the ergodic ones.

Exercise 3.14 (Unique Ergodicitiy of Isomorphic Systems)

(a) Let (K, T) and (L, S) be isomorphic topological systems. Prove that (K, T) is uniquely ergodic if and only if (L, S) is uniquely ergodic.
(b) Show that the shift $([0, 1), \alpha)$ and the rotation $(\mathbb{T}, a)$, $a := \mathrm{e}^{2\pi \mathrm{i}\alpha}$, are (topologically) isomorphic.
(c) Prove that the skew shift $([0, 1)^2, \widetilde{T_\alpha})$ and the skew rotation $(\mathbb{T}^2, T_a)$, $a := \mathrm{e}^{2\pi \mathrm{i}\alpha}$ (from Example 2.31) are (topologically) isomorphic, and therefore simultaneously uniquely ergodic or not.

Exercise 3.15 (Strong Mixing) Prove the implication (iii)$\Rightarrow$(iv) in Proposition 3.44.

Exercise 3.16 (Shifts Are Strongly Mixing) Prove that a shift (X, μ, T) with state space (Y, ν) is strongly mixing.

Exercise 3.17 (Not Weakly Mixing Systems) Show, by using the definition only, that rotations on finite groups and on $\mathbb{T}$ are not weakly mixing.

Exercise 3.18 (Exact Transformations) Prove that a non-invertible Bernoulli shift is exact and that an exact transformation is mixing.

Chapter 4
Some Fourier Analysis

In this chapter we study the Banach space $\mathrm{M}(\mathbb{T})$ of complex, Borel measures on the unit circle (endowed with the norm given by the total variation) in some detail. For this purpose we introduce the Fourier transform and the related Koopman–von Neumann lemma as fundamental and extremely powerful tools.

4.1 Fourier Transform

Recall that for $n \in \mathbb{Z}$ the function $\mathbf{z}^n : \mathbb{T} \to \mathbb{T}$ is defined by $z \mapsto z^n$ and the linear combinations of such functions, the trigonometric polynomials, form a dense subspace of $\mathrm{C}(\mathbb{T})$ (see Corollary 1.12), and hence of $\mathrm{L}^2(\mathbb{T}, \mu)$ for each positive measure $\mu \in \mathrm{M}(\mathbb{T})$. Recall also the following estimate for a measure $\mu \in \mathrm{M}(\mathbb{T})$ and $f \in \mathrm{L}^1(\mathbb{T}, |\mu|)$

$$\left| \int_{\mathbb{T}} f \, \mathrm{d}\mu \right| \le \int_{\mathbb{T}} |f| \, \mathrm{d}|\mu| = \|f\|_{\mathrm{L}^1(\mathbb{T},|\mu|)}.$$

Definition 4.1 Let $\mu \in \mathrm{M}(\mathbb{T})$, and let $n \in \mathbb{Z}$. We call

$$\hat{\mu}(n) := \int_{\mathbb{T}} \mathbf{z}^{-n} \, \mathrm{d}\mu$$

the nth **Fourier coefficient** (or nth **Fourier–Stieltjes coefficient**) of μ. The sequence $\hat{\mu} = (\hat{\mu}(n))_{n\in\mathbb{Z}}$ is called the **Fourier transform of** μ, and the mapping $\mathscr{F} : \mu \mapsto \hat{\mu}$ the **Fourier transform** (or **Fourier–Stieltjes transform**).

Proposition 4.2 *The Fourier transform* $\mathscr{F} : \mathrm{M}(\mathbb{T}) \to \ell^\infty(\mathbb{Z})$ *is an injective, linear contraction.*

T. Eisner, B. Farkas, *A Journey Through Ergodic Theorems*, Birkhäuser Advanced Texts Basler Lehrbücher, https://doi.org/10.1007/978-3-031-96506-7_4

Proof It is clear that $\mathscr{F} : \mathrm{M}(\mathbb{T}) \to \mathbb{C}^{\mathbb{Z}}$ is linear. For each $n \in \mathbb{Z}$ we have $|\hat{\mu}(n)| \leq \langle |\mathbf{z}^{-n}|, |\mu| \rangle \leq |\mu|(\mathbb{T}) = \|\mu\|$, so $\mathscr{F}$ maps into $\ell^\infty(\mathbb{Z})$ and it is a contraction. To show injectivity, suppose $\hat{\mu}(n) = 0$ for each $n \in \mathbb{Z}$. Then $\langle p, \mu \rangle = 0$ holds for every trigonometric polynomial $p \in \mathrm{P}(\mathbb{T})$. Since $\mathrm{P}(\mathbb{T})$ is dense in $\mathrm{C}(\mathbb{T})$, we obtain $\mu = 0$. ∎

We now relate the Fourier transform of measures to the Fourier transform of functions. If we identify $\mathrm{L}^1(\mathbb{T}, \mathrm{m})$ with a closed subspace of $\mathrm{M}(\mathbb{T})$ (under the isometry $J_{\mathrm{m}} : f \mapsto f \cdot \mathrm{m}$ from Proposition 1.26, where m is the Haar measure on $\mathbb{T}$) we get back the classical definition of the Fourier transform of L^1-functions given by

$$\hat{f}(n) = \int_{\mathbb{T}} f\mathbf{z}^{-n} \, \mathrm{dm} \text{ for } f \in \mathrm{L}^1(\mathbb{T}, \mathrm{m}) \text{ and } n \in \mathbb{Z},$$

or, in a more familiar way (after identifying $[0, 1)$ with addition mod 1 with $\mathbb{T}$, cf. Example 2.13),

$$\hat{f}(n) = \int_{[0,1]} f(s)\mathrm{e}^{-2\pi ins} \, \mathrm{d}s \text{ for } f \in \mathrm{L}^1([0, 1], \lambda) \text{ and } n \in \mathbb{Z}.$$

Thus Definition 4.1 consistently extends the notion of Fourier transform of functions. Since $(\mathbf{z}^n)_{n\in\mathbb{Z}}$ is an orthonormal basis in $\mathrm{L}^2(\mathbb{T}, \mathrm{m})$, we immediately obtain the following important fact, which we record here for completeness.

Proposition 4.3 (Parseval's Identity) *The Fourier transform*

$$\mathscr{F} : \mathrm{L}^2(\mathbb{T}, \mathrm{m}) \to \ell^2(\mathbb{Z})$$

is a unitary operator.

Next some basic computational rules for the Fourier transform come, see Exercise 4.1.

Proposition 4.4 (Properties of the Fourier Transform) *Let $\mu, \nu \in \mathrm{M}(\mathbb{T})$.*

(a) *For the rotation $\tau_a : \mathbb{T} \to \mathbb{T}$ by $a \in \mathbb{T}$ we have $\widehat{(\tau_a)_*\mu}(n) = a^{-n}\hat{\mu}(n)$ for every $n \in \mathbb{Z}$.*
(b) *For $m, n \in \mathbb{Z}$ we have $\widehat{\mathbf{z}^m \cdot \mu}(n) = \hat{\mu}(n - m)$.*
(c) *For $n \in \mathbb{Z}$ we have $\hat{\overline{\mu}}(n) = \overline{\hat{\mu}(-n)}$, where $\overline{\mu}$ is the complex conjugate of the measure μ.*

Our next aim is to study the asymptotic behavior of the sequence $(\hat{\mu}(n))_{n\in\mathbb{Z}}$. A measure $\mu \in \mathrm{M}(\mathbb{T})$ is called **Rajchman** if

$$\lim_{|n|\to\infty} \hat{\mu}(n) = 0.$$

For example, for $f \in \mathrm{L}^2(\mathbb{T}, \mathrm{m})$ the measure $f \cdot \mathrm{m}$ is Rajchman by, e.g., Parseval's identity. Rajchman measures form a closed subspace of $\mathrm{M}(\mathbb{T})$, see Exercise 4.2.

Proposition 4.5 *Let $\mu \in \mathrm{M}(\mathbb{T})$ be a Rajchman measure. Then every $\nu \in \mathrm{M}(\mathbb{T})$ with $\nu \ll \mu$ is a Rajchman measure.*

Proof Since μ is Rajchman, so is the measure $\mathbf{z}^k \cdot \mu$ for each fixed $k \in \mathbb{Z}$ (see Proposition 4.4 (b)). As a consequence, $p \cdot \mu$ is a Rajchman measure for each trigonometric polynomial $p \in \mathrm{P}(\mathbb{T})$. Since $\nu \ll |\mu|$, by the Radon–Nikodym theorem (Theorem 1.27), we can write $\nu = g \cdot |\mu|$ for some $g \in \mathrm{L}^1(\mathbb{T}, |\mu|)$. We have the **polar decomposition** $\mu = h \cdot |\mu|$ for some measurable $h : \mathbb{T} \to \mathbb{T}$, so that $\nu = g\overline{h} \cdot \mu$. For $f := g\overline{h}$ and for every trigonometric polynomial p we obtain

$$\begin{aligned}\limsup_{|n|\to\infty} \left|\mathscr{F}(f \cdot \mu)(n)\right| &\le \limsup_{|n|\to\infty} \left|\mathscr{F}((f-p) \cdot \mu)(n)\right| \le \|\mathscr{F}((f-p) \cdot \mu)\|_\infty \\ &\le \|f - p\|_{\mathrm{L}^1(\mathbb{T},|\mu|)}.\end{aligned}$$

With an appropriate choice of $p \in \mathrm{P}(\mathbb{T})$ the term $\|f - p\|_{\mathrm{L}^1(\mathbb{T},|\mu|)}$ can be made arbitrarily small. This proves the assertion. ∎

Since m is a Rajchman measure, we obtain the classical Riemann–Lebesgue Lemma stating that all measures $\mu \in \mathrm{M}(\mathbb{T})$ which are absolutely continuous with respect to m are Rajchman.

Theorem 4.6 (Riemann–Lebesgue Lemma) *For $f \in \mathrm{L}^1(\mathbb{T}, \mathrm{m})$ we have*

$$\lim_{|n|\to\infty} \hat{f}(n) = 0.$$

Remark 4.7 Since μ, $\overline{\mu}$ are equivalent measures, and since $\overline{\hat{\mu}(-n)} = \hat{\overline{\mu}}(n)$ for each $n \in \mathbb{Z}$, the proof of Proposition 4.5 yields (considering $\limsup_{n\to\infty}$ instead of $\limsup_{|n|\to\infty}$) that

$$\mu \text{ is Rajchman} \iff \lim_{n\to\infty} \hat{\mu}(n) = 0 \iff \lim_{n\to-\infty} \hat{\mu}(n) = 0.$$

4.2 Discrete and Continuous Measures

The Fourier transform of a measure $\mu \in \mathrm{M}(\mathbb{T})$ is very useful when studying qualitative properties of μ. The following is an elementary example for this.

Proposition 4.8 *For $\mu \in \mathrm{M}(\mathbb{T})$ we have*

$$\lim_{N\to\infty} \frac{1}{N} \sum_{n=0}^{N-1} \hat{\mu}(n) = \mu(\{1\}).$$

Proof The assertion follows from the formula $\sum_{n=0}^{N-1} z^n = \frac{z^N-1}{z-1}$ for $z \neq 1$, and from Lebesgue's dominated convergence theorem. ■

Therefore, the limit of arithmetic averages of the Fourier coefficients of μ tell us whether μ contains the Dirac measure δ_1 as a component. As we shall see shortly, with similar arguments we can see from the Fourier coefficients whether μ vanishes on all at most countable sets. We recall the name of this latter property.

Definition 4.9 Let $\mu \in \mathrm{M}(\mathbb{T})$ be a complex measure. A point $a \in \mathbb{T}$ is called an **atom** of μ if $\mu\{a\} \neq 0$ (from now on we use the abbreviation $\mu\{a\}$ for $\mu(\{a\})$). The measure μ is called

- **continuous** if it has no atoms, i.e., if μ vanishes on all at most countable sets.
- **discrete** if there exists an at most countable set $A \subset \mathbb{T}$ and a function $c : A \to \mathbb{C} \setminus \{0\}$ such that

$$\mu = \sum_{a \in A} c_a \delta_a.$$

  Here, δ_a is the Dirac measure at a, and the series here is unconditionally convergent in $\mathrm{M}(\mathbb{T})$.

A measure which is continuous and discrete at the same time is identically 0.

The following decomposition is classical and will be of extreme importance in the next chapter. The proof is left as Exercise 4.6.

Proposition 4.10 (Continuous–Discrete Decomposition)

(a) *The subsets*

$$\mathrm{M_c}(\mathbb{T}) := \{\mu : \mu \in \mathrm{M}(\mathbb{T}) \text{ is continuous}\},$$
$$\mathrm{M_d}(\mathbb{T}) := \{\mu : \mu \in \mathrm{M}(\mathbb{T}) \text{ is discrete}\}$$

are closed subspaces of $\mathrm{M}(\mathbb{T})$.

(b) *For each $\mu \in \mathrm{M}(\mathbb{T})$ the set of atoms of μ is at most countable and there is a unique continuous measure μ_c and a unique discrete measure μ_d such that $\mu = \mu_\mathrm{c} + \mu_\mathrm{d}$. Moreover, we have*

$$\|\mu_\mathrm{d}\| = |\mu_\mathrm{d}|(\mathbb{T}) = \sum_{a \in \mathbb{T}} |\mu\{a\}|.$$

The following result of Wiener connecting continuity of a measure μ to the asymptotic behavior of $\hat{\mu}$ has an elementary proof and, as we shall see, a number of important applications.

Proposition 4.11 (Wiener's Lemma) *For* $\mu \in \mathrm{M}(\mathbb{T})$

$$\lim_{N\to\infty} \frac{1}{N} \sum_{n=0}^{N-1} |\hat{\mu}(n)|^2 = \sum_{a\in\mathbb{T}} |\mu\{a\}|^2. \tag{4.1}$$

In particular, μ *is continuous if and only if*

$$\lim_{N\to\infty} \frac{1}{N} \sum_{n=0}^{N-1} |\hat{\mu}(n)|^2 = 0.$$

Proof Let $\Delta = \{(z, z) : z \in \mathbb{T}\} \subset \mathbb{T} \times \mathbb{T}$ be the diagonal. Fubini's theorem and Lebesgue's dominated convergence theorem imply

$$\begin{aligned}
\frac{1}{N} \sum_{n=0}^{N-1} |\hat{\mu}(n)|^2 &= \frac{1}{N} \sum_{n=0}^{N-1} \int_{\mathbb{T}} z^{-n} \,\mathrm{d}\mu(z) \int_{\mathbb{T}} \overline{w}^{-n} \,\mathrm{d}\overline{\mu}(w) \\
&= \int_{\mathbb{T}\times\mathbb{T}} \frac{1}{N} \sum_{n=0}^{N-1} (z\overline{w})^{-n} \,\mathrm{d}(\mu \times \overline{\mu})(z, w) \\
&= \int_{\Delta} \mathbf{1} \,\mathrm{d}(\mu \times \overline{\mu})(z, w) + \int_{\Delta^c} \frac{1}{N} \frac{(z\overline{w})^{-N} - 1}{z\overline{w} - 1} \,\mathrm{d}(\mu \times \overline{\mu})(z, w) \\
&\xrightarrow[N\to\infty]{} \int_{\mathbb{T}\times\mathbb{T}} \mathbf{1}_{\Delta} \,\mathrm{d}(\mu \times \overline{\mu})(z, w) = \int_{\mathbb{T}} \left(\int_{\mathbb{T}} \mathbf{1}_{\{w\}}(z) \,\mathrm{d}\mu(z) \right) \mathrm{d}\overline{\mu}(w) \\
&= \int_{\mathbb{T}} \mu\{w\} \,\mathrm{d}\overline{\mu}(w) = \sum_{a\in\mathbb{T}} |\mu\{a\}|^2.
\end{aligned}$$

The final assertion follows directly from (4.1). ■

Given $\nu \in \mathrm{M}(\mathbb{T})$ with $\nu \ll \mu$ let us denote by $\|\nu\|_2$ the $\mathrm{L}^2(|\mu|)$ norm of the Radon–Nikodym derivative of ν.

Proposition 4.12 (Wiener's Lemma, Uniform Version) *For each continuous measure* $\mu \in \mathrm{M}(\mathbb{T})$ *one has*

$$\lim_{N\to\infty} \sup_{\substack{\nu\ll\mu \\ \|\nu\|_2\le 1}} \frac{1}{N} \sum_{n=0}^{N-1} |\hat{\nu}(n)|^2 = 0.$$

In particular, for a continuous measure $\mu \in \mathrm{M}(\mathbb{T})$

$$\lim_{N\to\infty} \sup_{k\in\mathbb{Z}} \frac{1}{N} \sum_{n=0}^{N-1} |\hat{\mu}(n+k)|^2 = 0.$$

Proof Let $\nu \ll \mu$ and write $\nu = f \cdot \mu$ with the Radon–Nikodym derivative $f \in \mathrm{L}^2(\mu)$ satisfying $\|f\|_2 \leq 1$. Let $\Delta = \{(z, z) : z \in \mathbb{T}\} \subset \mathbb{T} \times \mathbb{T}$ be the diagonal. Fubini's theorem implies

$$\begin{aligned}
\frac{1}{N}\sum_{n=0}^{N-1} |\hat{\nu}(n)|^2 &= \frac{1}{N}\sum_{n=0}^{N-1} \int_{\mathbb{T}} z^{-n} \, \mathrm{d}\nu(z) \int_{\mathbb{T}} \overline{w}^{-n} \, \mathrm{d}\overline{\nu}(w) \\
&= \frac{1}{N}\sum_{n=0}^{N-1} \int_{\mathbb{T}} z^{-n} f(z) \, \mathrm{d}\mu(z) \int_{\mathbb{T}} \overline{w}^{-n}\overline{f(w)} \, \mathrm{d}\overline{\mu}(w) \\
&= \int_{\mathbb{T}\times\mathbb{T}} \frac{1}{N}\sum_{n=0}^{N-1} (z\overline{w})^{-n} f(z)\overline{f(w)} \, \mathrm{d}(\mu \times \overline{\mu})(z, w) \\
&= \int_{\Delta} f \otimes \overline{f} \, \mathrm{d}(\mu \times \overline{\mu}) + \int_{\Delta^c} \frac{1}{N} \frac{(z\overline{w})^{-N} - 1}{z\overline{w} - 1} f(z)\overline{f(w)} \, \mathrm{d}(\mu \times \overline{\mu})(z, w) \\
&= \int_{\Delta^c} \frac{1}{N} \frac{(z\overline{w})^{-N} - 1}{z\overline{w} - 1} f(z)\overline{f(w)} \, \mathrm{d}(\mu \times \overline{\mu})(z, w).
\end{aligned}$$

Now we can write by Cauchy–Schwarz that

$$\begin{aligned}
&\left| \int_{\Delta^c} \frac{1}{N} \frac{(z\overline{w})^{-N} - 1}{z\overline{w} - 1} f(z)\overline{f(w)} \, \mathrm{d}(\mu \times \overline{\mu})(z, w) \right|^2 \\
&\leq \left(\int_{\Delta^c} \left| \frac{1}{N} \frac{(z\overline{w})^{-N} - 1}{z\overline{w} - 1} f(z)\overline{f(w)} \right| \mathrm{d}\left|\mu \times \overline{\mu}\right|(z, w) \right)^2 \\
&\leq \int_{\Delta^c} \left| \frac{1}{N} \frac{(z\overline{w})^{-N} - 1}{z\overline{w} - 1} \right|^2 \mathrm{d}\left|\mu \times \overline{\mu}\right|(z, w) \cdot \int_{\mathbb{T}\times\mathbb{T}} |f(z)\overline{f(w)}|^2 \, \mathrm{d}\left|\mu \times \overline{\mu}\right|(z, w) \\
&= \int_{\Delta^c} \left| \frac{1}{N} \frac{(z\overline{w})^{-N} - 1}{z\overline{w} - 1} \right|^2 \mathrm{d}\left|\mu \times \overline{\mu}\right|(z, w) \cdot \|f\|_2^4 \\
&\leq \int_{\Delta^c} \left| \frac{1}{N} \frac{(z\overline{w})^{-N} - 1}{z\overline{w} - 1} \right|^2 \mathrm{d}\left|\mu \times \overline{\mu}\right|(z, w) \to 0,
\end{aligned}$$

by Lebesgue's theorem.

The last assertion follows by taking ν of the form $\mathbf{z}^k \cdot \mu$, $k \in \mathbb{Z}$. ■

A sequence $(a_n)_{n\in\mathbb{N}_0}$ in a Banach space E is called **Cesàro convergent** to $a \in E$ if the arithmetic averages, or **Cesàro averages**,

$$\frac{1}{N}\sum_{n=0}^{N-1} a_n \text{ converge to } a \text{ as } N \to \infty.$$

In this case a is called the **Cesàro limit** of the sequence. Thus for a continuous measure μ the sequence $(|\hat{\mu}(n)|^2)_{n\in\mathbb{N}_0}$ is Cesàro convergent to 0. Here are some elementary properties of Cesàro convergence, see Exercise 4.4.

Proposition 4.13 (Properties of the Cesàro Limit) *Let E be a Banach space and let $(a_n)_{n\in\mathbb{N}_0}$ be a sequence in E.*

(a) *Let $(b_n)_{n\in\mathbb{N}_0}$ be another sequence in E with $A := \{n : a_n = b_n\}$ having density 1, i.e., satisfying*

$$\mathrm{d}(A) := \lim_{N\to\infty} \frac{|A \cap [1, N]|}{N} = 1.$$

If the set $\{a_n, b_n : n \in \mathbb{N}_0 \text{ with } a_n \neq b_n\}$ is bounded, then

$$\lim_{N\to\infty} \frac{1}{N} \sum_{n=0}^{N-1} a_n = \lim_{N\to\infty} \frac{1}{N} \sum_{n=0}^{N-1} b_n,$$

meaning also that both of the above limits exist as soon as one of them does.

(b) *The sequence $(a_n)_{n\in\mathbb{N}_0}$ has at most one Cesàro limit.*

(c) *The Cesàro limit is shift invariant, i.e.,*

$$\lim_{N\to\infty} \frac{1}{N} \sum_{n=0}^{N-1} a_n = \lim_{N\to\infty} \frac{1}{N} \sum_{n=1}^{N} a_n$$

whenever one of these limits exists.

(d) *If $(a_n)_{n\in\mathbb{N}_0}$ is convergent to a, then it is Cesàro convergent to a.*

(e) *Suppose $(a_n)_{n\in\mathbb{N}_0}$ is a periodic sequence, i.e., satisfies $a_n = a_{n+m}$ for some $m \in \mathbb{N}$ and each $n \in \mathbb{N}_0$. Then $(a_n)_{n\in\mathbb{N}_0}$ is Cesàro convergent to*

$$\frac{1}{m} \sum_{n=0}^{m-1} a_n.$$

(f) *Suppose that sequences $(\lambda_n)_{n\in\mathbb{N}_0} \subset \mathbb{C}$ and $(a_n)_{n\in\mathbb{N}_0} \subset E$ are bounded, one of them is convergent, the other one Cesàro convergent with respective limits λ and a. Then $(\lambda_n a_n)_{n\in\mathbb{N}_0}$ is Cesàro convergent to λa.*

Remark 4.14 The Cesàro limit is not multiplicative, i.e.,

$$\lim_{N\to\infty} \frac{1}{N} \sum_{n=0}^{N-1} a_n b_n \neq \lim_{N\to\infty} \frac{1}{N} \sum_{n-0}^{N-1} a_n \cdot \lim_{N\to\infty} \frac{1}{N} \sum_{n=0}^{N-1} b_n$$

in general, as the example $(a_n) := (0, 1, 0, 1, \ldots)$ and $(b_n) := (1, 0, 1, 0 \ldots)$ in $\mathbb{C}$ shows.

For a Rajchman measure we have $|\hat{\mu}(n)|^2 \to 0$, so as a consequence of Propositions 4.13 (d) and 4.11 we see that every Rajchman measure $\mu \in \mathrm{M}(\mathbb{T})$ is continuous. However, not every continuous measure is Rajchman as the following example shows.

Example 4.15 (Cantor–Lebesgue Measure) Consider the ternary Cantor set $C \subset [0, 1]$ and the compact space $K := \{0, 1\}^{\mathbb{N}}$. Then

$$\theta : K \to C, \quad (x_n)_{n\in\mathbb{N}} \mapsto \sum_{n=1}^{\infty} \frac{2x_n}{3^n}$$

is a homeomorphism. Consider the Bernoulli shift $B(1/2, 1/2)$ on K with the corresponding measure μ. If $S : C \to C$ is given by $Sx = 3x - \lfloor 3x \rfloor$, then S is measurable, the push-forward measure $\theta_*\mu$ is S-invariant and θ is an isomorphism between the measure-preserving systems $B(1/2, 1/2)$ and $(C, \theta_*\mu, S)$. Now $\theta_*\mu$ can be extended to the Borel σ-algebra $\mathcal{B}([0, 1])$ by 0 on subsets of $[0, 1] \setminus C$. This extension is called the **Cantor–Lebesgue measure** and is denoted by μ_{CL}. By construction $\mathrm{supp}(\mu_{\mathrm{CL}}) = C$ and it is easy to see that μ_{CL} is a continuous measure. We can consider the push-forward of μ_{CL} under the map $s \mapsto \mathrm{e}^{2\pi \mathrm{i} s}$, which is a continuous measure on $\mathbb{T}$ also denoted by μ_{CL}. By Exercise 4.3 for each $n \in \mathbb{Z}$

$$\hat{\mu}_{\mathrm{CL}}(n) = \hat{\mu}_{\mathrm{CL}}(3n).$$

Since $\mu_{\mathrm{CL}} \neq \mathrm{m}$, there is $n \in \mathbb{Z} \setminus \{0\}$ such that $\hat{\mu}_{\mathrm{CL}}(n) \neq 0$. It follows that μ_{CL} is not a Rajchman measure.

For more information on Rajchman measures see, e.g., Lyons [763].

We finally present another consequence of the relation (4.1), being opposite to the case of continuous measures where the right-hand side of (4.1) is the minimal possible.

Every measure $\mu \in \mathrm{M}(\mathbb{T})$ satisfies

$$\lim_{N\to\infty} \frac{1}{N} \sum_{n=0}^{N-1} |\hat{\mu}(n)|^2 \leq \|\hat{\mu}\|_\infty^2 \leq \|\mu\|^2.$$

Thus the case when we have equalities here can be regarded as extremal.

Proposition 4.16 *For a measure $\mu \in \mathrm{M}(\mathbb{T})$ the following assertions are equivalent.*

(i) $\lim\limits_{N\to\infty} \frac{1}{N} \sum_{n=0}^{N-1} |\hat{\mu}(n)|^2 = \|\mu\|^2$.

(ii) *The support* $\mathrm{supp}(\mu)$ *has at most one element, i.e., μ is a scalar multiple of a Dirac measure.*

Proof (i)⇒(ii): The hypothesis and Proposition 4.11 yield

$$\|\mu\|^2 = \lim_{N\to\infty} \frac{1}{N}\sum_{n=0}^{N-1} |\hat{\mu}(n)|^2 = \sum_{a \text{ atom}} |\mu\{a\}|^2 \le \|\mu\| \sum_{a \text{ atom}} |\mu\{a\}| \le \|\mu\|^2,$$

implying $|\mu\{a\}|^2 = \|\mu\| \cdot |\mu\{a\}|$ or, equivalently, $|\mu\{a\}| \in \{0, \|\mu\|\}$ for every atom a. We conclude that μ is a scalar multiple of a Dirac measure.
(ii)⇒(i) follows from a simple computation for $\mu = c\delta_a$. ∎

4.3 The Koopman–von Neumann Lemma

Let $\mu \in \mathrm{M}(\mathbb{T})$ be a continuous measure. By Wiener's Lemma, it is intuitively clear that for a given number $r > 0$ the inequality $|\hat{\mu}(n)| > r$ cannot hold for n from a too large subset of $\mathbb{Z}$ or, equivalently, $\mathbb{N}_0$. To understand what "too large" means here, and to describe better in which sense the Fourier coefficients $\hat{\mu}(n)$ converge to 0, we introduce the following notions.

Definition 4.17

(a) The **upper density** $\overline{\mathrm{d}}(A)$ of a subset $A \subset \mathbb{N}_0$ is

$$\overline{\mathrm{d}}(A) := \limsup_{N\to\infty} \frac{|A \cap \{1, \dots, N\}|}{N} = \limsup_{N\to\infty} \frac{|A \cap \{0, \dots, N-1\}|}{N},$$

where $|B|$ denotes the number of elements in B (why does the second equality hold?). Analogously, the **lower density** $\underline{\mathrm{d}}(A)$ is defined by $\liminf$ replacing $\limsup$. Obviously, $0 \le \underline{\mathrm{d}}(A) \le \overline{\mathrm{d}}(A) \le 1$. If $\overline{\mathrm{d}}(A) = \underline{\mathrm{d}}(A)$, this common value (given by the limit) is called the **density** of A and denoted by $\mathrm{d}(A)$.

(b) We say that a sequence $(x_n)_{n\in\mathbb{N}}$ in a topological space Ω **converges in density** to $x \in \Omega$ if there is a subset $J = \{n_1 < n_2 < \cdots\}$ of $\mathbb{N}$ with density 1 such that $x_{n_j} \to x$ as $j \to \infty$. We denote this by writing

$$x = \mathrm{D}\text{-}\lim_{n\to\infty} x_n \text{ or } x_n \xrightarrow{\mathrm{D}} x.$$

Here are some essential properties.

Proposition 4.18

(a) *For each subset* $A \subset \mathbb{N}_0$ *we have* $\overline{\mathrm{d}}(A^c) = 1 - \underline{\mathrm{d}}(A)$.
(b) *For each pair of subsets* $A, B \subset \mathbb{N}_0$

$$\overline{\mathrm{d}}(A \cup B) \le \overline{\mathrm{d}}(A) + \overline{\mathrm{d}}(B).$$

(c) *For each pair of disjoint subsets* $A, B \subset \mathbb{N}_0$

$$\underline{\mathrm{d}}(A \cup B) \geq \underline{\mathrm{d}}(A) + \underline{\mathrm{d}}(B).$$

(d) *If* $\mathrm{d}(A) = 1$, *then for each* $B \subset \mathbb{N}_0$

$$\overline{\mathrm{d}}(A \cap B) = \overline{\mathrm{d}}(B) \text{ and } \underline{\mathrm{d}}(A \cap B) = \underline{\mathrm{d}}(B).$$

(e) *If* $A_1, \ldots, A_k \subset \mathbb{N}_0$ *satisfy* $\mathrm{d}(A_1) = \cdots = \mathrm{d}(A_k) = 1$, *then* $\mathrm{d}(A_1 \cap \cdots \cap A_k) = 1$.

(f) *If* $A_k \subset \mathbb{N}_0$ *and* $\mathrm{d}(A_k) = 1$ *for every* $k \in \mathbb{N}$, *then there is a subset* $A \subset \mathbb{N}$ *such that* $\mathrm{d}(A) = 1$ *and for each* $k \in \mathbb{N}$ *the set* $A \setminus A_k$ *is finite.*

Proof The proof of (a)–(e) is left as Exercise 4.8.
(f) We may suppose by (e) that $A_{k+1} \subset A_k$ for every $k \in \mathbb{N}$. By the hypothesis for each $k \in \mathbb{N}$ there exists $n_k \in \mathbb{N}$ such that

$$\frac{|A_k \cap \{1, \ldots, m\}|}{m} > 1 - \frac{1}{k} \text{ for every } m \geq n_k. \tag{4.2}$$

Moreover, we can assume that $(n_k)_{k \in \mathbb{N}}$ is strictly increasing. Define

$$A := \bigcap_{k \in \mathbb{N}} \Big(A_k \cup \{1, \ldots, n_k\} \Big).$$

Then, of course, $A \setminus A_k \subset \{1, \ldots, n_k\}$ and hence is finite for every $k \in \mathbb{N}$. We claim that $\underline{\mathrm{d}}(A) = 1$. Set $n_0 := 0$ and let $m \in \mathbb{N}$ be arbitrary. Take $k \in \mathbb{N}_0$ such that $n_k < m \leq n_{k+1}$. For $x \in A_k \cap \{1, \ldots, m\}$, by monotonicity, it follows that $x \in A_j \cap \{1, \ldots, m\}$ for each $j \in \{1, \ldots, k\}$. On the other hand, for $j > k$ we have $x \in \{1, \ldots, n_{k+1}\} \subset \{1, \ldots, n_j\}$. Altogether we obtain $x \in A \cap \{1, \ldots, m\}$ and thus by (4.2)

$$\frac{|A \cap \{1, \ldots, m\}|}{m} \geq \frac{|A_k \cap \{1, \ldots, m\}|}{m} \geq 1 - \frac{1}{k}.$$

By taking $m \to \infty$ this implies $\mathrm{d}(A) = 1$. ∎

The next very useful result relates convergence of Cesàro averages and convergence in density, see [659].

Proposition 4.19 (Koopman–von Neumann Lemma)

(a) *For a bounded sequence* $(a_n)_{n \in \mathbb{N}}$ *in* $[0, \infty)$ *the following are equivalent.*

(i) $\mathrm{D}\text{-}\lim\limits_{n \to \infty} a_n = 0$.

(ii) $\lim\limits_{N \to \infty} \frac{1}{N} \sum_{n=1}^{N} a_n = 0$.

(iii) $\lim_{N\to\infty} \frac{1}{N}\sum_{n=1}^{N} a_n^2 = 0.$

(b) *For a bounded sequence* $(b_n)_{n\in\mathbb{N}}$ *in* $(-\infty, 1]$ *the following are equivalent.*

(i) $\operatorname{D-lim}_{n\to\infty} b_n = 1.$

(ii) $\lim_{N\to\infty} \frac{1}{N}\sum_{n=1}^{N} b_n = 1.$

Moreover, if in addition $b_n \geq 0$ *for each* $n \in \mathbb{N}$*, then these assertions are equivalent to*

(iii) $\lim_{N\to\infty} \frac{1}{N}\sum_{n=1}^{N} b_n^2 = 1.$

Proof (a) We first prove the implication (i)⇒(ii). Let $J \subset \mathbb{N}$ have density 1 such that $\lim_{n\to\infty, n\in J} a_n = 0$. Take an arbitrary $\varepsilon > 0$ and $N \in \mathbb{N}$ such that for each $n \in J$ with $n \geq N$ we have $a_n \leq \varepsilon$. The following estimates hold for each such n:

$$\frac{1}{n}\sum_{j=1}^{n} a_j = \frac{1}{n}\sum_{j=1}^{N} a_j + \frac{1}{n}\sum_{j=N+1}^{n} a_j \leq \sup_{j\in\mathbb{N}} a_j \cdot \frac{N}{n} + \frac{1}{n}\sum_{\substack{j=N+1\\ j\notin J}}^{n} a_j + \frac{1}{n}\sum_{\substack{j=N+1\\ j\in J}}^{n} a_j$$

$$\leq \sup_{j\in\mathbb{N}} a_j \left(\frac{N}{n} + \frac{|\{1,\ldots,n\}\setminus J|}{n}\right) + \varepsilon.$$

Taking $\limsup_{n\to\infty}$ of both sides, we obtain by using Proposition 4.18 (a)

$$\limsup_{n\to\infty} \frac{1}{n}\sum_{j=1}^{n} a_j \leq \sup_{j\in\mathbb{N}} a_j \,\overline{\mathrm{d}}(\mathbb{N}\setminus J) + \varepsilon = \varepsilon.$$

Since $\varepsilon > 0$ was arbitrary, the assertion follows.
(ii)⇒(i): For $k \in \mathbb{N}$ let $B_k := \{n \in \mathbb{N} : a_n \geq 1/k\}$. Then

$$\frac{|B_k \cap \{1,\ldots,N\}|}{N} \leq \frac{1}{N}\sum_{n=1}^{N} k a_n \to 0 \text{ as } N \to \infty$$

by the hypothesis. This means $\mathrm{d}(B_k) = 0$ for each $k \in \mathbb{N}$. By Proposition 4.18 (f) (applied to $A_k = B_k^c$) there is $A \subset \mathbb{N}$ with density 1 such that $A \cap B_k$ is finite for each $k \in \mathbb{N}$, implying $\lim_{\substack{n\to\infty\\ n\in A}} a_n = 0$.
(ii)⇒(iii) follows by boundedness of (a_n) whereas the converse direction is a consequence of the Cauchy–Schwarz inequality (or the convexity of the square function $t \mapsto t^2$).
(b) follows from (a) by considering $a_n := 1 - b_n$. ■

Returning to Fourier coefficients of measures we can state the following as an immediate consequence of the previous discussion.

Proposition 4.20 *Let $\mu \in \mathrm{M}(\mathbb{T})$ be a probability measure. Then the following equivalences hold.*

(a) $\operatorname{D-lim}_{n\to\infty} |\hat{\mu}(n)| = 1 \iff$ *μ is a Dirac measure.*
(b) $\operatorname{D-lim}_{n\to\infty} |\hat{\mu}(n)| = 0 \iff$ *μ is continuous.*

We now introduce a class of measures which are continuous but not Rajchman.

Definition 4.21 A continuous probability measure μ on $\mathbb{T}$ is called **rigid along a subsequence** (n_k) **of** $\mathbb{N}$ if $\lim_{k\to\infty} \hat{\mu}(n_k) = 1$. A subsequence (n_k) of $\mathbb{N}$ is called a **rigidity sequence** if there exists a rigid measure along it.

It follows from Wiener's Lemma (Proposition 4.11) and the Koopman–von Neumann lemma (Proposition 4.19) that every rigidity sequence has density zero. For a characterization of rigidity sequences in terms of measure-preserving systems and more see Sect. 6.5. Here we just mention that rigidity sequences are difficult to characterize explicitly and for now just some classes of examples and counterexamples are known. For example, (2^n) is a rigidity sequence and (n^2), the sequence of primes (p_n) and (p_n^2) are not, see Eisner and Grivaux [343], Bergelson et al. [115], Fayad and Thouvenot [383], Cuny et al. [266], and Badea and Grivaux [71] as well as the references in Sect. 6.5 for these and more examples.

4.4 Positive Definite Sequences

In this section we characterize complex sequences which are the Fourier transform of a positive measure. Clearly, such sequences need to be bounded. The following definition gives another necessary property of such sequences.

Definition 4.22 A function $c : \mathbb{Z} \to \mathbb{C}$ (i.e., a sequence $(c(n))_{n\in\mathbb{Z}}$) is called **positive definite** if for all sequences $(x_n)_{n\in\mathbb{N}} \subset \mathbb{C}$ with finite support

$$\sum_{i,j\in\mathbb{N}} c(i-j)x_i\overline{x_j} \geq 0.$$

A sequence $(c(n))_{n\in\mathbb{Z}}$ is therefore positive definite if and only if for each $N \in \mathbb{N}$ the matrix $(c(i-j))_{i,j=1}^N$ is positive semi-definite. Some further elementary properties are listed in Exercise 4.13.

Example 4.23 The Fourier transform $\hat{\mu}$ of a positive measure $\mu \in \mathrm{M}(\mathbb{T})$ is positive definite. Indeed, for $(x_n)_{n\in\mathbb{N}} \subset \mathbb{C}$ with finite support

$$\sum_{i,j} \hat{\mu}(i-j) x_i \overline{x_j} = \sum_{i,j} \int_{\mathbb{T}} \mathbf{z}^{j-i} x_i \overline{x_j} \, \mathrm{d}\mu = \int_{\mathbb{T}} \sum_i \mathbf{z}^{-i} x_i \cdot \sum_j \overline{\mathbf{z}^{-j} x_j} \, \mathrm{d}\mu$$
$$= \int_{\mathbb{T}} \Big| \sum_k \mathbf{z}^k x_k \Big|^2 \, \mathrm{d}\mu \geq 0.$$

Our ultimate aim is to show that each positive definite sequence is the Fourier transform of a positive measure. For this we need some preparation, which takes up the rest of this section.

A class of positive definite sequences can be given using the convolution. The **convolution** of two sequences $a, b \in \ell^1(\mathbb{Z})$ is the sequence $a * b$ defined by

$$(a * b)(n) = \sum_{k \in \mathbb{Z}} a(n-k) b(k).$$

Note that for each $n \in \mathbb{Z}$ the previous series is absolutely convergent.

Example 4.24 For $a \in \ell^1(\mathbb{Z})$ let the sequence $\tilde{a}$ be defined by $\tilde{a}(j) := \overline{a}(-j)$ for $j \in \mathbb{Z}$. Then $a * \tilde{a}$ is positive definite. Indeed, by substituting $k = l - j$ and then by interchanging summation we obtain for $(x_n)_{n \in \mathbb{N}}$ with finite support

$$\sum_{i,j} (a * \tilde{a})(i-j) x_i \overline{x_j} = \sum_{i,j} \sum_k a(i-j-k) \overline{a(-k)} x_i \overline{x_j}$$
$$= \sum_{i,j} \sum_l a(i-l) x_i \overline{a(j-l)} \overline{x_j} = \sum_l \sum_i a(i-l) x_i \cdot \overline{\sum_j a(j-l) x_j} \geq 0.$$

As a next step we prove that the product of positive definite sequences is positive definite. For this purpose some facts from linear algebra need to be recalled. The **Hadamard product** $A \circ B$ of two complex matrices $A = (a_{ij})$ and $B = (b_{ij})$ of the same dimension is given by

$$(A \circ B)_{ij} = a_{ij} b_{ij}.$$

Note that

$$(A, B) \mapsto A \circ B \text{ is bilinear.}$$

The proof of the following result is left to the reader.

Lemma 4.25 *Let A, B be complex $n \times 1$ matrices, and let C, D be complex $1 \times k$ matrices. Then*

$$(A \circ B)(C \circ D) = (AC) \circ (BD).$$

Proposition 4.26 (Schur) *The Hadamard product of complex positive semi-definite quadratic matrices of the same dimension is positive semi-definite.*

Proof Let A, B be positive semi-definite $N \times N$ matrices, and let $a_1, \ldots, a_N \in \mathbb{C}^N$ and $b_1, \ldots, b_N \in \mathbb{C}^N$ be the pairwise orthogonal eigenvectors of A and B, respectively, with corresponding eigenvalues $\alpha_1, \ldots, \alpha_N \geq 0$ and $\beta_1, \ldots, \beta_N \geq 0$. We can write

$$A = \sum_{j=1}^{N} \alpha_j a_j a_j^\top \text{ and } B = \sum_{k=1}^{N} \beta_k b_k b_k^\top.$$

Therefore, by Lemma 4.25 and by bilinearity it follows that

$$A \circ B = \sum_{j,k=1}^{N} \alpha_j \beta_k (a_j a_j^\top) \circ (b_k b_k^\top) = \sum_{j,k=1}^{N} \alpha_j \beta_k (a_j \circ b_k)(a_j \circ b_k)^\top,$$

proving the statement. ■

An immediate consequence of Schur's theorem is the following fundamental result.

Proposition 4.27 *The coordinatewise product of positive definite sequences is again positive definite.*

We can now prove the promised result.

Theorem 4.28 (Bochner–Herglotz) *A sequence $c = (c(k))_{k \in \mathbb{Z}}$ in $\mathbb{C}$ is positive definite if and only if it is the Fourier transform of a positive measure μ on $\mathbb{T}$, i.e.,*

$$\hat{\mu}(k) = c(k) \text{ for each } k \in \mathbb{Z}.$$

Remark 4.29 Working on a problem of Carathéodory [222], see also [223], concerning positive harmonic functions on the unit disc, both Herglotz and Riesz discovered the crux of matter behind the above characterization independently, and published their result in 1911, see Herglotz [528] and Riesz [870]. While Herglotz formulated, in essence, the above result explicitly, Riesz' paper does not elaborate on the connection to Fourier coefficients directly. Note, however, that according to [528] Herglotz knew about Riesz' characterization, and he also mentioned several other papers from 1911, for example by Toeplitz [1003] and Fischer [390], which are connected to the problem of Carathéodory. Bochner has proved the analog of Theorem 4.28 for positive definite functions on the real line and on $\mathbb{R}^d$, see [167]. This explains the name under which this theorem is often cited.

Proof of Theorem 4.28 One implication was shown in Example 4.23. For the other one suppose that c is a positive definite sequence. For $n \in \mathbb{N}$ consider the characteristic sequences $\mathbf{1}_{[0,n)}$ and $\mathbf{1}_{(-n,0]}$ of the intervals $[0,n) := \{0,\dots,n-1\} \subset \mathbb{Z}$ and $(-n,0] = \{-n+1,\dots,0\}$, respectively. Then by Example 4.24 and Proposition 4.27 the sequence $a_n := \frac{1}{n}\mathbf{1}_{[0,n)} * \mathbf{1}_{-(n,0]} \cdot c$ is positive definite. For fixed $k \in \mathbb{Z}$ we clearly have

$$a_n(k) = c(k) \max\left\{1 - \tfrac{|k|}{n}, 0\right\} \to c(k) \text{ as } n \to \infty.$$

Moreover, $a_n = (a_n(k))_{k\in\mathbb{Z}}$ has support in $\{-n+1,\dots,0,\dots,n-1\}$, so

$$h_n := \sum_k a_n(k)\mathbf{z}^k$$

is a trigonometric polynomial. We first claim that it is positive. Let $n \in \mathbb{N}$ be fixed. For $z \in \mathbb{T}$ and $N \in \mathbb{N}$ with $N \geq n$ we have by positive definiteness

$$\begin{aligned} 0 &\leq \frac{1}{N}\sum_{i=1}^{N}\sum_{j=1}^{N} a_n(i-j)z^i\bar{z}^j = \frac{1}{N}\sum_{i=1}^{N}\sum_{j=1}^{N} a_n(i-j)z^{i-j} \\ &= \frac{1}{N}\sum_{k=-N}^{N} a_n(k)z^k\Big(N - |k|\Big) = \sum_{k=-n+1}^{n-1} a_n(k)z^k\Big(1 - \frac{|k|}{N}\Big), \end{aligned}$$

and here the right-hand side converges to $h_n(z)$ as $N \to \infty$.

We now consider the measures $\mu_n := h_n \cdot \mathrm{m}$. By Proposition 1.26

$$\|\mu_n\| = \int_{\mathbb{T}} |h_n| \,\mathrm{dm} = \int_{\mathbb{T}} h_n \,\mathrm{dm} = a_n(0) = c(0),$$

so, by Proposition 1.22 and Theorem 1.8(b), the sequence $(\mu_n)_{n\in\mathbb{N}}$ has a subsequence $(\mu_{n_j})_{j\in\mathbb{N}}$ with $\mu_{n_j} \to \mu$ in the weak* topology as $j \to \infty$ for some positive $\mu \in \mathrm{M}(\mathbb{T})$. For each fixed $k \in \mathbb{Z}$ we obtain

$$\begin{aligned} \hat{\mu}(k) &= \langle \mathbf{z}^{-k}, \mu\rangle = \lim_{j\to\infty} \langle \mathbf{z}^{-k}, \mu_{n_j}\rangle \\ &= \lim_{j\to\infty} \int_{\mathbb{T}} \mathbf{z}^{-k} h_{n_j} \,\mathrm{dm} = \lim_{j\to\infty} a_{n_j}(k) = c(k). \end{aligned}$$

The proof is complete. ■

The Bochner–Herglotz theorem will be used in the next chapter to construct scalar spectral measures.

Exercises

Exercise 4.1 (Properties of the Fourier Transform) Prove Proposition 4.4

Exercise 4.2 (Rajchman Measures) Prove that Rajchman measures form a closed subspace of $\mathrm{M}(\mathbb{T})$.

Exercise 4.3 (Cantor–Lebesgue Measure) Work out the details of Example 4.15.

Exercise 4.4 (Cesàro Limit) Prove Proposition 4.13.

Exercise 4.5 (Weighted Averages) Show the following generalization of Proposition 4.13(d). Let $(w_n) \subset [0, \infty)$ be a sequence of non-negative numbers, set $W_N := \sum_{n=1}^{N} w_n$ for $N \in \mathbb{N}$ and suppose that $\lim_{N\to\infty} W_N = \infty$. For a sequence (b_n) in $\mathbb{C}$ and for $b \in \mathbb{C}$ we have

$$\lim_{n\to\infty} b_n = b \quad \Longrightarrow \quad \lim_{N\to\infty} \frac{1}{W_N} \sum_{n=1}^{N} b_n w_n = b.$$

Exercise 4.6 (Discrete and Continuous Measures) Prove Proposition 4.10.

Exercise 4.7 (Convergence in Density)

(a) Prove that for $\lambda \in \mathbb{T}$ the limit

$$\operatorname*{D\text{-}lim}_{n\to\infty} \lambda^n$$

exists in $\mathbb{C}$ if and only if $\lambda = 1$.

(b) Let E be a normed space and let (a_n) be a bounded sequence in E. Then for every $l \in \mathbb{N}$

$$\operatorname*{D\text{-}lim}_{n\to\infty} a_n = 0 \quad \Longrightarrow \quad \operatorname*{D\text{-}lim}_{n\to\infty} a_{ln} = 0.$$

Exercise 4.8 (Upper and Lower Density) Show the assertions (a)–(e) from Proposition 4.18. Prove also that for each pair of real numbers $\alpha, \beta \in [0, 1]$ with $\alpha \leq \beta$ there is a set $A \subset \mathbb{N}_0$ such that $\underline{\mathrm{d}}(A) = \alpha$ and $\overline{\mathrm{d}}(A) = \beta$.

Exercise 4.9 (Positive Lower Density) Show the following assertions.

(a) A subsequence (k_n) of $\mathbb{N}$ has positive lower density if and only if it has sublinear growth, i.e., there exists $C > 0$ such that $k_n \leq Cn$ holds for every $n \in \mathbb{N}$.

(b) Let (a_n) be sequence in $[0, \infty)$ such that

$$\liminf_{N\to\infty} \frac{1}{N} \sum_{n=1}^{N} a_n > 0.$$

Then there is a $c > 0$ such that the set $E := \{n \in \mathbb{N} : a_n > c\}$ has positive lower density.

Exercise 4.10 (Density and Gaps) Prove that a syndetic set $A \subset \mathbb{N}$ has positive lower density. On the other hand, show that if A has positive upper density and $A = \{k_n : n \in \mathbb{N}\}$ for a subsequence $(k_n)_{n \in \mathbb{N}}$, then

$$\liminf_{n \to \infty}(k_{n+1} - k_n) < \infty.$$

Give an example of a syndetic set A with $\underline{\mathrm{d}}(A) < \overline{\mathrm{d}}(A)$.

Exercise 4.11 (Density of Polynomial Sequences) Compute the density of polynomial sequences, i.e., sequences of the form $(P(n))_{n \in \mathbb{N}}$ for a polynomial P with $P(\mathbb{N}) \subset \mathbb{N}_0$. (Note that such sequences are eventually strictly increasing and therefore can be identified, up to finitely many terms, with a subsequence of $\mathbb{N}_0$.)

Exercise 4.12 (Density of Primes)

(a) Prove that the set of primes $\mathbb{P}$ has density zero using the prime number theorem (see Theorem 10.21 below).
(b) Show the assertion in (a) using more elementary methods as follows. Denote by $\pi(x)$ the number of primes in $[0, x]$. Show first that for each $n \in \mathbb{N}$

$$n^{\pi(2n) - \pi(n)} \leq \binom{2n}{n} < 4^n.$$

Use this and a telescopic summation to estimate $\pi(2^k)$ and show that the density of $\mathbb{P}$ is 0, i.e.,

$$\lim_{n \to \infty} \frac{\pi(n)}{n} = 0.$$

Exercise 4.13 (Positive Definite Sequences) Show, using the definition only, that a positive definite sequence $c \in \mathbb{C}^{\mathbb{Z}}$ is bounded and satisfies $c(0) \geq 0$ and $|c(n)| \leq c(0)$ for each $n \in \mathbb{Z}$.

Exercise 4.14 (Summation by Parts) Let $(a_n)_{n \in \mathbb{N}}$ be as sequence in $\mathbb{C}$ and let $(b_n)_{n \in \mathbb{N}}$ be a sequence in a Banach space E. Denote $B_0 := 0$ and $B_n := \sum_{j=1}^{n} b_j$ for $n \in \mathbb{N}$. Prove that

$$\sum_{n=1}^{N} a_n b_n = \sum_{n=1}^{N-1} (a_n - a_{n+1}) B_n + a_N B_N.$$

holds for every $N \in \mathbb{N}$. (This is known as **summation by parts** and is a discrete analogue of integration by parts.)

Exercise 4.15 (Kronecker's Lemma) Let $(a_n)_{n\in\mathbb{N}}$ be a sequence in a Banach space E and $(b_n)_{n\in\mathbb{N}}$ be a sequence of real numbers with $0 < b_n < b_{n+1}$ for each $n \in \mathbb{N}$ and $\sup_{n\in\mathbb{N}} b_n = \infty$. Prove the following result of Kronecker [686]: If the series

$$\sum_{n=1}^{\infty} \frac{a_n}{b_n}$$

converges, then

$$\lim_{N\to\infty} \frac{1}{b_N} \sum_{n=1}^{N} a_n = 0.$$

In particular, if

$$\sum_{n=1}^{\infty} \frac{a_n}{n} \text{ converges, then } \quad \lim_{N\to\infty} \frac{1}{N} \sum_{n=1}^{N} a_n = 0.$$

(Hint: To prove the first statement use summation by parts.)

Chapter 5
The Spectral Theorem

Spectral measures and the spectral theorem are powerful tools to study unitary operators on Hilbert spaces via Fourier analysis. Recall that a linear operator $U : H \to K$ between two Hilbert spaces is called **unitary** if $UU^* = I_K$ and $U^*U = I_H$. A unitary operator U is surjective and preserves scalar products, so it is an isometry. Conversely, every surjective isometry between two Hilbert spaces preserves scalar products and hence is unitary. This can be easily seen from the following well-known fact.

Proposition 5.1 (Polarization Identity) *For a Hilbert space H, a sesquilinear form $b : H \times H \to \mathbb{C}$ satisfies*

$$b(x, y) = \frac{1}{4}\sum_{k=0}^{3} \mathrm{i}^k b(x + \mathrm{i}^k y, x + \mathrm{i}^k y) \textit{ for every } x, y \in H.$$

Relevant examples of unitary operators are provided by invertible measure-preserving systems. Indeed, their Koopman operator is unitary on the corresponding L^2-space by Proposition 2.27. Another important class of unitary operators is given in the next example, see Exercise 5.1.

Example 5.2

(1) Let (X, μ) be a measure space, and let $g : X \to \mathbb{T}$ be a measurable function. Then the multiplication operator $M_g : \mathrm{L}^2(X, \mu) \to \mathrm{L}^2(X, \mu)$, $M_g : f \mapsto gf$, is unitary.
(2) In particular, for every positive measure $\mu \in \mathrm{M}(\mathbb{T})$ the multiplication operator $M_{\mathbf{z}}$ is unitary on $\mathrm{L}^2(\mathbb{T}, \mu)$.

Our aim is to prove that the previous example represents the general case of unitary operators. For this purpose, to each unitary operator U we shall assign a family of measures on the spectrum of U such that the corresponding Fourier coefficients

T. Eisner, B. Farkas, *A Journey Through Ergodic Theorems*, Birkhäuser Advanced Texts Basler Lehrbücher, https://doi.org/10.1007/978-3-031-96506-7_5

completely describe the sesquilinear forms associated to the powers of U. The following elementary fact tells that these measures should be sought in $\mathrm{M}(\mathbb{T})$.

Proposition 5.3 *The spectrum of a unitary operator $U \in \mathscr{L}(H)$ is a subset of $\mathbb{T}$.*

Proof Since U is a contraction, its spectrum is contained in the closed unit disc: $\sigma(U) \subset \overline{\mathbb{D}} = \{z : |z| \leq 1\}$. Since U is invertible and its inverse U^{-1} is a contraction, $\sigma(U^{-1}) \subset \overline{\mathbb{D}}$. So by the spectral mapping theorem, Proposition 1.6, $\sigma(U) \subset \mathbb{C} \setminus \mathbb{D}$, and the inclusion $\sigma(U) \subset \mathbb{T}$ follows. ■

The following elementary result is recorded here for later reference.

Proposition 5.4 (Kernel and Range) *Let H, K be Hilbert spaces. For each $S \in \mathscr{L}(H, K)$ we have*

$$\ker(S) = \operatorname{ran}(S^*)^\perp \text{ and } \ker(S)^\perp = \overline{\operatorname{ran}(S^*)}.$$

Proof We have $x \in \operatorname{ran}(S^*)^\perp$ if and only if $0 = (x|S^*y) = (Sx|y)$ for every $y \in H$. But this holds if and only if $Sx = 0$. The first equality is proven, and the second one follows directly from the fact that $\overline{\operatorname{ran}(S^*)} = \operatorname{ran}(S^*)^{\perp\perp}$. ■

Let H be a Hilbert space and let $S \in \mathscr{L}(H)$. A closed subspace $F \subset H$ is **reducing** for S, or S-reducing, if it is S- and S^*-invariant.

Proposition 5.5 *Let $S \in \mathscr{L}(H)$ and let $F \subset H$ be a closed subspace. Then F is S^*-invariant if and only if $F^\perp$ is S-invariant.*

Proof Suppose F is S^*-invariant, and let $x \in F^\perp$. To prove $Sx \perp F$ take $y \in F$. Then we have $S^*y \in F$ and hence $(Sx|y) = (x|S^*y) = 0$. It follows that $Sx \perp F$. The converse implication can be proved analogously. ■

5.1 Scalar Spectral Measures

Recall from Chap. 4 that for $n \in \mathbb{Z}$ the nth Fourier coefficient $\hat{\mu}(n)$ of a measure $\mu \in \mathrm{M}(\mathbb{T})$ is

$$\hat{\mu}(n) = \int_{\mathbb{T}} \mathbf{z}^{-n} \, \mathrm{d}\mu.$$

Theorem 5.6 (Scalar Spectral Measures) *Let U be a unitary operator on a Hilbert space H. For each pair of vectors $x, y \in H$ there is a unique measure $\sigma_{x,y} \in \mathrm{M}(\mathbb{T})$ such that its Fourier coefficients satisfy*

$$\hat{\sigma}_{x,y}(n) = \int_{\mathbb{T}} \mathbf{z}^{-n} \, \mathrm{d}\sigma_{x,y} = (x|U^n y) \quad \textit{for every } n \in \mathbb{Z}.$$

Moreover, for every $x \in H$ the measure $\sigma_x := \sigma_{x,x}$ is positive.

Proof For given $x \in H$ define $c(n) := (x|U^n x)$ for $n \in \mathbb{Z}$. We show that $(c(n))_{n\in\mathbb{Z}}$ is a positive definite sequence. For each finitely supported sequence $(a_n)_{n\in\mathbb{N}}$ in $\mathbb{C}$

$$\sum_{i,j\in\mathbb{N}} c(i-j)a_i\overline{a_j} = \sum_{i,j\in\mathbb{N}} (U^j x|U^i x)a_i\overline{a_j} = \Big(\sum_{j\in\mathbb{N}} \overline{a_j}U^j x\Big|\sum_{i\in\mathbb{N}} \overline{a_i}U^i x\Big) \geq 0.$$

Therefore, by the Bochner–Herglotz Theorem, see Theorem 4.28, there is a positive measure $\sigma_{x,x} \in \mathrm{M}(\mathbb{T})$ with $c(n) = \hat{\sigma}_{x,x}(n)$ for each $n \in \mathbb{Z}$. For $y \in H$ we now set

$$\sigma_{x,y} := \frac{1}{4}\sum_{k=0}^{3} \mathrm{i}^k \sigma_{x+\mathrm{i}^k y}.$$

The linearity of $\mu \mapsto \langle \mathbf{z}^{-n}, \mu\rangle$ and the polarization identity (Proposition 5.1) applied to the sesquilinear form $(x, y) \mapsto (x|U^n y)$ yield $\hat{\sigma}_{x,y}(n) = (x|U^n y)$. Uniqueness follows from the injectivity of the Fourier transform, see Proposition 4.2. ∎

The measures $\sigma_{x,y} \in \mathrm{M}(\mathbb{T})$ from the previous proposition are called **scalar spectral measures** associated to U and will be studied in the following.

For a trigonometric polynomial $p \in \mathrm{P}(\mathbb{T})$ with

$$p = \sum_{j=-N}^{N} a_j \mathbf{z}^j$$

we set

$$p(U) := \sum_{j=-N}^{N} a_j U^j = \sum_{j=1}^{N} a_{-j}(U^*)^j + \sum_{j=0}^{N} a_j U^j.$$

Note that the mapping $\mathrm{P}(\mathbb{T}) \to \mathscr{L}(H)$, $p \mapsto p(U)$ is linear and multiplicative. In particular we have $(\mathbf{z}p)(U) = Up(U)$. For each pair $x, y \in H$ of vectors and each pair of trigonometric polynomials $p, q \subset \mathrm{P}(\mathbb{T})$ a short calculation yields

$$(p(U)x|q(U)y) = \int_{\mathbb{T}} p\overline{q}\,\mathrm{d}\sigma_{x,y}. \tag{5.1}$$

In the following we shorten $\mathrm{L}^2(\mathbb{T}, \sigma_x)$ to $\mathrm{L}^2(\sigma_x)$.

Proposition 5.7 (Properties of Scalar Spectral Measures) *Let $U \in \mathscr{L}(H)$ be a unitary operator with scalar spectral measures $(\sigma_{x,y})_{x,y\in H}$.*

(a) *The mapping*

$$H \times H \to \mathrm{M}(\mathbb{T}), \quad (x, y) \mapsto \sigma_{x,y}$$

is sesquilinear, and we have $\sigma_{x,y} = \overline{\sigma_{y,x}}$ *for every* $x, y \in H$. *Moreover,*

$$\|\sigma_x\| = \|x\|^2 \quad \textit{for each } x \in H.$$

(b) *For a trigonometric polynomial* $p \in \mathrm{P}(\mathbb{T})$

$$\|p(U)x\| = \|p\|_{\mathrm{L}^2(\sigma_x)} \quad \textit{for each } x \in H.$$

(c) *For all* $x, y \in H$ *and all* $p, q \in \mathrm{P}(\mathbb{T})$

$$\sigma_{p(U)x,q(U)y} = p\overline{q} \cdot \sigma_{x,y} \quad \textit{and } \sigma_{p(U)x} = |p|^2 \cdot \sigma_x.$$

(d) *For all* $x, y \in H$ *and all* $p, q \in \mathrm{P}(\mathbb{T})$

$$\left| \int_{\mathbb{T}} p\overline{q} \, \mathrm{d}\sigma_{x,y} \right| \le \|p\|_{\mathrm{L}^2(\sigma_x)} \|q\|_{\mathrm{L}^2(\sigma_y)}.$$

(e) *For each pair* $x, y \in H$

$$\|\sigma_{x,y}\| \le \|x\| \|y\|.$$

In particular, the mappings $H \times H \to \mathrm{M}(\mathbb{T})$, $(x, y) \mapsto \sigma_{x,y}$ *and* $H \to \mathrm{M}(\mathbb{T})$, $x \mapsto \sigma_x$ *are continuous.*

(f) *For each pair* $x, y \in H$ *there is* $h_{x,y} \in \mathrm{L}^2(\sigma_x)$ *such that* $\sigma_{x,y} = h_{x,y} \cdot \sigma_x$.

Proof (a) is clear from the definition of σ_x and from the fact that $\|\sigma_x\| = \langle \mathbf{z}^0, \sigma_x \rangle$, since σ_x is positive.
(b) For $x \in H$ and a trigonometric polynomial $p \in \mathrm{P}(\mathbb{T})$ we have by (5.1) that

$$\|p(U)x\|^2 = (p(U)x | p(U)x) = \int_{\mathbb{T}} |p|^2 \, \mathrm{d}\sigma_x.$$

(c) By (5.1) we obtain for $p, q \in \mathrm{P}(\mathbb{T})$ and for $n \in \mathbb{Z}$ that

$$\int_{\mathbb{T}} \mathbf{z}^n \, \mathrm{d}\sigma_{p(U)x,q(U)y} = (U^n p(U)x | q(U)y) = \int_{\mathbb{T}} \mathbf{z}^n p\overline{q} \, \mathrm{d}\sigma_{x,y}.$$

By the uniqueness of the scalar spectral measures we obtain $\sigma_{p(U)x,q(U)y} = p\overline{q} \cdot \sigma_{x,y}$, whence also the last assertion follows.
(d) By (b)

$$\left| \int_{\mathbb{T}} p\overline{q} \, \mathrm{d}\sigma_{x,y} \right| = |(p(U)x | q(U)y)| \le \|p(U)x\| \cdot \|q(U)y\| = \|p\|_{\mathrm{L}^2(\sigma_x)} \|q\|_{\mathrm{L}^2(\sigma_y)}.$$

(e) By the assertions in (a) and (d) we have

$$|\langle p, \sigma_{x,y}\rangle| \leq \|p\|_{\mathrm{L}^2(\sigma_x)}\|\mathbf{1}\|_{\mathrm{L}^2(\sigma_y)} \leq \|p\|_\infty \|x\|\|y\|.$$

Taking the supremum with respect to $p \in \mathrm{P}(\mathbb{T})$ with $\|p\|_\infty \leq 1$, we obtain the asserted inequality (use the density of $\mathrm{P}(\mathbb{T})$ in $\mathrm{C}(\mathbb{T})$, see Proposition 1.12 and Theorem 1.19).
(f) By part (d) the mapping $\mathrm{P}(\mathbb{T}) \to \mathbb{C}$, $p \mapsto \int_{\mathbb{T}} p \,\mathrm{d}\sigma_{x,y}$ is a continuous linear functional w.r.t. the $\mathrm{L}^2(\sigma_x)$-norm, hence can be uniquely extended to a continuous linear functional on $\mathrm{L}^2(\sigma_x)$ (recall that $\mathrm{P}(\mathbb{T})$ is dense in $\mathrm{L}^2(\sigma_x)$). The Riesz–Fréchet theorem yields the function $h_{x,y} \in \mathrm{L}^2(\sigma_x)$ with the required properties. ■

As a consequence, for each fixed, bounded and measurable $f : \mathbb{T} \to \mathbb{C}$ the mapping $b_f : H \times H \to \mathbb{C}$ defined by

$$b_f(x, y) := \int_{\mathbb{T}} f \,\mathrm{d}\sigma_{x,y}$$

is sesquilinear, conjugate symmetric and continuous. Indeed, continuity follows from

$$|b_f(x, y)| \leq \|f\|_\infty \|\sigma_{x,y}\| \leq \|f\|_\infty \|x\|\|y\|.$$

Continuous sesquilinear forms have the following property, which can be proven easily with the help of the Riesz–Fréchet theorem, see Exercise 5.3.

Proposition 5.8 *For Hilbert spaces H and K and a continuous sesquilinear form $b : H \times K \to \mathbb{C}$ there exists a unique operator $S \in \mathscr{L}(H, K)$ such that*

$$b(x, y) = (Sx|y)_K \textit{ for every } x \in H \textit{ and } y \in K.$$

This proposition, applied to b_f from above, yields a unique bounded linear operator, denoted by $f(U)$, such that

$$\int_{\mathbb{T}} f \,\mathrm{d}\sigma_{x,y} = b_f(x, y) = (f(U)x|y) \text{ for each } x, y \in H. \tag{5.2}$$

We now collect some properties of the operators $f(U)$ and the corresponding measures $\sigma_{f(U)x,g(U)y}$.

Proposition 5.9 *Let U be a unitary operator on a Hilbert space H with scalar spectral measures $(\sigma_{x,y})_{x,y\in H}$. For each pair $f, g : \mathbb{T} \to \mathbb{C}$ of bounded and measurable functions and each pair $x, y \in H$ of vectors we have*

(a) $\sigma_{f(U)x,g(U)y} = f\overline{g} \cdot \sigma_{x,y}$,
(b) $\left|\int_{\mathbb{T}} f\overline{g} \,\mathrm{d}\sigma_{x,y}\right| \leq \|f\|_{\mathrm{L}^2(\sigma_x)}\|g\|_{\mathrm{L}^2(\sigma_y)}$,
(c) $\|f(U)x\| = \|f\|_{\mathrm{L}^2(\sigma_x)}$.

Proof (a) For each $n \in \mathbb{Z}$ we have that

$$\hat{\sigma}_{f(U)x,y}(n) = (f(U)x|U^n y) = \int_{\mathbb{T}} f \, \mathrm{d}\sigma_{x,U^n y} = \int_{\mathbb{T}} f \mathbf{z}^{-n} \, \mathrm{d}\sigma_{x,y} = \widehat{f \cdot \sigma_{x,y}}(n).$$

Hence, by the injectivity of the Fourier transform, it follows that $\sigma_{f(U)x,y} = f \cdot \sigma_{x,y}$. By similar reasonings we obtain $\sigma_{f(U)x,g(U)y} = f\overline{g} \cdot \sigma_{x,y}$.
(b) Take trigonometric polynomials $p_n \in \mathrm{P}(\mathbb{T})$ with $p_n \to f$ in $\mathrm{L}^2(\sigma_x)$ and let $q \in \mathrm{P}(\mathbb{T})$. Then we have by Proposition 5.7 (d)

$$\left| \int_{\mathbb{T}} p_n q \, \mathrm{d}\sigma_{x,y} \right| \le \|p_n\|_{\mathrm{L}^2(\sigma_x)} \|q\|_{\mathrm{L}^2(\sigma_y)}.$$

The right-hand side converges to $\|f\|_{\mathrm{L}^2(\sigma_x)} \|q\|_{\mathrm{L}^2(\sigma_y)}$. For the left-hand side we have by Proposition 5.7 and by the continuity of the scalar product on $\mathrm{L}^2(\sigma_x)$ that

$$\left| \int_{\mathbb{T}} p_n q \, \mathrm{d}\sigma_{x,y} \right| = \left| \int_{\mathbb{T}} p_n q h_{x,y} \, \mathrm{d}\sigma_x \right| \to \left| \int_{\mathbb{T}} f q h_{x,y} \, \mathrm{d}\sigma_x \right|$$

as $n \to \infty$. Altogether, we obtain

$$\left| \int_{\mathbb{T}} f q \, \mathrm{d}\sigma_{x,y} \right| \le \|f\|_{\mathrm{L}^2(\sigma_x)} \|q\|_{\mathrm{L}^2(\sigma_y)}.$$

We repeat this argument for a sequence of trigonometric polynomials $(q_n)_{n \in \mathbb{N}}$ with $q_n \to \overline{g}$, and obtain the statement.
(c) By part (a) we have $\|f(U)x\|^2 = \int_{\mathbb{T}} |f|^2 \, \mathrm{d}\sigma_x = \|f\|^2_{\mathrm{L}^2(\sigma_x)}$. ■

5.2 Cyclic Subspaces

Given a unitary operator U on a Hilbert space H and a vector $x \in H$ we define

$$Z(x) := \overline{\mathrm{lin}}\{U^n x : n \in \mathbb{Z}\} = \overline{\{p(U)x : p \in \mathrm{P}(\mathbb{T})\}},$$

and call $Z(x)$ the **cyclic subspace** generated by x. Evidently, the cyclic subspace $Z(x)$ and its orthogonal complement $Z(x)^{\perp}$ both **reduce** U, i.e., they are invariant under U and U^*. It is easy to see that the restriction of U to reducing subspaces is again unitary. A vector $x \in H$ is called **cyclic** if $Z(x) = H$.

By Proposition 5.7(b), for fixed $x \in H$ the mapping $(\mathrm{P}(\mathbb{T}), \|\cdot\|_{\mathrm{L}^2(\sigma_x)}) \to Z(x)$, $p \mapsto p(U)x$ is isometric, so it has a unique extension W to $\mathrm{L}^2(\sigma_x)$, which is isometric and surjective onto $Z(x)$. We introduce the notation

$$f(U)x := Wf \in Z(x),$$

which is then consistent with the definition of $f(U)$ if f is bounded and measurable. Note that for $f \in \mathrm{L}^2(\sigma_x)$ there is, in general, no such bounded operator as $f(U)$.

Proposition 5.10 *Let $U \in \mathscr{L}(H)$ be a unitary operator on a Hilbert space H and fix $x \in H$. Then*

$$Z(x) = \{f(U)x : f \in \mathrm{L}^2(\sigma_x)\},$$

and the mapping $W : \mathrm{L}^2(\sigma_x) \to Z(x)$, $f \mapsto f(U)x$ is unitary and ***intertwines*** *the restricted operator $U|_{Z(x)}$ and the multiplication $M_{\mathbf{z}}$ by $\mathbf{z}$ on $\mathrm{L}^2(\sigma_x)$, i.e., $U|_{Z(x)}W = WM_{\mathbf{z}}$. In particular, $U|_{Z(x)}$ and $M_{\mathbf{z}}$ are* ***unitarily equivalent****.*

Proof That the mapping $\mathrm{P}(\mathbb{T}) \to \mathrm{lin}\{U^n x : n \in \mathbb{Z}\}$, $p \mapsto p(U)x$ intertwines the two mentioned operators is clear. By approximation, it follows that W is intertwining as well. Everything else has been already discussed in the preceding paragraph. ∎

Remark 5.11 In particular, since $\mathbf{1}(U) = I$, the relation $W\mathbf{1} = x$ holds.

The next result states that the scalar spectral measures are supported in the spectrum of the operator.

Corollary 5.12 *For each $x \in H$ we have* $\mathrm{supp}(\sigma_x) \subset \sigma(U)$. *If x is a cyclic vector, then* $\mathrm{supp}(\sigma_x) = \sigma(U)$.

Proof We have $\sigma(U|_{Z(x)}) \subset \sigma(U)$ since $Z(x)$ is a reducing subspace of U. On the other hand, Propositions 1.21 and 5.10 yield the equality $\mathrm{supp}(\sigma_x) = \sigma(U|_{Z(x)})$. (We use here the obvious fact that unitarily equivalent operators have the same spectrum). ∎

If U has a cyclic vector, then it is a direct consequence of the above that U is unitarily equivalent to a multiplication operator. More precisely, we have the following important result, which will be used frequently throughout this book.

Theorem 5.13 (Spectral Theorem for Operators with Cyclic Vectors, Multiplication Form) *Let $U \in \mathscr{L}(H)$ be a unitary operator with a cyclic vector $x \in H$. Then U is unitarily equivalent to a multiplication operator $M = M_{\mathbf{z}}$ on $\mathrm{L}^2(\mathbb{T}, \mu)$ with a positive Borel measure $\mu \in \mathrm{M}(\mathbb{T})$ with the vector x corresponding to* $\mathbf{1}$ *under this unitary equivalence. Moreover,* $\sigma(M) = \mathrm{supp}(\mu)$.

The general form of the spectral theorem in multiplication form is treated in Exercise 5.7. The importance of this form of the spectral theorem was emphasized by Halmos [515].

We now relate the cyclic subspaces and the properties of the scalar spectral measures.

Proposition 5.14 (Absolute Continuity and Spectral Measures) *Let $U \in \mathscr{L}(H)$ be a unitary operator.*

(a) *For each $x \in H$ and each $f \in \mathrm{L}^2(\sigma_x)$*

$$\sigma_{f(U)x} = |f|^2 \cdot \sigma_x.$$

(b) *For each $x \in H$ and for each positive measure $\mu \in \mathrm{M}(\mathbb{T})$*

$$\mu \ll \sigma_x \quad \Longleftrightarrow \quad \mu = \sigma_y \quad \text{for some } y \in Z(x).$$

(c) *For each $x, y \in H$*

$$\sigma_x \perp \sigma_y \quad \Longrightarrow \quad \sigma_{x,y} = 0 \quad \Longleftrightarrow \quad Z(x) \perp Z(y).$$

If $x, y \in Z(u)$ for some $u \in H$, then $\sigma_x \perp \sigma_y \Longleftrightarrow Z(x) \perp Z(y)$.

Proof (a) For $f \in \mathrm{P}(\mathbb{T})$ the statement is Proposition 5.7(c), the case of general $f \in \mathrm{L}^2(\sigma_x)$ follows by approximation. We leave the details to the reader.
(b) If $y \in Z(x)$, then there is $f \in \mathrm{L}^2(\sigma_x)$ such that $y = f(U)x$ (Proposition 5.10). So $\sigma_y \ll \sigma_x$ by part (a). Conversely, let the positive measure $\mu \in \mathrm{M}(\mathbb{T})$ be absolutely continuous with respect to σ_x. By the Radon–Nikodym theorem (Theorem 1.27) there is a positive $g \in \mathrm{L}^1(\sigma_x)$ such that $\mu = g \cdot \sigma_x$. Then for $f := \sqrt{g}$ we have $f \in \mathrm{L}^2(\sigma_x)$ and $\sigma_{f(U)x} = f^2 \cdot \sigma_x = g \cdot \sigma_x = \mu$ by (a).
(c) Since $\sigma_{x,y} \ll \sigma_x$ and $\sigma_{x,y} \ll \sigma_y$ by Proposition 5.7, the first implication follows from Lemma 1.25(c), (d). Suppose $\sigma_{x,y} = 0$. Then for each $p \in \mathrm{P}(\mathbb{T})$ we have $(p(U)y|x) = \langle p, \sigma_{x,y}\rangle = 0$, so $x \perp Z(y)$, implying $Z(x) \perp Z(y)$. On the other hand, if $Z(x) \perp Z(y)$, then $(p(U)x|y) = 0$ for each $p \in \mathrm{P}(\mathbb{T})$, showing $\sigma_{x,y} = 0$. Finally, suppose that $x, y \in Z(u)$ for some $u \in H$, and $\sigma_{x,y} = 0$. Then by Proposition 5.10 there are $f, g \in \mathrm{L}^2(\sigma_u)$ such that $x = f(U)u$ and $y = g(U)u$. By approximation, we conclude from Proposition 5.7(a) that $f\overline{g} \cdot \sigma_u = \sigma_{x,y} = 0$. Therefore, $f\overline{g} = 0$ in $\mathrm{L}^2(\sigma_u)$, so $\sigma_x \perp \sigma_y$ by part (a). ∎

5.3 Projections and Eigenvalues

In this section we study algebraic properties of the mapping $f \mapsto f(U)$, defined in (5.2) by Proposition 5.8, and introduce spectral projections. Let $\mathrm{BM}(\mathbb{T})$ denote the vector space of bounded and measurable $\mathbb{C}$-valued functions on $\mathbb{T}$, which becomes a Banach space with the supremum norm and a Banach algebra with the pointwise multiplication.

Theorem 5.15 (Bounded Borel Functional Calculus) *Let $U \in \mathscr{L}(H)$ be a unitary operator on a Hilbert space H. The mapping*

$$\Psi : \mathrm{BM}(\mathbb{T}) \to \mathscr{L}(H), \quad \Psi(f) := f(U)$$

has the following properties.

(a) $\Psi(\mathbf{z}) = U$, $\Psi(\mathbf{1}) = I$.
(b) *Ψ is linear, contractive, and satisfies $\Psi(f\overline{g}) = \Psi(f)\Psi(g)^*$ for each $f, g \in \mathrm{BM}(\mathbb{T})$.*
(c) *If $(f_n)_{n\in\mathbb{N}}$ is $\|\cdot\|_\infty$-bounded and pointwise convergent to f, then $\Psi(f_n)x \to \Psi(f)x$ as $n \to \infty$ for each $x \in H$.*
(d) *If $f \in \mathrm{BM}(\mathbb{T})$ is real-valued, then $\Psi(f)$ is self-adjoint.*

Proof (a) is clear.
(b) Linearity of Ψ is obvious. Note that, for $f \in \mathrm{BM}(\mathbb{T})$, by Propositions 5.7 and 5.9 we have that

$$\|\Psi(f)x\| = \|f\|_{\mathrm{L}^2(\sigma_x)} \le \|f\|_\infty \sqrt{\|\sigma_x\|} = \|f\|_\infty \|x\|$$

for every $x \in H$, showing that Ψ is a contraction. Let $f, g \in \mathrm{BM}(\mathbb{T})$. Then, by Proposition 5.9,

$$(g(U)^* f(U)x\,|\,y) = (f(U)x\,|\,g(U)y) = ((f\overline{g})(U)x\,|\,y) \text{ for every } x, y \in H.$$

Therefore the identity $\Psi(f\overline{g}) = \Psi(f)\Psi(g)^*$ follows.
(c) Take $(f_n)_{n\in\mathbb{N}}$ and f as in the assertion. Then for each $x \in H$

$$\|\Psi(f - f_n)x\|^2 = \int_{\mathbb{T}} |f - f_n|^2 \,\mathrm{d}\sigma_x \to 0$$

as $n \to \infty$ by Lebesgue's dominated convergence theorem.
(d) follows from (a) and (b): $\Psi(f) = \Psi(\overline{f}) = \Psi(f)^*$. ∎

For a Borel set $A \subset \mathbb{T}$ define $P_A := \mathbf{1}_A(U)$, i.e.,

$$\int_{\mathbb{T}} \mathbf{1}_A \,\mathrm{d}\sigma_{x,y} = (P_A x\,|\,y) \text{ for every } x, y \in H.$$

From the properties of the above Borel functional calculus one can easily deduce that each operator $P_A \in \mathscr{L}(H)$ is a projection (called **spectral projection**), i.e., $P_A P_A = P_A$, and that $P_{\mathbb{T}\setminus A} = I - P_A$. It also follows that P_A is an orthogonal projection and $x = P_A x$ holds if and only of $\sigma_x(\mathbb{T} \setminus A) = 0$. The mapping

$$\mathcal{B}(\mathbb{T}) \to \mathscr{L}(H), \quad A \mapsto P_A$$

is called the **projection valued spectral measure** of the operator U. See Exercise 5.9 also for some further properties.

Before considering the important case when A consists of just one point, we first need the following characterization.

Proposition 5.16 (Characterization of Eigenvalues via Scalar Spectral Measures) *Let $U \in \mathscr{L}(H)$ be a unitary operator with scalar spectral measures $(\sigma_{x,y})_{x,y\in H}$. Then for $x \in H$ and $\lambda \in \mathbb{T}$ the following assertions are equivalent.*

(i) $Ux = \lambda x$.
(ii) $\sigma_{x,y} = (x|y)\delta_\lambda$ *for each* $y \in H$.
(iii) $\sigma_x = \|x\|^2\delta_\lambda$.

Proof (i)⇒(ii): If $Ux = \lambda x$, then $U^n x = \lambda^n x$ for each $n \in \mathbb{Z}$. We then have for each $n \in \mathbb{Z}$ and each $y \in H$ that $\lambda^{-n}(x|y) = (U^{-n}x|y) = (x|U^n y) = \hat{\sigma}_{x,y}(n)$. It follows, by the injectivity of the Fourier transform, that $\sigma_{x,y} = (x|y)\delta_\lambda$. The implication (ii)⇒(iii) is trivial.
(iii)⇒(i): If $\sigma_x = \|x\|^2\delta_\lambda$, then by the definition of σ_x (see Theorem 5.6) we have $\lambda \in \mathbb{T}$ and

$$(Ux|x) = \int_{\mathbb{T}} \mathbf{z}\, \mathrm{d}\sigma_x = \lambda\|x\|^2.$$

The equality in the Cauchy–Schwarz inequality implies $Ux = cx$ for some $c \in \mathbb{T}$, and substituting Ux by cx in the above gives $c = \lambda$. ■

For each $\lambda \in \mathbb{T}$ we have the orthogonal projection $P_\lambda := P_{\{\lambda\}} := \mathbf{1}_{\{\lambda\}}(U)$.

Proposition 5.17 *For $\lambda \in \mathbb{T}$ the operator P_λ is the orthogonal projection onto* $\ker(\lambda - U)$.

Proof We have $(\lambda - \mathbf{z})\mathbf{1}_{\{\lambda\}} = 0$, so $(\lambda - U)P_\lambda = 0$, meaning $\operatorname{ran}(P_\lambda) \subset \ker(\lambda - U)$. If $x \in \ker(\lambda - U)$, then by Proposition 5.16

$$(P_\lambda x|x) = \int_{\mathbb{T}} \mathbf{1}_{\{\lambda\}}\, \mathrm{d}\sigma_x = \|x\|^2 = (x|x).$$

It follows $\|P_\lambda x - x\|^2 = (P_\lambda x - x|P_\lambda x - x) = (P_\lambda x - x|P_\lambda x) = 0$, i.e., $P_\lambda x = x$. The equality $\operatorname{ran}(P_\lambda) = \ker(\lambda - U)$ is proven. The orthogonality of the projection follows from the fact that P_λ is contractive by $\|P_\lambda\| \le \|\mathbf{1}_{\{\lambda\}}\|_\infty = 1$. ■

The following shows the connection between eigenvalues, scalar spectral measures and the spectral projections P_λ.

Proposition 5.18 *Let $U \in \mathscr{L}(H)$ be a unitary operator with scalar spectral measures $(\sigma_{x,y})_{x,y\in H}$. For $x \in H$ and $\lambda \in \mathbb{T}$ the following assertions are equivalent.*

(i) $\sigma_x\{\lambda\} > 0$.

(ii) $\|P_\lambda x\| > 0$.
(iii) $Z(x) \cap \ker(\lambda - U) \neq \{0\}$.
(iv) $(x|y) \neq 0$ *for some* $y \in \ker(\lambda - U)$.

Proof The equivalence (i)⇔(ii) follows from $\|P_\lambda x\|^2 = \sigma_x\{\lambda\}$. For the implication (ii)⇒(iii) note that $P_\lambda x = \mathbf{1}_{\{\lambda\}}(U)x \in Z(x) \cap \ker(\lambda - U)$ (by Propositions 5.10 and 5.17). Suppose y is as in (iv). Then $0 \neq (x|y) = (P_\lambda x|y)$, so $P_\lambda x \neq 0$, and that is (ii). To see the implication (iii)⇒(iv), notice that if $x \perp \ker(\lambda - U)$, then $Z(x) \perp \ker(\lambda - U)$. ∎

5.4 Decompositions and Ideals

We shall show that certain decompositions of $\mathrm{M}(\mathbb{T})$ lead to orthogonal decompositions of H into U- and U^*-invariant subspaces. Before that we need another short excursion to measure theory.

Let $\mathcal{X}$ be a σ-algebra over a set X, and consider the Banach space $\mathfrak{M}(X, \mathcal{X})$ of complex measures on $(X, \mathcal{X})$. A subspace $I \subset \mathfrak{M}(X, \mathcal{X})$ is called an (order) **ideal**, if $\mu \in I$, $\nu \in \mathfrak{M}(X, \mathcal{X})$ and $|\nu| \leq |\mu|$ imply $\nu \in I$.

Proposition 5.19 (Properties of Ideals)

(a) *For a closed subspace* $I \subset \mathfrak{M}(X, \mathcal{X})$ *the following assertions are equivalent.*

   (i) I *is an ideal.*
   (ii) *For each* $\mu, \nu \in \mathfrak{M}(X, \mathcal{X})$ *with* $\nu \ll \mu$ *and* $\mu \in I$ *also* $\nu \in I$ *holds.*

(b) *Let* I *and* J *be closed ideals of* $\mathfrak{M}(X, \mathcal{X})$. *Then* $I \cap J$ *is a closed ideal and*

$$I \cap J = \{0\} \quad \iff \quad \nu \perp \mu \text{ for each } \nu \in I \text{ and } \mu \in J.$$

(c) *Let* I *and* J *be closed ideals of* $\mathfrak{M}(X, \mathcal{X})$ *with* $I \cap J = \{0\}$. *Then* $I + J$ *is a closed ideal.*

Proof (a) Suppose (ii) and that $\mu \in I$, $|\nu| \leq |\mu|$. Then $|\mu| \in I$, and $\nu \ll |\mu|$, so $\nu \in I$ by the hypothesis.
Suppose (i), let $\mu \in I$, and let $\nu \ll \mu$. Then by the Radon–Nikodym theorem there is $f \in \mathrm{L}^1(|\mu|)$ such that $\nu = f \cdot |\mu|$. Consider $f_n := \operatorname{sign}(f)\min(|f|, n\mathbf{1})$, then $|f_n \cdot \mu| = |f_n| \cdot |\mu| \leq n|\mu|$, hence $f_n|\mu| \in I$ by the hypothesis. By Lebesgue's dominated convergence theorem $f_n \to f$ in $\mathrm{L}^1(|\mu|)$, so $f_n \cdot |\mu| \to f \cdot |\mu| = \nu$, and $\nu \in I$ follows, since I is closed.
(b) Suppose $I \cap J = \{0\}$, and let $\mu \in I$ and $\nu \in J$. Then by the Lebesgue decomposition (Theorem 1.27(a)), $\mu = \mu_a + \mu_s$ with $\mu_a \ll |\nu| \in J$ and $\mu_s \perp \nu$. Since also $\mu_a \ll \mu$, we obtain $\mu_a \in I \cap J$, hence $\mu_a = 0$ and $\mu = \mu_s \perp \nu$. The converse implication is even easier.
(c) Obviously $I+J$ is a linear subspace of $\mathfrak{M}(X, \mathcal{X})$. By part (b) and Lemma 1.25(e) for $\mu \in I$ and $\nu \in J$ we have $\|\mu+\nu\| = |\mu+\nu|(X) = |\mu|(X)+|\nu|(X) = \|\mu\|+\|\nu\|$.

Therefore, the projection

$$P_I : I + J \to I, \qquad \mu + \nu \mapsto \mu$$

is bounded, and by Proposition 1.4 the subspace $I + J$ is closed. To prove that $I + J$ is an ideal, take $\mu \in I$, $\nu \in J$ and let $\rho \in \mathfrak{M}(X, \mathcal{X})$ be such that $\rho \ll \mu + \nu$. Then by Lemma 1.25(f) there are $\mu', \nu' \in \mathfrak{M}(X, \mathcal{X})$ with $\mu' \ll \mu$, $\nu' \ll \nu$ and $\rho = \mu' + \nu' \in I + J$. The rest follows from (a). ■

We see that even though $\mathfrak{M}(X, \mathcal{X})$ is far from being a Hilbert space, it carries a structure that is reminiscent to that of a Hilbert space.

Example 5.20 We consider $\mathrm{M}(\mathbb{T}) = \mathfrak{M}(\mathbb{T}, \mathcal{B}(\mathbb{T}))$.

(1) The subspaces $\mathrm{M_d}(\mathbb{T})$ and $\mathrm{M_c}(\mathbb{T})$ are closed ideals, see Exercise 5.5.
(2) The subspace $\mathrm{M_r}(\mathbb{T})$ of Rajchman measures is a closed ideal, see Exercise 4.2 and Proposition 4.5.
(3) For a given $\mu \in \mathrm{M}(\mathbb{T})$ the subspaces

$$I(\mu) := \{\nu \in \mathrm{M}(\mathbb{T}) : \nu \ll \mu\} \text{ and } I(\mu)^{\perp} := \{\nu \in \mathrm{M}(\mathbb{T}) : \nu \perp \mu\}$$

are closed ideals, see Lemma 1.25 and Exercise 5.5. In particular, $\mathrm{L}^1(\mathbb{T}, \mathrm{m})$ is the smallest ideal that contains m (we use the identification from Proposition 1.26.)

Let $U \in \mathscr{L}(H)$ be a unitary operator on a Hilbert space H. We associate to each closed ideal I in $\mathrm{M}(\mathbb{T})$ a subspace in H by defining

$$H(I) := \{x \in H : \sigma_{x,y} \in I \text{ for all } y \in H\}.$$

Proposition 5.21 *For a closed ideal I in* $\mathrm{M}(\mathbb{T})$

$$H(I) = \big\{x \in H : \sigma_x \in I\big\}. \tag{5.3}$$

Moreover, $H(I)$ is a closed, U-reducing subspace of H. If J is another closed ideal in $\mathrm{M}(\mathbb{T})$ *with $I \cap J = \{0\}$, then*

$$H(I + J) = H(I) \oplus H(J)$$

as an orthogonal direct sum.

Proof To show (5.3) note that the inclusion "$\subset$" is trivial. For the converse inclusion take $x \in H$ such that $\sigma_x \in I$. Since $\sigma_{x,y} \ll \sigma_x$ for each $y \in H$ by Proposition 5.7 (f), we obtain by Proposition 5.19 that $\sigma_{x,y} \in I$, and the asserted equality is proven. Since $\sigma_{Ux,y} = \sigma_{x,U^*y}$ and $\sigma_{U^*x,y} = \sigma_{x,Uy}$ for every $x, y \in H$, the invariance

of $H(I)$ follows immediately. By Proposition 5.7(e) the mapping $x \mapsto \sigma_x$ is continuous, so $H(I)$ is closed by the closedness of I.
Let J be as in the assertion, then, by Proposition 5.19, $I + J$ is a closed ideal. Take $x \in H(I)$ and $y \in H(J)$. Then $\sigma_{x,y} \in I$ and $\sigma_{x,y} = \overline{\sigma_{y,x}} \in J$. From the hypothesis we obtain $\sigma_{x,y} = 0$, which yields $x \perp y$, by Proposition 5.14(c). Hence, $H(I) \perp H(J)$.
We finally prove that $H(I + J) = H(I) \oplus H(J)$, where only the inclusion "$\subset$" requires a proof. Take $z \in H(I + J)$, so $\sigma_z = \mu + \nu$ for some measures $\mu \in I$ and $\nu \in J$. We then have $\sigma_z = |\sigma_z| = |\mu + \nu| = |\mu| + |\nu|$ (since $\mu \perp \nu$), whence we conclude $\mu = |\mu| \geq 0$ and $\nu = |\nu| \geq 0$. Both measures μ and ν are absolutely continuous with respect to σ_z. By Proposition 5.14 there are $x, y \in Z(z)$ with $\sigma_x = \mu$ and $\sigma_y = \nu$. Take $f, g \in \mathrm{L}^2(\sigma_z)$ with $x = f(U)z$ and $y = g(U)z$. Then $\sigma_x = |f|^2 \cdot \sigma_z$ and $\sigma_y = |g|^2 \cdot \sigma_z$, and hence

$$\sigma_z = \mu + \nu = \sigma_x + \sigma_y = |f|^2 \cdot \sigma_z + |g|^2 \cdot \sigma_z = (|f|^2 + |g|^2) \cdot \sigma_z.$$

This yields that $|f|^2 + |g|^2 = \mathbf{1}$ holds σ_z-almost everywhere. Since $\sigma_x \perp \sigma_y$, we have $|f| \cdot |g| = 0$ and thus $|f| + |g| = \mathbf{1}$. Define $x' := |f|(U)z$ and $y' := |g|(U)z$. Then $\sigma_{x'} = |f|^2 \cdot \sigma_z = \sigma_x = \mu \in I$, so $x' \in H(I)$ (and the same arguments show $y' \in H(J)$). We conclude $z = \mathbf{1}(U)z = (|f| + |g|)(U)z = x' + y' \in H(I) + H(J)$. ■

In the next chapter we study concrete versions of such abstract decompositions.

5.5 Self-Adjoint Operators

We present here a version of the spectral theorem for bounded self-adjoint operators needed in Chap. 13 by deducing it from the unitary case. Furthermore, we refer, e.g., to [342, Sec. 18] for the spectral theorem in the case of normal operators, via a different approach.

We begin with some preparations. Let H be a Hilbert space and $T \in \mathscr{L}(H)$ a **self-adjoint** operator, i.e., satisfying $T = T^*$. For a self-adjoint operator $T \in \mathscr{L}(H)$ one has

$$(Tx|x) \in \mathbb{R} \text{ for every } x \in H.$$

It is also easy to see, by polarization, that this latter property in fact characterizes self-adjoint operators.

Lemma 5.22 *The spectrum $\sigma(T)$ of a self-adjoint operator $T \in \mathscr{L}(H)$ is contained in $\mathbb{R}$.*

Proof It suffices to prove that for $\alpha \in \mathbb{R} \setminus \{0\}$ one has $\alpha \mathrm{i} \in \rho(T)$ ($\rho(T)$ is the resolvent set of T), i.e., that $\alpha \mathrm{i} - T$ is bijective. For every $x \in H$ we can write

$$\begin{aligned}\|\alpha \mathrm{i}x - Tx\|^2 &= |\alpha|^2\|x\|^2 - 2\mathrm{Re}(\alpha\mathrm{i}(Tx|x)) + \|Tx\|^2 \\ &= |\alpha|^2\|x\|^2 + \|Tx\|^2 \ge |\alpha|^2\|x\|^2.\end{aligned} \tag{5.4}$$

This implies the injectivity of $\alpha\mathrm{i} - T$ for every $\alpha \in \mathbb{R} \setminus \{0\}$. By Proposition 5.4 we have that $\mathrm{ran}(\alpha\mathrm{i} - T)$ is dense in H. The estimate in (5.4) implies the closedness of $\mathrm{ran}(\alpha\mathrm{i} - T)$, and altogether $\alpha\mathrm{i} \notin \sigma(T)$. ■

Consider the rational function

$$C(t) := \frac{t - \mathrm{i}}{t + \mathrm{i}} = 1 - \frac{2\mathrm{i}}{t + \mathrm{i}},$$

which maps $\mathbb{R}$ into $\mathbb{T}$ injectively. As a matter of fact, $C : \mathbb{R} \to \mathbb{T} \setminus \{1\}$ is a homeomorphism with inverse

$$C^{-1}(w) = \mathrm{i}\frac{1 + w}{1 - w}.$$

By the previous result the **Cayley transform** $C(T) := (T - \mathrm{i})(T + \mathrm{i})^{-1}$ of the self-adjoint operator T is well-defined and a bounded linear operator on H. It is also easy to see that $U := C(T)$ is unitary. By the spectral mapping theorem, Proposition 1.6, we have

$$\sigma(C(T)) = C(\sigma(T)),$$

so, in particular $1 \in \rho(U)$. We can also conclude

$$T = \mathrm{i}(1 + U)(1 - U)^{-1}.$$

Consider the scalar spectral measures $\sigma_{x,y}$ corresponding to the unitary operator $U = C(T)$ and the family of the push-forward measures $\tau_{x,y} := C_*^{-1}\sigma_{x,y}$ ($x, y \in H$) under the Cayley transform $C : \mathbb{R} \to \mathbb{T} \setminus \{1\}$. For each $n \in \mathbb{N}_0$ we have by the continuous functional calculus

$$(T^n x|y) = ((C^{-1}(U))^n x|y) = \int_{\mathbb{T}} (C^{-1}(w))^n \, \mathrm{d}\sigma_{x,y}(w) = \int_{\mathbb{R}} t^n \, \mathrm{d}\tau_{x,y}(t). \tag{5.5}$$

Based on these identities we call $\tau_{x,y}$ **scalar spectral measures** for the bounded self-adjoint operator $T \in \mathscr{L}(H)$. As in Sect. 5.3 one can define the measurable functional calculus for T. Let $f : \mathbb{R} \to \mathbb{C}$ be a bounded, Borel measurable function, so $f \circ C^{-1} : \mathbb{T} \setminus \{1\} \to \mathbb{C}$ is bounded and measurable. We define $f(T) = f(C^{-1}(U))$ and using (5.2) obtain

$$(f(T)x|y) = \int_{\mathbb{T}} f \circ C^{-1} \, \mathrm{d}\sigma_{x,y} = \int_{\mathbb{R}} f \, \mathrm{d}\tau_{x,y}.$$

Note also that for every $x, y \in H$ we have $\operatorname{supp}(\tau_{x,y}) \subset \sigma(T)$. In particular, if $\sigma(T) \subset \{0\}$ then, by the equality in (5.5) we have $(Tx|y) = 0$ for every $x, y \in H$, so $T = 0$.

The properties of the mapping $f \mapsto f(T) \in \mathscr{L}(H)$ are summarized in the next theorem, with an easy but lengthy proof left to the reader.

Theorem 5.23 (Bounded Borel Functional Calculus) *Let $T \in \mathscr{L}(H)$ be a self-adjoint operator on a Hilbert space H. The mapping*

$$\Psi : \mathrm{BM}(\mathbb{R}) \to \mathscr{L}(H), \quad \Psi(f) := f(T).$$

has the following properties.

(a) $\Psi(\mathrm{id}) = T$, $\Psi(\mathbf{1}) = I$.
(b) *Ψ is linear, contractive, and satisfies $\Psi(f\overline{g}) = \Psi(f)\Psi(g)^*$ for each $f, g \in$* $\mathrm{BM}(\mathbb{R})$.
(c) *If $(f_n)_{n\in\mathbb{N}}$ is a $\|\cdot\|_\infty$-bounded and pointwise convergent sequence in* $\mathrm{BM}(\mathbb{R})$ *with limit f, then $\Psi(f_n)x \to \Psi(f)x$ as $n \to \infty$ for each $x \in H$.*
(d) *If $f \in$* $\mathrm{BM}(\mathbb{T})$ *is real-valued, then $\Psi(f)$ is self-adjoint.*

Analogously to unitary operators, for a self-adjoint operator T and a Borel set $A \subset \mathbb{R}$ the operator $P_A := \mathbf{1}_A(T)$ is an orthogonal projection, called **spectral projection** corresponding to $A \subset \mathbb{R}$. The equality $x = P_A x$ holds if and only if $\tau_{x,x}(\mathbb{R}\setminus A) = 0$. Moreover, $P_\lambda := P_{\{\lambda\}}$ is the orthogonal projection onto $\ker(\lambda - T)$. So, in particular, for different $\lambda, \mu \in \mathbb{C}$ we have $P_\lambda P_\mu = 0$ and the eigenspaces $\ker(\lambda - T)$ and $\ker(\mu - T)$ are orthogonal.

Theorem 5.24 (Spectral Theorem for Compact Self-Adjoint Operators) *Let $T \in \mathscr{L}(H)$ be a compact self-adjoint operator on a Hilbert space.*

(a) *The set $\sigma(T) \setminus \{0\}$ consists of eigenvalues only, the eigenspaces are pairwise orthogonal, each eigenspace corresponding to a non-zero eigenvalue is finite dimensional, the spectrum has 0 as the only possible accumulation point. Moreover,*

$$H = \bigoplus_{\lambda\in\sigma(T)} \ker(\lambda - T).$$

Therefore, there is an at most countable orthonormal system $(e_n)_{n\in N}$ of eigenvectors with corresponding non-zero eigenvalues $(\lambda_n)_{n\in N}$ such that

$$Tx = \sum_{n\in N} \lambda_n (x|e_n)e_n$$

(where the sum is unconditionally convergent). In particular, one has $\|T\| = \sup_{n\in N} |\lambda_n| = r(T)$, the spectral radius of T (with the convention $\sup\emptyset = 0$).

(b) *The operator T is Hilbert–Schmidt if and only if*

$$M^2 := \sum_{n\in N} |\lambda_n|^2 < \infty,$$

in which case for the Hilbert–Schmidt norm $\|T\|_{\mathrm{HS}} = M$.

(c) *Suppose T is a self-adjoint Hilbert–Schmidt operator. For $a > 0$ let $P = \mathbf{1}_{(-\infty,-a)\cup(a,\infty)}(T)$ be the corresponding spectral projection. Then we have*

$$\dim \operatorname{ran}(P) \leq \frac{\|T\|_{\mathrm{HS}}^2}{a^2}.$$

Proof (a) We sketch a proof for our case of self-adjoint operators, but remark that the assertions follow also from the Riesz–Schauder theory, see Theorem 1.7. Let $\lambda \in \sigma(T) \setminus \{0\}$. Since T is self-adjoint $\|(\lambda - T)^* x\|^2 = \|(\lambda - T)x\|^2$ for every $x \in H$ and thus

$$\ker(\lambda - T) = \ker(\lambda - T)^* = \operatorname{ran}(\lambda - T)^{\perp}.$$

So if $\lambda \notin P\sigma(T)$, then $\operatorname{ran}(\lambda - T)$ is dense, and the only possibility for $\lambda - T$ not being invertible is that $\operatorname{ran}(\lambda - T)$ is not closed. In this case there is a sequence (x_n) of unit vectors such that $(\lambda - T)x_n \to 0$, by compactness of T we may pass to a subsequence and then assume that $Tx_n \to y$. We conclude $x_n \to \frac{y}{\lambda}$, implying $Ty = \lambda y$ with $\|y\| = 1$, an impossibility if we assume $\lambda \notin P\sigma(T)$. We arrive at the inclusion $\sigma(T) \setminus \{0\} \subset P\sigma(T)$. For every $\lambda \neq 0$ we have that $\frac{1}{\lambda} T|_{\ker(\lambda-T)} = I_{\ker(\lambda-T)}$ is compact, so $\dim\ker(\lambda - T)$ must be finite. Suppose that there is a sequence (λ_n) of pairwise different scalars in $\sigma(T)$ with $r := \inf_n |\lambda_n| > 0$ and take a sequence (x_n) of unit vectors with $Tx_n = \lambda_n x_n$ for each $n \in \mathbb{N}$. Since for different n, m the vectors x_n, x_m are orthogonal, we have $\|Tx_n - Tx_m\|^2 = \|\lambda_n x_n - \lambda_m x_m\|^2 = |\lambda_n|^2 + |\lambda_m|^2 \geq 2r^2$, so (Tx_n) has no Cauchy subsequence, contradicting the compactness of T. It follows that $\sigma(T)$ is at most countable and can have only 0 as an accumulation point.

Define the closed T-reducing subspace

$$H_1 := \bigoplus_{\lambda\in\sigma(T)\setminus\{0\}} \ker(\lambda - T).$$

The restriction of T to $H_1^{\perp}$ is compact and self-adjoint with spectrum contained in $\{0\}$. It follows that this restriction is the 0 operator, and in fact $H_1^{\perp} = \ker(T)$ (see the argument preceding Theorem 5.23). For each $\lambda \in \sigma(T) \setminus \{0\}$ take a finite orthonormal basis in $\ker(\lambda - T)$. These altogether yield an orthonormal basis $(e_n)_{n\in N}$ in H_1 and the required orthonormal system in H.

(b) Let $(e_n)_{n\in N}$ be an orthonormal system as in (a) which we complete to an orthonormal basis $(e_n)_{n\in I}$ with $e_n \in \ker(T)$ for $n \in I \setminus N$. Then we have

$$\|T\|_{\mathrm{HS}}^2 = \sum_{n\in I} \|Te_n\|^2 = \sum_{n\in N} \|Te_n\|^2 = \sum_{n\in I} |\lambda_n|^2 = M^2,$$

proving (b).
(c) We notice that

$$Px = \sum_{n\in N, |\lambda_n|>a} (x|e_n)e_n.$$

Whence we conclude

$$\dim(\mathrm{ran}(P)) = \sum_{n\in N, |\lambda_n|>a} \|e_n\|^2 \le \frac{1}{a^2} \sum_{n\in N, |\lambda_n|>a} \|\lambda_n e_n\|^2 \le \frac{\|T\|_{\mathrm{HS}}^2}{a^2}.$$

The proof is complete. ■

5.6 Lebesgue Spectrum

Let $U \in \mathscr{L}(H)$ be a unitary operator on a (separable) Hilbert space. We say that U has **Rajchman spectrum** if there is a set $N \subset \mathbb{N}$ such that

$$H = \bigoplus_{m\in N} Z(x_m) \tag{5.6}$$

with σ_{x_m} being Rajchman measures, and U has **Lebesgue spectrum** if the σ_{x_m} are equivalent to the Haar measure m on $\mathbb{T}$.

Proposition 5.25 *A unitary operator $U \in \mathscr{L}(H)$ on a separable Hilbert space H is **weakly stable**, i.e.,*

$$(U^n x|y) \to 0 \textit{ as } n \to \infty \textit{ for all } x, y \in H$$

if and only if it has Rajchman spectrum. In particular, a unitary operator with Lebesgue spectrum is weakly stable.

The proof of this result is left as Exercise 5.12.

A unitary operator U with Lebesgue spectrum is said to have **countable Lebesgue spectrum** if N is countably infinite in the decomposition in (5.6).

The following is a characterization of this property, with a proof left as Exercise 5.13.

Proposition 5.26 *Let $U \in \mathscr{L}(H)$ be a unitary operator on a separable Hilbert space H.*

(a) *U has Lebesgue spectrum if and only if for every $x \in H$ with $\|x\| = 1$ the vectors $U^n x$, $n \in \mathbb{Z}$, form an orthonormal basis in the cyclic subspace $Z(x)$.*
(b) *The following assertions are equivalent:*

 (i) *U has countable Lebesgue spectrum.*
 (ii) *There is an orthonormal system $(x_m)_{m\in\mathbb{N}}$ in H such that*

$$H = \bigoplus_{m=1}^{\infty} Z(x_m).$$

 and for each $m \in \mathbb{N}$ the sequence $(U^k x_m)_{k\in\mathbb{Z}}$ is an orthonormal basis of the cyclic subspace $Z(x_m)$.
 (iii) *There is closed subspace $H_0 \subset H$ with $\dim(H_0) = \infty$ such that $H_0 \perp U^k H_0$ for all $k \in \mathbb{Z}$ and*

$$H = \bigoplus_{k\in\mathbb{Z}} U^k H_0.$$

An invertible measure-preserving system (X, μ, T) is said to have **Rajchman spectrum/Lebesgue spectrum/countable Lebesgue spectrum** if its Koopman operator U on $\mathrm{L}^2(X, \mu)$ restricted to the subspace

$$\mathrm{L}^2_0(X, \mu) := \mathbf{1}^\perp = \left\{ f \in \mathrm{L}^2(X, \mu) : \int_X f \,\mathrm{d}\mu = 0 \right\}$$

has the respective property.

The previous characterization can be used to show that non-trivial (two-sided) Bernoulli shifts have countable Lebesgue spectrum. Again, the proof is left to the reader, see Exercise 5.14.

Proposition 5.27 *A non-trivial, two-sided Bernoulli shift $B(p_0, \ldots, p_{k-1})$ has countable Lebesgue spectrum.*

We remark here that one has the following characterization of mixing properties.

Remark 5.28 (Mixing and Spectral Measures) An invertible measure-preserving system (X, μ, T) is strongly mixing if and only if its Koopman operator on $\mathrm{L}^2_0(X, \mu)$ has Rajchman spectrum (this is trivial in view of the above). In turn, the system is weakly mixing if and only if the spectral measures of the Koopman operator on $\mathrm{L}^2_0(X, \mu)$ are all continuous measures (such a unitary operator is said to have **continuous spectrum**). We shall return to these fact after we discussed spectral decomposition in Chap. 6, see Remark 6.31.

It turns out that one can find one single measure, which contains detailed information about the unitary operator, and which together with the so-called **multiplicity function** completely determines the operator (up to unitary equivalence).

Theorem 5.29 (Hahn–Hellinger) *Let $U \in \mathscr{L}(H)$ be a unitary operator on a separable Hilbert space H with associated family of scalar spectral measures $(\sigma_{x,y})_{x,y\in H}$ on $\mathbb{T}$.*

(a) *There is a positive measure $\sigma \in \mathrm{M}(\mathbb{T})$ with the following properties:*

(1) $\sigma_y \ll \sigma$ *for every* $y \in H$.
(2) *If* $0 \leq \nu \ll \sigma$, *then there is* $y \in H$ *with* $\sigma_y = \nu$.

The measure σ is, up to equivalence, unique with these properties. Moreover, $\mathrm{supp}(\sigma) = \sigma(U)$. *The equivalence class of σ is called the **maximal spectral type** of U.*

(b) *There is an at most countable decreasing sequence $(B_m)_{m\in N}$ of Borel sets in $\mathbb{T}$ and a sequence $(x_m)_{m\in N}$ of vectors of H such that*

$$H = \bigoplus_{m\in N} Z(x_m) \text{ and } \sigma_{x_m} = \mathbf{1}_{B_m}\sigma.$$

The function

$$M : \mathbb{T} \to \mathbb{N}_0 \cup \{\infty\}, \quad M := \sum_{m\in N} \mathbf{1}_{B_m}$$

*is called the **multiplicity function** of U.*

(c) *Two unitary operators U_1, U_2 on separable Hilbert spaces are unitarily equivalent if and only if they have the same maximal spectral type and multiplicity function.*

This result is true for normal operators instead of unitaries, see [342, Ch. 18] for some more details or the book by Nadkarni [813] (Chapter 1) for a complete proof; in this context also the Appendix of Parry's book [843] is recommended. That U has Rajchman spectrum means that its maximal spectral type consists of Rajchman measures. Whereas, countable Lebesgue spectrum is equivalent to the fact that the maximal spectral type is (equivalent to) m and the multiplicity function is constant infinity. Since non-trivial, invertible Bernoulli shifts have this property, the Koopman operators of any two such systems are unitarily equivalent. This means, by definition, that all these systems are **spectrally isomorphic** (however, not necessarily isomorphic as mentioned in Sect. 2.2).

Exercises

Exercise 5.1 (Multiplication Operators) Let (X, μ) be a σ-finite measure space. For a fixed measurable function $g : X \to \mathbb{C}$ consider the multiplication operator

$$M_g : \mathrm{L}^0(X, \mu) \to \mathrm{L}^0(X, \mu), \quad f \mapsto gf.$$

(a) Prove the following assertions.

- The operator M_g leaves $\mathrm{L}^2(X, \mu)$ invariant if and only if g is essentially bounded with respect to μ. In this case M_g is bounded with $\|M_g\| = \|g\|_\infty$.
- M_g is unitary if and only if $|g| = \mathbf{1}$ holds μ-almost everywhere.
- If g is (essentially) bounded, then the spectrum of M_g on $\mathrm{L}^2(X, \mu)$ satisfies $\sigma(M_g) = \operatorname{ess\,ran}(g)$, where the essential range of g is defined by

$$\operatorname{ess\,ran}(g) := \{c \in \mathbb{C} : \mu(g^{-1}(U)) > 0 \ \forall \text{ open set } U \subset \mathbb{C} \text{ with } c \in U\}.$$

(b) For unitary M_g determine the scalar spectral measures and the spectral projections P_A.
(c) Determine the bounded Borel calculus, i.e., the mapping Ψ from Theorem 5.15, in the case of unitary M_g.

Exercise 5.2 (Shift) Consider the Hilbert space $\ell^2(\mathbb{Z})$, and the shift operator $S : \ell^2(\mathbb{Z}) \to \ell^2(\mathbb{Z})$, $(x_n)_{n\in\mathbb{Z}} \to (x_{n+1})_{n\in\mathbb{Z}}$ thereon. Prove that S is a unitary operator and determine its scalar spectral measures.

Exercise 5.3 Prove Proposition 5.8.

Exercise 5.4 (Cyclic Vectors) Give an example of a unitary operator with (and without) a cyclic vector.

Exercise 5.5 (Ideals in $\mathrm{M}(\mathbb{T})$)

(a) Prove that $\mathrm{M_c}(\mathbb{T})$ and $\mathrm{M_d}(\mathbb{T})$ are closed ideals in $\mathrm{M}(\mathbb{T})$.
(b) Prove that $I(\mu)$ and $I(\mu)^\perp$ are closed ideals in $\mathrm{M}(\mathbb{T})$ for every $\mu \in \mathrm{M}(\mathbb{T})$.
(c) Let $M \subset \mathrm{M}(\mathbb{T})$ be a subset in $\mathrm{M}(\mathbb{T})$. Prove that

$$M^\perp := \{\nu : \nu \perp \mu \text{ for every } \mu \in M\}$$

is a closed ideal.

Exercise 5.6 (Orthogonal Decomposition into a Direct Sum of Cyclic Subspaces) Let $U \in \mathscr{L}(H)$ be a unitary operator on a separable Hilbert space H. Prove that there is an at most countable set N of unit vectors such that

$$H = \bigoplus_{x \in N} Z(x)$$

as an orthogonal direct sum.

Exercise 5.7 (Spectral Theorem, Multiplication Form) Let $U \in \mathscr{L}(H)$ be a unitary operator on a separable Hilbert space H. Using the result of the foregoing exercise, prove that there is a complete, separable, metric space Ω, a σ-finite, regular, positive, Borel measure μ on Ω, and a bounded measurable function $g : \Omega \to \mathbb{T}$ such that U is unitarily equivalent to the multiplication operator $M_g : \mathrm{L}^2(\Omega, \mu) \to \mathrm{L}^2(\Omega, \mu)$, $f \mapsto gf$.

Exercise 5.8 (Commutativity of the Borel Functional Calculus) Let U be a unitary operator on a Hilbert space H. Prove that for every bounded and measurable function $f : \mathbb{T} \to \mathbb{C}$ the operator $f(U)$ commutes with every bounded operator on H which commutes with U (and U^*).

Exercise 5.9 (Spectral Projections) Let U be a unitary operator on a Hilbert space H. Given a Borel set $B \in \mathcal{B}(\mathbb{T})$ we define $E(B) := P_B = \mathbf{1}_B(U)$. Prove the following assertions.

(a) Each operator $E(B)$, $B \in \mathcal{B}(\mathbb{T})$, is an orthogonal projection.
(b) $E(\mathbb{T}) = I$ and $E(\emptyset) = 0$.
(c) For every Borel set $B \subset \mathbb{T}$ with $B \cap \sigma(U) = \emptyset$ we have $E(B) = 0$.
(d) For a pairwise disjoint sequence $(B_n)_{n\in\mathbb{N}}$ in $\mathcal{B}(\mathbb{T})$,

$$E\Big(\bigcup_{n\in\mathbb{N}} B_n\Big)x = \sum_{n\in\mathbb{N}} E(B_n)x \text{ for all } x \in H.$$

(e) For $B, C \in \mathcal{B}(\mathbb{T})$ one has $E(B \cap C) = E(B)E(C)$.
(f) An operator $S \in \mathscr{L}(H)$ commutes with U (and U^*) if and only if

$$E(B)S = SE(B) \text{ for every } B \in \mathcal{B}(\mathbb{T}).$$

(g) For every $x, y \in H$ we have $(E(\cdot)x\,|\,y) = \sigma_{x,y}(\cdot)$ and

$$(U^n x\,|\,y) = \int_{\sigma(U)} \mathbf{z}^n \,\mathrm{d}(E(\cdot)x\,|\,y) \text{ for every } n \in \mathbb{Z}.$$

(h) If $B \in \mathcal{B}(\sigma(U))$ has non-empty relative interior in $\sigma(U)$, then $E(B) \neq 0$.

Exercise 5.10 (Weak Stability) Let $S \in \mathscr{L}(H)$ be a unitary operator on a Hilbert space, and let $x \in H$. Prove that $S^n x \to 0$ weakly if and only if $S^{*n}x \to 0$ weakly. Is this equivalence true for contractions or isometries?

Exercise 5.11 (Weak Stability) Let $D, D' \subset H$ be subsets in a Hilbert space H such that $\overline{\mathrm{lin}}(D) = H = \overline{\mathrm{lin}}(D')$. For $n \in \mathbb{N}$ let $S_n \in \mathscr{L}(H)$ be a contraction. Show that the following assertions are equivalent.

(i) $S_n x \to 0$ weakly for every $x \in H$.
(ii) $S_n x \to 0$ weakly for every $x \in D$.
(iii) $(S_n x | y) \to 0$ for every $x \in D$, $y \in D'$.

Exercise 5.12 (Rajchman Spectrum) Prove Proposition 5.25. (Hint: Use the result from the foregoing exercise.)

Exercise 5.13 (Lebesgue Spectrum) Prove Proposition 5.26.

Exercise 5.14 (Bernoulli Shifts and Countable Lebesgue Spectrum) Prove Proposition 5.27.

Exercise 5.15 Let $T \in \mathscr{L}(H)$ be a self-adjoint contraction on a Hilbert space H, and suppose that T is **positive** (semi-definite), i.e., satisfies

$$(Tx|x) \geq 0 \text{ for every } x \in H.$$

Prove that for every $x \in H$

$$T^n x \to P_{\mathrm{fix}\, T} x \text{ as } n \to \infty,$$

where $P_{\mathrm{fix}\, T}$ is the orthogonal projection onto $\mathrm{fix}\, T = \ker(I - T)$.

Exercise 5.16 Prove that for each self-adjoint contraction $T \in \mathscr{L}(H)$ on a Hilbert space and for each $x \in H$ the sequence $(T^{2n} x)_{n \in \mathbb{N}}$ converges.

Chapter 6
Decompositions in Hilbert Spaces

We now discuss three fundamental decompositions for linear contractions on Hilbert spaces which turn out to be important for ergodic theorems: the von Neumann decomposition, the rational spectrum decomposition and the Jacobs–de Leeuw–Glicksberg decomposition. These all share the common feature that in one of the components we collect eigenvectors of the operator. The idea behind this is that the action of the operator on eigenvectors is very simple, so we like to think of these components as *structured*. On our way to present these decompositions we also encounter the decomposition of Szőkefalvi-Nagy and Foiaş.

6.1 Von Neumann's Decomposition

Before we present the famous von Neumann decomposition due to F. Riesz [871], we briefly recall some elementary properties of contractions on Hilbert spaces in detail, because of their fundamental importance for our aims.

Proposition 6.1 (Eigenvectors of Contractions on Hilbert Spaces) *Let $S \in \mathscr{L}(H)$ be a contraction on a Hilbert space H. For $x \in H$ the following assertions are equivalent.*

(i) $Sx = x$.
(ii) $S^*x = x$.
(iii) $(Sx|x) = \|x\|^2$.

In particular, $\mathrm{fix}(S) = \mathrm{fix}(S^*)$. *In addition,* $\ker(\lambda - S) = \ker(\overline{\lambda} - S^*)$ *holds for each $\lambda \in \mathbb{T}$, and for different $\mu, \lambda \in \mathbb{T}$ we have* $\ker(\mu - S) \perp \ker(\lambda - S)$.

T. Eisner, B. Farkas, *A Journey Through Ergodic Theorems*, Birkhäuser Advanced Texts Basler Lehrbücher, https://doi.org/10.1007/978-3-031-96506-7_6

Proof It is clear that (i) or (ii) implies (iii). Supposing (iii) and using that S is a contraction we obtain

$$\|x - Sx\|^2 = \|x\|^2 - 2\mathrm{Re}(x|Sx) + \|Sx\|^2 = \|Sx\|^2 - \|x\|^2 \le 0.$$

This proves (i), while (ii) follows by symmetry, so the equivalence of the three assertions has been proved. For $\lambda \in \mathbb{T}$ the operator $\frac{1}{\lambda}S = \overline{\lambda}S$ is again a contraction. So, by the first part, $\ker(\lambda - S) = \mathrm{fix}(\overline{\lambda}S) = \mathrm{fix}(\lambda S^*) = \ker(\overline{\lambda} - S^*)$.
Let now $\mu, \nu \in \mathbb{T}$, $x \in \ker(\lambda - S)$ and $y \in \ker(\mu - S) = \ker(\overline{\mu} - S^*)$. Then

$$(\lambda - \mu)(x|y) = (Sx|y) - (x|S^*y) = (Sx|y) - (Sx|y) = 0.$$

If $\mu \neq \lambda$, then $(x|y) = 0$ must hold. ■

The next result yields, in particular, that angle-preserving linear transformations are precisely the linear isometries.

Corollary 6.2 *Let H, K be Hilbert spaces and $S \in \mathscr{L}(H, K)$ be a contraction. For $x \in H$ the following assertions are equivalent.*

(i) $\|Sx\| = \|x\|$.
(ii) $(Sx|Sy) = (x|y)$ *for every* $y \in H$.
(iii) $S^*Sx = x$.

Proof Only the implication (i)⇒(iii) requires a proof. If $\|Sx\| = \|x\|$, then $\|x\|^2 = (S^*Sx|x)$. Proposition 6.1, applied to the contraction S^*S, yields $S^*Sx = x$. ■

Recall from Chap. 5 that for a Hilbert space H and $S \in \mathscr{L}(H)$ a closed subspace F is **reducing** for S, or S-reducing, if it is S- and S^*-invariant. The following was also discussed in Chap. 5.

Remark 6.3 For a closed subspace $F \subset H$ the following equivalences hold:

$$F \text{ is reducing} \iff F^\perp \text{ is reducing} \iff F \text{ and } F^\perp \text{ are } S\text{-invariant}.$$

The following results was discovered by F. Riesz [871] in his proof of von Neumann's mean ergodic theorem.

Theorem 6.4 (Von Neumann's Decomposition for Contractions) *Let S be a contraction on a Hilbert space H. Then the orthogonal decomposition*

$$H = \mathrm{fix}(S) \oplus \overline{\mathrm{ran}(I - S)} \tag{6.1}$$

into closed reducing subspaces holds.

Proof We have $\mathrm{fix}(S) = \mathrm{fix}(S^*)$ by Proposition 6.1 and therefore $\mathrm{fix}(S)$ is S- and S^*-invariant and hence reducing, and so is its orthogonal complement by

Remark 6.3. By Proposition 5.4, applied to the operator $I - S$, we obtain

$$H = \mathrm{fix}(S) \oplus \mathrm{fix}(S)^{\perp} = \mathrm{fix}(S) \oplus \mathrm{fix}(S^*)^{\perp} = \mathrm{fix}(S) \oplus \overline{\mathrm{ran}(I - S)}.$$

The proof is complete. ■

Note that if $S \in \mathscr{L}(H)$ is a contraction, then so is S^* having the same von Neumann decomposition as S.

6.2 The Rational Spectrum Decomposition

Let $S \in \mathscr{L}(H)$ be a contraction on a Hilbert space H. One part in the von Neumann decomposition was $\mathrm{fix}(S)$ consisting of eigenvectors corresponding to the eigenvalue 1. We now make this part larger by collecting eigenvectors corresponding to all eigenvalues which are roots of unity.

Recall that $\lambda \in \mathbb{T}$ is called **rational** if there is $q \in \mathbb{N}$ with $\lambda^q = 1$, i.e., if λ is a root of unity, and otherwise **irrational**. Clearly, $\lambda \in \mathbb{T}$ is rational if and only if its argument is a rational multiple of π.

Definition 6.5 Let H be a Hilbert space and let $S \in \mathscr{L}(H)$. The subspace

$$H_{\mathrm{rat}} := \overline{\mathrm{lin}}\{x \in H : Sx = \lambda x \text{ for some rational } \lambda \in \mathbb{T}\}$$

is called the **rational spectrum component** of S.

Clearly, H_{rat} is a closed S-invariant subspace of H. Moreover, it has the following representation.

Lemma 6.6 (Representation of the Rational Spectrum Component) *For a Hilbert space H and $S \in \mathscr{L}(H)$ the following assertions hold.*

(a) $\mathrm{fix}(S) \subset \mathrm{fix}(S^2) \subset \mathrm{fix}(S^{2\cdot 3}) \subset \cdots \subset \mathrm{fix}(S^{n!}) \subset \cdots$.
(b) *If S is a contraction, then* $H_{\mathrm{rat}} = \overline{\bigcup_{k\in\mathbb{N}} \mathrm{fix}(S^k)}$.

Proof (a) follows from the fact that $k|l$ (k divides l) implies $\mathrm{fix}(S^k) \subset \mathrm{fix}(S^l)$.
(b) Observe first that by (a), $\bigcup_{k\in\mathbb{N}} \mathrm{fix}(S^k)$ is an S-invariant linear subspace of H. It is enough to prove

$$\mathrm{lin}\{y \in H : Sy = \lambda y \text{ for some } \lambda \in \mathbb{T} \text{ with } \lambda^k = 1\} = \mathrm{fix}(S^k) \tag{6.2}$$

for each $k \in \mathbb{N}$. The inclusion "$\subset$" is trivial.
To see the other inclusion "$\supset$" let $k \in \mathbb{N}$ and take $x \in \mathrm{fix}(S^k)$. Consider the subspace $E = \mathrm{lin}\{x, Sx, \ldots, S^{k-1}x\}$ which is then S-invariant. The operator $S|_E : E \to E$ is an invertible isometry, so a unitary operator, since S is a contraction and $(S|_E)^k = I$. Thus it possesses an orthonormal basis of eigenvectors, and again because of

$(S|_E)^k = I$, all the eigenvalues are k^{th} roots of unity. It follows that $x \in \text{lin}\{y \in H : Sy = \lambda y \text{ for some } \lambda \in \mathbb{T} \text{ with } \lambda^k = 1\}$. ∎

Now, we immediately obtain the following decomposition.

Proposition 6.7 (Rational Spectrum Decomposition) *Let $S \in \mathscr{L}(H)$ be a contraction on a Hilbert space H. Then the orthogonal decomposition*

$$H = H_{\text{rat}} \oplus \bigcap_{k \in \mathbb{N}} \overline{\text{ran}(I - S^k)}$$

into closed S-reducing subspaces holds.

Proof The von Neumann decomposition (Theorem 6.4) applied to the powers of S implies the orthogonal decompositions

$$H = \text{fix}(S^k) \oplus \overline{\text{ran}(I - S^k)} \text{ for all } k \in \mathbb{N}. \tag{6.3}$$

Thus by Lemma 6.6(a) we obtain the orthogonal decomposition

$$H = \overline{\bigcup_{k \in \mathbb{N}} \text{fix}(S^k)} \oplus \bigcap_{k \in \mathbb{N}} \overline{\text{ran}(I - S^k)},$$

where the components are S-reducing by Proposition 6.1 (see also Remark 6.3). The rest follows from Lemma 6.6(b). ∎

6.3 The Szőkefalvi-Nagy–Foiaş and Wold Decompositions

In this section we present a technique that allows us to apply the spectral theorem to study arbitrary contractions.

Proposition 6.8 *For a contraction $S \in \mathscr{L}(H)$ on a Hilbert space H define*

$$H_{\text{uni}} := \big\{x \in H : \|S^n x\| = \|S^{*n} x\| = \|x\| \textit{ for all } n \in \mathbb{N}\big\}.$$

Then H_{uni} is a closed, S-reducing subspace and the restriction of S to H_{uni} is unitary. Furthermore, H_{uni} is the largest closed, S-reducing subspace of H such that the restriction of S becomes unitary.

This result appeared first in [966] by Szőkefalvi-Nagy and Foiaş, the proof we present here is due to Foguel [392].

Proof Let $F \subset H$ be a closed, S-reducing subspace of H such that $S|_F$ is unitary. Then $S^*|_F = (S|_F)^* = (S|_F)^{-1}$ is also unitary. Therefore, the operators S^n and S^{*n} are isometries on F for each $n \in \mathbb{N}$. It follows that $F \subset H_{\text{uni}}$.

By Corollary 6.2 for each $n \in \mathbb{N}$ the identity $\|S^n x\| = \|x\|$ holds if and only if $S^{*n} S^n x = x$, and $\|S^{*n} x\| = \|x\|$ holds if and only if $S^n S^{*n} x = x$. Hence we obtain

$$H_{\text{uni}} = \{x \in H : S^{*n} S^n x = x = S^n S^{*n} x \text{ for every } n \in \mathbb{N}\},$$

or in other words

$$H_{\text{uni}} = \bigcap_{n \in \mathbb{N}} \Big(\text{fix}(S^{*n} S^n) \cap \text{fix}(S^n S^{*n}) \Big). \tag{6.4}$$

Now it is evident that H_{uni} is a closed subspace of H.
Next we show that H_{uni} is a reducing subspace. For $x \in H_{\text{uni}}$ and $n \in \mathbb{N}$ we have $\|S^n Sx\| = \|S^{n+1} x\| = \|x\| = \|Sx\|$. On the other hand $\|S^{*n} Sx\| = \|S^{*(n-1)} S^* Sx\| = \|S^{*(n-1)} x\| = \|x\| = \|Sx\|$. Altogether we conclude that $Sx \in H_{\text{uni}}$. By symmetry we also obtain $S^* x \in H_{\text{uni}}$. By $SS^* x = x = S^* Sx$ for $x \in H_{\text{uni}}$, both operators S and S^* are unitary on H_{uni}. ∎

For a given contraction $S \in \mathscr{L}(H)$ the subspace H_{uni} in the foregoing proposition is called the **unitary part** of H with respect to S. If we want to emphasize the corresponding operator, we write $H_{\text{uni}}(S)$. Its orthogonal complement

$$H_{\text{cnu}} := H_{\text{cnu}}(S) := H_{\text{uni}}(S)^{\perp}$$

is called the **completely non-unitary part** of H with respect to S. It is an S-reducing subspace of H and contains no non-trivial, closed, S-reducing subspace of H on which S acts as a unitary operator. The decomposition

$$H = H_{\text{uni}} \oplus H_{\text{cnu}} \tag{6.5}$$

into S-reducing subspaces is called the **Szőkefalvi-Nagy–Foiaş decomposition** corresponding to the contraction.

Remark 6.9 For a contraction S on a Hilbert space H we have $\ker(\lambda - S) \subset H_{\text{uni}}$ for every $\lambda \in \mathbb{T}$ by Proposition 6.1 and the definition of H_{uni}.

The next proposition is due to Foguel [392] and yields that on the completely non-unitary part the powers S^n converge weakly to 0.

Proposition 6.10 (Weak Stability on the Completely Non-unitary Subspace) *Let S be a contraction on a Hilbert space H. Then $S|_{H_{\text{cnu}}}$ is **weakly stable**, i.e.,*

$$(S^n x | y) \to 0 \text{ as } n \to \infty \text{ for every } x, y \in H_{\text{cnu}}.$$

Proof Let $u \in H$ be arbitrary and notice that $\lim_{n\to\infty} \|S^n u\|$ exists since S is a contraction. We have for $k \in \mathbb{N}$

$$\begin{aligned}\|S^{*k} S^k S^n u - S^n u\|^2 &= \|S^{*k} S^{n+k} u\|^2 - 2\|S^{n+k} u\|^2 + \|S^n u\|^2\\ &\le \|S^{n+k} u\|^2 - 2\|S^{n+k} u\|^2 + \|S^n u\|^2\\ &= \|S^n u\|^2 - \|S^{n+k} u\|^2 \to 0 \text{ as } n \to \infty.\end{aligned}$$

Moreover, for $n \ge k$

$$\begin{aligned}\|S^k S^{*k} S^n u - S^n u\|^2 &\le \|S^{*k} S^n u - S^{n-k} u\|^2\\ &= \|S^{*k} S^n u\|^2 - 2\|S^n u\|^2 + \|S^{n-k} u\|^2\\ &\le -\|S^n u\|^2 + \|S^{n-k} u\|^2 \to 0 \text{ as } n \to \infty.\end{aligned}$$

In particular, for each $k \in \mathbb{N}_0$ and every $u, v \in H$

$$\big((I - S^{*k} S^k) S^n u \big| v\big) \to 0 \quad \text{and} \quad \big((I - S^k S^{*k}) S^n u \big| v\big) \to 0 \text{ as } n \to \infty.$$

Hence for $D_k := \operatorname{ran}(I - S^{*k} S^k) \cup \operatorname{ran}(I - S^k S^{*k})$ we have

$$\big(S^n x \big| y\big) \to 0 \text{ for every } x \in H \text{ and } y \in D_k \text{ as } n \to \infty.$$

Interchanging S and S^* leads to

$$\big(S^n x \big| y\big) = \big(x \big| S^{*n} y\big) \to 0 \text{ for every } x \in D_k \text{ and } y \in H \text{ as } n \to \infty.$$

Therefore by a standard approximation argument

$$(S^n x | y) \to 0 \text{ for every } x, y \in \overline{\bigcup_{k\in\mathbb{N}} D_k} \text{ as } n \to \infty.$$

Since, by Theorem 5.4 and by (6.4),

$$\begin{aligned}\overline{\bigcup_{k\in\mathbb{N}} D_k} &= \overline{\bigcup_{k\in\mathbb{N}} \Big(\operatorname{ran}(I - S^k S^{*k}) \cup \operatorname{ran}(I - S^{*k} S^k)\Big)}\\ &= \Big(\bigcap_{k\in\mathbb{N}} \big(\operatorname{fix}(S^k S^{*k}) \cap \operatorname{fix}(S^{*k} S^k)\big)\Big)^{\perp} = H_{\mathrm{uni}}^{\perp} = H_{\mathrm{cnu}},\end{aligned}$$

the proof is complete. ■

The previous result can be used to show the following useful characterization of weak convergence to 0, we leave the proof as Exercise 6.7.

Proposition 6.11 *For $S \in \mathscr{L}(H)$ a contraction on a Hilbert space H and $x \in H$ the following assertions are equivalent.*

(i) $S^n x \to 0$ *weakly as* $n \to \infty$, *i.e.*, $(S^n x|y) \to 0$ *as* $n \to \infty$ *for each* $y \in H$.
(ii) $(S^n x|x) \to 0$ *as* $n \to \infty$.
(iii) $S^{*n} x \to 0$ *weakly as* $n \to \infty$.

For isometries more detailed information and a finer decomposition are available. For this, a basic building block is the **right shift** R on $\ell^2(\mathbb{N}_0)$ defined on the elements of the standard orthonormal basis $(e_n)_{n\in\mathbb{N}_0}$ by

$$Re_n := e_{n+1}.$$

Theorem 6.12 (Wold Decomposition) *Let S be an isometry on a Hilbert space H. Then*

$$H_{\text{uni}} = \bigcap_{n\in\mathbb{N}} \text{ran}(S^n),$$

and H_{cnu} can be written as an orthogonal sum

$$H_{\text{cnu}} = \bigoplus_{\alpha\in A} H_\alpha$$

for some index set A, where each H_α is a closed S-invariant subspace of H and $S : H_\alpha \to H_\alpha$ is unitarily equivalent to the right shift R on $\ell^2(\mathbb{N}_0)$.

Proof Recall that, since S^n is an isometry, we have $S^{*n}S^n = I$ and the subspace $\text{ran}(S^n)$ is closed. Then $F := \bigcap_{n\in\mathbb{N}} \text{ran}(S^n)$ is an S-invariant closed subspace, $S|_F : F \to F$ is surjective, and, by construction, F is the largest subspace of H with these properties. So we have $H_{\text{uni}} \subset F$, and since $S|_F$ is unitary, $F = H_{\text{uni}}$ follows.
For the proof of the decomposition of H_{cnu}, define for $n \in \mathbb{N}$ the subspace F_n as the orthogonal complement of $\text{ran}(S^n)$ in $\text{ran}(S^{n-1})$. By construction

$$\text{ran}(S^{n+1}) \oplus F_{n+1} = \text{ran}(S^n) = S\text{ran}(S^{n-1}) = S(\text{ran}(S^n) \oplus F_n).$$

Since S is an isometry, it preserves scalar products, and we obtain for $n \in \mathbb{N}$ that $S(\text{ran}(S^n) \oplus F_n) = \text{ran}(S^{n+1}) \oplus SF_n$ and $\text{ran}(S^{n+1}) \perp SF_n$, hence we must have $SF_n = F_{n+1}$.
The subspaces F_k for $k \in \mathbb{N}$ are pairwise orthogonal, and clearly

$$H_{\text{uni}} = \bigcap_{n\in\mathbb{N}} \text{ran}(S^n) \subset \Big(\bigoplus_{k\in\mathbb{N}} F_k\Big)^{\perp}.$$

On the other hand, if $x \perp F_k$ for all $k \in \mathbb{N}$, then $x \in \operatorname{ran}(S^n)$ for all $n \in \mathbb{N}$. Therefore

$$\Big(\bigoplus_{k\in\mathbb{N}} F_k\Big)^{\perp} \subset H_{\text{uni}}.$$

Altogether it follows that

$$H_{\text{cnu}} = \bigoplus_{k\in\mathbb{N}} F_k.$$

Finally, take an orthonormal basis $(e_\alpha)_{\alpha\in A}$ in F_1. Then $(S^k e_\alpha)_{\alpha\in A}$ is an orthonormal basis in $S^k F_1 = F_{k+1}$ for each $k \in \mathbb{N}$, and therefore with $H_\alpha := \overline{\operatorname{lin}}\{S^k e_\alpha : k \in \mathbb{N}_0\}$ we clearly have

$$\bigoplus_{\alpha\in A} H_\alpha = \bigoplus_{k\in\mathbb{N}} F_k = H_{\text{cnu}},$$

and $(S^k e_\alpha)_{k\in\mathbb{N}}$ is an orthonormal basis in H_α. The last statement about unitary equivalence follows immediately. ■

We finish this section with a famous result of Szőkefalvi-Nagy [964] (see also Shäffer [916] and Szőkefalvi-Nagy [965]) which we will need later. As other results of this section, it shows that understanding unitary operators helps a lot to understand arbitrary Hilbert space contractions.

Theorem 6.13 (Existence of Unitary Dilations) *Let $T \in \mathscr{L}(H)$ be a contraction on a Hilbert space H. Then there exist a Hilbert space K, a unitary operator $U \in \mathscr{L}(K)$ and an isometric embedding $J : H \to K$ such that for each $n \in \mathbb{N}_0$ one has $J^*U^nJ = T^n$. Moreover, the operator U leaves* $\operatorname{ran}(J)$ *invariant if and only if T is an isometry.*

A pair (J, U) with $J : H \to K$ a linear isometry and $U \in \mathscr{L}(K)$ with $J^*UJ = T$ is called a **dilation** of T. The previous result of Szőkefalvi-Nagy states that a Hilbert space contraction T has a unitary dilation (J, U) such (J, U^n) is a dilation of T^n for every $n \in \mathbb{N}_0$.

Proof First we assume that T is an isometry and consider the Wold decomposition

$$H_{\text{cnu}} = \bigoplus_{\alpha\in A} H_\alpha.$$

Consider also the right shift $\tilde{R}$ on $\ell^2(\mathbb{Z})$ and the natural embedding $\tilde{J} : \ell^2(\mathbb{N}_0) \to \ell^2(\mathbb{Z})$. Then $(\tilde{J}, \tilde{R}^n)$ is a dilation of R^n for each $n \in \mathbb{N}_0$. Now if $U_\alpha : H_\alpha \to \ell^2(\mathbb{N}_0)$ is a unitary intertwining R and $T|_{H_\alpha}$, then we can define the operator

$$J : H = H_{\text{uni}} \oplus \bigoplus_{\alpha\in A} H_\alpha \to H_{\text{uni}} \oplus \bigoplus_{\alpha\in A} \ell^2(\mathbb{Z}), \ \text{by } J := I \oplus \bigoplus_{\alpha\in A} \tilde{J}U_\alpha.$$

The operator U defined as

$$U : H_{\text{uni}} \oplus \bigoplus_{\alpha \in A} \ell^2(\mathbb{Z}) \to H_{\text{uni}} \oplus \bigoplus_{\alpha \in A} \ell^2(\mathbb{Z}), \quad U = T|_{H_{\text{uni}}} \oplus \bigoplus_{\alpha \in A} \tilde{R}$$

is clearly unitary, and it is also immediately seen that (J, U^n) is a dilation of T^n for each $n \in \mathbb{N}_0$. So the statement is proven if T is an isometry.
Next, we construct to an arbitrary contraction $T \in \mathscr{L}(H)$ Hilbert space isometries $J : H \to K$ and $S : K \to K$ such that (J, S^n) is a dilation of T^n for each $n \in \mathbb{N}_0$. Since T is a contraction and T^*T is self-adjoint, we have that $\sigma(I - T^*T) \subset [0, \infty)$. By the functional calculus in Sect. 5.5 there is a self-adjoint square root $(I - T^*T)^{1/2}$ of $I - T^*T$. Let $K := \ell^2(\mathbb{N}_0; H)$, the Hilbert space of square summable sequences in H, and for $(x_n)_{n \in \mathbb{N}_0}$ set

$$S(x_n)_{n \in \mathbb{N}_0} := (Tx_0, (I - T^*T)^{1/2}x_0, x_1, x_2, \dots).$$

It is easy to check that S is an isometry on K, and together with the natural embedding $J : H \to K$, $Jx = (x, 0, \dots)$ we have that (J, S^n) is a dilation of T^n for each $n \in \mathbb{N}_0$. Using the construction of a unitary dilation for isometries from first part of the proof finishes the argument.
To see the last statement, observe that $P := JJ^* : K \to K$ is an orthogonal projection, with $PUJ = JJ^*UJ = JT$. If T is an isometry, then PUJ is an isometry, so $PUJ = UJ$ must hold, and U leaves $\operatorname{ran}(J)$ invariant. Conversely, if U leaves $\operatorname{ran}(J)$ invariant, then for every $x \in H$ there is $y \in \operatorname{ran}(J)$ with $UJx = Jy$. Thus $\|x\| = \|y\|$ and $Tx = J^*UJx = J^*Jy = y$, since J is an isometry. This yields $\|Tx\| = \|y\| = \|x\|$ for every $x \in H$, i.e., T is an isometry. ∎

6.4 The Jacobs–de Leeuw–Glicksberg Decomposition

For a given contraction $S \in \mathscr{L}(H)$ on a Hilbert space we now enlarge the structured part H_{rat} by collecting all eigenvectors corresponding to unimodular eigenvalues. We first restrict ourselves to the case of unitary operators $U \in \mathscr{L}(H)$ with scalar spectral measures $(\sigma_{x,y})_{x,y \in H}$. According to Proposition 4.10, the space $\mathrm{M}(\mathbb{T})$ decomposes into the direct sum of the closed ideals $\mathrm{M}_{\mathrm{c}}(\mathbb{T})$ and $\mathrm{M}_{\mathrm{d}}(\mathbb{T})$, i.e.,

$$\mathrm{M}(\mathbb{T}) = \mathrm{M}_{\mathrm{d}}(\mathbb{T}) \oplus \mathrm{M}_{\mathrm{c}}(\mathbb{T}). \tag{6.6}$$

This induces, by Proposition 5.21, an orthogonal decomposition of the Hilbert space

$$H = H_{\mathrm{d}} \oplus H_{\mathrm{c}}$$

into the closed U- and U^*-invariant subspaces

$$H_{\mathrm{d}}(U) := \{x \in H : \sigma_x \text{ is discrete}\},$$
$$H_{\mathrm{c}}(U) := \{x \in H : \sigma_x \text{ is continuous}\}.$$

Next we give different descriptions of these parts. Recall that for every $a \in \mathbb{T}$, P_a is the orthogonal projection onto $\ker(a - U)$, see Proposition 5.17.

Proposition 6.14 (Structured Part for Unitary Operators) *For a unitary operator $U \in \mathscr{L}(H)$ we have*

$$H_{\mathrm{d}} = \overline{\mathrm{lin}}\{x \in H : x \textit{ is an eigenvector of } U\},$$

and for the orthogonal projection P_{d} onto H_{d} we have

$$P_{\mathrm{d}}x = \sum_{a \in \mathbb{T}} P_a x \textit{ for every } x \in H,$$

where the summands are pairwise orthogonal, at most countably many of them are non-zero and the series on the right-hand side converges unconditionally. Moreover, we have for each $x, y \in H$

$$\lim_{N\to\infty} \frac{1}{N} \sum_{n=0}^{N-1} |(U^n x|y)|^2 = \sum_{a \in \mathbb{T}} |(P_a x|y)|^2$$

and, in particular,

$$\lim_{N\to\infty} \frac{1}{N} \sum_{n=0}^{N-1} |(U^n x|x)|^2 = \sum_{a \in \mathbb{T}} \|P_a x\|^4 \le \|P_{\mathrm{d}} x\|^4. \tag{6.7}$$

Proof If $x \in H$ is an eigenvector of U, then by Proposition 5.16 we have $\sigma_x = \|x\|^2\delta_\lambda$, so that $\sigma_x \in \mathrm{M}_{\mathrm{d}}(\mathbb{T})$. On the other hand, if $\sigma_x \in \mathrm{M}_{\mathrm{d}}(\mathbb{T})$, then there is a countable set $A \subset \mathbb{T}$ such that $\sigma_x(A^c) = 0$. Since

$$(P_A x|x) = \int_{\mathbb{T}} \mathbf{1}_A \,\mathrm{d}\sigma_x = \int_{\mathbb{T}} \mathbf{1} \,\mathrm{d}\sigma_x = (x|x),$$

we obtain $P_A x = x$. By Proposition 5.17, $\mathbf{1}_{\{a\}}(U)P_A x = \mathbf{1}_{\{a\}}(U)x \in \ker(a - U)$. By Theorem 5.15 we obtain

$$x = P_A x = \sum_{a \in A} \mathbf{1}_{\{a\}}(U)x \in \overline{\mathrm{lin}}\{y \in H : y \text{ is an eigenvector of } U\}.$$

The description of H_{d} and the statement about the orthogonal projection P_{d} are proven, while the unconditional convergence of the series follows from Bessel's inequality.
For the last statement take $x, y \in H$. By Wiener's formula in Proposition 4.11

$$\lim_{N\to\infty}\frac{1}{N}\sum_{n=0}^{N-1}|(U^n x|y)|^2 = \lim_{N\to\infty}\frac{1}{N}\sum_{n=0}^{N-1}|\hat{\sigma}_{x,y}(n)|^2$$
$$= \sum_{a\in\mathbb{T}}|\sigma_{x,y}\{a\}|^2 = \sum_{a\in\mathbb{T}}|(P_a x|y)|^2,$$

where we have used Proposition 5.16 for the last equality. Taking $y := x$, (6.7) follows from Parseval's equality by

$$\sum_{a\in\mathbb{T}}\|P_a x\|^4 \le \Bigl(\sum_{a\in\mathbb{T}}\|P_a x\|^2\Bigr)^2 = \|P_{\mathrm{d}} x\|^4.$$

The proof is complete. ■

Remark 6.15 As can easily be seen from the proof, the inequality in (6.7) is strict as soon as $P_a x \neq 0$ and $P_b x \neq 0$ for some $a, b \in \mathbb{T}$ with $a \neq b$.

Next we turn our attention to the component H_{c} and first present an important lemma. The **(upper/lower) density of a subsequence** $(n_k)_{k\in\mathbb{N}}$ in $\mathbb{N}$ is the (upper/lower) density of the set $\{n_k : k \in \mathbb{N}\}$, see Sect. 4.3.

Lemma 6.16 *Let S be a contraction on a Hilbert space H. For $x \in H$ the following assertions are equivalent.*

(i) $\lim_{N\to\infty}\frac{1}{N}\sum_{n=0}^{N-1}|(S^n x|x)| = 0.$

(i') $\lim_{N\to\infty}\frac{1}{N}\sum_{n=0}^{N-1}|(S^n x|x)|^2 = 0.$

(ii) $\lim_{N\to\infty}\frac{1}{N}\sum_{n=0}^{N-1}|(S^n x|y)| = 0$ *for every* $y \in H$.

(ii') $\lim_{N\to\infty}\frac{1}{N}\sum_{n=0}^{N-1}|(S^n x|y)|^2 = 0$ *for every* $y \in H$.

(iii) *There is a subsequence $(n_j)_{j\in\mathbb{N}}$ of density 1 such that $S^{n_j}x \to 0$ weakly as $j \to \infty$.*

(iv) $$\lim_{N\to\infty}\sup_{y\in H,\|y\|\le 1}\frac{1}{N}\sum_{n=0}^{N-1}|(S^n x|y)| = 0.$$

(iv') $$\lim_{N\to\infty}\sup_{y\in H,\|y\|\le 1}\frac{1}{N}\sum_{n=0}^{N-1}|(S^n x|y)|^2 = 0.$$

(v) $\lim_{N-M\to\infty} \sup_{y\in H,\|y\|\le 1} \frac{1}{N-M} \sum_{n=M}^{N-1} |(S^n x|y)| = 0.$

(v') $\lim_{N-M\to\infty} \sup_{y\in H,\|y\|\le 1} \frac{1}{N-M} \sum_{n=M}^{N-1} |(S^n x|y)|^2 = 0.$

Proof The implications (v)⇒(iv)⇒(ii)⇒(i) are trivial while (i)⇒(i'), (ii)⇒(ii'), (iv)⇒(iv'), (v)⇒(v') and the converse implications are immediate consequences of the contractivity property and the Cauchy–Schwarz inequality, respectively (cf. the easy part of the Koopman–von Neumann lemma (Proposition 4.19)).
The implication (iii)⇒(ii) follows again from the Koopman–von Neumann lemma (Proposition 4.19).
(ii)⇒(iii): Consider the closed subspace $Y = \overline{\mathrm{lin}}\{S^n x : n \in \mathbb{N}_0\}$ and a dense subset $D = \{y_k : k \in \mathbb{N}\} \subset Y$. For $k \in \mathbb{N}$ let $A_k \subset \mathbb{N}$ be a subset with $\mathrm{d}(A_k) = 1$ and $\lim_{n\to\infty,\, n\in A_k}(S^n x|y_k) = 0$ (use the Koopman–von Neumann Lemma 4.19). Take $A \subset \mathbb{N}$ such that $\mathrm{d}(A) = 1$ and $A\setminus A_k$ is finite for every $k \in \mathbb{N}$ (use Proposition 4.18). Then clearly $\lim_{n\to\infty,\, n\in A}(S^n x|y_k) = 0$ for every $k \in \mathbb{N}$. By an approximation argument we conclude $\lim_{n\to\infty,\, n\in A}(S^n x|y) = 0$ for every $y \in Y$. If $y \in Y^\perp$, then $(S^n x|y) = 0$ holds for every $n \in \mathbb{N}$. Therefore we obtain $\lim_{n\to\infty,\, n\in A}(S^n x|y) = 0$ for every $y \in H$, and that was to be proved.

It remains to prove the implication (i')⇒(v'). We suppose first that S is unitary. Let $y \in H$ satisfy $\|y\| \le 1$, without loss of generality we can assume that y belongs to the cyclic subspace $Z(x)$. Then for the spectral measures we have $\sigma_{x,y} \ll \sigma_x$ with the Radon–Nikodym derivative $f \in \mathrm{L}^2(|\sigma_x|)$ satisfying $\|f\|_2 = \|y\|$, see Sect. 5.2. For any $M \in \mathbb{N}_0$ we have $|(S^{M+n}x|y)|^2 = |\hat{\sigma}_{x,S^{*M}y}(n)|^2$. So the uniform version of Wiener's lemma, Proposition 4.12, implies the assertion (v').
We turn to the general case of contractions. By the Szőkefalvi-Nagy dilation theorem, Theorem 6.13, there is a Hilbert space K, a unitary operator U on K and an isometric embedding $J : H \to K$ such that for each $n \in \mathbb{N}_0$ one has $J^*U^nJ = S^n$. If $x \in H$ is as in (i'), then

$$\begin{aligned}\lim_{N\to\infty} \frac{1}{N} \sum_{n=0}^{N-1} |(U^n Jx|Jx)|^2 &= \lim_{N\to\infty} \frac{1}{N} \sum_{n=0}^{N-1} |(J^*U^n Jx|x)|^2 \\ &= \lim_{N\to\infty} \frac{1}{N} \sum_{n=0}^{N-1} |(S^n x|x)|^2 = 0.\end{aligned}$$

So by the already established case of unitary operators

$$\lim_{N-M\to\infty} \sup_{y\in H,\|y\|\le 1} \frac{1}{N-M} \sum_{n=M}^{N-1} |(S^n x|y)|^2$$

$$= \lim_{N-M\to\infty} \sup_{y\in H,\|y\|\le 1} \frac{1}{N-M} \sum_{n=M}^{N-1} |(U^n Jx|Jy)|^2 = 0.$$

The proof is complete. ∎

Remark 6.17 Jones and Lin [585] proved the equivalence of (iv), (v) to the assertions (i), (ii), and (iii) in a much more general setting including power bounded operators on reflexive Banach spaces. A linear operator T on a Banach space E is called **power bounded** if $\sup_{n\in\mathbb{N}} \|T^n\| < \infty$.[1] The proof in [585] reduces the general case to the one of isometries on Hilbert spaces which they prove by different methods. Note that the equivalence of (i), (ii) to the orthogonality to H_{d} for Hilbert space isometries was done before by Neveu, see [681, Proposition 1.2] using our line of arguments with an alternative proof due to Lin [742] using Proposition 6.10.

Proposition 6.18 (Almost Weakly Stable Part for Unitary Operators) *Let $U \in \mathscr{L}(H)$ be a unitary operator on a Hilbert space. A vector $x \in H$ belongs to H_{c} if and only if*

$$\text{D-}\lim_{n\to\infty} U^n x = 0 \textit{ weakly.}$$

Proof By definition, $x \in H_{\mathrm{c}}$ if and only if the scalar spectral measure σ_x is continuous. By Wiener's lemma (Proposition 4.11) and by the Koopman–von Neumann Lemma 4.19 this holds if and only if

$$0 = \lim_{N\to\infty} \frac{1}{N} \sum_{n=1}^{N} |\hat{\sigma}_x(n)| = \lim_{N\to\infty} \frac{1}{N} \sum_{n=1}^{N} |(U^n x|x)|$$

Lemma 6.16 finishes the proof. ∎

Given a contraction $S \in \mathscr{L}(H)$ we define the closed subspace

$$H_{\mathrm{kr}} := H_{\mathrm{kr}}(S) := \overline{\mathrm{lin}}\{x \in H : Sx = \lambda x \text{ for some } \lambda \in \mathbb{T}\},$$

called the **Kronecker** (or **reversible**) **part**.

Remark 6.19 (S is Unitary on H_{kr}) It follows from Proposition 6.1 that H_{kr} is a reducing subspace and $S^*S = SS^* = I$ holds thereon, i.e., S is unitary on H_{kr}.

[1] For example, every contraction is power bounded.

We now come to the deepest result of this chapter, describing the orthogonal complement of the Kronecker part.

Theorem 6.20 (Jacobs–de Leeuw–Glicksberg Decomposition) *Let $S \in \mathscr{L}(H)$ be a contraction on a Hilbert space H. Then the orthogonal decomposition*

$$H = H_{\mathrm{kr}} \oplus H_{\mathrm{aws}}$$

into two closed S-reducing subspaces holds, where

$$H_{\mathrm{aws}} := H_{\mathrm{aws}}(S) := \bigl\{x \in H : \mathrm{D}\text{-}\lim_{n\to\infty} S^n x = 0 \textit{ weakly}\bigr\}.$$

The subspace H_{aws} is called **almost weakly stable part**. Vectors belonging to H_{aws} are called **almost weakly stable** (or **flight**) **vectors**, see Remark 6.27 below for one more name.

Proof Consider the Szőkefalvi-Nagy–Foiaş decomposition $H = H_{\mathrm{uni}} \oplus H_{\mathrm{cnu}}$ for the contraction S. By Remark 6.19 we have $H_{\mathrm{kr}} \subset H_{\mathrm{uni}}$. For the unitary operator $U := S|_{H_{\mathrm{uni}}}$ consider the discrete-continuous decomposition $H_{\mathrm{uni}} = H_{\mathrm{d}} \oplus H_{\mathrm{c}}$. By Proposition 6.14 we have $H_{\mathrm{kr}} = H_{\mathrm{d}}$. We set $H_0 := H_{\mathrm{c}} \oplus H_{\mathrm{cnu}}$. By construction $H_{\mathrm{kr}} \perp H_0$ and $H_{\mathrm{kr}} \oplus H_0 = H$. It remains to prove that $H_0 = H_{\mathrm{aws}}$. Decompose a given $x \in H_0$ as $x = x_{\mathrm{c}} + x_{\mathrm{cnu}}$ with $x_{\mathrm{cnu}} \in H_{\mathrm{cnu}}$, $x_{\mathrm{c}} \in H_{\mathrm{c}}$. For $y \in H$ we have

$$|(S^n x|y)| \le |(S^n x_{\mathrm{c}}|y)| + |(S^n x_{\mathrm{cnu}}|y)|,$$

where the last term tends to 0 for $n \to \infty$ by Foguel's result (Proposition 6.10). For the first term we have by Proposition 6.18

$$|(S^n x_{\mathrm{c}}|y)| \xrightarrow{\mathrm{D}} 0 \text{ as } n \to \infty,$$

where the density one sequence is independent of y. This proves $H_0 \subset H_{\mathrm{aws}}$. Conversely, suppose $x \in H_{\mathrm{aws}}$, and decompose $x = x_{\mathrm{kr}} + x_0$ with $x_{\mathrm{kr}} \in H_{\mathrm{kr}}$ and $x_0 \in H_0$. Then we have

$$(S^n x|x) = (S^n x_{\mathrm{kr}}|x) + (S^n x_0|x),$$

where the first and the last terms converge to 0 in density (for the last one we use the already proved inclusion $H_0 \subset H_{\mathrm{aws}}$). So we must have

$$0 = \lim_{N\to\infty} \frac{1}{N} \sum_{n=0}^{N-1} |(S^n x_{\mathrm{kr}}|x)|^2 = \lim_{N\to\infty} \frac{1}{N} \sum_{n=0}^{N-1} |(S^n x_{\mathrm{kr}}|x_{\mathrm{kr}})|^2 = \sum_{a\in\mathbb{T}} \|P_a x_{\mathrm{kr}}\|^4.$$

We conclude $\|x_{\mathrm{kr}}\|^2 = \sum_{a\in\mathbb{T}} \|P_a x_{\mathrm{kr}}\|^2 = 0$, and obtain $x = x_0$. The equality $H_0 = H_{\mathrm{aws}}$ is proven. ■

The following shows that one can replace D- lim in Theorem 6.20 by the limit along some (not necessarily large) subsequence of $\mathbb{N}$ in the definition of H_{aws}.

Proposition 6.21 (Almost Weakly Stable Part for Contractions) *For a contraction $S \in \mathscr{L}(H)$ on a Hilbert space H and for $x \in H$ the following assertions are equivalent.*

(i) *There is a subsequence $(n_j)_{j\in\mathbb{N}}$ in $\mathbb{N}$ such that $\lim_{j\to\infty} S^{n_j}x = 0$ weakly.*
(ii) *$x \in H_{\mathrm{aws}}$, i.e., there is a subsequence $(n_j)_{j\in\mathbb{N}}$ in $\mathbb{N}$ of density 1 such that $\lim_{j\to\infty} S^{n_j}x = 0$ weakly.*

Proof We only have to show the implication (i)⇒(ii). Suppose (i), let z be an eigenvector corresponding to a unimodular eigenvalue $\lambda \in \mathbb{T}$ and write $x = x_{\mathrm{kr}} + x_{\mathrm{aws}}$ with $x_{\mathrm{kr}} \in H_{\mathrm{kr}}$ and $x_{\mathrm{aws}} \in H_{\mathrm{aws}}$. Since by Proposition 6.1

$$\left|(S^{n_j}x|z)\right| = \left|(x|S^{*n_j}z)\right| = \left|\lambda^{n_j}(x|z)\right| = \left|(x|z)\right|$$

holds for every $j \in \mathbb{N}$, (i) implies $(x|z) = 0$. Thus $x \perp H_{\mathrm{kr}}$, i.e., $x_{\mathrm{kr}} = 0$ and therefore $x = x_{\mathrm{aws}} \in H_{\mathrm{aws}}$. ■

The following is an operator theoretic version of Wiener's Lemma, see Proposition 4.11.

Proposition 6.22 (Abstract Wiener Lemma) *For each contraction $S \in \mathscr{L}(H)$ on a Hilbert space H and each $x, y \in H$ we have*

$$\lim_{N\to\infty} \frac{1}{N}\sum_{n=0}^{N-1} |(S^n x|y)|^2 = \sum_{a\in\mathbb{T}} |(P_a x|y)|^2 \leq \|P_{\mathrm{kr}}x\|^2 \|P_{\mathrm{kr}}y\|^2,$$

where P_a denotes the orthogonal projection onto $\ker(a - S)$.

We leave the proof as Exercise 6.10.

We close this section with another characterization of the Kronecker part. A vector $x \in H$ with $S^{\mathbb{N}_0}x := \{S^n x : n \in \mathbb{N}_0\} \subset H$ being relatively compact is called **asymptotically almost periodic**. Define

$$H_{\mathrm{aap}} := \{x \in H : x \text{ is asymptotically almost periodic}\}.$$

It is Exercise 6.11 to show that for a contraction $S \in \mathscr{L}(H)$ the set H_{aap} is a closed, S-invariant subspace of H.

Proposition 6.23 (Characterization of the Kronecker Part for Isometries) *Let $S \in \mathscr{L}(H)$ be an isometry on a Hilbert space H. Then $H_{\mathrm{aap}} = H_{\mathrm{kr}}$, the Kronecker part. Moreover, for every $x \in H_{\mathrm{kr}}$ even the set*

$$S^{\mathbb{Z}}x := \{S^n x : n \in \mathbb{Z}\} \subset H$$

is relatively compact (recall that S is unitary on H_{kr}.)

Proof If $x \in \ker(a - S)$ for some $a \in \mathbb{T}$, then $S^{\mathbb{Z}}x = \{a^n x : n \in \mathbb{Z}\}$ is relatively compact. The inclusion $H_{\mathrm{kr}} \subset H_{\mathrm{aap}}$ and the last assertion follow immediately.

Let $x \in H_{\mathrm{aap}}$, and define $K := \overline{S^{\mathbb{N}_0}x}$, which is a compact set satisfying $SK \subset K$. We thus obtain a topological system (K, S). Let $z \in K$ be an almost periodic point[2] (which exists by Birkhoff's Theorem 3.10), take $\varepsilon > 0$, and consider the open ball $U = \mathrm{B}(z, \varepsilon)$. Then the set of return times $R_U(z) = \{n_1 < n_2 < \cdots\}$ of z to U is syndetic by the definition of almost periodicity (see Chap. 3). Let $m \in \mathbb{N}$ be such that $S^m x \in \mathrm{B}(z, \varepsilon)$. Since S is an isometry, we can write for each $k \in \mathbb{N}$

$$\begin{aligned}&\|S^{n_k}x - x\| = \|S^{n_k+m}x - S^m x\|\\&\le \|S^{n_k+m}x - S^{n_k}z\| + \|S^{n_k}z - z\| + \|z - S^m x\| < 2\|S^m x - z\| + \varepsilon < 3\varepsilon.\end{aligned}$$

Write $x = x_{\mathrm{kr}} + x_{\mathrm{aws}}$ with $x_{\mathrm{kr}} \in H_{\mathrm{kr}}$ and $x_{\mathrm{aws}} \in H_{\mathrm{aws}}$. By Proposition 6.21 there is a set J with density 1 such that $\lim_{n\to\infty, n\in J} S^n x_{\mathrm{aws}} = 0$ weakly. By Proposition 4.18 and Exercise 4.10 the set $J \cap R_U(z)$ has positive lower density, in particular $J \cap R_U(z) = \{m_1 < m_2 < \cdots\}$ is infinite. We conclude by orthogonality

$$\begin{aligned}9\varepsilon^2 &> \|S^{m_j}x - x\|^2 = \|S^{m_j}x_{\mathrm{kr}} - x_{\mathrm{kr}}\|^2 + \|S^{m_j}x_{\mathrm{aws}} - x_{\mathrm{aws}}\|^2\\&\ge \|S^{m_j}x_{\mathrm{aws}}\|^2 + \|x_{\mathrm{aws}}\|^2 - 2\mathrm{Re}(S^{m_j}x_{\mathrm{aws}}|x_{\mathrm{aws}}) \to 2\|x_{\mathrm{aws}}\|^2\end{aligned}$$

as $j \to \infty$. Since $\varepsilon > 0$ was arbitrary, we obtain $x_{\mathrm{aws}} = 0$, i.e., $x = x_{\mathrm{kr}} \in H_{\mathrm{kr}}$. ■

For contractions an additional component can appear in H_{aap}.

Theorem 6.24 (Relative Compactness of Orbits of Contractions) *Let $S \in \mathscr{L}(H)$ be a contraction on a Hilbert space H. Then one has the orthogonal decomposition*

$$H_{\mathrm{aap}} = H_{\mathrm{kr}} \oplus \big\{x : \lim_{n\to\infty} \|S^n x\| = 0\big\}. \tag{6.8}$$

Proof To show that the two subsets are orthogonal, let $x, y \in H$ satisfy $Sx = ax$ for some $a \in \mathbb{T}$ and $\lim_{n\to\infty} S^n y = 0$. Then we have by Proposition 6.1

$$|(x|y)| = |a^{-n}(x|y)| = |(S^{*n}x|y)| = |(x|S^n y)| \to 0 \text{ as } n \to \infty.$$

The orthogonality of the subspaces follows now by linearity and by the continuity of the scalar product. The inclusion "$\supset$" in (6.8) is clear. For the converse inclusion let $x \in H_{\mathrm{aap}}$ with $x \perp H_{\mathrm{kr}}$. Let P_{uni} be the orthogonal projection onto H_{uni}. Since $S^n P_{\mathrm{uni}}x = P_{\mathrm{uni}}S^n x$, we see that $P_{\mathrm{uni}}x \subset H_{\mathrm{aap}}$ and hence $P_{\mathrm{uni}}x \in H_{\mathrm{kr}}$ by Proposition 6.23, leading to $P_{\mathrm{uni}}x = 0$. We have shown $x \in H_{\mathrm{cnu}}$. Thus

[2] From the present proof it follows that every point is almost periodic in (K, S).

by Proposition 6.10, $\lim_{n\to\infty} S^n x = 0$ weakly. Since $x \in H_{\mathrm{aap}}$, there exists a subsequence $(n_j)_{j\in\mathbb{N}}$ of $\mathbb{N}$ and $z \in H$ such that $\lim_{j\to\infty} S^{n_j}x = z$. Uniqueness of the weak limit implies $z = 0$. Since this holds for every convergent subsequence of the orbit $(S^n x)_{n\in\mathbb{N}_0}$, we conclude $\lim_{n\to\infty} S^n x = 0$. ■

The Jacobs–de Leeuw–Glicksberg decomposition can be elaborated in the far more general setting of weakly compact operator semigroups on Banach spaces. Some elements of this theory, with much less elementary proofs, can be found, for instance, in [342, Ch. 16] or [337, Sec. I.1]. The original papers are due to Jacobs [572] and Glicksberg and de Leeuw [276, 277]. For a single power bounded operator on a reflexive Banach space there is a short and elegant proof by Jones and Lin, see [586].

6.5 Weak Mixing and Rigidity Revisited

We return now to measure-preserving systems, and based on Jacobs–de Leeuw–Glicksberg decomposition we present an equivalent definition of weak mixing.

Proposition 6.25 *For a measure-preserving system the following assertions are equivalent.*

(i) *(X, μ, T) is weakly mixing.*
(ii) *For each measurable set $A \subset X$ there is a sequence $(n_k)_{k\in\mathbb{N}}$ such that for every $g \in \mathrm{L}^1(X, \mu)$*

$$\langle T^{n_k}\mathbf{1}_A, g\rangle \to \mu(A)\langle \mathbf{1}, g\rangle.$$

(iii) *For each measurable set $A \subset X$ there is a sequence $(n'_k)_{k\in\mathbb{N}}$ of density 1 such that for each measurable set $B \subset X$*

$$\mu(T^{-n'_k}A \cap B) \to \mu(A)\mu(B).$$

Proof (i)⇒(ii) follows by a standard approximation argument, while (iii)⇒(i) is trivial.
(ii)⇒(iii): Let $A \subset X$ be a measurable set and let $(n_k)_{k\in\mathbb{N}}$ be such that

$$\langle T^{n_k}\mathbf{1}_A, g\rangle \to \mu(A)\langle \mathbf{1}, g\rangle$$

for each $g \in \mathrm{L}^1(X, \mu)$ (it is enough to require this for each $g \in \mathrm{L}^2(X, \mu)$). Then for $f := \mathbf{1}_A - \mu(A)\mathbf{1}$ we have

$$\langle T^{n_k} f, g\rangle = \langle T^{n_k}\mathbf{1}_A, g\rangle - \mu(A)\langle \mathbf{1}, g\rangle \to 0.$$

By Proposition 6.21 we obtain that f is almost weakly stable, and hence there is a sequence $(n'_k)_{k\in\mathbb{N}}$ with density 1 such that $\langle T^{n'_k} f, g\rangle \to 0$ for each $g \in \mathrm{L}^2(X,\mu)$. This implies

$$\mu(T^{-n'_k}A \cap B) = \langle T^{n'_k}\mathbf{1}_A, \mathbf{1}_B\rangle \to \mu(A)\mu(B),$$

as required. ■

The following charaterization of weak mixing via functions is analogous (up to assertion (v)) to the characterization of mixing given in Proposition 3.44. The proof is left as Exercise 6.17.

Proposition 6.26 (Characterization of Weak Mixing via Functions) *Let (X, μ, T) be a measure-preserving system with Koopman operator T on $\mathrm{L}^0(X,\mu)$. For $p, q \in [1,\infty]$ satisfying $\frac{1}{p} + \frac{1}{q} = 1$ let $D \subset \mathrm{L}^p(X,\mu)$ and $D' \subset \mathrm{L}^q(X,\mu)$ be subsets whose linear hull is dense in the respective spaces. Let finally $\mathcal{E}$ and $\mathcal{E}'$ be $\cap$-stable generators*[3] *of the underlying σ-algebra. Then the following assertions are equivalent.*

(i) *The system (X, μ, T) is weakly mixing.*
(ii) *For every $A \in \mathcal{E}$ and $B \in \mathcal{E}'$*

$$\operatorname*{D\text{-}lim}_{n\to\infty} \mu(T^{-n}A \cap B) = \mu(A)\mu(B).$$

(iii) *For every $f \in D$ and every $g \in D'$*

$$\operatorname*{D\text{-}lim}_{n\to\infty} \langle T^n f, g\rangle = \langle f, \mathbf{1}\rangle\langle \mathbf{1}, g\rangle. \tag{6.9}$$

(iv) *For every $f \in \mathrm{L}^p(X,\mu)$ and every $g \in \mathrm{L}^q(X,\mu)$*

$$\operatorname*{D\text{-}lim}_{n\to\infty} \langle T^n f, g\rangle = \langle f, \mathbf{1}\rangle\langle \mathbf{1}, g\rangle.$$

(v) *One has the orthogonal decomposition* $\mathrm{L}^2(X,\mu) = \mathbb{C}\mathbf{1} \oplus H_{\mathrm{aws}}$.

Remark 6.27 The equivalence of (i) and (v) in Proposition 6.26 motivates the following terminology which we will use later in Chap. 13. For a general measure-preserving system (X, μ, T) one calls a function $f \in \mathrm{L}^2(X,\mu)$ **weakly mixing** if it belongs to H_{aws}, i.e., satisfies the equivalent conditions of Lemma 6.16 for the corresponding Koopman operator. In particular, the Jacobs–de Leeuw–Glicksberg decomposition in Theorem 6.20 takes the form

$$\mathrm{L}^2(X,\mu) = \overline{\mathrm{lin}}\{f : Tf = \lambda f \text{ for a } \lambda \in \mathbb{T}\} \oplus \{f : f \text{ is weakly mixing}\}. \tag{6.10}$$

[3] Recall that $\mathcal{E}$ is $\cap$-stable if $A, B \in \mathcal{E}$ implies $A \cap B \in \mathcal{E}$.

In opposite to mixing, weak mixing is easy to check via the spectrum of the corresponding Koopman operator.

Proposition 6.28 (Spectral Characterization of Weak Mixing) *For a measure-preserving system (X, μ, T) with Koopman operator T on $H := \mathrm{L}^2(X, \mu)$, the following assertions are equivalent.*

(i) *(X, μ, T) is weakly mixing.*
(ii) *The Kronecker part of the Koopman operator is trivial, i.e., $H_{\mathrm{kr}} = \mathbb{C}\mathbf{1}$ and one has the decomposition*

$$\mathrm{L}^2(X, \mu) = \mathbb{C}\mathbf{1} \oplus \{f : f \textit{ is weakly mixing}\}. \tag{6.11}$$

(iii) *The only eigenvalue of T is 1 and it is a geometrically simple (i.e., $\dim(\mathrm{fix}(T)) = 1$).*

Proof (i)$\Rightarrow$(ii): Since weak mixing implies ergodicity by Remark 3.43, we have $\mathrm{fix}(T) = \mathbb{C}\mathbf{1}$ by Proposition 3.24. Let now $\lambda \in \mathbb{T}$ be an eigenvalue of T with eigenfunction $f \in H$. Since $(T^n f|f) = \lambda^n \|f\|_2^2$, Proposition 6.26(iv) applied to $g := \overline{f}$ yields that $\mathrm{D}\text{-}\lim_{n\to\infty}\lambda^n$ exists. This implies $\lambda = 1$ by the result in Exercise 4.7, and $H_{\mathrm{kr}} = \mathrm{fix}(T) = \mathbb{C}\mathbf{1}$ follows.
(ii)$\Rightarrow$(i) follows from the Jacobs–de Leeuw–Glicksberg decomposition, see Theorem 6.20, and the equivalence of (i) and (v) in Proposition 6.26.
The equivalence (ii)$\Leftrightarrow$(iii) is a direct consequence of Proposition 3.24. ■

Remark 6.29

(a) The following is yet another property that is equivalent to weak mixing of (X, μ, T) (which is based on the previous spectral characterization):

(i') For every $f \in \mathrm{L}^2(X, \mu)$ there is a subsequence (n_k) of $\mathbb{N}$ such that

$$\lim_{k\to\infty} (T^{n_k} f|f) = |(f|\mathbf{1})|^2$$

We leave the proof as Exercise 6.18.

(b) Using Proposition 6.28 and the Jacobs–de Leeuw–Glicksberg decomposition (Theorem 6.20) one obtains another equivalent condition for weak mixing:

(ii') For every $f \in \mathrm{L}^2(X, \mu)$ there is a subsequence (n_k) of $\mathbb{N}$ with density 1 such that

$$\lim_{k\to\infty} T^{n_k} f = (f|\mathbf{1})\mathbf{1} \text{ weakly,}$$

see Exercise 6.18.

(c) As Bergelson [100] observed, one can replace T^{-n} and T^n in Proposition 6.26 by T^{-n^2} and T^{n^2} and obtain a characterization of weak mixing using the sequence (n^2) of density zero. In particular, a measure-preserving system (X, μ, T) is weakly mixing if and only if for every measurable $A, B \subset X$ one has

$$\operatorname{D-lim}_{n\to\infty} \mu(T^{-n^2} A \cap B) = \mu(A)\mu(B),$$

cf. Kunszenti-Kovács [698, Thm. 1.1]. Moreover, one can replace (n^2) here by various other sequences such as the prime numbers or their squares. One implication is a direct corollary of (a) and for the converse one needs the notion of equidistribution. We leave the proof as Exercise 10.8 in Chap. 10, when more tools will be available.

Remark 6.30 It is easy to see that a measure-preserving system is totally ergodic if and only if $H_{\mathrm{rat}} = \mathbb{C}\mathbf{1}$, i.e., every eigenfunction of the Koopman operator on $\mathrm{L}^2(X, \mu)$ corresponding[4] to a rational[5] eigenvalue is constant, see Exercise 6.1. Since by Proposition 6.28 weakly mixing systems satisfy $H_{\mathrm{kr}} = \mathbb{C}\mathbf{1}$, i.e., every eigenfunction is constant, we see that weak mixing implies total ergodicity as asserted in Remark 3.43.

Remark 6.31 (Weak Mixing and Continuous Spectrum) An invertible measure-preserving system (X, μ, T) is weakly mixing if and only if the Koopman operator T on $\mathrm{L}^2_0(X, \mu) = \mathbf{1}^\perp$ has continuous spectrum, i.e., the spectral measures on $\mathrm{L}^2_0(X, \mu)$ are all continuous. This follows from the previous spectral characterization and the definition of scalar spectral measures.

We now briefly revisit rigidity sequences and begin with the following definition.

Definition 6.32 A measure-preserving system (X, μ, T) (and the transformation T) is called **rigid along a subsequence** (n_k) **of** $\mathbb{N}$ if its Koopman operator on $\mathrm{L}^2(X, \mu)$ satisfies $\lim_{k\to\infty} T^{n_k} = I$ in the strong operator topology i.e., $T^{n_k} f \to f$ for each $f \in \mathrm{L}^2(X, \mu)$ as $k \to \infty$. A measure-preserving system (and the corresponding transformation) is called **rigid** if it is rigid along some subsequence of $\mathbb{N}$.

Clearly, rigid transformations cannot be mixing whenever the underlying measure space (X, μ) is non-trivial in the sense that there exists a measurable $A \subset X$ with $\mu(A) \in (0, 1)$ (or, equivalently, the system is not Markov isomorphic to the one-point system.) However, they are often weakly mixing, see Remark 6.34.

Remark 6.33 Since for a Hilbert space contraction convergence of its powers to the identity in the weak and strong operator topology are equivalent, see, e.g., [337, Lem. II.5.11], one can replace "strong" by "weak" in the above definition. Moreover,

[4] By Proposition 3.29(a) one can replace $\mathrm{L}^2(X, \mu)$ by $\mathrm{L}^p(X, \mu)$ for every $p \in [1, \infty]$.

[5] As usual, we call $\lambda \in \mathbb{T}$ rational if $\lambda^q = 1$ holds for some $q \in \mathbb{N}$.

a measure-preserving system (X, μ, T) is rigid along (n_k) if and only if

$$\lim_{k\to\infty} \mu(T^{-n_k}A\triangle A) = 0 \quad \text{for each measurable } A \subset X.$$

(Note that this property was taken as original definition by Walters [1032].) We leave the proof as Exercise 6.19(a).

Remark 6.34 It is easy to show that every separable (i.e. the corresponding L^2 space is separable) measure-preserving system with **discrete spectrum** (i.e., with $H = H_{\mathrm{kr}}$ in the corresponding Jacobs–de Leeuw–Glicksberg decomposition) is rigid, see Walters [1032] and [337, Prop. IV.3.3]. Moreover, rigidity is in some sense in contrast to weak mixing. Indeed, by Remark 6.29(b) weak mixing of a measure-preserving system (X, μ, T) means that for every $f \in \mathrm{L}^2_0(X, \mu)$ the sequence $(T^n f)_{n\in\mathbb{N}}$ converges to 0 along a subsequence of $\mathbb{N}$ of density 1. However, if the system is in addition rigid, then for every $f \in \mathrm{L}^2(X, \mu)$ the sequence $(T^n f)_{n\in\mathbb{N}}$ still converges to f along some subsequence of $\mathbb{N}$, which then necessarily has density zero. Note that both weak mixing and rigidity[6] are "typical" properties in the Baire category sense, see the discussion after Remark 3.43, so a typical transformation is both rigid and weakly mixing. For (classes of) examples of such transformations, see, e.g., Walters [1032], Cornfeld et al. [260, Ch. 14], Veech [1017], Ferenczi et al. [385], Bergelson et al. [115], Eisner and Grivaux [343], Chaika [239], Adams [11]. We also mention here the following recent result of Ryzhikov [899] connecting rigidity and mixing along a sequence:[7] For every subsequence of $\mathbb{N}$ with density zero there exists a rigid measure-preserving system which is (strongly) mixing along this sequence, see also El Abdalaoui and Adams [354].

The following characterization seems to be well known.

Proposition 6.35 *A subsequence (n_k) of $\mathbb{N}$ is a rigidity sequence if and only if there exists a non-trivial[8] measure-preserving system which is weakly mixing and rigid along (n_k).*

Note that the "if" direction is easy and is left as Exercise 6.19(b). For the "only if" direction based on a construction of a stationary Gaussian process with a given spectral measure see, e.g., Bergelson et al. [115] and Eisner and Grivaux [343].

Proposition 6.35 in particular gives a way to generate (classes of) examples of rigidity sequences. Surprisingly, Adams [11] showed that in Proposition 6.35 one

---

[6] The mentioned result about rigidity is due to Katok [633] (see, e.g., Nadkarni [813, Ch. 8]) strengthening the result of Rokhlin [877] who showed that a "typical" measure-preserving system is not mixing.

[7] For a subsequence (n_k) of $\mathbb{N}$, a measure-preserving system (X, μ, T) is called **mixing along** (n_k) if for each pair of measurable sets $A, B \subset X$ one has $\lim_{k\to\infty} \mu(T^{-n_k}A \cap B) = \mu(A)\mu(B)$ or, equivalently, $\lim_{k\to\infty} \int_X T^{n_k} f \cdot g \,\mathrm{d}\mu = \int_X f \,\mathrm{d}\mu \cdot \int_X g \,\mathrm{d}\mu$ holds for every $f, g \in \mathrm{L}^2(X, \mu)$. Of course, every such system is weakly mixing.

[8] In the above sense, see the paragraph after Definition 6.32.

can replace weak mixing by just ergodicity. The "if" direction for irrational rotations on the unit circle (which are ergodic but not weakly mixing) was shown before by Fayad and Thouvenot [383], whereas Fayad and Kanigowski [381] showed that there exist rigidity sequences which are not rigidity sequences for irrational rotations.

For related results, examples and related notions (such as α-rigidity, uniform rigidity and IP-rigidity) see, e.g., Aaronson [2, §3], Glasner and Maon [471], Nadkarni [813, Ch. 8], Huang and Ye [564], James et al. [573], Eisner and Grivaux [343], Bergelson et al. [115], Grivaux [500], Yancey [1056, 1057], Aaronson et al. [5], Kunde [695, 696], Griesmer [498], Badea and Grivaux [71], Cai [214], Badea et al. [72], Luo and Zhao [762], El Abdalaoui and Adams [354], Ackelsberg [10].

We finish this chapter with a notion which is opposite to rigidity: A measure-preserving system is called **mildly mixing** if it has no non-constant rigid functions.[9] (As we will see in Exercise 8.13, (X, μ, T) is mildly mixing if and only if it has no non-trivial rigid factors.) This notion which is stronger than weak mixing and weaker than mixing was introduced by Walters [1032] and Furstenberg and Weiss [456], see also Aaronson et al. [6, 7], Furstenberg [446, Sec. 9.4–9.5], Schmidt and Walters [919], Schmidt [918], Bergelson [103], [107, Thm. 5.1], Zhang [1061], Bergelson and McCutcheon [139], Frączek and Lemańczyk [440, 441], Frączek et al. [442], Queffelec [860, Sec. 3.5], Kanigowski and Kułaga-Przymus [621], Robertson [873] for properties and examples of such systems.

Exercises

Exercise 6.1 (Total Ergodicity and Rational Spectrum) Show the characterization of total ergodicity stated in Remark 6.30.

Exercise 6.2 (Weakly Stable Part) Let $S \in \mathscr{L}(H)$ be a contraction on a Hilbert space. Define

$$H_{\mathrm{ws}} := \{x \in H : S^n x \to 0 \text{ weakly}\}.$$

Prove that H_{ws} is a closed S-invariant subspace of H. Is this subspace S-reducing?

Exercise 6.3 (Stability of Unitary Multiplication Operators) Let μ be a Borel probability measure on $\mathbb{T}$ and consider the unitary multiplication operator $U : f \mapsto \mathbf{z} \cdot f$ on $\mathrm{L}^2(\mathbb{T})$, cf. Proposition 1.21 and Example 5.2(2). Prove the following assertions:

[9] For a measure-preserving system (X, μ, T) and a subsequence (n_k) of $\mathbb{N}$, a function $f \in \mathrm{L}^2(X, \mu)$ is called **rigid along** (n_k) if $\lim_{k\to\infty} \|T^{n_k} f - f\|_2 = 0$. A function $f \in \mathrm{L}^2(X, \mu)$ is called **rigid** if it is rigid along some subsequence (n_k) of $\mathbb{N}$. Note that by Exercise 6.9, f is rigid if and only if $\limsup_{n\to\infty} |(T^n f \,|\, f)| = \|f\|_2^2$.

(a) U is almost weakly stable, i.e., $H = H_{\mathrm{aws}}$ if and only if μ is a continuous measure.
(b) U is weakly stable, i.e., $H = H_{\mathrm{ws}}$ if and only if μ is a Rajchman measure.

Exercise 6.4 (Isometric Part) Let $S \in \mathscr{L}(H)$ be a contraction on a Hilbert space. Define

$$H_{\mathrm{iso}} = \{x \in H : \|S^n x\| = \|x\| \text{ for every } n \in \mathbb{N}\}.$$

Prove that H_{iso} is an S-invariant closed subspace of H and that S acts isometrically on it. Show that if E is an S-invariant subspace on which S acts as an isometry, then $E \subset H_{\mathrm{iso}}$ and moreover that for each $x \in H_{\mathrm{iso}}^{\perp}$ we have $S^n x \to 0$ weakly as $n \to \infty$. Is H_{iso} S-reducing?

Exercise 6.5 (Unitary Dilation) Let $T \in \mathscr{L}(H)$ be a non-invertible isometry on a Hilbert space H, and let U be a unitary dilation defined on $K \supset H$. Prove that J^* is the orthogonal projection of K onto H, with J the identical embedding, and $U^n x = T^n x$ for each $x \in H$.

Exercise 6.6 (Norm Stable Part) Let $S \in \mathscr{L}(H)$ be a contraction on a Hilbert space. Define

$$H_{\mathrm{s}} := \{x \in H : S^n x \to 0 \text{ in norm}\}.$$

Prove that H_{s} is a closed S-invariant subspace of H. Is this subspace S-reducing?

Exercise 6.7 (Weak Convergence to 0) Show Proposition 6.11.

Exercise 6.8 For $m \in \mathbb{N}$ let $S_m \in \mathscr{L}(H)$ be contractions on a Hilbert space H, and let $x \in H$. Prove the equivalence of the following assertions.

(i) There is $\lambda \in \mathbb{T}$ and a subsequence (m_k) of $\mathbb{N}$ such that $S_{m_k} x \to \lambda x$ as $k \to \infty$.
(ii) $\limsup_{m\to\infty} |(S_m x|x)| = \|x\|^2$.

Exercise 6.9 Let $S \in \mathscr{L}(H)$ be a contraction on a Hilbert space, let $x \in H$. Prove the equivalence of the following assertions.

(i) $\liminf_{n\to\infty} \|S^n x - x\| = 0$.
(ii) $\limsup_{n\to\infty} \mathrm{Re}(S^n x|x) = \|x\|^2$.
(iii) $\limsup_{n\to\infty} |(S^n x|x)| = \|x\|^2$.

(Hint: Use the result in the foregoing exercise and the recurrence properties of torus rotations.)

Exercise 6.10 Prove Proposition 6.22.

Exercise 6.11 (Relatively Compact $\mathbb{N}_0$-Orbits) Let $S \in \mathscr{L}(H)$ be a contraction on a Hilbert space H. Prove that

$$H_{\mathrm{aap}}(S) = \big\{x \in H : \{S^n x : n \in \mathbb{N}_0\} \subset H \text{ is relatively compact}\big\}$$

is a closed, S-invariant subspace of H. Is this subspace S-reducing?

Exercise 6.12 (Relatively Compact $\mathbb{Z}$-Orbits) Let $U \in \mathscr{L}(H)$ be a unitary operator on a Hilbert space H. Show that

$$H_{\mathrm{ap}}(U) = \big\{x \in H : \{U^n x : n \in \mathbb{Z}\} \subset H \text{ is relatively compact}\big\}$$

is a closed, U-reducing subspace of H. Prove that in general $H_{\mathrm{ap}} \subsetneq H_{\mathrm{aap}}$, and give an example with equality here. (The vectors $x \in H_{\mathrm{ap}}(U)$ are called **almost periodic**.)

Exercise 6.13 (Almost Weakly Stable Part) Let $S \in \mathscr{L}(H)$ be a contraction on a Hilbert space H. Prove, only by using the definition involving D-lim, that $H_{\mathrm{aws}}(S)$ is a closed, S-invariant subspace of H. Show furthermore that for each $k \in \mathbb{N}$

$$H_{\mathrm{aws}}(S) = H_{\mathrm{aws}}(S^k).$$

Exercise 6.14 (Unitary Multiplication Operators) Let $\mu \in \mathrm{M}(\mathbb{T})$ be a positive measure. Consider the multiplication operator $M_{\mathbf{z}} : \mathrm{L}^2(\mathbb{T}, \mu) \to \mathrm{L}^2(\mathbb{T}, \mu)$. Determine the Kronecker and the almost weakly stable parts, and the corresponding orthogonal projections.

Exercise 6.15 (Multiplication Operators) Let (X, μ) be a finite measure space and let $g : X \to \overline{\mathbb{D}}$ be a measurable function, where $\overline{\mathbb{D}}$ is the closed unit disc in $\mathbb{C}$. Consider the multiplication operator $M_g : \mathrm{L}^2(X, \mu) \to \mathrm{L}^2(X, \mu)$, $f \mapsto gf$. Determine the unitary and the completely non-unitary parts, and the corresponding orthogonal projections.

Exercise 6.16 (Two-Sided Shift) Consider the left shift on $\ell^2(\mathbb{Z})$. Determine the Kronecker and the almost weakly stable parts.

Exercise 6.17 (Characterization of Weak Mixing) Prove Proposition 6.26 by utilizing Proposition 6.25 and the linearity of D-lim.

Exercise 6.18 (Weak Mixing and Subsequences) Prove the assertions in Remark 6.29 (a) and (b).

Exercise 6.19 (Rigidity and Rigidity Sequences)

(i) Show the assertion of Remark 6.33.
(ii) Prove the "if" direction in Proposition 6.35.

Part II
Classical Ergodic Theorems

Chapter 7
Classical Ergodic Theorems and More

We now come to the classical ergodic theorems due to von Neumann [1027] and Birkhoff [156] from 1931/1932, along with some of their extensions such as the uniform ergodic theorem for uniquely ergodic systems and the polynomial mean ergodic theorem due to Furstenberg from 1977. On the way we also present an important tool known as the van der Corput trick.

To begin with let S be a linear operator on a Banach space. For every $N \in \mathbb{N}$, the operator

$$A_N := \frac{1}{N} \sum_{n=0}^{N-1} S^n \tag{7.1}$$

is called the Nth **ergodic average** or **Cesàro average** of S. We are interested in the convergence (in various senses) of these ergodic averages as $N \to \infty$ and a description of their limit.

7.1 The Mean Ergodic Theorem

Let us begin with a simple observation. Each $\lambda \in \mathbb{T} \setminus \{1\}$ satisfies

$$\left| \frac{1}{N} \sum_{n=0}^{N-1} \lambda^n \right| = \left| \frac{\lambda^N - 1}{N(\lambda - 1)} \right| \leq \frac{2}{N|\lambda - 1|}.$$

T. Eisner, B. Farkas, *A Journey Through Ergodic Theorems*, Birkhäuser Advanced Texts Basler Lehrbücher, https://doi.org/10.1007/978-3-031-96506-7_7

On the other hand, $\frac{1}{N}\sum_{n=0}^{N-1} 1^n = 1$ for every $N \in \mathbb{N}$. Thus we have

$$\lim_{N\to\infty} \frac{1}{N}\sum_{n=0}^{N-1} \lambda^n = \begin{cases} 1, & \text{if } \lambda = 1, \\ 0, & \text{otherwise.} \end{cases}$$

Thus convergence holds in both cases $\lambda = 1$ and $\lambda \neq 1$, but with different limits and by different reasons. We shall see such a phenomenon several times later and even twice (or more often) in this chapter.

Let us recall from Chap. 6 von Neumann's decomposition for a Hilbert space contraction $S \in \mathscr{L}(H)$. Theorem 6.4 yields the orthogonal decomposition

$$H = \operatorname{fix}(S) \oplus \overline{\operatorname{ran}(I - S)}.$$

Evidently, for $x \in \operatorname{fix}(S)$ the ergodic averages satisfy $A_N x = x$. As a first step we characterize the elements of $\overline{\operatorname{ran}(I - S)}$.

Proposition 7.1 (Characterization of the Complement of the Fixed Component) *Let $S \in \mathscr{L}(H)$ be a contraction on a Hilbert space. Then $x \perp \operatorname{fix}(S)$, i.e., $x \in \overline{\operatorname{ran}(I - S)}$ holds if and only if*

$$\lim_{N\to\infty} \frac{1}{N}\sum_{n=0}^{N-1} S^n x = 0. \tag{7.2}$$

Proof For each $N \in \mathbb{N}$ we have

$$(I - S)\sum_{n=0}^{N-1} S^n = \sum_{n=0}^{N-1} S^n(I - S) = \sum_{n=0}^{N-1}(S^n - S^{n+1}) = I - S^N. \tag{7.3}$$

Since $\|S^N\| \leq 1$ for each $N \in \mathbb{N}$, we conclude that for $x = (I - S)y \in \operatorname{ran}(I - S)$

$$\left\| \frac{1}{N}\sum_{n=0}^{N-1} S^n x \right\| \leq \frac{2\|y\|}{N},$$

and (7.2) follows. Since all ergodic averages are again contractions, (7.2) holds also for all $x \in \overline{\operatorname{ran}(I - S)}$ by a standard approximation argument.

From (7.3) we conclude that

$$I - \frac{1}{N}\sum_{n=0}^{N-1} S^n = \frac{1}{N}\sum_{n=0}^{N-1}(I - S^n) = \frac{1}{N}\sum_{n=0}^{N-1}(I - S)\sum_{j=0}^{n-1} S^j.$$

Hence for each $x \in H$ we have

$$x = \frac{1}{N}\sum_{n=0}^{N-1} S^n x + \frac{1}{N}(I - S)\sum_{n=0}^{N-1}\sum_{j=0}^{n-1} S^j x.$$

Now if x satisfies (7.2), then

$$x = \lim_{N\to\infty} (I - S)\frac{1}{N}\sum_{n=0}^{N-1}\sum_{j=0}^{n-1} S^j x,$$

therefore $x \in \overline{\operatorname{ran}(I - S)}$. ■

Thus, we see that only the part $\operatorname{fix}(S)$ contributes non-zero values to the limit of ergodic averages, and the limit does not change if we project the vector onto $\operatorname{fix}(S)$. For this reason, $\operatorname{fix}(S)$ is called **characteristic** for the ergodic averages $(A_N)_{N\in\mathbb{N}}$, cf. Definition 9.23. We immediately obtain the first ergodic theorem which was originally formulated by von Neumann [1027] for the continuous time case (see also von Neumann [1028], Carleman [225, 226] and Hopf [541]). The proof we present is due to Riesz [871].

Theorem 7.2 (Von Neumann's Mean Ergodic Theorem for Contractions) *For a contraction $S \in \mathscr{L}(H)$ on a Hilbert space H*

$$\lim_{N\to\infty} \frac{1}{N}\sum_{n=0}^{N-1} S^n x = P_{\operatorname{fix}(S)}x \quad \textit{for each } x \in H, \tag{7.4}$$

where $P_{\operatorname{fix}(S)}$ is the orthogonal projection onto the fixed space $\operatorname{fix}(S)$.

Proof Convergence in (7.4) clearly holds for every $x \in \operatorname{fix}(S)$. By von Neumann's decomposition and linearity, it remains to show that (7.4) holds for every $x \in \overline{\operatorname{ran}(I - S)}$ with limit zero. But this is exactly Proposition 7.1. ■

By modifying the proof one sees that this assertion can be strengthened to the following uniform version, see Exercise 7.2. For a sequence (a_n) in a Banach space E and $a \in E$ we say that a is the **Banach limit** of (a_n), in notation

$$\lim_{N-M\to\infty} \frac{1}{N-M}\sum_{n=M}^{N-1} a_n = a$$

if for every $\varepsilon > 0$ there is $n_0 \in \mathbb{N}$ such that for every $N, M \in \mathbb{N}$ with $N - M \geq n_0$ one has

$$\left\| \frac{1}{N-M}\sum_{n=M}^{N-1} a_n - a \right\| < \varepsilon.$$

This is equivalent to requiring

$$\lim_{n\to\infty} \sup_{m\in\mathbb{N}} \left\| \frac{1}{n} \sum_{k=m}^{m+n-1} a_k - a \right\| = 0.$$

Proposition 7.3 (Mean Ergodic Theorem: Banach Limit Formulation) *For a contraction $S \in \mathscr{L}(H)$ on a Hilbert space H*

$$\lim_{N-M\to\infty} \frac{1}{N-M} \sum_{n=M}^{N-1} S^n x = P_{\mathrm{fix}(S)} x \quad \textit{for each } x \in H.$$

Another corollary characterizes ergodicity of a measure-preserving system in terms of the limit of ergodic averages.

Corollary 7.4 (Characterization of Ergodicity) *For a measure-preserving system (X, μ, T) and its Koopman operator T on $\mathrm{L}^2(X, \mu)$ the following assertions are equivalent.*

(i) *The measure-preserving system (X, μ, T) is ergodic.*
(ii) *For every $f \in \mathrm{L}^2(X, \mu)$ one has*

$$\lim_{N\to\infty} \frac{1}{N} \sum_{n=0}^{N-1} T^n f = \int_X f \,\mathrm{d}\mu \cdot \mathbf{1} \quad \textit{in } \mathrm{L}^2(X, \mu).$$

Proof If (ii) holds, then by the mean ergodic theorem every T-invariant function in $\mathrm{L}^2(X, \mu)$ is constant, therefore the system is ergodic.

Assume now that the system is ergodic and let $f \in \mathrm{L}^2(X, \mu)$. By the mean ergodic theorem, the limit of $\frac{1}{N} \sum_{n=0}^{N-1} T^n f$ exists and equals $P_{\mathrm{fix}(T)} f$ which is a constant function by ergodicity. Moreover, this constant function satisfies

$$(P_{\mathrm{fix}(T)} f \,|\, \mathbf{1}) = (f \,|\, \mathbf{1}) = \int_X f \,\mathrm{d}\mu,$$

implying (ii). ■

Remark 7.5 By Corollary 7.4 and Proposition 7.3 we see that in (ii) of Corollary 7.4 one can replace the Cesàro limit $\lim\limits_{N\to\infty} \frac{1}{N} \sum_{n=0}^{N-1}$ by the Banach limit $\lim_{N-M\to\infty} \frac{1}{N-M} \sum_{n=M}^{N-1}$, and the equivalence remains true.

The following strengthening of Poincaré's recurrence (Theorem 2.3) is an easy corollary of Proposition 7.3, see Hopf [542, §11] (the original proof of Khinchin [646] went along different lines).

Theorem 7.6 (Khinchin's Recurrence) *Let (X, μ, T) be a measure-preserving system and let $A \subset X$ be a measurable set. Then for every $\varepsilon > 0$ the set*

$$R_{\varepsilon,A} := R_\varepsilon := \{n \in \mathbb{N} : \mu(A \cap T^{-n}A) > \mu(A)^2 - \varepsilon\}$$

is syndetic.

Proof Let $A \subset X$ be measurable. Proposition 7.3 and Pythagoras's theorem imply

$$\lim_{N-M\to\infty} \frac{1}{N-M} \sum_{n=M}^{N-1} \mu(A \cap T^{-n}A) = \lim_{N-M\to\infty} \frac{1}{N-M} \sum_{n=M}^{N-1} (T^n \mathbf{1}_A | \mathbf{1}_A)$$
$$= (P_{\mathrm{fix}\, T} \mathbf{1}_A | \mathbf{1}_A) = \| P_{\mathrm{fix}\, T} \mathbf{1}_A \|^2 \geq \| P_{\mathbb{C}\mathbf{1}} \mathbf{1}_A \|^2 = \mu(A)^2. \tag{7.5}$$

Assume that the set R_ε is not syndetic for some $\varepsilon > 0$. Then there exists a family of intervals $([M_j, N_j])_{j\in\mathbb{N}}$ such that for every $j \in \mathbb{N}$ one has $N_j - M_j \geq j$ and $R_\varepsilon \cap [M_j, N_j] = \emptyset$, i.e., $\mu(A \cap T^{-n}A) \leq \mu(A)^2 - \varepsilon$ for every $n \in [M_j, N_j]$, contradicting (7.5). ∎

Remark 7.7

(a) Since mixing systems satisfy $\lim_{n\to\infty} \mu(A \cap T^{-n}A) = \mu(A)^2$ for all measurable sets $A \subset X$, this result is in a sense optimal.
(b) For more generalizations of Poincaré's and Khinchin's recurrence theorems and related results see, e.g., Bergelson [104, 106], Leibman [716], Bergelson and Downarowicz [116], Kuang and Ye [688], Boshernitzan et al. [174], Adams et al. [12]. For non-commutative versions see Niculescu and Ströh [825], Niculescu et al. [826]. Extensions concerning multiple recurrence will be discussed later, see Theorem 13.5 and Sect. 13.9.

7.2 Mean Ergodic Operators on Banach Spaces

We now turn to operators on general Banach spaces. An operator $S \in \mathscr{L}(E)$ on a Banach space E is called **mean ergodic** if the limit

$$Px := \lim_{N\to\infty} \frac{1}{N} \sum_{n=0}^{N-1} S^n x \text{ exists for all } x \in E.$$

In this case, P is a projection onto $\mathrm{fix}(S)$ called the **mean ergodic projection** of S (see Exercise 7.5 and Proposition 7.11).

Remark 7.8 (Koopman Operators are Mean Ergodic on L^p, $p < \infty$) Since the L^1-norm is dominated by the L^2-norm by the Cauchy–Schwarz inequality, the mean ergodic theorem and a standard approximation argument imply that the Koopman operator T for a measure-preserving system (X, μ, T) is mean ergodic also on $\mathrm{L}^1(X, \mu)$. By similar methods one can directly show that T is mean ergodic on every $\mathrm{L}^p(X, \mu)$, $p \in [1, \infty)$, see Exercise 7.3.

The following generalizes mean ergodicity of Koopman operators on L^1 to a large class of L^1-operators.

Proposition 7.9 (Dunford–Schwartz Operators) *Let (X, μ) be a probability space and let $S : \mathrm{L}^1(X, \mu) \to \mathrm{L}^1(X, \mu)$ be a linear operator which is contractive on L^1, leaves L^∞ invariant and is contractive for the L^∞-norm. Then S is mean ergodic on $\mathrm{L}^1(X, \mu)$.*

An operator which is simultaneously an L^1 and an L^∞ contraction, as in the assertion above, is called a **Dunford–Schwartz operator** over the measure space (X, μ). So the previous result can be rephrased as follows: Every Dunford–Schwartz operator over a probability space is mean ergodic on $\mathrm{L}^1(X, \mu)$.

Proof By the Riesz–Thorin interpolation theorem, see, e.g., Kadets' book [606, §14.3], $S : \mathrm{L}^1 \cap \mathrm{L}^\infty \to \mathrm{L}^1 \cap \mathrm{L}^\infty$ is a contraction for the L^p-norm for each $p \in [1, \infty]$. So in particular it is a contraction on the Hilbert space $\mathrm{L}^2(X, \mu)$, where its Cesàro averages converge by von Neumann's mean ergodic theorem in the L^2-norm, hence also in the L^1-norm by the Cauchy–Schwarz inequality. By the standard density argument the convergence follows on L^1 as well. ■

In fact, many more operators are mean ergodic than just contractions on Hilbert spaces or Dunford–Schwartz operators, see, e.g., [337, Sec. I.2.1] or [342, Ch. 8] and the references below in Sect. 7.8. An important class is that of power bounded operators on reflexive Banach spaces, see Lorch [757], Kakutani [610] and Yosida [1059]. In what follows we give a short proof of this fact and present some details from the rich theory of mean ergodic operators. We start with a simple observation. Recall that an operator $S \in \mathscr{L}(E)$ is called **power bounded**[1] if $\sup_{n\in\mathbb{N}} \|S^n\| < \infty$.

Proposition 7.10 *Let $S \in \mathscr{L}(E)$ be a power bounded operator on a Banach space E and consider its Cesàro means*

$$A_N := \frac{1}{N} \sum_{n=0}^{N-1} S^n.$$

For $x \in E$ the following assertions are equivalent.

(i) $\lim_{N\to\infty} A_N x = 0$.
(ii) *There is a subsequence $(N_k)_{k\in\mathbb{N}}$ such that $\lim_{k\to\infty} A_{N_k} x = 0$ weakly, i.e., for each $x' \in E'$ one has $\lim_{k\to\infty}\langle A_{N_k} x, x'\rangle = 0$.*

[1] Thus E can be equivalently renormed such that S becomes contractive (using the norm defined by $\|x\|_S := \sup_{n\in\mathbb{N}_0} \|S^n x\|$). Note that such renorming does not change convergence of sequences but can change certain properties of the space such as the property of being a Hilbert space.

(iii) $x \in \overline{\operatorname{ran}}(I - S)$.
(iv) *For each* $x' \in \operatorname{fix}(S')$ *one has* $x \in \ker(x')$.

Proof The implication (i)⇒(ii) is trivial. Suppose $A_{N_k}x \to 0$ as $k \to \infty$ weakly. Then, as in the proof of Proposition 7.1,

$$x - A_{N_k}x = \frac{1}{N_k}\sum_{n=0}^{N_k-1}(I - S^n)x = \frac{1}{N_k}(I - S)\sum_{n=0}^{N_k-1}\sum_{j=0}^{n-1}S^j x \in \operatorname{ran}(I - S).$$

It follows that x belongs to the weak closure of $\operatorname{ran}(I-S)$, hence to the norm closure $\overline{\operatorname{ran}}(I - S)$ by the Hahn–Banach theorem. This proves the implication (ii)⇒(iii).
(iii)⇒(iv): Let $x' \in \operatorname{fix}(S')$, so x' vanishes on $\operatorname{ran}(I - S)$ and hence also on $\overline{\operatorname{ran}}(I - S)$. It follows that $x \in \ker(x')$ holds whenever $x \in \overline{\operatorname{ran}}(I - S)$.
Implication (iv)⇒(iii) follows by the Hahn–Banach theorem.
(iii)⇒(i) can be proved by a telescopic sum argument together with the standard approximation argument as in the proof of Proposition 7.1. ■

Since for $x \in \operatorname{fix}(S)$ one has $A_N x = x$, from the above it follows immediately that $\operatorname{fix}(S) \cap \overline{\operatorname{ran}}(I - S) = \{0\}$ and that for each $x \in \operatorname{fix}(S) \oplus \overline{\operatorname{ran}}(I - S)$ the sequence $(A_N x)_{N\in\mathbb{N}}$ of the Cesàro means converges. It occurs that this subspace is already the maximal subspace on which the sequence of the Cesàro means converges.

Proposition 7.11 *Let E be a Banach space, let $S \in \mathscr{L}(E)$ be power bounded, and consider its Cesàro means $(A_N)_{N\in\mathbb{N}}$. Then the following assertions hold.*

(a) *The collection F of vectors $x \in E$ for which $(A_N x)_{N\in\mathbb{N}}$ converges is a closed and S-invariant subspace.*
(b) *The mapping $P : F \to E$, $Px := \lim_{n\to\infty} A_n x$ is bounded, linear, and maps into F.*
(c) *We have $SP = PS = P$ and $P^2 = P$ with* $\operatorname{ran}(P) = \operatorname{fix}(S)$, $\operatorname{ran}(I - P) = \ker(P) = \overline{\operatorname{ran}}(I - S)$.
(d) $F = \operatorname{fix}(S) \oplus \overline{\operatorname{ran}}(I - S)$.

We leave the proof as Exercise 7.6(a).

The following is a characterization[2] of power bounded mean ergodic operators.

Theorem 7.12 (Characterization of Mean Ergodicity) *Let $S \in \mathscr{L}(E)$ be power bounded on a Banach space E and let $(A_N)_{N\in\mathbb{N}}$ be the sequence of its Cesàro averages. Then the following assertions are equivalent.*

(i) *S is mean ergodic.*
(ii) *For every $x \in E$ there exists a subsequence (N_k) of $\mathbb{N}$ such that $(A_{N_k}x)$ converges weakly as $k \to \infty$.*

[2] The equivalences (i)⇔(ii)⇔(iii) go back to Yosida [1059] and Kakutani [610] while the equivalence (i)⇔(iv) was proven by Sine [950] for contractions (see also Sato [912] for a simple proof based on the Hahn–Banach theorem) and extended by Lloyd [755] to more general operators.

(iii) *The space E decomposes as the direct sum*

$$E = \operatorname{fix}(S) \oplus \overline{\operatorname{ran}}(I - S).$$

(iv) **(Sine's criterion)** *The fixed space* fix S *separates* fix S', *i.e., for each* $x' \in$ fix $S' \setminus \{0\}$ *there is* $x \in$ fix S *with* $\langle x, x' \rangle \neq 0$.

Proof (i)⇔(iii) follows from Proposition 7.11.
(iii)⇔(iv): By Proposition 7.11(d) $F = \operatorname{fix}(S) \oplus \overline{\operatorname{ran}}(I - S)$ and it is a closed subspace. Since for $x' \in E'$ the inclusion $\overline{\operatorname{ran}}(I - S) \subset \ker(x')$ is equivalent to $x' \in \operatorname{fix}(S')$, the assertions (iii) and (iv) are both equivalent to the fact that the only bounded linear functional which vanishes on F is 0.
The implication (i)⇒(ii) is trivial and we leave the proof of the implication (ii)⇒(iv) as Exercise 7.7. ■

Remark 7.13 One can add the following assertions to the list of equivalences in Theorem 7.12:

(i') The operator S is **weakly mean ergodic**, i.e., $(A_N x)_{N\in\mathbb{N}}$ is weakly convergent for each $x \in E$.
(ii') For every $x \in E$ the sequence $(A_N x)_{N\in\mathbb{N}}$ has a weak cluster point.

We leave the straightforward proof as Exercise 7.7, or refer, e.g., to [342, Thm. 8.20] or [337, Thm. I.2.9]).

Remark 7.14 In Theorem 7.12, the power boundedness of S can be replaced by a weaker assumption (which is clearly necessary for mean ergodicity, see Exercise 7.5), namely $\lim_{n\to\infty} \frac{\|S^n x\|}{n} = 0$ for every $x \in E$ and S is **Cesàro bounded**, i.e., the ergodic means (A_N) are bounded, see Krengel [683, Sec. 2.1].

Now we come to the next key lemma, promised already in Remark 2.34.

Proposition 7.15 *Let $S \in \mathscr{L}(E)$ be a power bounded operator on a Banach space E. Then* $\operatorname{fix}(S')$ *separates the points of* $\operatorname{fix}(S)$.

Proof Let $x \in \operatorname{fix} S \setminus \{0\}$, then by Proposition 7.10 $x \notin \overline{\operatorname{ran}}(I - S)$. By the Hahn–Banach Theorem 1.1 there is $x' \in E'$ vanishing on $\overline{\operatorname{ran}}(I - S)$, which is equivalent to $x' \in \operatorname{fix}(S')$, with $\langle x, x' \rangle \neq 0$. This proves the claim. ■

We can now prove the final result of this section:

Theorem 7.16 (Lorch–Kakutani–Yosida) *Every power bounded operator S on a reflexive Banach space E is mean ergodic.*

Proof Since E is reflexive, we may identify E and E'' via the canonical embedding, under which we can identify S with S'', see Sect. 1.1. Since $\operatorname{fix}(S'') = \operatorname{fix}(S)$ separates $\operatorname{fix}(S')$ by Proposition 7.15, we obtain the mean ergodicity of S by Sine's criterion, see Theorem 7.12. ■

Remark 7.17 In fact, Theorem 7.12 (more precisely, the equivalent condition (ii') from Remark 7.13) leads to a much larger[3] class of mean ergodic operators than just power bounded operators on reflexive spaces, namely operators with weakly relatively compact orbits.

Remark 7.18 If $S \in \mathscr{L}(E)$ is a power bounded, mean ergodic operator on a Banach space E, then

$$\lim_{N-M\to\infty} \frac{1}{N-M} \sum_{n=M}^{N-1} S^n x = Px \text{ for each } x \in E,$$

with mean ergodic projection $P \in \mathscr{L}(E)$. This follows from the decomposition in Theorem 7.12 (iii).

Remark 7.19 (Convergence Rate in the Mean Ergodic Theorem) An easy observation of Butzer and Westphal [212] states the equivalence

$$\lim_{N\to\infty} \sum_{n=0}^{N} S^n x = 0 \iff x = 0$$

for every linear bounded operator S on a Banach space E and every $x \in E$. Thus for mean ergodic operators one has by linearity

$$\frac{1}{N} \sum_{n=1}^{N} S^n x - Px = o(1/N) \iff x \in \operatorname{fix} S,$$

where P denotes the mean ergodic projection (see also Exercise 7.5). We leave the details as Exercise 7.6(b).

On the other hand, for power bounded operators on reflexive Banach spaces Browder [196] and Butzer and Westphal [212] showed the equivalence

$$\frac{1}{N} \sum_{n=1}^{N} S^n x - Px = O(1/N) \iff x \in \operatorname{fix} S \oplus \operatorname{ran}(I - S). \tag{7.6}$$

For a topological counterpart of this result see Remark 7.27 and for more on rates of convergence see Sect. 7.8 below.

[3] This class is indeed larger by the Banach–Alaoglu theorem.

7.3 Uniform Convergence for Uniquely Ergodic Systems

Recall from Chap. 3 that a topological system (K, T) is called uniquely ergodic if it has a unique invariant probability measure, which is then a fortiori ergodic. For such systems one has a stronger convergence of ergodic averages. Before discussing this we introduce a new notion.

Definition 7.20 Let (K, T) be a topological system. A sequence $(\mu_N)_{N\in\mathbb{N}}$ of measures in $\mathrm{M}(K)$ is called **asymptotically T-invariant** if

$$\lim_{N\to\infty} |\langle f \circ T, \mu_N\rangle - \langle f, \mu_N\rangle| = 0$$

holds for every $f \in \mathrm{C}(K)$.

The following easy but important lemma explains the value of such sequences.

Lemma 7.21 *Every weak* accumulation point of an asymptotically T-invariant sequence of measures is a T-invariant measure.*

The proof is left as Exercise 7.8.
A crucial step in the proof of the Krylov–Bogolyubov theorem (see Theorem 2.33) was to show that $\frac{1}{N}\sum_{n=0}^{N-1} \delta_{T^n x}$ defines an asymptotically T-invariant sequence of measures. The following is a slightly more general statement, which helps to produce invariant measures with particular properties. The proof is left as Exercise 7.9.

Lemma 7.22 (Examples of Asymptotically Invariant Sequences) *Let (K, T) be a topological system and let $(\nu_n)_{n\in\mathbb{N}}$ be a sequence of probability measures in $\mathrm{M}(K)$. Then the measures μ_N, $N \in \mathbb{N}$, defined by*

$$\mu_N := \frac{1}{N}\sum_{n=0}^{N-1} T_*^n \nu_N$$

form an asymptotically T-invariant sequence.

The following is a version of the pointwise ergodic theorem (coming in Sect. 7.6 below) for uniquely ergodic systems and continuous functions.

Theorem 7.23 (Uniform Convergence for Uniquely Ergodic Systems) *Let (K, T) be a uniquely ergodic topological system with (unique) invariant probability measure $\mu \in \mathrm{M}(K)$. Then for every $f \in \mathrm{C}(K)$*

$$\lim_{N\to\infty} \frac{1}{N}\sum_{n=0}^{N-1} T^n f = \int_K f \,\mathrm{d}\mu \cdot \mathbf{1} \text{ in the supremum norm.}$$

It is Exercise 7.12 to prove this theorem by *Sine's criterion* (equivalence (i)⇔(iv) in Theorem 7.12.)

Proof Observe that for $x \in K$ with

$$\mu_N := \frac{1}{N}\sum_{n=0}^{N-1} \delta_{T^n x} = \frac{1}{N}\sum_{n=0}^{N-1} T_*^n \delta_x$$

we have

$$\frac{1}{N}\sum_{n=0}^{N-1} f(T^n x) = \frac{1}{N}\sum_{n=0}^{N-1} \langle f, \delta_{T^n x}\rangle = \langle f, \mu_N\rangle.$$

By Lemma 7.22 the sequence $(\mu_N)_{N\in\mathbb{N}}$ of probability measures is asymptotically T-invariant. Thus every accumulation point of $(\mu_N)_{N\in\mathbb{N}}$ is an invariant probability measure (Lemma 7.21), which has to be equal to μ by unique ergodicity. Thus we have

$$\frac{1}{N}\sum_{n=0}^{N-1} f(T^n x) = \frac{1}{N}\sum_{n=0}^{N-1} \langle f, \delta_{T^n x}\rangle \to \langle f, \mu\rangle \text{ as } N \to \infty$$

for every $x \in K$.

It remains to show that this convergence is uniform in x. Assume the contrary, i.e., that there exists $f \in \mathrm{C}(K)$ and $\varepsilon > 0$ such that for every $N_0 \in \mathbb{N}$ there exist $N > N_0$ and $x_N \in K$ with

$$\left|\frac{1}{N}\sum_{n=0}^{N-1} f(T^n x_N) - \int_K f \,\mathrm{d}\mu\right| = \left|\langle f, \mu_N - \mu\rangle\right| \geq \varepsilon, \tag{7.7}$$

where $\mu_N := \frac{1}{N}\sum_{n=0}^{N-1} \delta_{T^n x_N}$. However, the sequence $(\mu_N)_{N\in\mathbb{N}}$ of probability measures is asymptotically invariant by Lemma 7.22 with the unique accumulation point μ, contradicting (7.7). ■

In fact, pointwise convergence of ergodic averages characterizes unique ergodicity.

Proposition 7.24 *Let (K, T) be a topological system and suppose that for each $f \in \mathrm{C}(K)$ there is a number $C_f \in \mathbb{C}$ such that*

$$\lim_{N\to\infty} \frac{1}{N}\sum_{n=0}^{N-1} f(T^n x) = C_f \text{ for each } x \in K.$$

Then (K, T) is uniquely ergodic.

Proof For $\mu \in \mathrm{M}_1(K,T)$ (μ an invariant probability measure), by the dominated convergence theorem, we conclude that

$$C_f = \int_K \lim_{N\to\infty} \frac{1}{N}\sum_{n=0}^{N-1} f(T^n x)\,\mathrm{d}\mu = \lim_{N\to\infty} \frac{1}{N}\sum_{n=0}^{N-1} \int_K f(T^n x)\,\mathrm{d}\mu = \langle f, \mu\rangle$$

for every $f \in \mathrm{C}(K)$. Thus any two T-invariant probability measures are equal, i.e., (K, T) is uniquely ergodic. ■

Corollary 7.25 (Characterization of Unique Ergodicity) *Let (K, T) be a topological system.*

(a) *(K, T) is uniquely ergodic if and only if the Koopman operator T on* $\mathrm{C}(K)$ *is mean ergodic with* $\mathrm{fix}(T) = \mathbb{C}\mathbf{1}$.
(b) *(K, T) is uniquely ergodic and the unique invariant probability measure μ has full support, i.e., satisfies* $\mathrm{supp}(\mu) = K$ *if and only if (K, T) is minimal and the Koopman operator is mean ergodic on* $\mathrm{C}(K)$.

The proof of this corollary is left as Exercise 7.11.

Via the previous results we see that whenever the Cesàro averages converge pointwise everywhere to constants, then the convergence is actually uniform. With the help of the characterization in Theorem 7.12 one can prove the following, more general and at first sight surprising fact.

Proposition 7.26 *Let (K, T) be a topological system with Koopman operator T on* $\mathrm{C}(K)$. *If for some $f \in \mathrm{C}(K)$ the Cesàro averages*

$$\frac{1}{N}\sum_{n=0}^{N-1} T^n f$$

converge pointwise everywhere *to a* continuous *function, then they converge uniformly.*

Proof By Proposition 7.10 it suffices to prove weak convergence of the sequence $(\frac{1}{N}\sum_{n=0}^{N-1} T^n f)_{N\in\mathbb{N}}$. By the Riesz–Markov–Kakutani Theorem 1.19 we identify $\mathrm{C}(K)'$ with the space $\mathrm{M}(K)$ of regular Borel measures. By assumption, there is $g \in \mathrm{C}(K)$ with $\frac{1}{N}\sum_{n=0}^{N-1} T^n f \to g$ pointwise everywhere as $N \to \infty$. By Lebesgue's theorem we conclude for every $\mu \in \mathrm{M}(K)$ that

$$\left|\int_K \left(\frac{1}{N}\sum_{n=0}^{N-1} T^n f - g\right)\mathrm{d}\mu\right| \le \int_K \left|\frac{1}{N}\sum_{n=0}^{N-1} T^n f - g\right|\mathrm{d}|\mu| \to 0$$

as $N \to \infty$, proving weak convergence (since g is continuous) and the assertion. ■

Remark 7.27 (Convergence Rate for Minimal Systems) A topological counterpart of (7.6) is a result of Gottschalk and Hedlund [479, Thm. 14.11] stating that for a minimal topological system (K, T) and $f \in \mathrm{C}(K)$

$$\|A_N f\|_\infty = O(1/N) \iff f \in (I - T)\mathrm{C}(K).$$

Lin and Sine [749] showed that the "only if" direction fails in general without the minimality assumption even for uniquely ergodic systems.

For a characterization of uniform convergence of ergodic averages for general (not necessarily uniquely ergodic) topological dynamical systems see Edeko [330, Thm. 2.15].

7.4 The Polynomial Mean Ergodic Theorem

In this section we study the L^2-mean convergence of so-called polynomial ergodic averages and prove the following result of Furstenberg [446, Sec. 3–4]. Recall the definition of H_{rat} from Sect. 6.2.

Theorem 7.28 (Polynomial Mean Ergodic Theorem for Isometries) *Let P be a polynomial with integer coefficients such that $P(\mathbb{N}_0) \subset \mathbb{N}_0$. For an isometry $S \in \mathscr{L}(H)$ on a Hilbert space H the* ***polynomial ergodic averages***

$$\frac{1}{N}\sum_{n=0}^{N-1} S^{P(n)}x \tag{7.8}$$

converge for every $x \in H$. Moreover, non-constant polynomials P satisfy

$$\lim_{N\to\infty} \frac{1}{N}\sum_{n=0}^{N-1} S^{P(n)}x = 0 \text{ if } x \perp H_{\mathrm{rat}}. \tag{7.9}$$

Of course, the assertion remains true if the condition $P(\mathbb{N}_0) \subset \mathbb{N}_0$ is relaxed to $P(\mathbb{N}_0 + m) \subset \mathbb{N}_0$ for some fixed $m \in \mathbb{N}$; in this case one has to start averaging from m and not from 0.

Remark 7.29

(a) Similarly to how the fixed space is characteristic for the averages in the mean ergodic theorem, we say that the rational spectrum component is *characteristic* for polynomial ergodic averages. This is justified by the fact that the limit of (7.8) remains unchanged when replacing x by the projection $P_{H_{\mathrm{rat}}}x$ onto the rational spectrum component H_{rat}.
(b) Following the terminology of Bergelson et al. [123], we say that P is an **integer polynomial** if P is a polynomial with $P(\mathbb{Z}) \subset \mathbb{Z}$. It is easy to extend

Theorem 7.28 to such polynomials mapping $\mathbb{N}_0$ to $\mathbb{N}_0$ (such as $P(n) = \frac{n(n+1)}{2}$), see Exercise 7.13.

To prove Theorem 7.28 we need the following technical but very useful result due to van der Corput which helps to control the norm of the averages of a sequence via the averages of the *"correlation"* between the members of the sequence, see Kuipers and Niederreiter [690, Lem. 3.1] and Bergelson [103, Eq. (1.5)].

Lemma 7.30 (Van der Corput's Inequality, Finitary Version) *Let H be a Hilbert space, let $N \in \mathbb{N}_0$, and let $u_0, \dots, u_{N-1} \in H$. Then for each $J \in \{1, \dots, N-1\}$*

$$\Big\|\sum_{n=0}^{N-1} u_n\Big\|^2 \le \frac{N+J}{J}\sum_{n=0}^{N-1}\|u_n\|^2 + \frac{2(N+J)}{J^2}\sum_{h=1}^{J-1}(J-h)\,\mathrm{Re}\sum_{n=0}^{N-h-1}(u_{n+h}|u_n).$$

Proof For the sake of convenience set $u_n := 0$ for $n \in \mathbb{Z}\setminus\{0, \dots, N-1\}$. Then we have for each $h \in \{0, \dots, J-1\}$

$$\sum_{n=0}^{N-1} u_n = \sum_{n=0}^{N+J-1} u_{n-h}.$$

Summing up these identities for $h = 0, \dots, J-1$ we obtain

$$J\sum_{n=0}^{N-1} u_n = \sum_{h=0}^{J-1}\sum_{n=0}^{N+J-1} u_{n-h} = \sum_{n=0}^{N+J-1}\sum_{h=0}^{J-1} u_{n-h}.$$

Therefore, by the Cauchy–Schwarz inequality (or the convexity of the function $t \mapsto t^2$), we can write

$$J^2\Big\|\sum_{n=0}^{N-1} u_n\Big\|^2 \le \Big(\sum_{n=0}^{N+J-1}\Big\|\sum_{h=0}^{J-1} u_{n-h}\Big\|\Big)^2 \le (N+J)\sum_{n=0}^{N+J-1}\Big\|\sum_{h=0}^{J-1} u_{n-h}\Big\|^2. \tag{7.10}$$

Next, we compute

$$\begin{aligned}\sum_{n=0}^{N+J-1}\Big\|\sum_{h=0}^{J-1} u_{n-h}\Big\|^2 &= \sum_{n=0}^{N+J-1}\sum_{h=0}^{J-1}\|u_{n-h}\|^2 + 2\mathrm{Re}\sum_{i=0}^{J-1}\sum_{j=0}^{i-1}\sum_{n=0}^{N+J-1}(u_{n-i}|u_{n-j})\\ &= J\sum_{n=0}^{N-1}\|u_n\|^2 + 2\mathrm{Re}\sum_{i=0}^{J-1}\sum_{j=0}^{i-1}\sum_{n=i}^{N+i-1}(u_{n-i}|u_{n-j})\\ &= J\sum_{n=0}^{N-1}\|u_n\|^2 + 2\mathrm{Re}\sum_{i=0}^{J-1}\sum_{n=0}^{N-1}\sum_{j=0}^{i-1}(u_n|u_{n+i-j}).\end{aligned}$$

After substituting $h := i - j$ and collecting the formally identical terms (of which, for every $h \in \{1, \dots, J-1\}$ there are as many as $J - h$) we can continue as

$$\begin{aligned}\sum_{n=0}^{N+J-1} \left\| \sum_{h=0}^{J-1} u_{n-h} \right\|^2 &= J \sum_{n=0}^{N-1} \|u_n\|^2 + 2\mathrm{Re} \sum_{h=1}^{J-1} (J-h) \sum_{n=0}^{N-1} (u_n | u_{n+h}) \\ &= J \sum_{n=0}^{N-1} \|u_n\|^2 + 2\mathrm{Re} \sum_{h=1}^{J-1} (J-h) \sum_{n=0}^{N-h-1} (u_n | u_{n+h}).\end{aligned}$$

Inserting this back into (7.10) and dividing by J^2 yield the assertion. ■

By passing to limits in the above van der Corput inequality we directly obtain the following powerful result, see Bergelson [103, Thm. 1.5] and Furstenberg [446, Lem. 4.9, 9.24].

Lemma 7.31 (Van der Corput's Lemma) *For a Hilbert space H and a bounded sequence $(u_n)_{n\in\mathbb{N}}$ in H define*

$$\gamma_h := \limsup_{N\to\infty} \frac{1}{N} \left| \sum_{n=0}^{N-1} (u_n | u_{n+h}) \right|, \qquad h \in \mathbb{N}.$$

Then the inequality

$$\limsup_{N\to\infty} \left\| \frac{1}{N} \sum_{n=0}^{N-1} u_n \right\|^2 \le 2 \liminf_{J\to\infty} \frac{1}{J} \sum_{h=0}^{J-1} \gamma_h$$

holds. In particular,

$$\liminf_{J\to\infty} \frac{1}{J} \sum_{h=0}^{J-1} \gamma_h = 0 \quad \Longrightarrow \quad \lim_{N\to\infty} \frac{1}{N} \sum_{n=0}^{N-1} u_n = 0.$$

Proof By the finitary version of van der Corput's inequality (Lemma 7.30) we can write

$$\frac{1}{N^2} \left\| \sum_{n=0}^{N-1} u_n \right\|^2 \le \frac{N+J}{JN^2} \sum_{n=0}^{N-1} \|u_n\|^2 + \frac{2(N+J)}{J^2N^2} \sum_{h=1}^{J-1} (J-h) \left| \sum_{n=0}^{N-h-1} (u_{n+h} | u_n) \right|. \tag{7.11}$$

Let $h \in \mathbb{N}$ be fixed. Then

$$\limsup_{N\to\infty} \frac{1}{N} \left| \sum_{n=0}^{N-h-1} (u_n | u_{n+h}) \right| = \limsup_{N\to\infty} \frac{1}{N} \left| \sum_{n=0}^{N-1} (u_n | u_{n+h}) \right| = \gamma_h.$$

Now, if $J \in \mathbb{N}$ is fixed, then for the second term on the right-hand side of (7.11)

$$\begin{aligned}\limsup_{N\to\infty} \frac{2(N+J)}{NJ^2}\sum_{h=1}^{J-1}(J-h)\frac{1}{N}\left|\sum_{n=0}^{N-h-1}(u_{n+h}|u_n)\right| &\leq \frac{2}{J^2}\sum_{h=1}^{J-1}(J-h)\gamma_h\\ &\leq \frac{2}{J}\sum_{h=0}^{J-1}\gamma_h,\end{aligned}$$

and for the first term on the right-hand side of (7.11)

$$\limsup_{N\to\infty} \frac{N+J}{JN^2}\sum_{n=0}^{N-1}\|u_n\|^2 \leq \frac{1}{J}\sup_{n\in\mathbb{N}_0}\|u_n\|^2.$$

Take now $\limsup_{N\to\infty}$ in the inequality (7.11), and use the previous two estimates, being true for every $J \in \mathbb{N}$, to conclude

$$\begin{aligned}\limsup_{N\to\infty} \frac{1}{N^2}\left\|\sum_{n=0}^{N-1} u_n\right\|^2 &\leq \lim_{J\to\infty}\frac{1}{J}\sup_{n\in\mathbb{N}_0}\|u_n\|^2 + 2\liminf_{J\to\infty}\frac{1}{J}\sum_{h=0}^{J-1}\gamma_h\\ &= 2\liminf_{J\to\infty}\frac{1}{J}\sum_{h=0}^{J-1}\gamma_h.\end{aligned}$$

The proof is complete. ∎

Remark 7.32 With a little bit more work one can prove that, using the above notation, the inequality

$$\limsup_{N\to\infty}\left\|\frac{1}{N}\sum_{n=0}^{N-1} u_n\right\|^2 \leq \limsup_{J\to\infty}\frac{1}{J}\sum_{h=0}^{J-1}\gamma_h$$

holds, see, e.g., [342, Lem. 9.28]. A dynamical proof via the mean ergodic theorem is presented in Edeko et al. [333]. For further extensions and modifications of the van der Corput lemma we refer to Bergelson [103, Thm. 3.3], Bergelson et al. [141, Lem. 4.2], Bergelson and Leibman [131, pp. 445–446], Bergelson and Lesigne [137], Kunszenti-Kovács [698], Bergelson and Moreira [142, Thm. 2.12], El Abdalaoui [353, Thm. 2.2].

We have now the tools to prove the announced polynomial mean ergodic theorem for isometries.

Proof of Theorem 7.28 We use the rational spectrum decomposition from Sect. 6.2 and define $A_N := \frac{1}{N}\sum_{n=0}^{N-1} S^{P(n)}$. Let $\lambda \in \mathbb{T}$ be rational, i.e., $\lambda^k = 1$ for some

$k \in \mathbb{N}$, and let $x \in H$ satisfy $Sx = \lambda x$. It is easy to see that

$$P(n+k) \equiv P(n) \bmod k$$

holds for every $n \in \mathbb{N}$. By the equality $S^{P(n)}x = \lambda^{P(n)}x$, the sequence $(S^{P(n)}x)_{n\in\mathbb{N}}$ is k-periodic and hence Cesàro convergent by Proposition 4.13(d). Since for each $N \in \mathbb{N}$ the operators A_N are contractions, by linearity we obtain that the sequence $(A_N x)_{N\in\mathbb{N}}$ converges for every $x \in H_{\mathrm{rat}}$.
Note that for constant polynomials convergence on H is clear. It remains to show (7.9) for non-constant P. We prove this by induction on the degree $d := \deg(P)$ of the polynomial P. For $d = 1$, i.e., when $P(n) = an + b$ for some $a, b \in \mathbb{N}_0$, $a \neq 0$ we have

$$\frac{1}{N}\sum_{n=0}^{N-1} S^{P(n)}x = S^b\Big(\frac{1}{N}\sum_{n=0}^{N-1}(S^a)^n x\Big).$$

Let $x \in H$ be such that $x \perp H_{\mathrm{rat}}$. Then, by Lemma 6.6, we have $x \perp \mathrm{fix}(S^a)$. Now the validity of (7.9) follows from Proposition 7.1 applied to S^a.
Let $d \geq 2$ and assume that (7.9) holds for all polynomials with integer coefficients mapping $\mathbb{N}_0$ to $\mathbb{N}_0$ of degree smaller than d. Let $x \perp H_{\mathrm{rat}}$ and let P be as in the assertion with $\deg(P) = d$. Consider $u_n := S^{P(n)}x$ for $n \in \mathbb{N}$, which form a bounded sequence in H. For $h \in \mathbb{N}$ define $Q_h(\cdot) := P(\cdot+h) - P(\cdot)$. Then each Q_h is a polynomial with integer coefficients and satisfies $\deg(Q_h) < \deg(P)$. Observe that for large enough h we have $Q_h(\mathbb{N}_0) \subset \mathbb{N}_0$. Since S is an isometry, it follows for such h that

$$(u_{n+h}|u_n) = (S^{P(n+h)}x|S^{P(n)}x) = (S^{P(n+h)-P(n)}x|x) = (S^{Q_h(n)}x|x).$$

By the induction hypothesis we conclude that

$$\gamma_h := \limsup_{N\to\infty}\Big|\frac{1}{N}\sum_{n=0}^{N-1}(u_{n+h}|u_n)\Big| = \limsup_{N\to\infty}\Big|\frac{1}{N}\sum_{n=0}^{N-1}(S^{Q_h(n)}x|x)\Big| = 0.$$

Therefore, (7.9) follows from the van der Corput Lemma 7.31. ∎

Remark 7.33 (Szőkefalvi-Nagy's Dilation Theorem) Using Szőkefalvi-Nagy's dilation theorem, Theorem 6.13, one can easily show that the assertions of Theorem 7.28 hold for all contractions on Hilbert spaces. A recent result of ter Elst and Müller [999] extends this to all power bounded operators on Hilbert spaces, see also [347]. The case of contractions on reflexive spaces (in particular on L^p-spaces) is still open.

Remark 7.34 (Uniform Version of Theorem 7.28) Using a version of the van der Corput lemma for the uniform Cesàro limit $\lim_{N-M\to\infty}\frac{1}{N-M}\sum_{n=M}^{N-1}$ instead of the Cesàro limit $\lim_{N\to\infty}\frac{1}{N}\sum_{n=1}^{N}$, see Bergelson [103, Thm. 3.3], the arguments above show that the assertion of Theorem 7.28 remains valid for the uniform Cesàro limit. In particular, non-constant polynomials P satisfy

$$\lim_{N-M\to\infty}\frac{1}{N-M}\sum_{n=M}^{N-1} S^{P(n)}x = 0 \text{ if } x \perp H_{\mathrm{rat}}. \tag{7.12}$$

We now present a polynomial generalization of Theorem 7.6 and follow the proof given in Bergelson [113, Sec. 11.3], see Bergelson et al. [117] for a stronger assertion and a different proof.

Theorem 7.35 (Khinchin's Recurrence for Polynomials) *Let (X, μ, T) be a measure-preserving system and $P \in \mathbb{Z}[\cdot]$ with $P(0) = 0$ and $P(\mathbb{N}_0) \subset \mathbb{N}_0$. Then for every measurable set $A \subset X$ and every $\varepsilon > 0$ the set*

$$R_\varepsilon := R_{\varepsilon,A} := \{n : \mu(A \cap T^{-P(n)}A) > \mu(A)^2 - \varepsilon\}$$

is syndetic.

Proof Let $A \subset X$ be a measurable set with $\mu(A) > 0$ (otherwise the assertion is trivial) and take $\varepsilon > 0$. Decompose $\mathbf{1}_A = f_{\mathrm{rat}} + f_{\mathrm{rest}}$ corresponding to the rational spectrum decomposition (Proposition 6.7). By the Pythagoras theorem we have

$$\|f_{\mathrm{rat}}\| = \|P_{H_{\mathrm{rat}}}\mathbf{1}_A\| \geq \|P_{\mathbb{C}\mathbf{1}}\mathbf{1}_A\| = \mu(A)$$

and in particular $f_{\mathrm{rat}} \neq 0$. Denote $f_k := P_{\mathrm{fix}\,T^k} f_{\mathrm{rat}} = P_{\mathrm{fix}\,T^k}\mathbf{1}_A$ for $k \in \mathbb{N}$. Then for each $k \in \mathbb{N}$ we have $f_k \geq 0$ since the mean ergodic projection of T^k is positive (i.e. maps positive function into positive ones). From Lemma 6.6 we obtain $f_{\mathrm{rat}} = \lim_{k\to\infty} f_{k!}$ in L^2, so we also have $f_{\mathrm{rat}} \geq 0$.

Let $k \in \mathbb{N}$ be such that $\|f_k - f_{\mathrm{rat}}\| \leq \frac{\varepsilon}{2\|f_{\mathrm{rat}}\|}$. By the assumption $P(0) = 0$ we see that $P(kn)$ is divisible by k and hence $T^{P(kn)} f_k = f_k$ for every $n \in \mathbb{N}$. Thus by Remark 7.34, the triangle inequality and the Pythagoras theorem we have

$$\begin{aligned}
&\lim_{N-M\to\infty}\frac{1}{N-M}\sum_{n=M}^{N-1}\mu(A \cap T^{-P(kn)}A)\\
&= \lim_{N-M\to\infty}\frac{1}{N-M}\sum_{n=M}^{N-1}(T^{P(kn)}\mathbf{1}_A|\mathbf{1}_A)\\
&= \lim_{N-M\to\infty}\frac{1}{N-M}\sum_{n=M}^{N-1}(T^{P(kn)}f_{\mathrm{rat}}|f_{\mathrm{rat}})
\end{aligned}$$

$$
\begin{aligned}
&\geq \lim_{N-M\to\infty} \frac{1}{N-M} \sum_{n=M}^{N-1} (T^{P(kn)} f_k | f_{\text{rat}}) - \frac{\varepsilon}{2\|f_{\text{rat}}\|} \|f_{\text{rat}}\| \\
&= \|f_k\|^2 - \frac{\varepsilon}{2} \geq \|P_{\mathbb{C}\mathbf{1}} \mathbf{1}_A\|^2 - \frac{\varepsilon}{2} = \mu(A)^2 - \frac{\varepsilon}{2}.
\end{aligned}
$$

So we see that the set $\{n : \mu(A \cap T^{-P(kn)} A) > \mu(A)^2 - \varepsilon\}$ is syndetic. Since every n in this set satisfies $kn \in R_\varepsilon$, the set R_ε is syndetic, too. ∎

Remark 7.36 The above proof is based on the fact that for every $\varepsilon > 0$ there exists $k \in \mathbb{N}$ (depending on ε and A) such that

$$
\lim_{N-M\to\infty} \frac{1}{N-M} \sum_{n=M}^{N-1} \mu(A \cap T^{-P(kn)} A) \geq \mu(A)^2 - \varepsilon,
$$

see also Bergelson and Leibman [134, Theorems 4.1 and 4.2] for a multiple version. Note that, unlike the linear case (see the proof of Theorem 7.6), the above inequality fails for $k = 1$ in general, see the example due to Bergelson [100] presented in Exercise 7.14.

7.5 Multiple Ergodic Theorems for Special Systems

We now return to measure-preserving systems and look at the following generalization of the mean ergodic theorem concerning so-called *multiple ergodic averages* for Koopman operators.

Theorem 7.37 (Host–Kra, Multiple Ergodic Theorem) *Let (X, μ, T) be an ergodic measure-preserving system, let $k \geq 1$, and let $f_1, \ldots, f_k \in \mathrm{L}^\infty(X, \mu)$. Then the limit of the multiple ergodic averages*

$$
\lim_{N\to\infty} \frac{1}{N} \sum_{n=0}^{N-1} \left(T^n f_1\right) \cdot \left(T^{2n} f_2\right) \cdots \left(T^k f_k\right) \tag{7.13}
$$

exists in $\mathrm{L}^2(X, \mu)$.

We shall discuss some ingredients from the proof of this result later in Chap. 13 for general ergodic measure-preserving systems. Here we only discuss two special classes, for which the proof is easy and short. The first such particular case is when the system is completely structured, meaning that the Kronecker part of the Koopman operator is the entire L^2-space. Recall from Remark 6.34 that such a system is termed as having **discrete spectrum**.

Proposition 7.38 (Multiple Ergodic Theorem for Systems with Discrete Spectrum) *Let (X, μ, T) be a measure-preserving system with discrete spectrum. Then for every $k \geq 1$ and every $f_1, \ldots, f_k \in \mathrm{L}^\infty(X, \mu)$ the limit in* (7.13) *exists in the* L^2*-norm.*

Proof First, let $f_1, \ldots, f_k \in \mathrm{L}^\infty(X, \mu)$ be eigenfunctions of the Koopman operator T. Then we have

$$\begin{aligned} &\frac{1}{N}\sum_{n=0}^{N-1} (T^n f_1) \cdot (T^{2n} f_2) \cdots (T^{kn} f_k) \\ &\qquad = \frac{1}{N}\sum_{n=0}^{N-1} (\lambda_1 \lambda_2^2 \cdots \lambda_k^k)^n \cdot f_1 \cdots f_k, \end{aligned}$$

and we have convergence in $\mathrm{L}^2(X, \mu)$ for $N \to \infty$ as asserted. By linearity we obtain the existence of the limit whenever $f_1, \ldots, f_k$ are linear combinations of L^∞-eigenfunctions of T. Since one can approximate every L^2-eigenfunction of T in L^2 by bounded eigenfunctions by Proposition 3.29, the assertion follows now by a standard approximation argument. ■

The other special case is when the system is weakly mixing. By Proposition 6.28 this is precisely the case when the Kronecker part of the Koopman operator is trivial, i.e., contains only constants.

Proposition 7.39 (Multiple Ergodic Theorem for Weakly Mixing Systems) *Let (X, μ, T) be a weakly mixing measure-preserving system. Then for every $k \geq 1$ and every $f_1, \ldots, f_k \in \mathrm{L}^\infty(X, \mu)$*

$$\lim_{N\to\infty} \frac{1}{N}\sum_{n=0}^{N-1} (T^n f_1) \cdot (T^{2n} f_2) \cdots (T^{kn} f_k) = \Big(\int_X f_1 \,\mathrm{d}\mu \cdots \int_X f_k \,\mathrm{d}\mu\Big) \cdot \mathbf{1}$$

holds in $\mathrm{L}^2(X, \mu)$.

Proof The proof is by induction on k. For $k = 1$ the assertion is just the mean ergodic theorem for ergodic systems (recall that every weakly mixing system is ergodic). Suppose now that $k \geq 2$ and the assertion is true for $k - 1$ functions, and let $f_1, \ldots, f_k \in \mathrm{L}^\infty(X, \mu)$. Decompose $f_k = P_{\mathrm{kr}} f_k + (I - P_{\mathrm{kr}}) f_k$. By linearity it suffices to handle the two cases $f_k = P_{\mathrm{kr}} f_k$ or $f_k = (I - P_{\mathrm{kr}}) f_k$ separately. Since the system is weakly mixing, P_{kr} is the orthogonal projection onto $\mathbb{C}\mathbf{1}$ by

Proposition 6.26. So if $f_k = P_{\mathrm{kr}} f_k \in H_{\mathrm{kr}}$, then

$$\begin{aligned}\frac{1}{N}\sum_{n=0}^{N-1}\big(T^n f_1\big)\cdot\big(T^{2n} f_2\big)\cdots\big(T^{(k-1)n} f_{k-1}\big)\cdot\big(T^{kn} f_k\big)\\ = \frac{1}{N}\sum_{n=0}^{N-1}\big(T^n f_1\big)\cdot\big(T^{2n} f_2\big)\cdots\big(T^{(k-1)n} f_{k-1}\big)\cdot f_k,\end{aligned}$$

and by the induction hypothesis we obtain

$$\begin{aligned}\lim_{N\to\infty}\frac{1}{N}\sum_{n=0}^{N-1}\big(T^n f_1\big)\cdot\big(T^{2n} f_2\big)\cdots\big(T^{kn} f_k\big)\\ = \Big(\int_X f_1\,\mathrm{d}\mu\cdots\int_X f_{k-1}\,\mathrm{d}\mu\Big)\cdot\int_X f_k\,\mathrm{d}\mu\cdot\mathbf{1}.\end{aligned}$$

Next, suppose that $f_k = (I - P_{\mathrm{kr}})f_k$, i.e., $f \in H_{\mathrm{aws}}$ (see Theorem 6.20). We apply the van der Corput trick (Lemma 7.31) to the functions $u_n := \big(T^n f_1\big)\cdot\big(T^{2n} f_2\big)\cdots\big(T^{kn} f_k\big)$. We obtain by the T-invariance of μ that

$$\begin{aligned}(u_n|u_{n+h}) &= \int_X \big(T^n f_1\cdot T^{2n} f_2\cdots T^{kn} f_k\big)\\ &\quad\cdot\big(T^{n+h}\overline{f_1}\cdot T^{2(n+h)}\overline{f_2}\cdots T^{k(n+h)}\overline{f_k}\big)\,\mathrm{d}\mu\\ &= \int_X \big(f_1\cdot T^n f_2\cdots T^{(k-1)n} f_k\big)\\ &\quad\cdot\big(T^h\overline{f_1}\cdot T^{n+2h}\overline{f_2}\cdots T^{(k-1)n+kh}\overline{f_k}\big)\,\mathrm{d}\mu\\ &= \int_X \big(f_1 T^h\overline{f_1}\big)\cdot T^n\big(f_2 T^{2h}\overline{f_2}\big)\cdots T^{(k-1)n}\big(f_k T^{kh}\overline{f_k}\big)\,\mathrm{d}\mu.\end{aligned}$$

By the continuity of the scalar product and by the induction hypothesis

$$\lim_{N\to\infty}\frac{1}{N}\sum_{n=0}^{N-1}(u_n|u_{n+h}) = \int_X f_1 T^h\overline{f_1}\,\mathrm{d}\mu\cdot\int_X f_2 T^{2h}\overline{f_2}\,\mathrm{d}\mu\cdots\int_X f_k T^{kh}\overline{f_k}\,\mathrm{d}\mu.$$

Whence we conclude that

$$
\begin{aligned}
&\limsup_{J\to\infty}\frac{1}{J}\sum_{h=0}^{J-1}\Big|\lim_{N\to\infty}\frac{1}{N}\sum_{n=0}^{N-1}(u_n|u_{n+h})\Big|\\
&=\limsup_{J\to\infty}\frac{1}{J}\sum_{h=0}^{J-1}\Big|\int_X f_1T^h\overline{f_1}\,\mathrm{d}\mu\cdot\int_X f_2T^{2h}\overline{f_2}\,\mathrm{d}\mu\cdots\int_X f_kT^{kh}\overline{f_k}\,\mathrm{d}\mu\Big|\\
&\le\|f_1\|_\infty^2\cdots\|f_{k-1}\|_\infty^2\limsup_{J\to\infty}\frac{1}{J}\sum_{h=0}^{J-1}\int_X|f_kT^{kh}\overline{f_k}|\,\mathrm{d}\mu=0,
\end{aligned}
$$

since $f_k\in H_{\mathrm{aws}}$ (use Theorem 6.20 and Lemma 4.19), and van der Corput's lemma finishes the proof. ■

7.6 The Pointwise Ergodic Theorem

We devote this section to the famous pointwise ergodic theorem due to Birkhoff [156].

Theorem 7.40 (Birkhoff's Pointwise Ergodic Theorem) *Let (X,μ,T) be a measure-preserving system. Then for every $f\in\mathrm{L}^1(X,\mu)$ the averages*

$$\frac{1}{N}\sum_{n=0}^{N-1}f(T^nx) \tag{7.14}$$

converge for μ-almost every $x\in X$. Moreover, the system (X,μ,T) is ergodic if and only if for every $f\in\mathrm{L}^1(X,\mu)$ the above limit equals $\int_X f\,\mathrm{d}\mu$ μ-almost everywhere.

The proof of the last part (characterization of ergodicity) is left as Exercise 7.17.

> We note here again that whenever we speak of pointwise almost everywhere convergence for a *"function"* $f\in\mathrm{L}^1$, we fix a representative of the equivalence class of f. We shall not stress this aspect anymore, unless it is really necessary.

Remark 7.41 Taking in particular $f := \mathbf{1}_A$ for a measurable set $A \subset X$ yields that for ergodic systems

$$\lim_{N\to\infty} \frac{|\{n \leq N : T^n x \in A\}|}{N} = \lim_{N\to\infty} \frac{1}{N} \sum_{n=0}^{N-1} \mathbf{1}_A(T^n x) = \int_X \mathbf{1}_A \,\mathrm{d}\mu = \mu(A),$$

i.e., time mean (on the left) equals space mean (on the right) for a.e. $x \in X$, providing the proof of the quasi-ergodic hypothesis for ergodic systems and almost every initial state.

Before proving Birkhoff's ergodic theorem, we present without proof two of its many extensions, and refer to Sect. 7.8 for further results. The first one concerns positive operators on reflexive L^p-spaces.

Theorem 7.42 (Akcoglu [17]) *Let S be a positive contraction on $\mathrm{L}^p(X, \mu)$ for a general measure space (X, μ) and $p \in (1, \infty)$. Then for every $f \in \mathrm{L}^p(X, \mu)$ (7.15) holds.*

For Dunford–Schwartz operators (which are contractions on $\mathrm{L}^p(X, \mu)$ for every $p \in [1, \infty]$, see the proof of Proposition 7.9) one has almost everywhere convergence in L^1, strengthening both Proposition 7.9 and Theorem 7.40.

Theorem 7.43 (Dunford–Schwartz [323]) *Let S be a Dunford–Schwartz operator on a general measure space (X, μ). Then for every $f \in \mathrm{L}^1(X, \mu)$ the averages*

$$\frac{1}{N} \sum_{n=1}^{N} S^n f \text{ converge } \mu\text{-almost everywhere.} \tag{7.15}$$

For further extensions and generalizations of Theorem 7.40 we refer to Sect. 7.8.

There are many, substantially different, proofs of the pointwise ergodic theorem. We present one which is relatively long but very structured and, most importantly, the developed techniques will be useful later. To show the main part of Theorem 7.40, we begin with the following observation. Define for the Koopman operator T on $\mathrm{L}^1(X, \mu)$ and for $N \in \mathbb{N}$

$$A_N := \frac{1}{N} \sum_{n=0}^{N-1} T^n,$$

and consider the subspace

$$D := \operatorname{fix}(T) \oplus (I - T)\mathrm{L}^\infty(X, \mu) \subset \mathrm{L}^1(X, \mu).$$

Then $D \cap \mathrm{L}^2(X, \mu)$ is dense in $\mathrm{L}^2(X, \mu)$ by von Neumann's decomposition (Theorem 6.4), and hence D is dense in $\mathrm{L}^1(X, \mu)$. For every $f \in D$ the ergodic averages converge in the (essential) supremum norm and hence almost everywhere.

Indeed, by the identity in (7.3) we have for each $f \in \mathrm{fix}(T)$, each $g = (I - T)h$ with $h \in \mathrm{L}^\infty(X, \mu)$ and almost every $x \in X$

$$\left|\frac{1}{N}\sum_{n=0}^{N-1}(T^n(f+g))(x) - f(x)\right| = \left|\frac{1}{N}\sum_{n=0}^{N-1}(T^n g)(x)\right| = \frac{1}{N}\left|h(x) - (T^N h)(x)\right| \leq \frac{2\|h\|_\infty}{N}$$

for all $N \in \mathbb{N}$.

Thus it remains to show that the set

$$F := \left\{f \in \mathrm{L}^1(X, \mu) : \lim_{N\to\infty} A_N f \text{ exists a.e.}\right\} \tag{7.16}$$

is closed in $\mathrm{L}^1(X, \mu)$.

To show this result for the almost everywhere convergence, we introduce the following notion. For a sequence $(S_N)_{N\in\mathbb{N}}$ of linear operators on $\mathrm{L}^1(X, \mu)$ the corresponding **maximal operator**[4] is defined by

$$S^* f := \sup_{N\in\mathbb{N}} |S_N f| \tag{7.17}$$

for $f \in \mathrm{L}^1(X, \mu)$, where the supremum is defined pointwise and may be infinity at some points.[5]

Although this operator is not linear, it has two important properties:

(1) $S^*(\alpha f) = |\alpha| \cdot S^* f$ for each $\alpha \in \mathbb{C}$ and $f \in \mathrm{L}^1(X, \mu)$,
(2) $S^*(f + g) \leq S^* f + S^* g$ for each $f, g \in \mathrm{L}^1(X, \mu)$,

see Exercise 7.16.

Definition 7.44 Let $(S_N)_{N\in\mathbb{N}}$ be a sequence of operators on $\mathrm{L}^1(X, \mu)$. We say that the maximal operator S^* (or the sequence $(S_N)_{N\in\mathbb{N}}$) satisfies a **(weak) maximal inequality** if there exists a function $c : (0, \infty) \to [0, \infty]$ with $\lim_{\lambda\to\infty} c(\lambda) = 0$ satisfying

$$\mu[S^* f > \lambda] \leq c(\lambda) \text{ for each } \lambda > 0,\ f \in \mathrm{L}^1(X, \mu) \text{ with } \|f\|_1 \leq 1. \tag{7.18}$$

The inequality $\|S^* f\|_1 \leq C\|f\|_1$ for some constant $C > 0$ and all $f \in \mathrm{L}^1(X, \mu)$ is called the (L^1-)**strong maximal inequality**.

---

[4] This operator goes back to Hardy and Littlewood [522] who in 1930 introduced it for translations.

[5] Ephremidze [371] showed that for an ergodic system (X, μ, T) on a complete non-atomic probability space and the averages (A_N) given by $A_N = \frac{1}{N}\sum_{n=1}^{N} T^n$, the maximal function $A^* f$ determines f uniquely, see also Jones [589].

The strong maximal inequality clearly implies the weak one for the function c given by $c(\lambda) = \frac{C}{\lambda}$.

The following shows the importance of maximal inequalities in the study of almost everywhere convergence of ergodic averages.

Proposition 7.45 (Banach's Principle) *Let $(S_N)_{N\in\mathbb{N}}$ be a sequence of linear operators on $\mathrm{L}^1(X,\mu)$ for some probability space (X,μ). If the corresponding maximal operator satisfies a maximal inequality, then the set*

$$F := \Big\{ f \in \mathrm{L}^1(X,\mu) : \lim_{N\to\infty} S_N f \text{ exists a.e.} \Big\}$$

is a closed linear subspace of $\mathrm{L}^1(X,\mu)$.

Proof It is clear that F is a linear subspace of $\mathrm{L}^1(X,\mu)$. To show the closedness, take $f \in \overline{F}$ and a sequence $(f_n)_{n\in\mathbb{N}}$ in F with $f_n \to f$ in $\mathrm{L}^1(X,\mu)$, where we can assume without loss of generality that $\|f_n - f\|_1 \neq 0$ for every $n \in \mathbb{N}$. We need to show that $(S_N f)_{N\in\mathbb{N}}$ is a.e. a Cauchy sequence in $\mathbb{C}$, i.e., that

$$h := \limsup_{k,l\to\infty} |S_k f - S_l f| = 0 \text{ almost everywhere.}$$

For $k, l, n \in \mathbb{N}$ we have

$$\begin{aligned} |S_k f - S_l f| &\le |S_k(f - f_n)| + |S_k f_n - S_l f_n| + |S_l(f_n - f)| \\ &\le 2S^*(f - f_n) + |S_k f_n - S_l f_n|. \end{aligned}$$

Since by the hypothesis $f_n \in F$ holds, we conclude

$$\begin{aligned} h = \limsup_{k,l\to\infty} |S_k f - S_l f| &\le 2S^*(f - f_n) + \limsup_{k,l\to\infty} |S_k f_n - S_l f_n| \\ &= 2S^*(f - f_n). \end{aligned}$$

Let $\lambda > 0$. By the above and the maximal inequality it follows that

$$\mu([h > 2\lambda]) \le \mu([S^*(f - f_n) > \lambda]) \le c\Big(\frac{\lambda}{\|f - f_n\|_1}\Big) \to 0 \text{ as } n \to \infty.$$

Thus $\mu[h > 2\lambda] = 0$ holds for every $\lambda > 0$, i.e., $h = 0$ almost everywhere. ∎

We now come back to the measure-preserving system (X,μ,T) with the ergodic averages $(A_N)_{N\in\mathbb{N}}$ of the Koopman operator T and shall prove that the corresponding maximal operator satisfies the maximal inequality with $c(\lambda) = \frac{1}{\lambda}$. The following result due to Yosida and Kakutani [1060] and Wiener [1045] (cf. the original proof of Birkhoff [156] as well as Khinchin [645], Kolmogorov [654] and Riesz [872]) is the key to achieve this.

Theorem 7.46 (Maximal Ergodic Theorem) *Let (X, μ, T) be a measure-preserving system with Koopman operator T on $\mathrm{L}^1(X, \mu)$. Define for $n \in \mathbb{N}$*

$$S_n := \sum_{j=0}^{n-1} T^j,$$

take $f \in \mathrm{L}^1(X, \mu)$ be real-valued and set $E_N := [S_n f > 0 \text{ for some } n \leq N]$. Then for every $N \in \mathbb{N}$

$$\int_{E_N} f \, \mathrm{d}\mu \geq 0.$$

As a consequence, for $E := [S_n f > 0 \text{ for some } n \in \mathbb{N}]$ we obtain

$$\int_E f \, \mathrm{d}\mu \geq 0. \tag{7.19}$$

Proof The last assertion follows from the first one and the facts that $E_N \subset E_{N+1}$ for each $N \in \mathbb{N}$ and $E = \bigcup_{N \in \mathbb{N}} E_N$.
To show the first assertion, set $f_0 := 0$, $f_n := S_{n-1} f$ for $n \in \mathbb{N}$, and for $N \in \mathbb{N}$ define the function

$$F_N := \max\{f_0, f_1, \ldots, f_N\}.$$

The functions F_N satisfy $TF_N + f \geq Tf_n + f = f_{n+1}$ for every $n \leq N$. This implies

$$TF_N + f \geq \max\{f_1, \ldots, f_N\}.$$

If $x \in E_N$, i.e., $F_N(x) > 0$, then

$$\max\{f_1(x), \ldots, f_N(x)\} = \max\{f_0(x), f_1(x), \ldots, f_N(x)\} = F_N(x)$$

and therefore it follows that

$$TF_N + f \geq F_N \text{ on } E_N.$$

Since $F_N \geq 0$, and hence $TF_N \geq 0$, and since $F_N = 0$ on $X \setminus E_N$, we conclude

$$\begin{aligned} \int_{E_N} f \, \mathrm{d}\mu &\geq \int_{E_N} F_N \, \mathrm{d}\mu - \int_{E_N} TF_N \, \mathrm{d}\mu = \int_X F_N \, \mathrm{d}\mu - \int_{E_N} TF_N \, \mathrm{d}\mu \\ &\geq \int_X F_N \, \mathrm{d}\mu - \int_X TF_N \, \mathrm{d}\mu = 0 \end{aligned}$$

by the measure-preserving property, proving the assertion. ∎

We can now prove the desired maximal inequality.

Corollary 7.47 (Maximal Inequality) *Let (X, μ, T) be a measure-preserving system with ergodic averages $(A_N)_{N\in\mathbb{N}}$ and the corresponding maximal operator A^* and $f \in \mathrm{L}^1(X, \mu)$. Then the inequality*

$$\mu([A^* f > \lambda]) \leq \frac{\|f\|_1}{\lambda} \text{ for all } \lambda > 0$$

holds. In particular, A^* satisfies the maximal inequality with $c(\lambda) := \frac{1}{\lambda}$. Moreover, if $f \in \mathrm{L}^p(X, \mu)$ and $p > 1$, then $A^* f \in \mathrm{L}^p(X, \mu)$, and

$$\|A^* f\|_p \leq \frac{p}{p-1}\|f\|_p.$$

The previous inequality on L^p is called **strong L^p-maximal inequality** (with constant $p/(p-1)$).

Proof Let $\lambda > 0$ and consider the real-valued function $g := |f| - \lambda$. Observe that

$$g + \cdots + T^{n-1}g > 0 \iff |f| + \cdots + T^{n-1}|f| > n\lambda \iff \frac{1}{n}\sum_{j=0}^{n-1} T^j|f| > \lambda.$$

Thus the maximal ergodic theorem (Theorem 7.46) implies for the set

$$E := \Big[\frac{1}{n}\sum_{j=0}^{n-1} T^j|f| > \lambda \text{ for some } n \in \mathbb{N}\Big]$$

that

$$\int_E (|f| \quad \lambda)\, \mathrm{d}\mu \geq 0.$$

Therefore we obtain

$$\begin{aligned}
\|f\|_1 &\geq \int_E |f|\, \mathrm{d}\mu \geq \lambda \cdot \mu\Big[\frac{1}{n}\sum_{j=0}^{n-1} T^j|f| > \lambda \text{ for some } n \in \mathbb{N}\Big] \\
&\geq \lambda \cdot \mu\Big[\Big|\frac{1}{n}\sum_{j=0}^{n-1} T^j f\Big| > \lambda \text{ for some } n \in \mathbb{N}\Big] \\
&= \lambda \cdot \mu([A^* f > \lambda]),
\end{aligned}$$

and the first part of the theorem is proved.

Since $0 \leq A^*f \leq A^*|f|$, it suffices to show the last assertion for $f \in \mathrm{L}^p(X,\mu)$ with $f \geq 0$. Recall from (1.3) that for $g \geq 0$ measurable and $q \in (0,\infty)$ one has

$$\|g\|_q^q = \int_0^\infty q t^{q-1} \mu([g > t])\, \mathrm{d}t. \tag{7.20}$$

We shall apply this first to the bounded functions $g_n = \min(A^*f, n\mathbf{1})$, $n \in \mathbb{N}$, which are thus in $\mathrm{L}^p(X,\mu)$. By the proof of the first part of the theorem $\lambda\mu([A^*f > \lambda]) \leq \int_{[A^*f>\lambda]} f\, \mathrm{d}\mu$ for each $\lambda > 0$. Now, for $\lambda > 0$ and $n \in \mathbb{N}$ we have $[g_n > \lambda] = [A^*f > \lambda]$ if $n > \lambda$ and $[g_n > \lambda] = \emptyset$ otherwise. It follows that $\lambda\mu([g_n > \lambda]) \leq \int_{[g_n>\lambda]} f\, \mathrm{d}\mu$ for every $n \in \mathbb{N}$ and $\lambda > 0$. For a fixed $n \in \mathbb{N}$ we obtain by (7.20) and Fubini's theorem

$$\begin{aligned}
\|g_n\|_p^p &= \int_0^\infty p\lambda^{p-1}\mu([g_n > \lambda])\, \mathrm{d}\lambda \leq p \int_0^\infty \lambda^{p-2} \int_{[g_n>\lambda]} f\, \mathrm{d}\mu\, \mathrm{d}\lambda \\
&= p \int_X f(x) \int_{[0,g_n(x))} \lambda^{p-2}\, \mathrm{d}\lambda\, \mathrm{d}\mu(x) \\
&= \frac{p}{p-1} \int_X f \cdot g_n^{p-1}\, \mathrm{d}\mu \leq \frac{p}{p-1} \|f\|_p \|g_n\|_p^{p-1},
\end{aligned}$$

where in the last step we used Hölder's inequality. This yields $\|g_n\|_p \leq \frac{p}{p-1}\|f\|_p$ and letting $n \to \infty$ finishes the proof, by the monotone convergence theorem. ■

We summarize now the proof of the pointwise ergodic theorem.

Proof of Theorem 7.40 By von Neumann's decomposition (Theorem 6.4) and by the proof of von Neumann's mean ergodic theorem, the ergodic averages converge a.e. on the dense subset

$$D := \mathrm{fix}(T) \oplus (I - T)\mathrm{L}^\infty(X,\mu)$$

of $\mathrm{L}^1(X,\mu)$. By the maximal inequality (Corollary 7.47) and Banach's principle (Proposition 7.45), the ergodic averages converge a.e. on the closure of D in $\mathrm{L}^1(X,\mu)$, i.e., on the whole $\mathrm{L}^1(X,\mu)$. ■

We remark that neither the previous nor other proofs of the pointwise ergodic theorem give a description of the points for which the convergence actually holds. For ergodic systems such points got their own name.

Definition 7.48 Let (X,μ,T) be an ergodic measure-preserving system and $f \in \mathrm{L}^1(X,\mu)$. We call $x \in X$ **generic** for f with respect to T if

$$\lim_{N\to\infty} \frac{1}{N} \sum_{n=0}^{N-1} f(T^n x) = \int_X f\, \mathrm{d}\mu.$$

Thus, the pointwise ergodic theorem says in particular that for ergodic systems and for a function $f \in \mathrm{L}^1(X, \mu)$, the generic points form a set of full measure. (Note that this set depends on the representative of the equivalence class of f.)

Remark 7.49 For a topological dynamical system (K, T), with K metrizable, and an ergodic measure on K there exists a set of full measure which is generic with respect to every function $f \in \mathrm{C}(K)$, see Exercise 7.18. Moreover, we saw in Sect. 7.3 that a topological system (K, T) is uniquely ergodic if and only if for each $f \in \mathrm{C}(K)$ and each $\mu \in \mathrm{M}_1(K, T)$ every point is generic.

Remark 7.50 (Pointwise Convergence of Polynomial Averages) It is a good exercise to go through the proof of Theorem 7.23 to check that it does not work for polynomial ergodic averages. Also the proof of the pointwise ergodic theorem presented above does not go through for polynomial ergodic averages. Indeed, already the first step fails, namely to find a good dense subset of functions for which a.e. convergence holds. We will discuss pointwise convergence of polynomial averages in more detail in Chap. 12 and Theorem 14.45 below.

Remark 7.51 (Pointwise Convergence, Banach Limit and (Tempered) Følner Sequences) In contrast to the mean convergence (see Proposition 7.3), the averages

$$\lim_{N-M\to\infty} \frac{1}{N-M} \sum_{n=M}^{N-1} T^n f$$

fail to converge almost everywhere in general. For example, for invertible measure-preserving systems on standard, non-atomic probability spaces (see Remark 8.48 below) Akcoglu and del Junco [21] showed that the averages

$$\lim_{N\to\infty} \frac{1}{\lfloor \sqrt{N} \rfloor} \sum_{n=N}^{N+\lfloor \sqrt{N} \rfloor-1} T^n f$$

do not converge a.e. for every ergodic, invertible, non-atomic measure-preserving system and some characteristic function f. For more examples and related results see, e.g., Emerson [364], del Junco and Rosenblatt [283], Schwartz [930], Bellow et al. [95], Jones and Wierdl [598], Butler and Rosenblatt [211]. There was a lot of effort to find subsets F_N of $\mathbb{N}$, $\mathbb{Z}$ (or, more generally, of a general locally compact amenable group G) such that the averages (or the analogous integral expressions, in the general case)

$$\lim_{N\to\infty} \frac{1}{|F_N|} \sum_{n\in F_N} T^n f$$

converge a.e. for every (invertible) measure-preserving system (with an action of G) and every L^1-function using the notion of a **Følner sequence**[6] with certain restrictions, see Chatard [241], Bewley [151], Tempelman [997], Greenleaf [494], Emerson [364], Greenleaf, Emerson [495], Ornstein and Weiss [837] and Tempelman [996, Sec. 5.6] for Shulman's result for L^2-functions from [938]. Finally, Lindenstrauss [751] extended Shulman's result to L^1-functions providing a pointwise ergodic theorem for locally compact second countable amenable groups along **tempered**[7] **Følner sequences**.

For an alternative proof of Lindenstrauss' theorem see Weiss [1040], cf. also Bowen and Nevo [189], whereas a Banach space valued version and a non-commutative version can be found in Pogorzelski [851, 852] and Cadilhac and Wang [213], respectively.

7.7 More on the Maximal Inequality

We now show that the validity of a maximal inequality is necessary for almost everywhere convergence, extending Banach's principle formulated in Proposition 7.45.

Theorem 7.52 *Let (X, μ) be a probability space and let $(S_N)_{N\in\mathbb{N}_0}$ be a sequence of linear bounded operators on $\mathrm{L}^1(X, \mu)$ with corresponding maximal operator S^*. Consider the following assertions.*

(i) *The limit $\lim_{N\to\infty} S_N f$ exists a.e. for every $f \in \mathrm{L}^1(X, \mu)$.*
(ii) *$S^* f$ is a.e. finite for every $f \in \mathrm{L}^1(X, \mu)$.*
(iii) *The operator S^* satisfies a maximal inequality.*
(iv) *The set*

$$F := \left\{ f \in \mathrm{L}^1(X, \mu) : \lim_{N\to\infty} S_N f \text{ exists a.e.} \right\}$$

is a closed linear subspace of $\mathrm{L}^1(X, \mu)$.

[6] For a locally compact second countable group G, a sequence (F_N) of compact subsets of G is called a Følner sequence if $|F_N| > 0$ for every $N \in \mathbb{N}$ and $\lim_{N\to\infty} \frac{|KF_N \triangle F_N|}{|F_N|} = 0$ holds for every compact non-empty set $K \subset G$, where $|\cdot|$ denotes the (left) Haar measure on G, see Sect. 11.2. The existence of a Følner sequence is one of many equivalent conditions to amenability for locally compact second countable groups.

[7] A Følner sequence (F_N) is called **tempered** if it satisfies the **Shulman condition** $\sup_{N\in\mathbb{N}} \frac{|\bigcup_{k<N} F_k^{-1} F_N|}{|F_N|} < \infty$. Lindenstrauss [751] also showed that every Følner sequence contains a tempered Følner subsequence, making his result general enough to cover all locally compact second countable amenable groups.

Then (i)⇒(ii)⇔(iii)⇒(iv). *Moreover,* (i)⇔(ii)⇔(iii)⇔(iv) *whenever there exists a dense subset D of* $\mathrm{L}^1(X, \mu)$ *such that the limit* $\lim_{N\to\infty} S_N f$ *exists a.e. for every* $f \in D$.

Proof The implication (i)⇒(ii) and the additional assertion are clear, whereas the implication (iii)⇒(iv) was proved in Proposition 7.45. The implication (iii)⇒(ii) is clear, thus it remains to show (ii)⇒(iii).

Assume (ii). Then for every $f \in \mathrm{L}^1(X, \mu)$ and every $\varepsilon > 0$ there exists $m \in \mathbb{N}$ such that $\mu([S^* f > m]) \leq \varepsilon$. Thus we have for every $\varepsilon > 0$

$$\begin{aligned}\mathrm{L}^1(X, \mu) &= \bigcup_{m=1}^{\infty} \{f \in \mathrm{L}^1(X, \mu) : \mu([S^* f > m]) \leq \varepsilon\} \\ &= \bigcup_{m=1}^{\infty} \bigcap_{n=1}^{\infty} \left\{ f \in \mathrm{L}^1(X, \mu) : \mu\Big(\Big[\sup_{1\leq N\leq n} |S_N f| > m\Big]\Big) \leq \varepsilon \right\} \\ &= \bigcup_{m=1}^{\infty} \bigcap_{n=1}^{\infty} M_{m,n},\end{aligned}$$

with $M_{m,n} := \{f \in \mathrm{L}^1(X, \mu) : \mu\Big(\big[\sup_{1\leq N\leq n} |S_N f| > m\big]\Big) \leq \varepsilon\}$. Since L^1-convergence implies convergence in measure (see Exercise 7.20), for each $n, m \in \mathbb{N}$ the set $M_{m,n}$ is closed in $\mathrm{L}^1(X, \mu)$, see Exercise 7.21. Thus, Baire's category theorem (Theorem 1.3) implies that for every $\varepsilon > 0$ there exist $m \in \mathbb{N}$, $\delta > 0$ and $f_0 \in \mathrm{L}^1(X, \mu)$ such that for every $f \in \mathrm{L}^1(X, \mu)$ with $\|f - f_0\|_1 \leq \delta$ one has $\mu([S^* f > m]) \leq \varepsilon$. This means that for every $g \in \mathrm{L}^1(X, \mu)$ with $\|g\|_1 \leq 1$

$$\mu([S^*(f_0 + \delta g) > m]) \leq \varepsilon.$$

Since, by the triangle inequality,

$$S^* g \leq \frac{1}{\delta}(S^*(f_0 + \delta g) + S^* f_0),$$

we obtain

$$\begin{aligned}\mu\left(\left[S^* g > \tfrac{2m}{\delta}\right]\right) &\leq \mu\left([S^*(f_0 + \delta g) > m]\right) + \mu\left([S^* f_0 > m]\right) \\ &\leq \varepsilon + \varepsilon = 2\varepsilon.\end{aligned}$$

For $\lambda > 0$ set

$$c(\lambda) := \sup_{\|g\|_1 \leq 1} \mu\left([S^* g > \lambda]\right) \in [0, \infty].$$

The function c converges monotonically to 0 by the above and satisfies by definition the maximal inequality (7.18), implying (iii). ∎

7.8 Some Further Developments

We mention here some related results and directions. For even more information we refer, e.g., to Krengel [683] and Chaps. 9, 12, 13, 15 and 16 below.

More on Mean Ergodic Operators

It is easy to see that every mean ergodic operator S on a Banach space E satisfies $\lim_{n\to\infty} \frac{\|S^n x\|}{n} = 0$ for every $x \in E$ and, in particular, $\sup_{n\in\mathbb{N}} \frac{\|S^n\|}{n} < \infty$. Moreover, every mean ergodic operator on a Banach space satisfies the von Neumann decomposition. (Both assertions are left as Exercise 7.5.) For examples of mean ergodic operators with sublinear and linear growth of powers see Hille [533], Kornfeld and Kosek [660], Tomilov and Zemánek [1004] and Kosek [662].

By Remark 7.17, every operator on a Banach space with relatively weakly compact orbits[8] is mean ergodic. As mentioned above, this class includes all power bounded operators on reflexive spaces. Note that mean ergodicity of power bounded operators characterizes reflexivity for spaces with a Schauder basis by Fonf et al. [396], cf. [397]. A related characterization of reflexivity via mean ergodic operators is due to Farkas and Kreidler [379].

Clearly, every mean ergodic operator is **Cesàro bounded**, i.e., the ergodic averages (7.1) are uniformly bounded. Assani [37] (see also Émilion [365, 366]) showed that the matrix $\begin{pmatrix} -1 & 2 \\ 0 & -1 \end{pmatrix}$ is Cesàro bounded, but not mean ergodic. Moreover, even weak convergence of ergodic averages is not sufficient for mean ergodicity even for Hilbert space operators, see Derriennic [297] and Tomilov and Zemánek [1004]. (Note that for power bounded operators weak and strong convergence of (7.1) coincide by Theorem 7.12.) On the other hand, Émilion [366] showed that every Cesàro bounded *positive* operator on a reflexive Banach lattice (for example L^p with $p \in (1, \infty)$) is mean ergodic, see also Derriennic [297].

For connections of mean ergodicity with Cesàro boundedness and other concepts of boundedness see Bermúdez et al. [150], Cohen et al. [250] and Bonilla and Müller [171].

Abstract mean ergodic theorems in the context of general operator semigroups were developed in Eberlein [329], Nagel [815] and Sato [911], see also Krengel [683, Ch. 2], Schreiber [923] and Gerlach and Kunze [465].

[8] Such operators are power bounded.

An operator S on a Banach space E is called **uniformly (mean) ergodic** if its ergodic averages converge in the operator norm. Analogously to Theorem 7.12, Dunford [320] showed that an operator S satisfying $\lim_{n\to\infty} \frac{\|S^n\|}{n} = 0$ is uniformly ergodic if and only if either $1 \in \rho(S)$ or 1 is a simple pole of the resolvent of S. Another equivalent condition is that $\operatorname{ran}(I - S)$ is closed and $E = \operatorname{fix} S \oplus \operatorname{ran}(I - S)$. Lin [743] proved that the closedness of $\operatorname{ran}(I - S)$ is already equivalent to uniform ergodicity. See Exercise 7.23 for a partial discussion of these statements. For positive operators[9] on L^p-spaces one can replace the assumption $\lim_{n\to\infty} \frac{\|S^n\|}{n} = 0$ by $r(S) \leq 1$, see Karlin [629]. For more information and related results and examples see[10] Lin [743–745], Axmann [70], Lloyd [756], Lotz [759], Derndinger et al. [290, Appendix W], Wacker [1030], Mbekhta and Zemánek [781], Yoshimoto [1058], Becker [87], Lin et al. [748], Beltrán-Meneu et al. [98], Mukhamedov and Al-Rawashdeh [809] and, e.g., [337, Sec. I.2.2] and Krengel [683, Sec. 2.2].

For a recent survey on operator (mean) ergodic theorems we refer to Cohen and Lin [252].

Note that for Koopman operators uniform mean ergodicity holds only in the (eventually) periodic case, see Exercise 7.24 and Hölz [540], where the latter reference treats both the measure-preserving and the topological case.

More Proofs of the Pointwise Ergodic Theorem

Chacon and Ornstein [238] introduced the so-called filling scheme approach to the maximal inequality and the pointwise ergodic theorem in the proof of their ratio ergodic theorem discussed below in this section, see also Garsia [461], Neveu [822], Derriennic [299].

Kamae [618] presented a proof of the pointwise ergodic theorem using non-standard analysis without using a maximal inequality. Katznelson and Weiss [635] adapted Kamae's argument to obtain a simple proof which uses neither non-standard analysis nor a maximal inequality. For a proof for standard probability spaces via the backward Vitali lemma see Rudolph [892, Sec. 3.2].

Simple proofs of the maximal ergodic theorem appeared in Garsia [461] and Keane and Petersen [637], where in [637] the authors prove a generalization of the maximal ergodic theorem which immediately yields the pointwise ergodic theorem. For one more proof of the maximal ergodic theorem in the ergodic case see Jones [587]. For a proof of Birkhoff's ergodic theorem via a covering argument using a version of the maximal ergodic theorem see Shields [936] and [937, Sec. I.3]. Bochi

[9] Note that an isometric lattice isomorphism on a Banach lattice is uniformly ergodic if and only if it is periodic, see, e.g., Derndinger et al. [290, Thm. W.12] or Derriennic [298]. In particular, this applies to Koopman operators of measure-preserving transformations, see also Exercise 7.24.

[10] This list is not complete.

[166] presented a proof of the maximal ergodic theorem and Birkhoff's ergodic theorem via a generalization of Rokhlin's lemma (see Theorem 3.23).

In his study of the polynomial ergodic theorem discussed in Sect. 12.4 below, Bourgain [183] gave a proof of the maximal inequality by a reduction to the shift on $\mathbb{Z}$ (known as Calderón's principle, see Exercise 12.9). Moreover, for L^2-functions he showed pointwise convergence via variational inequalities (which he proved using Fourier analytic methods) providing more quantitative information on convergence, see also Jones et al. [590] and Krause [670], [671, §3.3]. This approach was extended to commuting transformations by Jones et al. [596].

Beyond Birkhoff's Ergodic Theorem: Sign Change and Non-integrable Functions

Halász [511] showed that for every ergodic measure-preserving system and every integrable function with mean zero, the ergodic averages change sign (weakly) infinitely often a.e. For more results in this direction see, e.g., Marcus and Petersen [771], Petersen [848, Sec. 3.1] and Mondal et al. [801].

Of course, pointwise convergence of ergodic averages makes sense for arbitrary (not necessarily integrable) functions. Kesten and Tanny showed that for every ergodic measure-preserving system (X, μ, T) and every measurable function f on X, the averages $\frac{1}{N}\sum_{n=1}^{N} T^n f$ converge a.e. if and only if they are bounded a.e., see Kesten [642, Acknowledgement]. For further proofs of this result (also without the ergodicity assumption) see Aaronson [3, comment before Theorem C], Woś [1050], [1051, Thm. 2], [1052], Derriennic [299, 4(b)] and Hagelstein et al. [510]. For an example of a non-integrable function with a.e. convergent ergodic averages for a mixing system see Aaronson [1, Construction 2].

For more results on convergence (respectively, divergence) of ergodic averages for non-integrable measurable functions see [642], Aaronson [1, 3], Woś [1051, 1052], Broise et al. [195], Major [765], Buczolich [200, 202], Svetic [963], Derriennic [299], Hochman [538], Cuny and Derriennic [265].

We also mention here a result of Kornfeld [661] stating that for every topological system (K, T) such that T is *aperiodic*[11] with respect to some T-invariant Borel probability measure μ on K there exists $f \in \mathrm{C}(K)$ with $f = g - Tg$ for some measurable function g which is non-integrable with respect to μ. For irrational rotations on the unit circle this was proven by Anosov [32].

[11] For the definition of aperiodicity see the paragraph after Definition 12.33.

Pointwise, Ratio, Dominated and Subadditive Ergodic Theorems and More General Operators

The pointwise ergodic theorem was extended to Markov operators partially by Doob [312, 313] and Kakutani [612] and fully by Hopf [543]. Generalizing this further, Dunford and Schwartz [323] showed a.e. convergence of ergodic averages (7.14) for Dunford–Schwartz operators, see Theorem 7.43 above. For generalizations of Birkhoff's ergodic theorem for sequences of functions see Maker [766] and Breiman [192]. Stein [959] proved[12] a.e. convergence in L^p for $p > 1$ of powers (without averaging) of Dunford–Schwartz operators which are self-adjoint and positive semi-definite on L^2, see Burkholder and Chow [210] for the case $p = 2$, Rota [883] for an analogous result for doubly stochastic operators, Ornstein [833] for a counterexample for $p = 1$ and Bellow et al. [94] for an extension to normal Dunford–Schwartz operators with spectrum in a Stolz region. As a converse of Hopf's ergodic theorem, Helmberg [527] showed that if T is non-singular and conservative[13] on a probability space (X, μ), then almost everywhere convergence of Birkhoff's averages (7.14) for every $f \in L^\infty(X, \mu)$ implies the existence of an invariant probability measure equivalent to μ.

Chacon [235] constructed an example of a positive isometry on L^1 for which the ergodic averages diverge a.e.[14] for some bounded function. Ionescu Tulcea [570] showed that such counterexamples are typical in the Baire category sense among positive invertible isometries on L^1. (Note that positive contractions on L^p for $p > 1$ satisfy the pointwise ergodic theorem by Theorem 7.42 by Akcoglu [17], see Ionescu Tulcea [569] for the isometric case, de la Torre [275] for an extension to invertible doubly power bounded positive operators and Kan [620] for an extension to power bounded Lamperti operators,[15] see below.) Roth [884] presented the following complete characterization of positive contractions on L^1 satisfying the pointwise ergodic theorem as those for which there exists a strictly positive L^1-function g with $\limsup_{N\to\infty} \frac{1}{N}\sum_{n=1}^{N} T^n g < \infty$ a.e. The case of positive Cesàro power bounded operators was investigated in Derriennic and Lin [301], Émilion [367] and Martín-Reyes and De la Torre [774], see also Sato [910] and Martín-Reyes and de la Rosa [773].

---

[12] In particular, this implies a.e. convergence of $T^{2n} f$ as $n \to \infty$ for every $f \in L^p$, $p > 1$ and every self-adjoint Dunford–Schwartz operator T, cf. Exercise 5.16.

[13] A non-singular map $T : X \to X$ is called **conservative** if for each measurable $A \subset X$ one has $\mu(A \setminus \bigcup_{n=1}^{\infty} T^{-n} A) = 0$.

[14] By the **stochastic ergodic theorem** due to Krengel [679, Thm. 2.3], the ergodic averages converge stochastically for every contraction on L^1.

[15] A bounded linear operator on an L^p-space, $p \in [1, \infty)$, is called **Lamperti** if it separates supports, i.e., maps functions with disjoint supports to functions with disjoint supports. Lamperti operators include L^p-isometries for $p \neq 2$, positive isometries on L^2 and positive operators with positive inverse for every p, see Banach [74], Lamperti [703], Ionescu Tulcea [569] and Kan [620].

Without assuming positivity, every isometry on an L^p-space for $p \in (1, \infty)$, $p \neq 2$ satisfies the pointwise ergodic theorem as shown by Ionescu Tulcea [569]. A counterexample for $p = 2$ was observed by Akcoglu and Sucheston [27] based on Burkholder [209], where the corresponding operator is even unitary.[16] Moreover, the pointwise ergodic theorem fails in general for invertible doubly power bounded operators on L^p with $p \in (1, \infty)$, see Feder [384] and Assani [36, 37], see also Cohen [249].

Chacon and Ornstein [238] proved the following generalization of the pointwise ergodic theorem known as **ratio ergodic theorem** being for Koopman operators of transformations preserving a σ-finite measure due to Hopf [542, §14].

Theorem (Chacon, Ornstein) *Let S be a positive contraction on $\mathrm{L}^1(X, \mu)$ for a σ-finite measure space (X, μ). Then for every $f, g \in \mathrm{L}^1(X, \mu)$ with $g \geq 0$ the ratios*

$$\frac{f + Sf + \cdots + S^N f}{g + Sg + \cdots + S^N g} \tag{7.21}$$

converge almost everywhere as $N \to \infty$ on the set where $g + Sg + \cdots + S^N g$ eventually become strictly positive.

For a detailed proof, beside the original paper, we refer, e.g., to e.g., Garsia [463, Ch. 2]. For counterexamples in the case of positive contractions on L^p when $p \in (1, \infty)$ see Fong and Lin [400]. Neveu [820] and Chacon [233] showed that the limit equals $\frac{P_{\mathrm{fix}\,S} f}{P_{\mathrm{fix}\,S} g}$ (and in particular $\frac{\int f \,\mathrm{d}\mu}{\int g \,\mathrm{d}\mu}$ in the ergodic case), see also Krengel [683, Thm. 3.3.4].

For other proofs, generalizations and related results we refer to Hopf [544], Neveu [820, 822], Chacon [233, 234], Brunel [197, 198], Akcoglu [15, 16], Garsia [462], Bishop [159], Klimko and Sucheston [652], Akcoglu and Chacon [20], Akcoglu and Brunel [19], Roth [884, 885], Kamae and Keane [619], Zweimüller [1077, 1078], Rugh and Thomine [895].

We mention here the following general ergodic theorem of Dowker [316], and refer to Lin [746] for a short proof by a deduction from the Chacon–Ornstein theorem. For a version of Dowker's theorem from Lin [746] see Exercise 7.22.

Theorem (Dowker) *Let $(X, \mathcal{X}, \mu)$ be a σ-finite measure space, let $T : X \to X$ be bi-measurable, i.e., let T be measurable such that TB is measurable for each measurable B. Suppose*[17] *that $\mu \ll T_*\mu$, and define the measure ν by $\nu(A) := \mu(TA)$ on the sub-σ-algebra $\mathcal{Y} := \{T^{-1}B : B \in \mathcal{X}\}$. Since $\nu \ll \mu|_{\mathcal{Y}}$, we can take*

[16] This was strengthened by Paszkiewicz [844] who constructed a unitary operator U on an L^2-space such that for any subsequence (k_n) of $\mathbb{N}$ there exists $f \in \mathrm{L}^2$ such that the averages $\frac{1}{N}\sum_{n=1}^{N} U^{k_n} f$ fail to converge a.e.

[17] This property is termed as positively non-singular in [316].

the Radon–Nikodym derivative $w = \frac{\mathrm{d}\nu}{\mathrm{d}\mu}$, *being* $\mathcal{Y}$*-measurable. For* $n \in \mathbb{N}_0$ *define*

$$w_n(x) := \prod_{j=0}^{n-1} w(T^j x).$$

Let $f, g \in \mathrm{L}^1(X, \mu)$ *and let* $g(x) > 0$ *for almost every* $x \in X$. *Then the ratios*

$$\frac{\sum_{k=0}^{n} f(T^k x) w_k(x)}{\sum_{k=0}^{n} g(T^k x) w_k(x)}$$

converge almost everywhere to a finite limit.

One says that an operator S on an L^p-space with $p \in (1, \infty)$ satisfies the **dominated ergodic theorem in** L^p if the **strong maximal inequality**

$$\|S^* f\|_p \leq C_p \|f\|_p \text{ for all } f \in \mathrm{L}^p(X, \mu) \tag{7.22}$$

holds for some constant $C > 0$ and for the maximal operator S^* corresponding to the sequence $(\frac{1}{N}\sum_{n=1}^{N} S^n)_{N \in \mathbb{N}}$, cf. Corollary 7.47 above. This clearly implies the weak maximal inequality

$$\mu([S^* f > \lambda]) \leq \frac{C_p^p \|f\|_p^p}{\lambda^p}$$

for every $f \in \mathrm{L}^p(X, \mu)$ and, by an L^p-variation of Banach's principle (cf. Remark 12.19 below), the pointwise ergodic theorem[18] as soon as S is mean ergodic. The dominated ergodic theorem on L^p for every $p \in (1, \infty)$ was proved for Koopman operators by Yosida and Kakutani [1060], whereas Dunford and Schwartz [323] generalized it to Dunford–Schwartz operators. Akcoglu [17] proved the dominated ergodic theorem for positive contractions on L^p for $p \in (1, \infty)$ with $C_p = \frac{p}{p-1}$, which was proved for positive invertible isometries before by Ionescu Tulcea [569] and in another special case by Chacon and Olsen [237]. This was extended to invertible doubly power bounded positive[19] operators by de la Torre [275] and to power bounded Lamperti operators by Kan [620]. The case of positive Cesàro bounded operators was investigated in Émilion [367] and Martín-Reyes and de la Torre [774].

[18] In this case, the way of convergence in the mean ergodic theorem is as in the dominated convergence theorem, explaining the name "dominated ergodic theorem".

[19] The positivity cannot be dropped here since, as mentioned above, the pointwise ergodic theorem then fails in general by Feder [384] and Assani [36, 37], see also Cohen [249].

For $p = 1$, the strong maximal inequality (7.22) is false in general even for measure-preserving systems and their Koopman operators. Instead, one has the following **dominated ergodic theorem in** L^1 for a Koopman operator T on $\mathrm{L}^1(X, \mu)$ and $f \in \mathrm{L}^1(X, \mu)$, the maximal function satisfies $A^*f \in \mathrm{L}^1(X, \mu)$ if $f \in \mathrm{L}\log\mathrm{L}$, i.e., $\int_X |f| \log^+ |f| \,\mathrm{d}\mu < \infty$, where $\log^+ := \max\{\log, 0\}$, and in this case one has

$$\|A^*f\|_1 \leq 2\left(1 + \int_X |f| \log^+ |f| \,\mathrm{d}\mu\right).$$

This result is due to Wiener [1044] and Yosida and Kakutani [1060] for Koopman operators, with a generalization to Dunford–Schwartz operators due to Dunford and Schwartz [323]. Ornstein [835] showed that for an invertible ergodic measure-preserving system (X, μ, T) and $f \geq 0$, $f \in \mathrm{L}\log\mathrm{L}$ is also necessary for $A^*f \in \mathrm{L}^1(X, \mu)$, see Derriennic [292] for a different proof using a "reverse maximal inequality"[20] and an extension to ratio ergodic averages.

For a measure-preserving system (X, μ, T), a sequence (f_n) of measurable functions on X is called **subadditive with respect to** T if

$$f_{n+k} \leq f_n + f_k \circ T^n \qquad \forall n, k \in \mathbb{N}.$$

(This generalizes the classical additive sequence given by $f_N := f + Tf + \cdots + T^N f$.) Kingman [649] presented[21] the following **subadditive ergodic theorem**: For a measure-preserving system (X, μ, T) and a subadditive sequence (f_N) in $\mathrm{L}^1(X, \mu)$ with $\inf_{N\in\mathbb{N}} \frac{1}{N} \int_X f_N \,\mathrm{d}\mu > -\infty$, the averages $\frac{f_N}{N}$ converge a.e. to a T-invariant integrable function (which is constant whenever T is ergodic). He also presented a subadditive version of the maximal ergodic theorem. For alternative proofs of Kingman's subbaditive ergodic theorem and its various extensions see Kingman [650, 651] (with a proof by Burkholder in [650, pp. 900–902]), Hammersley [518], Derriennic [293, 296], Smythe [952], Ishitani [571], del Junco [281], Smeltzer [951], Akcoglu and Sucheston [28, 29], Nguyen [824], Fong [399], Ghoussoub and Steele [467], Akcoglu and Krengel [25], Derriennic and Krengel [300], Katznelson and Weiss [635], Neveu [823], Moulin Ollagnier [807], Ghoussoub [466], Liggett [740], Brogini Bratti [194], Émilion and Hachem [368], Çömez [228], Krengel [683, §5.1], Schürger [926–929], Levental [737], Steele [958], Silva and Thieullen [944], Gologan [472], Karlsson and Margulis [631], Çömez and Lin [230], Dooley et al. [315], Auffinger et al. [57], Karlsson [630], Bochi [166], Rugh and Thomine [895], Gouëzel and Karlsson [480] and Kazakevičius [636]. Numerous applications are discussed in, e.g., Kingman [650, 651], Hammersley [518], Steele [957], Smythe and Wierman [953, 954], Deken [280], Guivarch [501], Derriennic [295, 296],

[20] See also Sato [913].

[21] Kingman's formulation [649] is more general for stationary subadditive processes, see also Krengel [683, Thm. 5.3 and 5.4].

Liggett [740], Krengel [683, §5.1], Karlsson and Margulis [631], Hardcastle [520], Hardcastle and Khanin [521], Durrett [328, Ch. 6], Gouëzel and Karlsson [480] and Kazakevičius [636].

For the multiplicative ergodic theorem of Oseledeč and more (operator) ergodic theorems see, e.g., Krengel [683] and the references therein.

Rate of Convergence

Determining the speed (or rate) of convergence of ergodic averages has been an active area of research with many results. We briefly mention here just some of them (in addition to those in Remarks 7.19 and 7.27).

It is easy to see that there is no universal speed of mean convergence for measure-preserving systems, see, e.g., Derriennic [298, p. 144]: Assume that for some operator S on a Banach space E one has $\|\frac{1}{N}\sum_{n=1}^{N} S^n f - P_{\text{fix}\, S} f\| \leq a(N)$ for some sequence $(a(N))$ converging to zero and every $f \in E$ with $\|f\| \leq 1$. Then the principle of uniform boundedness implies that S is uniformly ergodic on E. As mentioned above in this section, for Koopman operators this implies periodicity of the corresponding measure-preserving system. For more on the non-existence of a universal speed of mean and pointwise convergence see Krengel [682], del Junco and Rosenblatt [283], Kakutani and Petersen [615], Liardet and Volný [739], Bayart et al. [85]. Thus one needs assumptions on the function f to make conclusions about the corresponding rate.

For a mean ergodic operator S, by Remark 7.14 and the triangle inequality it suffices to study convergence of ergodic averages to zero on $\overline{\text{ran}(I - S)}$. Butzer and Westphal [212] showed that for a power bounded operator S on a Banach space, the ergodic averages satisfy $\|A_N f\| = o(1/N)$ if and only if $f = 0$. Moreover, for reflexive spaces $\|A_N f\| = O(1/N)$ holds if and only if $f \in \text{ran}(I - S)$,[22] see Browder [196] and [212]. A topological counterpart of this is a result of Gottschalk and Hedlund [479, Thm. 14.11] stating that for a minimal topological system (K, T) and $f \in \mathrm{C}(K)$, one has $\|A_N f\|_\infty = O(1/N)$ if and only if $f \in (I - T)\mathrm{C}(K)$. Lin and Sine [749] showed that the "only if" direction fails in general without the minimality assumption even for uniquely ergodic systems. They also proved that for a contraction S on an L^1-space, $\|A_N f\|_1 = O(1/N)$ holds if and only if $f \in \text{ran}(I - S)$. Note that this equivalence fails for some power bounded operator on L^1 by Fonf et al. [398].

A very useful tool for proving $\lim_{N\to\infty} S_N f = 0$ a.e. for a family of operators (S_N) is to show mean convergence of $S_N f$ with rate. A basic observation here is that, for a sequence of L^2-functions (f_n), $\sum_{n=1}^{\infty} \|f_n\|_2 < \infty$ implies by the triangle inequality $\sum_{n=1}^{\infty} |f_n| \in \mathrm{L}^2$ which in turn implies $\lim_{n\to\infty} f_n = 0$ a.e. We will use this principle later to show pointwise convergence of weighted averages in

[22] Elements of $\text{ran}(I - S)$ are called **coboundaries** of S.

Sect. 16.2. Analogously, in order to obtain a rate for a.e. convergence one usually uses a (higher) speed of mean convergence.

As mentioned above, Bourgain [183] gave some quantitative information on the speed of a.e. convergence for L^2-functions by proving variational inequalities, see also Jones et al. [590, 596] and Krause [670].

For a sample of results on rates in the mean and pointwise ergodic theorem we refer to, e.g., Halász [511], Gaposhkin [460], Kachurovskiĭ [600], Derriennic and Lin [302], Cohen and Lin [251], O'Brien [828], Derriennic [298], Weber [1036], Wu [1053], Zhao and Woodroofe [1063], Cuny and Lin [267], Cuny [263, 264], Kachurovskiĭ and Sedalishchev [604, 605], Gomilko et al. [473], Kachurovskiĭ and Podvigin [601–603], cf. also Avigad et al. [68].

Multiparameter Ergodic Theorems

Dunford [321] showed norm and almost everywhere convergence of multiparameter ergodic averages

$$\frac{1}{N_1 \cdots N_d} \sum_{n_1=1}^{N_1} \cdots \sum_{n_d=1}^{N_d} T_1^{n_1} \cdots T_d^{n_d} f \tag{7.23}$$

as $\min\{N_1, \ldots, N_d\} \to \infty$ for every $d \in \mathbb{N}$, every injective measure-preserving (not necessarily commuting) transformations $T_1, \ldots, T_d$ on a probability space (X, μ) and every $f \in \mathrm{L}^p(X, \mu)$ for $p > 1$. Moreover, for $d = 2$ he showed a.e. convergence for every $f \in \mathrm{L}\log\mathrm{L}$. Independently, Zygmund [1079] proved convergence of the averages (7.23) (in the time continuous case) for $f \in \mathrm{L}(\log\mathrm{L})^{d-1}$ for all d. This was extended to certain groups of measure-preserving transformations by Calderón [215] (for $d = 2$) and to Dunford–Schwartz operators by Dunford and Schwartz [322] (for $f \in \mathrm{L}^p$, $p > 1$) and Fava [380] (for $f \in \mathrm{L}(\log\mathrm{L})^{d-1}$), see also Sucheston [961], Olsen [832] and Krengel [683, Sec. 6.1]. Oscillation inequalities for averages (7.23) in the commutative case were done by Jones et al. [595]. Note that convergence of (7.23) fails in general for $p = 1$ and $d \geq 2$ even for commuting transformations, see, e.g., Mirek et al. [797, Section 1.2].

Bourgain observed the polynomial version of the Dunford–Zygmund theorem on almost everywhere convergence of averages

$$\frac{1}{N_1 \cdots N_d} \sum_{n_1=1}^{N_1} \cdots \sum_{n_d=1}^{N_d} T_1^{P_1(n_1)} \cdots T_d^{P_d(n_d)} f$$

for polynomials $P_1, \ldots, P_d \in \mathbb{Z}[\cdot]$ with no constant term whenever the transformations $T_1, \ldots, T_d$ commute, but he never published his observation. For a proof of Bourgain's observation and further results including oscillation inequalities for multivariate polynomials see Mirek et al. [797], Bourgain et al. [187] and Mehlhop [787], where the last paper also treats averages along (multidimensional subsets of) primes.

Exercises

Exercise 7.1 (Mean Ergodic Theorem, Proof via the Spectral Theorem) Prove von Neumann's mean ergodic theorem for unitary operators by means of the spectral theorem. Restrict the investigation to a cyclic subspace.

Exercise 7.2 (Mean Ergodic Theorem) Prove Proposition 7.3.

Exercise 7.3 (Mean Ergodicity of Koopman Operators in L^p) Prove the assertion of Remark 7.8. (Hint: For $p > 2$ estimate the L^p-norm by L^2-norm for bounded functions.)

Exercise 7.4 Which of the following operators is mean ergodic on the given Banach space? Determine, where applicable, the mean ergodic projection.

(a) Left/right shift on $\ell^1(\mathbb{N})$.
(b) Left/right shift on the space $\mathrm{c}_0(\mathbb{N})$ of null-sequences (with supremum norm).
(c) An isometric multiplication operator on $\ell^1(\mathbb{N})$. (When is a multiplication operator isometric on $\ell^1(\mathbb{N})$?)
(d) An isometric multiplication operator on $\mathrm{c}_0(\mathbb{N})$ (with supremum norm). (When is a multiplication operator isometric on $\mathrm{c}_0(\mathbb{N})$?)

Exercise 7.5 (Mean Ergodic Operators) Let E be a Banach space and $S \in \mathscr{L}(E)$ be mean ergodic. Prove the following assertions.

(a) $\lim_{n\to\infty} \frac{\|S^n x\|}{n} = 0$ holds for every $x \in E$.
(b) The operator P defined by

$$Px := \lim_{N\to\infty} \frac{1}{N} \sum_{n=0}^{N} S^n x$$

is a bounded, linear projection with $\operatorname{ran}(P) = \operatorname{fix}(S)$ and $\ker(P) = \overline{\operatorname{ran}(I - S)}$.

Exercise 7.6

(a) Prove Proposition 7.11.
(b) Show the first two equivalences from Remark 7.19.

Exercise 7.7 Prove the equivalence of each of the assertions in Remark 7.13 to the mean ergodicity of the given operator S, and prove the missing equivalences in the proof of Theorem 7.12.

Exercise 7.8 (Asymptotically Invariant Sequences of Measures)

(a) Show that a topological system is uniquely ergodic if and only if every asymptotically T-invariant sequence of probability measures converges to an invariant probability measure.
(b) Prove the assertion of Lemma 7.21.

Exercise 7.9 (Asymptotic Invariance) Let (K, T) be a topological system and $(\nu_n)_{n\in\mathbb{N}}$ be a sequence of probability measures in $\mathrm{M}(K)$. Then the sequence $(\mu_N)_{N\in\mathbb{N}}$ of measures μ_N, $N \in \mathbb{N}$, defined by

$$\mu_N := \frac{1}{N}\sum_{n=0}^{N-1} T_*^n \nu_N,$$

is asymptotically T-invariant.

Exercise 7.10 (Uniform Convergence of Ergodic Averages)

(a) Does the statement of Proposition 7.24 remain true without the assumption that the everywhere limit of ergodic averages should be constant?
(b) Show that the assertion of Proposition 7.24 remains true if the ergodic averages converge everywhere to a constant function for every f in a dense subset of $\mathrm{C}(K)$.

Exercise 7.11 (Characterization of Unique Ergodicity) Prove Corollary 7.25.

Exercise 7.12 (Uniform Convergence via Sine's Criterion) Prove Theorem 7.23 by using Sine's criterion: A power bounded operator S on a Banach space is mean ergodic if and only if fix S separates fix S' (see Theorem 7.12).

Exercise 7.13 (Polynomial Mean Ergodic Theorem)

(a) Consider the Hilbert space $H = \ell^2(\mathbb{N})$ and a sequence $g = (g_n) \in \ell^\infty(\mathbb{N})$ with $|g_n| = 1$ for each $n \in \mathbb{N}$. Let $P \in \mathbb{Z}[\cdot]$ map $\mathbb{N}_0$ to $\mathbb{N}_0$ and let $S = M_g$ be the multiplication operator by g. Determine the limit in the polynomial mean ergodic theorem for the isometry S.
(b) Show that in Theorem 7.28 one can replace polynomials with integer coefficients by integer polynomials, see Remark 7.29(b). Determine the limit in the polynomial mean ergodic theorem for such polynomials and the multiplication operator from (a).

Exercise 7.14 (Polynomial Khinchin Recurrence) Consider the additive group $X = \mathbb{Z}/15\mathbb{Z}$ consisting of 15 points endowed with the normalized counting measure μ and the transformation T given by $x \mapsto x + 1$. Show that the set $A = \{0, 7\}$ satisfies

$$\lim_{N-M\to\infty} \frac{1}{N-M}\sum_{n=M}^{N-1} \mu(A \cap T^{n^2}A) = \frac{2}{225} < \frac{4}{225} = \mu(A)^2.$$

This shows the assertion in Remark 7.36.

Exercise 7.15 (Pointwise Convergence) Show that almost everywhere convergence on, say, $L^\infty([0,1],\lambda)$ is not induced by a topology. (Hint: Prove first that for a topological space X, a sequence $(x_n) \subset X$ converges to $x \in X$ if and only if every subsequence of (x_n) has a subsequence converging to x. Present finally a sequence of bounded measurable functions (f_n) on $[0,1]$ which diverges *everywhere* but has the property that every subsequence of (f_n) has a subsequence converging almost everywhere to 0.)

Exercise 7.16 (Maximal Operator) Let (X,μ) be a probability space and $(S_N)_{N\in\mathbb{N}}$ be a sequence of linear operators on $L^1(X,\mu)$. Show that the maximal operator S^* defined by (7.17) has the following two properties:

(1) $S^*(\alpha f) = |\alpha| \cdot S^* f$ for each $\alpha \in \mathbb{C}$ and $f \in \mathscr{L}^1(X,\mu)$,
(2) $S^*(f+g) \le S^* f + S^* g$ for each $f, g \in L^1(X,\mu)$.

Exercise 7.17 (Ergodicity and Pointwise a.e. Convergence) Deduce von Neumann's mean ergodic theorem for the Koopman operator of a measure-preserving system from Birkhoff's pointwise ergodic theorem. Prove that a measure-preserving system (X,μ,T) is ergodic if and only if the limit of ergodic averages equals $\int_X f \,\mathrm{d}\mu$ μ-almost everywhere for every $f \in L^1(X,\mu)$.

Exercise 7.18 (Generic Points in Topological Systems) Prove the assertion in Remark 7.49.

Exercise 7.19 (Markov's Inequality) Let (X,μ) be a measure space and $f \in L^0(X,\mu)$, $f \ge 0$. Prove that for every $\varepsilon > 0$ the following inequality holds

$$\mu([f \ge \varepsilon]) \le \frac{1}{\varepsilon} \int_X f \,\mathrm{d}\mu.$$

Exercise 7.20 (Convergence in Measure) Let (X,μ) be a probability space, let $p \in [1,\infty]$ and (g_j) be a sequence in $L^p(X,\mu)$ with $g_j \to g$ in L^p. Prove that for $\lambda > 0$ one has

$$\lim_{j\to\infty} \mu([|g_j - g| > \lambda]) = 0.$$

(Recall that this limit being 0 means, by definition, that (g_j) **converges in measure** (or in probability) to g.)

Exercise 7.21 (More on Maximal Inequality) Show that the sets $M_{m,n}$ presented in the proof of Theorem 7.52 are closed in $L^1(X,\mu)$.

Exercise 7.22 (A Version of Dowker's Ergodic Theorem) Let $(X,\mathcal{X},\mu)$ be a σ-finite measure space, let $T : X \to X$ be measurable such that $\mu \ll T_*\mu$, and set $\nu := T_*\mu$. Let $w = \frac{\mathrm{d}\nu}{\mathrm{d}\mu}$ be the Radon–Nikodym derivative of ν with respect to μ.

For $n \in \mathbb{N}_0$ define

$$w_n(x) := \prod_{j=0}^{n-1} w(T^j x) \quad (x \in X).$$

Deduce from the Chacon–Ornstein theorem, see Sect. 7.8, that for $f, g \in \mathrm{L}^1(X, \mu)$ with $g(x) > 0$ for almost every $x \in X$ the ratios

$$\frac{\sum_{k=0}^{n} f(T^k x) w_{k+1}(x)}{\sum_{k=0}^{n} g(T^k x) w_{k+1}(x)}$$

converge almost everywhere to a finite limit. (Hint: Consider the weighted Koopman operator S defined on $\mathrm{L}^1(X, \mu)$ via $Sf := w \cdot f \circ T$ and apply the Chacon–Ornstein theorem suitably.)

Exercise 7.23 (Uniformly Mean Ergodic Operators) Let S be a power bounded operator on a Banach space. Suppose that S is **uniformly mean ergodic**, i.e., the Cesàro averages

$$\frac{1}{N} \sum_{n=1}^{N} S^n$$

converge in the operator norm (then, of course, to the mean ergodic projection). Prove the following assertions.

(a) $I - S$ is invertible on $\mathrm{ran}(I - S)$.
(b) $\mathrm{ran}(I - S)$ is closed.
(c) $1 \in \rho(S) := \mathbb{C} \setminus \sigma(S)$ if and only if $\mathrm{fix}(S) = \{0\}$.
(d) Either $1 \in \rho(S)$, or 1 is isolated in $\sigma(S)$.

(Hint: Use Proposition 7.11.)

Exercise 7.24 (Uniform Mean Ergodicity and Koopman Operators) Let (X, μ, T) be an ergodic measure-preserving system. Show that the following assertions are equivalent.

(i) The Koopman operator T is uniformly mean ergodic on $\mathrm{L}^2(X, \mu)$.
(ii) 1 is isolated in $\sigma(T)$.
(iii) The Koopman operator T is periodic on $\mathrm{L}^2(X, \mu)$.

(Hint: Use the results in the foregoing exercise.) Note that this equivalence holds also without the ergodicity assumption, see Hölz [540].

Exercise 7.25 (Ergodic Hilbert Transform) The aim of this exercise is the proof of the following result of Campbell from [218] (see also [217]).

Theorem *Let $U \in \mathscr{L}(H)$ be a unitary operator on a Hilbert space H. For each $x \in H$ the limit*

$$\mathcal{H}x := \lim_{N\to\infty} \sum_{\substack{n=-N\\ n\neq 0}}^{N} \frac{U^n x}{n} \tag{7.24}$$

exists.[23]

The limit operator $\mathcal{H}$ is called the **ergodic Hilbert transform** of U.

(a) First prove convergence in (7.24) for $x \in \operatorname{fix}(U) + \operatorname{ran}(I - U)$.
(b) Show that for the partial sums

$$h_N(z) := \sum_{n=1}^{N} \frac{z^n - \overline{z}^n}{n}$$

one has

$$\sup_{N\in\mathbb{N}} \sup_{z\in\mathbb{T}} |h_N(z)| < \infty.$$

Hint: Use that for $z = \mathrm{e}^{\mathrm{i}t}$, $t \in [0, 2\pi]$

$$h_N(z) = 2\mathrm{i} \sum_{n=1}^{N} \frac{\sin(nt)}{n},$$

and relate this expression to

$$\frac{\sin(N+\frac{1}{2})t}{\sin(\frac{t}{2})} - 1 = \sum_{n=1}^{N} (z^n + \overline{z}^n) = 2\sum_{n=1}^{N} \cos(nt).$$

(c) Prove that the partial sums

$$\mathcal{H}_N := \sum_{\substack{n=-N\\ n\neq 0}}^{N} \frac{U^n}{n} = \sum_{n=1}^{N} \frac{U^n - U^{*n}}{n}$$

[23] In case U is the Koopman operator of an invertible measure-preserving system (X, μ, T) on $H = \mathrm{L}^2(X, \mu)$ the pointwise almost everywhere convergence in (7.24) was established (among other results) by Cotlar [261] and is known as Cotlar's theorem.

are uniformly bounded, i.e., $\sup_{N \in \mathbb{N}} \|\mathcal{H}_N\| < \infty$. Finish the proof of Campbell's theorem using the above together with the von Neumann decomposition (Theorem 6.4).

(d) Formulate and prove the analogue of Campbell's theorem for Hilbert space contractions.

Chapter 8
Factors of Measure-Preserving Systems

Factors of measure-preserving systems are important to understand the structures behind ergodic theorems. A fundamental tool for their study is provided by Markov operators which are briefly touched upon here, while a more systematic treatment can be found in [342, Ch. 13]. We further discuss a few important examples of factors: the fixed factor, the rational spectrum factor, the Kronecker factor and the Abramov factor in some detail, with a further characterization of weak mixing and a discussion of topological models.

8.1 Factors

We begin with some motivating examples and essential constructions.

Example 8.1 (Products) Let (X, μ, T) and (Y, ν, S) be two measure-preserving systems and consider the product probability space $(X \times Y, \mu \times \nu)$. The mapping

$$X \times Y \to X \times Y, \quad (x, y) \to (Tx, Sy)$$

is measure-preserving and yields the **product system** $(X \times Y, \mu \times \nu, T \times S)$ (see Lemma 2.6). Analogously, one defines the product of finitely many measure-preserving systems. Obviously, for $a \in \mathbb{T}^d$ the system $(\mathbb{T}^d, \mathrm{m}^d, a)$ is a product of d one-dimensional torus rotations.

Example 8.2 (Skew Product) Let (X, μ, T) be a measure-preserving system, let (Y, ν) be a probability space, and let $\rho : X \times Y \to Y$ be measurable (where $X \times Y$ is equipped with the product σ-algebra) such that $\rho_x : y \mapsto \rho(x, y)$ preserves ν for each $x \in X$. Define

$$T_\rho : X \times Y \to X \times Y, \quad T_\rho(x, y) := (Tx, \rho(x, y)).$$

T. Eisner, B. Farkas, *A Journey Through Ergodic Theorems*, Birkhäuser Advanced Texts Basler Lehrbücher, https://doi.org/10.1007/978-3-031-96506-7_8

It is easy to see that $(X \times Y, \mu \times \nu, S)$ is a measure-preserving system, called the ρ-**skew product**. If $Y = G$ is a compact group, $\nu = \mathrm{m}_G$, the Haar measure, and $\rho(x, \cdot)$ is a left rotation for each $x \in X$, then the skew product is called a **group extension**. A concrete example of a group extension is the skew product $(\mathbb{T}^2, \mathrm{m}^2, T_a)$ from Example 2.15 where $T_a(x, y) := (ax, xy)$ for a fixed $a \in \mathbb{T}$.

As in the topological situation (see Example 2.31.4), one defines the (measurable) **cocycle** associated to a skew product $(X \times Y, \mu \times \nu, T_\rho)$: We set

$$\widehat{\rho} : \mathbb{N}_0 \times X \to Y^Y, \qquad \widehat{\rho}(n, x) := \rho_{T^{n-1}(x)} \circ \cdots \circ \rho_{T(x)} \circ \rho_x .$$

(As in the topological situation, cf. Example 2.31(d), the mapping itself is sometimes referred to as the cocycle of the skew product.) Recall also the cocycle relation

$$\widehat{\rho}(0, x) = \mathrm{id}, \quad \widehat{\rho}(m + n, x) = \widehat{\rho}(m, T^n(x)) \circ \widehat{\rho}(n, x)$$

for every $n, m \in \mathbb{N}_0$, and $x \in X$.

The previous two examples share the common feature that the projection onto the first component respects the dynamics. We give a name to this phenomenon.

Definition 8.3 Let (X, μ, T) and (Y, ν, S) be two measure-preserving systems. The system (Y, ν, S) is called a **point factor** of (X, μ, T) (and (X, μ, T) is called a **point extension** of (Y, ν, S)) if there exists a measure-preserving map $\pi : X \to Y$ such that $S \circ \pi = \pi \circ T$ holds almost everywhere, i.e., such that the following diagram is commutative (π is then called a **point factor map**).

$$\begin{array}{ccc} X & \xrightarrow{T} & X \\ \downarrow{\scriptstyle \pi} & & \downarrow{\scriptstyle \pi} \\ Y & \xrightarrow{S} & Y \end{array}$$

Recall from Definition 2.9 that if π is essentially invertible, the systems are called **point isomorphic** and π is called a **point isomorphism of measure-preserving systems**.

Factors of topological dynamical systems are defined as follows. If (K, T) and (L, S) are topological systems and $\pi : K \to L$ is a continuous surjection with $\pi \circ T = S \circ \pi$, then (L, S) is called a (point) **factor** of (K, T), (L, S) an **extension** of (K, T) and π a (continuous) factor map.

Example 8.4 (Products, Skew Rotation and Group Extensions) Each of the systems (X, μ, T) and (Y, ν, S) is a factor of the product system $(X \times Y, \mu \times \nu, T \times S)$, and the factor map is given by the corresponding coordinate projections $\pi_1(x, y) := x$ or $\pi_2(x, y) := y$, respectively.

Analogously, the rotation system $(\mathbb{T}, \mathrm{m}, a)$ is a factor of the skew rotation from Example 2.15. Generally, a group extension as in Example 8.2 is indeed an extension of the original system (X, μ, T), explaining the origin of the name "group extension".

We now look at the properties of the Koopman operators induced by point factor maps. Recall the notation

$$\langle f, g \rangle = \int_X fg \, \mathrm{d}\mu \text{ for } f \in \mathrm{L}^p(X, \mu),\ \ g \in \mathrm{L}^q(X, \mu),$$

with $p, q \in [1, \infty]$, $\frac{1}{p} + \frac{1}{q} = 1$.

Proposition 8.5 (Koopman Operators of Point Factor Maps) *Let $\pi : (X, \mu, T) \to (Y, \nu, S)$ be a point factor map. The Koopman operator $S_\pi : \mathrm{L}^0(Y, \nu) \to \mathrm{L}^0(X, \mu)$ has the following properties.*

(a) $S_\pi f \geq 0$ *for each positive* $f \in \mathrm{L}^0(Y, \nu)$.
(b) $S_\pi(fg) = S_\pi(f) S_\pi(g)$ *for each pair* $f, g \in \mathrm{L}^0(Y, \nu)$.
(c) $S_\pi |f| = |S_\pi f|$ *for every* $f \in \mathrm{L}^0(Y, \nu)$.
(d) $S_\pi : \mathrm{L}^1(Y, \nu) \to \mathrm{L}^1(X, \mu)$ *is an isometry.*
(e) $S_\pi \mathbf{1} = \mathbf{1}$.
(f) $\langle S_\pi f, \mathbf{1} \rangle = \langle f, \mathbf{1} \rangle$ *for each* $f \in \mathrm{L}^1(Y, \nu)$, *i.e.,* $S_\pi' \mathbf{1} = \mathbf{1}$.
(g) S_π *intertwines the Koopman operators* T *and* S, *i.e.,* $S_\pi S = T S_\pi$.

We distill some operator theoretic definitions out of these properties.

Definition 8.6

(a) Let (X, μ), (Y, ν) be probability spaces. We call a linear operator $S : \mathrm{L}^1(Y, \nu) \to \mathrm{L}^1(X, \mu)$ a **Markov operator**[1] if

  (1) S is **positive**, i.e., $Sf \geq 0$ whenever $f \geq 0$.
  (2) $S\mathbf{1} = \mathbf{1}$.
  (3) $S'\mathbf{1} = \mathbf{1}$, i.e., $\langle Sf, \mathbf{1} \rangle = \langle f, \mathbf{1} \rangle$ for every $f \in \mathrm{L}^1(Y, \nu)$.

  A Markov operator S is called a **Markov homomorphism** (or **Markov embedding**) if $|Sf| = S|f|$ holds for every $f \in \mathrm{L}^1(Y, \nu)$, and a bijective Markov homomorphism is called a **Markov isomorphism**.

(b) We call a measure-preserving system (Y, ν, S) a **Markov factor** (or sometimes simply a **factor**) of (X, μ, T) if there is a Markov homomorphism $J : \mathrm{L}^1(Y, \nu) \to \mathrm{L}^1(X, \mu)$ intertwining the Koopman operators T and S, i.e., such that the following diagram is commutative.

$$\begin{array}{ccc} \mathrm{L}^1(Y, \nu) & \xrightarrow{\ S\ } & \mathrm{L}^1(Y, \nu) \\ {\scriptstyle J}\downarrow & & \downarrow{\scriptstyle J} \\ \mathrm{L}^1(X, \mu) & \xrightarrow{\ T\ } & \mathrm{L}^1(X, \mu) \end{array}$$

[1] In the literature the definition of a Markov operator often includes only the conditions that S is positive and $S'\mathbf{1} = \mathbf{1}$. In that setting the operators satisfying the extra condition $S\mathbf{1} = \mathbf{1}$, as here, should be called **bi-Markov**.

In this case, we also say that (X, μ, T) is a **Markov extension** (or sometimes simply an **extension**) of (Y, ν, S). We denote this by writing $J : (\mathrm{L}^1(Y, \nu), S) \to (\mathrm{L}^1(X, \mu), T)$ and say that J is a **Markov homomorphism** between the measure-preserving systems (Y, ν, S) and (X, μ, T).

(c) Two measure-preserving systems (X, μ, T) and (Y, ν, S) are called **Markov isomorphic** (or just **isomorphic**) if there is a Markov isomorphism $J : \mathrm{L}^1(Y, \nu) \to \mathrm{L}^1(X, \mu)$ intertwining the Koopman operators.

Remark 8.7

(a) Products of Markov operators are Markov.
(b) Every Markov homomorphism $S : \mathrm{L}^1(Y, \nu) \to \mathrm{L}^1(X, \mu)$ is an isometry by

$$\|Sf\|_1 = \int_X |Sf| \, \mathrm{d}\mu = \int_X S|f| \, \mathrm{d}\mu = \int_Y |f| \, \mathrm{d}\nu = \|f\|_1.$$

(c) By the above the Koopman operator $S := S_\pi$ of a point factor map π given by $Sf = f \circ \pi$ is a Markov homomorphism, and a point factor (Y, ν, S) is a (Markov) factor. In particular, point isomorphic systems (see Definition 2.9) are (Markov) isomorphic. Note that the converse fails in general, see Exercise 8.1.
(d) Recall that a measure-preserving system (X, μ, T) is invertible if its Koopman operator T is invertible. In this case T is a Markov isomorphism.

Remark 8.8 Point factors are defined via (essential) *surjections* $\pi : X \to Y$ respecting the structure of the probability spaces and the dynamics, which corresponds to *quotients*. In the above, one goes "up" to the level of functions to have *injections* $S : \mathrm{L}^1(Y, \nu) \to \mathrm{L}^1(X, \mu)$ respecting the structure of the L^1-spaces and the Koopman operators, which corresponds to *subspaces* (more precisely, sublattices in the terminology of Sect. 8.2 below).

Proposition 8.9 *A Markov factor (and in particular a point factor) of an ergodic/weakly mixing/strongly mixing measure-preserving system is ergodic/weakly mixing/strongly mixing.*

The proof is left as Exercise 8.2.

The following classical result shows that skew products from Example 8.2 represent a typical form of extensions.

Theorem 8.10 (Rokhlin Skew Product Theorem) *Let (X, μ, T) be an ergodic measure-preserving system, let (Y, ν, S) be a factor of (X, μ, T), and suppose that both systems are regular. Then (X, μ, T) is a ρ-skew product of (Y, ν, S) and a regular probability space with an appropriate ρ.*

A proof can be found in, e.g., [468, Thm. 3.18].

8.2 Markov Operators and Characterizations of Factors

Our next aim is to characterize Markov homomorphisms and factors in various, very different ways. First we collect some general properties of Markov operators. Even if not stated every time, we consider exclusively probability spaces (unless explicitly said otherwise).

Proposition 8.11 (Basic Properties of Positive/Markov Operators) *Let (X, μ) and (Y, ν) be probability spaces and let $S : \mathrm{L}^1(Y, \nu) \to \mathrm{L}^1(X, \mu)$ be a positive operator, i.e., satisfying $Sf \geq 0$ for each $f \geq 0$.*

(a) *S maps real-valued functions to real-valued functions, and* $\mathrm{Re}(Sf) = S\mathrm{Re}\, f$.
(b) *S respects conjugation, i.e., $S\overline{f} = \overline{Sf}$.*
(c) *If $g \leq f$, then $Sg \leq Sf$.*
(d) *$|Sf| \leq S|f|$ for every $f \in \mathrm{L}^1(Y, \nu)$.*
(e) *If $S\mathbf{1} \leq \mathbf{1}$, then it restricts to $S : \mathrm{L}^\infty(Y, \nu) \to \mathrm{L}^\infty(X, \mu)$ and this restriction is contractive with respect to the L^∞-norms.*
(f) *If $S'\mathbf{1} = \mathbf{1}$, then S is contractive with respect to the L^1-norms.*
(g) *If S is a Markov operator, then for each $p \in [1, \infty]$ and $f \in \mathrm{L}^p(Y, \nu)$ we have $\|Sf\|_p \leq \|f\|_p$.*
(h) *The adjoint operator S' is positive as well. Moreover, if $S\mathbf{1} \leq \mathbf{1}$ then S' extends to a positive contraction from $\mathrm{L}^1(X, \mu)$ to $\mathrm{L}^1(Y, \nu)$. If S is a Markov operator, then so is the extension $S' : \mathrm{L}^1(X, \mu) \to \mathrm{L}^1(Y, \nu)$.*

Given a Markov operator $S : \mathrm{L}^1(Y, \nu) \to \mathrm{L}^1(X, \mu)$, the extension $S' : \mathrm{L}^1(X, \mu) \to \mathrm{L}^1(Y, \nu)$ provided by the very last statement is called the **Markov adjoint** of S. Some basic properties of Markov adjoint operators are summarized in Exercise 8.6.

Proof The proof of the assertions (a)–(f) is left as Exercise 8.3, while (g) is handled in Exercise 8.5. To show (h), let $f \in \mathrm{L}^\infty(X, \mu)$ with $f \geq 0$ and take a measurable $A \subset Y$. We have by positivity of S

$$\int_A S'f \,\mathrm{d}\nu = \langle \mathbf{1}_A, S'f \rangle = \langle S\mathbf{1}_A, f \rangle \geq 0. \tag{8.1}$$

Since A was arbitrary, $S'f \geq 0$ follows.

Assume now $S\mathbf{1} \leq \mathbf{1}$ and let $f \in \mathrm{L}^\infty(X, \mu)$. Since S' is positive, we obtain

$$\|S'f\|_{\mathrm{L}^1} = \int_Y |S'f| \,\mathrm{d}\nu \leq \int_Y S'|f| \,\mathrm{d}\nu = \int_Y S\mathbf{1} \cdot |f| \,\mathrm{d}\nu \leq \|f\|_{\mathrm{L}^1}.$$

Thus S' extends to a contraction on $\mathrm{L}^\infty(X, \mu)$ which is positive by the computation (8.1) applied to positive $f \in \mathrm{L}^\infty(X, \mu)$. Assume finally that S is Markov. Then $S'\mathbf{1} = \mathbf{1}$ and we have already shown that S' extends to a positive operator on $\mathrm{L}^1(X, \mu)$ which we denote by S' again. By $S''\mathbf{1} = S\mathbf{1} = \mathbf{1}$ we see that this operator is Markov. ∎

Remark 8.12 (Perron–Frobenius Operators) Let T be a Koopman operator of a measure-preserving system (X, μ, T). Then by Exercise 8.6 we have $T'' = T$, so for each $f, g \in \mathrm{L}^\infty(X, \mu)$ it follows that

$$\int_X Tf \cdot g \, \mathrm{d}\mu = \int_X T'' f \cdot g \, \mathrm{d}\mu = \int_X f \cdot T' g \, \mathrm{d}\mu.$$

We conclude that $T' : \mathrm{L}^1(X, \mu) \to \mathrm{L}^1(X, \mu)$ is the Perron–Frobenius operator corresponding to the measure-preserving transformation T as discussed briefly for general non-singular transformations in Remark 2.28.

It is useful to look at the different algebraic structures on $\mathrm{L}^0(X, \mu)$ which are respected by Koopman operators. A subspace $E \subset \mathrm{L}^0(X, \mu)$ is called a (vector) **sublattice** if $f \in E$ implies $|f| \in E$, while a subspace E is a **subalgebra** in $\mathrm{L}^0(X, \mu)$ if it is closed under multiplication, and E is **conjugation invariant** if $\overline{f} \in E$ for every $f \in E$. For example, $\mathrm{L}^\infty(X, \mu)$ is a subalgebra and for each $p \in [1, \infty]$ the space $\mathrm{L}^p(X, \mu)$ is a sublattice in $\mathrm{L}^0(X, \mu)$. Note that a subspace E is conjugation invariant if and only if $\mathrm{Re} f \in E$ for every $f \in E$. Moreover, if $f, g \in E$ are real-valued and E is a sublattice, then

$$\max\{f, g\} = \tfrac{1}{2}(f + g + |f - g|) \in E \text{ and } \min\{f, g\} = \tfrac{1}{2}(f + g - |f - g|) \in E.$$

For more information on vector lattices see, e.g., Schaefer [915].

The following result connects these structures, a proof can be found, e.g., in [342, Thm. 7.23].

Proposition 8.13 (Sublattices vs. Subalgebras)

(a) *For a conjugation invariant, $\|\cdot\|_\infty$-closed subspace $E \subset \mathrm{L}^\infty(X, \mu)$ with $\mathbf{1} \in E$ the following are equivalent.*

  (i) *E is a subalgebra of $\mathrm{L}^\infty(X, \mu)$.*
  (ii) *E is a sublattice of $\mathrm{L}^\infty(X, \mu)$.*

(b) *For a subalgebra E as in* (a) *and a conjugation-preserving, linear operator $J : E \to \mathrm{L}^\infty(Y, \nu)$ with $J\mathbf{1} = \mathbf{1}$ the following assertions are equivalent.*

  (i) *J is **multiplicative**, i.e., $J(fg) = Jf \cdot Jg$ for each pair $f, g \in E$.*
  (ii) *J is a **lattice homomorphism**, i.e., $J|f| = |Jf|$ for each $f \in E$.*

The following shows that both multiplicativity and being an isometry for the L^1-norm characterize Markov homomorphisms.

Proposition 8.14 (Characterization of Markov Homomorphisms) *For a Markov operator $S : \mathrm{L}^1(Y, \nu) \to \mathrm{L}^1(X, \mu)$ the following assertions are equivalent.*

(i) *S is multiplicative on $\mathrm{L}^\infty(Y, \nu)$.*
(ii) *$S'S = I$, the identity on $\mathrm{L}^1(Y, \nu)$.*

(iii) *S is an isometry with respect to the* L^1*-norms.*
(iv) *S is a Markov homomorphism.*

Proof (i)⇒(ii): For $f, g \in \mathrm{L}^\infty(Y, \nu)$ we have $\langle Sf, Sg\rangle = \langle Sf\,Sg, \mathbf{1}\rangle = \langle S(fg), \mathbf{1}\rangle = \langle fg, \mathbf{1}\rangle = \langle f, g\rangle$. This implies that $S'Sf = f$ for $f \in \mathrm{L}^\infty(Y, \nu)$, and (ii) follows by density.
The implication (ii)⇒(iii) follows from the fact that the Markov operators S and S' are contractive for the L^1-norm. Indeed, $\|f\|_1 = \|S'Sf\|_1 \le \|Sf\|_1 \le \|f\|_1$.
(iii)⇒(iv): Let $f \in \mathrm{L}^1(Y, \nu)$. Since $|Sf| \le S|f|$ and S is isometric, we must have $|Sf| = S|f|$.
(iv)⇒(i) has been discussed in Proposition 8.13(b). ■

Remark 8.15

(a) Let S be a Markov homomorphism. Then S is isometric for the L^p-norms for each $p \in [1, \infty]$.
(b) If S is a Markov operator which is isometric for the L^2-norms (or for the L^p-norm for some $p \in [1, \infty)$), then S is a Markov homomorphism, see [342, Thm. 13.9].
(c) By (iv)⇒(i) in Proposition 8.14, every Markov homomorphism $S : \mathrm{L}^1(Y, \nu) \to \mathrm{L}^1(X, \mu)$ maps characteristic functions to characteristic functions, i.e., for every measurable set $A \subset Y$ there exists a measurable set $B \subset X$ with $J\mathbf{1}_A = \mathbf{1}_B$.

Proposition 8.16 (Sublattices in L^1 and Subalgebras of L^∞)

(a) *Let E be a closed, conjugation invariant sublattice in $\mathrm{L}^1(X, \mu)$ with $\mathbf{1} \in E$. Then*

$$F := \mathrm{L}^\infty(X, \mu) \cap E$$

is a closed, conjugation invariant subalgebra in $\mathrm{L}^\infty(X, \mu)$ which is dense in E.
(b) *Let $F \subset \mathrm{L}^\infty(X, \mu)$ be a conjugation invariant subalgebra with $\mathbf{1} \in F$. Then the L^1-closure E of F is a closed, conjugation invariant sublattice of $\mathrm{L}^1(X, \mu)$.*
(c) *Let E be a closed, conjugation invariant sublattice in $\mathrm{L}^1(X, \mathcal{X}, \mu)$ with $\mathbf{1} \in E$. Then there is a sub-σ-algebra $\mathcal{X}'$ of $\mathcal{X}$ such that $E = \mathrm{L}^1(X, \mathcal{X}', \mu)$.*

The property that $\mathbf{1}$ belongs to a sublattice/subalgebra will be expressed sometimes by saying that the sublattice/subalgebra is **unital**.

Proof (a) The subspace F is clearly a conjugation invariant sublattice, and it is closed in $\mathrm{L}^\infty(X, \mu)$. Indeed, if $\|f_n - f\|_\infty \to 0$ as $n \to \infty$ and $f_n \in F$, then $\|f_n - f\|_1 \le \|f_n - f\|_\infty \to 0$ and $f \in E$, since E is closed. Altogether we conclude that $f \in F$. Let $f \in E$, by conjugation invariance we may suppose that f is real-valued, and even that $f \ge 0$. Then $f_n := \min\{n\mathbf{1}, f\} \in F$ converges to f by Lebesgue's dominated convergence theorem, which proves the density of F in E.

(b) is clear by Proposition 8.13 applied to the $\|\cdot\|_\infty$-closure of F (note that the closure of a sublattice is a sublattice).
(c) Set $F := \mathrm{L}^\infty(X, \mu) \cap E$, which is then a closed subalgebra of $\mathrm{L}^\infty(X, \mu)$ by part (a). Define

$$\mathcal{X}' := \{A \in \mathcal{X} : \mathbf{1}_A \in E\} = \{A \in \mathcal{X} : \mathbf{1}_A \in F\}.$$

It is easy to see that $\mathcal{X}'$ is a sub-σ-algebra. The $\mathcal{X}'$-measurable simple functions belong to $F \subset E$, and are dense in $\mathrm{L}^1(X, \mathcal{X}', \mu)$. We conclude $\mathrm{L}^1(X, \mathcal{X}', \mu) \subset E$. For the converse inclusion take $f \in E$. Since E is conjugation invariant we may suppose that f is real-valued. Let $\alpha \in \mathbb{R}$, and define $g := \max\{f - \alpha\mathbf{1}, 0\}$ and $f_n := \min\{ng, \mathbf{1}\}$. Since E is a sublattice, we have $g, f_n \in E$. By the monotone convergence theorem we obtain that $\mathbf{1}_{[f>\alpha]} = \lim_{n\to\infty} f_n$ belongs to E. This, being true for each $\alpha \in \mathbb{R}$, implies that f is $\mathcal{X}'$-measurable. ∎

Remark 8.17 Of course, part (a) and (b) in the previous statement remain true if one replaces the space $\mathrm{L}^1(X, \mu)$ by any of the spaces $\mathrm{L}^p(X, \mu)$ with $p \in [1, \infty]$.

The central result of this section is the following characterization.

Theorem 8.18 (Characterization of Factors via Sub-σ-algebras) *For two measure-preserving systems (X, μ, T) and (Y, ν, S) the following assertions are equivalent.*

(i) *(Y, ν, S) is a Markov factor of (X, μ, T).*
(ii) *There exists a T-invariant sub-σ-algebra $\mathcal{X}'$ of $\mathcal{X}$ such that $(Y, \mathcal{Y}, \nu, S)$ and $(X, \mathcal{X}', \mu, T)$ are Markov isomorphic.*

The analogous result holds for point factors and point isomorphism.

Proof The implication (ii)⇒(i) is trivial.
(i)⇒(ii): Let $J : \mathrm{L}^1(Y, \nu) \to \mathrm{L}^1(X, \mu)$ be a Markov homomorphism. Then $E = \mathrm{ran}(J)$ is a closed, conjugation invariant sublattice of $\mathrm{L}^1(X, \mu)$ by Propositions 8.11 and 8.14. Since $JS = TJ$, we obtain that E is invariant under T. By Proposition 8.16 there is a sub-σ-algebra $\mathcal{X}'$ of $\mathcal{X}$ such that $E = \mathrm{L}^1(X, \mathcal{X}', \mu)$. Since E is T-invariant and $T\mathbf{1}_A = \mathbf{1}_{T^{-1}A}$ for each $A \in \mathcal{X}'$, we conclude that $\mathcal{X}'$ is T-invariant. Finally, $J : \mathrm{L}^1(Y, \mathcal{Y}, \nu) \to \mathrm{L}^1(X, \mathcal{X}', \mu)$ is a Markov isomorphism. The last assertion can be proved analogously. ∎

The following characterization of (Markov) factors via subalgebras of $\mathrm{L}^\infty(X, \mu)$ will be essential.

Proposition 8.19 (Characterization of Factors via Subalgebras) *Let (X, μ, T) be a measure-preserving system, and let $F \subset \mathrm{L}^\infty(X, \mu)$ be a conjugation invariant, T-invariant subalgebra with $\mathbf{1} \in F$. Then there is a T-invariant sub-σ-algebra $\mathcal{X}'$ of $\mathcal{X}$ such that $(X, \mathcal{X}', \mu, T)$ is a Markov factor of $(X, \mathcal{X}, \mu, T)$ such that F is dense in $\mathrm{L}^1(X, \mathcal{X}', \mu)$.*

Proof The L^1-closure E of F is a closed, conjugation invariant sublattice by Proposition 8.16(b), and there is a T-invariant sub-σ-algebra $\mathcal{X}'$ such that $E = \mathrm{L}^1(X, \mathcal{X}', \mu)$. Since F is T-invariant, so is E, and hence the identity $\mathrm{id} : \mathrm{L}^1(X, \mathcal{X}', \mu) \to \mathrm{L}^1(X, \mathcal{X}, \mu)$ provides a Markov homomorphism from $(X, \mathcal{X}', \mu, T)$ to $(X, \mathcal{X}, \mu, T)$. ∎

Remark 8.20 Of course, every Markov factor of $(X, \mathcal{X}, \mu, T)$ comes from the subalgebra $\mathrm{L}^\infty(X, \mathcal{X}', \mu)$ of $\mathrm{L}^\infty(X, \mu)$ for some T-invariant sub-σ-algebra $\mathcal{X}'$ of $\mathcal{X}$ by Theorem 8.18. This shows that every Markov factor arises in the way described in Proposition 8.19.

Remark 8.21 For a factor (Y, ν, S) of a measure-preserving system (X, μ, T) with the Markov embedding $J : \mathrm{L}^1(Y, \nu) \to \mathrm{L}^1(X, \mu)$ one has that $J\mathrm{L}^\infty(Y, \nu)$ is a conjugation invariant, T-invariant subalgebra of $\mathrm{L}^\infty(X, \mu)$ containing $\mathbf{1}$. On the other hand, as we saw in Proposition 8.19, every conjugation invariant, T-invariant subalgebra F with $\mathbf{1} \in F$ comes from a factor, i.e., is of the form $J\mathrm{L}^\infty(Y, \nu)$ for some factor (Y, ν, S) of (X, μ, T) with the corresponding Markov embedding $J : \mathrm{L}^1(Y, \nu) \to \mathrm{L}^1(X, \mu)$. In contrast to sub-$\sigma$-algebras, the factor (Y, ν, S) is not uniquely determined by the subalgebra $F = J\mathrm{L}^\infty(Y, \nu)$, but is unique up to (Markov) isomorphism.

Note, however, that different subalgebras may give rise to the same factor, actually what matters is the L^1-closure of the subalgebra which is a sublattice of $\mathrm{L}^1(X, \mu)$ by Proposition 8.16(b). The reason why one chooses here subalgebras of $\mathrm{L}^\infty(X, \mu)$ instead of sublattices of $\mathrm{L}^1(X, \mu)$ is that in a concrete situation the closedness of a certain family of functions under multiplication is often easer to check as we will see in Sect. 8.3 below.

Finally, one has a one-to-one correspondence between Markov factors and closed, T-invariant, conjugation invariant unital sublattices of L^1. We leave the proof as Exercise 8.8.

Proposition 8.22 (Characterization of Factors via Sublattices) *Let (X, μ, T) be a measure-preserving system with Koopman operator T on $\mathrm{L}^1(X, \mu)$, and let $E \subset \mathrm{L}^1(X, \mu)$ be a closed, T-invariant, conjugation invariant sublattice with $\mathbf{1} \in E$. Then there is a T-invariant sub-σ-algebra $\mathcal{X}'$ of $\mathcal{X}$ such that $(X, \mathcal{X}', \mu, T)$ is a Markov factor of $(X, \mathcal{X}, \mu, T)$ with $E = \mathrm{L}^1(X, \mathcal{X}', \mu)$.*

We now define projections onto a factor.

Definition 8.23 Let $J : (\mathrm{L}^1(Y, \nu), S) \to (\mathrm{L}^1(X, \mu), T)$ be a Markov homomorphism between the measure-preserving systems (Y, ν, S) and (X, μ, T), so (Y, ν, S) is a Markov factor of (X, μ, T). The operator $P := JJ' : \mathrm{L}^1(X, \mu) \to \mathrm{L}^1(X, \mu)$ is called the **projection onto the factor**.

Remark 8.24 In probability theory $P = JJ'$ is also called the conditional expectation operator. However, for us it will be more convenient to define the **conditional expectation operator** (associated with the factor) as $E_Y := J' : \mathrm{L}^1(Y, \nu) \to \mathrm{L}^1(X, \mu)$ to emphasize that E_Y acts between different spaces, see

Sect. 8.6 below. If $(Y, \mathcal{Y}, \nu) = (X, \mathcal{X}', \mu|_{\mathcal{X}'})$ for some sub-σ-algebra and if one *identifies* $\mathrm{L}^1(X, \mathcal{X}', \mu|_{\mathcal{X}'})$ with a sublattice of $\mathrm{L}^1(X, \mathcal{X}, \mu)$ via the identical embedding, then the two terms mean the same: $J' = E_Y = P : \mathrm{L}^1(X, \mathcal{X}, \mu) \to \mathrm{L}^1(X, \mathcal{X}, \mu)$ is the projection onto $\mathrm{L}^1(X, \mathcal{X}', \mu|_{\mathcal{X}'})$. This justifies the terminology used here.

Proposition 8.25 (Projections onto a Factor) *Let* $J : (\mathrm{L}^1(Y, \nu), S) \to (\mathrm{L}^1(X, \mu), T)$ *be a Markov homomorphism, and let* $P = JJ' : \mathrm{L}^1(X, \mu) \to \mathrm{L}^1(X, \mu)$. *Then the following assertions hold.*

(a) *P is a Markov operator.*
(b) *P is a projection.*
(c) $\mathrm{ran}(P) = \mathrm{ran}(J)$.
(d) *P leaves* $\mathrm{L}^2(X, \mu)$ *invariant, and when restricted to* $\mathrm{L}^2(X, \mu)$ *it is the orthogonal projection onto* $J(\mathrm{L}^2(Y, \nu))$.
(e) *For each* $f, g \in \mathrm{L}^\infty(X, \mu)$ *we have* $P(Pf \cdot g) = Pf \cdot Pg$.

Proof (a) P is a product of two Markov operators by Proposition 8.11(h), hence is itself a Markov operator.
(b) We have $P^2 = JJ'JJ' = JIJ' = JJ' = P$ by Proposition 8.14.
(c) The inclusion $\mathrm{ran}(P) \subset \mathrm{ran}(J)$ is clear. Since $PJ = JJ'J = J$ by Proposition 8.14, we obtain the inclusion $\mathrm{ran}(J) \subset \mathrm{ran}(P)$.
(d) By Proposition 8.11(g) the operator P is L^2-contractive and by (b) it is a projection, hence it is an orthogonal projection. Since J maps $\mathrm{L}^2(Y, \nu)$ to $\mathrm{L}^2(X, \mu)$ by Proposition 8.11(g), the assertion easily follows from (c).
(e) For $f, g, h \in \mathrm{L}^\infty(Y, \nu)$ we have $\langle J'(Jf \cdot g), h\rangle = \langle f J'g, h\rangle$ by Proposition 8.14. We thus obtain $J'(Jf \cdot g) = f \cdot J'g$. Now we can write $P(Pf \cdot g) = JJ'(JJ'f \cdot g) = J(J'f \cdot J'g) = JJ'f \cdot JJ'g = Pf \cdot Pg$. ■

Remark 8.26 We call a closed linear subspace Z of $\mathrm{L}^2(X, \mu)$ for a measure-preserving system (X, μ, T) **induced by a factor** if $Z = J\mathrm{L}^2(Y, \nu)$ for some factor (Y, ν, S) of (X, μ, T) with the corresponding Markov embedding $J : \mathrm{L}^1(Y, \nu) \to \mathrm{L}^1(X, \mu)$. (Recall that J as a Markov operator maps $\mathrm{L}^2(Y, \nu)$ to $\mathrm{L}^2(X, \mu)$, see Proposition 8.11(g).) In this case, the orthogonal projection onto Z is (the restriction of) $P_Z = JJ'$ from Proposition 8.25. By Proposition 8.11(g), this projection satisfies

$$\|P_Z f\|_\infty \leq \|f\|_\infty \text{ for all } f \in \mathrm{L}^\infty(X, \mu). \tag{8.2}$$

Moreover, one has

$$P_Z f > 0 \text{ whenever } f > 0, \tag{8.3}$$

where $f > 0$ means $f \geq 0$, $f \neq 0$ as an element of $\mathrm{L}^2(X, \mu)$. Indeed, let $f > 0$. Since P_Z is a positive operator we have $P_Z f \geq 0$. To show $P_Z f \neq 0$ observe

$$\int_X P_Z f \, \mathrm{d}\mu = \langle f, JJ'\mathbf{1} \rangle = \langle f, \mathbf{1} \rangle = \int_X f \, \mathrm{d}\mu. \tag{8.4}$$

Remark 8.27 As in Remark 8.21, a factor is not uniquely determined by the subspace it induces, but it is unique up to an isomorphism. When we define a factor via the subspace or the subalgebra of functions it induces we do not make precise which one we take. The reader can for example think of the factor coming from the corresponding (unique) sub-σ-algebra.

We thus obtain the following diagram.

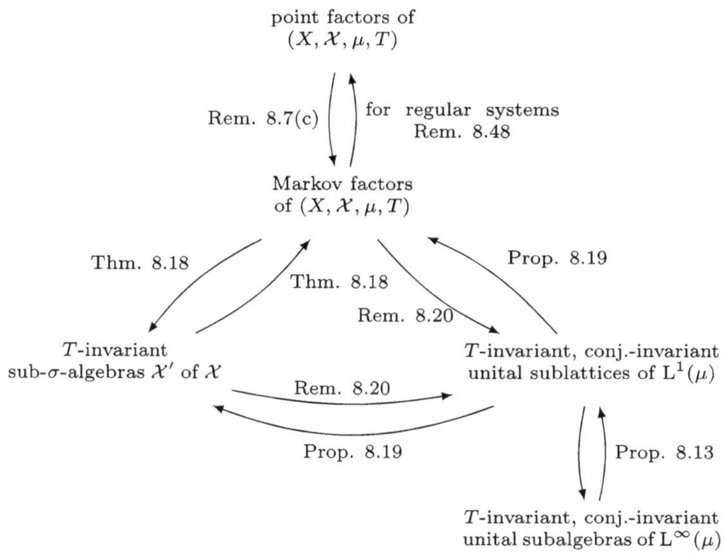

Definition 8.28 (Ordering of Factors) Let (Y, ν_Y, S) and (Z, ν_Z, S_Y) be two factors of a measure-preserving system (X, μ, T) with Markov embeddings J_Y and J_Z, respectively. We say that (Y, ν_Y, S_Y) is **smaller** than (Z, ν_Z, S_Z) if $\mathrm{ran}(J_Y) \subset \mathrm{ran}(J_Z)$. We write $(Y, \nu_Y, S) \leq (Z, \nu_Z, S_Y)$.

Remark 8.29

(1) The relation $\leq$ is transitive and reflexive on sets of factors of (X, μ, T), and is preserved under Markov isomorphisms. It is antisymmetric on the set of (Markov) isomorphism classes.
(2) The relation between two factors can be equivalently described by the inclusion of the corresponding sub-σ-algebras, by the inclusion of the corresponding unital subalgebras, or by the inclusion of the Hilbert spaces $J_Y \mathrm{L}^2(Y, \nu_Y) \subset J_Z \mathrm{L}^2(Z, \nu_Z)$.

Ordering of factors will play an essential role in the Furstenberg–Zimmer Structure Theorem 13.50.

8.3 Examples

We will now use the above characterizations to discuss the important factors mentioned at the beginning. We start with the two extremal cases.

Example 8.30 (The Whole System and the Trivial Factor) Clearly, every measure-preserving system (X, μ, T) is a factor of itself, inducing the whole space $\mathrm{L}^2(X, \mu)$. Moreover, every measure-preserving system has the trivial one-point system $(\{0\}, \delta_0, \mathrm{id})$ as a factor, which we call the **trivial factor** (being the smallest among the factors). The subspace induced by the trivial factor is $\mathbb{C}\mathbf{1}$, the subspace containing the constants only.

We now present classical factors coming from eigenfunctions of the Koopman operator.

Example 8.31 (Factors Arising from Eigenfunctions) For a subgroup $G \subset \mathbb{T}$ the set

$$F_G := \mathrm{lin}\{f \in \mathrm{L}^\infty(X, \mu) : Tf = \lambda f \text{ for some } \lambda \in G\}$$

is a T-invariant, conjugation invariant subalgebra of $\mathrm{L}^\infty(X, \mu)$ with $\mathbf{1} \in F_G$, hence gives rise to a factor of (X, μ, T) by Proposition 8.19. The subspace of $\mathrm{L}^2(X, \mu)$ induced by this factor is

$$\overline{\mathrm{lin}}\{f \in \mathrm{L}^2(X, \mu) : Tf = \lambda f \text{ for some } \lambda \in G\}.$$

The following already familiar factors are precisely of this type.

(a) The **fixed factor** $\mathrm{fix}(X, \mu, T)$ with $G = \{1\}$. The subspace of $\mathrm{L}^2(X, \mu)$ induced by this factor is $\mathrm{fix}(T) := \{f \in \mathrm{L}^2(X, \mu) : Tf = f\}$ and the projection onto this factor is the mean ergodic projection $P_{\mathrm{fix}} : \mathrm{L}^1(X, \mu) \to \mathrm{L}^1(X, \mu)$ (with range $\mathrm{fix}\, T$). The system is ergodic if and only if $\mathrm{fix}\, T$ contains constants only, in which case the fixed factor can be assumed to be trivial, i.e., the one-point system. In this case the corresponding T-invariant sub-σ-algebra $\mathcal{X}_{\mathrm{fix}}$ contains only sets of measure 1 and 0.

(b) The **rational spectrum factor** (or component) $\mathrm{rat}(X, \mu, T)$ with the group $G = \{a \in \mathbb{T} : a \text{ is rational}\}$. The L^2-closure of F_G, being the subspace induced by this factor, is the rational spectrum component H_{rat} of the Koopman operator on $H = \mathrm{L}^2(X, \mu)$ (see Sect. 6.2).

(c) The **Kronecker factor** $\mathrm{kr}(X, \mu, T)$ with $G = \mathbb{T}$. In this case, the Kronecker part H_{kr} associated to the Koopman operator T (see Sect. 6.4) is the closure of $F_{\mathbb{T}}$ in $H = \mathrm{L}^2(X, \mu)$ and the subspace induced by the Kronecker factor.

We now consider the extremal case when the Kronecker factor coincides with the whole system, i.e., when $(X, \mu, T) = \mathrm{kr}(X, \mu, T)$. Recall that such systems are said to have **discrete spectrum**, see Remark 6.34 and Sect. 7.5. (The other extremal case when the Kronecker factor is trivial will be discussed in Theorem 8.39 below.)

Theorem 8.32 (Halmos–von Neumann)

(a) *A compact Abelian group rotation system has discrete spectrum.*
(b) *An ergodic measure-preserving system with discrete spectrum is* (Markov) isomorphic *to an ergodic (necessarily Abelian), compact group rotation system.*

An effective way to prove this is to consider topological models as described below in Sect. 8.5 and prove the analogous statement first for topological systems. The proof can be found, e.g., in [336, Sec. 6.4] or [342, Ch. 17, Thm. 17.11], while the original paper, for the metrizable case, is due to Halmos and von Neumann [517]. Recently, Edeko [330, Thm. 3.10] showed that every, not necessarily ergodic, system with discrete spectrum is Markov isomorphic to a compact group rotation bundle, cf. also Kwiatkowski [701]. For an extension to the case of general group actions we refer to Kreidler and Hermle [529].

Next, we present factors of a similar but more general type due to Abramov [8].

Example 8.33 (Abramov Factors) Let (X, μ, T) be an ergodic measure-preserving system. We first define generalized eigenfunctions inductively. We call eigenfunctions of the Koopman operator T with modulus 1 generalized eigenfunction of order 1. We further call $f \in \mathrm{L}^\infty(X, \mu)$ a **generalized eigenfunction of order** k for $k \geq 2$ if $|f| = 1$ a.e. and $Tf = gf$ holds for some generalized eigenfunction g of order $k - 1$. Then for any given $k \in \mathbb{N}$ we define

$$F_k := \mathrm{lin}\{\text{generalized eigenfunctions of } T \text{ of order } \leq k\},$$

which is a T-invariant, conjugation invariant subalgebra of $\mathrm{L}^\infty(X, \mu)$ with $\mathbf{1} \in F_k$ (see Exercise 8.14), hence defines a factor of (X, μ, T) by Proposition 8.19, called the kth **Abramov factor** of (X, μ, T) denoted by $A_k := A_k(X, \mu, T)$. The **Abramov factor** is defined analogously starting with

$$F := \mathrm{lin}\{\text{generalized eigenfunctions of } T \text{ of some order}\}$$

and denoted by $A := A(X, \mu, T)$. Clearly, $A_1(X, \mu, T) = \mathrm{kr}(X, \mu, T)$ by Proposition 3.26.

If (X, μ, T) is ergodic and $A(X, \mu, T) = (X, \mu, T)$ holds, then the system (X, μ, T) has, similarly to the statement of the Halmos–von Neumann theorem, some group-theoretic structure, see Abramov [8] and Haase and Moriakov [509]. Systems satisfying $A(X, \mu, T) = (X, \mu, T)$ are said to have **quasi-discrete spectrum**.

We now turn to more concrete examples.

Example 8.34 (Skew Rotation) For given $a \in \mathbb{T}$ consider the skew rotation system $(\mathbb{T}^2, \mathrm{m}^2, T_a)$, where

$$T_a : \mathbb{T}^2 \to \mathbb{T}^2, \quad (x, y) \mapsto (ax, xy),$$

see Example 2.15. The rotation system $(\mathbb{T}, \mathrm{m}, a)$ is a factor of the skew shift $(\mathbb{T}^2, \mathrm{m}^2, T_a)$ as discussed in Example 8.4. Next we describe the fixed, Kronecker and Abramov factors. Let T denote the corresponding Koopman operator on $\mathrm{L}^2(\mathbb{T}^2)$, and consider the rotation system $(\mathbb{T}, \mathrm{m}, a)$ with Koopman operator L_a on $\mathrm{L}^2(\mathbb{T})$. Write

$$f = \sum_{n \in \mathbb{Z}^2} a_n \mathbf{z}^n$$

as an $\mathrm{L}^2(\mathbb{T}^2)$-convergent series, where $\mathbf{z}^n := \mathbf{z}_1^{n_1}\mathbf{z}_2^{n_2}$, $n = (n_1, n_2) \in \mathbb{Z}^2$ and $\mathbf{z}_1$ and $\mathbf{z}_2$ are given by $\mathbf{z}_1(x, y) := x$ and $\mathbf{z}_2(x, y) := y$. Then we have $T\mathbf{z}^n = (L_a\mathbf{z}_1^{n_1})\mathbf{z}_1^{n_2}\mathbf{z}_2^{n_2}$ and

$$Tf = \sum_{n \in \mathbb{Z}^2} a_n (L_a\mathbf{z}_1^{n_1})\mathbf{z}_1^{n_2}\mathbf{z}_2^{n_2}.$$

If $\lambda f = Tf$ for some $\lambda \in \mathbb{T}$, then

$$\sum_{n \in \mathbb{Z}^2} \lambda a_n \mathbf{z}_1^{n_1}\mathbf{z}_2^{n_2} = \lambda f = Tf = \sum_{n \in \mathbb{Z}^2} a_n (L_a\mathbf{z}_1^{n_1})\mathbf{z}_1^{n_2}\mathbf{z}_2^{n_2}.$$

Hence we must have $a_n a^{n_1}\mathbf{z}_1^{n_1+n_2}\mathbf{z}_2^{n_2} = \lambda a_n \mathbf{z}_1^{n_1}\mathbf{z}_2^{n_2}$ for each $n = (n_1, n_2) \in \mathbb{Z}^2$. For each fixed $n_2 \in \mathbb{Z}$ we obtain that

$$a_{(j,n_2)}a^j = \lambda a_{(j+n_2,n_2)} \text{ for every } j \in \mathbb{Z}.$$

Since $\sum_{j\in\mathbb{Z}} |a_{(j,n_2)}|^2 < \infty$, it follows that $a_{(j,n_2)} = 0$ for every $j \in \mathbb{Z}$ provided $n_2 \neq 0$. We thus conclude that

$$f = \sum_{m \in \mathbb{Z}} a_{(m,0)}\mathbf{z}_1^m.$$

This implies that the Kronecker factor is isomorphic to the rotation factor $(\mathbb{T}, \mathrm{m}, a)$.

Observe that, while the functions $\mathbf{z}_1^m$, $m \in \mathbb{Z}$, are eigenfunctions of T corresponding to the eigenvalues a^m, the monomials $\mathbf{z}_2^m$, $m \in \mathbb{Z}$, are generalized eigenfunctions of order 2, since

$$(T\mathbf{z}_2)(x, y) = \mathbf{z}_2(ax, xy) = \mathbf{z}_1\mathbf{z}_2(x, y).$$

Thus the Abramov factor in the ergodic case coincides with the whole system, i.e., the system has quasi-discrete spectrum.

Example 8.35 (Heisenberg Systems) Consider the set G of upper triangular 3×3 real matrices with ones on the diagonal, i.e.,

$$G = \left\{ \begin{pmatrix} 1 & x & z \\ 0 & 1 & y \\ 0 & 0 & 1 \end{pmatrix} : x, y, z \in \mathbb{R} \right\},$$

which is a group with the matrix multiplication (a subgroup of $\mathrm{GL}_3(\mathbb{R})$), called **Heisenberg group**. For $x, y, z \in \mathbb{R}$ we introduce the abbreviation

$$[x, y, z] := \begin{pmatrix} 1 & x & z \\ 0 & 1 & y \\ 0 & 0 & 1 \end{pmatrix},$$

and identify $[x, y, z] \in G$ with $(x, y, z) \in \mathbb{R}^3$. With the Euclidean topology of $\mathbb{R}^3$ the group G becomes a topological group. We can write

$$[x_0, y_0, z_0] \cdot [x, y, z] = [x + x_0, y + y_0, z + z_0 + x_0 y].$$

Consider the subgroup $H := \{[x, y, z] : x, y, z \in \mathbb{Z}\}$. Then H is closed and discrete in G. The set G/H of *left cosets* becomes a Hausdorff topological space if endowed with the quotient topology under the quotient map

$$q(g) := gH, \quad G \to G/H,$$

i.e., $A \subset G/H$ is open if and only if $q^{-1}(A)$ is open in G (so that q is continuous). The subset $A = \{[x, y, z] : x, y, z \in [0, 1)\}$ of G is a complete set of representatives (a so-called **fundamental domain**), i.e., contains from every left coset precisely one element. So $G = AH = \bigcup_{h \in H} Ah$ as a union of pairwise disjoint sets. Since $\overline{A}$ is compact in G, q is continuous and $q(\overline{A}) = G/H$, this latter set is compact. For a given $a = [\alpha, \beta, \gamma] \in G$ the left rotation by a on G/H is defined by $gH \mapsto agH$. In this way we obtain an invertible topological system, denoted by $(\mathbb{H}, a) = (\mathbb{H}, [\alpha, \beta, \gamma])$ and called the **Heisenberg system**. Systems of this kind will be discussed in some detail in Sect. 11.4.

Next we present a probability measure on $\mathbb{H}$ which is invariant under each left rotation by an element $g \in G$. After identifying G with $\mathbb{R}^3$ as above we can consider the three-dimensional Lebesgue measure λ on G, and multiplication by g clearly preserves λ. Let $A, B \subset \mathbb{R}^3$ be two Borel measurable complete sets of representatives of the left cosets in G/H. Note that $M \in \mathcal{B}(G/H)$ if and only if $q^{-1}(M)$ is Borel in G. We prove that for each Borel set $M \subset G/H$ we have $\lambda\bigl(A \cap q^{-1}(M)\bigr) = \lambda\bigl(B \cap q^{-1}(M)\bigr)$. Since $q^{-1}(M)h = q^{-1}(M)$ for every $h \in H$ and since λ is invariant under the left rotations by $g \in G$, we have the following chain of equalities

$$\begin{aligned}\lambda\big(A \cap q^{-1}(M)\big) &= \lambda\Big(A \cap q^{-1}(M) \cap \bigcup_{h\in H} Bh\Big) = \sum_{h\in H} \lambda\big(A \cap q^{-1}(M) \cap Bh\big)\\ &= \sum_{h\in H} \lambda\big(Ah^{-1} \cap q^{-1}(M)h^{-1} \cap B\big)\\ &= \sum_{h\in H} \lambda\big(Ah^{-1} \cap q^{-1}(M) \cap B\big)\\ &= \lambda\Big(\bigcup_{h\in H} Ah^{-1} \cap q^{-1}(M) \cap B\Big) = \lambda\big(q^{-1}(M) \cap B\big).\end{aligned}$$

For a Borel set $M \subset G/H$ define

$$\mathrm{m}_{\mathbb{H}}(M) := \lambda\big(A \cap q^{-1}(M)\big)$$

(note that, by the above, it is immaterial which complete, Borel set of representatives one chooses here). Then $\mathrm{m}_{\mathbb{H}}$ is a probability measure on $\mathcal{B}(G/H)$, and we claim that $\mathrm{m}_{\mathbb{H}}$ is invariant under every left rotation by $g \in G$. Indeed, for $g \in G$ the set $g^{-1}A$ is again a measurable complete set of representatives. Hence, by the invariance of λ and by the previously established fact we obtain that

$$\lambda\big(A \cap q^{-1}(gM)\big) = \lambda\big(A \cap gq^{-1}(M)\big) = \lambda\big(g^{-1}A \cap q^{-1}(M)\big) = \lambda\big(A \cap q^{-1}(M)\big).$$

The measure $\mathrm{m}_{\mathbb{H}}$ is also called the **Haar measure** on $\mathbb{H}$. For each $a \in G$, $(\mathbb{H}, \mathrm{m}_{\mathbb{H}}, a)$ is a measure-preserving system, as in the topological setting called a **Heisenberg system**, see also Sect. 11.4.

Example 8.36 (Rotation Factors of Heisenberg Systems) For every $\alpha, \beta, \gamma \in \mathbb{R}$ the two dimensional shift $([0, 1)^2, \lambda^2, (\alpha, \beta))$ is a factor of the Heisenberg system $(\mathbb{H}, \mathrm{m}_{\mathbb{H}}, [\alpha, \beta, \gamma])$ under the point factor map

$$[x, y, z]H \mapsto (x \bmod 1,\ y \bmod 1),$$

see Exercise 8.15.

Proposition 8.37 (The Kronecker Factor of Heisenberg Systems) *For $a = [\alpha, \beta, \gamma]$ with $\alpha = \beta = 0$, the Kronecker factor of $(\mathbb{H}, \mathrm{m}_{\mathbb{H}}, a)$ is the entire system. If $\alpha \neq 0$ or $\beta \neq 0$, then the Kronecker factor is the rotation factor from Example 8.36.*

Proof The center of the Heisenberg group G is

$$Z(G) := \big\{g \in G : gh = hg \text{ for all } h \in G\big\} = \big\{[0, 0, r] : r \in \mathbb{R}\big\},$$

see Exercise 8.15. For $g \in G$ let L_g be the Koopman operator on $\mathrm{L}^2(\mathbb{H})$ of the left rotation by g, and let $T = L_a$. It is easy to see that $F := \bigcap_{g \in Z(G)} \operatorname{fix}(L_g)$ consists precisely of functions $f \in \mathrm{L}^2(\mathbb{H})$ which depend on the first two coordinates only, see Exercise 8.15, i.e., F corresponds to the rotation factor from Example 8.36.
If $\alpha = \beta = 0$, then $a \in Z(G)$ and it is Exercise 8.15 to show that the Kronecker factor coincides with the system itself. In the following suppose that $(\alpha, \beta) \neq 0$. We start with the case $\beta \neq 0$.
For every $g, h \in G$ we have $L_{hg} = L_g L_h$ and $L_g^* = L_g^{-1} = L_{g^{-1}}$. To determine the Kronecker factor let $f \in \mathrm{L}^2(\mathbb{H})$ and $\lambda \in \mathbb{T}$ such that $Tf = \lambda f$. We prove that for every $h \in Z(G)$ one has $L_h f = f$, and this implies that $f \in F$, i.e., f belongs to the rotation factor by what has been said at the beginning of the proof.
For $x \in \mathbb{R}$ we set $g_x := [x, 0, 0] \in G$, and prove by induction that

$$a^n g_x a^{-n} = [x, 0, -n\beta x] \text{ for } n \in \mathbb{N}_0.$$

For a given $r \in \mathbb{R}$ and $n \in \mathbb{N}$ set $r_n := \frac{r}{-\beta n}$. Then $g_{r_n} \to [0, 0, 0]$ and $a^n g_{r_n} a^{-n} \to [0, 0, r]$ as $n \to \infty$. For each $g \in \mathrm{C}(\mathbb{H})$ we obtain

$$L_{g_{r_n}} g - T^{n*} L_{g_{r_n}} T^n g \to g - L_{[0,0,r]} g \text{ as } n \to \infty,$$

and by density this convergence extends to $g \in \mathrm{L}^2(\mathbb{H})$. For f as above we can write

$$\begin{aligned}
(f - L_{[0,0,r]} f \mid f) &= \lim_{n \to \infty} (L_{g_{r_n}} f - T^{n*} L_{g_{r_n}} T^n f \mid f) \\
&= \lim_{n \to \infty} \big((L_{g_{r_n}} f \mid f) - (L_{g_{r_n}} T^n f \mid T^n f) \big) \\
&= \lim_{n \to \infty} \big((L_{g_{r_n}} f \mid f) - |\lambda^n|^2 (L_{g_{r_n}} f \mid f) \big) = 0.
\end{aligned}$$

By Proposition 6.1 the equality $L_{[0,0,r]} f = f$ follows. The case when $\alpha \neq 0$ can be established analogously with the choice $g_x = [0, x, 0]$. Thus the Kronecker factor is a subset of F. For the converse inclusion see Exercise 8.15. ■

Remark 8.38 (Abramov Factors of Heisenberg Systems) Furstenberg showed in [448] that for the Heisenberg system in the ergodic case and for some choice of α, β, γ the Abramov and the Kronecker factors coincide (and are strictly smaller than the system itself).

8.4 The Kronecker Factor and Weak Mixing

In Sect. 6.5 we characterized weak mixing systems in terms of the Kronecker part of the Koopman operator on the Hilbert space L^2. Now that we understand the Kronecker part as a factor of the measure-preserving system, we repeat this characterization using the new terminology.

Theorem 8.39 (Characterization of Weak Mixing in Terms of the Kronecker Factor) *For a measure-preserving system (X, μ, T) with Koopman operator T on $\mathrm{L}^p(X, \mu)$, $p \in [1, \infty]$, the following assertions are equivalent.*

(i) *(X, μ, T) is weakly mixing.*
(ii) *The Kronecker factor is trivial, i.e., H_{kr} contains constant functions only.*
(iii) $\dim \operatorname{fix} T = 1$ *(i.e., (X, μ, T) is ergodic) and the point spectrum of the Koopman operator T satisfies $P\sigma(T) = \{1\}$.*

Proof The Koopman operator T leaves each L^p-space invariant and T is an isometry on each L^p-space, so $P\sigma(T) \subset \mathbb{T}$. Recall further that by Proposition 3.29, the set $P\sigma(T)$ of eigenvalues of T is independent of $p \in [1, \infty]$ and for a given $\lambda \in \mathbb{T}$ the space of L^∞-eigenfunctions of T is dense in the space of L^p-eigenfunctions (for every $p \in [1, \infty]$). It follows that it is enough to consider the case $p = 2$ which was established in Proposition 6.28. ∎

Example 8.40 (Group Rotations are not Weakly Mixing) Since for the torus rotation $(\mathbb{T}^d, \mathrm{m}^d, a)$ the Kronecker factor is the whole system, these systems are not weakly mixing. More generally, compact Abelian group rotation systems (G, m_G, a) with G having at least two elements are never weakly mixing since they have discrete spectrum by Theorem 8.32.

For a further remarkable characterization of weak mixing we need the following.

Remark 8.41 Let (X, μ, T), (Y, ν, S) be a measure-preserving systems. For $f \in \mathrm{L}^2(X, \mu)$ and $g \in \mathrm{L}^2(Y, \nu)$ define $f \otimes g : (x, y) \mapsto f(x)g(y)$. Then $f \otimes g \in \mathrm{L}^2(X \times Y, \mu \times \nu)$ and the linear hull of such functions is dense in $\mathrm{L}^2(X \times Y, \mu \times \nu)$. We have

$$(f \otimes g | u \otimes v)_{\mathrm{L}^2(X\times Y, \mu\times\nu)} = (f|u)_{\mathrm{L}^2(X,\mu)} (g|v)_{\mathrm{L}^2(Y,\nu)}.$$

The Koopman operator of the product system is denoted by $T \otimes S$ and we have $(T \otimes S)(f \otimes g) = Tf \otimes Sg$ for $f \in \mathrm{L}^2(X, \mu)$, $g \in \mathrm{L}^2(Y, \nu)$.

Proposition 8.42 (Product Characterization of Weak Mixing) *For a measure-preserving system (X, μ, T) the following assertions are equivalent.*

(i) *(X, μ, T) is weakly mixing.*
(ii) *$(X \times Y, \mu \times \nu, T \times S)$ is ergodic for every ergodic system (Y, ν, S).*
(iii) *$(X \times X, \mu \times \mu, T \times T)$ is ergodic.*
(iv) *$(X \times Y, \mu \times \nu, T \times S)$ is weakly mixing for every weakly mixing system (Y, ν, S).*
(v) *$(X \times X, \mu \times \mu, T \times T)$ is weakly mixing.*

Proof (i)⇒(ii): Let (Y, ν, S) be an ergodic system and take $f, u \in \mathrm{L}^2(X, \mu)$ and $g, v \in \mathrm{L}^2(Y, \nu)$. Then for the Koopman operator $T \otimes S$ we have

$$((T \otimes S)^n (f \otimes g)|u \otimes v) = (T^n f|u) \cdot (S^n g|v).$$

Since $(T^n f|u) \xrightarrow{\mathrm{D}} (f|\mathbf{1}_X)(\mathbf{1}_X|u)$ by Proposition 6.26, and since we have $\frac{1}{N}\sum_{n=0}^{N-1}(S^n g|v) \to (g|\mathbf{1}_Y)(\mathbf{1}_Y|v)$ by ergodicity and the mean ergodic theorem (Proposition 7.4), we conclude

$$\frac{1}{N}\sum_{n=0}^{N-1}((T\otimes S)^n(f\otimes g)|u\otimes v) \to (f|\mathbf{1}_X)(\mathbf{1}_X|u)(g|\mathbf{1}_Y)(\mathbf{1}_Y|v)$$

$$= (f\otimes g|\mathbf{1}_X\otimes\mathbf{1}_Y)(\mathbf{1}_X\otimes\mathbf{1}_Y|u\otimes v)$$

in the L^2-norm as $N\to\infty$ (cf. Proposition 4.13(e)). By Remark 8.41 a standard approximation argument yields

$$\lim_{N\to\infty}\frac{1}{N}\sum_{n=0}^{N-1}(T\otimes S)^n h = \int_{X\times Y} h\,\mathrm{d}(\mu\times\nu)\cdot\mathbf{1}_{X\times Y}$$

in the L^2-norm for every $h\in\mathrm{L}^2(X\times Y,\mu\times\nu)$, so Proposition 7.4 establishes the ergodicity of the product system.

(ii)⇒(iii): The one-point system $(\{0\},\delta_0,\mathrm{id})$ is ergodic and the product $(X\times\{0\},\mu\times\delta_0,T\times\mathrm{id})$ is isomorphic to (X,μ,T). We obtain from the hypothesis that the latter system is ergodic. Now the validity of (iii) follows also from the hypothesis (ii).

(iii)⇒(i): Let $f,g\in\mathrm{L}^2(X,\mu)$. By ergodicity of the product system and the mean ergodic theorem (see Proposition 7.4) we have

$$\lim_{N\to\infty}\frac{1}{N}\sum_{n=0}^{N-1}|(T^n f|g)|^2 = \lim_{N\to\infty}\frac{1}{N}\sum_{n=0}^{N-1}(T^n f|g)\overline{(T^n f|g)}$$

$$= \lim_{N\to\infty}\frac{1}{N}\sum_{n=0}^{N-1}((T\otimes T)^n(f\otimes\overline{f})|g\otimes\overline{g})$$

$$= (f\otimes\overline{f}|\mathbf{1}\otimes\mathbf{1})(\mathbf{1}\otimes\mathbf{1}|g\otimes\overline{g}) = |(f|\mathbf{1})|^2|(\mathbf{1}|g)|^2$$

in the L^2-norm. Thus (i) follows from Proposition 6.26 and the Koopman–von Neumann lemma (Proposition 4.19).

(ii)⇒(iv): Let (Y,ν,S) be a weakly mixing system, and let (Z,η,R) be an ergodic system. Then by the already proven implication (i)⇒(ii) the system $(Y\times Z,\nu\times\eta,S\times R)$ is ergodic, so by the hypothesis the system $((X\times(Y\times Z),\mu\times(\nu\times\eta),T\times(S\times R))$ is ergodic. Hence $(X\times Y,\mu\times\mu,T\times S)$ is weakly mixing by the already proven implication (ii)⇒(i).

(iv)⇒(i): Since the one-point system $(\{0\},\delta_0,\mathrm{id})$ is weakly mixing, the assertion is clear.

Finally, (iv) clearly implies (v), and the implication (v)⇒(iii) is trivial. ■

Remark 8.43 It follows from Proposition 8.42 that if (X, μ, T) is weakly mixing, then for any $k \in \mathbb{N}$ the kth product $T \times \cdots \times T$ (k times) is ergodic (in fact, weakly mixing). This is *not* true in context of infinite invariant measures, see Kakutani and Parry [614].

Remark 8.44 ((Weak) Mixing via Spectral Measures) Let (X, μ, T) be a measure-preserving system which is ergodic and invertible (this is only for simplifying things a bit), and consider its Koopman operator T on the Hilbert space $H = \mathrm{L}^2(X, \mu)$. The operator T is then unitary, due to the assumption of invertibility. The subspace

$$H_0 := \mathrm{fix}(T)^{\perp} = (\mathbb{C}\mathbf{1})^{\perp} = \Big\{f : \int_X f \,\mathrm{d}\mu = 0\Big\}$$

is closed and T-reducing, and T restricts to a unitary operator on H_0. Now it is easy to see that the system (X, μ, T) is

(a) weakly mixing if and only if for every $f \in H_0$ the scalar spectral measure σ_f is continuous, cf. Sect. 6.4.
(b) strongly mixing if and only if for every $f \in H_0$ the scalar spectral measure σ_f is a Rajchman measure, or in other words if T has Rajchman spectrum on H_0, cf. Sect. 5.6.

Finally, we present a remarkable property of weakly mixing systems discovered by Furstenberg during his study of multiple ergodic averages. A measure-preserving system is called **weakly mixing of order** $k \geq 2$ if for every choice $A_0, \ldots, A_{k-1} \subset X$ of measurable subsets

$$\text{D-}\lim_{n\to\infty} \mu(A_0 \cap T^{-n} A_1 \cap \cdots \cap T^{-(k-1)n} A_{k-1}) = \mu(A_0)\mu(A_1)\cdots\mu(A_{k-1}). \tag{8.5}$$

Clearly, weak mixing of order $k = 2$ is precisely weak mixing, and a kth-order weakly mixing system is weakly mixing of any order m with $m \leq k$, in particular, it is weak mixing. The proof of the following surprising result, see Furstenberg [446, Sec. 4.2] and also Furstenberg and Katznelson [450] and Furstenberg et al. [455], can be based on the previous characterization of weakly mixing systems (Proposition 8.42) and the corresponding multiple ergodic theorem for such systems (Proposition 7.39), we leave this as Exercise 8.17.

Proposition 8.45 (Furstenberg) *Every weakly mixing measure-preserving*[2] *system is weakly mixing of all orders.*

The motivation for the notion of weak mixing of all orders is the following property introduced by Rokhlin [878]: A measure-preserving system is called **(strongly) mixing of order** $k \geq 2$ if for every choice $A_0, \ldots, A_{k-1} \subset X$ of

[2] Recall that for us measure-preserving is synonymous to probability-preserving. The analogous statement to Furstenberg's theorem for infinite measure spaces is false.

measurable subsets

$$\lim_{n_1,\ldots,n_{k-1}\to\infty} \mu(A_0 \cap T^{-n_1}A_1 \cap \cdots \cap T^{-(n_1+\cdots+n_{k-1})}A_{k-1}) \\ = \mu(A_0)\mu(A_1)\cdots\mu(A_{k-1}).$$

Thus, strong mixing of order 2 is just the classical strong mixing. A famous open problem of Rokhlin is whether strong mixing implies strong mixing of all orders. (Even the case $k = 3$ and $(n_1, n_2) = (n, 2n)$ is still open.) A particular case has been settled by Host [546], see also Nadkarni [813, Ch. 10].

Theorem 8.46 (Host) *Let (X, μ, T) be an invertible, regular system. Assume that the maximal spectral type*[3] *of its Koopman operator on $\mathrm{L}^2(X, \mu)$ is singular. If the system is strongly mixing, then it is strongly mixing of all orders.*

For further classes of mixing transformations of all orders and more results see Sinai [948], Blum and Hanson [164], Totoki [1005], Ledrappier [713], Marcus [770], Kalikow [616], Rudolph [891], Schmidt and Ward [920], Ryzhikov [898], Starkov [956], Lemańczyk and Parreau [724, 725], de la Rue [273], Bashtanov [81], Fayad and Kanigowski [382], Gutman et al. [502], Bergelson and Zelada [148] as well as Ryzhikov [900, 901].

Remark 8.47 It is easy to see that a modification of weak mixing of order k in the sense of Rokhlin is false in general already for $k = 3$. Indeed, let (k_n) be a rigidity sequence[4] and let (X, μ, T) be a non-trivial weakly mixing system which is rigid along (k_n) (see Proposition 6.35). Take further a measurable $A \subset X$ with $\mu(A) \in (0, 1)$ and denote $B := X \setminus A$. Since both T^{k_n} and T^{2k_n} converge strongly to the identity operator, we have

$$\lim_{n\to\infty} \mu(T^{-k_n}A \cap T^{-2k_n}B) = \lim_{n\to\infty} (T^{k_n}\mathbf{1}_A | T^{2k_n}\mathbf{1}_B) \\ = (\mathbf{1}_A|\mathbf{1}_B) = \mu(A \cap B) = 0 \neq \mu(A)\mu(B).$$

8.5 Topological Models

The Krylov–Bogolyubov Theorem 2.33 allows to turn a topological dynamical system (K, T), generally in many ways, into a measure-preserving system (K, μ, T), with a T-invariant probability measure μ. Now it is natural to ask which measure-preserving systems arise this way. Here we briefly discuss some possible answers.

We first mention a class of measure spaces in which Markov homomorphisms are Koopman operators of point maps.

---

[3] Recall the definition in Theorem 5.29.

[4] Such as, for example, (2^n), recall Definition 4.21 and the text below it.

Remark 8.48

(a) A measurable space $(X, \mathcal{X})$ is called **standard Borel** if X is a complete, separable metric space and $\mathcal{X}$ is its Borel algebra. Note that there may be many different (completely metrizable, separable) *topologies* on X that have $\mathcal{X}$ as the Borel algebra. A fundamental result in descriptive set theory is that such spaces are, up to a measurable isomorphism, the at most countable discrete spaces or the unit interval $[0, 1]$, i.e., for such measurable spaces there is an invertible, measurable Borel map $\varphi : X \to Z$ with $Z \in \{[0, 1], \mathbb{N}, \{0, \ldots, n-1\} : n \in \mathbb{N}_0\}$ endowed with the Borel algebra, see [638, II. (15.6)] (we also note that inverse of a bijective Borel map between standard Borel spaces is automatically Borel measurable, i.e., a Borel automorphism, this is a result of **Lusin** and **Suslin**, see [638, II. (15.2)]).

(b) If $(X, \mathcal{X})$ and $(Y, \mathcal{Y})$ are standard Borel spaces and $f : X \to Y$ is measurable, then one can find a completely metrizable, separable topology on X which has $\mathcal{X}$ as Borel sets and makes $f : X \to Y$ continuous (where Y bears the initially fixed topology making it a standard Borel space), see [638, II.13.A] or [342, Sec. F.2].

(c) A probability space $(X, \mathcal{X}, \mu)$ is called a **standard Borel probability space** if the underlying measurable space is standard Borel. It is called a **standard Lebesgue space**, or a **standard space** for short, if it is measure-theoretically (i.e., point) isomorphic to a standard Borel space, i.e. there is a standard Borel probability space $(Z, \mathcal{Z}, \nu)$ and an essentially invertible measure-preserving map $\varphi : X \to Z$.

(d) For a standard Borel probability space $(X, \mathcal{X}, \mu)$ with μ being a continuous (i.e., non-atomic) measure there is a Borel measurable bijection $\varphi : X \to [0, 1]$ such that $\varphi_* \mu = \lambda$, the Lebesgue measure, see [638, II.17.F].

(e) A result of von Neumann is that Markov isomorphisms between standard probability spaces (more precisely between the corresponding L^1-spaces) are induced by point maps, see, [1026], [446, Ch. 5.6] (or [342, App. F]). Generalizations, as to under which conditions (Markov or Boolean) homomorphism are induced by point maps, can be found, e.g., in Maharam [764], Sikorski [941].

Next we discuss a method which can turn a measure-preserving system into a topological one with an invariant measure that the so arising measure-reserving systems are Markov isomorphic, and if we start with a standard system the resulting "model" will be compact metric.

Theorem 8.49 (Topological Models) *Let (X, μ, T) be a measure-preserving system.*

(a) *There is a topological system (K, S), i.e., a compact (Hausdorff) space K and a continuous mapping $S : K \to K$, and there is an invariant measure $\nu \in \mathrm{M}_1(K)$ such that the measure-preserving systems (X, μ, T) and (K, ν, S) are Markov isomorphic.*

(b) *If* $\mathrm{L}^1(X, \mu)$ *is separable,*[5] *then* K *in* (a) *can be taken to be metrizable.*

The measure-preserving system (K, ν, S) is called a **topological model** of (X, μ, T). We only indicate the main steps of the proof, for details we refer, e.g., to [342, Ch. 12].

Sketch of Proof (a) Consider the Koopman operator T on $\mathrm{L}^1(X, \mu)$ and take a unital, conjugation invariant and T-invariant subalgebra of $\mathrm{L}^\infty(X, \mu)$ which is dense in $\mathrm{L}^1(X, \mu)$ (for instance $\mathrm{L}^\infty(X, \mu)$ itself is a possible choice). The L^∞-closure A of this subalgebra is still unital, conjugation invariant and T-invariant. Now the Gelfand–Naimark theorem[6] tells that there is a compact (Hausdorff) space K, an isometric isomorphism $\Phi : (A, \|\cdot\|_{\mathrm{L}^\infty}) \to (\mathrm{C}(K), \|\cdot\|_{\mathrm{C}(K)})$, which is unital (i.e., $\Phi(\mathbf{1}_X) = \mathbf{1}_K$), multiplicative, and conjugation-preserving. Since A is $\|\cdot\|_{\mathrm{L}^\infty}$-closed every $f \in A$ with $f \geq 0$ (such as elements $f = |g|^2$, $g \in A$) has a positive square root in A. It follows that for $f \geq 0$, $f \in A$ one has $\Phi(f) = \Phi(f^{1/2}f^{1/2}) = \Phi(f^{1/2})\overline{\Phi(f^{1/2})} = |\Phi(f^{1/2})|^2 \geq 0$, i.e., Φ is a positive operator. The very same argument yields that $\Phi^{-1} : \mathrm{C}(K) \to A$ is a positive operator. By the Riesz–Markov–Kakutani representation Theorem 1.19 we have for the adjoint $\Phi^{-1'} : A' \to \mathrm{M}(K)$. Since the positive functional $\varphi : f \mapsto \langle f, \mathbf{1}\rangle$ is in A' we can set $\nu := \Phi^{-1'}\varphi$, defining a (regular) Borel probability measure on K. Finally, we define the linear operator $S := \Phi T \Phi^{-1}$, which is then unital, multiplicative and conjugation-preserving. Recurring to the Gelfand–Naimark theory again,[7] we see that S is induced by a continuous map on K (which we denote also by S, for consistency reasons). It is easy to check that Φ extends to a Markov isomorphism between $\mathrm{L}^1(X, \mu)$ and $\mathrm{L}^1(K, \nu)$ (use the denseness of A), which intertwines the Koopman operators T and S, i.e., the systems are Markov isomorphic.
(b) If $\mathrm{L}^1(X, \mu)$ is separable, then it has a countable dense subset comprised of essentially bounded functions. It follows that the subalgebra A in (a) can be taken to be separable (in the L^∞-norm). So finally the arising $\mathrm{C}(K)$ space will be separable and it follows that K is metrizable. ∎

The topological model arising from the subalgebra $A = \mathrm{L}^\infty(X, \mu)$ as above is called the **Stone model** of the measure-preserving system (X, μ, T), it is separable only if $\mathrm{L}^\infty(X, \mu)$ is finite dimensional. In general, there may be several different other models. Note that models of invertible systems are invertible. Moreover, if the original system is ergodic/weakly mixing/strongly mixing so is each of its topological models. Thus all these properties are preserved by switching to topological models, a useful feature that will be utilized later. It is a deep theorem of Jewett and Krieger that an (invertible, standard) ergodic measure-preserving system

---

[5] This is equivalent to the fact that the σ-algebra $\mathcal{X}$ is generated by an at most countable family of sets together with the measurable null sets.

[6] The original paper is [464]; a detailed proof with application to topological models can be found, e.g., in [342, Ch. 4].

[7] See, e.g., Theorem 4.13 in [342].

(X, μ, T) has a uniquely ergodic (and minimal) topological model (with strictly positive invariant measure and the underlying compact space being a Cantor set). This was first proved for weakly mixing systems by Jewett in [581], and the general case was then stablished by Krieger in [685], with a couple of other proofs by Bellow and Furstenberg [92], Hansel and Raoult [519], Denker [289]. We note that the Stone model is uniquely ergodic if and only if $\mathrm{L}^\infty(X, \mu)$ is finite dimensional. (This is so because unique ergodicity implies that the corresponding Koopman operator is mean ergodic on $\mathrm{C}(K)$ by Corollary 7.25, so the original Koopman operator T is mean ergodic on $\mathrm{L}^\infty(X, \mu)$; it can be shown that this property implies the finite dimensionality of $\mathrm{L}^\infty(X, \mu)$, see, e.g., [342, Sec.12.5]).

Remark 8.50 (Standard vs. Regular)

(a) Furstenberg [446, Def. 5.5] calls a measure space $(X, \mathcal{X}, \mu)$ **regular** if X is compact metric with $\mathcal{X}$ being the Borel algebra and μ a (regular) Borel probability measure. Evidently, regular probability spaces are standard. Furstenberg calls measure spaces **equivalent** if the corresponding L^1-spaces are Markov isomorphic.
(b) Furstenberg proved that any separable system is equivalent to a regular one, see, e.g., Furstenberg [446, Ch. 5.2, 5.6] or e.g., [342, Thm. 12.2]. In our terminology, this means that such systems have a metrizable topological model (this is the statement of Theorem 8.49).
(c) If $(X, \mathcal{X}, \mu)$ is any probability space then it has a (compact) *topological model* in the above sense and the probability space $(X, \mathcal{X}, \mu)$ has a regular (metric) model if and only if $\mathrm{L}^1(X, \mathcal{X}, \mu)$ is separable. By the above mentioned result of von Neumann (see Remark 8.48(e)) a standard probability space has a regular topological model to which it is point isomorphic.

In many of the arguments in the rest of this book it is enough to restrict attention to orbits $\{T^{n_1} f_1, \dots, T^{n_k} f_k : n_1, \dots, n_k \in \mathbb{N}_0\}$ of finitely many functions, which are then contained in a separable subalgebra, so give rise to a factor that has a regular model. That is why it is also sufficient to prove many of the results for regular systems only, not pursuing the greatest possible generality.

Topological models can be also used to construct invertible extensions.

Proposition 8.51 (Existence of Invertible Extensions)

(a) *Every measure-preserving system is a (Markov) factor of an invertible one.*
(b) *For each measure-preserving system (X, μ, T) there is an invertible extension $(\tilde{X}, \tilde{\mu}, \tilde{T})$ which is minimal in the following sense. If (Z, ν, S) is an invertible extension of (X, μ, T) with Markov homomorphism $J_Z : (\mathrm{L}^1(X, \mu), T) \to (\mathrm{L}^1(Z, \nu), S)$, then J_Z factorizes through the Markov homomorphism $J_{\tilde{X}} : (\mathrm{L}^1(X, \mu), T) \to (\mathrm{L}^1(\tilde{X}, \tilde{\mu}), \tilde{T})$, i.e., there exists a Markov homomorphism*

J such that the following diagram is commutative:

$$\begin{array}{ccc} (\mathrm{L}^1(\tilde{X},\tilde{\mu}),\tilde{T}) & & \\ \uparrow{\scriptstyle J_{\tilde{X}}} & \searrow{\scriptstyle J} & \\ (\mathrm{L}^1(X,\mu),T) & \xrightarrow{J_Z} & (\mathrm{L}^1(Z,\nu),S) \end{array}$$

In particular, $(\tilde{X}, \tilde{\mu}, \tilde{T})$ is a factor of (Z, ν, S). Any two such minimal invertible extensions of (X, μ, T) are isomorphic.

(c) *If (X, μ, T) is regular (standard), then the minimal invertible extension can be chosen to be regular, too.*

(d) *A measure-preserving system is ergodic/weakly mixing if and only if its minimal invertible extension has the corresponding property.*

We note that the invertible point extension, called **natural extension**, constructed in Exercise 2.8 is "the" minimal invertible extension of a measure-preserving system (X, μ, T) with $TX = X$. The general case can be then handled by taking a compact topological model (K, μ_K, T_K) and a surjective subsystem therein living on $K_0 := \bigcap_{n\in\mathbb{N}_0} T_K^n K$, with finally considering the natural extension of this "subsystem". We leave it as Exercise 8.11 to work out the details, and refer to [342, Sec. 13.5] for a comprehensive treatment.

Note further that the Koopman operator on L^2 of an invertible extension of a measure-preserving system is a unitary dilation (in the sense of Theorem 6.13) of the Koopman operator on L^2 of the original system.

8.6 Conditional Expectation and Disintegration of Measures

The notions and results of this section will be heavily used in Chaps. 9, 13 and 15.

Let (X, μ) and (Y, ν) be two probability spaces and let $J : \mathrm{L}^1(Y, \nu) \to \mathrm{L}^1(X, \mu)$ be a Markov homomorphism. For example, J can be induced by a measure-preserving map $\pi : X \to Y$ via $Jf = f \circ \pi$ as a Koopman operator. Consider the Markov adjoint operator $E_Y := J' : \mathrm{L}^1(X, \mu) \to \mathrm{L}^1(Y, \mu)$ which is a Markov operator between these spaces (see Proposition 8.11(h)) and hence has all properties listed in Proposition 8.11. We call E_Y the **conditional expectation operator** and present here some of its further properties needed later, see Remark 8.24 concerning this terminology.

Proposition 8.52 (Properties of Conditional Expectations) *Let $J : \mathrm{L}^1(Y, \nu) \to \mathrm{L}^1(X, \mu)$ be a Markov homomorphism between two probability spaces with corresponding conditional expectation operator $E_Y : \mathrm{L}^1(X, \mu) \to \mathrm{L}^1(Y, \nu)$. Then the following assertions hold.*

(a) *$E_Y(Jf) = f$ holds for every $f \in \mathrm{L}^1(Y, \nu)$. In particular, $E_Y(\mathbf{1}_{\pi^{-1}(A)}) = \mathbf{1}_A$ for every measurable $A \subset Y$ if J is induced by a point factor map π.*

(b) $\int_X f \,\mathrm{d}\mu = \int_Y E_Y f \,\mathrm{d}\nu$ *holds for every* $f \in \mathrm{L}^1(X, \mu)$.
(c) *(Product rule)* $E_Y(Jg \cdot f) = gE_Y f$ *holds for every* $g \in \mathrm{L}^\infty(Y, \nu)$ *and every* $f \in \mathrm{L}^1(X, \mu)$.

Proof (a) follows from $E_Y J = J'J = I$ on $\mathrm{L}^1(Y, \nu)$, see Proposition 8.14.
(b) By $J\mathbf{1} = \mathbf{1}$ we have $\int_Y J'f \,\mathrm{d}\nu = \langle J'f, \mathbf{1}\rangle = \langle f, J\mathbf{1}\rangle = \int_X f \,\mathrm{d}\mu$.
(c) Let $A \subset Y$ be measurable. Since by Proposition 8.14 J is multiplicative on bounded functions,

$$\int_A E_Y(Jg \cdot f)\,\mathrm{d}\nu = \int_Y \mathbf{1}_A E_Y(Jg \cdot f)\,\mathrm{d}\nu = \int_X J\mathbf{1}_A \cdot Jg \cdot f \,\mathrm{d}\mu$$
$$= \int_X J(\mathbf{1}_A g) \cdot f \,\mathrm{d}\mu = \int_Y \mathbf{1}_A g E_Y f \,\mathrm{d}\nu = \int_A g E_Y f \,\mathrm{d}\nu.$$

Since A was arbitrary, $E_Y(Jg \cdot f) = gE_Y f$ follows. ∎

Remark 8.53 The properties (b) and (c) characterize E_Y uniquely, i.e., if $E : \mathrm{L}^1(X, \mu) \to \mathrm{L}^1(Y, \nu)$ is a linear operator satisfying (b) and (c), then $E = E_Y$. Indeed, let $f \in \mathrm{L}^1(X, \mu)$ and consider a ν-measurable set $A \subset Y$. By assumption we have

$$\int_A Ef \,\mathrm{d}\nu = \int_Y \mathbf{1}_A Ef \,\mathrm{d}\nu = \int_Y E(J\mathbf{1}_A f)\,\mathrm{d}\nu = \int_X J\mathbf{1}_A \cdot f \,\mathrm{d}\mu$$
$$= \int_Y \mathbf{1}_A J'f \,\mathrm{d}\nu = \int_A E_Y f \,\mathrm{d}\nu.$$

Since A was arbitrary $Ef = E_Y f$ follows.

Remark 8.54 As in Remark 8.26, the conditional expectation satisfies

$$\|E_Y f\|_\infty \le \|f\|_\infty \text{ for all } f \in \mathrm{L}^\infty(X, \mu), \tag{8.6}$$

and, by Proposition 8.52(b) (or (8.3)),

$$E_Y f > 0 \text{ whenever } f > 0, \tag{8.7}$$

where $f > 0$ again means $f \ge 0$, $f \ne 0$ as an element of $\mathrm{L}^1(X, \mu)$.

Remark 8.55 Let (Y, ν, S) be a factor of a measure-preserving system (X, μ, T) with corresponding Markov embedding J. We show that $S'E_Y T = E_Y$ and

$$E_Y T = SE_Y \text{ if } (Y, \nu, S) \text{ is invertible.} \tag{8.8}$$

Indeed, by the definition of a factor we have $JS = TJ$ and hence $S'E_Y = E_Y T'$. Since T is an isometry, it satisfies $T'T = I$ implying $S'E_Y T = E_Y$. Multiplying by S on the left proves $E_Y T = S E_Y$ in the case when S is invertible, i.e., when $SS' = I$.

Remark 8.56 (Extension to Positive Measurable Functions)

(a) Let $T : \mathrm{L}^1(Y, \nu) \to \mathrm{L}^1(X, \mu)$ be a positive (bounded) linear operator. We extend T to positive measurable functions as follows. Let $g \in \mathrm{L}^0(Y, \nu)$ be positive, i.e., $g \geq 0$. Take an arbitrary sequence (g_N) in $\mathrm{L}^1(Y, \nu)$ of positive functions which converge increasingly to g, for example given by $g_N := g\mathbf{1}_{[g \leq N]}$. Then (Tg_N) is an increasing sequence in $\mathrm{L}^1(X, \mu)$, hence converges pointwise a.e. to a measurable function which we denote by Tg (of course, the limit can be ∞ at some points). It can be shown that the limit is independent of the choice of the approximating sequence, the mapping $g \mapsto Tg$ is hence well-defined, order-preserving and positive homogeneous, see Exercise 8.18.

(a) In Chap. 13 we will need the following stronger version of Proposition 8.52(c):

(c') $E_Y(Jg \cdot f) = gE_Y f$ holds for every $f \in \mathrm{L}^1(X, \mu)$ and every $g \in \mathrm{L}^0(Y, \nu)$ with $g \geq 0$ and $Jg \cdot f \in \mathrm{L}^1(X, \mu)$. Also $E_Y(Jg \cdot f) = gE_Y f$ holds whenever $f, g \geq 0$ are measurable.

Here, Jg is defined as described in (a), i.e., Jg is the pointwise limit of (Jg_N) with $0 \leq g_N \nearrow g$ and $g_N \in \mathrm{L}^1(Y, \nu)$, with $E_Y f$ and $E_Y(Jg \cdot f)$ defined accordingly in the second statement. We first prove the first part of (c'). By (c) in Proposition 8.52 we have

$$E_Y(Jg_N \cdot f) = g_N E_Y f \text{ for all } N \in \mathbb{N}. \tag{8.9}$$

Since E_Y is a contraction w.r.t. the L^1-norm,

$$\|E_Y(Jg_N \cdot f) - E_Y(Jg \cdot f)\|_{\mathrm{L}^1(Y,\nu)} \leq \|Jg_N \cdot f - Jg \cdot f\|_{\mathrm{L}^1(X,\mu)}$$

which converges to 0 as $N \to \infty$ by $|Jg_N \cdot f - Jg \cdot f| \leq Jg \cdot |f| = |Jg \cdot f|$ and by Lebesgue's dominated convergence theorem. Thus the left-hand side of (8.9) converges in $\mathrm{L}^1(Y, \nu)$ as $N \to \infty$ to $E_Y(Jg \cdot f)$. By (8.9) we have $|E_Y(Jg_N \cdot f)| = g_N |E_Y f| \nearrow g|E_Y f|$, thus by the monotone convergence theorem and by what has been said above, we can conclude that $g|E_Y f|$ belongs to $\mathrm{L}^1(Y, \nu)$. Since $|g_N E_Y f| \leq g|E_Y f|$, Lebesgue's dominated convergence theorem implies that the right-hand side of (8.9) converges in $\mathrm{L}^1(Y, \nu)$ to $gE_Y f$, finishing the proof of (c') in the first case. Now suppose that $f, g \geq 0$ are measurable functions. Take $f_N \in \mathrm{L}^\infty(X, \mu)$, $f_N \geq 0$ with $f_N \nearrow f$ and similarly (g_N). Then we have pointwise ν-a.e. $Jg_N \cdot f_N \nearrow Jg \cdot f$, so

$$E_Y(Jg \cdot f) = \lim_{N\to\infty} E_Y(Jg_N \cdot f_N) = \lim_{N\to\infty} Jg_N E_Y f_N = JgE_Y f.$$

We now introduce the disintegration of a measure following Furstenberg [445] and Tao [977].

Theorem 8.57 (Disintegration of Measures) *Let $J : \mathrm{L}^1(Y,\nu) \to \mathrm{L}^1(X,\mu)$ be a Markov homomorphism between two probability spaces with corresponding conditional expectation operator $E_Y : \mathrm{L}^1(X,\mu) \to \mathrm{L}^1(Y,\nu)$ and assume that (X,μ) is compact metric with μ being a (regular) Borel measure. Then there exists a measurable*[8] *map $y \mapsto \mu_y$, $Y \mapsto \mathrm{M}_1(X)$ with the following properties.*

(a) *For every $f \in \mathrm{L}^1(X,\mu)$ one has $f \in \mathrm{L}^1(X,\mu_y)$ for ν-a.e. $y \in Y$ with*

$$(E_Y f)(y) = \int_X f \, \mathrm{d}\mu_y \quad \text{for } \nu\text{-a.e. } y \in Y \tag{8.10}$$

and[9]

$$\int_Y \Big(\int_X f \, \mathrm{d}\mu_y \Big) \mathrm{d}\nu(y) = \int_X f \, \mathrm{d}\mu. \tag{8.11}$$

In particular, each μ-null set is a μ_y-null set for almost every $y \in Y$. This map is essentially unique in the sense that every other map $y \mapsto \mu'_y \in \mathrm{M}_1(X)$ with the same properties satisfies $\mu_y = \mu'_y$ for ν-a.e. $y \in Y$.

(b) *For every $g \in \mathrm{L}^\infty(Y,\nu)$ one has for ν-a.e. $y \in Y$*

$$Jg = g(y) \cdot \mathbf{1} \quad \mu_y\text{-a.e.} \tag{8.12}$$

In particular, if J is induced by a measure-preserving map $\pi : X \to Y$ via $Jf = f \circ \pi$, then for ν-a.e. $y \in Y$

$$\mu_{\pi(x)} = \mu_y \text{ for } \mu_y\text{-a.e. } x \in X. \tag{8.13}$$

(c) *Suppose that (Y,ν,S) is an invertible factor of a measure-preserving system (X,μ,T) and the embedding J is induced by a measure-preserving map $\pi : X \to Y$ via $Jf = f \circ \pi$. Then*

$$T_*\mu_y = \mu_{Sy} \text{ for } \nu\text{-a.e. } y \in Y. \tag{8.14}$$

Proof (a) One can find a countable set of linearly independent functions which contains $\mathbf{1}$ and has dense $(\mathbb{Q}+\mathrm{i}\mathbb{Q})$-linear hull D in $\mathrm{C}(X)$. Since D is countable one can find a $(\mathbb{Q}+\mathrm{i}\mathbb{Q})$-*linear* mapping E from D to Borel measurable functions on Y such that $Eg = E_Y g$ holds ν-a.e. for each $g \in D$. By countability there is a set $A \subset Y$ with full measure such that for each $y \in A$ and each $g \in D$

[8] Here *measurability* means that $y \mapsto \langle f, \mu_y \rangle \in \mathbb{C}$ is measurable for each $f \in \mathrm{C}(X)$.

[9] One often shortens (8.11) as $\mu = \int_Y \mu_y \, \mathrm{d}\nu(y)$.

$$|Eg(y)| \leq \|g\|_\infty.$$

One easily shows that for each $y \in A$ the mapping $L_y : D \to \mathbb{C}$, $g \mapsto Eg(y)$ extends to a continuous $\mathbb{C}$-linear functional on $\mathrm{C}(X)$, still denoted by L_y, with $\|L_y\| \leq 1$ (and also that $y \mapsto L_y g$ is measurable for every $g \in \mathrm{C}(X)$). Since $L_y \mathbf{1} = 1$, we even have $\|L_y\| = 1$. By the Riesz–Markov–Kakutani Theorem 1.19 there exists a Borel measure $\mu_y \in \mathrm{M}(X)$ with

$$L_y g = \int_X g \, \mathrm{d}\mu_y$$

for each $g \in \mathrm{C}(X)$, and $\|\mu_y\| = |\mu_y|(X) = 1$, so $|\mu_y|$ is a probability measure. The continuous $\mathbb{R}$-linear functional Re L_y is represented by the signed measure Re μ_y, and we have $\|\mathrm{Re}\, \mu_y\| = 1$. It follows that Re μ_y is a probability measure, and then from $\|\mathrm{Re}\, \mu_y\| = \|\mu_y\| = \mu_y(X) = 1$ that $\mu_y = \mathrm{Re}\, \mu_y$, so μ_y itself is a probability measure. For $y \notin A$ take $\mu_y = \mu$ (or any other fixed probability measure). For each $f \in D$ and $y \in A$ we have

$$\int_Y \int_X f \, \mathrm{d}\mu_y \, \mathrm{d}\nu(y) = \int_Y Ef \, \mathrm{d}\nu = \int_Y E_Y f \, \mathrm{d}\nu = \int_X f \, \mathrm{d}\mu$$

so (8.10) and (8.11) hold. Now if $f \in \mathrm{C}(X)$ is arbitrary and $f_n \in D$ satisfy $f_n \to f$ in the sup-norm, then $E_Y f_n \to E_Y f$ in $\mathrm{L}^\infty(Y, \nu)$, and we conclude that $E_Y f(y) = \int_X f \, \mathrm{d}\mu_y$ holds for ν-almost every $y \in Y$, and that (8.10) and (8.11) hold as well. Standard approximation argument based on monotone convergence implies that if $f_n \geq 0$ are measurable functions on X satisfying (8.10) and (8.11) and $f_n \leq f_{n+1}$ for each $n \in \mathbb{N}$, then (8.10) and (8.11) hold for $f = \lim_{n\to\infty} f_n$. Consider

$$\mathcal{A} := \{B \subset X : B \text{ Borel s.t. (8.10) and (8.11) hold for } f = \mathbf{1}_B\}.$$

Since bounded measurable functions satisfying (8.10) and (8.11) form a vector space, by the above approximation argument, $\mathcal{A}$ is a σ-algebra. Moreover, by the approximation argument and Urysohn's lemma (or a simple direct construction using the metric) $\mathcal{A}$ contains all open sets and hence all Borel sets. Thus each bounded measurable function f and hence each positive measurable function satisfies (8.10) and (8.11). By linearity we also obtain that each integrable function f satisfies (8.10) and (8.11). It also follows that if B is a μ-null set, then it is also a μ_y-null set for ν-almost every $y \in Y$.

We turn to the proof of uniqueness. Let $y \mapsto \mu'_y$ be another map with the same properties as $y \mapsto \mu_y$. Then by (8.10) for every open set $G \subset X$ there exists $A_G \subset Y$ with full measure such that for every $y \in A_G$

$$\mu'_y(G) = \int_X \mathbf{1}_G \, \mathrm{d}\mu'_y = \int_X \mathbf{1}_G \, \mathrm{d}\mu_y = \mu_y(G).$$

Take a countable basis $\mathcal{B}$ for open sets, set $M := \bigcap_{G\in\mathcal{B}} A_G$ and take $y \in M$. Then by the above $\mu'_y(B) = \mu_y(B)$ holds for every open set B and hence for every Borel set, implying $\mu'_y = \mu_y$.
(b) To show (8.12) take $g \in \mathrm{L}^\infty(Y, \nu)$. By (a) and Proposition 8.52(a) we have $\int_X (Jg)\, \mathrm{d}\mu_y = E_Y(Jg)(y) = g(y)$ for ν-a.e. $y \in Y$. Since by Proposition 8.14 $|Jg|^2 = J(|g|^2)$, this implies

$$\begin{aligned}
&\int_Y \int_X |(Jg)(x) - g(y)|^2 \,\mathrm{d}\mu_y(x)\,\mathrm{d}\nu(y)\\
&\quad= \int_Y \int_X \Big(|Jg|^2(x) - 2\mathrm{Re}\,((Jg)(x)\overline{g(y)}) + |g(y)|^2\Big)\,\mathrm{d}\mu_y(x)\,\mathrm{d}\nu(y)\\
&\quad= \int_Y \int_X \Big(|g(y)|^2 - 2\mathrm{Re}\,(g(y)\overline{g(y)}) + |g(y)|^2\Big)\,\mathrm{d}\mu_y(x)\,\mathrm{d}\nu(y) = 0.
\end{aligned}$$

So $(Jg)(x) = g(y)$ for ν-a.e. $y \in Y$ and μ_y-a.e. $x \in X$.

We now assume that J is induced by a measure-preserving map $\pi : X \to Y$ via $Jf = f \circ \pi$ and prove (8.13). By separability of $\mathrm{C}(X)$ it suffices to show that for each fixed $f \in \mathrm{C}(X)$ there is a set A of full measure with $\int_X f\,\mathrm{d}\mu_{\pi(x)} = \int_X f\,\mathrm{d}\mu_y$ for μ_y-almost every x and every $y \in A$. Take $f \in \mathrm{C}(X)$ and set $g := E_Y f \in \mathrm{L}^\infty(Y, \nu)$. By (8.10) and (8.12) there is a set of full measure $A \subset Y$ such that for every $y \in A$ there is $B_y \subset Y$ of full measure with $(Jg)(x) = g(y) = \int_X f\,\mathrm{d}\mu_y$ for every $x \in B_y$ and for every $y \in A$. On the other hand, there is $B \subset X$ with $\mu(B) = 1$ and $(Jg)(x) = g(\pi(x))$ for every $x \in B$. Moreover, $g(\pi(x)) = \int_X f\,\mathrm{d}\mu_{\pi(x)}$ holds for every $x \in \pi^{-1}(A)$. Since $\pi^{-1}(A) \cap B$ has full measure, there is $C \subset Y$ with full measure such that for every $y \in C$ one has $\mu_y(\pi^{-1}(A) \cap B) = 1$. If $y \in A \cap C$ (this set has full ν-measure), then $\mu_y(\pi^{-1}(A) \cap B \cap B_y) = 1$. Take $x \in \pi^{-1}(A) \cap B \cap B_y$, then

$$\int_X f\,\mathrm{d}\mu_{\pi(x)} = g(\pi(x)) = (Jg)(x) = g(y) = \int_X f\,\mathrm{d}\mu_y.$$

(c) Again by separability of $\mathrm{C}(X)$ it suffices to show that for a fixed $f \in \mathrm{C}(X)$ one has $\int_X f\,\mathrm{d}(T_*\mu_y) = \int_X f\,\mathrm{d}\mu_{Sy}$ for ν-a.e. $y \in Y$. This, however, follows from (8.10) and (8.8) by

$$\begin{aligned}
\int_X f\,\mathrm{d}(T_*\mu_y) &= \int_X Tf\,\mathrm{d}\mu_y = (E_Y Tf)(y) = (SE_Y f)(y)\\
&= (E_Y f)(Sy) = \int_X f\,\mathrm{d}\mu_{Sy}
\end{aligned}$$

for ν-a.e. $y \in Y$. ■

Example 8.58 (Product Measure) When $X = Y \times Z$, $\mu = \nu \times \nu_Z$ and $J : \mathrm{L}^1(Y, \nu) \to \mathrm{L}^1(X, \mu)$, $Jf = f \otimes \mathbf{1}$, then the disintegration measures satisfy $\mu_y(B) = \nu_Z(B_y)$, $B \subset X$ measurable with $B_y = \{z \in Z : (y, z) \in B\}$, and one rediscovers Fubini's theorem, see Exercise 8.19.

Example 8.59 (Disc and Torus) Consider the unit disc $\overline{\mathbb{D}} := \{z \in \mathbb{C} : |z| \le 1\}$ with the two-dimensional Lebesgue measure λ^2 normalized such that $\lambda^2(\overline{\mathbb{D}}) = 1$, the unit interval $[0, 1]$ with the one-dimensional Lebesgue measure λ^1 and the torus $\mathbb{T}$ with the Haar measure m. Define the linear operator

$$J : \mathrm{L}^1([0, 1], \lambda^1) \to \mathrm{L}^1(\overline{\mathbb{D}}, \lambda^2), \quad (Jf)(z) := f(|z|),$$

which is easily seen to be a Markov embedding between these spaces. For the corresponding disintegration measures m_t, $t \in (0, 1]$, we have $\mathrm{m}_t(B) = \mathrm{m}(\mathbb{T} \cap t^{-1}B)$ for Borel sets $B \subset \overline{\mathbb{D}}$. Similarly, one can define the Markov embedding

$$\tilde{J} : \mathrm{L}^1(\mathbb{T}, \mathrm{m}) \to \mathrm{L}^1(\overline{\mathbb{D}}, \lambda^2), \quad (\tilde{J}f)(z) := f\left(\tfrac{z}{|z|}\right).$$

Here the corresponding disintegration measures λ_z^2, $z \in \mathbb{T}$ are given by $\lambda_z^2(B) = \lambda^2(B \cap [0, z])$, for B Borel, where $[0, z]$ denotes the line segment connecting 0 and z. We leave the details as Exercise 8.20.

Remark 8.60 Let (Y, ν, S) be a factor of a regular measure-preserving system (X, μ, T) with the Markov embedding $J : \mathrm{L}^1(Y, \nu) \to \mathrm{L}^1(X, \mu)$ given by $Jf = f \circ \pi$ for some measure-preserving point map $\pi : X \to Y$ and suppose that (Y, ν) is regular as well (or at least point isomorphic to such a system). We show that for ν-a.e. $y \in Y$ the measure μ_y is supported in the y-fiber $X_y := \pi^{-1}(\{y\})$ in the sense $\mu_y(X_y) = 1$. Indeed, let (A_n) be an at most countable set of subsets of Y generating (together with the measurable null sets) the σ-algebra in Y. Denote $f_n := J\mathbf{1}_{A_n} = \mathbf{1}_{\pi^{-1}(A_n)}$ for every n. Take M to be the set of full measure in Y such that (8.10) holds for every function f_n and let $y \in M$. We show that $\mu_y(X_y) = 1$. Let (n_j) be a subsequence of $\mathbb{N}$ such that $\{y\} = \bigcap_j A_{n_j}$. It suffices to show $\mu_y(\pi^{-1}(A_{n_j})) = 1$ for every j. But the definition of M implies by Proposition 8.52(a) that for every j

$$\int_X f_{n_j} \, \mathrm{d}\mu_y = (E_Y f_{n_j})(y) = \mathbf{1}_{A_{n_j}}(y) = 1$$

finishing the argument.

Disintegrating with respect to the fixed factor is an important special case and has its own name.

Proposition 8.61 (Ergodic Decomposition) *Let (X, μ, T) be a regular measure-preserving system and let (Y, ν, id) be its fixed factor. Then the disintegration* *measures μ_y in Theorem 8.57 are ergodic T-invariant probability measures on X for ν-a.e. $y \in Y$.*

Proof The T-invariance, i.e., $T_*\mu_y = \mu_y$ for almost every y, follows from Theorem 8.57(c) since $S = \mathrm{id}$ is, of course, invertible.

To show ergodicity, let f be a bounded measurable function on X. By Birkhoff's pointwise ergodic theorem and the standard approximation argument it is enough to show that for ν-a.e. $y \in Y$ the ergodic averages $\frac{1}{N}\sum_{n=1}^{N} T^n f$ converge pointwise μ_y-a.e. to $\int_X f \, \mathrm{d}\mu_y$. We know that by Birkhoff's and von Neumann's ergodic theorems these averages converge pointwise to $P_{\mathrm{fix}} f$ outside a set of μ-measure zero. This set has μ_y-measure zero for ν-a.e. $y \in Y$ by Theorem 8.57. So it remains to show that for ν-a.e. $y \in Y$ one has $P_{\mathrm{fix}} f = \int_X f \, \mathrm{d}\mu_y$ μ_y-a.e. But this is true since by (8.12) and Theorem 8.57 for ν-a.e. $y \in Y$

$$P_{\mathrm{fix}} f = J E_Y f = (E_Y f)(y) = \int_X f \, \mathrm{d}\mu_y \quad \mu_y\text{-a.e.},$$

where the first equality holds by Proposition 8.25(d). ■

Exercises

Exercise 8.1 (Markov vs. Point Factors/Isomorphic Systems) Present examples of Markov factors which are not point factors and Markov isomorphic systems which are not point isomorphic. (Hint: Take finite systems with appropriate σ-algebras of measurable sets.)

Exercise 8.2 (Preservation of Ergodicity, Unique Ergodicity, Transitivity and Minimality) Show the following assertions.

(a) Markov (and point) factors of ergodic/weakly mixing/strongly mixing systems are ergodic/weakly mixing/strongly mixing.
(b) Factors of uniquely ergodic topological systems are uniquely ergodic, and the same holds for transitivity and minimality.

Exercise 8.3 (Positive Operators) Prove (a)–(f) of Proposition 8.11.

Exercise 8.4 (Jensen's Inequality) Let (X, μ), (Y, ν) be probability spaces, let $S : \mathrm{L}^1(Y, \nu) \to \mathrm{L}^1(X, \mu)$ be a Markov operator, and let $g : [0, \infty) \to \mathbb{R}$ be a convex function. Prove that for every positive $f \in \mathrm{L}^1(Y, \nu)$ such that $g \circ f \in \mathrm{L}^1(Y, \nu)$ we have $g \circ Sf \le S(g \circ f)$.

Exercise 8.5 (L^p-contractivity of Markov Operators) Use the result of the foregoing exercise to prove that a Markov operator $S : \mathrm{L}^1(Y, \nu) \to \mathrm{L}^1(X, \mu)$ respects the L^p-spaces and is contractive for the L^p-norms for $p \in [1, \infty]$.

Exercise 8.6 (Markov Adjoint) Let $S : \mathrm{L}^1(Y, \nu) \to \mathrm{L}^1(X, \mu)$ and $R : \mathrm{L}^1(X, \mu) \to \mathrm{L}^1(Z, \lambda)$ be Markov operators. Show the following assertions.

(a) $S'' = S$.
(b) $S'|_{\mathrm{L}^2(X,\mu)} = (S|_{\mathrm{L}^2(Y,\nu)})^*$.
(c) RS is a Markov operator.
(d) $(RS)' = S'R'$.

Exercise 8.7 (Transition Probabilities) Let $(X, \mathcal{X})$ and $(Y, \mathcal{Y})$ be measurable spaces. A mapping $P : Y \times \mathcal{X} \to \mathbb{R}$ is called **Markov transition probability kernel** (or **Markov kernel** for short) if it satisfies the following:

(1) For each $y \in Y$ the mapping $A \mapsto P(y, A)$ is a probability measure on $(X, \mathcal{X})$.
(2) For each $A \in \mathcal{X}$ the mapping $y \mapsto P(y, A)$ is measurable on $(Y, \mathcal{Y})$.

Suppose that $(X, \mathcal{X}, \mu)$ and $(Y, \mathcal{Y}, \nu)$ are probability spaces and let P be a Markov kernel. Prove the following assertions.

(a) If, for $f \in \mathrm{L}^1(Y, \nu)$ and $A \in \mathcal{X}$, we set

$$(\widetilde{S}f)(A) := \int_Y P(\cdot, A) f \,\mathrm{d}\nu,$$

then $\widetilde{S}f$ defines a finite (complex) measure on $(X, \mathcal{X})$. The linear operator $\widetilde{S} : \mathrm{L}^1(Y, \nu) \to \mathfrak{M}(X, \mathcal{X})$ is continuous.

(b) Suppose that the Markov kernel P satisfies

$$P(y, \cdot) \ll \mu \text{ for } \nu\text{-almost every } y \in Y.$$

Then for $f \in \mathrm{L}^1(Y, \nu)$ the measure $\widetilde{S}f$ is absolutely continuous with respect to μ. The Radon–Nikodym derivative

$$Sf := \frac{\mathrm{d}\widetilde{S}f}{\mathrm{d}\mu}$$

defines a positive operator $S : \mathrm{L}^1(Y, \nu) \to \mathrm{L}^1(X, \mu)$ for which $S'\mathbf{1} = \mathbf{1}$. If $\widetilde{S}\mathbf{1}_Y = \mu$, then S is a Markov operator.

Exercise 8.8 (Characterization of Factors via Sublattices) Prove Proposition 8.22.

Exercise 8.9 (Projection Onto a Factor) For a given Markov homomorphism $J : (\mathrm{L}^1(Y, \nu), S) \to (\mathrm{L}^1(X, \mu), T)$ and for $P := JJ'$ prove the following assertions.

(a) $P'' = P$.
(b) $PTP = TP$.
(c) If $SJ' = J'T$, then $PT = TP$.

Exercise 8.10 (The Compact Factor) Let (X, μ, T) be a measure-preserving system with Koopman operator T. A function $f \in \mathrm{L}^2(X, \mu)$ is called **compact** (or **asymptotically almost periodic** with respect to T) if the orbit

$$\{T^n f : n \in \mathbb{N}_0\}$$

is relatively compact in $\mathrm{L}^2(X, \mu)$, see also the definition of asymptotical almost periodicity in Sect. 6.4. Prove that

$$E := \{f \in \mathrm{L}^2(X, \mu) : f \text{ is a compact function}\}$$

is a closed, T-invariant and conjugation invariant sublattice in $\mathrm{L}^2(X, \mu)$ with $\mathbf{1} \in E$, and the set $F := \mathrm{L}^\infty(X, \mu) \cap E$ is a closed subalgebra yielding the Kronecker factor (cf. Exercise 6.11).

Exercise 8.11 (Invertible Extensions) Prove Proposition 8.51.

Exercise 8.12 (Weakly Mixing Functions) Let (X, μ, T) be a measure-preserving system and denote by $W := H_{\mathrm{aws}}$ the set of all weakly mixing functions. Show that

$$W \cap \mathrm{L}^\infty(X, \mu) \text{ is dense in } W.$$

(Hint: Use the Jacobs–de Leeuw–Glicksberg decomposition (6.10) and Example 8.31(c)).)

Exercise 8.13 (Rigid Functions Build a Factor) Let (X, μ, T) be a measure-preserving system. Recall that a function $f \in \mathrm{L}^2(X, \mu)$ is called **rigid along** $(n_k)_{k\in\mathbb{N}}$ if $\lim_{k\to\infty} \|T^{n_k} f - f\|_2 = 0$. Show that rigid functions along a fixed subsequence $(n_k)_{k\in\mathbb{N}}$ of $\mathbb{N}$ induce a factor. Describe the corresponding sub-σ-algebra of sets.

Exercise 8.14 (Abramov Factors) Let (X, μ, T) be a measure-preserving system. Prove the following assertion for the sets A and A_k from Example 8.33.

(a) The set A and for each $k \in \mathbb{N}$ the sets A_k are closed, conjugation invariant and T-invariant subalgebras of $\mathrm{L}^\infty(X, \mu)$.
(b) $A_k \subset A_{k+1} \subset A$ for each $k \in \mathbb{N}$.

Exercise 8.15 (Heisenberg Systems) For $\alpha, \beta, \gamma \in \mathbb{R}$ consider the Heisenberg system $(\mathbb{H}, \mathrm{m}_{\mathbb{H}}, a)$ with $a := [\alpha, \beta, \gamma]$.

(a) Prove that as in Example 8.36

$$[x, y, z]\Pi \mapsto (x \bmod 1, y \bmod 1)$$

is a point factor map onto the two-dimensional shift $([0, 1)^2, \lambda^2, (\alpha, \beta))$.
(b) Determine the center of the Heisenberg group.

(c) Show that for $f \in \mathrm{L}^2(\mathbb{H})$ one has $f \in \bigcap_{g \in Z(G)} \operatorname{fix}(L_g)$ if and only if f depends on the first two coordinates only. Show further that every such f belongs to the Kronecker factor of the Heisenberg system.

(d) Prove that in the case $\alpha = \beta = 0$, the system $(\mathbb{H}, \mu, a)$ coincides with its Kronecker factor, i.e., has discrete spectrum. (Hint: use Exercise 8.10.)

Exercise 8.16 (Affine Endomorphisms on $\mathbb{T}$) For $a \in \mathbb{T}$ and $k \in \mathbb{N}$ define $T : \mathbb{T} \to \mathbb{T}$ by $Tx = ax^k$. Prove the following assertions.

(a) $(\mathbb{T}, \mathrm{m}, T)$ is a measure-preserving system.
(b) $(\mathbb{T}, \mathrm{m}, T)$ is ergodic if and only if $k \geq 2$ or a is irrational.
(c) For $k \geq 2$ the Kronecker factor of $(\mathbb{T}, \mathrm{m}, T)$ is trivial, i.e., consists of constants.

Exercise 8.17 (Weak Mixing of All Orders) Prove that every weakly mixing system (X, μ, T) is weakly mixing of all orders. (Hint: Use Proposition 7.39 for the product system $(X \times X, \mu \times \mu, T \times T)$.)

Exercise 8.18 (Extension of Positive Operators to Positive Measurable Functions) Let $T : \mathrm{L}^1(Y, \nu) \to \mathrm{L}^1(X, \mu)$ be a positive bounded linear operator.

(a) For $g \in \mathrm{L}^0(Y, \nu)$, $g \geq 0$ take a sequence of positive functions $g_N \in \mathrm{L}^1(Y, \nu)$ with $g_N \nearrow g$ pointwise almost everywhere. Prove that the limit $\lim_{N\to\infty} Tg_N$ exists almost everywhere, and defines (a possibly $[0, \infty]$-valued) measurable function, the limit is independent of the choice of the sequence (g_N). We denote it by Tg.

(b) Prove that $g \mapsto Tg$, defined on positive measurable functions, extends T and is positive homogeneous, additive, and order-preserving.

Exercise 8.19 (Conditional Expectation and Disintegration) Determine the conditional expectation operator and the disintegration measures μ_y defined in Theorem 8.57 in the product setting of Example 8.58.

Exercise 8.20 (Disintegration over the Disc) Work out the details of Example 8.59.

Chapter 9
First Applications of Ergodic Theorems

In this chapter we present some applications of ergodic theorems and related techniques in different areas of mathematics.

9.1 Applications in Probability

Let $(\Omega, \mathbb{P})$ be a probability space and for $j \in \mathbb{N}$ let $f_j : \Omega \to \mathbb{R}$ be measurable functions, i.e., random variables. Recall that the random variables f_j are called **independent** if for every $m \in \mathbb{N}$ and all Borel sets $B_1, \dots, B_m \subset \mathbb{R}$ one has

$$\mathbb{P}\Big(\bigcap_{j=1}^{m} f_{n_j}^{-1} B_j\Big) = \prod_{j=1}^{m} \mathbb{P}(f_{n_j}^{-1} B_j)$$

for every choice of different $n_1, \dots, n_m \in \mathbb{N}$. This will be assumed in the following. Suppose further that the random variables are **identically distributed**, i.e., that $f_{j*}\mathbb{P} = f_{i*}\mathbb{P}$ holds for each $i, j \in \mathbb{N}$ (recall that $f_{j*}\mathbb{P}$ is the push-forward measure defined by $f_{j*}\mathbb{P}(B) = \mathbb{P}(f_j^{-1} B))$. This provides a special case of stationary stochastic processes from Example 2.22. We set $\nu := f_{1*}\mathbb{P}$, the common distribution of the random variables.

Consider the one-sided shift system with state space $(\mathbb{R}, \nu)$, i.e., the measure-preserving system $(X, \mathcal{X}, \mu, T)$ with $X = \mathbb{R}^{\mathbb{N}}$, the product σ-algebra $\mathcal{X} = \mathcal{B}(\mathbb{R}^{\mathbb{N}})$, the product measure μ on $\mathcal{X}$ with ν in each component, and the shift T defined as $T(x_n)_{n\in\mathbb{N}} = (x_{n+1})_{n\in\mathbb{N}}$. This system is ergodic by Example 3.18. As in Example 2.22, consider the mapping

$$\theta : \Omega \to X, \quad \theta(\omega) := (f_j(\omega))_{j\in\mathbb{N}},$$

satisfying $\theta_*\mathbb{P} = \mu$ since the random variables are independent. The Koopman operator $S_\theta : \mathrm{L}^1(X, \mu) \to \mathrm{L}^1(\Omega, \mathbb{P})$ is a Markov homomorphism, so we have

T. Eisner, B. Farkas, *A Journey Through Ergodic Theorems*, Birkhäuser Advanced Texts Basler Lehrbücher, https://doi.org/10.1007/978-3-031-96506-7_9

$\mathbb{E}(S_\theta g) = \int_X g \, \mathrm{d}\mu$ for every $g \in \mathrm{L}^1(X, \mu)$, where $\mathbb{E}(h) = \int_\Omega h \, \mathrm{d}\mathbb{P}$ is the **expectation** of the random variable h.

We first show how the mean ergodic theorem from Sect. 7.1 can be used to show the weak law of large numbers.

Theorem 9.1 (Weak Law of Large Numbers) *Let $f_n \in \mathrm{L}^1(\Omega, \mathbb{P})$, $n \in \mathbb{N}$, be identically distributed, independent, real random variables. Then for every $\varepsilon > 0$*

$$\mathbb{P}\Big[\Big|\frac{1}{N}\sum_{n=1}^{N} f_n - \mathbb{E}(f_1)\Big| > \varepsilon\Big] \to 0 \quad \text{as } N \to \infty.$$

Proof Let (X, μ, T) be as above and let $g : X \to \mathbb{R}$ be the projection onto the first component, then $g \in \mathrm{L}^1(X, \mu)$. Since T is mean ergodic on $\mathrm{L}^1(X, \mu)$ by Remark 7.8, from ergodicity (see Proposition 3.24) we conclude

$$\lim_{N\to\infty} \frac{1}{N}\sum_{n=1}^{N} T^n g = \int_X g \, \mathrm{d}\mu \cdot \mathbf{1} \text{ in } \mathrm{L}^1(X, \mu).$$

This yields

$$\lim_{N\to\infty} \mathbb{E}\Big|\frac{1}{N}\sum_{n=1}^{N} f_n - \mathbb{E}(f_1)\Big| = \lim_{N\to\infty} \Big\|S_\theta\Big(\frac{1}{N}\sum_{n=1}^{N} T^n g - \int_X g \, \mathrm{d}\mu \cdot \mathbf{1}\Big)\Big\|_{\mathrm{L}^1(\Omega,\mathbb{P})} = 0.$$

It is clear (see Exercise 7.19) that

$$\mathbb{P}\Big[\Big|\frac{1}{N}\sum_{n=1}^{N} f_n - \mathbb{E}(f_1)\Big| > \varepsilon\Big] \le \frac{1}{\varepsilon}\mathbb{E}\Big|\frac{1}{N}\sum_{n=1}^{N} f_n - \mathbb{E}(f_1)\Big|.$$

The proof is hence complete. ∎

With structurally the same proof it is possible to strengthen this result. We first need a lemma for that.

Lemma 9.2 *Let (X, μ), (Y, ν) be probability spaces and let $\theta : X \to Y$ be a measure-preserving mapping. Suppose that functions $g_n \in \mathrm{L}^0(Y, \nu)$, $n \in \mathbb{N}$, converge ν-almost everywhere to $g \in \mathrm{L}^0(Y, \nu)$. Then $S_\theta g_n \to S_\theta g$ almost everywhere with respect to μ.*

Proof We can suppose without loss of generality that the functions g_n are real-valued. Moreover, note that for each sequence of real-valued functions $(h_n)_{n\in\mathbb{N}}$ in $\mathrm{L}^0(Y, \nu)$ we have

$$S_\theta \sup_{n\in\mathbb{N}} h_n = \sup_{n\in\mathbb{N}} S_\theta h_n \text{ and } S_\theta \inf_{n\in\mathbb{N}} h_n = \inf_{n\in\mathbb{N}} S_\theta h_n.$$

As a consequence

$$S_\theta \limsup_{n\to\infty} g_n = \limsup_{n\to\infty} S_\theta g_n \text{ and } S_\theta \liminf_{n\to\infty} g_n = \liminf_{n\to\infty} S_\theta g_n.$$

The proof is complete. ■

We are now ready to deduce the strong law of large numbers from the pointwise ergodic theorem from Sect. 7.6.

Theorem 9.3 (Kolmogorov's Strong Law of Large Numbers [655]) *For $n \in \mathbb{N}$ let $f_n \in \mathrm{L}^1(\Omega, \mathbb{P})$ be identically distributed, independent real random variables. Then*

$$\frac{1}{N}\sum_{n=1}^{N} f_n \to \mathbb{E}(f_1) \quad \mathbb{P}\text{-almost surely as } N \to \infty.$$

Proof The proof is almost the same as for the weak law of large numbers using the pointwise ergodic theorem (Theorem 7.40). With the notations of that proof we obtain

$$\frac{1}{N}\sum_{n=1}^{N} T^n g \to \int_X g \,\mathrm{d}\mu \cdot \mathbf{1} \quad \mu\text{-almost everywhere as } N \to \infty.$$

By Lemma 9.2 applied to S_θ we conclude

$$\frac{1}{N}\sum_{n=1}^{N} f_n = S_\theta \frac{1}{N}\sum_{n=1}^{N} T^n g \to \mathbb{E}(f_1) \quad \mathbb{P}\text{-almost everywhere as } N \to \infty.$$

The proof is complete. ■

The previous result can be generalized to stationary sequences, the proof being exactly the same as the previous one (recall from Example 2.21 that for a stationary process $(f_n)_{n\in\mathbb{N}_0}$ there is a probability-preserving transformation T with $f_n = f_0 \circ T^n$ for every $n \in \mathbb{N}$). We leave the details to the reader, or refer to Doob [314, Thm. X.2.1].

Theorem 9.4 (Strong Law of Large Numbers for Stationary Sequences) *Let $(f_n)_{n\in\mathbb{N}_0}$ be a stationary sequence on a regular probability space $(\Omega, \mathbb{P})$ with finite expectation. Then*

$$\frac{1}{N}\sum_{n=1}^{N} f_n \to E_{\mathcal{Y}}(f_0) \quad \mathbb{P}\text{-almost surely as } N \to \infty,$$

where $\mathcal{Y}$ is the σ-algebra of invariant sets corresponding to the measure-preserving transformation with $f_n = f_0 \circ T^n$ for every $n \in \mathbb{N}$.

Still before Birkhoff and von Neumann the following version of the strong law, which is not an application of ergodic theorems, was known to A. Rajchman [863]. Recall that two random variables f, g in L^1 are called **uncorrelated** if $\mathbb{E}(fg) = \mathbb{E}(f)\mathbb{E}(g)$, which is the case, for example, if they are independent.

Theorem 9.5 (Rajchman's Strong Law of Large Numbers) *Let (Ω, P) be a probability space. For $n \in \mathbb{N}$ let $f_n \in \mathrm{L}^2(\Omega, \mathbb{P})$ be uncorrelated real random variables with $\sup_{n\in\mathbb{N}} \|f_n - \mathbb{E} f_n\|_2 < \infty$. Then*

$$\frac{1}{N}\sum_{n=1}^{N}(f_n - \mathbb{E}(f_n)) \to 0 \quad \mathbb{P}\text{-almost surely as } N \to \infty.$$

The proof of this result relies on a property of bounded, orthogonal sequences in L^2 discovered by Banach [73] for $\mathrm{L}^2[0, 1]$, and independently by Rajchman [863] for the case of general measure spaces. We have learned the proof from an unpublished note of D. Werner [1041], which is based on the following lemma, see also Krengel's book [684, Satz 12.4].

Lemma 9.6 *Let (f_n) be a sequence in $\mathrm{L}^1(X)$ such that for every $\varepsilon > 0$*

$$\sum_{n=1}^{\infty} \mu[|f_n| > \varepsilon] < \infty.$$

Then $f_n \to 0$ holds for μ-almost every $x \in X$.

Proof We need to prove $\limsup_{n\to\infty} |f_n| = 0$ almost everywhere. For $\varepsilon > 0$ we have

$$[\limsup_{n\to\infty} |f_n| > \varepsilon] = \bigcap_{n=1}^{\infty}\bigcup_{k\geq n}[|f_k| > \varepsilon],$$

so by the first Borel–Cantelli lemma, see Exercise 2.9, we conclude from the assumption that $\mu[\limsup_{n\to\infty} |f_n| > \varepsilon] = 0$. ∎

Now we can state and prove the next proposition, which when applied to the pairwise orthogonal functions $f_n - \mathbb{E} f_n$, yields Rajchman's strong law of large numbers.

Proposition 9.7 (Banach–Rajchman Lemma) *Let (f_n) be a bounded sequence of pairwise orthogonal functions in $\mathrm{L}^2(X, \mu)$. Then*

$$\frac{1}{N}\sum_{n=1}^{N} f_n \to 0 \quad \mu\text{-almost everywhere and in } \mathrm{L}^2.$$

Proof To show convergence in L^2 is an easy exercise. To show almost everywhere convergence, set $A_n := \frac{1}{n}\sum_{j=1}^n f_j$. By orthogonality we have

$$\mu[|A_{n^2}| > \varepsilon] \leq \frac{1}{\varepsilon^2}\int_X |A_{n^2}|^2 \,\mathrm{d}\mu = \frac{1}{n^4\varepsilon^2}\sum_{j=1}^{n^2} \|f_j\|_2^2 \leq \frac{\sup_{j\in\mathbb{N}}\|f_j\|_2^2}{n^2\varepsilon^2},$$

so

$$\sum_{n=1}^{\infty} \mu[|A_{n^2}| > \varepsilon] < \infty,$$

and Lemma 9.6 implies $A_{n^2} \to 0$ almost everywhere.
Next, let $n \in \mathbb{N}$ and take $m := m_n \in \mathbb{N}$ such that $m^2 \leq n < (m+1)^2$ and write

$$A_n = \frac{nA_n - m^2 A_{m^2}}{n} + \frac{m^2 A_{m^2}}{n}.$$

If $n \to \infty$, then $m = m_n \to \infty$, so by the already proved first part $A_{m^2} \to 0$ almost everywhere, and by $m^2/n \leq 1$ also $m^2 A_{m^2}/n \to 0$ almost everywhere. We show, by using again Lemma 9.6, that $(nA_n - m^2 A_{m^2})/n \to 0$ almost everywhere. This is done by the following calculation for $\varepsilon > 0$, where we also utilize orthogonality:

$$\begin{aligned}\mu[|\tfrac{nA_n - m^2 A_{m^2}}{n}| > \varepsilon] &\leq \frac{1}{\varepsilon^2 n^2}\sum_{j=m^2+1}^{n} \|f_j\|_2^2 \\ &\leq \frac{(m+1)^2 - 1 - m^2}{\varepsilon^2 n^2}\sup_{j\in\mathbb{N}}\|f_j\|_2^2 \leq \frac{2\sqrt{n}}{\varepsilon^2 n^2}\sup_{j\in\mathbb{N}}\|f_j\|_2^2,\end{aligned}$$

implying

$$\sum_{n=1}^{\infty} \mu[|\tfrac{nA_n - m_n^2 A_{m_n^2}}{n}| > \varepsilon] < \infty,$$

and hence finishing the proof by Lemma 9.6. ■

This result will be utilized later for studying more involved pointwise ergodic theorems.

Remark 9.8 (The Menchoff–Rademacher Theorem) The Banach–Rajchman lemma, even in a strengthened form, can be obtained from deeper results in harmonic analysis, for example as follows. We recall the **Menchoff–Rademacher theorem**, see [1080, Thm. XIII.10.22], the original papers are Menchoff [788] and Rademacher [861].

Let $(f_n)_{n\in\mathbb{N}}$ be an orthonormal system in $\mathrm{L}^2(X,\mu)$ and let $(a_n\log(n))_{n\in\mathbb{N}}\in\ell^2(\mathbb{N})$. Then the series

$$\sum_{n=1}^{\infty} a_n f_n$$

converges μ-almost everywhere (and in L^2).

Now as in the Banach–Rajchman lemma, let $(f_n)_{n\in\mathbb{N}}$ be a bounded sequence of pairwise orthogonal functions and let $\gamma > 1/2$. Since $(\|f_n\|_2\log(n)/n^\gamma)_{n\in\mathbb{N}}\in\ell^2(\mathbb{N})$, by the above mentioned Menchoff–Rademacher theorem, the corresponding series

$$\sum_{n=1}^{\infty}\frac{f_n}{n^\gamma}$$

converges almost everywhere. By Kronecker's lemma (see Exercise 4.15) we obtain

$$\frac{1}{N^\gamma}\sum_{n=1}^{N} f_n \to 0 \text{ almost everywhere as } N\to\infty$$

for every $\gamma > 1/2$, being a strengthened form of the Banach–Rajchman lemma (Proposition 9.7). It follows from the law of iterated logarithm, see, e.g., Billingsley [153, Thm. 9.5] that one cannot take $\gamma = 1/2$ here.

9.2 Applications in Number Theory

Next we come to more surprising applications in number theory. The following property of real numbers was introduced by Borel [172] in 1909.

Definition 9.9 Let $x\in\mathbb{R}$ and $b\in\{2,3,\ldots\}$. Denote by $N_b(a,n)$ the number of appearances of the digit a in the first n positions after the "decimal point[1]" of x in base b. The number x is called **simply normal in base** b if

$$\lim_{n\to\infty}\frac{N_b(a,n)}{n}=\frac{1}{b} \text{ for each } a\in\{0,\ldots,b-1\},$$

and **simply normal** if x is simply normal in every base. Analogously, x is called **normal in base** b if for every finite word $S := a_1\ldots a_d\in\{0,1,\ldots,b-1\}^d$, the

[1] Only in base $b = 10$ this is a *decimal* point, we keep however this terminology.

number $N_b(S, n)$ of appearances of S among the first n positions after the decimal point of x in base b satisfies

$$\lim_{n\to\infty} \frac{N_b(S,n)}{n} = \frac{1}{b^d},$$

and **normal** if x is normal in every base.

For example, the number 0.01234567890123456789... is simply normal in base 10 and is not normal in base 2 (or any other base) (Exercise 9.1). The so-called Champernowne number 0.1234567891011121314... is normal in base 10 but it is unknown whether it is normal in other bases.

Only a few examples of normal numbers are known, see Sierpiński [940] and Becher and Figueira [86]. Even for the numbers $\sqrt{2}$, π and e it is not known whether they are normal. (In fact, it is not even known whether every digit appears in the decimal representation of these numbers infinitely often). Nevertheless, the following result shows that normal numbers are the rule and not the exception.

Theorem 9.10 (Borel) *Lebesgue almost every real number is normal.*

We present here an ergodic theoretic proof.

Proof Since the set of possible bases is countable, it is enough to show that for a fixed base b, almost every real number is normal in base b. Similarly, there are countably many finite words, so it suffices to prove that every fixed finite word $S = a_1 \dots a_d \in \{0, \dots, b-1\}^d$ has the asymptotic frequency $1/b^d$ of occurrence in almost every $y \in [0, 1]$. Consider the Bernoulli shift $B(1/b, \dots, 1/b)$ on $X := \{0, \dots, b-1\}^{\mathbb{N}}$. By Proposition 3.18, it is ergodic with respect to the product measure μ. Let $S = a_1 \dots a_d$ be a finite word in base b. Applying the pointwise ergodic theorem to the cylinder set $A := \{(x_1, x_2, \dots) : x_1 = a_1, \dots, x_d = a_d\}$ we obtain

$$\lim_{N\to\infty} \frac{1}{N} \sum_{n=0}^{N-1} \mathbf{1}_A(T^n x) = \mu(A) = \frac{1}{b^d}$$

for almost every $x \in X$. The term $\frac{1}{N}\sum_{n=0}^{N-1} \mathbf{1}_A(T^n x)$ on left-hand side equals the number of appearances of S in the first $N+d$ positions of x divided by N. Now, the mapping

$$\theta : X \to [0, 1], \quad (x_n)_{n\in\mathbb{N}} \mapsto \sum_{n=1}^{\infty} \frac{x_n}{b^n}$$

is measure-preserving (if $[0, 1]$ is endowed with the Lebesgue measure) and even essentially invertible, with essential inverse mapping y to the expansion $(y_n)_{n\in\mathbb{N}}$ of y in base b. An application of Lemma 9.2 finishes the proof. ∎

We now turn to a property of sequences of numbers which we will explore in more detail in the next chapter. At this point we only describe a relation between a number theoretic and an ergodic theoretic property.

Recall that the rotation system $(\mathbb{T}, a)$ is uniquely ergodic for irrational $a \in \mathbb{T}$ (see Theorem 3.40). Endow $[0, 1)$ with the topology that makes $\theta : t \mapsto \mathrm{e}^{2\pi i t}$ a homeomorphism between $[0, 1)$ and $\mathbb{T}$. If $a = \mathrm{e}^{2\pi i\alpha}$, we obtain a topological system $([0, 1), \alpha)$, with the continuous mapping $t \mapsto \{t + \alpha\}$. Recall the notation $\{x\} := x - \lfloor x \rfloor$ for the **fractional part** of $x \in \mathbb{R}$. The Lebesgue measure λ is an invariant measure for this system (see Example 2.13). The translation system $([0, 1), \lambda, \alpha)$ is isomorphic to $(\mathbb{T}, \mathrm{m}, a)$ via the previous homeomorphism. It is Exercise 3.14 to show that $([0, 1), \alpha)$ is uniquely ergodic if and only if $(\mathbb{T}, a)$ is uniquely ergodic.

Remark 9.11 We note that the isomorphism θ being a homeomorphism is vital. If (K, μ, T) and (L, ν, S) are isomorphic measure-preserving systems with (K, T) and (L, S) being topological systems, it is not true in general that (K, T) and (L, S) are simultaneously uniquely ergodic. In fact, one has the following result, called the **Jewett–Krieger theorem** [581], [685]: For every ergodic, invertible, standard measure-preserving system (X, μ, T) there is a uniquely ergodic topological system (L, S) with (unique) invariant measure ν such that (X, μ, T) and (L, ν, S) are isomorphic, and ν satisfies $\mathrm{supp}(\nu) = L$, see Sect. 8.5, where topological models were discussed briefly.

By unique ergodicity of the translation system $([0, 1), \lambda, \alpha)$ for irrational α, Theorem 7.23 implies the following.

Proposition 9.12 *For a given irrational number $\alpha \in \mathbb{R}$ consider the sequence $(\{n\alpha\})_{n\in\mathbb{N}_0}$. For each continuous function $f : [0, 1] \to \mathbb{C}$ with $f(0) = f(1)$ one has*

$$\int_0^1 f(s)\,\mathrm{d}s = \lim_{N\to\infty} \frac{1}{N} \sum_{n=0}^{N-1} f(\{n\alpha\}).$$

Note that one can omit here the condition $f(0) = f(1)$ and even replace continuity by Riemann-integrability, see Exercise 9.2. This phenomenon will be discussed later in Sect. 10.1.

We now replace the rotation on $\mathbb{T}$ by the skew rotation on $\mathbb{T}^2$ given by $T_a(x, y) := (ax, xy)$, see Example 2.15. Recall also the additive variant, the skew shift, defined on $[0, 1)^2$ by

$$\widetilde{T}_\alpha : [0, 1)^2 \to [0, 1)^2, \quad (x, y) \mapsto (\{x + \alpha\}, \{x + y\}).$$

By endowing $[0, 1)$ with the topology as above, this mapping becomes continuous, giving rise to the **topological skew shift** system $([0, 1)^2, \widetilde{T}_\alpha))$, which is uniquely ergodic if and only if the corresponding skew shift $(\mathbb{T}^2, T_a)$, $a = \mathrm{e}^{2\pi i\alpha}$, is uniquely ergodic, see Exercise 3.14.

Proposition 9.13 *The skew rotation $(\mathbb{T}^2, \mathrm{m}^2, T_a)$ is uniquely ergodic if and only if $a \in \mathbb{T}$ is irrational (i.e., not a root of unity). Equivalently, the skew shift $([0, 1)^2, \lambda^2, \widetilde{T}_\alpha)$ is uniquely ergodic if and only if α is irrational.*

The "only if" part is left as Exercise 9.5. One can prove the "if" part by ergodic theoretic arguments in a far more general situation, and the previous statement becomes a special case of the following result of Furstenberg [443], see also [342, Sec. 10.4 and Ch. 15, Supp.] for a proof.

Theorem 9.14 (Furstenberg's Skew Product Theorem) *Let (K, T) be a uniquely ergodic topological system with corresponding T-invariant probability measure μ. Let further $\alpha : K \to G$ be a continuous mapping, where G is a compact group. Consider the topological group extension system $(K \times G, \widetilde{S}_a)$ with $\widetilde{S}_a(x, g) = (Tx, \alpha(x)g)$, which has $\mu \otimes \mathrm{m}_G$ as an invariant measure. If $\mu \otimes \mathrm{m}_G$ is an ergodic measure, then $(K \times G, \widetilde{S}_a)$ is uniquely ergodic.*

This directly implies unique ergodicity of $(\mathbb{T}^2, \mathrm{m}^2, T_a)$ if a is irrational, i.e., the "if" part of Proposition 9.13. However, we give a different, number theoretic proof of it in the next chapter, see Proposition 10.20. Here we only describe what this unique ergodicity property yields in number theory in the spirit of Proposition 9.12. The following is a preparatory result with the proof left as Exercise 9.5.

Proposition 9.15 (Cocycle for the Skew Shift) *For each $(x, y) \in \mathbb{T}^2$ we have*

$$T_a^n(x, y) = \left(a^n x, a^{\frac{n(n-1)}{2}} x^n y\right)$$

and for the skew shift $([0, 1)^2, \lambda^2, \widetilde{T}_\alpha)$ and for every $(s, t) \in [0, 1)^2$

$$\widetilde{T}_\alpha^n(s, t) = \left(\{n\alpha + s\}, \{\tfrac{n(n-1)}{2}\alpha + ns + t\}\right).$$

Thus Proposition 9.13 implies the following result.

Proposition 9.16 *Let $\alpha \in \mathbb{R}$ be an irrational number and consider the sequence $(\{\frac{n(n-1)}{2}\alpha\})_{n\in\mathbb{N}_0}$. For each continuous function $f : [0, 1] \to \mathbb{C}$ with $f(0) = f(1)$ one has*

$$\int_0^1 f(s)\,\mathrm{d}s = \lim_{N\to\infty} \frac{1}{N} \sum_{n=0}^{N-1} f\left(\{\tfrac{n(n-1)}{2}\alpha\}\right).$$

Proof Apply Propositions 9.13, 9.15 and Theorem 7.23 to the continuous function $g : (x, y) \mapsto y$. ∎

Analogously to Proposition 9.12, see Exercise 9.2, the assertion of Proposition 9.16 holds for all Riemann-integrable functions $f : [0, 1] \to \mathbb{C}$.

The property, as described in Propositions 9.12 and 9.16, of the sequences $(\{n\alpha\})_{n\in\mathbb{N}_0}$ and $(\{\frac{n(n-1)}{2}\alpha\})_{n\in\mathbb{N}_0}$ means that they are *equidistributed* in $[0, 1)$. A more detailed study of equidistribution will be carried out in the next chapter.

9.3 Applications in Combinatorics

We now show how ergodic theory can help to solve combinatorial problems in number theory and encounter the first example of the following phenomenon: If a subset $E \subset \mathbb{N}$ is large (in some appropriate sense), then it contains some structured subsets. For us "largeness" will be quantified by the upper density defined by

$$\overline{\mathrm{d}}(E) = \limsup_{N\to\infty} \frac{|E \cap \{1, \ldots, N\}|}{N}$$

(see Sect. 4.3).

Polynomial Configurations in Large Sets

In the first example, a structured subset means a polynomial progression, and we show how ergodic theorems help to obtain such polynomial patterns in large sets of integers.

Theorem 9.17 (Furstenberg–Sárközy) *Let $E \subset \mathbb{N}$ have positive upper density and let $P \in \mathbb{Z}[\cdot]$ satisfy $P(0) = 0$. Then there exist $a, n \in \mathbb{N}$ with $a, a + P(n) \in E$.*

Remark 9.18 The following easy example shows that the condition $P(0) = 0$ cannot be dropped. Let $P(x) := 2x + 1$ and $E := 2\mathbb{N} + 1$ be the set of all odd numbers. It is clear that E has no progressions of the form $a, a + 2n + 1$.

For the polynomial $P(n) = n^2$ Sárközy [907] proved this with number theoretic arguments, while Furstenberg [445] gave an ergodic theoretic proof by first showing the following property of measure-theoretic dynamical systems. The general case is settled in Sárközy [908] and Furstenberg [446].

Theorem 9.19 (Polynomial Recurrence) *Let (X, μ, T) be a measure-preserving system, let $f \in \mathrm{L}^\infty(X, \mu)$ satisfy $f > 0$ (meaning $f \geq 0$ and $f \neq 0$) and take $P \in \mathbb{Z}[\cdot]$ with $P : \mathbb{N}_0 \to \mathbb{N}_0$ and $P(0) = 0$. Then*

$$\lim_{N\to\infty} \frac{1}{N} \sum_{n=0}^{N-1} \int_X f \cdot T^{P(n)} f \,\mathrm{d}\mu = \Big(f \,\Big|\, \lim_{N\to\infty} \frac{1}{N} \sum_{n=0}^{N-1} T^{P(n)} f \Big) > 0. \tag{9.1}$$

In particular, for every $A \subset X$ with $\mu(A) > 0$ one has

$$\mu\big(A \cap T^{-P(n)} A\big) > 0 \quad \textit{for some } n \in \mathbb{N}. \tag{9.2}$$

Remark 9.20

(a) Recall that the limit on the right-hand side of (9.1) exists in $\mathrm{L}^2(X, \mu)$ by the polynomial mean ergodic theorem (Theorem 7.28). Thus, (9.1) means that this limit has non-trivial correlation with f.
(b) Like the polynomial mean ergodic theorem, see Remark 7.29(b), both the Furstenberg–Sárközy Theorem 9.19 and the polynomial recurrence (Theorem 9.17) hold if one replaces $P \in \mathbb{Z}[\cdot]$ by the weaker assumption that P is an *integer polynomial*. We leave the details as Exercise 9.8.

The following abstract bridge between ergodic theory and combinatorics leads to the proof of the Furstenberg–Sárközy theorem and will be used later in the context of arithmetic progressions and multiple recurrence.

Theorem 9.21 (Furstenberg's Correspondence Principle) *Let $k \in \mathbb{N}$ and $s_1, \dots, s_k : \mathbb{N} \to \mathbb{Z}$. If for every invertible, ergodic, regular measure-preserving system (X, μ, T) and for every $A \subset X$ with $\mu(A) > 0$ there is $n \in \mathbb{N}$ such that*

$$\mu\Big(A \cap T^{-s_1(n)}(A) \cap \cdots \cap T^{-s_k(n)}(A)\Big) > 0,$$

then for every subset $E \subset \mathbb{N}$ with $\overline{\mathrm{d}}(E) > 0$ there exist $a, n \in \mathbb{N}$ such that $a, a + s_1(n), \dots, a + s_k(n) \in E$.

Proof Let $E \subset \mathbb{N}$ have upper density $\overline{\mathrm{d}}(E) > 0$. We construct a corresponding invertible, ergodic, regular measure-preserving system (X, μ, T) and a set $A \subset X$ with $\mu(A) > 0$ transferring the combinatorial problem into the realm of ergodic theory.

Let $K := \{0, 1\}^{\mathbb{Z}}$ and consider the shift (K, T) from Example 2.31. Then of course $\mathbf{1}_E \in K$. Define

$$X := \overline{\mathrm{orb}}(\mathbf{1}_E) = \overline{\{T^i \mathbf{1}_E : i \in \mathbb{Z}\}},$$

which is a T- and T^{-1}-invariant, compact set consisting of 0-1-sequences only, so $T : X \to X$ is bijective. Consider the cylinder set $A \subset X$ given by

$$A := \{(t_j)_{j \in \mathbb{Z}} \in X : t_0 = 1\}.$$

The set A has the following properties, whose proof we leave as Exercise 9.7.

(1) A is closed and open in X.
(2) The equivalence

$$T^m \mathbf{1}_E \in A \quad \Longleftrightarrow \quad m \in E$$

holds for each $m \in \mathbb{N}$.
(3) Let $n \in \mathbb{N}$ be fixed. The existence of $a \in \mathbb{N}$ with $a, a+s_1(n), \dots, a+s_k(n) \in E$ is equivalent to $A \cap T^{-s_1(n)} A \cap \cdots \cap T^{-s_k(n)} A \neq \emptyset$.

As a consequence, the proof will be complete if we can find an ergodic measure μ on X with $\mu(A \cap T^{-s_1(n)} A \cap \cdots \cap T^{-s_k(n)} A) > 0$ for some $n \in \mathbb{N}$. Thus, under the assumption of Theorem 9.21, it suffices to find a T-invariant, ergodic probability measure μ on X with $\mu(A) > 0$.

The construction of such a measure μ is similar to the proof of the Krylov–Bogolyubov Theorem 2.33. This is also the first place where we use the assumption on E. Let $(N_j)_{j\in\mathbb{N}}$ be a subsequence of $\mathbb{N}$ such that

$$\lim_{j\to\infty} \frac{|E \cap \{1, \ldots, N_j\}|}{N_j} = \overline{\mathrm{d}}(E) > 0.$$

We let $\mu_j := \frac{1}{N_j} \sum_{n=1}^{N_j} \delta_{T^n \mathbf{1}_E}$. Then by (2) above we conclude

$$\begin{aligned} \frac{|E \cap \{1, \ldots, N_j\}|}{N_j} &= \frac{1}{N_j} \sum_{n=1}^{N_j} \mathbf{1}_E(n) = \frac{1}{N_j} \sum_{n=1}^{N_j} \mathbf{1}_A(T^n \mathbf{1}_E) \\ &= \frac{1}{N_j} \sum_{n=1}^{N_j} \delta_{T^n \mathbf{1}_E}(A) = \mu_j(A). \end{aligned}$$

Since $(\mu_j)_{j\in\mathbb{N}}$ is a sequence of probability measures on the compact (metric) space X, by the Banach–Alaoglu theorem it has a weak* convergent subsequence with limit $\tilde{\mu}$ which is a probability measure. The proof that $\tilde{\mu}$ is a T-invariant measure on X with $\tilde{\mu}(A) = \overline{\mathrm{d}}(E) > 0$ is again left as Exercise 9.7. By Proposition 3.34(c) (and since $\mathbf{1}_A$ is continuous), one can even find an ergodic measure μ with $\mu(A) > 0$. The proof is complete. ∎

Remark 9.22 (Furstenberg's Correspondence and its Converse) Furstenberg's construction from the above proof provides the following (non-ergodic) variant of Furstenberg's correspondence principle, cf., e.g., Bergelson [102] and Kra [669]: For every $E \subset \mathbb{N}$ there exists a measure-preserving system (X, μ, T), a measurable set $A \subset X$ with $\mu(A) = \overline{\mathrm{d}}(E)$ and a subsequence (N_j) of $\mathbb{N}$ such that for every $k, n_1, \ldots, n_k \in \mathbb{N}$ one has

$$\begin{aligned} &\mu(A \cap T^{-n_1} A \cap \ldots \cap T^{-n_k} A) \\ &\quad = \lim_{j\to\infty} \frac{|E \cap (E - n_1) \cap \ldots \cap (E - n_k) \cap \{1, \ldots, N_j\}|}{N_j}. \end{aligned} \tag{9.3}$$

We leave the details as Exercise 9.7(d). Recently, Fish and Skinner [391] proved the following converse statement: For every measure-preserving system (X, μ, T) and every measurable $A \subset X$ there exists $E \subset \mathbb{N}$ such that for every $k \in \mathbb{N}_0$ and every $n_1, \ldots, n_k \in \mathbb{N}$

$$\mu(A \cap T^{-n_1} A \cap \ldots \cap T^{-n_k} A) = \mathrm{d}(E \cap (E - n_1) \cap \ldots \cap (E - n_k)),$$

i.e., $\mu(A) = \mathrm{d}(E)$ and (9.3) holds for $N_j = j$.

If we take $s_1 = P$ a polynomial with integer coefficients in the correspondence principle (Theorem 9.21) we see that for a proof of Theorem 9.17 it suffices to prove Theorem 9.19.

Proof of Theorem 9.19 Decompose f as $f = g + h$ according to the rational spectrum decomposition (Proposition 6.7) applied to $H := \mathrm{L}^2(X, \mu)$ and the Koopman operator T on H with $g \in H_{\mathrm{rat}}$ and $h \perp H_{\mathrm{rat}}$. By the polynomial mean ergodic theorem for contractions (Theorem 7.28), one has

$$\lim_{N\to\infty} \frac{1}{N} \sum_{n=0}^{N-1} T^{P(n)} h = 0$$

and therefore by orthogonality

$$\Big(f \,\Big|\, \lim_{N\to\infty} \frac{1}{N} \sum_{n=1}^{N} T^{P(n)} f\Big) = \Big(g \,\Big|\, \lim_{N\to\infty} \frac{1}{N} \sum_{n=1}^{N} T^{P(n)} g\Big).$$

Since the projection P_{rat} onto the rational spectrum component is a restriction of a Markov operator to L^2 (see Proposition 8.25 and Example 8.31(b)), we obtain that $g = P_{\mathrm{rat}} f \geq 0$, $g \in \mathrm{L}^\infty(X, \mu)$ and $\int_X g \,\mathrm{d}\mu = \int_X f \,\mathrm{d}\mu > 0$. Thus we can assume without loss of generality that $f = g \in H_{\mathrm{rat}}$ and $\|f\|_\infty = 1$. Recall from Lemma 6.6 that

$$H_{\mathrm{rat}} = \overline{\bigcup_{k\in\mathbb{N}} \mathrm{fix}(T^k)}. \tag{9.4}$$

Suppose first $f \in \mathrm{fix}(T^k)$ for some $k \in \mathbb{N}$. Since $P(0) = 0$, we conclude that k divides $P(k), P(2k), \ldots$, thus $(T^{P(kn)} f \,|\, f) = \|f\|_2^2$ for every $n \in \mathbb{N}$. Therefore by positivity

$$\lim_{N\to\infty} \frac{1}{kN+1} \sum_{n=0}^{kN} (T^{P(n)} f \,|\, f) \geq \limsup_{N\to\infty} \frac{1}{kN+1} \sum_{n=1}^{N} \|f\|_2^2 = \frac{1}{k} \|f\|_2^2 > 0.$$

Let now $f \in H_{\mathrm{rat}}$ be arbitrary and set $d := \int_X f \,\mathrm{d}\mu > 0$. By (9.4), there exists $k \in \mathbb{N}$ such that $f_k := P_{\mathrm{fix}(T^k)} f$ satisfies $\|f - f_k\|_2 < d^2/4$. Recall that $P_{\mathrm{fix}(T^k)}$ is also a restriction of a Markov operator to L^2, so $\|f_k\|_\infty \leq \|f\|_\infty = 1$ and $\int_X f_k \,\mathrm{d}\mu = d$. Observe that

$$\begin{aligned} |(T^n f \,|\, f) - (T^n f_k \,|\, f_k)| &= |(T^n f_k \,|\, f - f_k) + (T^n (f - f_k) \,|\, f)| \\ &\leq 2\|f - f_k\|_2 < \frac{d^2}{2} \end{aligned}$$

for each $n \in \mathbb{N}$. In particular, we have

$$|(T^{P(kn)} f|f) - (T^{P(kn)} f_k|f_k)| < \frac{d^2}{2}.$$

Since $P(0) = 0$ implies $T^{P(kn)} f_k = f_k$ (since $f_k \in \operatorname{fix}(T^k)$) the Cauchy–Schwarz inequality implies

$$(T^{P(kn)} f|f) \geq \|f_k\|_2^2 - \frac{d^2}{2} \geq (f_k|\mathbf{1})^2 - \frac{d^2}{2} = d^2 - \frac{d^2}{2} = \frac{d^2}{2} \text{ for each } n \in \mathbb{N}.$$

Hence we conclude

$$\begin{aligned} \lim_{N\to\infty} \frac{1}{N} \sum_{n=1}^{N} (T^{P(n)} f|f) &= \lim_{N\to\infty} \frac{1}{kN} \sum_{n=1}^{kN} (T^{P(n)} f|f) \\ &\geq \frac{1}{k} \limsup_{N\to\infty} \frac{1}{N} \sum_{n=1}^{N} (T^{P(nk)} f|f) \geq \frac{d^2}{2k}. \end{aligned}$$

The proof is complete. ∎

We now introduce the notion of *characteristic factor* going back to Furstenberg and Weiss [458].

Definition 9.23 Let $k \in \mathbb{N}$ and $a_1, \ldots, a_k : \mathbb{N}_0 \to \mathbb{N}_0$. Let further (X, μ, T) be a measure-preserving system and $Z \subset \mathrm{L}^2(X, \mu)$ be a closed linear subspace which is induced by a factor (Y, ν, S). We call both the subspace Z and the factor (Y, ν, S) **characteristic for the averages**

$$\frac{1}{N} \sum_{n=0}^{N-1} T^{a_1(n)} f_1 \cdot T^{a_2(n)} f_2 \cdots T^{a_k(n)} f_k, \quad N \in \mathbb{N}, \tag{9.5}$$

if for every $f_1, \ldots, f_k \in \mathrm{L}^\infty(X, \mu)$[2]

$$\lim_{N\to\infty} \left\| \frac{1}{N} \sum_{n=0}^{N-1} T^{a_1(n)} f_1 \cdots T^{a_k(n)} f_k - \frac{1}{N} \sum_{n=0}^{N-1} T^{a_1(n)} P_Z f_1 \cdots T^{a_k(n)} P_Z f_k \right\|_2 = 0, \tag{9.6}$$

[2] One takes bounded functions here to be sure that the product $T^{a_1(n)} f_1 \cdot T^{a_2(n)} f_2 \cdots T^{a_k(n)} f_k$ is well defined as an element of $\mathrm{L}^2(X, \mu)$. For $k = 1$ one can just take $f_1 \in \mathrm{L}^2(X, \mu)$ and, more generally, define characteristic subspaces for any contraction on a Hilbert space.

where P_Z denotes the orthogonal projection onto Z. (Recall that by (8.2) one has $P_Z f_1, \dots, P_Z f_k \in \mathrm{L}^\infty(X, \mu)$ so the products on the right-hand side in (9.6) are well defined as elements of $\mathrm{L}^2(X, \mu)$.)

In particular, both L^2-convergence and the limit of the averages (9.5) remain unchanged if one replaces the functions $f_1, \dots, f_k$ by their projections onto Z. This allows to reduce the complexity of proving the convergence and identifying properties of the limit of (9.5) for example to study recurrence.

Remark 9.24

(a) If Z is characteristic for the averages (9.5), then the L^2-limit of (9.5) is zero whenever at least one of the functions $f_1, \dots, f_k$ is orthogonal to Z. Moreover, by linearity this property characterizes subspaces which are characteristic and induced by a factor (since in this case $P_Z f_1, \dots, P_Z f_k \in \mathrm{L}^\infty(X, \mu)$ by (8.2)).

(b) If Z is induced by a factor (Y, ν, S) with the corresponding Markov factor map J and the conditional expectation operator E_Y, then by $P_Z = JE_Y$ and the properties of J (9.6) is equivalent to

$$\lim_{N\to\infty} \left\| \frac{1}{N}\sum_{n=0}^{N-1} T^{a_1(n)} f_1 \cdots T^{a_k(n)} f_k - J\frac{1}{N}\sum_{n=0}^{N-1} S^{a_1(n)} E_Y f_1 \cdots S^{a_k(n)} E_Y f_k \right\|_2 = 0.$$

This allows to replace (X, μ, T) by its factor (Y, ν, S) in the study of convergence and recurrence.

(c) Characteristic subspaces and characteristic factors are not unique. Indeed, the whole space or system, respectively, is always characteristic for each kind of averages, and any larger subspace or larger factor of a characteristic one is again characteristic. One therefore looks for characteristic subspaces and/or factors which are as small as possible in order to gain as much as possible from reducing to this subspace and/or factor.

Example 9.25 (Characteristic Factor for Polynomial Averages) By the polynomial mean ergodic theorem for contractions (Theorem 7.28) and Remark 9.24(a) we see that the subspace H_{rat} and the corresponding rational spectrum factor (see Example 8.31(b)) are characteristic for polynomial averages

$$\lim_{N\to\infty} \frac{1}{N}\sum_{n=0}^{N-1} T^{P(n)} f$$

for any non-constant $P \in \mathbb{Z}[\cdot]$ with $P : \mathbb{N}_0 \to \mathbb{N}_0$.

We will see more examples of characteristic factors in the rest of this section and in Chaps. 13 and 15.

Remark 9.26 A set $S \subset \mathbb{Z}$ is called a **set of recurrence** if for every invertible measure-preserving system (X, μ, T) and every measurable set $A \subset X$ with $\mu(A) > 0$ there exists $n \in S$ with $\mu(A \cap T^{-n}A) > 0$. By Furstenberg's correspondence principle, S is a set of recurrence if and only if[3] it is **intersective**, i.e., for every set $E \subset \mathbb{N}$ with positive upper density there exists $n \in S \setminus \{0\}$ such that $a, a + n \in E$ for some $a \in \mathbb{N}$ (or, equivalently, n belongs to the difference set $E - E$).

Thus Poincaré's recurrence theorem (Theorem 2.3) states that $\mathbb{N}$ is a set of recurrence. As discussed in Remark 9.18, $2\mathbb{N} + 1$ is not a set of recurrence. By polynomial recurrence (Theorem 9.19), the set $P(\mathbb{N})$ is a set of recurrence for every $P \in \mathbb{Z}[\cdot]$ with $P(0) = 0$. Bergelson and Håland [118] extended this to so-called *generalized polynomials*.[4] For more examples and related notions such as sets of strong recurrence and sets of uniform recurrence see Bergelson [101], Bergelson and Hindman [122], Bergelson and Håland [118], Forrest [402], McCutcheon [782], Bergelson et al. [117], Nair and Zaris [819], Frantzikinakis and Wierdl [432], Frantzikinakis [407], Fan and Schneider [374], Bergelson et al. [125, 127, 144], Griesmer [496, 498, 499]. Sets of multiple recurrence will be discussed in Sect. 13.9.

Arithmetic Progressions of Length 3

We continue with another application to combinatorics. Here a structured subset of a set in $\mathbb{N}$ means an arithmetic progression of length 3, and the goal of this section is to give an ergodic theoretic proof due to Furstenberg [445] of the following result.

Theorem 9.27 (Roth [886]) *Let $E \subset \mathbb{N}$ have positive upper density. Then there exist $a, n \in \mathbb{N}$ such that $a, a + n, a + 2n \in E$.*

We begin with some observations. Let (X, μ, T) be a measure-preserving system and consider the Jacobs–de Leeuw–Glicksberg decomposition $H = H_{\mathrm{kr}} \oplus H_{\mathrm{aws}}$ of the Hilbert space $H = \mathrm{L}^2(X, \mu)$ with respect to the Koopman operator, see Theorem 6.20. Let P_{kr} be the orthogonal projection onto the Kronecker part H_{kr}, and let P_{aws} be the complementary projection onto H_{aws}. By Example 8.31 P_{kr} is the (restriction of the) Markov projection onto the Kronecker factor. The following shows that for ergodic systems the subspace H_{kr} and the Kronecker factor are characteristic for double ergodic averages $\frac{1}{N}\sum_{n=1}^{N} T^n f \cdot T^{2n} g$, see Definition 9.23.

Proposition 9.28 (Kronecker Factor is Characteristic for Double Convergence) *Let (X, μ, T) be an ergodic measure-preserving system and let $f, g \in \mathrm{L}^\infty(X, \mu)$.*

[3] The "only if" direction holds by Theorem 9.21 whereas the converse follows from the full version of Furstenberg's correspondence principle formulated below in Sect. 13.1.

[4] In their terminology, generalized polynomials are functions obtained from the classical polynomials using the operations of addition, multiplication and taking the integer part (such as $f(t) = \alpha t\{\beta t^2\}$ for $\alpha, \beta \in \mathbb{R}$, where $\{\cdot\}$ denotes the fractional part).

If f or g belongs to H_{aws}, then

$$\lim_{N\to\infty} \frac{1}{N} \sum_{n=0}^{N-1} T^n f \cdot T^{2n} g = 0. \tag{9.7}$$

As a consequence, for each $f, g \in \mathrm{L}^\infty(X, \mu)$

$$\lim_{N\to\infty} \frac{1}{N} \sum_{n=0}^{N-1} T^n f \cdot T^{2n} g = \lim_{N\to\infty} \frac{1}{N} \sum_{n=0}^{N-1} T^n P_{\mathrm{kr}} f \cdot T^{2n} P_{\mathrm{kr}} g.$$

The proof we present is based on the van der Corput Lemma 7.31 and goes back to Furstenberg [446, Sec. 4.2], see also Furstenberg et al. [455, Sec. 3] and Bergelson [103], [110, Sec. 4.2.3].

Proof Set $u_n := T^n f \cdot T^{2n} g \in \mathrm{L}^2(X, \mu)$ for $n \in \mathbb{N}$ and let $h \in \mathbb{N}$. Since T is measure-preserving, we have

$$\begin{aligned}(u_n | u_{n+h}) &= \int_X T^n f \cdot T^{2n} g \cdot T^{n+h} \overline{f} \cdot T^{2n+2h} \overline{g} \, \mathrm{d}\mu \\ &= \int_X f \cdot T^h \overline{f} \cdot T^n (g \cdot T^{2h} \overline{g}) \, \mathrm{d}\mu.\end{aligned}$$

We thus obtain by the ergodicity of T (see Proposition 7.4) that

$$\begin{aligned}\gamma_h := \lim_{N\to\infty} \left| \frac{1}{N} \sum_{n=0}^{N-1} (u_n | u_{n+h}) \right| &= \left| \int_X f \cdot T^h \overline{f} \lim_{N\to\infty} \frac{1}{N} \sum_{n=0}^{N-1} T^n (g \cdot T^{2h} \overline{g}) \, \mathrm{d}\mu \right| \\ &= \left| \int_X f \cdot T^h \overline{f} \, \mathrm{d}\mu \right| \cdot \left| \int_X g \cdot T^{2h} \overline{g} \, \mathrm{d}\mu \right| \\ &= |(T^h f | f)| \cdot |(T^{2h} g | g)|.\end{aligned}$$

Therefore, if $g \in H_{\mathrm{aws}}(T)$, we can write

$$0 \le \limsup_{J\to\infty} \frac{1}{J} \sum_{h=0}^{J-1} \gamma_h \le \limsup_{J\to\infty} \frac{1}{J} \sum_{h=0}^{J-1} |(T^{2h} g | g)| \cdot \|f\|_\infty^2.$$

By Exercise 6.13 $H_{\mathrm{aws}}(T^2) = H_{\mathrm{aws}}(T)$, so $g \in H_{\mathrm{aws}}(T^2)$. By Proposition 6.18 and Lemma 6.16 we conclude now

$$\limsup_{J\to\infty} \frac{1}{J} \sum_{h=0}^{J-1} |(T^{2h} g | g)| = 0.$$

Van der Corput's Lemma 7.31 implies (9.7), provided $g \in H_{\mathrm{aws}}(T)$. The case when $f \in H_{\mathrm{aws}}$ can be proved analogously.
The final assertion follows by linearity, see Remark 9.24(a). ∎

One has the following analogue of the polynomial convergence presented in Theorem 7.28.

Theorem 9.29 (Double Convergence) *Let (X, μ, T) be an ergodic measure-preserving system and let $f, g \in \mathrm{L}^\infty(X, \mu)$. Then the limit*

$$\lim_{N\to\infty} \frac{1}{N} \sum_{n=0}^{N-1} T^n f \cdot T^{2n} g \text{ exists in the } \mathrm{L}^2\text{-norm.} \tag{9.8}$$

Proof Let $g \in \mathrm{L}^\infty(X, \mu)$ be fixed and consider for $N \in \mathbb{N}$ the linear operators $(S_N)_{N\in\mathbb{N}}$ on $\mathrm{L}^2(X, \mu)$ given by

$$S_N f := \frac{1}{N} \sum_{n=0}^{N-1} T^n f \cdot T^{2n} g$$

whose operator norms are uniformly bounded by $\|g\|_\infty$. If $f \in \ker(\lambda - T)$ for some $\lambda \in \mathbb{T}$, then

$$S_N f = \frac{1}{N} \sum_{n=0}^{N-1} (\lambda T^2)^n g \cdot f.$$

Since λT^2 is a contraction, it is mean ergodic (see Theorem 7.2), and the sequence $(S_N f)_{N\in\mathbb{N}}$ is convergent. By linearity and a standard approximation argument, we obtain the convergence on $H_{\mathrm{kr}} = \overline{\mathrm{lin}}\{f : Tf = \lambda f \text{ for some } \lambda \in \mathbb{T}\}$. The general case follows from Proposition 9.28 (recall that both f_{kr} and f_{aws} are bounded since P_{kr} is a Markov projection, see Proposition 8.11(e)). ∎

The following gives more information on the limit in (9.8).

Theorem 9.30 (Double Recurrence) *Let (X, μ, T) be an ergodic measure-preserving system and $f \in \mathrm{L}^\infty(X, \mu)$ satisfy $f > 0$. Then*

$$\lim_{N\to\infty} \frac{1}{N} \sum_{n=0}^{N-1} \int_X f \cdot T^n f \cdot T^{2n} f \,\mathrm{d}\mu = \Big(f \,\Big|\, \lim_{N\to\infty} \frac{1}{N} \sum_{n=0}^{N-1} T^n f \cdot T^{2n} f\Big) > 0.$$

In particular, for every $A \subset X$ with $\mu(A) > 0$

$$\mu(A \cap T^{-n} A \cap T^{-2n} A) > 0 \quad \textit{for some } n \in \mathbb{N}.$$

Remark 9.31 Double convergence in (9.8) and double recurrence hold also without the ergodicity assumption. This follows from first reducing a general measure-preserving system to a regular one and then using the ergodic decomposition discussed in Sect. 8.6. We leave the details as Exercise 9.9.

The following special case of Theorem 9.21 builds a bridge between Roth's Theorem 9.27 and double recurrence.

Proposition 9.32 (Correspondence Principle for 3-Term Progressions) *Assume that for every ergodic measure-preserving system* (X, μ, T) *and* $f \in \mathrm{L}^\infty(X, \mu)$ *with* $f > 0$

$$\lim_{N\to\infty} \frac{1}{N} \sum_{n=0}^{N-1} \int_X f \cdot T^n f \cdot T^{2n} f \,\mathrm{d}\mu = \Big(f \,\Big|\, \lim_{N\to\infty} \frac{1}{N} \sum_{n=0}^{N-1} T^n f \cdot T^{2n} f\Big) > 0. \tag{9.9}$$

Then every set $E \subset \mathbb{N}$ *with positive upper density contains arithmetic progressions of length* 3.

Thus to show Roth's Theorem 9.27 it remains to prove Theorem 9.30.

Proof of Theorem 9.30 By Theorem 9.29 we know that the limit

$$\lim_{N\to\infty} \frac{1}{N} \sum_{n=0}^{N-1} T^n f \cdot T^{2n} f$$

exists, implying the existence of the limits in (9.9). By (8.2) and (8.3) $P_{\mathrm{kr}} f \in \mathrm{L}^\infty(X, \mu)$ and $P_{\mathrm{kr}} f > 0$. Moreover, by the orthogonality of H_{aws} and H_{kr} and by Proposition 9.28 we obtain

$$\lim_{N\to\infty} \frac{1}{N} \sum_{n=0}^{N-1} \int_X f \cdot T^n f \cdot T^{2n} f \,\mathrm{d}\mu = \lim_{N\to\infty} \frac{1}{N} \sum_{n=0}^{N-1} \int_X P_{\mathrm{kr}} f \cdot T^n P_{\mathrm{kr}} f \cdot T^{2n} P_{\mathrm{kr}} f \,\mathrm{d}\mu.$$

Hence we may assume that $f \in \mathrm{L}^\infty(X, \mu) \cap H_{\mathrm{kr}}$, and without loss of generality also that $\|f\|_\infty \leq 1$. Let $0 < \varepsilon < \int_X f^3 \,\mathrm{d}\mu$. Define $K := \overline{T^{\mathbb{N}_0} f}$, which is a compact subset of $\mathrm{L}^2(X, \mu)$ by Proposition 6.23 and satisfies $TK = K$. We thus obtain a forward transitive topological system (K, T) which is easily seen to be minimal, see Exercise 9.3. By Proposition 3.6 every point, in particular $f \in K$, is almost periodic. For $U := \mathrm{B}(f, \varepsilon/3)$ the set of return times $R_U(f)$ is syndetic by the definition of almost periodicity (see Sect. 3.1). We have for each $n \in R_U(f)$ that

$$\|T^n f - f\|_2 < \frac{\varepsilon}{3} \quad \text{and}$$

$$\|T^{2n} f - f\|_2 \leq \|T^{2n} f - T^n f\|_2 + \|T^n f - f\|_2 < \frac{2\varepsilon}{3}.$$

From this we conclude

$$\begin{aligned}
&\left|\int_X f\cdot T^n f\cdot T^{2n} f\,\mathrm{d}\mu-\int_X f^3\,\mathrm{d}\mu\right|\\
&\le\left|\int_X f\cdot T^n f\cdot T^{2n} f\,\mathrm{d}\mu-\int_X f^2\cdot T^{2n} f\,\mathrm{d}\mu\right|+\left|\int_X f^2\cdot T^{2n} f\,\mathrm{d}\mu-\int_X f^3\,\mathrm{d}\mu\right|\\
&\le\int_X f\cdot|T^n f-f|\cdot T^{2n} f\,\mathrm{d}\mu+\int_X f^2\cdot|T^{2n} f-f|\,\mathrm{d}\mu\\
&\le\|f\|_\infty^2\|T^n f-f\|_2+\|f\|_\infty^2\|T^{2n} f-f\|_2<\varepsilon
\end{aligned}$$

for each $n\in R_U(f)$. As a consequence

$$\int_X f\cdot T^n f\cdot T^{2n} f\,\mathrm{d}\mu>\int_X f^3\,\mathrm{d}\mu-\varepsilon$$

for every such n. Since a syndetic set has positive lower density $d>0$ (if the length of the gaps is bounded by ℓ, then the lower density is $\ge\frac{1}{\ell}$, see Exercise 4.10), we conclude (see Exercise 9.4)

$$\liminf_{N\to\infty}\frac{1}{N}\sum_{n=0}^{N-1}\int_X f\cdot T^n f\cdot T^{2n} f\,\mathrm{d}\mu>d\cdot\Big(\int_X f^3\,\mathrm{d}\mu-\varepsilon\Big)>0.$$

The proof is complete. ∎

Remark 9.33 For generalizations of Roth theorem to groups and related results see Bergelson et al. [141], Bergelson and Leibman [131], Bergelson and McCutcheon [140], Solymosi [955], Zorin-Kranich and Chu [248], Durcik et al. [326].

The following generalization of Roth's theorem due to Szemerédi [970] was first proved using complex and ingenious arguments from combinatorics. Furstenberg [445] discovered an ergodic theoretic proof, and he established his correspondence principle precisely for this purpose.

Theorem 9.34 (Szemerédi) *A subset of the natural numbers with positive upper density contains arbitrarily long arithmetic progressions.*

We have proved the existence of three-terms arithmetic progressions, i.e., the result of Roth. The existence of longer progressions is much more difficult and will be discussed in detail in Chap. 13.

Exercises

Exercise 9.1 (Normal Numbers) Show that the number

$$0.01234567890123456789\ldots0123456789\ldots$$

(repeated blocks) is simply normal in base 10 but not normal in base 2.

Exercise 9.2 Prove that the assertion of Proposition 9.12 holds for every Riemann-integrable function $f : [0, 1] \to \mathbb{C}$. (Hint: Show it first for real-valued functions using an appropriate approximation.)

Exercise 9.3 (Isometric Systems) A topological system (K, T) is called **isometric** if the topology of K comes from a metric d satisfying $d(Tx, Ty) = d(x, y)$ for every pair $x, y \in K$. Prove that in an isometric system the notions of forward topological transitivity and minimality coincide.

Exercise 9.4 (Limit Inferior and Lower Density) Let $(a_n)_{n\in\mathbb{N}}$ be a sequence of positive numbers such that for some $\varepsilon > 0$ the set $A := \{n : a_n > \varepsilon\}$ has positive lower density. Show that

$$\liminf_{N\to\infty} \frac{1}{N} \sum_{n=1}^{N} a_n \geq \varepsilon \cdot \underline{\mathrm{d}}(A).$$

Exercise 9.5 (Skew Rotation/Shift)

(a) Show the "only if" part of Proposition 9.13.
(b) Prove Proposition 9.15.

Exercise 9.6 (Polynomial Averages for Totally Ergodic Systems) Let $P \in \mathbb{Z}[\cdot]$ be non-constant with $P(\mathbb{N}_0) \subset \mathbb{N}_0$ and $P(0) = 0$. Show that the trivial factor (with Markov projection denoted by $P_{\mathbb{C}\mathbf{1}}$) is characteristic for the convergence of the corresponding polynomial averages for totally ergodic systems, i.e., for every totally ergodic system (X, μ, T) and every $f \in \mathrm{L}^2(X, \mu)$ one has

$$\lim_{N\to\infty} \frac{1}{N} \sum_{n=0}^{N-1} T^{P(n)} f = \lim_{N\to\infty} \frac{1}{N} \sum_{n=0}^{N-1} T^{P(n)} P_{\mathbb{C}\mathbf{1}} f = \int_X f \, \mathrm{d}\mu \cdot \mathbf{1}.$$

Extend this to integer polynomials, cf. Remark 7.29(b) and Exercise 7.13(b).

Exercise 9.7 (Furstenberg's Correspondence Principle)

(a) Show that the set A in the proof of Theorem 9.21 has the properties (1)–(3) given there.
(b) Prove that the measure $\tilde{\mu}$ constructed in the proof of Theorem 9.21 is a T-invariant probability measure on X with $\tilde{\mu}(A) = \overline{\mathrm{d}}(E) > 0$.

(c) Show by modifying the construction that in Theorem 9.21 the upper density $\overline{\mathrm{d}}(E)$ can be replaced by the **upper Banach density** defined by

$$\overline{\mathrm{Bd}}(E) := \limsup_{N\to\infty} \sup_{M\in\mathbb{N}} \frac{|E \cap \{1+M, \ldots, N+M\}|}{N}.$$

(d) Prove the variant of Furstenberg's correspondence principle from (the first part of) Remark 9.22.

Exercise 9.8 (Polynomial Recurrence and Integer Polynomials) Extend the Furstenberg–Sárközy Theorem 9.19 and the polynomial recurrence (Theorem 9.17) to the class of integer polynomials, see Remark 9.20(b).

Exercise 9.9 (Double Convergence and Recurrence for General Systems) Work out the details from Remark 9.31.

Chapter 10
Equidistribution

As a preparation for the later chapters we discuss a classical number theoretic property which is closely related to ergodic theorems. Here we just touch this many-sided topic presenting only notions and results which we will need later. For more information see, e.g., Kuipers and Niederreiter [690].

A sequence $(s_n)_{n\in\mathbb{N}}$ is called **equidistributed** (or **uniformly distributed**) **in** $[0, 1)$ if for every interval $(a, b) \subset [0, 1)$,

$$\lim_{N\to\infty} \frac{|\{n \leq N : s_n \in (a, b)\}|}{N} = b - a. \tag{10.1}$$

One can replace open intervals (a, b) in the above by closed, or other types of intervals, still obtaining the same notion. A sequence $(s_n)_{n\in\mathbb{N}}$ in $\mathbb{R}$ is called **equidistributed** (or **uniformly distributed**) **modulo** 1 if the sequence $(\{s_n\})_{n\in\mathbb{N}}$ is equidistributed in $[0, 1)$. (Recall that $\{x\} = x - \lfloor x \rfloor$ is the fractional part of x.)

Next we define equidistributed sequences in $\mathbb{T}$ and call a connected subset $I \subset \mathbb{T}$ an **interval**. Consider the continuous mapping

$$e : \mathbb{R} \to \mathbb{T}, \quad t \mapsto \mathrm{e}^{2\pi i t}.$$

Then $I \subset \mathbb{T}$ is an interval if and only if $e^{-1}(I) = J + 2\pi\mathbb{Z}$ for some interval $J \subset \mathbb{R}$. Now, a sequence $(a_n)_{n\in\mathbb{N}}$ in $\mathbb{T}$ is called **equidistributed** (or **uniformly distributed**) **in** $\mathbb{T}$ if for every interval $I \subset \mathbb{T}$

$$\lim_{N\to\infty} \frac{|\{n \leq N : a_n \in I\}|}{N} = \mathrm{m}(I),$$

where m denotes the Haar measure on $\mathbb{T}$ (i.e., the normalized arclength measure).

Remark 10.1 It is easy to see that a sequence $(s_n)_{n\in\mathbb{N}}$ in $\mathbb{R}$ is equidistributed modulo 1 if and only if the sequence $(e(s_n))_{n\in\mathbb{N}}$ is equidistributed in $\mathbb{T}$, see

T. Eisner, B. Farkas, *A Journey Through Ergodic Theorems*, Birkhäuser Advanced Texts Basler Lehrbücher, https://doi.org/10.1007/978-3-031-96506-7_10

Exercise 10.1. Furthermore, we note that an equidistributed sequence is clearly dense in $\mathbb{T}$, $[0, 1)$ or in $\mathbb{R} \bmod 1$, respectively.

Remark 10.2 Observe that a sequence $(a_n)_{n\in\mathbb{N}}$ is equidistributed in $\mathbb{T}$ if and only if for each interval $I \subset \mathbb{T}$ we have

$$\mathrm{m}(I) = \int_{\mathbb{T}} \mathbf{1}_I \,\mathrm{dm} = \lim_{N\to\infty} \frac{1}{N} \sum_{n=1}^{N} \mathbf{1}_I(a_n).$$

The second equality motivates the characterization of equidistribution presented in the following section.

10.1 Weyl's Equidistribution Criterion

We present a classical criterion for equidistribution of sequences due to H. Weyl [1043]. While his original formulation was for equidistribution mod 1, we need the following considerations on the torus. A function $f : \mathbb{T} \to \mathbb{C}$ is called **Riemann integrable** if $f \circ e$ is Riemann integrable on $[0, 1]$. In this case we call

$$\int_0^1 f(e(t))\,\mathrm{d}t$$

the Riemann integral of f on $\mathbb{T}$, and we have $\int_{\mathbb{T}} f \,\mathrm{dm} = \int_0^1 f(e(t))\,\mathrm{d}t$.

Theorem 10.3 (Weyl's Equidistribution Criterion) *For a given sequence $(a_n)_{n\in\mathbb{N}}$ in $\mathbb{T}$ the following assertions are equivalent.*

(i) *The sequence $(a_n)_{n\in\mathbb{N}}$ is equidistributed in $\mathbb{T}$.*
(ii) *For every Riemann integrable function $f : \mathbb{T} \to \mathbb{C}$ one has*

$$\lim_{N\to\infty} \frac{1}{N} \sum_{n=1}^{N} f(a_n) = \int_{\mathbb{T}} f \,\mathrm{dm}. \tag{10.2}$$

(iii) *Every $f \in \mathrm{C}(\mathbb{T})$ satisfies (10.2); in other words, the measures $\frac{1}{N}\sum_{n=1}^{N} \delta_{a_n}$ converge to the Haar measure* m *in the weak* topology.*
(iv) *For every $\ell \in \mathbb{Z} \setminus \{0\}$ we have*

$$\lim_{N\to\infty} \frac{1}{N} \sum_{n=1}^{N} a_n^{\ell} = 0.$$

Analogous assertions hold for equidistribution in $[0, 1)$ *or modulo* 1, *respectively. In particular, a sequence* $(s_n)_{n\in\mathbb{N}}$ *is equidistributed modulo* 1 *if and only if for every* $\ell \in \mathbb{Z} \setminus \{0\}$

$$\lim_{N\to\infty} \frac{1}{N} \sum_{n=1}^{N} e(\ell s_n) = 0.$$

Proof We only prove the statement concerning equidistribution on $\mathbb{T}$, the rest is left as Exercise 10.1.

The implications (ii)⇒(iii)⇒(iv) are clear (the last one relies on the fact that the functions $\mathbf{z}^\ell : z \mapsto z^\ell$ have integral zero for $\ell \in \mathbb{Z} \setminus \{0\}$). Thus it remains to show the implications (iv)⇒(iii)⇒(i)⇒(ii).

(iv)⇒(iii): By assumption, (10.2) holds for all trigonometric monomials $f = \mathbf{z}^\ell$ with $\ell \in \mathbb{Z} \setminus \{0\}$, and it is immediate that it also holds for $f = \mathbf{1} = \mathbf{z}^0$. Thus, by linearity, (10.2) holds for every trigonometric polynomial. Since these are dense in $\mathrm{C}(\mathbb{T})$ (see Proposition 1.12) and since the operators $S_N : f \mapsto \frac{1}{N}\sum_{n=1}^{N} f(a_n)$, $N \in \mathbb{N}$, are uniformly bounded, (iii) follows by a standard approximation argument.

(iii)⇒(i): Let $I \subset \mathbb{T}$ be an interval. We have to prove (10.2) for the function $f = \mathbf{1}_I$ (see Remark 10.2). Let $\varepsilon > 0$. It is easy to construct $g, h \in \mathrm{C}(\mathbb{T})$ with $g \le \mathbf{1}_I \le h$ and $\int_{\mathbb{T}} (h - g)\, \mathrm{dm} < \varepsilon$ (take, for instance, suitable piecewise linear approximations of $\mathbf{1}_I$). Observe that $0 \le \int_{\mathbb{T}} (h - \mathbf{1}_I)\, \mathrm{dm} < \varepsilon$ and $0 \le \int_{\mathbb{T}} (\mathbf{1}_I - g)\, \mathrm{dm} < \varepsilon$, and therefore

$$\begin{aligned} \frac{1}{N}\sum_{n=1}^{N} g(a_n) - \int_{\mathbb{T}} g\, \mathrm{dm} - \varepsilon &< \frac{1}{N}\sum_{n=1}^{N} \mathbf{1}_I(a_n) - \int_{\mathbb{T}} \mathbf{1}_I\, \mathrm{dm} \\ &< \frac{1}{N}\sum_{n=1}^{N} h(a_n) - \int_{\mathbb{T}} h\, \mathrm{dm} + \varepsilon. \end{aligned}$$

Now, (iii) applied to g and h yield that for sufficiently large N

$$-2\varepsilon < \frac{1}{N}\sum_{n=1}^{N} \mathbf{1}_I(a_n) - \int_{\mathbb{T}} \mathbf{1}_I\, \mathrm{dm} < 2\varepsilon$$

proving the validity of (10.2) for $f = \mathbf{1}_I$.

(i)⇒(ii): By assumption, (10.2) holds for functions of the form $\mathbf{1}_I$ and hence for all step functions which are (finite) linear combinations of such characteristic functions. Since for every Riemann integrable function $f : \mathbb{T} \to \mathbb{R}$ and every $\varepsilon > 0$ there are step functions g, h on $\mathbb{T}$ with $g \le f \le h$ and $\int_{\mathbb{T}} (h - g)\, \mathrm{dm} < \varepsilon$, the same argument as in the proof of (iii)⇒(i) above finishes the proof (it suffices to prove (ii) for real-valued functions). ∎

Example 10.4 (Kronecker) For every irrational $\lambda \in \mathbb{T}$ the sequence $(\lambda^n)_{n\in\mathbb{N}}$ is equidistributed in $\mathbb{T}$. This is essentially contained in Proposition 9.12, but also follows directly from Weyl's equidistribution criterion (Theorem 10.3). Indeed, take $\ell \in \mathbb{Z} \setminus \{0\}$ and observe

$$\lim_{N\to\infty} \frac{1}{N} \sum_{n=1}^{N} \lambda^{\ell n} = \lim_{N\to\infty} \frac{\lambda^\ell}{N} \cdot \frac{1-\lambda^{\ell N}}{1-\lambda^\ell} = 0,$$

implying the equidistribution property. In the additive setting this means that the sequence $(n\alpha)_{n\in\mathbb{N}}$ is equidistributed modulo 1 for every irrational $\alpha \in \mathbb{R}$.

The following gives a (theoretical) way to produce equidistributed sequences due to Weyl [1043]. We follow here the exposition by Rosenblatt and Wierdl [881, Thm. 2.22], see also Kuipers and Niederreiter [690, Thm. 1.4.1].

Proposition 10.5 *Let $(k_n)_{n\in\mathbb{N}}$ be a strictly monotone sequence in $\mathbb{Z}$. Then for* m-*almost every $\lambda \in \mathbb{T}$ the sequence $(\lambda^{k_n})_{n\in\mathbb{N}}$ is equidistributed.*

Proof The functions f_n on $\mathbb{T}$ given by $f_n(z) := z^{k_n}$ for $n \in \mathbb{N}$ are pairwise orthogonal in $\mathrm{L}^2(\mathbb{T}, \mathrm{m})$ and satisfy $\|f_n\|_2 = 1$. The Banach–Rajchman Lemma 9.7 implies that for m-almost every $\lambda \in \mathbb{T}$ we have

$$0 = \lim_{N\to\infty} \frac{1}{N} \sum_{n=1}^{N} f_n(\lambda) = \lim_{N\to\infty} \frac{1}{N} \sum_{n=1}^{N} \lambda^{k_n}.$$

To show the equidistribution property, take $\ell \in \mathbb{Z} \setminus \{0\}$. Since $(\ell k_n)_{n\in\mathbb{N}}$ is strictly monotone, we can replace k_n by ℓk_n in the above and obtain that

$$\lim_{N\to\infty} \frac{1}{N} \sum_{n=1}^{N} \lambda^{\ell k_n} = 0 \text{ for m-almost every } \lambda \in \mathbb{T}.$$

Since a countable intersection of full measure sets still has full measure, Weyl's equidistribution criterion (Theorem 10.3) finishes the proof. ∎

More (and rather involved) examples of equidistributed sequences will be discussed in Sect. 10.4 below. On the other hand, it is easy to construct artificial examples of sequences which are not equidistributed, even if dense. The following is a non-trivial and *natural* example of such a sequence involving the natural logarithm log.

Example 10.6 Let $a_n := e(\log(n))$. Then $(a_n)_{n\in\mathbb{N}}$ is not equidistributed in $\mathbb{T}$ or, equivalently, the sequence $(\log(n))_{n\in\mathbb{N}}$ is not equidistributed modulo 1. To see this latter property we take the interval $I = (0, 1/2)$ and compute how often $\{\log(n)\}$ belongs to it as n ranges over $\mathbb{N}$. Let $N \in \mathbb{N}$ and $m \in \{0, \dots, N-1\}$. We have

$$\log(n) \in (m, m+1/2) \quad \Longleftrightarrow \quad n \in (\mathrm{e}^m, \mathrm{e}^{1/2}\mathrm{e}^m).$$

Thus,

$$\left|\left\{n \le \left\lfloor e^N \right\rfloor : \{\log(n)\} \in I\right\}\right| \le \sum_{m=0}^{N-1} \left((e^{1/2} - 1)e^m + 1\right)$$
$$= \frac{(e^{1/2} - 1)(e^N - 1)}{e - 1} + N,$$

implying

$$\lim_{N\to\infty} \frac{\left|\left\{n \le \left\lfloor e^N \right\rfloor : \{\log(n)\} \in I\right\}\right|}{\left\lfloor e^N \right\rfloor} \le \frac{e^{1/2} - 1}{e - 1} = \frac{1}{e^{1/2} + 1} < \frac{1}{2}.$$

Thus (10.1) fails and $(\log(n))_{n\in\mathbb{N}}$ is not equidistributed modulo 1. On the other hand, it is easy to see that $(a_n)_{n\in\mathbb{N}}$ is dense in $\mathbb{T}$ using that the sequence $\log(n)$ goes monotonically to infinity with $\log(n+1)-\log(n)$ tending to zero, see Exercise 10.2.

Remark 10.7 The function log is an example of a **logarithmico-exponential function** (i.e., a real function constructed by using finitely many combinations of usual arithmetic symbols and the function symbols exp and log) of subpolynomial growth. For such logarithmico-exponential functions f there is a complete characterization of the equidistribution property of the sequence $(e(f(n)))_{n\in\mathbb{N}}$ in terms of the growth of f, see Boshernitzan [177]. As an example we mention that $(e(n^\alpha))_{n\in\mathbb{N}}$ is equidistributed for every $\alpha > 0$.

10.2 Multidimensional Equidistribution

Condition (iii) in Theorem 10.3 motivates the following generalization to sequences in compact metric spaces.

Definition 10.8 Let K be a compact metric space with a Borel probability measure μ. Then a sequence $(a_n)_{n\in\mathbb{N}}$ in K is called **equidistributed** in K (with respect to the reference measure μ) if for every $f \in \mathrm{C}(K)$

$$\lim_{N\to\infty} \frac{1}{N} \sum_{n=1}^{N} f(a_n) = \int_{\mathbb{T}} f \, d\mu,$$

i.e., if the measures $\frac{1}{N}\sum_{n=1}^{N} \delta_{a_n}$ converge to μ in the weak* topology.

Remark 10.9 If μ has full support, then every equidistributed sequence $(a_n)_{n\in\mathbb{N}}$ in K with respect to μ is dense in K, see Exercise 10.3.

For $K = \mathbb{T}^d$ we take $\mu = \mathrm{m}^d$ the normalized Haar measure as the reference measure. The following criterion is a multidimensional analogue of Theorem 10.3 (again due to Weyl [1043]) which we state multiplicatively only and leave the proof as Exercise 10.4.

Theorem 10.10 (Weyl's Equidistribution Criterion for $\mathbb{T}^d$) *For a sequence $(a_n)_{n\in\mathbb{N}}$ in $\mathbb{T}^d$ the following assertions are equivalent.*

(i) *The sequence $(a_n)_{n\in\mathbb{N}}$ is equidistributed in $\mathbb{T}^d$.*
(ii) *For every $\ell \in \mathbb{Z}^d \setminus \{0\}$*

$$\lim_{N\to\infty} \frac{1}{N} \sum_{n=1}^{N} a_n^{\ell} = 0,$$

where the multi-index notation $(x_1, \ldots, x_d)^{\ell} := x_1^{\ell_1} \cdot \ldots \cdot x_d^{\ell_d}$ is used.

The following is a multidimensional version of Example 10.4.

Example 10.11 (Kronecker) For $\lambda \in \mathbb{T}^d$ and $n \in \mathbb{N}$ consider $a_n := \lambda^n = (\lambda_1^n, \ldots, \lambda_d^n)$. Then $(a_n)_{n\in\mathbb{N}}$ is equidistributed if and only if the corresponding topological system $(\mathbb{T}^d, \lambda)$ is minimal, i.e., if and only if $\lambda \in \mathbb{T}^d$ is **irrational** in the sense that for every $\ell \in \mathbb{Z}^d$

$$\lambda^{\ell} = (1, \ldots, 1) \quad \Longrightarrow \quad \ell = (0, \ldots, 0) \tag{10.3}$$

(cf. Theorem 3.40). The proof is left as Exercise 10.5. Note that the additive version of the irrationality condition for $\alpha \in \mathbb{R}^d$ is just linear independence of the coordinates over $\mathbb{Q}$, or, equivalently, for every $\ell \in \mathbb{Z}^d$

$$\ell_1\alpha_1 + \cdots + \ell_d\alpha_d = 0 \quad \Longrightarrow \quad \ell = (0, \ldots, 0). \tag{10.4}$$

Remark 10.12 The irrationality of λ implies irrationality of every coordinate of λ, but the converse implication is not true (cf. Theorem 3.40). Take for example λ whose coordinates are irrational and all equal. Although, by Example 10.11, the sequence $(\lambda^n)_{n\in\mathbb{N}}$ for such λ is not equidistributed in $\mathbb{T}^d$ (if $d \geq 2$), it is equidistributed in the diagonal $\{(z, \ldots, z) : z \in \mathbb{T}\}$ with respect to the induced (1-dimensional normalized Lebesgue, or Haar) measure on it (by Example 10.4). In general, if λ is rational in the above sense, i.e., $\lambda^l = (1, \ldots, 1)$ for some $l \neq (0, \ldots, 0)$, then the sequence $(\lambda^n)_{n\in\mathbb{N}}$ is equidistributed in a finite union of translated subtori. For this and more see, e.g., Tao [980, Prop. 1.1.5], [972].

For more on equidistribution as well as on related notions such as well-distribution, total equidistribution, δ-equidistribution see, e.g., Kuipers and Niederreiter [690] and Tao [980, Ch. 1].

10.3 Equidistribution of Polynomial Sequences

In Chap. 7 we have discussed a powerful tool developed by the number theorist J. van der Corput to estimate exponential sums. We recall Lemma 7.31 in the special case of scalar-valued sequences.

Lemma 10.13 (Van der Corput's Lemma) *Let $(u_n)_{n\in\mathbb{N}}$ be a bounded sequence in $\mathbb{C}$ and define for $h \in \mathbb{N}$*

$$\gamma_h := \limsup_{N\to\infty} \frac{1}{N}\left|\sum_{n=1}^{N} u_{n+h}\overline{u_n}\right|.$$

Then

$$\liminf_{J\to\infty} \frac{1}{J}\sum_{h=1}^{J} \gamma_h = 0 \quad\Longrightarrow\quad \lim_{N\to\infty} \frac{1}{N}\sum_{n=1}^{N} u_n = 0.$$

Based on this we can immediately deduce the following important corollary.

Corollary 10.14 *Let $(a_n)_{n\in\mathbb{N}}$ be a sequence in the compact group $\mathbb{T}^d$ such that for every $h \in \mathbb{N}$ the sequence $(a_{n+h}\overline{a_n})_{n\in\mathbb{N}}$ is equidistributed in $\mathbb{T}^d$. Then $(a_n)_{n\in\mathbb{N}}$ itself is equidistributed in $\mathbb{T}^d$.*

Proof By Theorem 10.10 it suffices to show that for every $\ell \in \mathbb{Z}^d \setminus \{0\}$

$$\lim_{N\to\infty} \frac{1}{N}\sum_{n=1}^{N} a_n^{\ell} = 0. \tag{10.5}$$

Take $\ell \in \mathbb{Z}^d \setminus \{0\}$ and define $u_n := a_n^{\ell}$. For $h \in \mathbb{N}$ we observe that, by the hypothesis and by Theorem 10.10,

$$\lim_{N\to\infty} \frac{1}{N}\sum_{n=1}^{N} u_{n+h}\overline{u_n} = \lim_{N\to\infty} \frac{1}{N}\sum_{n=1}^{N} (a_{n+h}\overline{a_n})^{\ell} = 0.$$

Thus (10.5) follows from Lemma 10.13. ∎

Remark 10.15 The sequences $(a_{n+h}\overline{a_n})_{n\in\mathbb{N}}$ are often called **(multiplicative) discrete derivatives** of $(a_n)_{n\in\mathbb{N}}$. This terminology becomes apparent when writing a_n as $e(s_n)$, so $a_{n+h}\overline{a_n} = e(s_{n+h} - s_n)$. The additive version of Corollary 10.14, where the terms $a_{n+h}\overline{a_n}$ are replaced by $s_{n+h} - s_n$, is called **van der Corput's difference theorem**, see Kuipers and Niederreiter [690, Thm. 3.1].

Recall that, by Proposition 10.5, for every subsequence $(k_n)_{n\in\mathbb{N}}$ of $\mathbb{N}$ the sequence $(\lambda^{k_n})_{n\in\mathbb{N}}$ is equidistributed in $\mathbb{T}$ for $\mathrm{m}_{\mathbb{T}}$ almost every $\lambda \in \mathbb{T}$. However, it is sometimes important to know for which λ exactly equidistribution occurs. Our aim in the rest of this chapter is to show that whenever $(k_n)_{n\in\mathbb{N}}$ is produced by polynomials, by primes or by polynomials of primes, one has equidistribution for every irrational λ. This will be important later.

We first need a simple fact whose proof we leave as Exercise 10.6.

Lemma 10.16 *Let $m \in \mathbb{N}$. For $\alpha \in \mathbb{R}$ and a sequence $(a_n)_{n\in\mathbb{N}}$ in $\mathbb{R}$ which is bounded from below*

$$\liminf_{N\to\infty} \frac{1}{N}\sum_{n=1}^{N} a_n \geq \alpha \quad \Longleftrightarrow \quad \liminf_{N\to\infty} \frac{1}{mN}\sum_{n=1}^{mN} a_n \geq \alpha.$$

For $\alpha \in \mathbb{C}$ and a bounded sequence $(a_n)_{n\in\mathbb{N}}$ in $\mathbb{C}$

$$\lim_{N\to\infty} \frac{1}{N}\sum_{n=1}^{N} a_n = \alpha \quad \Longleftrightarrow \quad \lim_{N\to\infty} \frac{1}{mN}\sum_{n=1}^{mN} a_n = \alpha.$$

The following is a polynomial version of Example 10.4. The proof we present is due to van der Corput [1012].

Theorem 10.17 (Weyl, Equidistribution of Polynomial Sequences) *Let P be a real polynomial, $P(x) = a_d x^d + \cdots + a_1 x + a_0$, such that at least one of the coefficients $a_1, \ldots, a_d$ is irrational. Then the sequence $(e(P(n)))_{n\in\mathbb{N}}$ is equidistributed in $\mathbb{T}$.*

Proof We proceed by induction on the degree d of P. If $d = 1$, the assertion follows from Example 10.4. Suppose that $d \geq 2$ and the assertion holds for polynomials with degree smaller than d.

We first consider the case when the leading coefficient a_d is rational and take $m \in \mathbb{N}$ such that $ma_d \in \mathbb{Z}$. Observe that for $\ell \in \mathbb{Z}\setminus\{0\}$ we have

$$\begin{aligned}\frac{1}{N}\sum_{n=1}^{mN} e(\ell P(n)) = \frac{1}{N}\sum_{n=1}^{N} e(\ell P(mn)) + \frac{1}{N}\sum_{n=1}^{N} e(\ell P(mn-1)) + \cdots \\ + \frac{1}{N}\sum_{n=1}^{N} e(\ell P(mn-m+1)).\end{aligned}$$

The polynomials $Q_{\ell,j}$ defined as $Q_{\ell,j}(x) := \ell P(mx - j) - \ell m^d a_d x^d$, $j \in \{0, \ldots, m-1\}$, are of degree at most $d-1$. They have at least one irrational coefficient (different from the constant term) and $e(Q_{\ell,j}(n)) = e(\ell P(mn - j))$ for each $j \in \{0, \ldots, m-1\}$ and $n \in \mathbb{N}$ since $ma_d \subset \mathbb{Z}$. Hence the equidistribution of $(e(P(n)))_{n\in\mathbb{N}}$ follows from the induction hypothesis, Lemma 10.16 and Theorem 10.3.

Suppose now that the leading coefficient of P is irrational. Define $u_n := e(P(n))$ and observe that

$$u_{n+h}\overline{u_n} = e(P(n+h) - P(n)) = e(Q_h(n))$$

for every $h \in \mathbb{N}$, where Q_h is the polynomial defined as $Q_h(x) := P(x+h) - P(x)$. It is easy to see that $\deg(Q_h) = \deg(P) - 1 = d - 1$ and that the leading coefficient of Q_h is irrational as well. The induction hypothesis together with Corollary 10.14 finishes the argument. ■

Remark 10.18 It is clear that equidistribution fails if all coefficients of P, with the possible exception of the constant term, are rational, see Exercise 10.7.

Analogously one can prove the multidimensional case (we leave the details to the reader).

Proposition 10.19 (Equidistribution of Multivariate Polynomials) *Let $P \in \mathbb{R}^d[\cdot]$ be a multivariate polynomial with at least one irrational (in the sense of* (10.4)*) coefficient different from the constant term. Then the sequence $(e(P(n)))_{n\in\mathbb{N}}$ is equidistributed in $\mathbb{T}^d$.*

In Sect. 9.2 we briefly mentioned how equidistribution is connected to unique ergodicity of certain group extensions. Here we illustrate by an easy example how equidistribution can be used in ergodic theory, providing the promised proof of the "if" part of Proposition 9.13.

Proposition 10.20 *If $a \in \mathbb{T}$ is irrational, then the skew shift $(\mathbb{T}^2, T_a)$ is uniquely ergodic.*

Proof First of all we recall that $(\mathbb{T}^2, \mathrm{m}^2, T_a)$ is ergodic (see Example 8.34). By Corollary 7.25 we therefore need to show

$$\lim_{N\to\infty} \frac{1}{N} \sum_{n=0}^{N-1} T_a^n f = \int_{\mathbb{T}} f \,\mathrm{dm} \cdot \mathbf{1} \tag{10.6}$$

for every $f \in \mathrm{C}(\mathbb{T}^2)$ (pointwise or uniformly, all the same here, see Sect. 7.3). We first consider trigonometric monomials $f = \mathbf{z}^m$, $m = (m_1, m_2) \in \mathbb{Z}^2$, where $\mathbf{z}^m = \mathbf{z}_1^{m_1}\mathbf{z}_2^{m_2}$ for $\mathbf{z}_1$ and $\mathbf{z}_2$ given by $\mathbf{z}_1(x, y) := x$ and $\mathbf{z}_2(x, y) := y$. For such f we have by Proposition 9.15

$$\frac{1}{N} \sum_{n=0}^{N-1} T_a^n \mathbf{z}^m = \frac{1}{N} \sum_{n=0}^{N-1} a^{m_1 n} \mathbf{z}_1^{m_1} \cdot a^{m_2 \frac{n(n-1)}{2}} \mathbf{z}_1^{n m_2} \mathbf{z}_2^{m_2}.$$

By the equidistribution theorem for polynomials (Theorem 10.17) and Weyl's equidistribution criterion (Theorem 10.3) we have for $x, y \in \mathbb{T}$ and $m \neq (0, 0)$

$$\lim_{N\to\infty} \frac{1}{N} \sum_{n=0}^{N-1} T_a^n \mathbf{z}^m(x, y) = \lim_{N\to\infty} \frac{1}{N} \sum_{n=0}^{N-1} a^{m_1 n} a^{m_2 \frac{n(n-1)}{2}} x^{nm_2} x^{m_1} y^{m_2} = 0.$$

Since (10.6) is trivial for $m = (0, 0)$, we obtain (10.6) for every trigonometric monomial, and then by linearity for every trigonometric polynomial. A standard density argument proves (10.6) for every $f \in \mathrm{C}(\mathbb{T}^2)$. ∎

10.4 Equidistribution of Primes and Polynomials of Primes

Up to now the book was more or less self-contained. This changes now dramatically since we turn our attention to prime numbers and need deep results from number theory. These will be used as a black box, and we only prepare here some important tools for the next chapters.

Denote by $\mathbb{P}$ the set of primes and list them increasingly in the sequence $(p_n)_{n\in\mathbb{N}}$, i.e., $p_1 = 2$, $p_2 = 3$ etc. Denote by $\pi : \mathbb{R} \to \mathbb{N}_0$ the prime counting function given by

$$\pi(x) := |\{n : p_n \leq x\}|.$$

The following classical result was proved independently by Hadamard and de la Vallée Poussin in 1896. For details see, e.g., Jameson [574, Sec. 3.4].

Theorem 10.21 (Prime Number Theorem) *The function π satisfies*

$$\lim_{x\to\infty} \frac{\pi(x)}{\frac{x}{\log(x)}} = 1.$$

To average over primes we shall need the following **modified von Mangoldt function** Λ' defined as

$$\Lambda'(n) := \log(n)\mathbf{1}_{\mathbb{P}}(n) = \begin{cases} \log(n), & n \in \mathbb{P}, \\ 0, & \text{otherwise}. \end{cases}$$

It involves a simple computation to show that the prime number theorem is equivalent to the following asymptotics for the function Λ'

$$\lim_{N\to\infty} \frac{1}{N} \sum_{n=1}^{N} \Lambda'(n) = 1, \tag{10.7}$$

see Exercise 10.9.

Notation 10.22 In what follows, we use p to denote a prime number without writing explicitly $p \in \mathbb{P}$.

The following easy but very useful result shows one important property of the function Λ': Averaging along the primes is the same as averaging with the weight Λ' along the natural numbers. Note that the boundedness assumption can be removed as observed by Wierdl [1047], cf. Exercise 12.8(b).

Lemma 10.23 *For a bounded sequence $(b_n)_{n\in\mathbb{N}}$ in $\mathbb{C}$ we have*

$$\lim_{N\to\infty} \left| \frac{1}{N} \sum_{n=1}^{N} \Lambda'(n) b_n - \frac{1}{\pi(N)} \sum_{p\le N} b_p \right| = 0. \tag{10.8}$$

Proof Assume without loss of generality that $|b_n| \le 1$ for every $n \in \mathbb{N}$. For $N \in \mathbb{N}$ we have by the prime number theorem (Theorem 10.21)

$$\begin{aligned}
\limsup_{N\to\infty} \left| \frac{1}{\pi(N)} \sum_{p\le N} b_p - \frac{1}{N} \sum_{n=1}^{N} \Lambda'(n) b_n \right| &= \limsup_{N\to\infty} \left| \frac{1}{N} \sum_{p\le N} (\log(N) - \log(p)) b_p \right| \\
&\le \lim_{N\to\infty} \left(\frac{\log(N)\cdot\pi(N)}{N} - \frac{1}{N} \sum_{p\le N} \log(p) \right) = 1 - \lim_{N\to\infty} \frac{1}{N} \sum_{n=1}^{N} \Lambda'(n) = 0,
\end{aligned}$$

where we used (10.7) for the last equality. ■

We now need some further notation. For $w \in \mathbb{N}$ define

$$W = W_w := \prod_{p\le w} p.$$

For $r < W$ coprime to W consider the **modified Λ'-function** defined as

$$\Lambda'_{r,w}(n) := \frac{\varphi(W)}{W} \Lambda'(Wn + r), \tag{10.9}$$

where the **Euler totient function** φ is defined by

$$\varphi(n) := |\{a < n : \ a, n \text{ are coprime}\}|.$$

We mention the following refinement of the prime number theorem being a qualitative version of the famous Siegel–Walfisz theorem, see, e.g., Jameson [574, Sec. 4.4].

Theorem 10.24 (Prime Number Theorem in Arithmetic Progressions) *Let $a, q \in \mathbb{N}$ be coprime and denote by $\pi_{a,q}(x)$ the number of primes not exceeding x which are congruent to a modulo q. Then*

$$\lim_{x\to\infty} \frac{\pi_{a,q}(x)}{\frac{x}{\log(x)}} = \frac{1}{\varphi(q)}.$$

This means that the primes are asymptotically equally distributed between the residue classes mod q. Again one can prove that this implies

$$\lim_{N\to\infty} \frac{1}{N}\sum_{n=1}^{N} \Lambda'_{r,w}(n) = 1.$$

The following is a special case of a deep result of Green and Tao [486, Prop. 10.2], see [340, Cor. 2.2], which we will use as black box. (Note that Green–Tao's result was formulated for Lipschitz nilsequences corresponding to connected, simply connected groups. We will define nilsequences in Sect. 14.1 and see that every polynomial sequence of the form $(e(P(n)))$ for $P \in \mathbb{R}[\cdot]$ is such a nilsequence, see in Example 14.13(c) and Remark 14.14.)

Theorem 10.25 (Green–Tao) *For every real polynomial P*

$$\lim_{w\to\infty} \limsup_{N\to\infty} \max_{r<W,(r,W)=1} \left| \frac{1}{N}\sum_{n=1}^{N} (\Lambda'_{r,w}(n) - 1)e(P(n)) \right| = 0. \tag{10.10}$$

We note that the result of Green and Tao in [486] is conditional to the so-called Möbius-Nilsequence-Conjecture which was settled by them later in [488].

The following restricts the sequence $\mathbb{N}$ in Example 10.4 to the primes and polynomials of primes. We will use the foregoing deep theorem of Green and Tao to reduce equidistribution of polynomials of primes to equidistribution of polynomials.

Theorem 10.26 (Equidistribution of (Polynomials) of Primes) *For every real polynomial P with at least one irrational non-constant coefficient the sequence $(P(p_n))_{n\in\mathbb{N}}$ is equidistributed modulo 1. In the multiplicative setting, for every such P the sequence $(e(P(p_n)))_{n\in\mathbb{N}}$ is equidistributed in $\mathbb{T}$.*

Proof Let P be an arbitrary real polynomial with at least one irrational non-constant coefficient. By Weyl's equidistribution criterion (Theorem 10.3) it is enough to show

$$\lim_{N\to\infty}\frac{1}{N}\sum_{n=1}^{N} e(P(p_n)) = 0.$$

Define $a_n := \Lambda'(n)e(P(n))$. By Lemma 10.23 it suffices to show

$$\lim_{N\to\infty}\frac{1}{N}\sum_{n=1}^{N} a_n = 0. \tag{10.11}$$

Since for $n > 1$ we have $a_n \neq 0$ if and only if $n \in \mathbb{P}$, Lemma 10.16 implies for every ω and $W = W_\omega$

$$\lim_{N\to\infty}\frac{1}{N}\sum_{n=1}^{N} a_n = \lim_{N\to\infty}\frac{1}{WN}\sum_{n=1}^{WN} a_n = \frac{1}{W}\sum_{r<W,\,(r,W)=1}\lim_{N\to\infty}\frac{1}{N}\sum_{n=1}^{N} a_{Wn+r}.$$

Thus, taking $\lim_{\omega\to\infty}$ and using Theorem 10.25 leads to

$$\begin{aligned}\lim_{N\to\infty}\frac{1}{N}\sum_{n=1}^{N} a_n &= \lim_{\omega\to\infty}\frac{1}{W}\sum_{r<W,\,(r,W)=1}\lim_{N\to\infty}\frac{1}{N}\sum_{n=1}^{N} a_{Wn+r}\\ &= \lim_{\omega\to\infty}\frac{1}{\varphi(W)}\sum_{r<W,\,(r,W)=1}\lim_{N\to\infty}\frac{1}{N}\sum_{n=1}^{N}\Lambda'_{r,\omega}(n)e(P(Wn+r))\\ &= \lim_{\omega\to\infty}\frac{1}{\varphi(W)}\sum_{r<W,\,(r,W)=1}\lim_{N\to\infty}\frac{1}{N}\sum_{n=1}^{N} e(P(Wn+r)).\end{aligned} \tag{10.12}$$

Since $P(W\cdot{+r})$ are again real polynomials with at least one irrational non-constant coefficient, the sequences $(P(Wn+r))_{n\in\mathbb{N}}$ are all equidistributed modulo 1 by Theorem 10.17 and hence the right-hand side of (10.12) equals zero by Weyl's equidistribution criterion (Theorem 10.3). Thus (10.11) is proved, and the proof is complete. ■

Of course, the above proof is so short because it uses the deep and powerful Theorem 10.25. The original proof of Theorem 10.26 in some special cases[1] is

---

[1] The estimation of the exponential sums corresponding to P given by $P(n) = n$ was the key step in Vinogradov's proof of the ternary Goldbach problem for sufficiently large integers, see, e.g., Vaughan [1015, Ch. 3].

due to Vinogradov [1022–1025] and uses the classical Hardy–Littlewood method (see, for example, Vaughan [1015]). Using Vinogradov's approach, Rhin [869] proved Theorem 10.26 together with the converse direction, namely that for a real polynomial P, the sequence $(P(p_n))_{n\in\mathbb{N}}$ is equidistributed mod 1 if and only if at least one of the non-constant coefficients of P is irrational. (Note that by Weyl's Theorem 10.17 this is the case if and only if the sequence $(P(n))_{n\in\mathbb{N}}$ is equidistributed modulo 1.) For yet another, shorter proof and more equidistribution results along primes see Bergelson et al. [126]. In particular, they generalize Rhin's result showing that for every Hardy field function (e.g., a logarithmico-exponential function) f of subpolynomial growth, $(f(p_n))_{n\in\mathbb{N}}$ is equidistributed modulo 1 if and only if $(f(n))_{n\in\mathbb{N}}$ is so.

Exercises

Exercise 10.1 (Equidistribution modulo 1)

(a) Prove that a sequence $(s_n)_{n\in\mathbb{N}}$ in $\mathbb{R}$ is equidistributed modulo 1 if and only if the sequence $(e(s_n))_{n\in\mathbb{N}}$ is equidistributed in $\mathbb{T}$.
(b) Formulate analogous conditions to the ones in Theorem 10.3 for equidistribution modulo 1 and prove the equivalence of them.

Exercise 10.2 Show that the set $\{\{\log(n)\} : n \in \mathbb{N}\}$ is dense in $[0, 1)$ (cf. Example 10.6).

Exercise 10.3 (Equidistribution in Compact Metric Spaces)

(a) Let K be a compact metric space and let $(a_n)_{n\in\mathbb{N}}$ be a sequence in K. Prove that if $(a_n)_{n\in\mathbb{N}}$ is equidistributed with respect to a measure with full support, then $(a_n)_{n\in\mathbb{N}}$ is dense.
(b) Construct a dense sequence in $K := \{0, 1\}$ which is not equidistributed with respect to any probability measure on K.

Exercise 10.4 (Multidimensional Weyl Criterion) Prove Theorem 10.10.

Exercise 10.5 (Equidistribution in $\mathbb{T}^d$) Work out the details of Example 10.11.

Exercise 10.6 (Cesàro Convergence) Prove Lemma 10.16.

Exercise 10.7 (Equidistribution of Polynomials) Show that the sequence $(e(P(n)))_{n\in\mathbb{N}}$ is not equidistributed if all coefficients of the polynomial P are rational, with the possible exception of the constant term. This shows that the irrationality assumption in Theorem 10.17 is necessary.

Exercise 10.8 (Bergelson, Weak Mixing along Subsequences and Equidistribution) Prove the assertion of Remark 6.29(c) and show in particular the following characterization of weak mixing: A measure-preserving system (X, μ, T) is weakly mixing if and only if for every measurable $A, B \subset X$ one has

$$\operatorname{D-}\lim_{n\to\infty} \mu(A \cap T^{-n^2} B) = \mu(A)\mu(B).$$

(Hint: For the "if" implication use the spectral theorem.)

Show further that one can replace (n^2) by every subsequence $(k_n) \subset \mathbb{N}$ such that (λ^{k_n}) is equidistributed in $\mathbb{T}$ for every irrational $\lambda \in \mathbb{T}$, e.g., by (p_n) or (p_n^2) (see Theorem 10.26).

Exercise 10.9 (Prime Number Theorem and the Modified von Mangoldt Function) Prove the equivalence

$$\lim_{n\to\infty} \frac{\pi(n)\log(n)}{n} = 1 \quad \Longleftrightarrow \quad \lim_{N\to\infty} \frac{1}{N}\sum_{n=1}^{N} \Lambda'(n) = 1.$$

(Hint: Use summation by parts.)

Chapter 11
Groups, Semigroups and Homogeneous Spaces

Topological groups and semigroups are fundamental objects in ergodic theory and this chapter collects some very basic facts about them, with the aim to set up the stage for *nilsystems*, which will be discussed in Chap. 14.

We first concentrate on groups and assume familiarity with the basic notions of group theory, such as the notion of a group itself, subgroups, conjugation, normal (invariant) subgroups, etc., as well as some fundamental results such as the isomorphism and homomorphism theorems (see e.g., Lang [706, Ch. I]). Here we focus, and that also very briefly, only on the new aspects which arise in connection with an additional topology. Books to be recommended here are Deitmar and Echterhoff [279], Rudin [888], Montgomery and Zippin [803], Hewitt and Ross [530, 531], Folland [395], and Hofmann and Morris [539].

11.1 Topological Groups

Group rotations are important examples of measure-preserving systems and have been mentioned briefly in Example 2.14. Here we provide some more detail about them and about topological groups in general. These will be crucial in Chap. 14. Recall that a group G endowed with a topology is a **topological group** if its multiplication and the inverse mapping

$$\mathrm{mul} : G \times G \to G, \quad (a, b) \mapsto ab \quad \text{and} \quad \mathrm{inv} : G \to G, \quad a \mapsto a^{-1},$$

are continuous. Left and right multiplications τ_g and ρ_g as well as the conjugation $\varphi_g := \tau_{g^{-1}} \circ \rho_g = \rho_g \circ \tau_{g^{-1}}$ by a fixed element $g \in G$ are therefore homeomorphisms. The identity (unit element) in the group G is denoted by id_G, often abbreviated by id. We use the following notation for given subsets $A, B \subset G$ of

T. Eisner, B. Farkas, *A Journey Through Ergodic Theorems*, Birkhäuser Advanced Texts Basler Lehrbücher, https://doi.org/10.1007/978-3-031-96506-7_11

a group G and an element $g \in G$:

$$A^{-1} := \{a^{-1} : a \in A\},$$
$$Ag := \{ag : a \in A\} \quad \text{and} \quad gB = \{gb : b \in B\},$$
$$AB := \{ab : a \in A, b \in B\},$$
$$A^1 := A,\ A^2 := A^1A = AA^1, \dots, A^{n+1} := A^nA = AA^n \quad (n \in \mathbb{N}).$$

A subset $A \subset G$ is called **symmetric** if $A^{-1} = A$.

Direct products and subgroups of topological groups are again topological groups (with the product topology and the subspace topology, respectively). Recall that a mapping $\varphi : G \to H$ between two groups is a **(group) homomorphism** if it is multiplicative, i.e., for each pair $g, h \in G$ one has $\varphi(gh) = \varphi(g)\varphi(h)$ (it then follows that $\varphi(\mathrm{id}_G) = \mathrm{id}_H$ and that $\varphi(g^{-1}) = \varphi(g)^{-1}$ for every $g \in G$). A conjugation φ_g is a homomorphism, and in fact an **isomorphism**, i.e., has an inverse (which is then automatically a homomorphism). An isomorphism of topological groups is an isomorphism of the (underlying algebraic) groups which is also a homeomorphism.

Proposition 11.1 *Let G be a Hausdorff topological group.*

(a) *If $U \subset G$ is an open neighborhood of id_G, then there is an open, symmetric neighborhood V of id_G with $VV \subset U$.*
(b) *The closure $\overline{H}$ of a subgroup $H \leq G$ is a subgroup of G.*
(c) *If $U \subset G$ is open and $A \subset G$ is arbitrary, then AU and UA are open.*
(d) *If $K \subset G$ is compact and $A \subset G$ is closed, then AK and KA are closed. If A is also compact, then AK and KA are compact, too.*

We leave the proof of these assertions as Exercise 11.1.

Proposition 11.2 *Let G, H be topological groups.*

(a) *A homomorphism $\varphi : G \to H$ is continuous if and only if it is continuous at id_G (or at any other element).*
(b) *If $\varphi : G \to H$ is a continuous group homomorphism and H is Hausdorff, then the* **kernel** $\ker(\varphi) := \varphi^{-1}(\mathrm{id}_H)$ *is a closed, normal subgroup of G.*
(c) *If N is a normal subgroup of G, then so is its closure $\overline{N}$.*

We leave the proof of this proposition as Exercise 11.2.

Let G be a group and $H \leq G$ be a subgroup. Recall that different sets of the form aH (called **left cosets**) are pairwise disjoint, and thus yield a partition of G. We define $G/H := \{gH : g \in G\}$, the set of left cosets of H. Suppose that G is a Hausdorff topological group, and consider the quotient mapping

$$q : G \to G/H, \quad g \mapsto gH.$$

Call a set $U \subset G/H$ open if $q^{-1}(U) = \bigcup_{gH \in U} gH$ is open in G. This defines a topology on G/H, the so-called **quotient topology**, which renders q continuous (it is the finest topology with this property). Moreover, the quotient mapping q is open, i.e., maps open sets into open sets. The quotient topology is Hausdorff if and only if H is closed (see Montgomery and Zippin [803, I.1.16]).

If N is a normal subgroup of G, then the multiplication

$$gN \cdot hN := ghN$$

and the quotient topology render G/N a topological group, called the **quotient** (or factor) **group**. The quotient map q is a (continuous and open) surjective homomorphism.

Subgroups and Connectedness

A non-empty subset H of a group is a subgroup if and only if $H^{-1}H \subset H$ (indeed, from this property it follows that $\mathrm{id}_G \in H$, $H^{-1} \subset H$, $H = H^{-1}$, and finally $HH \subset H$). Recall that a subset $A \subset X$ of a topological space is called **connected** if $A \subset A_0 \cup A_1$ with $A_0, A_1 \subset X$ open and $A \cap A_0 \cap A_1 = \emptyset$ implies $A_0 \cap A = \emptyset$ or $A_1 \cap A = \emptyset$ (equivalently, one could replace "open" by "closed" here). If X is connected then the only subsets of X that are both open and closed are $\emptyset$ and X. A continuous image and the closure of a connected set is connected. If $\mathcal{A}$ is a collection of connected subsets of X with the property that for each $A, B \in \mathcal{A}$ one has $A \cap B \neq \emptyset$, then $\bigcup_{A \in \mathcal{A}} A$ is connected. A **connected component** of a topological space is a connected subset which is maximal with respect to inclusion. Connected components are closed and yield a partition of the topological space, homeomorphisms permute connected components.

Proposition 11.3 *Let G be a topological group.*

(a) *Let $H \leq G$ be an open subgroup. Then H is closed and moreover, for every set $S \subset G$, the sets HS and SH are (open and) closed.*
(b) *Let U be a symmetric, open neighborhood of* id_G. *Then*

$$H := \bigcup_{n=1}^{\infty} U^n$$

is an open and closed subgroup.
(c) *If G is connected and U is a neighborhood of* id_G, *then*

$$G = \bigcup_{n=1}^{\infty} U^n.$$

The proof is left as Exercise 11.3. In the following we use the notation $\langle A \rangle$ for the subgroup generated by a $A \subset G$ in a given group G, i.e., the smallest subgroup containing A.

Proposition 11.4 *Let G be a Hausdorff topological group, and let G° be the connected component of the identity* $\mathrm{id} \in G$.

(a) *The component G° is a closed normal subgroup of G. Each connected component of G has the form gG° for some $g \in G$.*
(b) *Suppose G is locally connected (i.e., each open neighborhood of* id *contains a connected open neighborhood of* id*). Then for every set $S \subset G$, the set $G^\circ S = SG^\circ$ and the subgroup $\langle G^\circ, S \rangle$ generated by G° and S are (open and) closed.*
(c) *Let $H \leq G$ be a subgroup of G. Then $H^\circ \subset G^\circ$, and one has $H^\circ = G^\circ$ if and only if H is open.*

We leave the proof as Exercise 11.4.

Metrizability

For Hausdorff topological groups first countability is sufficient for metrizability, and that even by a special metric as shown by the next result.

Proposition 11.5 (Birkhoff and Kakutani) *Let G be a Hausdorff topological group such that the identity* id_G *has a countable neighborhood basis. Then G is metrizable and, moreover, there is a right invariant metric on G which induces the topology of G. If G is additionally compact, then there is an invariant metric inducing its topology.*

For a proof we refer to Montgomery and Zippin [803, I.1.22], see also the original papers by Birkhoff [154] and by Kakutani [609].

Locally Compact Groups

Recall that a Hausdorff topological space[1] X is called **locally compact** if every point in X has a compact neighborhood. By the Hausdorff property each point in X has a neighborhood basis consisting of compact sets. A **locally compact group** is a (Hausdorff) topological group whose topology is locally compact. On a locally compact space there are sufficiently many continuous functions. In fact Urysohn's Lemma 1.13 tells that for a compact set $K \subset X$ in a locally compact space X there

[1] We include here the Hausdorff property for the sake of convenience.

is $f \in \mathrm{C_c}(X)$ with $f|_K = \mathbf{1}$. Recall that $\mathrm{C_c}(X)$ denotes the $\mathbb{C}$-vector space of complex-valued continuous functions with compact support.

If G is a locally compact group and H is a subgroup, then the coset space G/H is locally compact. Indeed, if K is a compact neighbourhood of id_G, then under the quotient mapping $q(K)$ is a compact neighborhood of $H \in G/H$. By homogeneity each point in G/H has a compact neighborhood. If N is a closed normal subgroup of a locally compact group, then G/N is Hausdorff and a locally compact group.

Proposition 11.6 (Koopman Representation) *Let G be a locally compact group.*

(a) *Every function $f \in \mathrm{C_c}(G)$ is uniformly continuous, meaning that for every $\varepsilon > 0$ there is a neighborhood U of id_G such that for every $g, h \in G$ with $g^{-1}h \in U$ one has*

$$|f(g) - f(h)| < \varepsilon.$$

(b) *For $g \in G$ let L_g denote the Koopman operator of the left rotation by g. For each $f \in \mathrm{C_c}(G)$ the mapping*

$$G \to \mathrm{C_c}(G), \quad g \mapsto L_g f$$

is continuous with respect to the supremum norm on $\mathrm{C_c}(G)$.

For a proof we refer to Deitmar and Echterhoff [279, Lem. 1.3.6] (or [342, Prop. 10.11] for the case of compact groups).

Compact Group Rotations

If $a \in G$ is a fixed element, then we can consider the left rotation by a on G and on G/H:

$$g \mapsto ag, \quad \tau_a : gH \mapsto agH,$$

both mappings being continuous. If G is compact and H is closed, then G/H is Hausdorff and compact, and we obtain two *compact* topological systems (G, a) and $(G/H, \tau_a)$, where the latter is abbreviated also as $(G/H, a)$. The quotient mapping $q : G \to G/H$ intertwines the left rotations by a, i.e., $q(ag) = aq(g)$ for each $a \in G$.

Proposition 11.7 (Mean Ergodicity of Compact Group Rotations) *Let G be a compact topological group and let $a \in G$. Then the Koopman operator L_a of the* *left rotation by a is mean ergodic on* $\mathrm{C}(G)$*, i.e., for every* $f \in \mathrm{C}(G)$ *the Cesàro averages*

$$\frac{1}{N}\sum_{n=1}^{N} L_a^n f$$

converge in the supremum norm.

Proof For each $f \in \mathrm{C}(G)$ the mapping

$$G \to \mathrm{C}(G), \quad g \mapsto L_g f$$

is continuous by Proposition 11.6, so by compactness of G, the orbit $\{L_a^n f : n \in \mathbb{N}_0\}$ is relatively compact in $\mathrm{C}(G)$. The same follows for the convex hull $\mathrm{conv}(\{L_a^n f : n \in \mathbb{N}_0\})$, so in particular, the sequence $(A_N f)$ of Cesàro averages possesses a convergent subsequence. By Theorem 7.12 the mean ergodicity of L_a follows. ■

Corollary 11.8 (Uniform Convergence of Multiple Ergodic Averages) *Let* (G, a) *be a compact group rotation system with Koopman operator* L_a *and let* $k \in \mathbb{N}$*. For every* $f_1, \ldots, f_k \in \mathrm{C}(G)$ *the multiple ergodic averages*

$$\frac{1}{N}\sum_{n=1}^{N} L_a^n f_1 \cdot L_a^{2n} f_2 \cdots L_a^{kn} f_k \tag{11.1}$$

converge uniformly.

Proof Consider the compact group rotation system $(G^k, (a, a^2, \ldots, a^k))$ which is mean ergodic on $\mathrm{C}(G^k)$ by Proposition 11.7. So for the function $F = f_1 \otimes f_2 \otimes \cdots \otimes f_k$ the Cesàro averages

$$\frac{1}{N}\sum_{n=1}^{N} L^n_{(a,a^2,\ldots,a^k)} F = \frac{1}{N}\sum_{n=1}^{N} L_a^n f_1 \otimes L_a^{2n} f_2 \otimes \cdots \otimes L_a^{kn} f_k$$

converge uniformly on G^k, in particular, on the diagonal $\{(g, g, \ldots, g) : g \in G\}$, which is the assertion. ■

11.2 Haar Measure on Locally Compact Groups

Rotations on compact groups (endowed with their unique Haar probability measures) are one of the chief examples of measure-preserving systems. Here we discuss (left and right) invariant measures in a greater generality.

Let X be a locally compact space and recall that $\mathrm{C_c}(X)$ denotes the $\mathbb{C}$-vector space of complex-valued, continuous functions with compact support. A linear functional $\varphi : \mathrm{C_c}(X) \to \mathbb{C}$ is called **positive** if $\varphi(f) \geq 0$ for every $f \in \mathrm{C_c}(X)$ with $f \geq 0$. For example, if μ is a (positive) Radon measure, then

$$\varphi_\mu(f) = \int_X f \, \mathrm{d}\mu \quad (f \in \mathrm{C_c}(X))$$

defines a positive linear functional on $\mathrm{C_c}(X)$. By the Riesz–Markov–Kakutani representation theorem (see Theorem 1.23) each positive linear functional on $\mathrm{C_c}(X)$ arises in this way.

Let now G be a locally compact group. Then for each $a \in G$ the left and right multiplications $g \mapsto ag$, $g \mapsto ga$ are homeomorphisms. We denote the corresponding Koopman operators on the space $\mathrm{C_c}(G)$ of continuous functions by L_a and R_a, respectively. A non-zero (positive) Radon measure m on G is called **left-invariant** (or a **left Haar measure**) if $\varphi_\mathrm{m} \circ L_a = \varphi_\mathrm{m}$ for every $a \in G$, and analogously **right-invariant** (or a **right Haar measure**) if $\varphi_\mathrm{m} \circ R_a = \varphi_\mathrm{m}$ for every $a \in G$. The corresponding functionals φ_m are called left (respectively, right) **Haar integrals**. Let $\mathrm{inv} : G \to G$, $g \mapsto g^{-1}$ be the inversion mapping. Then for a left Haar measure m the image measure $\mathrm{inv}_*\mathrm{m}$ is a right Haar measure (and vice versa).

Example 11.9

(1) If G is a discrete group, then the counting measure m is a left and right invariant measure.
(2) For the torus $\mathbb{T}$ with the Euclidean topology, the arclength measure m is (left and right) invariant.
(3) If m_1 and m_2 are left Haar measures on G_1 and G_2, respectively, then the product measure $\mathrm{m}_1 \times \mathrm{m}_2$ is a left Haar measure on $G_1 \times G_2$. This, of course, extends to finite direct products.
(4) For the multiplicative group $(0, \infty)$ (with the Euclidean topology) the functional φ defined via

$$\varphi(f) := \int_0^\infty \frac{f(x)}{x} \, \mathrm{d}x \quad (f \in \mathrm{C_c}((0, \infty))$$

is a left and right Haar integral, see Exercise 11.12.
(5) Consider the group $G = \mathrm{GL}_n(\mathbb{R})$ of invertible $n \times n$ matrices with the topology inherited from the Euclidean topology on $\mathbb{R}^{n^2}$ (here and below we identify a matrix with an element of $\mathbb{R}^{n^2}$). The functional φ defined via

$$\varphi(f) = \int_G \frac{f(A)}{|\det(A)|^n} \, \mathrm{d}\lambda^{n^2}(A) \quad (f \in \mathrm{C_c}(G))$$

is a (left and right) Haar integral, see Exercise 11.13.

Proposition 11.10 *Let G be a locally compact group, and let* m *be a left Haar measure. Then* $\mathrm{supp}(\mathrm{m}) = G$*, or equivalently, the corresponding left Haar integral* $\varphi = \varphi_{\mathrm{m}}$ *is strictly positive, i.e., for every* $f \in \mathrm{C_c}(G)$ *with* $f \geq 0$ *and* $f \neq 0$ *one has*

$$\int_G f \,\mathrm{dm} = \varphi_{\mathrm{m}}(f) > 0.$$

The proof of this result is left as Exercise 11.11.

Now it is a result of fundamental importance that locally compact groups carry a left Haar measure (and then, by what is said above, also a right Haar measure).

Proposition 11.11 *Let G be a locally compact group. Then there is a left Haar measure* m *and any other left Haar measure on G is a constant multiple (by a strictly positive factor) of* m.

For a proof we refer, e.g., to Deitmar and Echterhoff [279, Sec. 1.3] or to Nachbin [812, Sec. II.8, II.9].

Since for each $a \in G$ and a left Haar integral φ_{m} also $\varphi_{\mathrm{m}} \circ R_a$ is a left Haar integral, by the previous result there is (a unique) $\Delta_{\mathrm{m}}(a) > 0$ with $\Delta_{\mathrm{m}}(a)\varphi_{\mathrm{m}} \circ R_a = \varphi_{\mathrm{m}}$. The function $\Delta := \Delta_G : G \to (0, \infty)$, $a \mapsto \Delta_{\mathrm{m}}(a)$, is independent of the chosen left Haar measure m and is called the **modular function** of the group G.

Proposition 11.12 *Let G be a locally compact group with modular function Δ and a left Haar measure* m.

(a) *For each* $f \in \mathrm{C_c}(G)$ *and* $g \in G$ *we have*

$$\int_G R_g f \,\mathrm{dm} = \Delta(g)^{-1} \int_G f \,\mathrm{dm} = \Delta(g)^{-1} \int_G L_g f \,\mathrm{dm}.$$

(b) *For the inversion mapping* $\mathrm{inv} : G \to G$, $g \mapsto g^{-1}$ *and for each* $f \in \mathrm{C_c}(G)$ *we have*

$$\int_G f \circ \mathrm{inv} \,\mathrm{dm} = \int_G f \cdot \Delta \circ \mathrm{inv} \,\mathrm{dm}.$$

(c) $\Delta : G \to (0, \infty)$ *is a continuous homomorphism.*

We leave the proof as Exercise 11.10 (or refer to Deitmar and Echterhoff [279, Thm. 1.4.1]). A locally compact group G is called **unimodular** if $\Delta = \mathbf{1}$, which holds if and only if there is a left Haar measure which is also right invariant (in which case every left Haar measure has this property). A **Haar measure** is, by definition, simultaneously left and right invariant. A locally compact group has a Haar measure if and only if it is unimodular. Discrete groups (multiples of the counting measure are the Haar measures), Abelian groups are (obviously) such, just as *nilpotent groups*, to be discussed in Chap. 14 below (see Nachbin [812, Ch. 2, Prop.25]). It turns out that compact groups have this property, too.

A locally compact group is said to be **σ-compact** if it is a union of countably many compact sets. In this case any (left, right) Haar measure is σ-finite.

In what follows we write $\mathrm{L}^p(G)$ for $\mathrm{L}^p(G, \mathrm{m})$, where m is a given left (or right) Haar measure on G.

Compact Groups

Let G be a compact group. Suppose that ν is a left invariant and μ is a right invariant probability measure on G. For each $f \in \mathrm{C}(G) = \mathrm{C_c}(G)$ we have

$$\begin{aligned}\int_G f(x)\,\mathrm{d}\mu(x) &= \int_G\int_G f(x)\,\mathrm{d}\mu(x)\,\mathrm{d}\nu(y) = \int_G\int_G f(xy)\,\mathrm{d}\mu(x)\,\mathrm{d}\nu(y)\\ &= \int_G\int_G f(xy)\,\mathrm{d}\nu(y)\,\mathrm{d}\mu(x) = \int_G\int_G f(y)\,\mathrm{d}\nu(y)\,\mathrm{d}\mu(x)\\ &= \int_G f(y)\,\mathrm{d}\nu(y),\end{aligned}$$

whence it follows that $\mu = \nu$. This implies that if there is a left invariant probability measure on G, then it is unique and, it is also right invariant (and the unique probability measure with this property). It is not hard to prove that a locally compact group with left Haar measure m is compact if and only if $\mathrm{m}(G) < \infty$ (see, e.g., [279, Prop. 1.4.5]).

Theorem 11.13 (Haar Measure on Compact Groups) *On every compact group G there is a unique left invariant Borel probability measure. Every left Haar measure is finite and also right invariant and inversion invariant.*

On compact groups the existence of a left Haar measure is somewhat easier to prove than in the general case of locally compact groups. By virtue of the Riesz–Markov–Kakutani theorem it suffices to prove that

$$\bigcap_{g\in G} \mathrm{fix}(L'_g)$$

contains a non-zero (positive) measure, where L_g is the Koopman operator of the left multiplication τ_g and L'_g is its adjoint. So appropriate fixed point theorems yield a left Haar measure. In the Abelian case, the Markov–Kakutani fixed point theorem, see [611, Thm. 1] and [772], does this, while for the general case of compact groups the Ryll-Nardzewski fixed point theorem, see [896], is directly applicable, see [254, §11]. There are other proofs which rely on fixed point theorems (e.g., of Kakutani [611, Thm. 2]) and on the Banach space duality between $\mathrm{C}(K)$ and $\mathrm{M}(K)$, see, e.g., [890, pp. 128–132], [606, Sec. 15.3.3], or [342, App. G.4].

11.3 Some Fourier Analysis

In this section we restrict our attention to locally compact, Abelian groups, and show that one can carry out Fourier-analytic calculations similarly to the case of $\mathbb{T}^d$ as done in Chap. 4.

Definition 11.14 Let G be a (locally) compact Abelian group. A continuous homomorphism $\chi : G \to \mathbb{T}$ is called a **character** of G.

Denote by $\hat{G}$ the set of all characters of G. The pointwise multiplication turns $\hat{G}$ into an Abelian group with identity element $\mathbf{1}$, the constant 1 character, while the inverse of $\chi \in \hat{G}$ is $\chi^{-1} = \overline{\chi}$. The group $\hat{G}$ is called the **dual group** (or **character group**) of G.

Proposition 11.15 *Let G and H be locally compact Abelian groups.*

(a) *Isomorphic topological groups have isomorphic dual groups.*
(b) *The dual group of $H \times G$ is isomorphic to the direct product $\hat{H} \times \hat{G}$ of the respective dual groups.*
(c) *If N is a closed, normal subgroup of G and $H = G/N$, then $\hat{H}$ is isomorphic to a subgroup of $\hat{G}$.*

We leave the proof as Exercise 11.5.

Example 11.16 (Finite Groups)

(a) If $G = \mathbb{Z}_m = \langle a \rangle$ is a cyclic group of order m, then $\hat{G}$ is isomorphic to $\mathbb{Z}_m$. More precisely, for each mth root of unity $\lambda \in \mathbb{T}$
$$a^n \mapsto \lambda^n$$
defines a character on G, and these mappings constitute the dual group.
(b) If G is a finite Abelian group (considered with the discrete topology), then it is isomorphic to a direct product of finite cyclic groups. So G and $\hat{G}$ are isomorphic groups.

Example 11.17 (The Additive Groups $\mathbb{R}^d$) The dual group of $\mathbb{R}^d$ is isomorphic to $\mathbb{R}^d$. More precisely, for $x, y \in \mathbb{R}^d$ let us set $e_y(x) := \mathrm{e}^{2\pi \mathrm{i}(x|y)}$. Then
$$\hat{G} = \{e_y : y \in \mathbb{R}^d\}$$
and the mapping $y \mapsto e_y$ is the required isomorphism, see Exercise 11.8.

Example 11.18 (Torus) The dual group of $\mathbb{T}^d$ is isomorphic to $\mathbb{Z}^d$. More precisely, we have
$$\hat{G} = \{\mathbf{z}^m : m \in \mathbb{Z}^d\},$$

with $\mathbf{z}^m : (z_1, \ldots, z_d) \mapsto z_1^{m_1} \cdots z_d^{m_d}$, and $m \mapsto \mathbf{z}^m$ is an isomorphism, see Exercise 11.8.

If G is a locally compact Abelian group, then endowed with the so-called **compact open topology** (or the topology of uniform convergence on compact sets) the dual group $\hat{G}$ becomes also a locally compact Abelian group. In the previous examples the group isomorphisms are, in fact, topological ones, i.e., also homeomorphisms.

Remark 11.19

(1) If G is compact, then $\hat{G}$ is endowed with the subspace topology inherited from $\mathrm{C}(G)$. In this case $\hat{G}$ is discrete.
(2) If G is discrete, then the compact subsets of G are the finite ones, and $\hat{G}$ is endowed with the pointwise topology, i.e., the subspace topology inherited from $\mathbb{T}^G$. In this case, $\hat{G}$ is compact.
(3) One can form the bidual group $\hat{\hat{G}}$, and **Pontrjagin's duality theorem** states that the mapping

$$G \to \hat{\hat{G}}, \quad g \mapsto \delta_g, \text{ the evaluation at } g \text{ given by } \delta_g(\chi) = \chi(g),$$

is an isomorphism of topological groups (here the proof of surjectivity presents the greatest challenge), see, e.g., to Deitmar and Echterhoff [279, Ch. 3] or Hewitt and Ross [530, Ch. Six].
(4) If G is a locally compact Abelian group which is compactly generated (i.e., $G = \langle K \rangle$ for a compact subset $K \subset G$), then it is isomorphic as topological group to a Cartesian product $\mathbb{R}^d \times \mathbb{Z}^m \times H$, where $d, m \in \mathbb{N}_0$ and H is a compact group, see, e.g., Deitmar and Echterhoff [279, Ch. 4] or Hewitt and Ross [530, Sec. 9].

In the rest of this subsection we concentrate on compact (Abelian) groups G. In this case, each character $\chi \subset \hat{G}$ belongs to $\mathrm{L}^2(G)$. As usual, on a compact group we consider the (unique) normalized Haar measure m, i.e. $\mathrm{m}(G) = 1$.

Proposition 11.20 (Orthonormality of Characters) *Let G be a compact Abelian group with normalized Haar measure* m, *and take* $\chi, \psi \in \hat{G}$. *Then*

$$(\chi|\psi)_{\mathrm{L}^2(G)} = \int_G \chi\overline{\psi}\,\mathrm{dm} = \begin{cases} 0 & \text{if } \chi \neq \psi, \\ 1 & \text{if } \chi = \psi. \end{cases}$$

Proof Since for $\chi, \psi \in \hat{G}$ also $\overline{\chi}\psi$ is a character, it is enough to show $(\chi|\mathbf{1}) = 0$ for $\chi \neq \mathbf{1}$. Take $\chi \in \hat{G}$ and $g \in G$. Then, by the invariance of m

$$\chi(g)(\chi|\mathbf{1}) = \int_G \chi(g)\chi\,\mathrm{dm} = \int_G \chi(g\cdot)\,\mathrm{dm} = \int_G \chi\,\mathrm{dm} = (\chi|\mathbf{1}).$$

If $(\chi|\mathbf{1}) \neq 0$, then $\chi(g) = 1$ holds for each $g \in G$, i.e., $\chi = \mathbf{1}$. ■

On a locally compact Abelian group there are sufficiently many characters to distinguish the group elements.

Proposition 11.21 (Characters Separate the Points) *Let G be a locally compact Abelian group. For each $g, h \in G$ with $g \neq h$, there is $\chi \in \hat{G}$ with $\chi(g) \neq \chi(h)$.*

The proof of this result is non-trivial. The case of discrete groups is much easier and can be found, e.g., in [342, Ch. 14., Suppl.]. For the general case one needs to develop representation theory for locally compact groups and some basic theory of Banach- and C^*-algebras, see, e.g., Deitmar and Echterhoff [279] or Hewitt and Ross [530, Sec. 22]. In combination with the Stone–Weierstraß theorem this yields the next important result.

Corollary 11.22 *Let G be a compact Abelian group with dual group $\hat{G}$. Then* $\operatorname{lin} \hat{G}$ *is dense in both* $\mathrm{C}(G)$ *and in* $\mathrm{L}^2(G)$.

Proof Consider the conjugation invariant subalgebra $A = \operatorname{lin} \hat{G}$ of $\mathrm{C}(G)$, which by Proposition 11.21 separates the points of G. So, by the Stone–Weierstraß Theorem 1.11 A is dense in $\mathrm{C}(G)$ (and then also in $\mathrm{L}^2(G)$). ∎

As an immediate consequence we obtain the following version of a Fourier-series expansion (recall that different characters are orthogonal by Proposition 11.20).

Proposition 11.23 (Fourier Series Expansion) *Let G be a compact Abelian group. The characters of G form an orthonormal basis in $\mathrm{L}^2(G)$, i.e., for each $f \in \mathrm{L}^2(G)$*

$$f = \sum_{\chi \in \hat{G}} (f|\chi)_{\mathrm{L}^2(G)} \chi,$$

where the summands are pairwise orthogonal (hence the sum is unconditionally convergent in $\mathrm{L}^2(G)$).

Remark 11.24 If G is a compact, metrizable group, then $\mathrm{C}(G)$ and $\mathrm{L}^2(G)$ are separable and $\hat{G}$ is at most countable. So if $\hat{G}$ is countably infinite and $(\chi_m)_{m \in \mathbb{N}_0}$ is any enumeration of $\hat{G}$, then for every $f \in \mathrm{L}^2(G)$ one can write

$$f = \sum_{m=0}^{\infty} (f|\chi_m) \chi_m .$$

Hence we can speak of an unconditionally convergent series.

From Proposition 11.7 we already know that the Koopman operator of a compact group rotation system (G, a) is mean ergodic on $\mathrm{C}(G)$, but for Abelian groups this also follows with Fourier series techniques, and we can describe ergodic properties in a manner similar to the proof of Proposition 3.25.

Proposition 11.25 *Let (G, a) be an Abelian group rotation system with Koopman operator L_a on $\mathrm{C}(G)$. Then L_a is mean ergodic on $\mathrm{C}(G)$, i.e., for each $f \in \mathrm{C}(G)$ the limit*

$$\lim_{N\to\infty} \frac{1}{N} \sum_{n=0}^{N-1} L_a^n f \tag{11.2}$$

exists in $\mathrm{C}(G)$ (with respect to the supremum norm). If (G, a) is two-sided topologically transitive, then the limit in (11.2) equals

$$\int_G f \,\mathrm{dm} \cdot \mathbf{1},$$

and, in particular, (G, m, a) is ergodic.

Proof Since the closed linear hull of $\hat{G}$ is $\mathrm{C}(G)$ and the Cesàro averages of the Koopman operator L_a are uniformly bounded, it suffices to establish convergence of the Cesàro averages for $f = \chi \in \hat{G}$. Since $L_a^n \chi = \chi(a)^n \chi$ for each $n \in \mathbb{Z}$, the existence of the limit in (11.2) follows at once (for the supremum norm). Now if (G, a) is two-sided topologically transitive, then $\mathrm{fix}(L_a) = \mathbb{C}\mathbf{1}$ is one-dimensional (in $\mathrm{C}(G)$, see Exercise 3.4), and the limit in (11.2) is a constant function. By integrating with respect to m, we conclude the assertion. The last statement concerning ergodicity follows by a standard density argument. ∎

Theorem 11.26 (Ergodicity and Minimality of Group Rotations) *Let G be a compact (not assumed to be Abelian) group and $a \in G$. Then for the rotation system (G, a) the following assertions are equivalent.*

(i) *(G, a) is minimal.*
(ii) *The set $\{a^n : n \in \mathbb{N}_0\}$ is dense in G.*
(iii) *(G, a) is topologically transitive.*
(iv) *(G, a) is topologically two-sided transitive.*
(v) *(G, m, a) is ergodic.*
(vi) *(G, a) is uniquely ergodic.*

If either of these condition holds, then G is Abelian and the Haar measure m is the unique invariant probability measure for (G, a).

Proof The implications (i)⇒(ii)⇒(iii)⇒(iv) are trivial (and were discussed partly in Proposition 3.2).
(iv)⇒(v): If $\{a^n g : n \in \mathbb{Z}\}$ is dense in G for some $g \in G$, then $\{a^n : n \in \mathbb{Z}\}$ is dense in G, and G is obviously Abelian. Now, Proposition 11.25 gives the ergodicity of (G, m, a).
(v)⇒(vi): We prove that m is the unique invariant probability measure for (G, a). Since, by Proposition 11.7, L_a is mean ergodic on $\mathrm{C}(G)$, ergodicity of (G, m, a)

implies that

$$\lim_{N\to\infty}\frac{1}{N}\sum_{n=1}^{N}L_a^n f = \int_G f\,\mathrm{dm}\cdot\mathbf{1}\quad\text{uniformly}$$

for every $f \in \mathrm{C}(G)$, see Proposition 3.24. Integrating this equality with respect to any invariant probability measure ν for (G, a), we conclude

$$\int_G f\,\mathrm{dm} = \int_G f\,\mathrm{d}\nu\quad\text{for every } f \in \mathrm{C}(G).$$

This yields $\mathrm{m} = \nu$.
Implication (vi)⇒(i) follows from Corollary 7.25 and the fact that the Haar measure m has full support. ■

11.4 Homogeneous Spaces

Let G be a Hausdorff topological group and X a Hausdorff topological space. Let $S : G \times X \to X$ be a continuous mapping such that for every $g \in G$ the mapping $S_g : x \mapsto S(g, x)$ is a homeomorphism of X and such that $g \mapsto S_g$ is a group homomorphism. Then we call (X, G, S) a G-**space** and S a G-**action**, we write gx for $S(g, x)$ if the action S is understood, and use the abbreviation (X, G) for (X, G, S). Two G-spaces (X, G) and (Y, G) are **isomorphic** if there is a homeomorphism $\Phi : Y \to X$ which commutes with the G-actions, i.e. for each $x \in X$ and $a \in G$ one has $\Phi(ax) = a\Phi(x)$. If the action of G on X is **transitive**, i.e., for each pair of points $x, y \in X$ there is $g \in G$ with $gx = y$ (or equivalently $g^{-1}y = x$), then we call (X, G) a **homogeneous** G-**space**. (This property is preserved by isomorphisms.)

Important examples of homogeneous spaces are provided by cosets spaces.

Example 11.27 (Coset Spaces) Let G be a Hausdorff topological group and let H be a closed subgroup of G. Consider the quotient mapping $q : G \to G/H$, which is continuous and open if G/H is endowed with the quotient topology. The group G acts on G/H by left multiplications:

$$G/H \to G/H,\quad gH \mapsto agH.$$

In this way we obtain the G-space $(G/H, G)$, which is easily seen to be homogeneous. If G is connected, so is G/H (this follows from the continuity of the quotient mapping).

Proposition 11.28 (Homogeneous Spaces vs. Coset Spaces) *Let (X, G) be a homogeneous G-space and let $x \in X$ be an arbitrary point. Then $H := \mathrm{Stab}(x) := \{g \in G : gx = x\}$ is a closed subgroup, called the* ***stabilizer*** *of x, and the mapping*

$$\Phi : G/H \to X, \quad \Phi(gH) := gx,$$

is well defined, bijective and continuous. The G-action commutes with Φ. Furthermore, the mapping Φ is a homeomorphism if (and only if) the mapping $G \to X$, $g \mapsto gx$ is open. In this case $(G/H, G)$ and (X, G) are (topologically) isomorphic G-spaces. This is so, in particular, if G is compact.

Proof The set $H := \mathrm{Stab}_G(x)$ is easily seen to be a closed subgroup of G. For $g \in G$ and $h \in H$ we have $ghx = gx$, so Φ is indeed well defined. It is plain to see that the mapping $\Phi : G/H \to X$ is injective, and since G acts transitively, it is even surjective. The inverse of Φ is given by $\Phi^{-1}(gx) = gH$. We prove that Φ is continuous. Take $g \in G$ and an open subset $U \subset X$ containing gx. Since the action is continuous, there is an open set $V \subset G$ with $g \in V$ such that $Vx = \{hx : h \in V\} \subset U$. Since the quotient mapping $q : G \to G/H$ is open (see Sect. 11.1), the set $q(V)$ is open in G/H, and $\Phi(q(V)) \subset U$. So Φ is continuous. (If G is compact, so is G/H and Φ is a homeomorphism.) For each $a, g \in G$ we have $\Phi(agH) = (ag)x = a\Phi(gH)$, so the G-action commutes with Φ.

Finally, suppose that the mapping $g \mapsto gx$ is open. Let $U \subset G/H$ be an open neighborhood of gH. Then, by the continuity of the quotient mapping $q : G \to G/H$, the pre-image $q^{-1}(U) \subset G$ is open, and by assumption so is $q^{-1}(U)x = \Phi(U)$. The proof is complete. ∎

Under some particular, but quite general, conditions on G and X one can verify the openness of the relevant mapping in the previous proposition, and hence arrive at the following result of Arens and see [34].

Theorem 11.29 (Arens) *Let G be a locally compact group, let $G_1 \leq G$ be an open subgroup containing G^0 with G/G_1 at most countable and such that G_1/G^0 is compact. Let (X, G) be a homogeneous G-space, where X is locally compact. Then for each $x \in X$ the G-spaces (X, G) and $(G/\mathrm{Stab}(x), G)$ are isomorphic (under $g\mathrm{Stab}(x) \mapsto gx$).*

A proof can be found also in Montgomery and Zippin [803, I.2.13].

Metrizability of a topology group is carried over to its coset spaces.

Proposition 11.30 *If G is a metrizable topological group and H is a closed subgroup, then the left coset space G/H is metrizable.*

Proof Let d be a compatible metric on G, which can be taken to be right-invariant by the Birkhoff–Kakutani Theorem 11.5. For $a, b \in G$ set

$$\rho(aH, bH) := \inf_{g,h \in H} d(ah, bg).$$

Then $\rho : G/H \times G/H \to [0, \infty)$ is symmetric. If $\rho(aH, bH) = 0$, then there are $h_n, g_n \in H$ with $d(ag_n, bh_n) \to 0$, so $d(a, bh_n g_n^{-1}) \to 0$ and $h_n g_n^{-1} \to b^{-1}a$. Since H is closed, it follows that $b^{-1}a \in H$, hence $aH = bH$. The function ρ satisfies the triangle inequality by Exercise 11.15.

It remains to prove that ρ is compatible with the topology of G/H. Since the quotient topology on G/H is the finest one making the quotient mapping $q : a \mapsto aH$ continuous and since q is clearly a contraction for the metric considered here, any ρ-open set is open in the quotient topology (in G/H). Conversely, let U be an open subset in G/H and $aH \in U$. Then $V := q^{-1}(U) \subset G$ is an open set containing a so there is $\varepsilon > 0$ such that $\mathrm{B}_d(a, \varepsilon) \subset V$. Now if $bH \in \mathrm{B}_\rho(aH, \varepsilon)$, then there is $h \in H$ with $bh \in \mathrm{B}_d(a, \varepsilon)$. It follows that $bH \in q(V) = U$. This proves that U is open for the topology induced by ρ. We conclude that ρ is compatible with the quotient topology on G/H. ■

Remark 11.31 In general there is no invariant metric on the homogeneous space G/H, even if the latter is compact, because G need not act equicontinuously on G/H, cf. Exercise 11.24.

Compact Homogeneous Spaces

Let G be a Hausdorff topological group and $\Gamma \le G$ be a discrete subgroup which is **cocompact**, i.e., G/Γ is compact; such a subgroup Γ is often referred to as a **uniform lattice** in G (we remark that a discrete subgroup of a Hausdorff topological group is closed). Then $(G/\Gamma, a)$ becomes a topological system for each fixed $a \in G$, called a **homogeneous rotation system**.

Example 11.32

(a) The torus rotation $\mathbb{T}^d$ is a compact homogeneous space via the trivial choice $G = \mathbb{T}^d$ and $\Gamma = \{\mathrm{id}\} \le G$. More interesting possibilities are available, such as $G = \mathbb{R}^d$ and $\Gamma = \mathbb{Z}^d$.

(b) Recall from Example 8.35 the Heisenberg group G of upper triangular 3×3 real matrices with ones on the diagonal, i.e.,

$$G = \left\{ \begin{pmatrix} 1 & x & z \\ 0 & 1 & y \\ 0 & 0 & 1 \end{pmatrix} : x, y, z \in \mathbb{R} \right\},$$

which is a locally compact group (bearing the Euclidean topology). The subgroup $\Gamma := \{[x, y, z] : x, y, z \in \mathbb{Z}\}$ is a uniform lattice, and the left coset space $\mathbb{H} = G/\Gamma$ is a compact homogeneous space, giving rise to the Heisenberg systems $(\mathbb{H}, a)$. More examples of this kind will be discussed in Chap. 14.

Proposition 11.33 *Let G be a Hausdorff topological group and $\Gamma \le G$ a closed subgroup. If there is a compact subset $K \subset G$ such that $G = K\Gamma$, then the homogeneous space G/Γ is compact, i.e., Γ is cocompact.*

The proof is left as Exercise 11.9.

Haar Measure on Quotients

Let G be a locally compact group and let H be a closed subgroup of G. Then G/H is a locally compact space.[2] For each $a \in G$ we denote by L_a the Koopman operator of the (left) multiplication τ_a by a on G/H, i.e., for $f \in \mathrm{C_c}(G/H)$

$$(L_a f)(xH) := f(axH) \quad (x \in G).$$

Then L_a is a linear operator on $\mathrm{C_c}(G/H)$ (and a lattice and algebra homomorphism).

A non-zero (positive) Radon measure m on G/H which is invariant under all left multiplications τ_a is called a **(left) Haar measure.**[3] We discuss the construction of such a left Haar measure building on the left Haar measures on the groups G and H, and prove the following result with an argumentation following Deitmar and Echterhoff [279, Sec. 1.5].

Theorem 11.34 *Let G be a locally compact group, and let $H \leq G$ be a closed subgroup of G. There exists a left Haar measure* m *on G/H if and only if the modular function Δ_H is the restriction of the modular function Δ_G to H. Such a measure is unique up to scalar factors, and for fixed Haar measures m_H, m_G on H and G, respectively, there is a unique Haar measure* m *on G/H which satisfies*

$$\int_G f \,\mathrm{dm}_G = \int_{G/H} \int_H f(xy) \,\mathrm{dm}_H(y) \,\mathrm{dm}(xH) \quad (f \in \mathrm{C_c}(G)). \tag{11.3}$$

The identity (11.3) is called **Weil's formula** (or **quotient integral formula**). Here, one identifies functions on G/H with functions on G that are constant on cosets gH, i.e., we write h also for $h \circ q$, where $h : G/H \to \mathbb{C}$ and $q : G \to G/H$ is the quotient mapping. In fact the second integral on the right-hand side of (11.3) yields exactly such functions (see below).

Before presenting the proof of this theorem let us draw the following important consequence (recall that discrete groups are unimodular).

Corollary 11.35 *Let G be a unimodular, locally compact group and let $\Gamma \leq G$ be a discrete, cocompact subgroup of G. Then there is a unique probability Haar measure on G/Γ.*

---

[2] It is important to note that G/H is *not* a group unless H is a normal subgroup.

[3] There is no need to add the word "left", because, in general, it is only meaningful to consider left invariance on the left coset space; for emphasis we write occasionally "left" even in this situation.

As a matter of fact, one can prove a slight strengthening of this result, namely, one can drop the assumption of the unimodularity of G, see Exercise 11.17 (or [342, Thm. 5.18]).

Proposition 11.36 *Let G be a locally compact group and let $\Gamma \leq G$ be a discrete, cocompact subgroup of G. Then G is unimodular and there exists a unique probability Haar measure on G/Γ.*

It is also worth making the next couple of remarks.

Remark 11.37

(a) Let $G = H_1 \times H_2$ with H_1, H_2 locally compact groups. Then one has $\mathrm{m}_G = \mathrm{m}_{H_1} \times \mathrm{m}_{H_2}$ for the corresponding left Haar measures, and the left Haar measure on G/H_1 is m_{H_2} (after identifying H_2 and G/H_1).
(b) With some more effort one can extend the validity of Weil's formula (11.3) to $f \in \mathrm{L}^1(G)$, by utilizing the denseness of $\mathrm{C_c}(G)$ in $\mathrm{L}^1(G)$ and by an approximation argument. The ambitious reader is invited to try this, or look up the proof in [279, proof of Thm. 1.5.2].
(c) Suppose that H is discrete and countable (with counting measure m_H) and that A is a measurable complete set of representatives for the left cosets (so $h_1 A \cap h_2 A = \emptyset$ for $h_1, h_2 \in A$ if $h_1^{-1} h_2 \notin H$, and $AH = G$). Weil's formula (11.3) yields for a Borel set $M \subset G/H$ and $f = \mathbf{1}_{A \cap q^{-1}(M)}$ (again $q : H \to G/H$ denotes the quotient map) that

$$\begin{aligned}
\mathrm{m}_G(A \cap q^{-1}(M)) &= \int_{G/H} \sum_{y \in H} \mathbf{1}_{A \cap q^{-1}(M)}(xy) \,\mathrm{dm}(xH) \\
&= \int_{G/H} \sum_{y \in H} \mathbf{1}_{Ay^{-1} \cap q^{-1}(M) y^{-1}}(x) \,\mathrm{dm}(xH) \\
&= \int_{G/H} \sum_{y \in H} \mathbf{1}_{Ay^{-1} \cap q^{-1}(M)}(x) \,\mathrm{dm}(xH) \\
&= \int_{G/H} \mathbf{1}_{q^{-1}(M)}(x) \,\mathrm{dm}(xH) \\
&= \int_{G/H} \mathbf{1}_M(xH) \,\mathrm{dm}(xH) = \mathrm{m}(M),
\end{aligned}$$

a formula that we already saw in the special case of Heisenberg systems, see Example 8.35.

To prove Theorem 11.34 we first need to give meaning to the second integral on the right-hand side of (11.3). For $f \in \mathrm{C_c}(G)$ let us define

$$(\Phi f)(x) := \int_H f(xy) \,\mathrm{dm}_H(y) \quad (x \in G),$$

where the integral exists since $y \mapsto f(xy)$ is continuous and has compact support. This integral operator will be important in the construction of the Haar measure on G/H. In fact, Φf is constant on the cosets xH, so one can think of Φf as a function on G/H, an identification that will be used without further mention.

Clearly, Φ commutes with each of the Koopman operators L_a, $a \in G$, and a short calculation shows for $f \in \mathrm{C_c}(G)$ and $g \in \mathrm{C_c}(G/H)$ that

$$\Phi(fg) = \Phi(f)g. \tag{11.4}$$

We leave the proof of this and some parts of the next lemma as Exercise 11.16.

Lemma 11.38 *For $f \in \mathrm{C_c}(G)$ the function Φf is continuous on G/H and has compact support, i.e., $\Phi f \in \mathrm{C_c}(G/H)$. The mapping $\Phi : \mathrm{C_c}(G) \to \mathrm{C_c}(G/H)$ is a positive, surjective linear operator.*

Proof We only prove surjectivity and leave the rest of the proof to the reader as Exercise 11.16. Consider the quotient mapping $q : G \to G/H$ and let $h \in \mathrm{C_c}(G/H)$. For each $y \in \operatorname{supp}(h)$ take $x_y \in q^{-1}(\{y\})$ and an open neighborhood U_y of x_y with $\overline{U_y}$ compact. Then the sets $q(U_y)$, $y \in \operatorname{supp}(h)$, yield an open covering of $\operatorname{supp}(h)$, so by compactness we can find a finite sub-covering $\operatorname{supp}(h) \subset q(U_{y_1}) \cup \dots \cup q(U_{y_N})$. The set $K := \overline{U_{y_1}} \cup \dots \cup \overline{U_{y_N}}$ is compact and satisfies $\operatorname{supp}(h) \subset q(K)$. By Urysohn's Lemma 1.13, we find a continuous function $\psi \in \mathrm{C_c}(G)$ with $\psi \geq 0$ and $\psi|_K = \mathbf{1}$. For $y \in \operatorname{supp}(h)$ there is $x \in K$ with $y = q(x)$ and we conclude from $\psi(x) = 1$, $\psi \geq 0$ and the continuity of ψ that

$$(\Phi\psi)(x) = \int_H \psi(xv)\,\mathrm{dm}_H(v) > 0$$

(recall that m_H has full support). Thus we can define the function $f : G \to \mathbb{C}$ by the following formula:

$$f(z) := \begin{cases} \frac{h(q(z))\psi(z)}{(\Phi\psi)(z)} & \text{if } h(q(z)) \neq 0, \\ 0 & \text{if } h(q(z)) = 0. \end{cases}$$

Then $f \in \mathrm{C_c}(G)$ and by (11.4) we conclude that $\Phi f = \Phi(\psi) \cdot h/\Phi(\psi) = h$. ■

Proof of Theorem 11.34 Suppose that there is a Haar measure m on G/H. Take a left Haar measure m_H on H and the corresponding integral operator Φ as above. Define the positive linear functional φ on $\mathrm{C_c}(G/H)$ by

$$\varphi(f) = \int_{G/H} \Phi f\,\mathrm{dm} \quad (f \in \mathrm{C_c}(G)),$$

which is easily seen to be left invariant and non-zero, i.e., a left Haar integral. As discussed in Sect. 11.2, there is a corresponding left Haar measure m_G on G given by

$$\int_{G/H} \Phi f \,\mathrm{dm} = \varphi(f) = \int_G f \,\mathrm{dm}_G \quad (f \in \mathrm{C_c}(G)). \tag{11.5}$$

For fixed $h \in H$ and $f \in \mathrm{C_c}(G)$ we can write

$$\begin{aligned}
\Delta_G(h) \int_G f \,\mathrm{dm}_G &= \int_G f(xh^{-1}) \,\mathrm{dm}_G(x) = \int_G R_{h^{-1}} f \,\mathrm{dm}_G \\
&= \int_{G/H} \Phi(R_{h^{-1}} f) \,\mathrm{dm} \\
&= \int_{G/H} \int_H f(xyh^{-1}) \,\mathrm{dm}_H(y) \,\mathrm{dm}(xH) \\
&= \Delta_H(h) \int_{G/H} \int_H f(xy) \,\mathrm{dm}_H(y) \,\mathrm{dm}(xH) \\
&= \Delta_H(h) \int_{G/H} \Phi f \,\mathrm{dm} \\
&= \Delta_H(h) \int_G f \,\mathrm{dm}_G.
\end{aligned}$$

This being true for every $f \in \mathrm{C_c}(G)$ and $h \in H$ implies $\Delta_G(h) = \Delta_H(h)$ for every $h \in H$.

Conversely, suppose that Δ_H is the restriction of Δ_G to H. Take the Haar measures m_G and m_H on G and H, respectively. The idea is to define an invariant integral via (11.5) by exploiting that Φ is surjective. To see that such a definition is meaningful we need to establish that for $f \in \ker \Phi$ one has $\int_G f \,\mathrm{dm}_G = 0$. So take $f \in \ker \Phi$ and $\psi \in \mathrm{C_c}(G)$ such that $(\Phi\psi)|_{\mathrm{supp}(f)} = \mathbf{1}_{\mathrm{supp}(f)}$ (use Urysohn's lemma and the surjectivity of Φ). We then have

$$0 = \int_G \Phi f(x) \cdot \psi(x) \,\mathrm{dm}_G(x) = \int_G \int_H f(xy)\psi(x) \,\mathrm{dm}_H(y) \,\mathrm{dm}_G(x)$$

and continue by Fubini's theorem

$$\begin{aligned}
&= \int_H \int_G f(xy)\psi(x) \,\mathrm{dm}_G(x) \,\mathrm{dm}_H(y) \\
&= \int_H \Delta_G(y^{-1}) \int_G f(x)\psi(xy^{-1}) \,\mathrm{dm}_G(x) \,\mathrm{dm}_H(y)
\end{aligned}$$

$$= \int_H \Delta_H(y^{-1}) \int_G f(x)\psi(xy^{-1}) \, \mathrm{d}m_G(x) \, \mathrm{d}m_H(y)$$

$$= \int_G f(x) \int_H \Delta_H(y^{-1})\psi(xy^{-1}) \, \mathrm{d}m_H(y) \, \mathrm{d}m_G(x),$$

and by Proposition 11.12(b) we continue

$$= \int_G f(x) \int_H \psi(xy) \, \mathrm{d}m_H(y) \, \mathrm{d}m_G(x)$$

$$= \int_G f \cdot \Phi\psi \, \mathrm{d}m_G = \int_{\mathrm{supp}(f)} f \cdot \Phi\psi \, \mathrm{d}m_G = \int_G f \, \mathrm{d}m_G.$$

So for an arbitrary $g \in \mathrm{C_c}(G/H)$ and any $f \in \mathrm{C_c}(G)$ with $\Phi f = g$ we can set

$$\varphi(g) := \int_G f \, \mathrm{d}m_G,$$

yielding a well-defined, non-zero, positive functional on $\mathrm{C_c}(G/H)$, which is easily seen to be left invariant. By the Riesz–Markov–Kakutani Theorem 1.23 we obtain the required left Haar measure on G/H. The above arguments also yield the validity of formula (11.3), and the uniqueness up to scalar multiples. ∎

11.5 Distality and Semigroups

On a metrizable compact group G there is an invariant metric d, so for a group rotation (G, a) distinct points $x, y \in G$ keep their positive distance under the dynamics: $d(a^n x, a^n y) = d(x, y) > 0$ for each $n \in \mathbb{N}_0$. In this section we look at a generalization of this phenomenon, and following the works of Auslander [58] and Ellis [361], we show how abstract (right-topological) semigroups can be used in the study of topological dynamical systems (and refer to Hindman and Strauss [536] for more information about these objects).

Definition 11.39 Given a topological system (K, T) two points $x, y \in K$ are called **proximal** if there is $z \in K$ such that for every open neighborhood U of z there is $n \in \mathbb{N}_0$ with $T^n(x), T^n(y) \in U$. The system (K, T) is called **distal** if from the proximality of $x, y \in K$ it always follows that $x = y$.

Remark 11.40 (Distality in the Metric Case) Let (K, d) be a metric space and $T : K \to K$ a continuous map. Then two points $x, y \in K$ are proximal if and only if

$$\inf_{n \in \mathbb{N}} d(T^n(x), T^n(y)) = 0.$$

As a consequence, the system (K, T) is distal if an only if for each pair $x, y \in K$ with $x \neq y$

$$\inf_{n\in\mathbb{N}} d(T^n(x), T^n(y)) > 0.$$

The proof is left as Exercise 11.18.

Proposition 11.41 (Proximality for the Shift) *Consider the (one-sided) shift system (K, T), $K = L^{\mathbb{N}_0}$ from Example 2.31. Then two points $x, y \in K$ are proximal if and only if for each $m \in \mathbb{N}$ there is $n \in \mathbb{N}$ such that*

$$x(n) = y(n),\ x(n+1) = y(n+1), \ldots, x(n+m) = y(n+m).$$

In particular, if L has more than one element, (K, T) is not distal.

We leave the proof of this result as Exercise 11.19.

Proposition 11.42 (Distality of Group Rotations) *Every compact group rotation (G, a) is distal.*

Proof Suppose the contrary, so let $g, h \in G$ be different proximal points and $z \in G$ witnessing this. Let U be an open neighborhood of id_G such that $Ug^{-1} \cap Uh^{-1} = \emptyset$. Since zU is an open neighborhood of z, by proximality there is $n \in \mathbb{N}_0$ such that $a^n g, a^n h \in zU$. It follows that $z^{-1}a^n \in Ug^{-1} \cap Uh^{-1}$, a contradiction. ∎

Example 11.43

(1) If (K, T) is a topological system and K is metrizable with compatible metric d such that $d(T(x), T(y)) = d(x, y)$ for every $x, y \in K$, then T is called an **isometry**, d a **T-invariant metric** and (K, T) an **isometric system**, cf. the example at the beginning of this section. (It can be proved elementarily that isometric systems must be invertible, but this will follow from the results below, see Theorem 11.49.) Isometric systems are evidently distal. In general, for a not necessarily metrizable space K, the system is called **pseudoisometric** if the topology is induced by T-invariant semimetrics (often called pseudometrics). For invertible systems (K, T) this is equivalent to the **equicontinuity** of $\{T^n : n \in \mathbb{Z}\}$ and (via the Arzelà–Ascoli Theorem) to the fact that the Koopman operator T has relatively compact orbits $\{T^n f : n \in \mathbb{Z}\}$ for every $f \in \mathrm{C}(K)$. Recall that $\{T^n : n \in \mathbb{Z}\}$ is **equicontinuous** if for every continuous semimetric d on K, every $x \in K$ and every $\varepsilon > 0$ there is a neighborhood U of x such that for every $y \in U$ and every $n \in \mathbb{Z}$ one has $d(T^n(x), T^n(y)) < \varepsilon$, cf. also Exercise 11.22. We refer to [529] for details (in fact, in a more general situation of group actions).

(2) Consider the closed unit disc $\overline{\mathbb{D}} = \{z \in \mathbb{C} : |z| \leq 1\}$ and the multiplication τ_a by $a \in \mathbb{T}$. The system $(\overline{\mathbb{D}}, \tau_a)$ is trivially isometric, hence distal. Each subset $K_r = \{z \in \overline{\mathbb{D}} : |z| = r\}$ $(r \in [0, 1])$ is τ_a-invariant, and if a is not a root of

unity, they are minimal subsystems. In this case, but also when a is a root of unity, $\overline{\mathbb{D}}$ is the disjoint union of its minimal subsystems.

(3) The skew rotation $(\mathbb{T}^2, a)$ from Example 2.31.6 and the Heisenberg system $(\mathbb{H}, a)$ from Example 8.35 are distal. This can be proved easily, but follows from more general facts in the coming Chap. 14 (both systems are examples of so-called **nilsystems**).

Every group rotation system (G, a) is equicontinuous (basically by Proposition 11.6). But not all homogeneous systems enjoy this property, for example, skew rotations are not equicontinuous, see Exercise 11.24.

Let (K, T) be a topological system. Define

$$\mathcal{E} := \mathcal{E}(T) := \overline{\{T^n : n \in \mathbb{N}_0\}}^{K^K}$$

with closure in the product topology of K^K (i.e., the topology of pointwise convergence of mappings $K \to K$). Then $\mathcal{E}$ is a compact subset of K^K with respect to this topology (by Tychonoff's Theorem).

Theorem 11.44 *Let (K, T) be a topological system. Then $\mathcal{E}(T)$ is a semigroup, i.e., for $R, S \in \mathcal{E}$ we have $RS \in \mathcal{E}$. Furthermore the mapping $R \mapsto RS$ is continuous on $\mathcal{E}$ for each fixed $S \in \mathcal{E}$, and $R \mapsto SR$ is continuous if and only if $S : K \to K$ is continuous.*

The proof is left as Exercise 11.20. The semigroup $\mathcal{E}(T)$ is called the **Ellis semigroup** (or **enveloping semigroup**) of (K, T), see Ellis and Gottschalk [363]. The first continuity property in the previous statement is expressed by saying that $\mathcal{E}(T)$ is a *right-topological semigroup*. Another result of Ellis states that such semigroups, if being compact, contain idempotent elements, see Ellis [362].

Theorem 11.45 (Ellis' Theorem on Idempotents) *Let $\mathcal{S}$ be a compact* right-topological semigroup, *i.e., $\mathcal{S}$ is a compact (Hausdorff) space and a semigroup such that $r \mapsto rs$ is continuous on $\mathcal{S}$ for each fixed $s \in \mathcal{S}$. Then there is an idempotent $p \in \mathcal{S}$, i.e., an element with $p = p^2$.*

The proof is left as Exercise 11.21.

Proposition 11.46 *Let (K, T) be a topological system. Two points $x, y \in K$ are proximal if and only if there is an idempotent $P \in \mathcal{E}(T)$ with $P(x) = P(y)$.*

Proof Let $x, y \in K$ and $z \in K$ be a point witnessing proximality. So for every U open neighbourhood of z there is $n := n_U \in \mathbb{N}_0$ with $T^n(x), T^n(y) \in U$. By compactness it follows that there is $S \in \mathcal{E}(T)$ with $S(x) = S(y)$. Define $\mathcal{S} := \{R \in \mathcal{E}(T) : R(x) = R(y)\}$. Then $\mathcal{S}$ is a non-empty, compact subsemigroup of $\mathcal{E}(T)$, and right-topological by Theorem 11.44, hence contains an idempotent P by Theorem 11.45. One of the implications is thus proved, while the other is clear. ∎

This characterization of proximal pairs, in combination with the existence of idempotent elements, is the key to the results below. For example, for each $x \in K$

and each idempotent $P \in \mathcal{E}(T)$ (and there is at least one) the points x and $P(x)$ are proximal, since $P(x) = P(P(x))$. The following refines this observation to the effect that the proximal point to x can be chosen to be almost periodic.

Theorem 11.47 (Auslander–Ellis) *Let (K, T) be a topological system, let $x \in K$ and let $M \subset \overline{\mathrm{orb}_+(x)}$ be an invariant set. Then there is an almost periodic point $y \in M$ such that x and y are proximal.*

Proof By Theorem 11.44 the Ellis semigroup $\mathcal{E}(T)$ is a compact right-topological semigroup. Let $x \in K$ be arbitrary and consider the orbit closure $L := \overline{\mathrm{orb}_+(x)} = \overline{\{T^n x : n \in \mathbb{N}_0\}}$. Take a minimal subsystem (M, T) of (L, T) (see Birkhoff's Theorem 3.10) and define $\mathcal{S} := \{S \in \mathcal{E} : S(x) \in M\}$. Since $L = \{S(x) : S \in \mathcal{E}(T)\}$, $\mathcal{S}$ is non-empty, and in fact a compact subsemigroup of $\mathcal{E}$ (see Exercise 11.20 for details).
As a compact right-topological semigroup, $\mathcal{S}$ contains an idempotent element $P = P^2 \in \mathcal{S}$ by Theorem 11.45 of Ellis. With $y := P(x) \in M$ we have $P(x) = P^2(x) = P(y)$, and this implies that x and y are proximal (see Proposition 11.46). Since $y = P(x) \in M$, and (M, T) is minimal, we conclude that y is an almost periodic point, see Proposition 3.6. The proof is complete. ∎

Utilizing this we can generalize the result in Exercise 9.3 about isometric systems.

Theorem 11.48 (Ellis) *A distal system is minimal if and only if it is topologically transitive. Moreover, each distal topological system (K, T) is the disjoint union of its minimal subsystems.*

Proof Let (K, T) be a distal system. Take $x \in K$ arbitrarily. Then there is an almost periodic point $y \in K$ which is proximal to x, see Theorem 11.47. Since the system is distal, we must have $x = y$, so x is an almost periodic point, i.e., contained in a minimal subsystem of (K, T). This proves then the first assertion, while the second one follows also immediately, since different minimal subsystems are disjoint (see Exercise 3.2). ∎

In [444] Furstenberg proved a more detailed structure theorem for distal systems, see also [468, Ch. 10] and [332]. Finally, we can relate the Ellis semigroup and the distality of a topological system in a fundamental manner.

Theorem 11.49 (Ellis [361]) *A topological system (K, T) is distal if and only if* id *is the unique idempotent in the Ellis semigroup $\mathcal{E}(T)$. This is the case if and only if $\mathcal{E}(T)$ is a group (with unit element* id*).*

Proof The Ellis semigroup contains id by definition. Suppose the unique idempotent in $\mathcal{E}(T)$ to be the identity $\mathrm{id} : K \to K$. This certainly holds if $\mathcal{E}(T)$ is a group. By Proposition 11.46 it follows that (K, T) is distal.
Conversely, suppose that (K, T) is distal. Let $P \in \mathcal{E}(T)$ be an idempotent. Since for each $x \in K$ the points x and $P(x)$ are proximal (see Proposition 11.46), we obtain $x = P(x)$, and $P = \mathrm{id}$ follows. For $S \in \mathcal{E}(T)$ consider the closed (compact) subsemigroup $\mathcal{E}(T)S$, which contains by Theorem 11.45 the unique idempotent id.

So there is $S' \in \mathcal{E}(T)$ with $S'S = \mathrm{id}$. Similarly, there is $S'' \in \mathcal{E}(T)$ with $S''S' = \mathrm{id}$. It follows that $S'' = S''S'S = S$. So S' is the inverse of S, and $\mathcal{E}(T)$ is a group. ∎

As a consequence we obtain that a distal system is in fact invertible.

Hindman's Theorem

In this section we give an example of how methods from topological dynamics can be used to obtain statements of combinatorial nature: Hindman's theorem, see Hindman [535]. The presentation here follows Furstenberg and Weiss [457].

Theorem 11.50 (Hindman, Arithmetic Formulation) *Let $r \in \mathbb{N}$ and $\mathbb{N} = A_1 \cup \cdots \cup A_r$ with some pairwise disjoint $A_1, \ldots, A_r \subset \mathbb{N}$. Then there is $j \in \{1, \ldots, r\}$ such that A_j contains a strictly increasing sequence $(n_k)_{k\in\mathbb{N}}$ such that every finite sum composed of different elements of the sequence belongs to A_j.*

Proof We shall use the Auslander–Ellis theorem, Theorem 11.47, stating that in a topological system every point is proximal to some almost periodic point. Now, consider the alphabet $L := \{1, \ldots, r\}$ and the shift system (K, T), $K = L^{\mathbb{N}_0}$. Define the point $x \in K$ by setting $x_0 = 1$ and

$$x_n = j \quad \text{if } n \in A_j.$$

(Such a j is uniquely determined by assumption.) Then, by what is said above, there is an almost periodic $y \in K$ which is proximal to x. Since y is almost periodic every finite block of it occurs infinitely often in y with bounded gaps, see Exercise 3.8. By Proposition 11.41 it follows that the finite blocks of y occur also in x infinitely often at the same position as in y. Set $j := y_0$. Then the block (j) occurs at the same position $n_1 \in \mathbb{N}$ at x and y, i.e., $x_{n_1} = j = y_{n_1}$. Next, consider the starting block $(y_0, y_1, \ldots, y_{n_1})$ of y, which occurs also at $n_2 > n_1$ in x and y, i.e.,

$$x_{n_2} = y_{n_2} = y(0),\ x_{n_2+1} = y_{n_2+1} = y(1), \ldots, x_{n_2+n_1} = y_{n_2+n_1} = y(n_1).$$

To sum up we have $x_{n_1} = j$, $x_{n_2} = j$, $x_{n_1+n_2} = j$, i.e., $n_1, n_2, n_1 + n_2 \in A_j$. Carrying on this procedure recursively and in the kth step considering the starting block $(y_0, y_1, \ldots, y_{n_1+n_2+\cdots+n_k})$ of y one can finish the proof easily. ∎

Given a (strictly increasing) subsequence $(n_k)_{k\in\mathbb{N}}$ of $\mathbb{N}$ one defines

$$\mathrm{FS}(n_k) := \Big\{ \sum_{j\in F} n_j : F \subset \mathbb{N} \text{ finite and non-empty} \Big\},$$

and calls $\mathrm{FS}(n_k)$ an **IP-set**,[4] see, e.g., [446, Def. 2.3] and [112]. The previous theorem of Hindman thus tells that if one partitions $\mathbb{N}$ into a finite number of sets, then one of these sets contains an IP-set, in particular, it bears some arithmetic structure. (Sometimes the term "IP-set" refers to sets that contain $\mathrm{FS}(n_k)$ for some infinite subsequence $(n_k)_{k\in\mathbb{N}}$.)

By using binary expansion of natural numbers, Hindman's theorem can be reformulated as a statement about finite sets, which we present next. (For details see, e.g., [457], [112] (or [342, Sec. 19.4]).

Let $\mathcal{F} := \mathcal{F}(\mathbb{N}_0)$ denote the set of finite, non-empty subsets of $\mathbb{N}_0$. For a given sequence $(\alpha_k)_{k\in\mathbb{N}}$ in $\mathcal{F}$ of pairwise disjoint sets we define

$$\mathrm{FU}(\alpha_j) := \Big\{\bigcup_{j\in F}\alpha_j : F\subset\mathbb{N} \text{ finite and non-empty}\Big\}\subset\mathcal{F}.$$

Theorem 11.51 (Hindman, Set-theoretic Version) *Let $\mathcal{F} = \mathcal{A}_1\cup\mathcal{A}_2\cup\ldots\cup\mathcal{A}_r$ with $\mathcal{A}_1,\ldots,\mathcal{A}_r$ pairwise disjoint. Then there is $j_0\in\{1,\ldots,r\}$ and a sequence $(\alpha_k)_{k\in\mathbb{N}}$ in $\mathcal{F}$ of pairwise disjoint finite sets such that $\mathrm{FU}(\alpha_k)\subset\mathcal{A}_{j_0}$. The sequence $(\alpha_k)_{k\in\mathbb{N}}$ can be chosen such that $\max\alpha_k < \min\alpha_{k+1}$ for all $j\in\mathbb{N}$.*

We close this chapter by the following characterization of distal systems, highlighting again the relevance of IP-sets in dynamics, see Furstenberg [449].

Theorem 11.52 *A topological system (K, T) is distal if and only if for each $x\in K$ and each open neighborhood U of x the set*

$$R_U(x) := \{n\in\mathbb{N} : T^n(x)\in U\}$$

of return times is an IP-set, i.e., intersects each IP-subset of $\mathbb{N}$.*

Exercises

Exercise 11.1 Prove Proposition 11.1.

Exercise 11.2 Show Proposition 11.2.

Exercise 11.3 Prove Proposition 11.3.

Exercise 11.4 Show Proposition 11.4.

Exercise 11.5 Prove Proposition 11.15.

Exercise 11.6 (Connectedness in Groups) For a topological group G prove the following assertions.

[4] The word "IP" refers to "Infinite Parallelepiped" or "IdemPotent".

(a) If G is connected and H is a subgroup of G, then the commutator subgroup $[G, H]$ is also connected.
(b) Denote by G° the connected component of G containing id. Every connected component of G is of the form $G^\circ h$ and hG° for some $h \in G$. In particular, $G^\circ h = hG^\circ$ for every $h \in G$, i.e., G° is normal.

Exercise 11.7 (Locally Connected Groups) Let G be a topological group which is **locally connected** (i.e., each open neighborhood of the identity id contains a connected, open neighborhood of id). For G°, the connected component containing id, show the following assertions.

(a) G° is an open subgroup of G.
(b) For every set $S \subset G$, the set $G^\circ S = SG^\circ$ is closed. (Hint: Use (a) and show $SG^\circ = G \setminus \cup_{g \notin SG^\circ} gG^\circ$.)
(c) For every set $S \subset G$, the subgroup $H := \langle G^\circ, S\rangle$ generated by G° and S is automatically closed.
(d) Assume that G is compact. Then $G = \cup_{j=1}^m g_j G^\circ$ for some $m \in \mathbb{N}$ and $g_1, \dots, g_m \in G$. In particular, G has finitely many connected components.

Exercise 11.8 Prove the assertions in Examples 11.17 and 11.18 concerning the dual group of $\mathbb{R}^d$ and $\mathbb{T}^d$, respectively.

Exercise 11.9 Show Proposition 11.33.

Exercise 11.10 (Modular Function) Prove Proposition 11.12.

Exercise 11.11 (Haar Integral) Prove that a left Haar integral φ on a locally compact group is strictly positive, i.e., for each $f \in \mathrm{C_c}(G)$ with $f \geq 0$ and $f \neq 0$, one has $\varphi(f) > 0$.

Exercise 11.12 Show that for the multiplicative group $(0, \infty)$ (with the Euclidean topology)

$$\varphi(f) := \int_0^\infty \frac{f(x)}{x}\,\mathrm{d}x \quad (f \in \mathrm{C_c}((0, \infty)))$$

defines a left and right Haar integral.

Exercise 11.13 Prove that for the group $G = \mathrm{GL}_n(\mathbb{R})$ the formula

$$\varphi(f) = \int_G \frac{f(A)}{|\det(A)|^n}\,\mathrm{d}\lambda^{n^2}(A) \quad (f \in \mathrm{C_c}(G))$$

defines a Haar integral.

Exercise 11.14 (Haar Measure on Discrete Groups) Prove that a locally compact group G with identity id and left Haar measure m is discrete if an only if $\mathrm{m}(\{\mathrm{id}\}) > 0$.

Exercise 11.15 Fill in the missing details in the proof of Proposition 11.30.

Exercise 11.16 Prove Lemma 11.38 and the results in the paragraph preceding it.

Exercise 11.17 Show Proposition 11.36.

Exercise 11.18 Prove the characterization of distality for compact metric systems as given in Remark 11.40.

Exercise 11.19 Prove Proposition 11.41.

Exercise 11.20 (Ellis' Theorem) This exercise furnishes the details for the proofs of Theorems 11.44 and 11.47. Let (K, T) be a topological system.

(a) Prove that

$$\mathcal{E} := \mathcal{E}(T) := \overline{\{T^n : n \in \mathbb{N}_0\}}^{K^K}$$

is a semigroup, i.e., for $R, S \in \mathcal{E}$ we have $RS \in \mathcal{E}$. Show further that the mapping $R \mapsto RS$ is continuous on $\mathcal{E}$ for each given $S \in \mathcal{E}$, and that $R \mapsto SR$ is continuous if and only if $S : K \to K$ is continuous.

(b) For $x \in K$ prove that its orbit closure is $\{S(x) : S \in \mathcal{E}\}$. Show that a compact subset $M \subset K$ is T -invariant if and only if it is invariant under each $S \in \mathcal{E}$.

(c) Let $M \subset K$ be a compact T-invariant subset of the orbit closure of a point $x \in K$. Building on (b) prove that $\mathcal{S} := \{S \in \mathcal{E} : S(x) \in M\}$ is non-empty and a compact subsemigroup of $\mathcal{E}$ (i.e., $R, S \in \mathcal{S}$ implies $RS \in \mathcal{S}$). Show further that $M = \{S(x) : S \in \mathcal{S}\}$.

Exercise 11.21 (Ellis' Theorem on Idempotents) Let $\mathcal{S}$ be a compact right-topological semigroup. Prove that there is an idempotent $p \in \mathcal{S}$, i.e., an element with $p = p^2$, following the next steps:

(a) Prove that $\mathcal{S}x$ is a closed subsemigroup for each $x \in \mathcal{S}$.

(b) Utilize Zorn's lemma to find a non-empty closed subsemigroup H in $\mathcal{S}$ which is minimal with respect to inclusion. Prove that for each $e \in H$ one has $He = H$.

(c) Prove that for each $e \in H$ the set $\{s : s \in H \text{ and } se = e\}$ is a closed subsemigroup of H, and conclude that $e^2 = e$.

Exercise 11.22 (Equicontinuous Systems) Let (K, T) be an invertible topological system and let K be metrizable with compatible metric d. We call (K, T) (uniformly) **equicontinuous**[5] if for every $\varepsilon > 0$ there is $\delta > 0$ such that

$$d(x, y) < \delta \quad \Longrightarrow \text{ for each } n \in \mathbb{Z} \text{ one has } d(T^n(x), T^n(y)) < \varepsilon.$$

Prove that $d^*(x, y) := \sup_{n \in \mathbb{Z}} d(T^n(x), T^n(y))$ defines a compatible metric on K which renders T isometric.

---

[5] Since K is compact and metrizable, this notion of equicontinuity is the same as the one in Example 11.43.

Exercise 11.23 (Distality and Group Extensions) Prove that a group extension $(K \times G, T_\rho)$ of a distal system (K, T) is distal. (Hint: Modify the proof of Proposition 11.42 suitably.)

Exercise 11.24 (Skew Rotations are not Equicontinuous) Show that skew rotations $(\mathbb{T}^2, a)$ are not equicontinuous.

Part III
More Ergodic Theorems

Chapter 12
Subsequential Ergodic Theorems

We now discuss generalizations of the classical ergodic theorems in the spirit of the polynomial mean ergodic theorem (Theorem 7.28). Suppose that, for a given subsequence $(k_n)_{n\in\mathbb{N}}$ of $\mathbb{N}$, we are interested in how systems evolve along the times $(k_n)_{n\in\mathbb{N}}$ and ignore all other times and are looking at the averages of the form

$$\frac{1}{N}\sum_{n=1}^{N} T^{k_n} f.$$

For example, we could be interested in the states on Wednesdays only (that would mean a periodic sequence $(k_n)_{n\in\mathbb{N}}$), or the orbit along the days that are numbered by the primes $(p_n)_{n\in\mathbb{N}}$ or a certain random sequence of times. Moreover, we distinguish between convergence in mean and pointwise convergence in L^p.

From now on it is convenient to switch from averages of the form $\frac{1}{N}\sum_{n=0}^{N-1}$ to $\frac{1}{N}\sum_{n=1}^{N}$ in the generalizations of the classical ergodic theorems. This, of course, changes neither convergence nor the limit of the averages.

Throughout this chapter, we understand under a subsequence $(k_n)_{n\in\mathbb{N}}$ of $\mathbb{N}$ an *eventually* strictly increasing sequence $(k_n)_{n\in\mathbb{N}}$ in $\mathbb{N}$ (i.e., the sequence is strictly increasing from an index on). Moreover, we allow that finitely many k_n do not belong to $\mathbb{N}$. If in this way finitely many summands are not defined (such as negative powers of non-invertible operators) we can ignore them or replace by, say, 0 (or any other value) since they do not contribute to the limit (or convergence) anyway.

12.1 Mixing Systems and Subsequential Averages

It is natural to ask for which measure-preserving systems the mean ergodic theorem holds for every subsequence $(k_n)_{n\in\mathbb{N}}$ with the same limit. A result of Blum and Hanson [163] gives an answer to this.

T. Eisner, B. Farkas, *A Journey Through Ergodic Theorems*, Birkhäuser Advanced Texts Basler Lehrbücher, https://doi.org/10.1007/978-3-031-96506-7_12

Theorem 12.1 (Blum–Hanson) *Let (X, μ, T) be a measure-preserving system with Koopman operator T on $\mathrm{L}^2(X, \mu)$. The system is strongly mixing if and only if for every subsequence $(k_n)_{n\in\mathbb{N}}$ of $\mathbb{N}$ and every $f \in \mathrm{L}^2(X, \mu)$*

$$\lim_{N\to\infty} \frac{1}{N} \sum_{n=1}^{N} T^{k_n} f = \Bigl(\int_X f \,\mathrm{d}\mu\Bigr) \cdot \mathbf{1}. \tag{12.1}$$

To prove this result, we shall extract the following operator-theoretic essence, discovered by Jones and Kuftinec in [583] and Akcoglu and Sucheston in [26]. Theorem 12.1 is a direct corollary of it and the characterization of mixing given in Proposition 3.44.

Theorem 12.2 (Jones–Kuftinec, Akcoglu–Sucheston) *Let $T \in \mathscr{L}(H)$ be a contraction on a Hilbert space H. For $x, x_0 \in H$ the following assertions are equivalent:*

(i) *The sequence $(T^n x)_{n\in\mathbb{N}}$ is weakly convergent to x_0.*
(ii) *For every subsequence $(k_n)_{n\in\mathbb{N}}$ of $\mathbb{N}$ one has*

$$\frac{1}{N} \sum_{n=1}^{N} T^{k_n} x \to x_0 \text{ in the norm of } H \text{ as } N \to \infty.$$

We remark that, in this case, one must have $x_0 = P_{\text{fix}}\, x$ (with P_{fix} being the mean ergodic projection). The proof we present here is due to Jones and Kuftinec, see [583]. Note that they did not use dilations, but rather proved the statement directly for contractions.

Proof (i)$\Rightarrow$(ii): Suppose that $(T^n x)_{n\in\mathbb{N}}$ converges weakly to x_0. Then, as said above, $x_0 = P_{\text{fix}}\, x$. After replacing x by $x - P_{\text{fix}}\, x$, we may assume that $P_{\text{fix}}\, x = 0$ is the weak limit of $(T^n x)_{n\in\mathbb{N}}$. Take an arbitrary subsequence $(k_n)_{n\in\mathbb{N}}$ of $\mathbb{N}$ for which we shall show

$$\frac{1}{N} \sum_{n=1}^{N} T^{k_n} x \to 0 \text{ as } N \to \infty. \tag{12.2}$$

We first prove (12.2) in the case when T is an isometry. Let $\varepsilon > 0$ be given. By assumption, there is $n_0 \in \mathbb{N}$ such that for every $n \geq n_0$ one has $|(T^n x|x)| < \varepsilon$. For $N \geq n_0$ we have

$$\Bigl\|\frac{1}{N} \sum_{n=1}^{N} T^{k_n} x\Bigr\|^2 = \frac{1}{N^2} \sum_{n,m=1}^{N} (T^{k_m} x | T^{k_n} x)$$

$$
\begin{aligned}
&= \frac{1}{N^2}\sum_{n=1}^{N} \|T^{k_n}x\|^2 + \frac{2}{N^2}\sum_{\substack{n,m=1\\ n<m}}^{N} \mathrm{Re}(T^{k_m}x\,|\,T^{k_n}x)\\
&= \frac{\|x\|^2}{N} + \frac{2}{N^2}\sum_{\substack{n,m=1\\ n<m}}^{N} \mathrm{Re}(T^{k_m-k_n}x\,|\,x)\\
&= \frac{\|x\|^2}{N} + \frac{2}{N^2}\sum_{\substack{n,m=1\\ 0<m-n<n_0}}^{N} \mathrm{Re}(T^{k_m-k_n}x\,|\,x) + \frac{2}{N^2}\sum_{\substack{n,m=1\\ m-n\geq n_0}}^{N} \mathrm{Re}(T^{k_m-k_n}x\,|\,x)\\
&\leq \frac{\|x\|^2}{N} + \frac{2n_0\|x\|^2}{N} + 2\varepsilon,
\end{aligned}
$$

since in the case $m-n \geq n_0$ one has $k_m - k_n \geq n_0$, so $|(T^{k_m-k_n}x\,|\,x)| < \varepsilon$. Letting $N \to \infty$, we obtain

$$
\limsup_{N\to\infty} \left\| \frac{1}{N}\sum_{n=1}^{N} T^{k_n}x \right\|^2 \leq 2\varepsilon
$$

for every $\varepsilon > 0$, so (12.2) is established, provided T is an isometry.

Next we consider the case of a general Hilbert space contraction $T \in \mathscr{L}(H)$. Take a unitary dilation (S, J) of T such that (S^n, J) is a dilation of T^n for each $n \in \mathbb{N}_0$, see Theorem 6.13. If $(T^n x)$ converges weakly to 0, then $(S^n Jx\,|\,Jx) \to 0$ as $n \to \infty$ and hence, by Exercise 6.7, $(S^n Jx)$ converges weakly to 0. So for each subsequence $(k_n)_{n\in\mathbb{N}}$ of $\mathbb{N}$ we have, by the above case of isometries, that

$$
\lim_{N\to\infty} \frac{1}{N}\sum_{n-1}^{N} S^{k_n}Jx = 0.
$$

But this implies

$$
\lim_{N\to\infty} \frac{1}{N}\sum_{n=1}^{N} T^{k_n}x = J^* \lim_{N\to\infty} \frac{1}{N}\sum_{n=1}^{N} S^{k_n}Jx = 0,
$$

and that is (12.2).

(ii)$\Rightarrow$(i): Suppose that even though (ii) holds, there is $y \in H$ such that $((T^n x\,|\,y))_{n\in\mathbb{N}}$ is not convergent. By Exercise 12.1 there is a subsequence $(k_n)_{n\in\mathbb{N}}$ such that $((T^{k_n}x\,|\,y))_{n\in\mathbb{N}}$ is Cesàro divergent, and a contradiction with (ii) ensues. So $((T^n x\,|\,y))_{n\in\mathbb{N}}$ converges for every $y \in H$ and by the Riesz–Fréchet representation theorem there exists $z \in H$ with $\lim_{n\to\infty} T^n x = z$ weakly. By (ii) we have $z = x_0$ finishing the proof. ∎

The previous theorem yields also a characterization of weak stability for contractions. Recall that for a contraction $T \in \mathscr{L}(H)$ the weakly stable part $H_{\mathrm{ws}}(T)$ consists of $x \in H$ with $T^n x \to 0$ weakly as $n \to \infty$, cf. Exercise 6.2. The Kronecker part $H_{\mathrm{kr}}(T)$ and the almost weakly stable part $H_{\mathrm{aws}}(T)$ were introduced in Sect. 6.4.

Corollary 12.3 *For a Hilbert space contraction $T \in \mathscr{L}(H)$ the following are equivalent.*

(i) *One has $H_{\mathrm{kr}}(T) = \operatorname{fix} T$ and $H_{\mathrm{ws}}(T) = H_{\mathrm{aws}}(T)$, i.e., $H = \operatorname{fix} T \oplus H_{\mathrm{ws}}(T)$.*
(ii) *For every $x \in H$ one has $T^n x \to P_{\operatorname{fix} T} x$ weakly as $n \to \infty$.*
(iii) *For every subsequence $(k_n)_{n\in\mathbb{N}}$ and $x \in H$ we have*

$$\lim_{N\to\infty} \frac{1}{N} \sum_{n=1}^{N} T^{k_n} x = P_{\operatorname{fix} T} x.$$

Proof The second assertion in (i) is a consequence of the Jacobs–de Leeuw–Glicksberg decomposition (Theorem 6.20). Now, (i)⇔(ii) is clear whereas (ii)⇔(iii) follows directly from Theorem 12.2. ∎

By a result of Müller and Tomilov [810] the above statements do not extend to the case of general power bounded operators on Hilbert spaces.

The following variant of Corollary 12.3 will help to characterize weak mixing.

Theorem 12.4 *For a Hilbert space contraction $T \in \mathscr{L}(H)$ the following are equivalent:*

(i) *One has $H_{\mathrm{kr}}(T) = \operatorname{fix} T$, i.e., $H = \operatorname{fix} T \oplus H_{\mathrm{aws}}(T)$.*
(ii) *For every $x \in H$ there is a subsequence $(m_n)_{n\in\mathbb{N}}$ of $\mathbb{N}$ such that $T^{m_n} x \to P_{\operatorname{fix} T} x$ weakly as $n \to \infty$.*
(iii) *For every subsequence $(k_n)_{n\in\mathbb{N}}$ with positive lower density and for every $x \in H$ we have*

$$\lim_{N\to\infty} \frac{1}{N} \sum_{n=1}^{N} T^{k_n} x = P_{\operatorname{fix} T} x.$$

Proof The second assertion in (i) is again a consequence of the Jacobs–de Leeuw–Glicksberg decomposition (Theorem 6.20) while the equivalence (i)⇔(ii) was shown in Proposition 6.21. Suppose (i) and take $x \in H$ which we write as $x = x_1 + x_0$ with $x_1 = P_{\operatorname{fix} T} x \in \operatorname{fix} T$ and $x_0 \in H_{\mathrm{aws}}$. Let (k_n) be any subsequence of $\mathbb{N}$ with positive lower density. Since $x_0 \in H_{\mathrm{aws}}$, there is a subsequence (m_n) of density 1 with $T^{m_n}x_0 \to 0$ as $n \to \infty$. Set $M := \{m_n : n \in \mathbb{N}\}$. Then we have

$$0 = \lim_{N\to\infty} \frac{1}{N} \sum_{\substack{n=1\\ k_n \in M}}^{N} T^{k_n} x_0 = \lim_{N\to\infty} \frac{1}{N} \sum_{n=1}^{N} T^{k_n} x_0,$$

so altogether we obtain

$$\lim_{N\to\infty} \frac{1}{N} \sum_{n=1}^{N} T^{k_n} x = P_{\mathrm{fix}\,T} x + \lim_{N\to\infty} \frac{1}{N} \sum_{n=1}^{N} T^{k_n} x_0 = P_{\mathrm{fix}\,T} x,$$

and (iii) is proved.
(iii)⇒(i): Suppose (iii) holds and let $x \in H$, $\lambda \in \mathbb{T}$ satisfy $Tx = \lambda x$. Take $r \in (0, 1)$ and consider the open neighborhood $U := \mathrm{B}(\lambda, r) \cap \mathbb{T}$ of λ in T. Then the set of return times $R_U(x) := \{k \in \mathbb{N} : \lambda^k \in U\} \subset \mathbb{N}$ is syndetic (see Proposition 3.5 and Proposition 3.3 for irrational λ whereas the assertion is clear if λ is rational), so it has positive lower density (see Exercise 4.10). The corresponding sequence $(k_n)_{n\in\mathbb{N}}$ of return times satisfies $\lambda^{k_n} \in U$ for every $n \in \mathbb{N}$. It follows that for every $N \in \mathbb{N}$

$$\Big\| \frac{1}{N} \sum_{n=1}^{N} T^{k_n} x \Big\| = \Big| \frac{1}{N} \sum_{n=1}^{N} \lambda^{k_n} \Big| \cdot \|x\| \geq (1 - r)\|x\|.$$

Assumption (iii) implies then $\|P_{\mathrm{fix}}\, x\| \geq (1 - r)\|x\|$. Taking $r \to 0$ we obtain $\|P_{\mathrm{fix}\,T} x\| = \|x\|$ and hence $x \in \mathrm{fix}\, T$ since $P_{\mathrm{fix}\,T}$ is an orthogonal projection. Altogether $H_{\mathrm{fix}\,T} = H_{\mathrm{kr}}$ follows. ■

This result was proved by Jones [584] even in the more general setting of Banach space contractions. He also gave the following immediate corollary, a characterization of weakly mixing systems being an analogue of the Blum–Hanson theorem.

Corollary 12.5 (Jones) *Let (X, μ, T) be a measure-preserving system with Koopman operator T on $f \in \mathrm{L}^2(X, \mu)$. The system is weakly mixing if and only if for every subsequence $(k_n)_{n\in\mathbb{N}}$ of $\mathbb{N}$ with positive lower density and every $f \in \mathrm{L}^2(X, \mu)$*

$$\lim_{N\to\infty} \frac{1}{N} \sum_{n=1}^{N} T^{k_n} f = \Big(\int_X f \,\mathrm{d}\mu \Big) \cdot \mathbf{1}.$$

Finally, we turn to pointwise convergence. For a particular class of invertible measure-preserving systems including two-sided Bernoulli shifts, namely those having Lebesgue spectrum (see Sect. 5.6), *any* subsequence will be suitable also for the pointwise ergodic theorem in the following sense.

Proposition 12.6 *Let (X, μ, T) be an invertible measure-preserving system with Lebesgue spectrum. Then there is a dense subspace $D \subset \mathrm{L}^2(X, \mu)$ such that for every $f \in D$ and every subsequence $(k_n)_{n\in\mathbb{N}}$ of $\mathbb{N}$ one has*

$$\lim_{N\to\infty} \frac{1}{N}\sum_{n=1}^{N} T^{k_n} f = \int_X f \, \mathrm{d}\mu \cdot \mathbf{1}$$

μ-almost everywhere.

Proof By Proposition 5.26(a), Lebesgue spectrum implies the orthogonality of $(T^n f)_{n\in\mathbb{Z}}$ for any $f \in \mathrm{L}^2(X)$. Let $(e_m)_{m\in N}$ be an orthonormal set such that $\mathbf{1}$, $T^k e_m$ for $k \in \mathbb{Z}$, $m \in N$ form an orthonormal basis in $\mathrm{L}^2(X, \mu)$. Take $f \in \mathrm{lin}\{\mathbf{1}, T^k e_m : k \in \mathbb{Z},\ m \in N\}$, i.e.,

$$f = c\mathbf{1} + \sum_{m\in F, |k|\le M} a_{mk} T^k e_m$$

for some coefficients a_{mk}, $M \in \mathbb{N}$, and $F \subset N$ with $|F| \le M$; by orthogonality $c = \int_X f \, \mathrm{d}\mu$. Then

$$\frac{1}{N}\sum_{n=1}^{N} T^{k_n} f = c\mathbf{1} + \sum_{m\in F, |k|\le M} a_{mk} \frac{1}{N}\sum_{n=1}^{N} T^{k+k_n} e_m$$

which by the Banach–Rajchman lemma (Proposition 9.7) converges to $c\mathbf{1}$ almost everywhere. ■

For the subsequential version of the pointwise ergodic theorem one needs besides a dense set of good functions also Banach's principle, which we discuss in Sect. 12.3 below. First we study subsequences that are good for the mean ergodic theorem and *every* measure-preserving system.

12.2 Mean Good Sequences

For a subsequential version of the mean ergodic theorem we introduce the following notion.

Definition 12.7 A subsequence $(k_n)_{n\in\mathbb{N}}$ of $\mathbb{N}$ is called a **good subsequence for the mean ergodic theorem** (or **mean good**, or just **good**) if for every invertible measure-preserving system (X, μ, T) and every $f \in \mathrm{L}^2(X, \mu)$ the averages

$$\frac{1}{N}\sum_{n=1}^{N} T^{k_n} f \tag{12.3}$$

converge in $L^2(X, \mu)$. A subsequence which is not mean good is called **mean bad** (or **bad for the mean ergodic theorem**, or just **bad**).

Note that the assumption of invertibility, implying that the corresponding Koopman operator on L^2 is unitary, is technical and can be dropped, see Remark 12.10 below.

Example 12.8 (Affine Sequences) A first easy example of a good subsequence is an affine subsequence of the form $k_n := an + b$ for a fixed $a \in \mathbb{N}$, $b \in \mathbb{Z}$. This follows from the mean ergodic theorem and the identity $T^{k_n} f = (T^a)^n (T^b f)$ if $b \geq 0$ and $T^{k_n} f = (T^a)^{n-n_0}(T^{n_0 a+b} f)$ for an appropriate n_0 and $n \geq n_0$ if $b < 0$.

It is easy to characterize good subsequences as follows, see Blum and Eisenberg [160] and, e.g., Berend et al. [99].

Proposition 12.9 (Characterization of Good Sequences) *For a subsequence $(k_n)_{n\in\mathbb{N}}$ of $\mathbb{N}$ the following assertions are equivalent.*

(i) *$(k_n)_{n\in\mathbb{N}}$ is a good subsequence for the mean ergodic theorem.*
(ii) *For every unitary operator S on a Hilbert space H and every $x \in H$ the averages*

$$\frac{1}{N}\sum_{n=1}^{N} S^{k_n} x \text{ converge.} \tag{12.4}$$

(iii) *For every $\lambda \in \mathbb{T}$ the limit*

$$d(\lambda) := \lim_{N\to\infty} \frac{1}{N}\sum_{n=1}^{N} \lambda^{k_n} \text{ exists.}$$

When either of these conditions holds, for every unitary operator S on a Hilbert space H and every $x \in H$

$$\lim_{N\to\infty} \frac{1}{N}\sum_{n=1}^{N} S^{k_n} x = \sum_{\lambda\in\mathbb{T}} d(\lambda) P_\lambda x, \tag{12.5}$$

where $P_\lambda = P_{\ker(\lambda - S)}$ denotes the orthogonal projection onto $\ker(\lambda - S)$.

Remark 12.10 For $\lambda_1, \lambda_2 \in \mathbb{T}$ with $\lambda_1 \neq \lambda_2$ we have $\operatorname{ran}(P_{\lambda_1}) \perp \operatorname{ran}(P_{\lambda_2})$ (Proposition 6.1), so the sum in (12.5) contains at most countably many non-zero terms and the series converges unconditionally (use Bessel's inequality). Moreover, using Szőkefalvi-Nagy's dilation theorem (Theorem 6.13) one can add the following assertion to the previous list of equivalent conditions, cf. Lin et al. [747]:

(ii') *For every contraction S on a Hilbert space H and every $x \in H$* (12.4) *holds.*

This shows in particular that one can drop the assumption of invertibility in Definition 12.7. The formula in (12.5) remains also true, see [342, Ch. 21]. One can also show the equivalence (ii)⇔(ii') and the validity of the limit formula (12.5) for Hilbert space contractions using the decomposition (6.5) into the unitary and the completely non-unitary part, Proposition 6.10 and Theorem 12.2. We leave the details as Exercise 12.2.

Proof The implication (ii)⇒(i) is trivial since the Koopman operator of an invertible measure-preserving system is unitary.
(i)⇒(iii): For $\lambda \in \mathbb{T}$ consider the measure-preserving system $(\mathbb{T}, \mathrm{m}, \lambda)$ with Koopman operator T on $\mathrm{L}^2(\mathbb{T}, \mathrm{m})$ and the monomial $\mathbf{z}$ given by $z \mapsto z$. Then $T^n\mathbf{z} = \lambda^n\mathbf{z}$ and the convergence of (12.3) in $\mathrm{L}^2(\mathbb{T}, \mathrm{m})$ implies the existence of $d(\lambda)$.
(iii)⇒(ii) Let S be a unitary operator on a Hilbert space H and take $x \in H$. By restricting to the cyclic subspace $Z(x) = \overline{\mathrm{lin}}\{S^n x : n \in \mathbb{Z}\}$ we can assume that x is a cyclic vector. By the spectral theorem (Theorem 5.13) we can moreover assume that S is the multiplication $M_\mathbf{z} : f \mapsto \mathbf{z} \cdot f$ on $\mathrm{L}^2(\mathbb{T}, \mu)$ for some probability measure μ on $\mathbb{T}$ and $x = \mathbf{1}$. Now, the identity $(M_\mathbf{z}^n\mathbf{1})(\lambda) = \lambda^n$ and the hypothesis that $(k_n)_{n\in\mathbb{N}}$ is good imply, by Lebesgue's dominated convergence theorem, the L^2-convergence of

$$\frac{1}{N}\sum_{n=1}^{N} M_\mathbf{z}^{k_n}\mathbf{1}. \tag{12.6}$$

The equivalence of the assertions is hence proved.
We now turn to the last assertion and stay, without loss of generality, in the setting of $S = M_\mathbf{z}$ on $\mathrm{L}^2(\mathbb{T}, \mu)$ as above. The limit of (12.6) at the point λ equals $\lim_{N\to\infty} \frac{1}{N}\sum_{n=1}^{N} \lambda^{k_n} = d(\lambda)$. It is easy to see, by using Proposition 5.16, that $P_\lambda\mathbf{1} = \mathbf{1}_{\{\lambda\}}$ holds for every $\lambda \in \mathbb{T}$ and thus we have

$$\begin{aligned} \lim_{N\to\infty} \frac{1}{N}\sum_{n=1}^{N} M_\mathbf{z}^{k_n}\mathbf{1} &= \sum_{\lambda:\, d(\lambda)\neq 0} d(\lambda)\mathbf{1}_{\{\lambda\}} \\ &= \sum_{\lambda \text{ atom of } \mu} d(\lambda)\mathbf{1}_{\{\lambda\}} = \sum_{\lambda\in P\sigma(M_\mathbf{z})} d(\lambda)P_\lambda\mathbf{1}. \end{aligned}$$

By Lebesgue's dominated convergence theorem we obtain (12.5). ■

Remark 12.11

(a) By the equidistribution result in Proposition 10.5 and Weyl's criterion (Theorem 10.3), for every subsequence $(k_n)_{n\in\mathbb{N}}$ of $\mathbb{N}$ one has $d(\lambda) = 0$ for $\mathrm{m}_\mathbb{T}$-almost every $\lambda \in \mathbb{T}$.
(b) If $(k_n)_{n\in\mathbb{N}}$ is good, then $d : \mathbb{T} \to \overline{\mathbb{D}} = \{z \in \mathbb{C} : |z| \le 1\}$ is a Borel measurable function as pointwise limit of a sequence of continuous functions. Moreover, in many cases the set $\{\lambda \in \mathbb{T} : d(\lambda) \neq 0\}$ (called the **spectrum of the sequence**

(k_n)) is countable. In addition to the examples below, we refer to Rosenblatt [880, Cor. 43] for the case of sequences with positive lower density and to Cuny et al. [266, Cor. 2.13] for the case of good subsequences with positive upper density. Very recently, Cuny and Parreau [268] showed that not every good subsequence has countable spectrum. (Cf. the opposite situation for the weighted analogue, see Proposition 16.3 and the references after it.) Note that by (a) the spectrum of a good subsequence always has Lebesgue measure zero.

Before turning to examples let us mention the following property of the function d, see [266, Sec. 2] for further properties.

Proposition 12.12 *Let $(k_n)_{n\in\mathbb{N}}$ be good. Then $G := \{\lambda : |d(\lambda)| = 1\}$ is a finite (cyclic) subgroup of $\mathbb{T}$.*

Proof We clearly have $d(1) = 1$ and $d(\overline{\lambda}) = \overline{d(\lambda)}$. We first show that $\operatorname{D-lim}_{n\to\infty} \lambda^{k_n} = d(\lambda)$ holds whenever $\lambda \in G$. Indeed, take $b_n := \operatorname{Re}(\overline{d(\lambda)}\lambda^{k_n})$ and apply the Koopman–von Neumann lemma (Proposition 4.19(b)) to conclude $\operatorname{D-lim}_{n\to\infty} \operatorname{Re}(\overline{d(\lambda)}\lambda^{k_n}) = 1$. Since $|\overline{d(\lambda)}\lambda^{k_n}| \le 1$, we also obtain $\operatorname{D-lim}_{n\to\infty}\overline{d(\lambda)}\lambda^{k_n} = 1$, and that was to be proved. Altogether we conclude that G is a subgroup of $\mathbb{T}$, and as such, it is either finite or dense in $\mathbb{T}$ (see Exercise 3.3). Assume, with the aim of arriving at a contradiction, that G is dense. By Remark 12.11(a) $d = 0$ almost everywhere with respect to m, and we conclude that d is nowhere continuous. This is a contradiction since d is a pointwise limit of continuous functions, hence has many continuity points, see, e.g., Kechris [638, Thm. 8.38]. ■

Next we turn to more sophisticated examples than affine sequences from Example 12.8.

Example 12.13 (Polynomial Sequences) By Proposition 7.28, for every $P \in \mathbb{Z}[\cdot]$ with positive leading coefficient the sequence $(P(n))_{n\in\mathbb{N}}$ is a good sequence for the mean ergodic theorem. We give here another proof for integer polynomials and calculate the limit. Let $P \in \mathbb{Q}[\cdot]$ be a polynomial of degree at least 1 such that $P(\mathbb{N}_0) \subset \mathbb{N}_0$.

For irrational $\lambda \in \mathbb{T}$ we have by Weyl's equidistribution theorem for polynomial sequences (Theorem 10.17)

$$d(\lambda) = \lim_{N\to\infty} \frac{1}{N} \sum_{n=1}^{N} \lambda^{P(n)} = 0.$$

Next, assume that $\lambda \in \mathbb{T}$ is a root of unity, i.e., $\lambda^q = 1$ for some $q \in \mathbb{N}$, and assume for simplicity (and without loss of generality) that $qP \in \mathbb{Z}[\cdot]$. Observe that for every $n \in \mathbb{N}$

$$P(n+q) - P(n) \equiv 0 \bmod q.$$

For $N \in \mathbb{N}$ write $N = mq + r$ with $r \in \{0, \dots, q-1\}$. We then have

$$\begin{aligned}\frac{1}{N}\sum_{n=1}^{N}\lambda^{P(n)} &= \frac{1}{N}\sum_{n=1}^{mq}\lambda^{P(n)} + \frac{1}{N}\sum_{n=mq+1}^{N}\lambda^{P(n)} \\ &= \frac{mq}{N}\frac{1}{q}\sum_{n=1}^{q}\lambda^{P(n)} + \frac{1}{N}\sum_{n=mq+1}^{N}\lambda^{P(n)} \to \frac{1}{q}\sum_{n=1}^{q}\lambda^{P(n)}\end{aligned}$$

as $N \to \infty$. Thus we see that

$$d(\lambda) = \lim_{N\to\infty}\frac{1}{N}\sum_{n=1}^{N}\lambda^{k_n} = \begin{cases}\frac{1}{q}\sum\limits_{n=1}^{q}\lambda^{P(n)}, & \lambda^q = 1 \text{ and } qP \in \mathbb{Z}[\cdot], \\ 0, & \lambda \text{ not a root of unity.}\end{cases} \tag{12.7}$$

This shows by Proposition 12.9 that the subsequence $(P(n))_{n\in\mathbb{N}}$ is good with the limit of subsequential ergodic averages given by (12.5) and (12.7) (note that by (12.7) this sequence has countable spectrum). In particular, by Remark 6.30 for totally ergodic systems[1] the limit of the ergodic averages for f along $(P(n))$ equals $\int_X f \,\mathrm{d}\mu$ (which we already saw in Exercise 9.6).

Note that the limit of polynomial averages $\frac{1}{N}\sum_{n=1}^{N} T^{P(n)}$ is in general not a projection. (This can easily be seen by using (12.7) for $P(n) := n^2$ and a rotation on a set of three points.) See Kunszenti-Kovács [697] and Kunszenti-Kovács et al. [699] for more information on this limit.

Example 12.14 (Primes and Polynomials of Primes) As before, we denote by p_n the nth prime number. Let P be an integer polynomial (i.e., $P \in \mathbb{Q}[\cdot]$ with $P(\mathbb{Z}) \subset \mathbb{Z}$) with positive leading coefficient. By using Proposition 12.9 we show that the sequence $(P(p_n))_{n\in\mathbb{N}}$ is good with countable spectrum and determine the limit of the corresponding ergodic averages in (12.5), i.e., the limit function d.

For irrational $\lambda \in \mathbb{T}$ we have by Theorem 10.26, more precisely by its proof,

$$\lim_{N\to\infty}\frac{1}{N}\sum_{n=1}^{N}\lambda^{P(p_n)} = 0$$

(the sequence $(\lambda^{P(p_n)})_{n\in\mathbb{N}}$ is equidistributed in $\mathbb{T}$ by Weyl's criterion, Theorem 10.3). Assume now that λ is a root of unity, say $\lambda^q = 1$ for some $q \in \mathbb{N}$, and assume without loss of generality again that $qP \in \mathbb{Z}[\cdot]$. It remains to show that the limit $d(\lambda) := \lim_{N\to\infty}\frac{1}{N}\sum_{n=1}^{N}\lambda^{P(p_n)}$ exists and to determine it.

[1] Recall that a measure-preserving system (X, μ, T) is **totally ergodic** if for each $k \in \mathbb{N}$ the kth iterate system (X, μ, T^k) is ergodic.

In what follows we will again denote prime numbers by p without saying it explicitly every time. Recall that by the prime number theorem in arithmetic progressions (Theorem 10.24), for every $a \in \{1, \ldots, q\}$ with $(a, q) = 1$ (i.e., a and q coprime)

$$\lim_{N\to\infty} \frac{|\{p \le N : p \equiv a \bmod q\}|}{N/\log(N)} = \frac{1}{\varphi(q)}, \tag{12.8}$$

where φ is the Euler totient function given by $\varphi(q) := |\{n < q : (n, q) = 1\}|$. Since by assumption $P(n + q) \equiv P(n) \bmod q$ holds for every $n \in \mathbb{N}$ and $\lambda^q = 1$, we obtain for $N \in \mathbb{N}$

$$\begin{aligned}\sum_{p\le N} \lambda^{P(p)} &= \sum_{a<q,\,(a,q)=1}\ \sum_{p\le N,\ p\equiv a \bmod q} \lambda^{P(p)} = \sum_{a<q,(a,q)=1}\ \sum_{p\le N,\ p\equiv a \bmod q} \lambda^{P(a)} \\ &= \sum_{a<q,(a,q)=1} \lambda^{P(a)} |\{p \le N : p \equiv a \bmod q\}|.\end{aligned}$$

Dividing by $\pi(N)$ (the number of primes $\le N$, see Sect. 10.4) and using (12.8) and the prime number theorem (Theorem 10.21) lead to

$$d(\lambda) = \lim_{N\to\infty} \frac{1}{\pi(N)} \sum_{p\le N} \lambda^{P(p)} = \frac{1}{\varphi(q)} \sum_{a<q,\,(a,q)=1} \lambda^{P(a)}. \tag{12.9}$$

We have shown that $(P(p_n))_{n\in\mathbb{N}}$ is a good sequence with countable spectrum and the limit of the corresponding ergodic averages is given by (12.5) and (12.9). Again, for totally ergodic systems the limit of the ergodic averages for f along $(P(p_n))$ equals $\int_X f \,\mathrm{d}\mu$.

It is not hard to construct a *bad* subsequence. Indeed, take for instance $k_n := 2n + a_n$ with $a_n \in \{0, 1\}$ such that the sequence $((-1)^{k_n})_{n\in\mathbb{N}} = ((-1)^{a_n})_{n\in\mathbb{N}}$ is Cesàro divergent, so (k_n) is bad for the mean ergodic theorem. In this way, it is possible to destroy goodness of sequences by the addition of a_n with $a_n \in \{0, 1\}$.

This shows that it is not the growth alone that determines whether a subsequence is good. However, growth can be responsible for the badness of a sequence as the following example shows.

Example 12.15 (Lacunary Sequences) A subsequence $(k_n)_{n\in\mathbb{N}}$ of $\mathbb{N}$ is called **lacunary** if there is $q > 1$ such that $k_{n+1} \ge qk_n$ holds for every $n \in \mathbb{N}$. In [90] Bellow showed that every lacunary sequence is bad. Following Rosenblatt and Wierdl [881] we present here a proof of this fact for lacunary sequences provided $q \ge 8$.

Suppose that a subsequence $(k_n)_{n\in\mathbb{N}}$ of $\mathbb{N}$ satisfies $k_{n+1} \ge 8k_n$ for every $n \in \mathbb{N}$. By Proposition 12.9 we need to find $\lambda = \mathrm{e}^{2\pi i\alpha} \in \mathbb{T}$ ($\alpha \in [0, 1]$) such that the sequence $(\frac{1}{N}\sum_{n=1}^{N} \lambda^{k_n})_{n\in\mathbb{N}}$ diverges.

We first show that for every given sequence $(I_n)_{n\in\mathbb{N}}$ of closed intervals in $[0, 1]$, each of them having length at least $1/4$, there exists $\alpha \in [0, 1]$ with $\{k_n\alpha\} := k_n\alpha - \lfloor k_n\alpha\rfloor \in I_n$ for every $n \in \mathbb{N}$ or, equivalently,

$$\alpha \in \bigcap_{n=1}^{\infty}(I_n + \mathbb{Z})/k_n. \tag{12.10}$$

We define a new sequence $(J_n)_{n\in\mathbb{N}}$ of closed intervals and a sequence $(m_n)_{n\in\mathbb{N}}$ in $\mathbb{Z}$ recursively, and start with $J_1 := I_1$ and $m_1 := 1$. Since $k_2 \geq 8$ and since the length of I_1 is at least $1/4$, there exists $m_2 \in \mathbb{Z}$ such that $[\frac{m_2}{k_2}, \frac{m_2+1}{k_2}] \subset I_1$. Thus we have

$$J_2 := (I_2 + m_2)/k_2 \subset \left[\tfrac{m_2}{k_2}, \tfrac{m_2+1}{k_2}\right] \subset J_1.$$

Since the length of J_2 is at least $\frac{1}{4k_2}$ and $k_3 \geq 8k_2$, there exists $m_3 \in \mathbb{N}$ with $[\frac{m_3}{k_3}, \frac{m_3+1}{k_3}] \subset J_2$, and we have $J_3 := (I_3 + m_3)/k_3 \subset J_2$, etc. Thus we obtain a sequence of closed, nested, non-empty subintervals $(J_n)_{n\in\mathbb{N}}$ of $[0, 1]$ with length not exceeding $\frac{1}{8^k}$. Denote by α the unique number in $[0, 1]$ with $\{\alpha\} := \bigcap_{n=1}^{\infty} J_n$. By construction we have $\alpha \in (I_n + m_n)/k_n$ for every $n \in \mathbb{N}$, proving (12.10).

The proof that $(k_n)_{n\in\mathbb{N}}$ is bad is completed by carefully choosing the intervals $I_1, I_2, \ldots$, each of them with length at least $1/4$. The details are left as Exercise 12.3.

Example 12.16 (Ergodic Sequences) A subsequence $(k_n)_{n\in\mathbb{N}}$ in $\mathbb{N}$ is called **ergodic** if $d(\lambda) = 0$ for every $\lambda \neq 1$, i.e., if $d = \mathbf{1}_{\{1\}}$. By Proposition 12.9, a sequence (k_n) is ergodic if and only if

$$\lim_{N\to\infty} \frac{1}{N}\sum_{n=1}^{N} S^{k_n}x = P_{\mathrm{fix}\, S}x \tag{12.11}$$

holds for every unitary operator S on a Hilbert space H and every $x \in H$, i.e., the assertion of the mean ergodic theorem holds along this sequence. Moreover, as Berend et al. [99] observed, (12.11) holds for every ergodic sequence (k_n), every asymptotically almost periodic operator[2] S on a Banach space E and every $x \in E$, see Exercise 12.4. Note that a sequence (k_n) is ergodic whenever (λ^{k_n}) is equidistributed in $\mathbb{T}$ for every irrational $\lambda \in \mathbb{T}$ and for every rational $\lambda \neq 1$ the sequence (λ^{k_n}) is equidistributed on the pertinent finite set of points (in the natural sense).

Clearly, $(n)_{n\in\mathbb{N}}$ is ergodic while $(2n)_{n\in\mathbb{N}}$ is not. Moreover, as we saw above, integer polynomials not of the form $p(n) = n + a$, lacunary sequences (which are not even good), primes and integer polynomials of primes are also not ergodic.

[2] An operator is called **asymptotically almost periodic** if every vector is asymptotically almost periodic with respect to it, see the definition before Proposition 6.23.

For examples of ergodic sequences see, e.g., Veech [1016], Blum and Cogburn [162], Boshernitzan [173], Ajtai et al. [14], Rosenblatt and Wierdl [881, Ch. 2], Boshernitzan et al. [175], Cuny et al. [266],[3] and Donoso et al. [310].

12.3 Pointwise Convergence and Weak Maximal Inequality

The study of pointwise convergence of subsequential averages is much more difficult than that of mean convergence. We begin with some general principles and connect almost everywhere convergence to a weak maximal inequality.

We first recall maximal operators from Sect. 7.6 and extend the notion of a weak maximal inequality from there to the L^p-setting. For a sequence $(S_N)_{N\in\mathbb{N}}$ of linear operators on $\mathrm{L}^1(X,\mu)$ the corresponding **maximal operator** is defined for $f \in \mathrm{L}^1(X,\mu)$ by

$$S^* f := \sup_{N\in\mathbb{N}} |S_N f|.$$

This (nonlinear) operator has the following properties.

(1) $S^*(\alpha f) = |\alpha| \cdot S^* f$ for each $\alpha \in \mathbb{C}$ and $f \in \mathrm{L}^1(X,\mu)$,
(2) $S^*(f+g) \le S^* f + S^* g$ for each $f, g \in \mathrm{L}^1(X,\mu)$.

Definition 12.17 Let $p \in [1,\infty)$ and (X,μ) be a probability space. We say that a sequence (S_N) of operators on $\mathrm{L}^p(X,\mu)$ satisfies a **weak (p,p)-maximal inequality** if there exists a constant $C > 0$ such that for the corresponding maximal operator S^*

$$\mu[S^* f > \lambda] \le \frac{C}{\lambda^p}\|f\|_p^p \qquad \text{for all } \lambda > 0 \tag{12.12}$$

holds for every $f \in \mathrm{L}^p(X,\mu)$. We say that a Koopman operator T (or the system (X,μ,T)) satisfies a **weak (p,p)-maximal inequality along (k_n)** if the sequence of the subsequential ergodic means

$$S_N := \frac{1}{N}\sum_{n=1}^{N} T^{k_n}$$

satisfy a weak (p,p)-maximal inequality.

Remark 12.18 Clearly (12.12) implies $S^* f < \infty$ a.e.

[3] In this paper, there is a gap in the example on return times sequences along polynomials on page 356.

Remark 12.19 (Banach's Principle in L^p) An easy modification of Proposition 7.45 gives the following: If a sequence (S_N) of linear bounded operators on $\mathrm{L}^p(X,\mu)$ satisfies a weak (p,p)-maximal inequality, then the set of functions $f \in \mathrm{L}^p(X,\mu)$ such that $(S_N f)$ converges a.e. is closed in $\mathrm{L}^p(X,\mu)$. We leave the proof as Exercise 12.10.

The following general principle connects almost everywhere convergence of averages (12.3) for L^p-functions to a weak (p,p)-maximal inequality, cf. Theorem 7.52.

Proposition 12.20 (Sawyer) *Let (X,μ,T) be an ergodic measure-preserving system, $p \in [1,\infty)$ and let (S_N) be a sequence of positive operators on $\mathrm{L}^p(X,\mu)$ commuting with T. Assume that for the maximal operator S^* the inequality $S^* f < \infty$ holds on a set of positive measure for every $f \in \mathrm{L}^p(X,\mu)$. Then (S_N) satisfies a weak (p,p)-maximal inequality.*

For a proof we refer to Garsia [463, Sec. 1.3] or the original work by Sawyer [914, Thm. 2 and its proof].

Remark 12.21 In particular, by Proposition 12.20 and Remark 12.18, for T and (S_N) as in Proposition 12.20, for $p \in [1,\infty)$ the following assertions are equivalent:

(i) (S_N) satisfies a weak (p,p)-maximal inequality.
(ii) $S^* f < \infty$ holds on a set of positive measure for every $f \in \mathrm{L}^p(X,\mu)$.
(iii) $S^* f < \infty$ holds a.e. for every $f \in \mathrm{L}^p(X,\mu)$.

Remark 12.22 One can reformulate the assertion of Proposition 12.20 as follows: If T does not satisfy a weak (p,p)-maximal inequality, then there exists a function $f \in \mathrm{L}^p(X,\mu)$ with $S^* f = \infty$ a.e., in particular $S_N f$ diverges a.e. In fact more holds, namely that the set of such functions is residual in the Baire category sense, see Sawyer [914, Thm. 4].

Note that, unlike in the classical case of the sequence (n) (see the discussion at the beginning of Sect. 7.6), for general measure-preserving systems and for many sequences like polynomials or primes there is a priori no good dense set of functions for which one has a.e. convergence of the subsequential averages (12.3).

12.4 Pointwise Good Sequences

We now turn to the study of pointwise convergence of subsequential ergodic averages, which is way more difficult than the case of mean convergence.

Definition 12.23 For $p \in [1,\infty]$, a subsequence $(k_n)_{n\in\mathbb{N}}$ in $\mathbb{N}$ is called a **good subsequence for the pointwise ergodic theorem in** L^p (or **pointwise good in** L^p) if for every invertible measure-preserving system (X,μ,T) and every $f \in \mathrm{L}^p(X,\mu)$ the subsequential averages (12.3) converge almost everywhere.

Remark 12.24 The assumption of invertibility in the above definition is technical and can be removed, see Exercise 12.6.

Every pointwise good sequence in L^2 is good. This follows from Proposition 12.9 by considering rotations on the unit circle. The converse is not true, see Boshernitzan et al. [175] (an explicit example is $(\lfloor a_n \rfloor)$ for $a_n := \sqrt{3}n^3 + \log n \cdot \log\log n$). Moreover, Bellow [91] constructed for any $p > 1$ a pointwise good sequence in L^p which is not pointwise good in L^q for every $q \in [1, p)$, see also Reinhold-Larsson [867, 868] and Rosenblatt and Wierdl [881, Ch. VII].

Contrary to the case of (mean) good sequences, there is no useful characterization of pointwise good sequences. We present here some known (classes of) examples and begin with sequences of positive density.

Definition 12.25 For a measure-preserving system (X, μ, T), a measurable subset $A \subset X$ and $x \in X$ consider the set

$$R_A(x) := \{n \in \mathbb{N} : T^n x \in A\}$$

of return times of x to A. We call the corresponding subsequence $(r_{A,x}(n))$ of $\mathbb{N}$ (where $r_{A,x}(n)$ is the nth visit time of x to A) **return times sequence**.

It is easy to see that for ergodic systems and almost every $x \in X$, the density of the set $R_A(x)$ and hence the corresponding subsequence $(r_{A,x}(n))$ exists and equals $\mu(A)$, see Exercise 12.7(a). This allows to treat ergodic averages along return times sequences via studying corresponding weighted ergodic averages whenever $\mu(A) > 0$ by the following observation due to Ryll-Nardzewski [897]. Weighted ergodic averages will be discussed in Chap. 16 in some more detail.

Remark 12.26 (Connection between Weighted and Subsequential Ergodic Theorems for Subsequences of Positive Density[4]) Suppose that a subsequence $(k_n)_{n\in\mathbb{N}}$ of $\mathbb{N}$ has positive density, i.e., the set $A := \{k_n : n \in \mathbb{N}\}$ satisfies

$$\delta := \lim_{N\to\infty} \frac{|\{1, 2, \dots, N\} \cap A|}{N} > 0.$$

Define $a_n := \mathbf{1}_A(n)$ for $n \in \mathbb{N}$. It is Exercise 12.7(b) to show that for every Banach space E and every sequence $(f_n)_{n\in\mathbb{N}}$ in E the *subsequential averages* $\frac{1}{N}\sum_{n=1}^N f_{k_n}$ converge if and only if the *weighted averages* $\frac{1}{N}\sum_{n=1}^N a_n f_n$ converge, and in this case the equality

$$\lim_{N\to\infty} \frac{1}{N}\sum_{n=1}^N f_{k_n} = \frac{1}{\delta} \cdot \lim_{N\to\infty} \frac{1}{N}\sum_{n=1}^N a_n f_n$$

holds.

[4] see Bellow and Jones [97, Prop. 1.7] for the scalar case.

Using this correspondence and the deep return times theorem of Bourgain on weighted averages, see Theorem 16.30 below, one has the following rich class of examples of pointwise good weights.

Example 12.27 (Almost All Return Times Sequences Are Pointwise Good in L^1) Let (X, μ, T) be an ergodic measure-preserving system and let $A \subset X$ satisfy $\mu(A) > 0$. Then for μ-a.e. $x \in X$ the return times sequence $(r_{A,x}(n))$ is pointwise good in L^1. We leave as Exercise 12.7(c) to deduce this from Remark 12.26 and Theorem 16.30. Note that we will not prove Theorem 16.30 but its "mean" version showing that for μ-a.e. $x \in X$ the return times sequence $(r_{A,x}(n))$ is good, see Example 16.29.

We now turn our attention to subsequences having density zero and revisit arithmetic sequences from the above. The following famous result of Bourgain shows that polynomial sequences are pointwise good in L^p for every $p > 1$ answering a question of Bellow and Furstenberg. (For the case $p = 1$ see Theorem 12.38 below.)

Theorem 12.28 (Polynomial Ergodic Theorem, Bourgain [181, 183]) *Let (X, μ, T) be an invertible measure-preserving system, $p > 1$ and $f \in \mathrm{L}^p(X, \mu)$. Then for every polynomial $P \in \mathbb{Z}[\cdot]$ the polynomial ergodic averages*

$$\frac{1}{N}\sum_{n=1}^{N} T^{P(n)} f \tag{12.13}$$

converge almost everywhere.

Of course, the limit of averages (12.13) is the same as for the mean convergence, see (12.5) and (12.7). Bourgain [183] also showed that sequences of the form $(\lfloor P(n) \rfloor)$ for a real polynomial $P \in \mathbb{R}[\cdot]$ are pointwise good in L^∞, see Krause [673] for the corresponding result in L^p for every $p > \frac{2d}{d+1}$, where $d = \deg P$.

Remark 12.29 Bourgain's proof is based on the observation that pointwise convergence of ergodic averages along a subsequence follows[5] from the corresponding maximal inequality and certain oscillation inequalities. These inequalities, being quantitative and finite in nature (i.e., involving finitely many elements of the orbit), can be reduced via the so-called Calderón principle[6] to the corresponding inequalities for the shift model $(\mathbb{Z}, +1)$.[7] The main work is to prove these inequalities for the shift model and polynomial sequences which is done by deep and technically

[5] In particular, this led to a new proof of Birkhoff's ergodic theorem providing some quantitative information about the speed of convergence, see [179, Sec. 2] and [183, Sec. 2].

[6] See Calderón [216] for the original version of this principle in the setting of flows. We illustrate the version used by Bourgain in the context of the (strong) maximal inequality in Exercise 12.9.

[7] Note that this system does not have a finite invariant measure and one is no more in the ergodic theoretic setting.

difficult harmonic analysis arguments involving maximal inequalities for certain Fourier multipliers combined with a major arc description of the exponential sums $\sum_{n=1}^{N} e^{2\pi i P(n)}$ from the Hardy–Littlewood circle method.

For excellent expositions of (a modification of) Bourgain's proof for L^2 see Thouvenot [1001], Rosenblatt and Wierdl [881], and Quas and Wierdl [110, Appendix B], where the latter one uses a simplification of the key step in Bourgain's proof due to Lacey [702]. We also refer to Krause [672], [671, Ch. 4, 7] for some ideas behind Bourgain's proof, cf. also Tao [986].

A modification of Bourgain's argument via a more detailed analysis of the corresponding variational operators was done by Krause [670], see also Zorin-Kranich [1072]. A higher dimensional counterpart of Bourgain's result via variational inequalities was established by Mirek and Trojan [798] and Mirek et al. [794, 795], see also Krause [671, Ch. 8–9]. For more generalizations and related results see Sects. 12.6, 15.9 and 16.6.

The following shows that the primes are also pointwise good in L^p for every $p > 1$. This was proved in a similar way by Bourgain [179] for $p = 2$ (where he mentioned that his arguments work for $p > \frac{1+\sqrt{3}}{2}$) and extended to all $p > 1$ by Wierdl [1047].

Theorem 12.30 (Ergodic Theorem along Primes) *Let (X, μ, T) be an (invertible) measure-preserving system, $p > 1$ and $f \in L^p(X, \mu)$. Then the ergodic averages along primes*

$$\frac{1}{N}\sum_{n=1}^{N} T^{p_n} f \tag{12.14}$$

converge almost everywhere.

Again, the limit of averages (12.14) is the same as for the mean convergence, see (12.5) and (12.9) with $P := \mathrm{id}$.

Remark 12.31 Analogously to Remark 12.26, Wierdl's proof of Theorem 12.30 begins with the observation that convergence of averages along primes (12.14) is equivalent to convergence of weighted *linear* ergodic averages

$$\frac{1}{N}\sum_{n=1}^{N} \Lambda'(n) T^n f \tag{12.15}$$

for the (unbounded) modified von Mangoldt function Λ' (see Sect. 10.4), see Exercise 12.8(b) based on Wierdl [1047, pp. 316–317] (for bounded functions this was done in Lemma 10.23). Then Wierdl proves a maximal inequality in L^p for $p > 1$ for the weighted averages to extend pointwise convergence from $p = 2$ (done by Bourgain in [179]) to $p > 1$ by extending Bourgain's method from [179] using more refined results from number theory.

Note that Bourgain's proof [183] of the polynomial ergodic theorem (Theorem 12.28) appeared one year later. He mentions there that his arguments from that paper also work to prove Theorem 12.30, giving an alternative proof to the result of Wierdl. We again refer to the exposition by Thouvenot [1001] for the case $p = 2$.

A more detailed analysis of the corresponding variational operators was carried out by Zorin-Kranich [1072], see also Mirek et al. [799]. For an extension of Theorem 12.30 to functions belonging to the Orlicz space $\mathrm{L}(\log \mathrm{L})^2(\log\log \mathrm{L})$ see Trojan [1007].

Theorem 12.30 was extended to powers of primes and L^2-functions by Wierdl [1048] and rediscovered by Nair [817]. This was generalized further by Nair [818] who showed that for every invertible measure-preserving system (X, μ, T), every $p > 1$, every $f \in \mathrm{L}^p(X, \mu)$ and every polynomial $P \in \mathbb{Z}[\cdot]$ the averages

$$\frac{1}{N}\sum_{n=1}^{N} T^{P(p_n)} f \tag{12.16}$$

converge almost everywhere. Mirek [791] presented a pointwise ergodic theorem for certain thin subsets of primes, see also Trojan [1008].

For more examples of pointwise good sequences see Sect. 12.6 below.

Remark 12.32 Contrary to uniform convergence of the classical ergodic averages for uniquely ergodic systems and continuous functions (Theorem 7.23), there are counterexamples for pointwise convergence of averages along any sequence of Banach density zero (such as non-linear polynomials and polynomials of primes) for such systems and such functions, see Pavlov [846] and Lima [741].

12.5 Universally Bad Sequences

We now discuss sequences for which pointwise convergence of averages (12.3) fails dramatically. The following definition was inspired by Krengel [680], who constructed the first example of such sequences.

Definition 12.33 For $p \geq 1$, a subsequence (k_n) of $\mathbb{N}$ is called **universally bad in** $\mathbf{L}^p$ if for every ergodic invertible measure-preserving system (X, μ, T) on a non-atomic standard probability space there exists $f \in \mathrm{L}^p(X, \mu)$ such that the averages (12.3) diverge on a set of positive measure.

Note that some authors use a formally stronger definition where the word "ergodic" is replaced by "aperiodic", see, e.g., Bellow [89, 90] and Bellow and Losert [97]. A standard measure-preserving system (X, μ, T) is called **aperiodic measure-preserving system** if the set of all periodic points of T (i.e., all $x \in X$ such that there exists $d \in \mathbb{N}$ with $T^d x = x$) has measure zero. One can show that every ergodic system over a non-atomic standard space is aperiodic. Note further that often more holds than just divergence of the averages (12.3) on a set of positive measure, see Theorem 12.36 below. Moreover, being universally bad is strictly stronger than being pointwise bad, see Remark 12.37 below.

For a useful reformulation of this property we first need some preparation.

Lemma 12.34 (Halmos' Conjugacy Lemma) *Let (X, μ) be a non-atomic standard probability space and let T be an aperiodic (e.g., ergodic), invertible measure-preserving transformation on (X, μ). Then every other invertible measure-preserving transformation $\tilde{T}$ on (X, μ) can be approximated by elements of the conjugacy class of T, i.e., there exists a sequence (S_j) of invertible measure-preserving transformations on (X, μ) such that the corresponding Koopman operators on $\mathrm{L}^p(X, \mu)$ for $p \in [1, \infty)$ (denoted by the same letters) satisfy*

$$\lim_{j\to\infty} S_j T S_j^{-1} f = \tilde{T} f \tag{12.17}$$

for each $f \in \mathrm{L}^p(X, \mu)$.

For a proof, using the aperiodic variant of Rokhlin's lemma (cf. Theorem 3.23) we refer to [514, p. 77]. With the help of this lemma one can prove the following general principle discovered by Conze, see [255, Thm. 1].

Proposition 12.35 (Conze's Principle) *Let (k_n) be a subsequence in $\mathbb{N}$ and $p \in [1, \infty)$. Assume that there exists an ergodic (or, more generally, aperiodic) invertible measure-preserving system on a non-atomic standard probability space satisfying a weak (p, p)-maximal inequality along (k_n) with some constant $C > 0$. Then every invertible measure-preserving system on a non-atomic standard probability space satisfies the weak (p, p)-maximal inequality along (k_n) with the same constant C.*

We adapt the argument in Rosenblatt and Wierdl [881, Proof of Thm. 5.9] based on the original work by Conze [255], cf. [205, Proof of Thm. 3.3]. Recall that all non-atomic standard probability spaces are point isomorphic to each other (and to $([0, 1], \lambda)$), see Remark 8.48(d). So if the assumption of Conze's principle holds on one non-atomic standard probability space, then it holds for each of them.

Proof Let (X, μ) be a non-atomic standard probability space and let T be an aperiodic measure-preserving transformation on (X, μ) subject to the weak (p, p)-maximal inequality along (k_n) with constant $C > 0$. Let $\tilde{T}$ be another invertible measure-preserving transformation on (X, μ), see the above discussion. By Halmos' conjugacy lemma (Lemma 12.34) we have a sequence (S_j) of invertible measure-preserving transformations on (X, μ) such that (12.17) holds, so for every $n \in \mathbb{N}$ we have $\lim_{j\to\infty} S_j T^n S_j^{-1} = \tilde{T}^n$ in the strong operator topology.

We have to show that the maximal operator $\tilde{T}^*$ corresponding to the subsequential averages of $\tilde{T}$ satisfies

$$\mu[\tilde{T}^* f > \lambda] \leq \frac{C}{\lambda^p} \|f\|_p^p \quad \text{for all } \lambda > 0, \text{ for every } f \in \mathrm{L}^p(X, \mu).$$

Fix $f \in \mathrm{L}^p(X, \mu)$, $\lambda > 0$ and $M \in \mathbb{N}$. By monotonicity it suffices to show

$$\mu\Big[\sup_{1\le N\le M}\Big|\frac{1}{N}\sum_{n=1}^{N}\tilde{T}^{k_n}f\Big| > \lambda\Big] \le \frac{C}{\lambda^p}\|f\|_p^p.$$

Observe that $\frac{1}{N}\sum_{n=1}^{N} S_j T^{k_n} S_j^{-1} f$ converges to $\frac{1}{N}\sum_{n=1}^{N}\tilde{T}^{k_n} f$ in $\mathrm{L}^p(X, \mu)$ as $j \to \infty$ for every $N \in \mathbb{N}$, and one also has convergence of $\sup_{1\le N\le M}$ of the absolute value of these functions. Since L^p-convergence implies convergence in measure, every sequence (g_j) converging to g in $\mathrm{L}^p(X, \mu)$ satisfies $\lim_{j\to\infty}\mu[|g_j| > \lambda] = \mu[|g| > \lambda]$, see Exercise 12.13. Thus it suffices to show for a fix $j \in \mathbb{N}$

$$\mu\Big[\sup_{1\le N\le M}\Big|\frac{1}{N}\sum_{n=1}^{N} S_j T^{k_n} S_j^{-1} f\Big| > \lambda\Big] \le \frac{C}{\lambda^p}\|f\|_p^p. \tag{12.18}$$

By the assumption on T and monotonicity we have

$$\mu\Big[\sup_{1\le N\le M}\Big|\frac{1}{N}\sum_{n=1}^{N} T^{k_n} g\Big| > \lambda\Big] \le \frac{C}{\lambda^p}\|g\|_p^p,$$

for every $g \in \mathrm{L}^p(X, \mu)$ and every $\lambda > 0$. Taking $g := S_j^{-1} f$ and using the fact that S_j preserves μ implies (12.18) completing the proof. ∎

We can now characterize universally bad sequences in terms of (the failure of) a weak (p, p)-maximal inequality. This follows from the results in Conze [255], see, e.g., Bellow [90, p. 75] and Rosenblatt and Wierdl [881, p. 114].

Theorem 12.36 *Let (k_n) be a subsequence of $\mathbb{N}$ and $p \in [1, \infty)$. The following assertions are equivalent.*

(i) *(k_n) is universally bad in L^p.*
(ii) *There exists an invertible measure-preserving system on a non-atomic standard probability space which does not satisfy a weak (p, p)-maximal inequality along (k_n).*

In this case for every ergodic measure-preserving system (X, μ, T) on a non-atomic standard probability space there exists $f \in \mathrm{L}^p(X, \mu)$ such that the subsequential averages (12.3) *are unbounded almost everywhere.*

Note that using Remark 12.22 one sees that the set of such f is residual in the Baire category sense.

Proof If (ii) holds, then by Conze's principle (Proposition 12.35) every ergodic invertible measure-preserving system (X, μ, T) on a non-atomic standard probability space does not satisfy a weak (p, p)-maximal inequality along (k_n). By Sawyer's result (Proposition 12.20), there exists $f \in L^p(X, \mu)$ such that the averages (12.3) are unbounded a.e. In particular, both (i) and the supplementary assertion hold.

It remains to show (i)⇒(ii). Suppose (i) holds while (ii) is false. To achieve a contradiction, by Banach's principle (Remark 12.19), it suffices to find an invertible ergodic measure-preserving system (X, μ, T) on a non-atomic standard probability space such that the averages (12.3) converge a.e. for every f from a dense subset of $L^p(X, \mu)$. As a concrete example consider the two-sided Bernoulli shift $B(1/2, 1/2)$ (which is an invertible ergodic system with (countable) Lebesgue spectrum, see Proposition 5.27). The linear hull D of characteristic functions $\mathbf{1}_A$ of cylinder sets is contained in L^∞ and is dense in L^p for any p. Let $f = \mathbf{1}_A$ with A a cylinder set depending only on the coordinates, say, $1, \ldots, m$, i.e., $A = \cdots \times \{0, 1\} \times A_1 \times A_2 \times \cdots \times A_m \times \{0, 1\} \times \cdots$. For fixed $j \in \{1, \ldots, m\}$ the functions $T^{k_n} f - \mu(A)\mathbf{1}$ with $n \equiv j \bmod m$ are pairwise orthogonal to each other. Writing

$$\frac{1}{N}\sum_{n=1}^{N} T^{k_n} f = \frac{1}{m}\Big(\frac{m}{N} \sum_{n\le N, n\equiv 1 \bmod m} T^{k_n} f + \cdots + \frac{m}{N} \sum_{n\le N, n\equiv m \bmod m} T^{k_n} f\Big),$$

the Banach–Rajchman lemma (Proposition 9.7) implies that

$$\frac{1}{N}\sum_{n=1}^{N} T^{k_n} f \to \mu(A)$$

almost everywhere. It also follows that for every $f \in D$ one has almost everywhere convergence of the subsequential averages $\frac{1}{N}\sum_{n=1}^{N} T^{k_n} f$ to the mean of f. The proof is complete. ■

Remark 12.37 As Conze [255, Lem. 4] observed, Theorem 12.36 easily implies that universally bad sequences must have lower density zero, see Exercise 12.12. Combined with Conze [255, Thm. 5] where he constructed a pointwise bad sequence with positive density, this shows that not all pointwise bad sequences are universally bad.

In 2010 Buczolich and Mauldin [207] finally answered the question of Bellow, Furstenberg and Bourgain whether Theorem 12.28 holds for $p = 1$ in the negative for the sequence of squares. They used Theorem 12.36 and constructed for every $C > 0$ a *periodic* shift on $[0, 1]$ endowed with the Lebesgue measure and a *bounded* function f on $[0, 1]$ for which (12.12) fails, see also Buczolich and Mauldin [206] and Buczolich [202]. LaVictoire [708] modified their method to handle general monomials and primes.

Theorem 12.38 (Buczolich and Mauldin [207] and LaVictoire [708]) *Let (k_n) be either a sequence (n^d) for some $d \geq 2$ or the sequence of primes (p_n). Then (k_n) is universally bad in L^1.*

Moreover, Buczolich [203] showed that sequences of the form $(P(n))$ for $P \in \mathbb{Z}[\cdot]$ with $\deg P = 2$ are also universally bad in L^1. The case of higher degree polynomials (except for monomials) is still open.

In the next section we list some more examples of universally bad sequences.

12.6 Further Directions

We now indicate some further research directions, for more see Sects. 14.4, 15.9 and 16.6.

More Good and Universally Bad Sequences, Strong Sweeping Out

Bourgain [183] showed that subsequences of the form $\lfloor P(n) \rfloor$ for real polynomials P are L^p-pointwise good for every $p > 1$. Good, ergodic, L^2-pointwise good and L^2-pointwise bad sequences of the form $(\lfloor a_n \rfloor)$, where (a_n) is a logarithmico-exponential function of subpolynomial growth, were studied in Boshernitzan and Wierdl [176] and Boshernitzan et al. [175]. While they provide a complete characterization of good sequences in terms of asymptotic growth of the function, for pointwise goodness and pointwise badness sufficient conditions are presented. For example, $(\lfloor n + \log n \rfloor)$ and sequences of the form $\lfloor n^\alpha \rfloor$ are mean good while $(\lfloor n^2 + \log n \rfloor)$ is not. Urban and Zienkiewicz [1010] showed that sequences of the form $\lfloor n^\alpha \rfloor$, $\alpha \in (1, 1001/1000)$, are L^1-pointwise good, which was extended by Mirek [792] to $\alpha \in (1, 30/29)$. See also Bergelson et al. [114], Jones and Wierdl [597] and Mondal [800] for a related (negative) results concerning sequences of the form (n^α) in the continuous setting.

First examples of pointwise good sequences in L^1 were so-called *uniform sequences* introduced by Brunel and Keane [199], see also Ryll-Nardzewski [897] and Bellow and Losert [97]. (These sequences are certain return times sequences and have therefore positive density.) Another class of examples consists of certain *block sequences*, see Bellow and Losert [96, 97]. For more examples of pointwise good sequences see Blum and Reich [165], Bellow and Losert [96], Bourgain [180], Bellow [91], Rosenblatt and Wierdl [881, Ch. III-IV], Lesigne et al. [736], Jones and Wierdl [598], Schneider [921], Jones et al. [591], Buczolich [201], La Victoire [707], Frantzikinakis et al. [421], Mirek [792], Trojan [1009], Daskalakis [271], cf. also Reich [866] and Cuny and Weber [269] (combined with Remark 12.26).

In [178] Bourgain provides a necessary condition for a.e. convergence in L^p for $p \in [1, \infty]$ known as *Bourgain's entropy criterion*, see also Jones [588] and Weber [1037]. It helps showing that a given sequence is pointwise bad, see, e.g., Rosenblatt [879] and Rosenblatt and Wierdl [881, Ch. VI]. More results on divergence of ergodic averages along subsequences can be found in Bellow and Losert [97], Akcoglu et al. [24] and Quas and Wierdl [859] as well as following references.

Bellow [90] showed that every lacunary sequence is universally bad in L^1, see also Rosenblatt [879] for the L^∞-version. Demeter [284] studied universally bad sequences in terms of the growth of the best constants in (finite versions of) the weak maximal inequality. For more examples of universally bad sequences see Conze [255], Rosenblatt [879], Rosenblatt and Wierdl [881, Ch. V], Jones and Wierdl [597] as well as sequences with the following stronger property.

A subsequence (k_n) is called **strongly sweeping out** if for every ergodic measure-preserving system (X, μ, T) on a non-atomic standard probability space and every $\varepsilon > 0$ there exists a set $A \subset X$ with $\mu(A) < \varepsilon$ such that

$$\liminf_{N\to\infty} \frac{1}{N}\sum_{n=1}^{N} T^{k_n}\mathbf{1}_A = 0 \text{ a.e.}, \quad \limsup_{N\to\infty} \frac{1}{N}\sum_{n=1}^{N} T^{k_n}\mathbf{1}_A = 1 \text{ a.e.}$$

(Some authors use "aperiodic" instead of "ergodic" here.) Clearly, every strongly sweeping out sequence is universally bad for every p. Rosenblatt [879] proved that the converse is false, see also Demeter [284, Proposition 4.3]. The first example of a universally bad sequence due to Krengel [680] is actually strongly sweeping out. Also every lacunary sequence is strongly sweeping out, see Akcoglu et al. [22] and Akcoglu et al. [18]. For the strong sweeping out property of sublacunary sequences see Mondal et al. [802]. Another example is a (non-lacunary) sequence $(2^n 3^m)$, see [18].

For more examples and related notions and results see del Junco and Rosenblatt [283, Thm. 1.3], Bellow and Losert [97], Rosenblatt [879], Rosenblatt and Wierdl [881, Ch. V], Akcoglu et al. [18], Losert [758], Bellow and Jones [93], Biró [158], Jones [588], Akcoglu and Jones [23], Akcoglu et al. [24], Quas and Wierdl [859], Parrish [839], and Cuny et al. [266].

Everywhere Convergence of Subsequential Averages

Analogously to the pointwise ergodic theorem, see Sect. 7.3, it is natural to ask for which topological systems for example the averages (12.13) and (12.14) converge *everywhere* for continuous functions. This is the case for rotations on the unit circle, see Sect. 12.2, and, more generally, for nilsystems discussed in Chap. 14, see Theorem 14.45 and Remark 14.48 below, Green and Tao [488, Thm. 7.1] and [340]. For more examples of **systems satisfying the prime number theorem** (i.e., "good" systems for everywhere convergence of averages along primes) we refer to Bourgain [185], Ferenczi and Mauduit [388], Kanigowski et al. [624], see also Fraczek et al. [439]. For an example of a uniquely ergodic system with strong properties not satisfying the prime number theorem see Kanigowski et al. [624]. For an example of a uniquely ergodic system with other strong properties for which convergence of polynomial ergodic averages fails on a residual set of points for some continuous functions see Pavlov [846].

Note that by Exercise 12.8(b) the study of systems which satisfy the dynamical version of the prime number theorem is equivalent to the study of systems for which the weighted averages (12.15) converge everywhere. This is closely related to Sarnak's conjecture discussed in Chap. 17.

Of course, the question of a.e. and everywhere convergence of the averages (12.3) makes sense also for sequences $(k_n) \subset \mathbb{N}$ which are *not subsequences* of $\mathbb{N}$. Surprisingly, Bergelson and Richter [145] showed that the classical number theoretic sequences (Ω_n) and (ω_n) are good for everywhere convergence for uniquely ergodic systems and continuous functions. Here Ω_n and ω_n denote the number of all prime factors of n counted with or without multiplicity,[8] respectively. Loyd [760] showed further the strong sweeping out property of (Ω_n), cf. also Loyd and Mondal [761]. For a related result see Donoso et al. [309].

Subsequential Averages and Mixing Properties

Recall that the Blum–Hanson result (Theorem 12.1) characterizes mixing in terms of norm convergence of subsequential ergodic averages in L^2 for every subsequence (k_n). Friedman and Ornstein [436] constructed a mixing transformation (X, μ, T), a subsequence (k_n) of $\mathbb{N}$ and $f \in L^1(X, \mu, T)$ such that the subsequential averages (12.3) diverge a.e almost everywhere. This was strengthened by Krengel [680] who, as mentioned above, constructed the first strongly sweeping out sequence.

For a modulated version of the Blum–Hanson theorem for L^1-functions see Brunel and Keane [199].

More General Operators

As for the classical ergodic averages, it is a natural question to consider convergence of subsequential averages (12.3) for more general operators. Here, one can again study norm convergence for Banach space operators and pointwise convergence for operators on L^p-spaces.

We discussed the Blum–Hanson characterization of mixing and its generalization to Hilbert space contractions by Jones and Kuftinec [583] and Akcoglu and Sucheston [26], as well as the analogous characterization of weak mixing and its Banach space version by Jones [584]. Berend et al. [99] showed that ergodic sequences with positive lower density are mean good for operators with relatively weakly compact orbits, while thin ergodic sequences are not.

---

[8] These two sequences are highly oscillating.

Olsen [829] extended Akcoglu's pointwise ergodic theorem for positive contractions on L^p, $p \in (1, \infty)$, to the above mentioned uniform subsequences introduced by Brunel and Keane [199].

Baxter and Olsen [84] showed that every subsequence of positive density which is pointwise good is also pointwise good for all Dunford–Schwartz operators. This was extended to arbitrary sequences which are L^p-pointwise good for all $p > 1$ by Jones and Olsen [593]. They also showed that the block sequences considered by Bellow and Losert [97] are pointwise good for power bounded Lamperti operators and operators dominated by positive contractions on L^p for $p > 1$. Moreover Olsen [830] showed that uniform sequences are L^p-pointwise good for every Lamperti contraction on L^p for $p > 1$. Connections to Wiener's lemma along subsequences can be found in Cuny et al. [266], see also Exercise 12.5.

Extending the just mentioned result of Jones and Olsen [593], Jones et al. [594] showed[9] using variational inequalities that ergodic averages along squares and primes converge a.e. for every positive contraction and every quasi-isometric Lamperti operator on L^p for $p > 1$. (Note that positive isometries appear already in Bourgain [180, 183].) See Cohen [249] (with an appendix by Cuny) for an example of a doubly power bounded operator on L^p, $p \in (1, \infty)$ with $p \neq 2$, such that the averages along primes or squares (or any other subsequence of $\mathbb{N}$) are a.e. unbounded for some $f \in \mathrm{L}^p$. Note that norm convergence of the averages along primes holds for every invertible power bounded operator on L^p with $p \in (1, \infty)$, see Eisner and Lin [346] (with an appendix by Wierdl).

Ter Elst and Müller [999] showed by an adaptation of the van der Corput lemma that polynomials are mean good for all power bounded operators on Hilbert spaces. This was extended by Eisner and Müller [347] to a large class of subsequences including the integer part of most subpolynomial logarithmico-exponential functions under the assumption that the operator has no unimodular eigenvalues (except 1).

Discrete Analogues in Harmonic Analysis

Bourgain's ideas behind Theorem 12.28 together with ideas of Stein et al. [566, 960] on the discrete singular Radon transform gave birth to a new branch of **discrete harmonic analysis**, "a rapidly developing field of mathematics that fuses together classical Fourier analysis, probability theory, ergodic theory, analytic number theory, and additive combinatorics in new and interesting ways."[10] For the history and a very good exposition see Krause's book [671]. For a sample of results which shed new light on (subsequential) pointwise ergodic theorems we refer to Ionescu

[9] See also the text after the proof of Theorem 5.2 and Appendix A by Wierdl in Eisner and Lin [346].

[10] Terence Tao, review on Krause's book [671], see https://bookstore.ams.org/gsm-224.

et al. [565], Mirek [791, 793], Mirek and Trojan [798], Mirek et al. [799], Mirek et al. [794, 795], Mirek et al. [796], Mirek et al. [797], Krause et al. [675], Bourgain et al. [187], Krause et al. [676], and Krause [673, 674].

Exercises

Exercise 12.1 (Cesàro Divergence of Subsequences) Let $(c_n)_{n\in\mathbb{N}_0}$ be a divergent sequence in $\mathbb{C}$. Prove that there is a subsequence $(k_n)_{n\in\mathbb{N}}$ of $\mathbb{N}$ such that the Cesàro averages

$$\frac{1}{N}\sum_{n=1}^{N} c_{k_n}$$

diverge.

Exercise 12.2 (Good Subsequences) Prove the equivalence (ii)⇔(ii') in Proposition 12.9 and Remark 12.10, respectively, in two different ways indicated in Remark 12.10.

Exercise 12.3 (Lacunary Sequences)

(a) Show that the Fibonacci numbers form a lacunary sequence, and hence are bad for the mean ergodic theorem.
(b) Work out the details of the construction at the end of Example 12.15.

Exercise 12.4 (Ergodic Sequences)

(a) Let S be an asymptotically almost periodic operator[11] on a Banach space E. We note that for such operators we have the following Jacobs–de Leeuw–Glicksberg decomposition

$$E = \overline{\mathrm{lin}}\{x \in E : Tx = \lambda x \text{ for a } \lambda \in \mathbb{T}\} \oplus \{x \in E : \|T^n x\| \to 0\}.$$

Show that (12.11) holds for every ergodic sequence (k_n) and every $x \in E$. (Note that $P_{\mathrm{fix}\,S}$ is defined here as the mean ergodic projection which exists by Theorem 7.12.) Does this property characterize ergodic sequences?
(b) Prove that every subsequence $(k_n)_{n\in\mathbb{N}}$ with density 1 is ergodic. Show further that every such (k_n) is pointwise good in L^∞.

[11] An operator S on a Banach space E is called **asymptotically almost periodic** if it has relatively compact orbits, i.e., for every $x \in X$ the set $S^{\mathbb{N}_0}x := \{S^n x : n \in \mathbb{N}_0\}$ is relatively compact in E, cf. the definition before Proposition 6.23.

Exercise 12.5 (Wiener's Lemma along Subsequences) Let $(k_n)_{n\in\mathbb{N}}$ be a good subsequence of $\mathbb{N}$ with the corresponding limit function d. Prove that every complex Borel measure $\mu \in \mathrm{M}(\mathbb{T})$ satisfies

$$\lim_{N\to\infty} \frac{1}{N}\sum_{n=1}^{N} |\hat{\mu}(k_n)|^2 = \int_{\mathbb{T}^2} d(\lambda_1\overline{\lambda_2})\, \mathrm{d}(\mu \times \overline{\mu})(\lambda_1, \lambda_2).$$

Suppose further that the spectrum $\Lambda := [d \neq 0] \subset \mathbb{T}$ is at most countable (which is true, i.e., for polynomial sequences, primes and polynomials of primes as shown above). Show that in this case every $\mu \in \mathrm{M}(\mathbb{T})$ satisfies

$$\lim_{N\to\infty} \frac{1}{N}\sum_{n=1}^{N} |\hat{\mu}(k_n)|^2 = \sum_{\lambda\in\Lambda} d(\lambda) \sum_{a \text{ atom}} \overline{\mu}(\{a\overline{\lambda}\})\mu(\{a\}).$$

Conclude that if μ is continuous, then

$$\mathrm{D}\text{-}\lim_{n\to\infty} \hat{\mu}(k_n) = 0.$$

This is a subsequential version of Wiener's lemma (Proposition 4.11). Moreover, show that the assertion of Wiener's lemma holds along (k_n), i.e.,

$$\lim_{N\to\infty} \frac{1}{N}\sum_{n=1}^{N} |\hat{\mu}(k_n)|^2 = \sum_{a \text{ atom}} |\mu\{a\}|^2$$

holds for every $\mu \in \mathrm{M}(\mathbb{T})$ if and only if (k_n) is ergodic.

Exercise 12.6 (Pointwise Good Subsequences) Show that the assumption of invertibility can be removed from the definition of pointwise good subsequences. (Hint: Use invertible extensions from Proposition 8.51 and the fact that a sequence (f_n) of integrable functions on a measure space (X, μ) converges a.e. if and only if $\int_X \limsup_{n,m\to\infty} |f_n - f_m|\, \mathrm{d}\mu = 0$.)

Exercise 12.7 (Return Times Sequences, Subsequential Versus Weighted Averages)

(a) Let (X, μ, T) be a measure-preserving system and let $A \subset X$ be measurable. Prove that for μ-a.e. $x \in X$ the density of the return times set $R_A(x)$, and hence of the return times sequence $(r_{x,A}(n))$, exists and equals $\mu(A)$.
(b) Show the assertion of Remark 12.26.
(c) Deduce the assertion of Example 12.27 from Remark 12.26 and Theorem 16.30.

Exercise 12.8 (Averages along Primes vs. Weighted Averages)

(a) For $n \in \mathbb{N}$ let $v_n, w_n > 0$ be given and set

$$V_N := \sum_{n=1}^{N} v_n \text{ and } W_N := \sum_{n=1}^{N} w_n.$$

Suppose that $\lim_{N\to\infty} W_N = \infty$ and $\lim_{N\to\infty} V_N = \infty$ and that one of the following conditions is satisfied:

(i) For each $n \in \mathbb{N}$ one has $\dfrac{v_n}{w_n} \geq \dfrac{v_{n+1}}{w_{n+1}}$.
(ii) For each $n \in \mathbb{N}$ one has $\dfrac{v_n}{w_n} \leq \dfrac{v_{n+1}}{w_{n+1}}$ and

$$\sup_{n\in\mathbb{N}} \frac{v_n W_n}{w_n V_n} < \infty.$$

Prove that for every sequence $(b_n) \subset \mathbb{C}$ and every $b \in \mathbb{C}$

$$\lim_{N\to\infty} \frac{1}{W_N} \sum_{n=1}^{N} w_n b_n \to b$$

implies

$$\lim_{N\to\infty} \frac{1}{V_N} \sum_{n=1}^{N} v_n b_n = b.$$

(Hint: Using summation by parts, see Exercise 4.14, show first

$$\frac{1}{V_N} \sum_{n=1}^{N-1} \left(\frac{v_n}{w_n} - \frac{v_{n+1}}{w_{n+1}} \right) W_n + \frac{v_N W_N}{w_N V_N} = 1 \tag{12.19}$$

and use summation by parts again.)

(b) Applying (a) to $v_n := 1$, $w_n := \log n$ and $v_n := \log n$, $w_n := 1$, respectively, show that for every sequence $(b_n) \subset \mathbb{C}$ and every $b \in \mathbb{C}$ one has the equivalence

$$\lim_{N\to\infty} \frac{1}{N} \sum_{p\leq N} b_p = b \quad \iff \quad \lim_{N\to\infty} \frac{1}{N} \sum_{n=1}^{N} \Lambda'(n) b_n = b,$$

where we use Notation 10.22. This generalizes Lemma 10.23 to arbitrary, not necessarily bounded sequences. (Hint: Apply (a) to the sequence (b_{p_n}) and use (10.7).)

Exercise 12.9 (Calderón's Principle) Prove the following version of Calderón's principle for the maximal inequality along a subsequence.

(a) Let (k_n) be a subsequence of $\mathbb{N}$ and $p \in (1, \infty]$. Assume that the system $(\mathbb{N}, +)$ together with the left shift satisfies

$$\left\| \sup_{N \in \mathbb{N}} \left| \frac{1}{N} \sum_{n=1}^{N} \varphi(k_n + \cdot) \right| \right\|_p \leq C \|\varphi\|_p$$

for every $\varphi : \mathbb{N} \to \mathbb{C}$ with finite support. Then for every measure-preserving system (X, μ, T) the strong maximal inequality

$$\|S^* f\|_p \leq C \|f\|_p$$

holds for every $f \in \mathrm{L}^p(X, \mu)$, where S^* is the maximal operator corresponding to the subsequential averages $S_N := \frac{1}{N} \sum_{n=1}^{N} T^{k_n}$. (Hint: Replace first $\sup_{N \in \mathbb{N}}$ by $\sup_{N \leq M}$ in the definition of S^* for a fixed $M \in \mathbb{N}$ and consider φ defined by

$$\varphi(j) := \begin{cases} f(T^j x), & j \leq J, \\ 0, & \text{otherwise} \end{cases}$$

for an appropriate J.)

(b) Formulate and prove the corresponding statement for the weak maximal inequality in L^1.

Exercise 12.10 (Banach's Principle) Prove the assertion in Remark 12.19.

Exercise 12.11 (Pointwise Convergence Versus Weak Maximal Inequality) Formulate and prove an L^p-analogue of Theorem 7.52.

Exercise 12.12 (Weak Maximal Inequality and Density) Let (k_n) be a subsequence of $\mathbb{N}$ with positive lower density and $p \in [1, \infty)$. Show that there exists $C > 0$ such that the weak (p, p)-maximal inequality (12.12) is satisfied for every measure-preserving system if $p = 1$ and for every ergodic measure-preserving system if $p \in (1, \infty)$. (Hint: Use the result in Exercise 4.9.)

Exercise 12.13 Let (X, μ) be a probability space, let $p \in [1, \infty)$ and (g_j) be a sequence in $\mathrm{L}^p(X, \mu)$ with $g_j \to g$ in L^p. Prove that for $\lambda > 0$ one has

$$\lim_{j \to \infty} \mu[|g_j| > \lambda] = \mu[|g| > \lambda].$$

Chapter 13
Multiple Recurrence

We now continue the study of the interplay between ergodic theory and combinatorics discovered by Furstenberg [445]. As we saw via Furstenberg's correspondence principle in Sect. 9.3, polynomial and double recurrence provide an ergodic theoretic proof for the existence of polynomial configurations (the Furstenberg–Sárközy Theorem 9.17) and progressions of length three (Roth's Theorem 9.27) in large sets of integers, respectively. In this chapter we study multiple recurrence leading to an ergodic theoretic approach to Szemerédi's Theorem 9.34.

We discuss four proofs of multiple recurrence in this book, two in this chapter and two more in Chap. 15. We first present the classical proof due to Furstenberg and Katznelson [450], see also Furstenberg et al. [455], based on the Furstenberg–Zimmer structure theorem (Theorem 13.50 below) following the abstract approach by Tao [977, Ch. 2]. Here, we use a stronger statement than the dichotomy between compact and weakly mixing extensions (Proposition 13.49 below) in one of the steps, namely the conditional Jacobs–de Leeuw–Glicksberg decomposition (Theorem 13.24 below). Then we very briefly discuss a proof based on the original approach of Furstenberg [445] using distal factors of finite order in Sect. 13.8. Finally, two proofs using Host–Kra–Ziegler factors (or, respectively, subspaces) and their relation to Furstenberg's distal factors of finite order follow in Chap. 15.

13.1 Motivation: Szemerédi's Theorem

The study of multiple recurrence takes its origin in the famous combinatorial result on the existence of arithmetic progressions in large sets of integers known as Szemerédi's theorem. Recall that an **arithmetic progression** of length $k+1$ is a set of integers of the form

$$a, a+n, \dots, a+kn.$$

T. Eisner, B. Farkas, *A Journey Through Ergodic Theorems*, Birkhäuser Advanced Texts Basler Lehrbücher, https://doi.org/10.1007/978-3-031-96506-7_13

Here, $a \in \mathbb{N}$ is called the **start** of the progression and $n \in \mathbb{N}$ is called the **step**. We will be interested in the following property.

Definition 13.1 Let $k \in \mathbb{N}$, $k \geq 2$. We say that a set $E \subset \mathbb{N}$ has

- **property (AP$_k$)** if E contains arithmetic progressions of length $k+1$, i.e.,

$$\exists a, n \in \mathbb{N} \ \text{ with } \ a, a+n, \ldots, a+kn \in E.$$

- **property (AP)** if it has property (AP$_k$) for every $k \in \mathbb{N}$, $k \geq 2$, i.e.,

$$\forall k \in \mathbb{N}\ \exists a, n \in \mathbb{N} \ \text{ with } \ a, a+n, \ldots, a+kn \in E.$$

We now recall Szemerédi's (1975) Theorem 9.34 which, roughly speaking, says that every asymptotically large (i.e., not vanishing) set of natural numbers contains very structured subsets. (Recall that the upper density of a set $E \subset \mathbb{N}$ is defined by $\overline{\mathrm{d}}(E) = \limsup_{N\to\infty} \frac{|E\cap\{1,\ldots,N\}|}{N}$.)

Theorem 13.2 (Szemerédi [970]) *Every set $E \subset \mathbb{N}$ with $\overline{\mathrm{d}}(E) > 0$ has property (AP).*

Two years after Szemerédi's highly involved combinatorial proof, Furstenberg [445] published an ergodic theoretic proof of this result using the following property of measure-preserving systems. We recall that $f > 0$ means $f \geq 0$ and $f \not\equiv 0$ (not identically 0 in $\mathrm{L}^0(X, \mu)$).

Definition 13.3 Let $k \in \mathbb{N}$. We say that a measure-preserving system (X, μ, T) satisfies

- **multiple recurrence of order** k, shortly (**MR$_k$**), if for every $f \in \mathrm{L}^\infty(X, \mu)$ with $f > 0$

$$\liminf_{N\to\infty} \frac{1}{N} \sum_{n=0}^{N-1} \int_X f \cdot T^n f \cdot T^{2n} f \cdots T^{kn} f \,\mathrm{d}\mu > 0. \tag{13.1}$$

- **multiple recurrence**, shortly (**MR**), if it satisfies multiple recurrence of order k for every $k \in \mathbb{N}$.

For f of the form $f = \mathbf{1}_A$ for some $A \subset X$ with $\mu(A) > 0$ inequality (13.1) means that

$$\liminf_{N\to\infty} \frac{1}{N} \sum_{n=0}^{N-1} \mu(A \cap T^{-n} A \cap \ldots \cap T^{-kn} A) > 0,$$

implying also that the set of multiple return times

$$\{n \in \mathbb{N} : \mu(A \cap T^{-n}A \cap \ldots \cap T^{-kn}A) > 0\}$$

has positive lower density, see Exercise 4.9(b). This explains the name "multiple recurrence".

The bridge between multiple recurrence results and the existence of arithmetic progressions (or other patterns) in large sets of integers is Furstenberg's correspondence principle, which was already discussed in some detail in Chap. 9. For the sake of completeness we present here both directions for the setting of arithmetic progressions.

Furstenberg's Correspondence Principle: Both Directions

Let $k \in \mathbb{N}$. Each of the following statements implies the others, cf. Remark 9.22:

(i) **(Furstenberg's k-recurrence)** Every measure-preserving system satisfies (MR_k).

(i') **(Furstenberg's k-recurrence for ergodic, invertible, regular[1] systems)** Every ergodic, invertible, regular measure-preserving system satisfies (MR_k).

(ii) For every ergodic, invertible, regular measure-preserving system (X, μ, T) and every $f \in \mathrm{L}^\infty(X, \mu)$ with $f > 0$

$$\limsup_{N\to\infty} \frac{1}{N} \sum_{n=1}^{N} \int_X f \cdot T^n f \cdots T^{kn} f \,\mathrm{d}\mu > 0. \tag{13.2}$$

(iii) **(Furstenberg's k-recurrence for sets, weak version).** For every measure-preserving system (X, μ, T) for every $A \subset X$ with $\mu(A) > 0$ there is $n \in \mathbb{N}$ such that

$$\mu\Big(A \cap T^{-n}A \cap \cdots \cap T^{-kn}A\Big) > 0.$$

(iii') **(Furstenberg's k-recurrence for sets in ergodic, invertible, regular systems, weak version).** For every ergodic, invertible, regular measure-preserving system (X, μ, T) and every $A \subset X$ with $\mu(A) > 0$ there is $n \in \mathbb{N}$ such that

$$\mu\Big(A \cap T^{-n}A \cap \cdots \cap T^{-kn}A\Big) > 0.$$

[1] I.e., X is a compact metric space and μ is a regular Borel probability measure.

(iv) **(Szemerédi's Theorem for $(k+1)$-term arithmetic progressions).** Every subset $E \subset \mathbb{N}$ with upper density $\overline{d}(E) > 0$ has property (AP_k).

(v) **(Szemerédi's Theorem for Banach density for $(k+1)$-term arithmetic progressions).** Every subset $E \subset \mathbb{N}$ with **upper Banach density**

$$\overline{\mathrm{Bd}}(E) := \limsup_{N\to\infty} \sup_{M\in\mathbb{N}} \frac{|E \cap \{1+M, \ldots, N+M\}|}{N} > 0$$

has property (AP_k).

(vi) **(Szemerédi's Theorem for $(k+1)$-term arithmetic progressions, finitary version).** For every $\varepsilon > 0$ there is $N = N(\varepsilon, k) \in \mathbb{N}$ such that whenever $\ell > N$ and $E \subset \{1, \ldots, \ell\}$ has at least $\varepsilon\ell$ elements, then E contains a $(k+1)$-term arithmetic progression.

(vii) **(Szemerédi's Theorem for Banach density for $(k+1)$-term arithmetic progressions, finitary version).** For every $\varepsilon > 0$ there is $N = N(\varepsilon, k) \in \mathbb{N}$ such that whenever $E \subset \mathbb{N}$ is contained in an interval of length $\ell \geq N$ and has at least $\varepsilon\ell$ elements, then E contains a $(k+1)$-term arithmetic progression.

Note that the implications (i)⇒(i')⇒(ii)⇒(iii'); (i)⇒(iii)⇒(iii'); (v)⇒(iv); (vii)⇒(vi) are trivial and the implications (vi)⇒(iv) and (vii)⇒(v) are left as Exercise 13.1. Implication (iii')⇒(iv) follows from the correspondence principle as presented in Theorem 9.21 for the choice $s_1(n) = n, \ldots, s_k(n) = kn$; for implication (iii')⇒(v) (proved in Furstenberg's original paper [445]) see Exercise 9.7(c). The finitary version of Szemerédi's theorem for Banach density (vii) (and hence also (vi)) can be deduced from (iv) (and hence also from (v)), see Furstenberg et al. [455, p. 529] or, e.g., [342, Sec. 20.2]. The finitary version can be used to prove multiple recurrence (see, e.g., [455], [977, §2.10] or [342, Sec. 20.2]), thus establishing implications (vii)⇒(iii) and (vi)⇒(iii). How (iii) implies (i) (and we close all implication chains) is described in Tao [977, §2.10] (see also [342, Sec. 20.2]), the arguments there are based on a technique of Varnavides [1014], used for the proof of a finitary version of Roth's theorem (Fig. 13.1). We refer to [124] for further aspects of the correspondence principle.

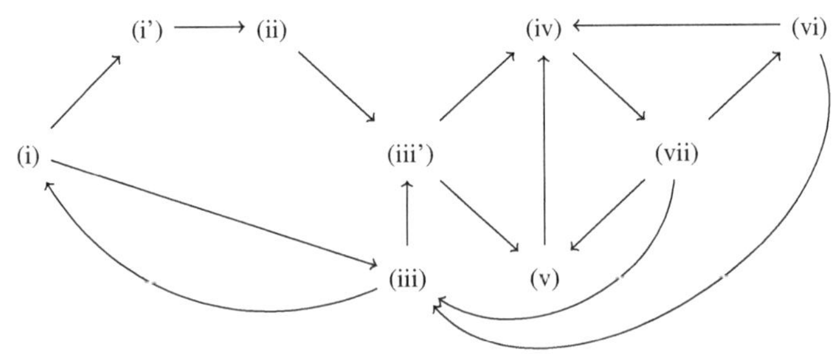

Fig. 13.1 Tracking down the implications in Furstenberg's correspondence principle

Remark 13.4 Furstenberg [445] showed the following stronger version of (i):

(i*) **(Furstenberg's recurrence theorem, uniform version).** For every measure-preserving system (X, μ, T) and every $f \in \mathrm{L}^\infty(X, \mu)$ with $f > 0$

$$\liminf_{N-M\to\infty} \frac{1}{N-M} \sum_{n=M+1}^{N} \int_X f \cdot T^n f \cdots T^{kn} f \,\mathrm{d}\mu > 0.$$

It is easy to see that this implies the following uniform version of (iii):

(iii*) For every measure-preserving system (X, μ, T) and every $A \subset X$ with $\mu(A) > 0$ the set

$$\Big\{n \in \mathbb{N} : \mu\Big(A \cap T^{-n}A \cap \cdots \cap T^{-kn}A\Big) > 0\Big\}$$

is syndetic.

We leave the proof of this implication as Exercise 13.15.

We note that by a result of Host and Kra, mentioned in Theorem 7.37 and discussed in Chap. 15 below, the liminf in (13.1) and the limsup (13.2) are equal, i.e., one has actually a limit. Note that the existence of the limit or, more generally, convergence of multiple averages

$$\frac{1}{N}\sum_{n=1}^{N} T^n f_1 \cdots T^{kn} f_k \tag{13.3}$$

in L^2 is much harder to prove than multiple recurrence (and was discovered much later). For details see Chap. 15 below.

We will not need here the full strength of the equivalences above and just use the implication (iii')⇒(iv) proved in Theorem 9.21. So to complete the ergodic theoretic proof of Szemerédi's theorem we are left with establishing the following result of Furstenberg from 1977.

Theorem 13.5 (Furstenberg, Multiple Recurrence for Ergodic Systems) *Every ergodic measure-preserving system (X, μ, T) has property (MR).*

Szemerédi's theorem had a direct precursor found by B.L. van der Waerden in 1927, see [1013].

Theorem 13.6 (van der Waerden) *For every finite partition $\mathbb{N} = C_1 \cup \ldots \cup C_m$ there exists $j \in \{1, \ldots, m\}$ such that C_j has property (AP).*

In other words, if $\mathbb{N}$ is colored with m colors, then at least one color class contains arbitrarily long arithmetic progressions.

Amazingly, just as Szemerédi's theorem has its *measure-preserving* dynamical systems counterpart (as we saw above), van der Waerden's theorem can be translated into the language of *topological dynamics*, see Furstenberg [445] and Furstenberg and Weiss [457]. We will not prove van der Waerden's theorem here and refer to [457] or, e.g., [342, Ch. 19]. It is Exercise 13.2 to show that van der Waerden's theorem is equivalent to the following finitary version of it.

Theorem 13.7 (van der Waerden, Finitary Version) *For every $m, k \in \mathbb{N}$ there exists $N \in \mathbb{N}$ such that for every finite partition $\{1, \ldots, N\} = C_1 \cup \ldots \cup C_m$ there exists $j \in \{1, \ldots, m\}$ such that C_j has property (AP_k).*

In other words, for every $m, k \in \mathbb{N}$ there exists $N \in \mathbb{N}$ such that for every coloring of $\{1, \ldots, N\}$ with m colors there exists a monochromatic arithmetic progression of length $k + 1$ in $\{1, \ldots, N\}$.

Of course, van der Waerden's theorem follows from Szemerédi's (by the subadditivity of the upper density, see Proposition 4.18) but we will utilize van der Waerden's result in the proofs of Furstenberg's multiple recurrence, hence of Szemerédi's theorem, discussed in this book. We will need van der Waerden's theorem in one crucial step used in all four proofs we discuss, namely to show that compact extensions preserve multiple recurrence (Proposition 13.44 below). For a proof of this step without using van der Waerden's theorem see Furstenberg et al. [455] or, e.g., Einsiedler and Ward [336, Sec. 7.9.2]. Note that, in opposite to Szemerédi's combinatorial proof [970] of Szemerédi's theorem, the original argument of Furstenberg [445] did not use van der Waerden's result either.

Remark 13.8 Khinchin [643] called van der Waerden's theorem one of three pearls of number theory. This theorem (together with Schur's result [925] on the existence of monochromatic patterns of $\mathbb{N}$ of the form $x, y, x + y$) is a classical example of a result in **(partition) Ramsey theory** which studies properties that are preserved by an element of an arbitrary finite partition, i.e., properties (P) such that as soon as an object A satisfies (P) and $A = C_1 \cup \ldots \cup C_m$ for some $m \in \mathbb{N}$ and $C_1, \ldots, C_m \subset A$, then at least one C_j satisfies (P), too (see, e.g., Graham and Butler [483]). Analogously, Szemerédi's theorem is a classical result in **density Ramsey theory** which studies properties which are preserved by "large" subsets. For connection of Ramsey theory to ergodic theory and topological dynamics, so-called **ergodic Ramsey theory** initiated by Furstenberg's proof of Szemerédi's theorem, see, e.g., Furstenberg and Weiss [457], Furstenberg [446], de Vries [278], Bergelson [102, 104, 105, 107, 109–112], Furstenberg and Katznelson [453, 454], Bergelson and Hindman [120, 121], Bergelson and Leibman [129, 130], McCutcheon [783], Bergelson and Moreira [143], and Moreira [805].

13.2 Single and Double Recurrence Revisited

Let (X, μ, T) be a measure-preserving system. We start with discussing single recurrence[2]

$$\lim_{N\to\infty} \frac{1}{N} \sum_{n=1}^{N} (T^n f | f) > 0 \text{ for all } f \in \mathrm{L}^2(X, \mu) \text{ with } f > 0 \tag{13.4}$$

by recalling and refining the proof of single convergence, i.e., the mean ergodic theorem (Theorem 7.2). We do this to extract an argument that would work for the general case. Recall the orthogonal von Neumann decomposition (6.1)

$$\mathrm{L}^2(X) = \mathrm{fix}(T) \oplus \overline{\mathrm{ran}(I - T)}. \tag{13.5}$$

Take $f \in \mathrm{L}^2(X, \mu)$ with $f > 0$ and decompose $f = f_{\mathrm{fix}} + f_{\mathrm{rest}}$ with f_{fix} being the orthogonal projection onto the $\mathrm{fix}(T)$ and f_{rest} onto $\overline{\mathrm{ran}(I - T)}$. As in the proof of the mean ergodic theorem (Theorem 7.2) one has $\frac{1}{N}\sum_{n=1}^{N} T^n f_{\mathrm{rest}} \to 0$ (as $N \to \infty$) by a telescopic sum argument, i.e., the second component of the right-hand side in (13.5) does not contribute to the limit in (13.4). Moreover, we clearly have $T^n f_{\mathrm{fix}} = f_{\mathrm{fix}}$. Thus by orthogonality

$$\lim_{N\to\infty} \frac{1}{N} \sum_{n=1}^{N} (T^n f | f) = \lim_{N\to\infty} \frac{1}{N} \sum_{n=1}^{N} (T^n f_{\mathrm{fix}} | f_{\mathrm{fix}}) = \| f_{\mathrm{fix}} \|^2_{\mathrm{L}^2(X,\mu)},$$

and it remains to show that $f_{\mathrm{fix}} \neq 0$ (which is clear for ergodic systems by $f_{\mathrm{fix}} = \int_X f \,\mathrm{d}\mu \cdot \mathbf{1}$). There are two ways to see this. The first one is to use the fact that $\mathrm{fix}(T)$ is induced by a factor, the fixed factor, and use the property (8.3) of projections onto factors. The second, more direct way is to use the mean ergodic theorem for $f_{\mathrm{fix}} = \lim_{N\to\infty} \frac{1}{N}\sum_{n=1}^{N} T^n f$ and the T-invariance of μ for $\int_X f_{\mathrm{fix}} \,\mathrm{d}\mu = \int_X f \,\mathrm{d}\mu > 0$.

To summarize, the fixed factor is *characteristic* (cf. Definition 9.23) for the classical ergodic averages $\frac{1}{N}\sum_{n=1}^{N} T^n f$. This together with (8.3) allows to replace f by f_{fix}, i.e., to replace the original system by its fixed factor. For the fixed factor (13.4) is clear due to the very structured behavior of the orbits (being one point).

Analogously we proceeded[3] for double convergence and recurrence for ergodic systems in Sect. 9.3, see Theorems 9.29 and 9.30. There the proof of both assertions was based on the Jacobs–de Leeuw–Glicksberg decomposition

$$\mathrm{L}^2(X, \mu) = \overline{\mathrm{lin}}\{f : Tf = \lambda f \text{ for some } \lambda \in \mathbb{T}\} \oplus \{f \text{ weakly mixing}\}, \tag{13.6}$$

[2] Since there are no multiple averages here, we can take $f \in \mathrm{L}^2(X, \mu)$ instead of $f \in \mathrm{L}^\infty(X, \mu)$.

[3] The arguments there were formulated for the averages $\frac{1}{N}\sum_{n=0}^{N-1}$ which can clearly be replaced by $\frac{1}{N}\sum_{n=1}^{N}$.

see (6.10). We showed in Proposition 9.28 that the Kronecker factor inducing H_{kr} is characteristic for double averages

$$\frac{1}{N}\sum_{n=1}^{N} T^n f \cdot T^{2n} g,$$

allowing (using (8.2) and (8.3) for double recurrence) to assume without loss of generality that our system coincides with its Kronecker factor, i.e., $\mathrm{L}^2(X,\mu) = H_{\mathrm{kr}}$. Double convergence followed immediately by reducing the question to eigenfunctions, i.e., to an elementary property of complex numbers. Double recurrence was showed by an argument using compactness of orbits and almost periodicity.

In both cases, we have used a decomposition into a part induced by a factor which is characteristic for the averages and into its orthogonal complement. The behavior of the orbits on the factor was analyzed using the structured behavior of the Koopman operator on this factor, leading finally to recurrence/convergence. This will be the general strategy to prove multiple recurrence (and then multiple convergence discussed in Chap. 15).

We will see three generalizations of the decomposition (13.6) for higher k below, these are the conditional Jacobs–de Leeuw–Glicksberg decomposition (Theorem 13.24) giving rise to the most known proof of multiple recurrence due to Furstenberg and Katznelson [450] and Furstenberg et al. [455], its version (13.52) for the distal factor of order $k-1$ in the spirit of the original proof by Furstenberg [445] and the Host–Kra decomposition (15.14) introduced in their proof of multiple convergence leading to more proofs of multiple recurrence, see Sects. 15.3 and 15.8 below.

13.3 Conditional Setting

We now introduce the abstract conditional approach due to Tao [977, Ch. 2] inspired by Furstenberg [445], Furstenberg and Katznelson [450] and Furstenberg et al. [455].[4] Here, L^∞-functions on the factor play the role of scalars.

Let (X,μ,T) be a measure-preserving system and let (Y,ν,S) be one of its factors via a Markov homomorphism $J : \mathrm{L}^1(Y,\nu) \to \mathrm{L}^1(X,\mu)$ (given by $Jf = f \circ \pi$ in the case of a point factor map $\pi : X \to Y$). Recall that the conditional expectation operator is defined as the Markov adjoint $E_Y := J' : \mathrm{L}^1(X,\mu) \to \mathrm{L}^1(Y,\nu)$, see Chap. 8.

[4] Furstenberg's proof as well as the other proofs by Furstenberg, Katznelson and Ornstein use disintegration as an important tool and hence need the regularity (or, equivalently, separability) assumption on the system. Tao's abstract approach is based on the proof by Furstenberg and Katznelson [450] but avoids disintegration completely and is valid for general systems. We will use disintegration and hence the regularity assumption only once in the proof of the conditional Jacobs–de Leeuw–Glicksberg decomposition (Theorem 13.24).

Notation 13.9 Throughout this chapter, we write $\mathrm{L}^p(X)$ for $\mathrm{L}^p(X,\mu)$ and $\mathrm{L}^p(Y)$ for $\mathrm{L}^p(Y,\nu)$, and also identify $c \in \mathrm{L}^\infty(Y)$ with its copy $Jc \in \mathrm{L}^\infty(X)$ to justify products cf with $c \in \mathrm{L}^\infty(Y)$, $f \in \mathrm{L}^p(X)$.

Definition 13.10 (Conditional Scalar Product and Conditional L^2-space) Let (X,μ,T) be a measure-preserving system and let (Y,ν,S) be one of its factors. For $f, g \in \mathrm{L}^2(X)$ we call the function

$$\langle f, g\rangle_{\mathrm{L}^2(X|Y)} := E_Y(f \cdot \overline{g}) \in \mathrm{L}^1(Y)$$

the **conditional scalar product** of f and g w.r.t. Y and the function

$$\|f\|_{\mathrm{L}^2(X|Y)} := \langle f, f\rangle^{\frac{1}{2}}_{\mathrm{L}^2(X|Y)} = (E_Y(|f|^2))^{\frac{1}{2}} \in \mathrm{L}^2(Y)$$

the **conditional norm** of f w.r.t. Y. Moreover, we define the **conditional L^2-space** by

$$\mathrm{L}^2(X|Y) := \left\{f \in \mathrm{L}^2(X) : \|f\|_{\mathrm{L}^2(X|Y)} \in \mathrm{L}^\infty(Y)\right\}.$$

It is plain to see that the conditional scalar product is sesquilinear.

Notation 13.11 We abbreviate $\|\cdot\|_{\mathrm{L}^{2,\infty}(X|Y)} := \|\|\cdot\|_{\mathrm{L}^2(X|Y)}\|_{\mathrm{L}^\infty(Y)}$.

Let us collect some of the most essential properties of these "conditional objects".

Remark 13.12 (a) For regular systems, by (8.10) one has the following representation using the disintegration measures for ν-a.e. $y \in Y$ (depending on f and g)

$$\langle f, g\rangle_{\mathrm{L}^2(X|Y)}(y) = \langle f, g\rangle_{\mathrm{L}^2(X,\mu_y)}, \quad \|f\|_{\mathrm{L}^2(X|Y)}(y) = \|f\|_{\mathrm{L}^2(X,\mu_y)}.$$

(b) For the trivial one-point factor the conditional scalar product and the conditional norm coincide with the classical, non-conditional ones. Moreover, for every factor one has by Proposition 8.52(b) for every $f \in \mathrm{L}^2(X)$

$$\|\|f\|_{\mathrm{L}^2(X|Y)}\|_{\mathrm{L}^2(Y)} = \|(E_Y(|f|^2))^{\frac{1}{2}}\|_{\mathrm{L}^2(Y)} = \|f\|_{\mathrm{L}^2(X)}. \tag{13.7}$$

(c) Also by Proposition 8.52(b) one has the implication

$$\langle f, g\rangle_{\mathrm{L}^2(X|Y)} = 0 \quad \Longrightarrow \quad f \perp g \text{ in } \mathrm{L}^2(X). \tag{13.8}$$

(d) By Proposition 8.52(c)

$$\|cf\|_{\mathrm{L}^2(X|Y)} = |c| \cdot \|f\|_{\mathrm{L}^2(X|Y)}, \tag{13.9}$$

and as a consequence

$$\|cf\|_{\mathrm{L}^{2,\infty}(X|Y)} \le \|c\|_{\mathrm{L}^\infty(Y)} \|f\|_{\mathrm{L}^{2,\infty}(X|Y)} \tag{13.10}$$

holds for all $c \in \mathrm{L}^\infty(Y)$ and $f \in \mathrm{L}^2(X|Y)$.

(e) Since E_Y is a contraction w.r.t. the L^∞-norm, one has

$$\|f\|_{\mathrm{L}^{2,\infty}(X|Y)} \le \|f\|_{\mathrm{L}^\infty(X)} \tag{13.11}$$

for every $f \in \mathrm{L}^\infty(X)$. Thus the space $\mathrm{L}^2(X|Y)$ satisfies

$$\mathrm{L}^\infty(X) \subset \mathrm{L}^2(X|Y) \subset \mathrm{L}^2(X).$$

Moreover, $\mathrm{L}^2(X|Y)$ is a linear subspace of $\mathrm{L}^2(X)$ and actually a **module over** $\mathrm{L}^\infty(Y)$ (a submodule of $\mathrm{L}^2(X)$) meaning that $c_1 f_1 + c_2 f_2 \in \mathrm{L}^2(X|Y)$ for every $f_1, f_2 \in \mathrm{L}^2(X|Y)$ and every $c_1, c_2 \in \mathrm{L}^\infty(Y)$.

(f) If (Y, ν, S) is invertible, then T is a contraction w.r.t. $\|\cdot\|_{\mathrm{L}^{2,\infty}(X|Y)}$ and hence $\mathrm{L}^2(X|Y)$ is T-invariant. This follows from (8.8) by

$$\langle Tf, Tg\rangle_{\mathrm{L}^2(X|Y)} = E_Y(T(f\overline{g})) = SE_Y(f\overline{g}) = S\langle f, g\rangle_{\mathrm{L}^2(X|Y)}. \tag{13.12}$$

(g) For every $f \in \mathrm{L}^2(X)$ the functions $f\mathbf{1}_{[E_Y(|f|^2)\le m]}$ belong to $\mathrm{L}^2(X|Y)$ for every $m > 0$ and converge to f in $\mathrm{L}^2(X)$ as $m \to \infty$.

(h) Every $f \in \mathrm{L}^2(X)$ satisfies $f = \mathbf{1}_{[\|f\|_{\mathrm{L}^2(X|Y)} \neq 0]} f$ by (13.7).

Remark 13.13

(a) For $f, g \in \mathrm{L}^2(X|Y)$ the following **conditional Cauchy–Schwarz inequality** holds:

$$|\langle f, g\rangle_{\mathrm{L}^2(X|Y)}| \le \|f\|_{\mathrm{L}^2(X|Y)} \cdot \|g\|_{\mathrm{L}^2(X|Y)} \quad \nu\text{-a.e.} \tag{13.13}$$

We leave the proof as Exercise 13.5. This implies in particular the **conditional triangle inequality**

$$\|f + g\|_{\mathrm{L}^2(X|Y)} \le \|f\|_{\mathrm{L}^2(X|Y)} + \|g\|_{\mathrm{L}^2(X|Y)} \quad \nu\text{-a.e.} \tag{13.14}$$

(b) Note that $\|\cdot\|_{L^{2,\infty}(X|Y)}$ is a norm on $L^2(X|Y)$ with respect to which $L^2(X|Y)$ becomes a Banach space, see Exercise 13.5 (or Tao [977, Ex. 2.13.2]). Thus one can refer to $L^2(X|Y)$ as a Hilbert module[5] over $L^\infty(Y)$.

Definition 13.14 A **finitely generated sub-$L^\infty(Y)$-module** of $L^2(X|Y)$ is a subset of the form

$$M(f_1, \ldots, f_d) := \{c_1 f_1 + \cdots + c_d f_d : c_1, \ldots, c_d \in L^\infty(Y)\}$$

for some $d \in \mathbb{N}$ and $f_1, \ldots, f_d \in L^2(X|Y)$. A **finitely generated module zonotope** is a set of the form

$$\{c_1 f_1 + \cdots + c_d f_d : c_j \in L^\infty(Y) \text{ with } \|c_j\|_{L^\infty(Y)} \le 1\ \forall j \in \{1, \ldots, d\}\}$$

for some $d \in \mathbb{N}$ and $f_1, \ldots, f_d \in L^2(X|Y)$. We call thereby $\{f_1, \ldots, f_d\}$ a **generating set** for the sub-$L^\infty(Y)$-module or the module zonotope, respectively. In these cases we shall also say that a **rank** of a finitely generated module (or, respectively, zonotope) is at most d.

We now define conditional analogues of eigenfunctions, asymptotic almost periodicity and weak mixing. (See Exercise 8.10 for the definition of asymptotically almost periodic functions.)

Definition 13.15 Let (Y, ν, S) be a factor of a measure-preserving system (X, μ, T). A function $f \in L^2(X|Y)$ is called

- **conditionally almost periodic** (w.r.t. Y) if for every $\varepsilon > 0$ there exists a finitely generated module zonotope Z such that for every $n \in \mathbb{N}_0$ there exists $g \in Z$ with $\|T^n f - g\|_{L^{2,\infty}(X|Y)} < \varepsilon$. We denote the space of all such functions by $A(X|Y)$.
- **conditional eigenfunction**[6] (w.r.t. Y) if it (and hence its orbit) is contained in a finitely generated T-invariant sub-$L^\infty(Y)$-module. We denote the space of all conditional eigenfunctions by $E(X|Y)$.

A function $f \in L^2(X)$ is called **conditionally weakly mixing** if

$$\operatorname*{D-lim}_{n\to\infty} \|\langle T^n f, f\rangle_{L^2(X|Y)}\|_{L^1(Y)} = 0, \tag{13.15}$$

where D- lim denotes limit in density, see Definition 4.17(b) (for the basic properties of D- lim see Sect. 4.3). We denote the set of conditionally weakly mixing functions in $L^2(X)$ relative to Y by $W(X|Y)$.

[5] For an approach via **Hilbert–Kaplansky modules** see Edeko et al. [331].

[6] This definition goes back to Zimmer [1066] and Furstenberg and Weiss [458] (and is similar to Furstenberg's definition of generalized eigenfunctions from [445] for ergodic systems, see the last assertion of Lemma 13.21 below).

Remark 13.16

(a) Since E_Y is a positive operator, by Proposition 8.52(b) one has for every $n \in \mathbb{N}$ and $f, g \in \mathrm{L}^2(X)$

$$\begin{aligned}\|\langle T^n f, g\rangle_{\mathrm{L}^2(X|Y)}\|_{\mathrm{L}^1(Y)} &= \int_Y |E_Y(T^n f \cdot \overline{g})| \,\mathrm{d}\nu \leq \int_Y E_Y|T^n f \cdot \overline{g}| \,\mathrm{d}\nu \\ &= \|T^n f \cdot \overline{g}\|_{\mathrm{L}^1(X)} \leq \|f\|_{\mathrm{L}^2(X)}\|g\|_{\mathrm{L}^2(X)}. \end{aligned} \tag{13.16}$$

In particular, the sequence $(\|\langle T^n f, g\rangle_{\mathrm{L}^2(X|Y)}\|_{\mathrm{L}^1(Y)})$ is bounded. By the Koopman–von Neumann lemma (Proposition 4.19), we see that $f \in \mathrm{L}^2(X)$ is conditionally weakly mixing if and only if

$$\lim_{N\to\infty} \frac{1}{N}\sum_{n=1}^{N} \|\langle T^n f, f\rangle_{\mathrm{L}^2(X|Y)}\|_{\mathrm{L}^1(Y)} = 0$$

or, equivalently, if

$$\lim_{N\to\infty} \frac{1}{N}\sum_{n=1}^{N} \|\langle T^n f, f\rangle_{\mathrm{L}^2(X|Y)}\|^2_{\mathrm{L}^1(Y)} = 0.$$

(b) Since by Proposition 8.52(b)

$$\begin{aligned}|\langle T^n f, f\rangle| &= \left|\int_Y E_Y(T^n f \cdot \overline{f}) \,\mathrm{d}\nu\right| \leq \int_Y |E_Y(T^n f \cdot \overline{f})| \,\mathrm{d}\nu \\ &= \|\langle T^n f, f\rangle_{\mathrm{L}^2(X|Y)}\|_{\mathrm{L}^1(Y)}\end{aligned}$$

we see that every conditionally weakly mixing function is (unconditionally) weakly mixing.

(c) If (Y, ν, S) is invertible, then $A(X|Y)$ is closed w.r.t. $\|\cdot\|_{\mathrm{L}^{2,\infty}(X|Y)}$ by Remark 13.12(f).

Remark 13.17 We use the definition of conditional weak mixing from Zorin-Kranich [1073, Sec. 8]. Tao [977, Sec. 2.14.1] called a function $f \in \mathrm{L}^2(X|Y)$ conditionally weakly mixing if

$$\lim_{N\to\infty} \frac{1}{N}\sum_{n=1}^{N} \|\langle T^n f, f\rangle_{\mathrm{L}^2(X|Y)}\|^2_{\mathrm{L}^2(Y)} = 0. \tag{13.17}$$

It is easy to see that the two definitions are equivalent for functions in $\mathrm{L}^2(X|Y)$, see Exercise 13.4(c).

Example 13.18

(a) Every $f \in \mathrm{L}^\infty(Y)$ is a conditional eigenfunction and conditionally almost periodic (with the module zonotope and the T-invariant sub-L^∞-module generated by $\mathbf{1}$, respectively).
(b) For the one-point factor Y (corresponding to the subalgebra $\mathbb{C}\mathbf{1}$ of $\mathrm{L}^\infty(X)$) the definition of conditional weak mixing and conditional almost periodicity coincides with the unconditional ones, respectively (we leave the proof as Exercise 13.4). However the definition of a conditional eigenfunction is different due to requiring finitely many generators instead of a single one.
(c) More generally, let (X, μ, T) be a product system, i.e., $X = Y \times \tilde{Y}$, $\mu = \nu \times \tilde{\nu}$ and $T = S \times \tilde{S}$ for some measure-preserving system $(\tilde{Y}, \tilde{\nu}, \tilde{S})$. Consider $\tilde{f} \in \mathrm{L}^2(Y)$. Then the function f defined by $f(y, \tilde{y}) := \tilde{f}(\tilde{y})$ is conditionally weakly mixing if and only if $\tilde{f}$ is weakly mixing in $(\tilde{Y}, \tilde{\nu}, \tilde{S})$. Analogously, f is conditionally almost periodic if and only if $\tilde{f}$ is almost periodic. We again leave the proof as Exercise 13.4.

Example 13.19 (Generalized Eigenfunctions) Every eigenfunction of the Koopman operator T is clearly a conditional eigenfunction and conditionally almost periodic (with the corresponding rank one sub-L^∞-module and module zonotope, respectively) with respect to both the trivial one-point factor and the fixed factor. Moreover, if (X, μ, T) is ergodic, then every generalized eigenfunction of order ≤ 2 (see Example 8.33) is a conditional eigenfunction and conditionally almost periodic (again with rank one sub-L^∞-module and zonotope, respectively) with respect to the Kronecker factor. Analogously, for every $k \in \mathbb{N}$ each generalized eigenfunction of order $\leq k + 1$ is conditionally almost periodic with respect to the Abramov factor $A_k(X, \mu, T)$. We leave the details as Exercise 13.10.

While it follows directly from the definition that $A(X|Y)$ and $E(X|Y)$ are linear subspaces of $\mathrm{L}^2(X)$, this is not as immediate for the subset $W(X|Y)$. To prove this we need the following lemma.

Lemma 13.20 *Let (X, μ, T) be a measure-preserving system with an invertible factor (Y, ν, S).*

(a) *For every $f \in W(X|Y)$ and every $g \in \mathrm{L}^2(X)$*

$$\text{D-}\lim_n \|\langle T^n f, g\rangle_{\mathrm{L}^2(X|Y)}\|_{\mathrm{L}^1(Y)} = 0. \tag{13.18}$$

(b) *The space $W(X|Y)$ is a closed linear subspace of $\mathrm{L}^2(X)$.*

Proof (a) We begin with the following observation regarding approximation. Let $f \in W(X|Y)$. By (13.16) and Remark 13.12(g) we can assume by approximating g by $g_m := g \cdot \mathbf{1}_{[E_Y(|g|^2)\leq m]}$ and f by $f_m := f \cdot \mathbf{1}_{[E_Y(|f|^2)\leq m]}$, $m \in \mathbb{N}$, that $f, g \in \mathrm{L}^2(X|Y)$. Note that $f_m \in W(X|Y)$ for every $m \in \mathbb{N}$. We thus have

$\langle T^n f, g\rangle_{\mathrm{L}^2(X|Y)} \in \mathrm{L}^\infty(Y)$ for each $n \in \mathbb{N}_0$ and can write by Proposition 8.52(b)

$$\begin{aligned}\|\langle T^n f, g\rangle_{\mathrm{L}^2(X|Y)}\|_{\mathrm{L}^1(Y)}^2 \leq \|\langle T^n f, g\rangle_{\mathrm{L}^2(X|Y)}\|_{\mathrm{L}^2(Y)}^2 &= \int_Y \left|\langle T^n f, g\rangle_{\mathrm{L}^2(X|Y)}\right|^2 \,\mathrm{d}\nu \\ &= \int_Y \langle\langle g, T^n f\rangle_{\mathrm{L}^2(X|Y)} T^n f, g\rangle_{\mathrm{L}^2(X|Y)} \,\mathrm{d}\nu \\ &= (\langle g, T^n f\rangle_{\mathrm{L}^2(X|Y)} T^n f | g)_{\mathrm{L}^2(X)}.\end{aligned}$$

We set $u_n := \langle g, T^n f\rangle_{\mathrm{L}^2(X|Y)} T^n f$ and calculate by Proposition 8.52(b),(c)

$$\begin{aligned}|(u_n|u_{n+h})_{\mathrm{L}^2(X)}| &= \left|\int_X \langle g, T^n f\rangle_{\mathrm{L}^2(X|Y)} T^n f \cdot \langle T^{n+h} f, g\rangle_{\mathrm{L}^2(X|Y)} T^{n+h}\overline{f} \,\mathrm{d}\mu\right| \\ &= \left|\int_Y \langle g, T^n f\rangle_{\mathrm{L}^2(X|Y)} \langle T^{n+h} f, g\rangle_{\mathrm{L}^2(X|Y)} E_Y(T^n f T^{n+h}\overline{f}) \,\mathrm{d}\nu\right| \\ &\leq \int_Y \left|\langle g, T^n f\rangle_{\mathrm{L}^2(X|Y)} \langle T^{n+h} f, g\rangle_{\mathrm{L}^2(X|Y)} E_Y(T^n f T^{n+h}\overline{f})\right| \,\mathrm{d}\nu,\end{aligned}$$

and continue by the conditional Cauchy–Schwarz inequality (13.13) and Remark 13.12(f)

$$\begin{aligned}&\leq \|g\|_{\mathrm{L}^{2,\infty}(X|Y)}^2 \|f\|_{\mathrm{L}^{2,\infty}(X|Y)}^2 \int_Y |E_Y(T^n f T^{n+h}\overline{f})| \,\mathrm{d}\nu \\ &= \|g\|_{\mathrm{L}^{2,\infty}(X|Y)}^2 \|f\|_{\mathrm{L}^{2,\infty}(X|Y)}^2 \|S\langle f, T^h f\rangle_{\mathrm{L}^2(X|Y)}\|_{\mathrm{L}^1(Y)} \\ &= \|g\|_{\mathrm{L}^{2,\infty}(X|Y)}^2 \|f\|_{\mathrm{L}^{2,\infty}(X|Y)}^2 \|\langle f, T^h f\rangle_{\mathrm{L}^2(X|Y)}\|_{\mathrm{L}^1(Y)}.\end{aligned}$$

Since $f \in W(X|Y)$, we conclude by Remark 13.16(a)

$$\limsup_{N\to\infty} \frac{1}{N} \sum_{n=1}^{N} |(u_n|u_{n+h})_{\mathrm{L}^2(X)}| = 0.$$

The van der Corput lemma (Lemma 7.31) and the Koopman–von Neumann lemma (Proposition 4.19) finish the proof of assertion (a), while (b) follows easily from (a) and (13.16), see Exercise 13.6. ∎

We now turn our attention to conditional eigenfunctions and observe first a basic property of finitely generated T-invariant sub-L^∞-modules, cf. Furstenberg [445, §6], Zorin-Kranich [1067], Durcik et al. [326, Appendix D] and Edeko et al. [331, Sec. 2].

Lemma 13.21 (Existence of Conditionally Orthonormal Generating Sets) *Let (Y, ν, S) be a (Markov) invertible factor of a measure-preserving system (X, μ, T). Then for every finitely generated T-invariant sub-$\mathrm{L}^\infty(Y)$-module $M = M(f_1, \dots, f_d)$ there exist $g_1, \dots, g_d \in \mathrm{L}^2(X|Y)$ with the following properties.*

(a) *$\langle g_j, g_k\rangle_{\mathrm{L}^2(X|Y)}$ vanishes ν-a.e. whenever $j \neq k$ and is ν-a.e. a 0-1-valued function on Y whenever $j = k$ for every $j, k \in \{1, \dots, d\}$. In particular, every $g \in M(g_1, \dots, g_d)$ has the representation*

$$g = \sum_{j=1}^{d} \langle g, g_j\rangle_{\mathrm{L}^2(X|Y)} g_j. \tag{13.19}$$

(c) *$M \subset M(g_1, \dots, g_d)$.*
(d) *$M(g_1, \dots, g_d)$ is T-invariant.*
(e) *The function $h := \sum_{j=1}^{d} |g_j|^2 \in \mathrm{L}^1(X)$ is T-invariant.*

If (Y, ν, S) is ergodic, one can additionally achieve that there exists $d' \in \{0, \dots, d\}$ such that $\|g_j\|_{\mathrm{L}^2(X|Y)} = 1$ ν-a.e. for every $j \in \{1, \dots, d'\}$ and $\|g_j\|_{\mathrm{L}^2(X|Y)} = 0$ ν-a.e. for every $j \in \{d' + 1, \dots, d\}$.

Proof We adapt the Gram–Schmidt orthogonalization argument to the conditional setting. Consider the function $q : [0, \infty) \to (0, \infty)$ given by

$$q(s) := \begin{cases} \frac{1}{s}, & s \neq 0, \\ 1, & s = 0, \end{cases}$$

and notice that $|1/q(s)| \leq \max\{s, 1\}$. Define $g_1, \dots, g_d$ recursively as follows. We set $\tilde{g_1} := f_1$, $g_1 := q(\|f_1\|_{\mathrm{L}^2(X|Y)}) f_1$ and

$$\tilde{g_j} := f_j - \sum_{l=1}^{j-1} \langle f_j, g_l\rangle_{\mathrm{L}^2(X|Y)} g_l, \quad g_j := q(\|\tilde{g_j}\|_{\mathrm{L}^2(X|Y)}) \tilde{g_j}, \quad j \in \{2, \dots, d\}.$$

By Remark 8.56(b)

$$\langle g_1, g_1\rangle_{\mathrm{L}^2(X|Y)} = E_Y(q(\|f_1\|^2_{\mathrm{L}^2(X|Y)})|f_1|^2) = q(\|f_1\|^2_{\mathrm{L}^2(X|Y)})\|f_1\|^2_{\mathrm{L}^2(X|Y)},$$

so $\langle g_1, g_1\rangle_{\mathrm{L}^2(X|Y)}$ is indeed 0-1-valued, and from (13.7) we conclude that $g_1 \in \mathrm{L}^2(X)$ and $g_1 \in \mathrm{L}^2(X|Y)$.

Suppose that the first part of (a) is proved for all $j, k < i$ for some $i \in \{2, \dots, d\}$. Take $j < i$ and notice that by the conditional Cauchy–Schwarz inequality (13.13) one has $\langle f_i, g_k\rangle_{L^2(X|Y)} \in L^\infty(Y)$ for each $k \le i-1$, thus $\tilde{g}_i \in L^2(X)$ and we can write

$$\begin{aligned}\langle \tilde{g}_i, g_j\rangle_{L^2(X|Y)} &= \langle f_i, g_j\rangle_{L^2(X|Y)} - \sum_{k=1}^{i-1}\langle f_i, g_k\rangle_{L^2(X|Y)}\langle g_k, g_j\rangle_{L^2(X|Y)}\\ &= \langle f_i, g_j\rangle_{L^2(X|Y)} - \langle f_i, g_j\rangle_{L^2(X|Y)}\langle g_j, g_j\rangle_{L^2(X|Y)},\end{aligned}$$

where also the induction hypothesis was used. By the induction hypothesis and again by the conditional Cauchy–Schwarz inequality (13.13) it follows that $\langle \tilde{g}_i, g_j\rangle_{L^2(X|Y)} = 0$ ν-a.e. We abbreviate $g := q(\|\tilde{g}_i\|_{L^2(X|Y)})$. Then by Remark 8.56, the conditional Cauchy–Schwarz inequality and by the induction hypothesis

$$E_Y(Jg\cdot|\tilde{g}_i g_j|) = g\cdot\langle|\tilde{g}_i|, |g_j|\rangle_{L^2(X|Y)} \le 1.$$

This implies $Jg\cdot|\tilde{g}_i g_j| \in L^1(X)$, so again by Remark 8.56(b) we conclude $\langle g_i, g_j\rangle_{L^2(X|Y)} = g\langle\tilde{g}_i, g_j\rangle_{L^2(X|Y)} = 0$ ν-a.e. Analogously to the case of g_1 one proves that $\langle g_i, g_i\rangle_{L^2(X|Y)}$ is 0-1-valued, $g_i \in L^2(X)$ and $g_i \in L^2(X|Y)$. By induction the proof of the first part of (a) is complete.

To see (13.19) take $g \in M(g_1, \ldots, g_d)$ and write

$$g = \sum_{j=1}^{d} c_j g_j \text{ for some } c_1, \ldots, c_d \in L^\infty(Y).$$

By Remark 13.12(h) we have

$$g = \sum_{j=1}^{d} c_j \mathbf{1}_{[\|g_j\|_{L^2(X|Y)}\neq 0]} g_j. \tag{13.20}$$

Take the conditional scalar product with g_i in (13.20) to conclude $\langle g, g_i\rangle_{L^2(X|Y)} = c_i\mathbf{1}_{[\|g_i\|_{L^2(X|Y)}\neq 0]}\langle g_i, g_i\rangle_{L^2(X|Y)}$. Since $\langle g_i, g_i\rangle_{L^2(X|Y)}$ is 0-1-valued, we obtain $\langle g, g_i\rangle_{L^2(X|Y)} = c_i\mathbf{1}_{[\|g_i\|_{L^2(X|Y)}\neq 0]}$ implying by (13.20) the validity of (13.19).
To show (b) take $j \in \{1, \ldots, d\}$. By definition of g_j we have

$$f_j = c_j g_j + \sum_{l=1}^{j-1}\langle f_j, g_l\rangle_{L^2(X|Y)} g_l$$

(using the convention $\sum_{l=1}^{0} := 0$) with

$$\|c_j\|_{L^\infty(Y)} \le \max\{\|\tilde{g}_j\|_{L^{2,\infty}(X|Y)}, 1\}.$$

Since the coefficients $\langle f_j, g_l\rangle_{\mathrm{L}^2(X|Y)}$ belong to $\mathrm{L}^\infty(Y)$ (by the conditional Cauchy–Schwarz inequality (13.13)), the assertion follows.
(c) We need to show $Tg_j \in M(g_1, \dots, g_d)$ for each $j \in \{1, \dots, d\}$, which we prove by induction on j. For $j = 1$ we have $T\tilde{g}_1 = Tf_1 \in M(g_1, \dots, g_d)$ by (b), so it follows that $Tg_1 \in S(q(\|f_1\|_{\mathrm{L}^2(X|Y)}))M(g_1, \dots, g_d)$, i.e.,

$$Tg_1 = S(q(\|f_1\|_{\mathrm{L}^2(X|Y)}))\sum_{k=1}^{d} a_{1,k}g_k \text{ for some } a_{1,1}, \dots, a_{1,d} \in \mathrm{L}^\infty(Y).$$

Again by Remark 13.12(h) applied to the functions g_k we can replace $a_{1,k}$ by $a_{1,k}\mathbf{1}_{[\|g_k\|_{\mathrm{L}^2(X|Y)}\neq 0]}$, i.e., assume without loss of generality that the coefficients satisfy

$$a_{1,k} = 0 \text{ whenever } \|g_k\|_{\mathrm{L}^2(X|Y)} = 0 \quad (k \in \{1, \dots, d\}). \tag{13.21}$$

Observe now that by linearity, (a) and Remark 8.56(b) again

$$\begin{aligned} S\langle g_1, g_1\rangle_{\mathrm{L}^2(X|Y)} &\overset{(8.8)}{=} \langle Tg_1, Tg_1\rangle_{\mathrm{L}^2(X|Y)} \\ &= S(q(\|f_1\|_{\mathrm{L}^2(X|Y)}))^2 \sum_{k,k'=1}^{d} a_{1,k}\overline{a_{1,k'}}\langle g_k, g_{k'}\rangle_{\mathrm{L}^2(X|Y)} \\ &\overset{(13.21)}{=} \sum_{k=1}^{d} |S(q(\|f_1\|_{\mathrm{L}^2(X|Y)}))a_{1,k}|^2. \end{aligned} \tag{13.22}$$

The left-hand side is 0-1-valued ν-a.e. by (a) and the definition of the Koopman operator, so we conclude that $|S(q(\|f_1\|_{\mathrm{L}^2(X|Y)}))a_{1,k}|^2 \leq 1$ holds ν-a.e. for every $k \in \{1, \dots, d\}$, and it follows that $Tg_1 \in M(g_1, \dots, g_d)$.

We now assume that $Tg_1, \dots, Tg_{j-1} \in M(g_1, \dots, g_d)$ and show $Tg_j \in M(g_1, \dots, g_d)$. By (b), the T-invariance of M and the definition of $\tilde{g}_j$ there exist some functions $(b_{j,k})_{j,k=1}^{d}$ in $\mathrm{L}^\infty(Y)$ with

$$T\tilde{g}_j = \sum_{k=1}^{d} b_{j,k}g_k - \sum_{l=1}^{j-1} S\langle f_j, g_l\rangle_{\mathrm{L}^2(X|Y)}Tg_l \in M(g_1, \dots, g_d).$$

We thus conclude $Tg_j \in S(q(\|\tilde{g}_j\|_{\mathrm{L}^2(X|Y)}))M(g_1, \dots, g_d)$, and as above for $j = 1$, we, in fact, obtain $Tg_j \in M(g_1, \dots, g_d)$.
(d) By (c) there exist $a_{j,k} \in \mathrm{L}^\infty(Y)$, $j, k \in \{1, \dots, d\}$, such that

$$Tg_j = \sum_{k=1}^{d} a_{j,k}g_k \text{ for all } j \in \{1, \dots, d\}, \tag{13.23}$$

and, as before in (13.21), we can also assume that $a_{j,k}(y) = 0$ for every $j \in \{1, \dots, d\}$ whenever $\|g_k\|_{L^2(X|Y)}(y) = 0$. Set $A = (a_{j,k})$ (being a $d \times d$ matrix with entries being elements of $L^\infty(Y)$). We have by (a) for every $j, j' \in \{1, \dots, d\}$

$$S\langle g_j, g_{j'}\rangle_{L^2(X|Y)} = \langle Tg_j, Tg_{j'}\rangle_{L^2(X|Y)} = \sum_{k,k'=1}^{d} a_{j,k}\overline{a_{j',k'}}\langle g_k, g_{k'}\rangle_{L^2(X|Y)}$$

$$= \sum_{k=1}^{d} a_{j,k}\overline{a_{j',k}} = (AA^*)_{j,j'}. \tag{13.24}$$

By (a) and the definition of the Koopman operator, ν-a.e. the left-hand side vanishes for $j \neq j'$ and takes 0-1-values for $j = j'$. Thus we obtain $0 \leq AA^* \leq I$ entrywise ν-a.e. In particular, for ν-a.e. $y \in Y$ the ℓ^2-operator norm of A satisfies $\|A\| = \|A^*\| \leq 1$, i.e., A acts as a contraction on $\mathbb{C}^d$ w.r.t. the Euclidean norm (for ν-a.e. $y \in Y$). Now consider $h := \sum_{j=1}^{d} |g_j|^2 \in L^1(X)$ and denote $\vec{g} := (g_1, \dots, g_d)^\top \in (L^2(X|Y))^d$. We can rewrite (13.23) as $T\vec{g} = A\vec{g}$, where T acts on each of the components of $\vec{g}$. Taking pointwise the Euclidean norm in $\mathbb{C}^d$ leads to

$$0 \leq Th = \sum_{j=1}^{d} |Tg_j|^2 \leq \|A\|^2 \sum_{j=1}^{d} |g_j|^2 \leq \sum_{j=1}^{d} |g_j|^2 = h.$$

Since T is an isometry on $L^1(X)$, we see that $Th = h$ proving (d).

Before we turn to the proof of the final statement we show that we can even achieve

$$\mathrm{supp}_Y g_1 \supset \mathrm{supp}_Y g_2 \supset \cdots \supset \mathrm{supp}_Y g_d, \tag{13.25}$$

where $\mathrm{supp}_Y g_j := [\|g_j\|_{L^2(X|Y)} \neq 0] = [\|g_j\|_{L^2(X|Y)} = 1]$ (up to a null set) for every j, still keeping the validity of (a)–(d). Indeed, observe that replacing g_1 by $g_1' := g_1 + \mathbf{1}_{Y \setminus \mathrm{supp}_Y g_1} g_2$ and g_2 by $g_2' := \mathbf{1}_{\mathrm{supp}_Y g_1} g_2$ increases the support of g_1 and decreases the support of g_2 without changing the properties (a)–(d). After applying this procedure with $g_2, \dots, g_d$ we can assume that $\mathrm{supp}_Y g_1 = \bigcup_{j=1}^{d} \mathrm{supp}_Y g_j$. Analogously we maximize $\mathrm{supp}_Y g_2$ by comparing g_2 to $g_3, \dots, g_d$ etc. After finitely many steps we arrive at (13.25).

Assume finally that (Y, ν, S) is ergodic. Then by (d) and (8.8) we see that $E_Y h = \sum_{j=1}^{d} \|g_j\|_{L^2(X|Y)}^2$ is S-invariant and hence constant. By (a) and (13.25) this implies $\nu(\mathrm{supp}_Y g_j) \in \{0, 1\}$ for every j and hence the existence of $d' \leq d$ with the desired properties. ∎

Remark 13.22 Lemma 13.21(d) shows that for ergodic systems (X, μ, T) every conditional eigenfunction is bounded, since so is the function h being constant in this case.

An immediate corollary of Lemma 13.21 is the following.

Lemma 13.23 (Conditional Eigenfunctions are Conditionally Almost Periodic) *Let (Y, ν, S) be a invertible (Markov) factor of a measure-preserving system (X, μ, T). Then $E(X|Y) \subset A(X|Y)$.*

Proof Let $f \in E(X|Y)$ and assume without loss of generality $\|f\|_{\mathrm{L}^2(X|Y)} \leq 1$. Let $g_1, \dots, g_d$ be a conditionally orthonormal generating set of a T-invariant sub-L^∞-module containing the orbit of f given by Lemma 13.21 and let $n \in \mathbb{N}$. By representation (13.19) we have

$$T^n f = \sum_{j=1}^{k} \langle T^n f, g_j \rangle_{\mathrm{L}^2(X|Y)} g_j$$

with coefficients $\langle T^n f, g_j \rangle_{\mathrm{L}^2(X|Y)}$ being bounded by 1 by the conditional Cauchy–Schwarz inequality (13.13) and Remark 13.12(f). Thus we have found a finitely generated module zonotope containing the orbit of f. ■

13.4 Conditional Jacobs–de Leeuw–Glicksberg Decomposition

The aim of this subsection is to prove the following conditional version of Theorem 6.20. Note that for the first proof of multiple recurrence presented in this chapter we will not need the full strength of this decomposition but rather a version of the classical dichotomy result being its direct corollary, see Proposition 13.49 below. We will use this stronger version for the third and fourth proof later in Sect. 15.8. Moreover, we will employ the equality (13.26) in the definition of the distal factors of finite order used in the second proof (which, however, can be omitted by a direct argument mentioned in Remark 13.39).

Theorem 13.24 (Conditional Jacobs de Leeuw–Glicksberg Decomposition) *Let (X, μ, T) be a measure-preserving system with invertible (Markov) factor (Y, ν, S). Then*

$$\overline{A(X|Y)} = \overline{E(X|Y)} \tag{13.26}$$

(with closure in $\mathrm{L}^2(X)$) and one has the orthogonal decomposition

$$\mathrm{L}^2(X) = \overline{A(X|Y)} \oplus W(X|Y) = \overline{E(X|Y)} \oplus W(X|Y).$$

Remark 13.25 The decomposition $\mathrm{L}^2(X) = \overline{A(X|Y)} \oplus W(X|Y)$ for general systems is due to Tao[7] [977, Sec. 2.14.2], cf. Furstenberg and Katznelson [450,

[7] Tao's argument does not use the separability assumption and is valid for general systems.

Prop. 2.2] and Furstenberg et al. [455, Thm. 10.1] for regular systems. The equality $\overline{A(X|Y)} = \overline{E(X|Y)}$ and therewith the decomposition $\mathrm{L}^2(X) = \overline{E(X|Y)} \oplus W(X|Y)$ was proven explicitly by Zorin-Kranich [1067] for regular systems with ergodic factor Y, see Austin et al. [67, Prop. 3.6] for the corresponding dichotomy for von Neumann dynamical systems. The general version, in the context of arbitrary group actions, is due to Jamneshan [576] and Edeko et al. [331]. Note that both Jamneshan [576] and Edeko et al. [331] work with slightly different notions of submodules.

We will prove Theorem 13.24 in several steps and begin with the orthogonality of the corresponding subspaces, see Tao [977, Lem. 2.14.2].

Lemma 13.26 (Orthogonality of $A(X|Y)$ and $W(X|Y)$) *Let (X, μ, T) be a measure-preserving system with an invertible (Markov) factor (Y, ν, S). Then for every $f \in W(X|Y)$ and $g \in A(X|Y)$ one has $\langle f, g\rangle_{\mathrm{L}^2(X|Y)} = 0$. In particular,*

$$A(X|Y) \perp W(X|Y) \text{ in } \mathrm{L}^2(X).$$

Proof The last assertion follows from Remark 13.12(c), provided that we already know the first one. Let now $f \in W(X|Y)$ and $g \in A(X|Y)$. We will show $\|\langle f, g\rangle_{\mathrm{L}^2(X|Y)}\|_{\mathrm{L}^1(Y)} = 0$.

Let $\varepsilon > 0$. Since $g \in A(X|Y)$, there exist $k \in \mathbb{N}$, $g_1, \dots, g_k \in \mathrm{L}^2(X|Y)$, $(a_{n,j})_{n\in\mathbb{N}, 1\le j\le k}$ in $\mathrm{L}^\infty(Y)$ with $\|a_{n,j}\|_{\mathrm{L}^\infty(Y)} \le 1$ and $(r_n)_{n\in\mathbb{N}} \subset \mathrm{L}^2(X|Y)$ with $\|r_n\|_{\mathrm{L}^{2,\infty}(X|Y)} \le \varepsilon$ such that

$$T^n g = \sum_{j=1}^{k} a_{n,j} g_j + r_n \text{ for each } n \in \mathbb{N}_0.$$

Since Y is an invertible factor of X, using (13.12), the conditional Cauchy–Schwarz inequality (13.13), (13.7) and the mean ergodic theorem yield

$$\begin{aligned}
\|\langle f, g\rangle_{\mathrm{L}^2(X|Y)}\|_{\mathrm{L}^1(Y)} &= \lim_{N\to\infty} \frac{1}{N}\sum_{n=1}^{N} \|\langle T^n f, T^n g\rangle_{\mathrm{L}^2(X|Y)}\|_{\mathrm{L}^1(Y)} \\
&\le \limsup_{N\to\infty} \frac{1}{N}\sum_{n=1}^{N}\Big(\sum_{j=1}^{k} \|\overline{a_{n,j}}\langle T^n f, g_j\rangle_{\mathrm{L}^2(X|Y)}\|_{\mathrm{L}^1(Y)} + \|\langle T^n f, r_n\rangle_{\mathrm{L}^2(X|Y)}\|_{\mathrm{L}^1(Y)}\Big) \\
&\le \limsup_{N\to\infty} \frac{1}{N}\sum_{n=1}^{N}\Big(\sum_{j=1}^{k} \|\langle T^n f, g_j\rangle_{\mathrm{L}^2(X|Y)}\|_{\mathrm{L}^1(Y)} + \varepsilon\big\| \|T^n f\|_{\mathrm{L}^2(X|Y)}\big\|_{\mathrm{L}^1(Y)}\Big) \\
&\le \sum_{j=1}^{k} \limsup_{N\to\infty} \frac{1}{N}\sum_{n=1}^{N} \|\langle T^n f, g_j\rangle_{\mathrm{L}^2(X|Y)}\|_{\mathrm{L}^1(Y)} + \varepsilon\|f\|_{\mathrm{L}^2(X)}.
\end{aligned}$$

By Lemma 13.20, the first term on the right-hand side equals zero. Taking $\varepsilon \to 0$ leads to $\|\langle f, g\rangle_{L^2(X|Y)}\|_{L^1(Y)} = 0$. ■

To proceed we need some preparations and introduce first the notion of conditional products.

Definition 13.27 Let (X, μ, T) be a regular measure-preserving system with factor (Y, ν, S). Consider the space $X \times X$ endowed with the product σ-algebra and the measure $\mu \times_Y \mu$ defined by

$$\int_{X\times X} f \otimes g \,\mathrm{d}(\mu \times_Y \mu) := \int_Y E_Y f \cdot E_Y g \,\mathrm{d}\nu \quad (f, g \geq 0 \text{ measurable}). \tag{13.27}$$

The existence and the uniqueness of this measure follows from the Carathéodory extension theorem as described below in Proposition 13.28. It is easy to see using (8.8) that this measure is $T \times T$-invariant whenever the factor (Y, ν, S) is invertible. In this case, the measure-preserving system $(X \times X, \mu \times_Y \mu, T \times T)$ is called the **conditional product** (or **relatively independent joining** or **fibered product**) of (X, μ, T) with itself with respect to Y.

Proposition 13.28 *Let (X, μ, T) be a regular measure-preserving system with factor (Y, ν, S). Then there exists a unique (regular) Borel probability measure $\mu \times_Y \mu$ on $X \times X$ with*

$$(\mu \times_Y \mu)(A \times B) = \int_Y E_Y \mathbf{1}_A \cdot E_Y \mathbf{1}_B \,\mathrm{d}\nu \ (A, B \subset X \text{ measurable}).$$

This measure satisfies (13.27).

Proof We define $\mu \times_Y \mu$ on measurable cylinder sets $A \times B$ by

$$(\mu \times_Y \mu)(A \times B) = \int_Y E_Y \mathbf{1}_A \cdot E_Y \mathbf{1}_B \,\mathrm{d}\nu.$$

We first observe that by positivity of the conditional expectation one has $0 \leq E_Y \mathbf{1}_A \leq 1$ implying by Hölder's inequality and Proposition 8.52(b)

$$(\mu \times_Y \mu)(A \times B) \leq \sqrt{\mu(A)\mu(B)}. \tag{13.28}$$

It is easy to check that $\mu \times_Y \mu$ can be extended to the semi-ring

$$\mathscr{C} := \left\{ M : M = \bigcup_{j=1}^{m} A_j \times B_j \right\}$$

of finite unions of measurable cylinder sets, where it becomes *finitely* additive. Thus it is also monotone and subadditive. It remains to prove its σ-subadditivity

in order to apply Carathéodory's extension theorem (Corollary 1.17). (Indeed, the σ-superadditivity on disjoint unions is a direct consequence of monotonicity and finite additivity.)
Let

$$A \times B \subset \bigcup_{n \in \mathbb{N}} A_n \times B_n$$

and let $\varepsilon > 0$. Take $A' \subset A$ and $B' \subset B$ compact sets with

$$\mu(A \setminus A') \leq \tfrac{\varepsilon^2}{4} \text{ and } \mu(B \setminus B') \leq \tfrac{\varepsilon^2}{4}.$$

Then we have by (13.28)

$$\begin{aligned}
&\int_Y E_Y \mathbf{1}_A \cdot E_Y \mathbf{1}_B \, \mathrm{d}\nu = \\
&\quad \int_Y E_Y \mathbf{1}_{A'} \cdot E_Y \mathbf{1}_{B'} \, \mathrm{d}\nu + \int_Y E_Y \mathbf{1}_{A \setminus A'} \cdot E_Y \mathbf{1}_{B'} \, \mathrm{d}\nu + \int_Y E_Y \mathbf{1}_A \cdot E_Y \mathbf{1}_{B \setminus B'} \, \mathrm{d}\nu \\
&\leq \int_Y E_Y \mathbf{1}_{A'} \cdot E_Y \mathbf{1}_{B'} \, \mathrm{d}\nu + \tfrac{\varepsilon}{2} + \tfrac{\varepsilon}{2} = \int_Y E_Y \mathbf{1}_{A'} \cdot E_Y \mathbf{1}_{B'} \, \mathrm{d}\nu + \varepsilon.
\end{aligned}$$

Analogously, again using (13.28), for each $n \in \mathbb{N}$ we can take U_n, V_n open with $A_n \subset U_n$, $B_n \subset V_n$ such that

$$\int_Y E_Y \mathbf{1}_{U_n} \cdot E_Y \mathbf{1}_{V_n} \, \mathrm{d}\nu \leq \int_Y E_Y \mathbf{1}_{A_n} \cdot E_Y \mathbf{1}_{B_n} \, \mathrm{d}\nu + \frac{\varepsilon}{2^n}.$$

Since $A' \times B' \subset \bigcup_{n \in \mathbb{N}} U_n \times V_n$, by compactness we can pass to a finite subcover

$$A' \times B' \subset \bigcup_{n=1}^{N} U_n \times V_n.$$

By the finite subadditivity we conclude

$$\begin{aligned}
\int_Y E_Y \mathbf{1}_{A'} \cdot E_Y \mathbf{1}_{B'} \, \mathrm{d}\nu &\leq \sum_{n=1}^{N} \int_Y E_Y \mathbf{1}_{U_n} \cdot E_Y \mathbf{1}_{V_n} \, \mathrm{d}\nu \\
&\leq \sum_{n=1}^{\infty} \int_Y E_Y \mathbf{1}_{A_n} \cdot E_Y \mathbf{1}_{B_n} \, \mathrm{d}\nu + \varepsilon.
\end{aligned}$$

Altogether we obtain

$$\int_Y E_Y \mathbf{1}_A \cdot E_Y \mathbf{1}_B \, \mathrm{d}\nu \leq \sum_{n=1}^{\infty} \int_Y E_Y \mathbf{1}_{A_n} \cdot E_Y \mathbf{1}_{B_n} \, \mathrm{d}\nu + 2\varepsilon$$

implying the σ-subadditivity of $\mu \times_Y \mu$ on $\mathscr{C}$. The first statement of the proposition follows now from the Carathéodory extension theorem (Corollary 1.17) whereas the second one follows then by using bilinearity and approximation. ■

Remark 13.29 When referring to the conditional product measure $\mu \times_Y \mu$ as above a tacit assumption will be that the considered measure-preserving system (X, μ, T) is regular (even if this condition is not stated explicitly every time). This assures, by the above, the existence of the conditional product measure *on the product space* $X \times X$. The restriction to regular systems is sufficient for our purposes. Note, however, that one can circumvent this restriction of regularity by making more use of topological models and Markov isomorphisms.

We will also encounter conditional products in Chap. 15 and refer to Glasner [468] for a systematic study of joinings and their role in modern ergodic theory. Here we just list the properties that will be needed later.

Remark 13.30 (Properties of Conditional Products)

(a) We leave as Exercise 13.7 to show that (Y, ν, S) is a factor of $(X \times X, \mu \times_Y \mu, T \times T)$ (for example via the Markov embedding given as $\tilde{J}f := Jf \otimes \mathbf{1}$ for the corresponding Markov embedding $J : \mathrm{L}^1(Y) \to \mathrm{L}^1(X)$) with the conditional expectation $\tilde{E}_Y$ given by

$$\tilde{E}_Y(f \otimes g) = E_Y f \cdot E_Y g \quad \nu\text{-a.e.} \tag{13.29}$$

(b) Let (X, μ, T) be regular and let $\pi : X \to Y$ be the corresponding point factor map, see Remark 8.48(c). Then the measure $\mu \times_Y \mu$ has the representation

$$\mu \times_Y \mu = \int_Y \mu_y \times \mu_y \, \mathrm{d}\nu(y),$$

where $\mu = \int_Y \mu_y \, \mathrm{d}\nu(y)$ is a disintegration of μ given by Theorem 8.57. Moreover, the measure $\mu \times_Y \mu$ is supported on the set

$$X \times_Y X := \{(x_1, x_2) \in X \times X : \pi(x_1) = \pi(x_2)\}$$

in the sense that $(\mu \times_Y \mu)(X \times_Y X) = 1$. In particular, $(X \times X, \mu \times_Y \mu, T \times T)$ is isomorphic to the system $(X \times_Y X, \mu \times_Y \mu, T \times T)$. We leave the details again as Exercise 13.7. This provides an equivalent definition of conditional products via disintegration originally introduced by Furstenberg [445] (and explaining the alternative name “fibered product”).

Inspired by the above remark we will shorten the notation for the conditional product to $X \times_Y X$.

We are now able to prove Theorem 13.24 and present the proof given by Zorin-Kranich [1067], cf. Furstenberg and Katznelson [450, Prop. 2.2] and Tao [977, Sec. 2.14.2].

Proof of Theorem 13.24 By Lemmata 13.26 and 13.23 it suffices to show that every $f \in \mathrm{L}^2(X)$ with $f \notin W(X|Y)$ correlates with a conditional eigenfunction, i.e., $f \not\perp E(X|Y)$.

Step 0: Reduction to regular systems. Assume that we know the assertion for regular systems. Let (X, μ, T) be a general measure-preserving system with invertible Markov factor (Y, ν, S) and $f \in \mathrm{L}^2(X)$ with $f \notin W(X|Y)$, i.e., satisfying

$$\limsup_{N\to\infty} \frac{1}{N}\sum_{n=1}^{N} \|\langle T^n f, f\rangle_{\mathrm{L}^2(X|Y)}\|_{\mathrm{L}^1(Y)} > 0 \tag{13.30}$$

by Remark 13.16(a). Consider now the Markov factor $\tilde{X}$ of X generated by the functions $T^n f, T^n f\cdot\overline{f}, n \in \mathbb{N}_0$ (see Sect. 8.2, in particular Proposition 8.22) and the corresponding Markov invertible subfactor $\tilde{Y}$ of Y generated by $E_Y(T^n f), E_Y(T^n f\cdot\overline{f}), n \in \mathbb{N}_0$. To simplify the notation we will view $\mathrm{L}^p(\tilde{X})$ as a subset of $\mathrm{L}^p(X)$ and $\mathrm{L}^p(\tilde{Y})$ as a subset of $\mathrm{L}^p(Y)$ for every $p \in [1,\infty]$ (cf. Theorem 8.18). In particular, by (13.30) we have $f \notin W(\tilde{X}|\tilde{Y})$. Since $\tilde{X}$ is separable and hence Markov isomorphic to a regular system by Remark 8.50, by assumption there exists a conditional eigenfunction $g \in \mathrm{L}^2(\tilde{X}|\tilde{Y})$ with $\langle f, g\rangle_{\mathrm{L}^2(\tilde{X})} \neq 0$. By the above identification $\mathrm{L}^2(\tilde{X}) \subset \mathrm{L}^2(X)$ we can view g as an element of $\mathrm{L}^2(X)$ and $\langle f, g\rangle_{\mathrm{L}^2(X)} \neq 0$ holds. Moreover, $E_Y(|g|^2) \in \mathrm{L}^\infty(Y)$ implying $g \in \mathrm{L}^2(X|Y)$. Thus it suffices to show that g is a conditional eigenfunction w.r.t. X and Y.

Since g is a conditional eigenfunction w.r.t. $\tilde{X}$ and $\tilde{Y}$, there exists a T-invariant $\mathrm{L}^\infty(\tilde{Y})$-module

$$\tilde{M} := \mathrm{L}^\infty(\tilde{Y})h_1 + \cdots + \mathrm{L}^\infty(\tilde{Y})h_l$$

for some $l \in \mathbb{N}$ and $h_1, \ldots, h_l \in \mathrm{L}^2(\tilde{X}|\tilde{Y})$ containing g. Consider the corresponding $\mathrm{L}^\infty(Y)$-module

$$M := \mathrm{L}^\infty(Y)h_1 + \cdots + \mathrm{L}^\infty(Y)h_l.$$

By $Th_j \in \tilde{M} \subset M$ for every $j \in \{1, \ldots, l\}$ we see that M is T-invariant. Moreover, M clearly contains g implying that g is a conditional eigenfunction w.r.t. X and Y finishing the reduction.

Thus we can assume without loss of generality that (X, μ, T) is regular, and in fact that X is compact metric. Suppose first that $f \in \mathrm{L}^2(X|Y)$, and let $\mu = \int_Y \mathrm{d}\mu_y$ be a disintegration of μ w.r.t. Y, see Sect. 8.6.

Step 1: Case $f \in \mathrm{L}^2(X|Y)$, construction of the operator U. By the mean ergodic theorem (Theorem 7.2) applied to the conditional product there exists a $T \times T$-invariant function $H \subset \mathrm{L}^2(X \times_Y X)$ with

$$\lim_{N\to\infty} \frac{1}{N}\sum_{n=1}^{N} T^n\overline{f} \otimes T^n f = H \text{ in } \mathrm{L}^2(X \times_Y X).$$

By (13.29) and (8.8) we have

$$\begin{aligned}(\tilde{E}_Y|T^n\overline{f}\otimes T^n f|^2)^{1/2} &= (\tilde{E}_Y(|T^n f|^2\otimes|T^n f|^2)^{1/2}\\ &= E_Y(T^n|f|^2) = S^n E_Y(|f|^2)\end{aligned}$$

implying $H \in \mathrm{L}^2(X \times_Y X|Y)$ with

$$\|H\|_{\mathrm{L}^{2,\infty}(X\times_Y X|Y)} \le \|f\|^2_{\mathrm{L}^{2,\infty}(X|Y)} =: C.$$

By the definition of H and by Remark 13.30

$$\lim_{N\to\infty}\int_Y\int_{X\times X}\left|\frac{1}{N}\sum_{n=1}^{N} T^n\overline{f}\otimes T^n f - H\right|^2 \mathrm{d}(\mu_y\times\mu_y)\,\mathrm{d}\nu(y) = 0,$$

so there exists a subsequence (N_j) in $\mathbb{N}$ such that

$$\lim_{j\to\infty}\frac{1}{N_j}\sum_{n=1}^{N_j} T^n\overline{f}\otimes T^n f = H \text{ in } \mathrm{L}^2(X\times X, \mu_y\times\mu_y) \tag{13.31}$$

holds for ν-a.e. $y \in Y$. This implies that $\overline{H(x_1, x_2)} = H(x_2, x_1)$ holds for $(\mu \times_Y \mu)$-a.e. (x_1, x_2) and

$$\int_{X\times X}|H|^2\,\mathrm{d}(\mu_y\times\mu_y) \le C^2 \text{ for } \nu\text{-a.e. } y\in Y. \tag{13.32}$$

For $y \in Y$ define an operator U_y on $\mathrm{L}^2(X, \mu_y)$ by

$$(U_y g)(x) := \int_X H(x', x)g(x')\,\mathrm{d}\mu_y(x').$$

By (13.32), for ν-a.e. $y \in Y$ the operator U_y is a self-adjoint Hilbert–Schmidt operator with Hilbert–Schmidt norm bounded by C. Moreover, one can show using approximation by continuous functions and the properties of the disintegration (Theorem 8.57) that the map $(x, y) \mapsto (U_y g)(x)$ is jointly measurable.

Define now an operator U on $\mathrm{L}^2(X)$ by

$$(Ug)(x) := (U_{\pi(x)}g)(x),$$

where $\pi : X \to Y$ is the corresponding point factor map (see Remark 8.48(c)). We first show that

$$Ug = U_y g \text{ in } \mathrm{L}^2(X, \mu_y) \text{ for } \nu\text{-a.e. } y \in Y. \tag{13.33}$$

Indeed, by (8.13) we have for every $g \in \mathrm{L}^2(X)$ and for ν-a.e. $y \in Y$

$$\begin{aligned}
&\int_X |Ug - U_y g|^2 \, \mathrm{d}\mu_y \\
&= \int_X \left| \int_X H(x', x) g(x') \, \mathrm{d}\mu_{\pi(x)}(x') \right. \\
&\quad \left. - \int_X H(x', x) g(x') \, \mathrm{d}\mu_y(x') \right|^2 \mathrm{d}\mu_y(x) = 0,
\end{aligned}$$

implying $Ug = U_y g$ in $\mathrm{L}^2(X, \mu_y)$. We next show that U is a bounded linear operator on $\mathrm{L}^2(X)$ with norm bounded by C. Indeed, by (8.11), (13.33) and since the norm of a Hilbert–Schmidt operator is bounded by its Hilbert–Schmidt norm, we have for $g \in \mathrm{L}^2(X)$

$$\begin{aligned}
\|Ug\|^2_{\mathrm{L}^2(X)} &= \int_Y \int_X |(Ug)(x)|^2 \, \mathrm{d}\mu_y(x) \, \mathrm{d}\nu(y) \\
&= \int_Y \int_X |(U_y g)(x)|^2 \, \mathrm{d}\mu_y(x) \, \mathrm{d}\nu(y) \\
&\leq \int_Y C^2 \int_X |g(x)|^2 \, \mathrm{d}\mu_y(x) \, \mathrm{d}\nu(y) = C^2 \|g\|^2_{\mathrm{L}^2(X)}.
\end{aligned}$$

By the $T \times T$-invariance of H and by (8.14), U commutes with T by

$$\begin{aligned}
(UTg)(x) = (U_{\pi(x)} Tg)(x) &= \int_X H(x', x) g(Tx') \, \mathrm{d}\mu_{\pi(x)}(x') \\
&= \int_X H(Tx', Tx) g(Tx') \, \mathrm{d}\mu_{\pi(x)}(x') \\
&= \int_X H(z, Tx) g(z) \, \mathrm{d}\mu_{\pi(Tx)}(z) \\
&= (Ug)(Tx) = (TUg)(x).
\end{aligned}$$

We next show

$$(Uf | f)_{\mathrm{L}^2(X)} > 0. \tag{13.34}$$

Indeed, by (8.13), (8.10), (13.31) and by the assumption on f

$$\begin{aligned}
(Uf | f)_{\mathrm{L}^2(X)} &= \int_X \int_X H(x', x) f(x') \, \mathrm{d}\mu_{\pi(x)}(x') \overline{f}(x) \, \mathrm{d}\mu(x) \\
&= \int_Y \int_X \int_X H(x', x) f(x') \overline{f}(x) \, \mathrm{d}\mu_{\pi(x)}(x') \, \mathrm{d}\mu_y(x) \, \mathrm{d}\nu(y)
\end{aligned}$$

$$= \int_Y \int_X \int_X H(x', x) f(x') \overline{f}(x) \, d\mu_y(x') \, d\mu_y(x) \, d\nu(y)$$

$$= \lim_{N\to\infty} \frac{1}{N} \sum_{n=1}^{N} \int_Y \int_X \int_X T^n \overline{f}(x') T^n f(x) f(x') \overline{f}(x)$$

$$d\mu_y(x') \, d\mu_y(x) \, d\nu(y)$$

$$= \lim_{N\to\infty} \frac{1}{N} \sum_{n=1}^{N} \int_Y \left| \int_X T^n f \cdot \overline{f} \, d\mu_y \right|^2 d\nu(y)$$

$$= \lim_{N\to\infty} \frac{1}{N} \sum_{n=1}^{N} \int_Y |E_Y(T^n f \cdot \overline{f})|^2 \, d\nu$$

$$= \lim_{N\to\infty} \frac{1}{N} \sum_{n=1}^{N} \| \langle T^n f, f \rangle_{L^2(X|Y)} \|^2_{L^2(Y)} > 0.$$

We finally observe that U is self-adjoint since by (13.33) we have

$$(Ug|h)_{L^2(X)} = \int_Y \int_X Ug(x)\overline{h}(x) \, d\mu_y(x) \, d\nu(y) = \int_Y (U_y g|h)_{L^2(X,\mu_y)} \, d\nu(y)$$

and since U_y is self-adjoint on $L^2(X, \mu_y)$ for ν-a.e. $y \in Y$.
Step 2: Case $f \in L^2(X|Y)$, construction of the conditional eigenfunction. To construct the desired conditional eigenfunction we will use spectral theory and functional calculus for self-adjoint Hilbert–Schmidt operators, see Sect. 5.5.

For $a > 0$ denote $r_a := \mathbf{1}_{(-\infty,-a)\cup(a,\infty)}$ and define the operator $r_a(U)$ via measurable functional calculus for self-adjoint operators. Consider further the function q on $\mathbb{R}$ defined by

$$q(s) := \begin{cases} s, & |s| \le C+1, \\ 0 & \text{otherwise}. \end{cases}$$

Since $\|U\| \le C$, we have $q(U) = U$ and we observe that $r_a q \to q$ uniformly as $a \to 0$, so by Theorem 5.23 $(r_a q)(U) \to U$ strongly as $a \to 0$. In particular, by (13.34) there exists $a > 0$ such that $(r_a(U)Uf\,|\,f)_{L^2(X)} > 0$, i.e., $r_a(U)Uf$ correlates with f. We fix such an a, set $r = r_a$ and show next that $r(U)Uf$ is a conditional eigenfunction.

For $g \in L^2(X)$ we first show

$$r(U)g = r(U_y)g \text{ in } L^2(X, \mu_y) \text{ for } \nu\text{-a.e. } y \in Y. \tag{13.35}$$

Consider a sequence of polynomials (p_n) on $\mathbb{R}$ with $p_n(-a) = 0 = p_n(a)$ converging to r pointwise and boundedly on $[-C, C]$ and uniformly on $[-C, -a -$

$\varepsilon] \cup [-a, a] \cup [a+\varepsilon, C]$ for every $\varepsilon > 0$. (This is clearly possible by the Weierstrass approximation theorem applied to appropriate continuous functions approximating r pointwise.) By (13.33) we have

$$p_n(U)g = p_n(U_y)g \text{ in } \mathrm{L}^2(X, \mu_y) \text{ for every } n \in \mathbb{N} \text{ and } \nu\text{-a.e. } y \in Y. \tag{13.36}$$

Take such a y. The left-hand side of (13.36) converges to $r(U)g$ in $\mathrm{L}^2(X, \mu)$ and the right-hand side converges to $r(U_y)g$ in $\mathrm{L}^2(X, \mu_y)$. Passing to a subsequence we have μ-a.e. convergence of the left-hand side which, by disintegration, implies μ_y-a.e. convergence for ν-a.e. $y \in Y$. This shows (13.35).

Recall from Step 1 that for ν-a.e. $y \in Y$ the operator U_y is self-adjoint and Hilbert–Schmidt on $\mathrm{L}^2(X, \mu_y)$ with its Hilbert–Schmidt norm bounded by C, in particular it is compact. Take such a y. By Theorem 5.23 $Q_y := r(U_y)$ is the orthogonal projection onto the closed linear span of the eigenfunctions of U_y belonging to the (finitely many) eigenvalues lying in $(-\infty, -a) \cup (a, \infty)$. And by Theorem 5.24 the projection Q_y on $\mathrm{L}^2(X, \mu_y)$ has rank at most $\frac{C^2}{a^2}$.

We now check that $r(U)Uf \in \mathrm{L}^2(X|Y)$. For ν-a.e. $y \in Y$ we have by (8.10), (13.35), (13.33) and since the operator norm is bounded by the Hilbert–Schmidt norm

$$\begin{aligned} \|r(U)Uf\|^2_{\mathrm{L}^2(X|Y)}(y) &= \int_X |r(U)Uf|^2 \,\mathrm{d}\mu_y = \|Q_y Uf\|^2_{\mathrm{L}^2(X,\mu_y)} \\ &\le \|Uf\|^2_{\mathrm{L}^2(X,\mu_y)} = \|U_y f\|^2_{\mathrm{L}^2(X,\mu_y)} \\ &\le C^2 \|f\|^2_{\mathrm{L}^2(X,\mu_y)} = C^2 \|f\|^2_{\mathrm{L}^2(X|Y)}(y). \end{aligned} \tag{13.37}$$

As f belongs to $\mathrm{L}^2(X|Y)$, so does $r(U)Uf$.

It remains to show that the orbit O of $r(U)Uf$ is contained in a finitely generated T-invariant sub-$\mathrm{L}^\infty(Y)$-module. Since T commutes with U we have

$$O = \{r(U)UT^n f : n \in \mathbb{N}_0\}.$$

Similarly to the calculation in (13.37), using (8.14), we see that this set is bounded in $\mathrm{L}^2(X|Y)$. Moreover, since O is countable, using (13.35) and since for ν-a.e. $y \in Y$ every $\mathrm{L}^\infty(Y)$-function is constant in $\mathrm{L}^2(X, \mu_y)$ by (8.12), the (automatically T-invariant) sub-$\mathrm{L}^\infty(Y)$-module generated by O is contained in the range of Q_y for ν-a.e. $y \in Y$. Adapting the Gram–Schmidt procedure as in the proof of Lemma 13.21 one can construct a conditionally orthogonal (possibly infinite) set of functions (h_j) in $\mathrm{L}^2(X|Y)$ with $\|h_j\|_{\mathrm{L}^2(X|Y)}$ being 0-1-valued for every j such that for every $n \in \mathbb{N}$ the sub-L^∞-module generated by $h_1, \ldots, h_n$ is T-invariant and contains the sub L^∞-module generated by the first n elements of the orbit O. In particular, for every j the function Th_j belongs to the sub-L^∞-module generated by $h_1, \ldots, h_j$. Since the set (h_j) is orthogonal in $\mathrm{L}^2(X, \mu_y)$ and since $\dim \operatorname{ran} Q_y \le \frac{C^2}{a^2}$ for ν-a.e. $y \in Y$, we see that for ν-a.e. $y \in Y$ at most $\frac{C^2}{a^2}$-many of h_j are non-zero in $\mathrm{L}^2(X, \mu_y)$.

We now construct a finitely generated T-invariant sub-L^∞-module which contains every h_j and hence the orbit O. Consider the functions on X given by

$$H_n := \sum_j h_j \mathbf{1}_{A_{j,n}}, \quad n \in \mathbb{N},$$

where the sets $A_{j,n}$ are defined as

$$A_{j,n} := \Big\{ y : \sum_{m=1}^{j} \|h_m\|_{\mathrm{L}^2(X|Y)}(y) = n,\ \|h_j\|_{\mathrm{L}^2(X|Y)}(y) = 1 \Big\}.$$

By construction, for every fixed $n \in \mathbb{N}$ these sets are pairwise disjoint, and so they are for every fixed j. In particular, using the notation $\operatorname{supp}_Y g := \{y : \|g\|_{\mathrm{L}^2(X|Y)}(y) \neq 0\}$,

$$H_1 = h_1 \mathbf{1}_{\operatorname{supp}_Y h_1} + h_2 \mathbf{1}_{\operatorname{supp}_Y h_2 \setminus \operatorname{supp}_Y h_1} + h_3 \mathbf{1}_{\operatorname{supp}_Y h_3 \setminus (\operatorname{supp}_Y h_1 \cup \operatorname{supp}_Y h_2)} + \cdots,$$

i.e., in each fiber H_1 equals h_j for the smallest j (depending on the fiber) for which $\|h_j\|_{\mathrm{L}^2(X|Y)}$ is non-zero, and equals zero if there is no such j. Analogously, H_2 in each fiber equals h_j for the second smallest j (again depending on the fiber) for which $\|h_j\|_{\mathrm{L}^2(X|Y)}$ is non-zero, and equals zero if there is no such j, etc. We see that each H_n belongs to $\mathrm{L}^2(X|Y)$ (with $\|H_n\|_{\mathrm{L}^2(X|Y)}$ being again 0-1-valued) and $H_n = 0$ for every $n > \frac{C^2}{a^2}$. Denote by N the greatest index such that $H_N \neq 0$. By construction

$$h_j = \sum_{n=1}^{N} H_n \mathbf{1}_{A_{j,n}} \quad \text{for every } j, \tag{13.38}$$

therefore each h_j (and hence the orbit O) is contained in the sub-L^∞-module generated by the functions $H_1, \dots, H_N$. Thus it remains to show that this sub-L^∞-module is T-invariant.

For each j denote by $a_{j,1}, \dots, a_{j,j}$ the functions in $\mathrm{L}^\infty(Y)$ with $Th_j = a_{j,1}h_1 + \cdots + a_{j,j}h_j$. Note that the conditional orthogonality of (h_j) implies that for every j and $m \le j$ one has $a_{j,m} = \langle Th_j, h_m \rangle_{\mathrm{L}^2(X|Y)}$ (one can choose $a_{j,m}$ in such a way), and hence by the conditional Cauchy–Schwarz inequality (13.13) and (8.14)

$$\|a_{j,m}\|_{\mathrm{L}^\infty(Y)} \le 1 \text{ for all } j \text{ and } m \le j. \tag{13.39}$$

For a fixed n we conclude from (13.38) that

$$TH_n = \sum_j Th_j \cdot \mathbf{1}_{S^{-1}A_{j,n}} = \sum_j \mathbf{1}_{S^{-1}A_{j,n}} \sum_{m=1}^{j} a_{j,m} \sum_{n=1}^{N} H_n \mathbf{1}_{A_{m,n}} =: \sum_{n=1}^{N} c_n H_n$$

with coefficients (being functions on Y)

$$c_n := \sum_j \mathbf{1}_{S^{-1}A_{j,n}} \sum_{m=1}^{j} a_{j,m} \mathbf{1}_{A_{m,n}}.$$

Thus it remains to show $c_n \in \mathrm{L}^\infty(Y)$ for every $n \in \{1, \dots, N\}$. Fix such an n. Since the sets $S^{-1}A_{j,n}$ are disjoint, we have

$$\|c_n\|_{\mathrm{L}^\infty(Y)} \le \sup_j \left\| \sum_{m=1}^{j} a_{j,m} \mathbf{1}_{A_{m,n}} \right\|_{\mathrm{L}^\infty(Y)}.$$

Since also the sets $A_{m,n}$ are disjoint, by (13.39) we obtain

$$\|c_n\|_{\mathrm{L}^\infty(Y)} \le \sup_j \max_{m \le j} \|a_{j,m}\|_{\mathrm{L}^\infty(Y)} \le 1.$$

This concludes Step 2 and hence finishes the proof for the case $f \in \mathrm{L}^2(X|Y)$.
Step 3: The general case. Let now $f \in \mathrm{L}^2(X) \setminus W(X|Y)$. Consider for $a > 0$ the measurable set

$$M_a := \{y \in Y : E_Y(|f|^2)(y) \le a\}.$$

Observe that $\mathbf{1}_{M_a} f \to f$ in $\mathrm{L}^2(X)$ as $a \to \infty$ and $\mathbf{1}_{M_a} f \in \mathrm{L}^2(X|Y)$ for each $a > 0$. Lemma 13.20(b) implies that there exists $a > 0$ such that $\mathbf{1}_{M_a} f \notin W(X|Y)$, hence by the above there exists a conditional eigenfunction g which correlates with $\mathbf{1}_{M_a} f$. Denote by $g_1, \dots, g_k$ the functions which generate a T-invariant sub-L^∞-module containing the orbit of g. Then the same sub-L^∞-module contains the orbit of $\mathbf{1}_{M_a} g$, i.e., $\mathbf{1}_{M_a} g$ is a conditional eigenfunction, too. Moreover, by

$$\langle f, \mathbf{1}_{M_a} g\rangle_{\mathrm{L}^2(X)} = \langle \mathbf{1}_{M_a} f, g\rangle_{\mathrm{L}^2(X)} > 0$$

$\mathbf{1}_{M_a} g$ correlates with f. The proof is complete. ∎

13.5 Weakly Mixing Extensions

In this section we consider a special case when the factor (Y, ν, S) is such that the part $\overline{A(X|Y)}$ is the smallest possible (i.e., equals $\mathrm{L}^2(Y)$ or, more precisely, $J\mathrm{L}^2(Y)$ for the corresponding Markov embedding J), and show that (MR) is preserved by such extensions (X, μ, T).

Definition 13.31 Let (Y, ν, S) be a factor of a measure-preserving system (X, μ, T). We say that $f \in \mathrm{L}^2(X)$ has **conditional** (or **relative**) **mean zero** if $E_Y f = 0$ holds.

Remark 13.32 By Proposition 8.52(c) we see that f has conditional mean zero if and only if

$$\langle f, g\rangle_{\mathrm{L}^2(X|Y)} = \overline{g} E_Y f = 0 \text{ for every } g \in \mathrm{L}^\infty(Y).$$

Assume that (Y, ν, S) is invertible. Since $W(X|Y)$ is conditionally orthogonal to $A(X|Y) \supset \mathrm{L}^\infty(Y)$ by Lemma 13.26, every conditionally weakly mixing function has conditional mean zero. Note that the converse is false. Indeed, as we will see later in Example 13.43, the only conditionally weakly mixing function for the skew rotation $(\mathbb{T}^2, \mathrm{m}^2, T_a)$ (see Example 8.34) relative to its Kronecker factor is constant zero by Lemma 13.26, while there are many functions with conditional mean zero.

Definition 13.33 We say that a measure-preserving system (X, μ, T) is a **weakly mixing extension** of its factor (Y, ν, S) if every $f \in \mathrm{L}^2(X)$ with conditional mean zero is conditionally weakly mixing.

Remark 13.34 Since every $f \in \mathrm{L}^2(X)$ decomposes[8] into the sum of $E_Y f$ (which belongs to $\mathrm{L}^2(Y)$ by Proposition 8.11(g)) and $f - E_Y f$ which has conditional mean zero, (X, μ, T) is a weakly mixing extension of (Y, ν, S) if and only if

$$\mathrm{L}^2(X) = \mathrm{L}^2(Y) + W(X|Y). \tag{13.40}$$

Moreover, this sum is direct and orthogonal if (Y, ν, S) is invertible by Lemma 13.26. This gives a simple form of the decomposition given by Theorem 13.24 (without the regularity assumption) for weakly mixing extensions.

Remark 13.35

(a) We call a measure-preserving system (X, μ, T) an **ergodic extension** of its factor (Y, ν, S) if every T-invariant $f \in \mathrm{L}^1(X)$ is (up to a null set an image of) a function on Y. Analogously to the product characterization of weak mixing given in Proposition 8.42, (X, μ, T) is a weakly mixing extension of (Y, ν, S) if and only if the conditional product system $X \times_Y X$ is an ergodic extension

[8] Recall that we identify functions in $\mathrm{L}^2(Y)$ with their images under J being functions in $\mathrm{L}^2(X)$.

of (Y, ν, S),[9] see Exercise 13.9(a). Furthermore, for ergodic systems (X, μ, T) this is equivalent to the property that the conditional product system $X \times_Y X$ is ergodic,[10] see Exercise 13.9(b).

(b) As in the situation of classical weak mixing, see Sect. 3.6, a typical (in the Baire category sense) extension is weakly mixing for non-atomic measures, see Schnurr [922] and Glasner and Weiss [470].

To show that weakly mixing extensions preserve (MR) we first need the following comparison result for multiple ergodic averages.

Proposition 13.36 (Characteristic Factor for Weakly Mixing Extensions) *Let (X, μ, T) be a weakly mixing extension of an invertible measure-preserving system (Y, ν, S) with Markov embedding $J : \mathrm{L}^1(Y) \to \mathrm{L}^1(X)$ and the Markov projection $P_Y = JE_Y$ onto (Y, ν, S). Then the factor (Y, ν, S) is* characteristic for multiple ergodic averages (13.3), *i.e., for every $k \in \mathbb{N}$ and every $f_1, \dots, f_k \in \mathrm{L}^\infty(X)$*

$$\lim_{N\to\infty} \left\| \frac{1}{N}\sum_{n=1}^{N} T^n f_1 \cdots T^{kn} f_k - \frac{1}{N}\sum_{n=1}^{N} T^n P_Y f_1 \cdots T^{kn} P_Y f_k \right\|_2 = 0. \qquad (13.41)$$

In particular, if at least one f_j has conditional mean zero, then

$$\lim_{N\to\infty} \frac{1}{N}\sum_{n=1}^{N} T^n f_1 \cdots T^{kn} f_k = 0 \qquad (13.42)$$

in $\mathrm{L}^2(X)$.

Proof [11] We only have to prove the first assertion. The proof goes by induction on k. For $k = 1$, by the mean ergodic theorem the averages

$$\frac{1}{N}\sum_{n=1}^{N}(T^n f_1 - T^n P_Y f_1) = \frac{1}{N}\sum_{n=1}^{N} T^n (f_1 - P_Y f_1)$$

converge in $\mathrm{L}^2(X)$ to $P_{\mathrm{fix}}(f_1 - P_Y f_1)$. But $P_{\mathrm{fix}}(f_1 - P_Y f_1) = 0$ since by the definition of a weak mixing extension the function $f_1 - P_Y f_1$ is conditionally weakly mixing and hence orthogonal to all eigenfunctions of T by Example 13.19 and Lemma 13.26.

[9] Weakly mixing extensions were originally introduced by Furstenberg and Katznelson [450] via (a variant of) this equivalent definition, see also Furstenberg [446, Def. 6.1].

[10] This is how Furstenberg et al. [455] defined weakly mixing extensions.

[11] The original proof by Furstenberg et al. [455], see also Furstenberg and Katznelson [450], does not use the van der Corput trick explicitly. Its first usage in this context is due to Bergelson (see also Furstenberg [446, Lem. 4.9]) who introduced it in [103] and whose proof in the weakly mixing case is mimicked here.

Suppose now that $k \geq 2$ and the assertion holds for $k-1$. Let $f_1, \dots, f_k \in \mathrm{L}^\infty(X)$ and assume without loss of generality that $\|f\|_\infty \leq 1$ for every $j \in \{1, \dots, k\}$. Moreover, by

$$\begin{aligned}
&T^n f_1 \cdot T^{2n} f_2 \cdots T^{kn} f_k - T^n(P_Y f_1) \cdots T^{kn}(P_Y f_k) \\
&\quad = T^n(f_1 - P_Y f_1) \cdot T^{2n} f_2 \cdots T^{kn} f_k + \cdots \\
&\qquad + T^n P_Y f_1 \cdots T^{(k-1)n} P_Y f_{k-1} \cdot T^{kn}(f_k - P_Y f_k)
\end{aligned} \tag{13.43}$$

we can assume that there exists $l \in \{1, \dots, k\}$ such that f_l has conditional mean zero, and hence is conditionally weakly mixing (by the definition of a weakly mixing extension).

We use the van der Corput trick and denote for $n \in \mathbb{N}$

$$u_n := T^n f_1 \cdot T^{2n} f_2 \cdots T^{kn} f_k.$$

By the T-invariance of μ

$$\begin{aligned}
(u_n|u_{n+h}) &= \int_X \left(T^n f_1 \cdot T^{2n} f_2 \cdots T^{kn} f_k\right) \cdot \left(T^{n+h}\overline{f_1} \cdot T^{2(n+h)}\overline{f_2} \cdots T^{k(n+h)}\overline{f_k}\right) \mathrm{d}\mu \\
&= \int_X \left(f_1 \cdot T^n f_2 \cdots T^{(k-1)n} f_k\right) \cdot \left(T^h\overline{f_1} \cdot T^{n+2h}\overline{f_2} \cdots T^{(k-1)n+kh}\overline{f_k}\right) \mathrm{d}\mu \\
&= \int_X \left(f_1 T^h\overline{f_1}\right) \cdot T^n\left(f_2 T^{2h}\overline{f_2}\right) \cdots T^{(k-1)n}\left(f_k T^{kh}\overline{f_k}\right) \mathrm{d}\mu.
\end{aligned}$$

By the inductive hypothesis applied to the functions $f_2 T^{2h}\overline{f_2}, \dots, f_k T^{kh}\overline{f_k}$ and the Cauchy–Schwarz inequality we have for every $h \in \mathbb{N}$

$$\begin{aligned}
\gamma_h &:= \limsup_{N\to\infty} \left| \frac{1}{N} \sum_{n=0}^{N-1} (u_n|u_{n+h}) \right| \\
&= \limsup_{N\to\infty} \left| \int_X \left(f_1 T^h\overline{f_1}\right) \cdot \frac{1}{N} \sum_{n=0}^{N-1} T^n\left(f_2 T^{2h}\overline{f_2}\right) \cdots T^{(k-1)n}\left(f_k T^{kh}\overline{f_k}\right) \mathrm{d}\mu \right| \\
&= \limsup_{N\to\infty} \left| \int_X \left(f_1 T^h\overline{f_1}\right) \cdot \frac{1}{N} \sum_{n=0}^{N-1} T^n\left(P_Y(f_2 T^{2h}\overline{f_2})\right) \cdots \right. \\
&\qquad \left. \cdots T^{(k-1)n}\left(P_Y(f_k T^{kh}\overline{f_k})\right) \mathrm{d}\mu \right| \\
&= \limsup_{N\to\infty} \left| \frac{1}{N} \sum_{n=0}^{N-1} \int_X P_Y(f_1 T^h\overline{f_1}) \cdot T^n\left(P_Y(f_2 T^{2h}\overline{f_2})\right) \cdots \right. \\
&\qquad \left. \cdots T^{(k-1)n}\left(P_Y(f_k T^{kh}\overline{f_k})\right) \mathrm{d}\mu \right|,
\end{aligned}$$

where we used for the last equality that P_Y is an orthogonal projection on $\mathrm{L}^2(X)$, see Proposition 8.25(d). Since we assumed $\|f_j\|_\infty \le 1$ for every $j \in \{1, \dots, k\}$ and $P_Y = JE_Y$ is a contraction w.r.t. the $\|\cdot\|_\infty$-norm, we can conclude by the T-invariance of μ that

$$\gamma_h \le \|P_Y(f_l T^{lh}\overline{f_l})\|_{\mathrm{L}^1(X)} = \|\langle T^{lh} f_l, f_l\rangle_{\mathrm{L}^2(X|Y)}\|_{\mathrm{L}^1(Y)}.$$

Thus we obtain

$$\frac{1}{H}\sum_{h=1}^{H} \gamma_h \le \frac{1}{H}\sum_{h=1}^{H} \|\langle T^{lh} f_l, f_l\rangle_{\mathrm{L}^2(X|Y)}\|_{\mathrm{L}^1(Y)}$$

which goes to zero as $H \to \infty$ since f_l is conditionally weakly mixing, see Exercise 4.7(b). The van der Corput Lemma 7.31 implies $\lim_{N\to\infty} \frac{1}{N}\sum_{n=1}^{N} u_n = 0$ in $\mathrm{L}^2(X)$ as desired. ■

We now obtain the following property of weakly mixing extensions as a direct corollary.

Proposition 13.37 (Weakly Mixing Extensions Preserve (MR)) *Let (X, μ, T) be a weakly mixing extension of an invertible measure-preserving system (Y, ν, S) and let $k \in \mathbb{N}$. If (Y, ν, S) satisfies (MR$_k$), then so does (X, μ, T).*

Proof Let $k \in \mathbb{N}$ and assume that (Y, ν, S) satisfies (MR$_k$). Take $f \in \mathrm{L}^\infty(X)$ with $f > 0$. We have to show (13.1). By Proposition 13.36 and (8.8)

$$\begin{aligned}
&\liminf_{N\to\infty} \frac{1}{N}\sum_{n=1}^{N} \int_X f \cdot T^n f \cdots T^{kn} f \,\mathrm{d}\mu \\
&= \liminf_{N\to\infty} \frac{1}{N}\sum_{n=1}^{N} \int_X f \cdot J(S^n(E_Y f) \cdots S^{kn}(E_Y f)) \,\mathrm{d}\mu \\
&= \liminf_{N\to\infty} \frac{1}{N}\sum_{n=1}^{N} \int_Y E_Y f \cdot S^n(E_Y f) \cdots S^{kn}(E_Y f) \,\mathrm{d}\nu, \qquad (13.44)
\end{aligned}$$

where we used that $P_Y = JE_Y$ is an orthogonal projection onto $J(\mathrm{L}^2(Y))$ by Proposition 8.25(d) and $J'\mathbf{1} = \mathbf{1}$. Since $E_Y f \in \mathrm{L}^\infty(Y)$ with $E_Y f > 0$ by Remark 8.54 and (Y, ν, S) satisfies (MR$_k$), the right-hand side of (13.44) is greater than zero, implying (13.1). ■

13.6 Compact Extensions

We now consider the opposite case to weakly mixing extensions (see Sect. 13.5), namely when the part $\overline{A(X|Y)}$ is the largest possible, and show that also in this case (MR) is preserved by such extensions.

We begin with the following crucial property of conditional eigenfunctions which together with (13.26) implies that $\overline{A(X|Y)}$ defines a factor.

Lemma 13.38 (Conditional Eigenfunctions Define a Factor) *Let (X, μ, T) be a measure-preserving system with an invertible (Markov) factor (Y, ν, S). Then the set $E(X|Y) \cap \mathrm{L}^\infty(X)$ is a conjugation-invariant, T-invariant subalgebra of $\mathrm{L}^\infty(X)$ containing $\mathbf{1}$ which is dense in $E(X|Y)$ w.r.t. the $\mathrm{L}^2(X)$-norm. In particular, $\overline{E(X|Y)}$ is induced by a factor of (X, μ, T). Moreover, this factor is (Markov) invertible whenever (Y, ν, S) is ergodic.*

Proof The second statement follows from Proposition 8.19 whereas it is clear that $E(X|Y) \cap \mathrm{L}^\infty(X)$ is a conjugation-invariant, T-invariant subalgebra of $\mathrm{L}^\infty(X)$ containing $\mathbf{1}$. Thus for the first statement it remains to show that $E(X|Y) \cap \mathrm{L}^\infty(X)$ is dense in $E(X|Y)$.

Let $f \in E(X|Y)$ and take a conditionally orthonormal generating set $\{g_1, \ldots, g_d\}$ for the T-invariant sub-$\mathrm{L}^\infty(Y)$-module containing the orbit of f given by Lemma 13.21. By Lemma 13.21(d) the function $h := \sum_{j=1}^d |g_j|^2 \in \mathrm{L}^1(X)$ is T-invariant, implying that for every $N \in \mathbb{N}$ the function $\mathbf{1}_{[h\le N]}$ is also T-invariant. Therefore the bounded functions $g_1\mathbf{1}_{[h\le N]}, \ldots, g_d\mathbf{1}_{[h\le N]}$ generate a T-invariant sub-L^∞-module containing $f\mathbf{1}_{[h\le N]}$. So the functions $f\mathbf{1}_{[h\le N]}$ are bounded conditional eigenfunctions converging to f in $\mathrm{L}^2(X)$ as $N \to \infty$, proving the first statement.

Assume now that (Y, ν, S) is ergodic and denote by T the Koopman operator on $\overline{E(X|Y)}$. We construct T^{-1} as follows. Let M be a T-invariant finitely generated sub-L^∞-module and take a conditionally orthonormal set $\{g_1, \ldots, g_r\}$ of generators of M with $\|g_j\|_{\mathrm{L}^2(X|Y)} = 1$ ν-a.e. for every $j \in \{1, \ldots, r\}$, see Lemma 13.21. Using the notation from the proof of Lemma 13.21 we have $T\vec{g} = A\vec{g}$, where $AA^* = I$ by (13.24). In particular, for ν-a.e. $y \in Y$ the $r \times r$ matrix A is surjective and hence invertible with $A^{-1} = A^*$, i.e., A is unitary. Thus we can now define T^{-1} on M by $T^{-1}\vec{g} = A^{-1}\vec{g}$ which is a contraction on M w.r.t. the $\mathrm{L}^2(X)$-norm. Since every two finitely generated T-invariant sub-L^∞-modules are included into a third one and the union of such submodules is dense in $\overline{E(X|Y)}$, T^{-1} extends to a contraction on $\overline{E(X|Y)}$ being the inverse of T. ∎

Remark 13.39 Using (13.26), Lemma 13.38 implies that $\overline{A(X|Y)}$ is also induced by a factor. This can also be shown directly, see [341, Lem. 5.6] (based on Zorin-Kranich [1067, Lem. 8,9]) for ergodic systems and Tao [977, Ex. 2.13.6] for the general case.

We now define compact extensions.

Definition 13.40 We call (X, μ, T) a **compact extension** of its factor (Y, ν, S) if $\overline{A(X|Y)} = \mathrm{L}^2(X)$.

If (Y, ν, S) is (Markov) invertible, then by Theorem 13.24 (X, μ, T) is a compact extension of (Y, ν, S) if and only if $\overline{E(X|Y)} = \mathrm{L}^2(X)$ and if and only if $W(X|Y) = \{0\}$.

Remark 13.41 (Consistency with the Furstenberg–Katznelson Definition) Several equivalent definitions similar to the one here appeared one year after Furstenberg's original proof in Furstenberg and Katznelson [450, Sec. 2] in the context of commuting transformations and multidimensional Szemerédi's theorem, with one more equivalent definition added by Furstenberg [446, p. 131]. As one of these equivalent definitions, they call an extension *compact* if there exists a dense set D of $\mathrm{L}^2(X)$ such that for every $f \in D$ and every $\delta > 0$ there exist finitely many functions $f_1, \dots, f_k \in \mathrm{L}^2(X)$ such that for every $n \in \mathbb{N}_0$

$$\min_{j\in\{1,\dots,k\}} \|T^n f - f_j\|_{\mathrm{L}^2(X|Y)} < \delta \quad \nu\text{-a.e.} \tag{13.45}$$

We show that the set of all functions satisfying the above condition equals $A(X|Y)$. It is clear that every such function is conditionally almost periodic (take the module zonotope generated by $f_1, \dots, f_k$). Conversely, let f be conditionally almost periodic. Take $\delta > 0$ and denote by $g_1, \dots, g_m$ finitely many functions generating the module zonotope Z corresponding to $\delta/2$. Denote $c := \max_{j\in\{1,\dots,m\}} \|g_j\|_{\mathrm{L}^{2,\infty}(X|Y)}$ which we may assume to be non-zero. Taking all linear combinations of $g_1, \dots, g_m$ with coefficients from a finite set of complex numbers which is $\frac{\delta}{2mc}$-dense in the unit disc we obtain the desired set of functions (f_j). Indeed, by the conditional triangle inequality (13.14) for every $g \in Z$ and ν-a.e. $y \in Y$ there exists j such that $\|g - f_j\|_{\mathrm{L}^2(X|Y)}(y) < \delta/2$, leading to (13.45).

Remark 13.42 (Consistency with Tao's Definition) Inspired by one of the equivalent definitions by Furstenberg and Katznelson [450, Sec. 2], Tao [977, Def. 2.13.7] called (X, μ, T) a *compact extension* of (Y, ν, S) if every $f \in \mathrm{L}^2(X|Y)$ is **conditionally almost periodic in measure**, i.e., for every $\varepsilon > 0$ there exists a measurable set $M \subset Y$ with $\nu(M) \geq 1 - \varepsilon$ such that the function $\mathbf{1}_M f$ is conditionally almost periodic. Let us denote the set of all functions in $\mathrm{L}^2(X)$ which are conditionally almost periodic in measure by $AM(X|Y)$. We show that if (Y, ν, S) is invertible then

$$AM(X|Y) = \overline{A(X|Y)} \tag{13.46}$$

holds[12] and thus the two definitions are equivalent. Clearly, every function which is conditionally almost periodic in measure belongs to $\overline{A(X|Y)}$. To show the converse let $\varepsilon > 0$, $f \in \overline{A(X|Y)}$ and take $(f_n) \subset A(X|Y)$ with $\lim_{n\to\infty} f_n = f$ in $\mathrm{L}^2(X)$. By Proposition 8.52(b) this means

$$\int_Y E_Y(|f_n - f|^2)\,\mathrm{d}\nu = \|f_n - f\|^2_{\mathrm{L}^2(X)} \to 0 \text{ as } n \to \infty.$$

Passing to a subsequence if necessary we can assume that $E_Y(|f_n - f|^2) \to 0$ as $n \to \infty$ holds ν-a.e. By Egorov's theorem there exists a measurable set $M \subset Y$ with $\nu(M) > 1 - \varepsilon$ such that $\mathbf{1}_M E_Y(|f_n - f|^2) \to 0$ uniformly as $n \to \infty$. By Proposition 8.52(c) this implies

$$\|\mathbf{1}_M f_n - \mathbf{1}_M f\|_{\mathrm{L}^{2,\infty}(X|Y)} \to 0 \text{ as } n \to \infty.$$

Since for every n the function $\mathbf{1}_M f_n$ is conditionally almost periodic (w.r.t. the same module zonotopes as f_n), so is $\mathbf{1}_M f$ by Remark 13.16(c).

Example 13.43 (Kronecker and Abramov Factors) By Example 13.19 and Lemma 13.23, the Kronecker factor is always a compact extension of both the trivial one-point factor and the fixed factor. The same also holds for the rational spectrum factor. Moreover, if (X, μ, T) is ergodic, then the factor $A_{k+1}(X, \mu, T)$ (see Example 8.33) is a compact extension of $A_k(X, \mu, T)$ for every $k \in \mathbb{N}$. This shows that a skew rotation $(\mathbb{T}^2, \mathrm{m}^2, T_a)$ discussed in Example 8.34 is a compact extension of the torus rotation $(\mathbb{T}, \mathrm{m}, a)$ for irrational a, whereas the case of rational a can be treated analogously. We leave the details as Exercise 13.10.

The following property of compact extensions is the heart of the classical proof by Furstenberg and Katznelson [450], see also Furstenberg et al. [455, Thm. 9.1] and Tao [977, Thm. 2.13.11]. Our presentation combines[13] the expositions in Tao [977, Sec. 2.13] and Jamneshan et al. [577, Sec. 10.3].

Proposition 13.44 (Compact Extensions Preserve (MR)) *Let (X, μ, T) be a compact extension of an invertible measure-preserving system (Y, ν, S). If (Y, ν, S) satisfies (MR), then so does (X, μ, T).*

Proof Let $f \in \mathrm{L}^\infty(X)$ with $f > 0$ and $k \in \mathbb{N}$. We have to show (13.1).
Step 1: Reductions. We first show that we can assume without loss of generality that f is of the form $f = \mathbf{1}_A$ for some measurable $A \subset X$ and that $f \in A(X|Y)$. By assumption there is $c > 0$ and a set $A \subset X$ of positive measure with $f \geq c$ for μ-a.e. $x \in A$, i.e., $f \geq c\mathbf{1}_A$. Since the expression $\int_X f \cdot Tf \cdots T^{kn} f\,\mathrm{d}\mu$ is monotone in f, we can assume without loss of generality that $f = \mathbf{1}_A$. Further, by (13.46)

[12] See, e.g., Zorin-Kranich [1067], cf. Furstenberg and Katznelson [450, Thm. 2.1] and Furstenberg [446, Thm. 6.13].

[13] We thank Henrik Kreidler for his insight regarding a measurability issue.

and the definition of compact extensions there exists $B \subset Y$ with $\nu(B) > 1 - \mu(A)$ such that $g := \mathbf{1}_B f$ is conditionally almost periodic. (Recall that $J\mathbf{1}_B$ is again a characteristic function by Remark 8.15(c).) In particular we have $0 \neq g \leq f$. Thus, indeed, we can assume without loss of generality that $f = g \in A(X|Y)$ and, by making A smaller if necessary, that $f = \mathbf{1}_A$.
Step 2: Using conditional almost periodicity. Denote $\delta := \int_X f \,\mathrm{d}\mu = \mu(A) > 0$. Define the set

$$E := \{y \in Y : (E_Y f)(y) > \delta/2\}.$$

By $\delta = \int_X f \,\mathrm{d}\mu = \int_Y E_Y f \,\mathrm{d}\nu \leq \nu(E) + \delta/2$ we have $\nu(E) \geq \delta/2$.

Let $\varepsilon > 0$ satisfy $\varepsilon < \frac{\delta}{16k}$. By conditional almost periodicity of f we can find $d \in \mathbb{N}$, $f_1, \ldots, f_d \in \mathrm{L}^2(X|Y)$ and $c_{1,n}, \ldots, c_{d,n} \in \mathrm{L}^\infty(Y)$ with $\|c_{j,n}\|_{\mathrm{L}^\infty(Y)} \leq 1$ for each $j \in \{1, \ldots, d\}$ and $n \in \mathbb{N}_0$ such that

$$T^n f = c_{1,n} f_1 + \cdots + c_{d,n} f_d + r_n$$

for every $n \in \mathbb{N}_0$, where the error term $r_n \in \mathrm{L}^2(X|Y)$ satisfies $\|r_n\|_{\mathrm{L}^2(X|Y)} < \varepsilon$ ν-a.e. for every $n \in \mathbb{N}$. By doubling d if necessary we can assume without loss of generality that each f_j is real-valued, and that $c_{j,n}$ are real-valued, too. Moreover, by replacing $c_{j,n}(y)$ by the nearest integer multiple of ε/d (which can clearly be achieved in a measurable way) we can further assume that the values of each function $c_{j,n}$ are multiples of ε/d with a somewhat larger error term so that

$$\|T^n f - c_{1,n} f_1 - \ldots - c_{d,n} f_d\|_{\mathrm{L}^2(X|Y)}(y) < 2\varepsilon \tag{13.47}$$

holds for every $n \in \mathbb{N}_0$ and ν-a.e. $y \in Y$. Denote by Λ the set of all integer multiples of ε/d whose absolute value is less than or equal to one.
Step 3: Using van der Waerden's theorem. Take $y \in Y$ satisfying (13.47) for every $n \in \mathbb{N}$. Fix $n \in \mathbb{N}$ and let $K \in \mathbb{N}$ to be chosen directly next. For every $a \in \{0, \ldots, K-1\}$ the d-tuple

$$C_{an}(y) := (c_{1,an}(y), \cdots, c_{d,an}(y))$$

takes at most $|\Lambda|$-many values, leading to a finite partition of $\{0, \ldots, K-1\}$. By the finitary version of van der Waerden's theorem (see Theorem 13.7), there exists K, $a(y)$ and $r(y)$ (depending on k, d and ε) such that the arithmetic progression $a(y), a(y) + r(y), \ldots, a(y) + (k-1)r(y) \in \{0, \ldots, K-1\}$ of length k belongs to the same element of the partition, i.e.,

$$C_{a(y)n}(y) = C_{(a(y)+r(y))n}(y) = \cdots = C_{(a(y)+(k-1)r(y))n}(y).$$

From this, using the conditional triangle inequality (13.14) and (13.9) in Remark 13.12 we conclude that for each fixed $n \in \mathbb{N}$ the measurable sets

$$D_{n,a,r,\lambda_1,\ldots,\lambda_d} := \bigcap_{j=0}^{k-1} \left\{ y : \|T^{(a+jr)n} f - \lambda_1 f_1 - \ldots - \lambda_d f_d\|_{L^2(X|Y)}(y) < 2\varepsilon \right\}$$

cover Y up to a null set, where $(a, r) \in \mathbb{N}^2$ satisfy $a+(k-1)r \leq K$ and $\lambda_1, \ldots, \lambda_d \in \Lambda$.

Step 4: (MR)-property of the factor and partitioning sets of "good" y. Denote

$$B_n := E \cap S^{-n}E \cap \ldots \cap S^{-(K-1)n}E \subset Y$$

for $n \in \mathbb{N}$. Since (Y, ν, S) has (MR) and $\mathbf{1}_E > 0$, we have

$$\liminf_{N\to\infty} \frac{1}{N} \sum_{n=1}^{N} \nu(B_n) = \liminf_{N\to\infty} \frac{1}{N} \sum_{n=1}^{N} \int_Y \mathbf{1}_E \cdot S^n \mathbf{1}_E \cdots S^{(K-1)n} \mathbf{1}_E \, d\nu > 0.$$

By Exercise 4.9(b) there exists a $c > 0$ and a set $M \subset \mathbb{N}$ of positive lower density $\eta > 0$ such that

$$\nu(B_n) > c \text{ for all } n \in M.$$

For such n, since the sets $D_{n,a,r,\lambda_1,\ldots,\lambda_d}$ from Step 3 cover Y up to a null set and since there are at most K possible values of such a and r, there exist a subset $\tilde{B}_n \subset B_n$ with

$$\nu(\tilde{B}_n) \geq \frac{\nu(B_n)}{K^2|\Lambda|^d} \geq \frac{c}{K^2|\Lambda|^d} \tag{13.48}$$

such that $\tilde{B}_n \subset D_{n,a,r,\lambda_1,\ldots,\lambda_d}$ for some $a = a(n), r = r(n) \leq K$ and $\lambda_1 = \lambda_1(n), \ldots, \lambda_d = \lambda_d(n) \in \Lambda$. Moreover, by the definition of B_n for every $y \in \tilde{B}_n$ we have $S^{bn}y \in E$ for every $b \in \{0, \ldots, K-1\}$, i.e., by (8.8)

$$(E_Y T^{bn} f)(y) = (E_Y f)(S^{bn} y) > \delta/2. \tag{13.49}$$

Step 5: Final estimate. Take $n \in M$ and let $a, r, \lambda_1, \ldots, \lambda_d$ be as in Step 4 (depending on n). Then we have by $\tilde{B}_n \subset D_{n,a,r,\lambda_1,\ldots,\lambda_d}$ and the conditional triangle inequality (13.14)

$$\|T^{(a+jr)n} f - T^{an} f\|_{L^2(X|Y)}(y) < 4\varepsilon$$

for almost every $y \in \tilde{B}_n$ and every $j \in \{1, \dots, k-1\}$. By $\|f\|_{L^\infty(X)} = 1$ and using the conditional Cauchy–Schwarz inequality (13.13) and the conditional triangle inequality (13.14) together with (13.11) several times, one easily obtains

$$\|T^{an} f \cdot T^{(a+r)n} f \cdots T^{(a+(k-1)r)n} f - (T^{an} f)^k\|_{L^2(X|Y)}(y) < 4(k-1)\varepsilon.$$

The conditional Cauchy–Schwarz inequality again applied to the conditional scalar product of $\mathbf{1}$ with the difference under the norm sign on the left-hand side above and (13.49) imply

$$\begin{aligned} E_Y(T^{an} f \cdot T^{(a+r)n} f \cdots T^{(a+(k-1)r)n} f)(y) &\geq (E_Y(T^{an} f)^k)(y) - 4(k-1)\varepsilon \\ &= (E_Y T^{an} f)(y) - 4(k-1)\varepsilon > \delta/2 - 4(k-1)\varepsilon > \delta/4. \end{aligned}$$

Integrating in $y \in \tilde{B}_n$ and using shift invariance of μ, Proposition 8.52(b) and (13.48) leads to

$$\begin{aligned} \int_X f \cdot T^{rn} f \cdots T^{(k-1)rn} f \, \mathrm{d}\mu &= \int_Y E_Y(T^{an} f \cdot T^{(a+r)n} f \cdots T^{(a+(k-1)r)n} f) \, \mathrm{d}\nu \\ &\geq \nu(\tilde{B}_n)\delta/4 \geq \frac{c\delta}{4K^2|\Lambda|^d}, \end{aligned}$$

where $r = r(n) \in \{1, \dots, K\}$ depends on n. Let now $N \in \mathbb{N}$ and consider $n \in M \cap \{1, \dots, N\}$. Then $r(n)n \leq KN$ and for every number in $\{1, \dots, KN\}$ there are at most K possibilities (if at all) to be written in the form $r(n)n$. Thus we have

$$\frac{1}{NK} \sum_{n=1}^{NK} \int_X f \cdot T^n f \cdots T^{(k-1)n} f \, \mathrm{d}\mu \geq \frac{c\delta}{4K^3|\Lambda|^d} \cdot \frac{\eta}{2}$$

for all N large enough (recall that $\eta > 0$ is the lower density of M). The desired property (13.1) follows by Lemma 10.16. ■

Remark 13.45

(a) The above proof shows that, for a given k, a compact extension satisfies (MR_k) using the (MR_K)-property of the factor for a much larger K (coming from van der Waerden's theorem).

(b) Furstenberg et al. [455, Sec. 9] gave an alternative proof of Proposition 13.44 without using van der Waerden's theorem, see also Einsiedler and Ward [336, Sec. 7.9.2]. This argument proves the (MR_k)-property of a compact extension given the (MR_k)-property of the factor.

13.7 The Furstenberg–Zimmer Structure Theorem

To present the famous Furstenberg–Zimmer structure theorem and the final argument of the classical proof of multiple recurrence for ergodic systems we need the following concept.

Definition 13.46 Let $(Y_\beta, \nu_\beta, S_\beta)$ be a totally ordered family of factors of a measure-preserving system (X, μ, T) with corresponding Markov embeddings $J_\beta : \mathrm{L}^1(Y_\beta) \to \mathrm{L}^1(X)$ (recall that the ordering is by the inclusion between $\mathrm{ran}(J_\beta)$). A factor (Y, ν, S) of (X, μ, T) with corresponding Markov embedding $J : \mathrm{L}^1(Y) \to \mathrm{L}^1(X)$ is called the **inverse limit** of $(Y_\beta, \nu_\beta, S_\beta)$ if $J\mathrm{L}^1(Y) = \overline{\cup_\beta J_\beta \mathrm{L}^1(Y_\beta)}^{\|\cdot\|_1}$ in $\mathrm{L}^1(X)$.

Remark 13.47 The factor (Y, ν, S) of (X, μ, T) is an inverse limit of the chain of factors $(Y_\beta, \nu_\beta, S_\beta)$ (with corresponding embeddings J, J_β) if and only if $J\mathrm{L}^2(Y) = \overline{\cup_\beta J_\beta \mathrm{L}^2(Y_\beta)}^{\|\cdot\|_2}$, and this is also equivalent to the fact that the sub-σ-algebra corresponding to (Y, ν, S) is generated by the union of the (totally ordered) sub-σ-algebras corresponding to $(Y_\beta, \nu_\beta, S_\beta)$. The proof is left as Exercise 13.11.

Remark 13.48 It is easy to see that the inverse limit of (Markov) invertible factors is itself (Markov) invertible, see Exercise 13.12.

We will further need the following dichotomy result being a direct corollary of Theorem 13.24 and Lemma 13.38. For the original proof of this dichotomy see Furstenberg and Katznelson [450, Prop. 2.2], Furstenberg et al. [455, Thm. 10.1] and for a proof for not necessarily separable systems see Tao [977, Thm. 2.14.9].

Proposition 13.49 (Dichotomy Between Compact and Weakly Mixing Extensions) *Let (X, μ, T) be an ergodic measure-preserving system with a (Markov) invertible factor (Y, ν, S). Then exactly one of the following assertions hold (Fig. 13.2).*

(a) *(X, μ, T) is a weakly mixing extension of (Y, ν, S).*
(b) *There exists an intermediate (Markov) invertible factor (Z, ν_Z, S_Z) of (X, μ, T) strictly larger than (Y, ν, S) which is a compact extension of (Y, ν, S).*

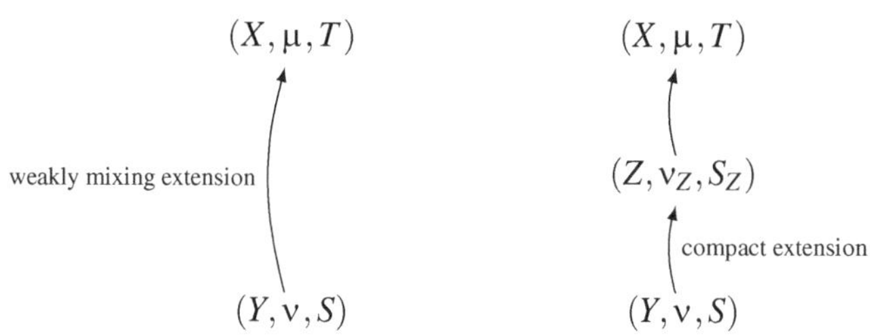

Fig. 13.2 The alternative: compact vs. weakly mixing extensions

Proof By Remark 13.34, (a) holds if and only if the orthogonal decomposition $\mathrm{L}^2(X) = \mathrm{L}^2(Y) \oplus W(X|Y)$ holds. Assume that (a) does not hold. Then by Theorem 13.24 $\overline{A(X|Y)} = \overline{E(X|Y)}$ must be strictly larger than $\mathrm{L}^2(Y)$ and we can take the corresponding factor (see Lemma 13.38) as (Z, ν_Z, S_Z). ∎

The following structure theorem is due to Furstenberg [445], [446, Ch. 6] and Zimmer [1066] for regular systems and to Tao[14] [977, Thm. 2.15.1].

Theorem 13.50 (Furstenberg–Zimmer Structure Theorem) *Let (X, μ, T) be an ergodic measure-preserving system. Then there exists an ordinal α and a family of (Markov) invertible factors $(Y_\beta, \nu_\beta, S_\beta)_{\beta \le \alpha}$ such that the following assertions hold.*

(a) *Y_0 is a point.*
(b) *$(Y_{\beta+1}, \nu_{\beta+1}, S_{\beta+1})$ is a compact extension of $(Y_\beta, \nu_\beta, S_\beta)$ for every successor ordinal $\beta + 1 \le \alpha$.*
(c) *$(Y_\beta, \nu_\beta, S_\beta)$ is an inverse limit of the family $(Y_\gamma, \nu_\gamma, S_\gamma)_{\gamma<\beta}$ for every limit ordinal $\beta \le \alpha$.*
(d) *(X, μ, T) is a weakly mixing extension of $(Y_\alpha, \nu_\alpha, S_\alpha)$.*

Moreover, if (X, μ, T) is regular, then α can be taken to be a countable ordinal.

$$Y_0 \xleftarrow{\text{cpt}} \cdots \xleftarrow{\text{cpt}} Y_n \xleftarrow{\text{cpt}} \cdots \qquad Y_\omega \xleftarrow{\text{cpt}} Y_{\omega+1} \xleftarrow{\text{cpt}} \cdots \qquad Y_\beta \xleftarrow{\text{cpt}} Y_{\beta+1} \xleftarrow{\text{cpt}} \cdots \qquad Y_\alpha \xleftarrow{\text{wm}} X$$

The family of factors $(Y_\beta, \nu_\beta, S_\beta)_{\beta \le \alpha}$ is called the **Furstenberg–Zimmer tower**, α is called its **height** and the factor $(Y_\alpha, \nu_\alpha, S_\alpha)$ is called the **maximal distal factor** of (X, μ, T). Beleznay and Foreman [88] showed that every countable ordinal can appear as the (smallest) height of a Furstenberg–Zimmer tower, see also Avigad and Towsner [69].

Proof If (X, μ, T) is weakly mixing, then with $\alpha := 0$ we are done. Otherwise, take the Kronecker factor (Y_1, ν_1, S_1) which is (Markov) invertible by Remark 6.19. If for some ordinal β the (Markov) invertible factors $(Y_\gamma, \nu_\gamma, S_\gamma)$ for each $\gamma < \beta$ (unequal to (X, μ, T)) have been found, then we proceed as follows. If β is a limit ordinal, then we set $(Y_\beta, \nu_\beta, S_\beta)$ as the inverse limit of $(Y_\gamma, \nu_\gamma, S_\gamma)$ for $\gamma < \beta$, see Remark 13.48. Now, if (X, μ, T) is a weakly mixing extension of $(Y_\beta, \nu_\beta, S_\beta)$, then we are done by the choice $\alpha := \beta$. If $\beta = \gamma + 1$ is a successor and (X, μ, T) is a weakly mixing extension of $(Y_\gamma, \nu_\gamma, S_\gamma)$, then we set $\alpha := \gamma$ and we are done. Otherwise, take a compact extension $(Y_\beta, \nu_\beta, S_\beta)$ of (X, μ, T) relative to $(Y_\gamma, \nu_\gamma, S_\gamma)$ (possible by Proposition 13.49). If (X, μ, T) is a weakly mixing extension of $(Y_\beta, \nu_\beta, S_\beta)$, then we are done. Otherwise, we continue the procedure, which terminates at some ordinal. Moreover, if (X, μ, T) is regular then this ordinal is countable since the construction yields a well-ordered family of closed subspaces $\mathrm{L}^2(Y_\gamma, \nu_\gamma)$ in the separable Hilbert space $\mathrm{L}^2(X, \mu)$. ∎

[14] As mentioned above, Tao's argument does not need the separability assumption and is valid for the general case.

Thus the last ingredient needed for the proof of multiple recurrence is the preservation of (MR) by chains of factors. We follow the proof in Tao [977, §2.15] to establish this.

Proposition 13.51 (Inverse Limits Preserve (MR)) *Let $k \in \mathbb{N}$ and let a measure-preserving system (Y, ν, S) be the inverse limit of an increasing family of (Markov) invertible factors $(Y_\beta, \nu_\beta, S_\beta)$. If every $(Y_\beta, \nu_\beta, S_\beta)$ satisfies (MR$_k$), then so does (Y, ν, S).*

Proof Take $f \in \mathrm{L}^\infty(Y)$ with $f > 0$, $\|f\|_\infty \leq 1$ and $c := \int_Y f \,\mathrm{d}\nu > 0$. We need to prove that $\int_Y f \cdot S^n f \cdots S^{kn} f \,\mathrm{d}\nu \geq \delta > 0$ for some $\delta > 0$ and for many $n \in \mathbb{N}$ (having positive lower density). The rough idea is the following. If f is bounded away from 0 on a large set, say, if for some $m \in \mathbb{N}$ we have $\nu([f > \frac{1}{m}]) > 1 - \frac{1}{k+1} + \frac{1}{m}$, then one can find a $\delta > 0$ such that the required estimate holds for every $n \in \mathbb{N}$. We will obtain a relativized version of the previous lower bound for the measure by projecting f to a suitable factor $(Y_\beta, \nu_\beta, S_\beta)$, and then make use of the property (MR$_k$) on that factor.
Step 1: Choice of β. Let us denote by $J_\beta : \mathrm{L}^1(Y_\beta) \to \mathrm{L}^1(Y)$, $E_\beta : \mathrm{L}^1(Y) \to \mathrm{L}^1(Y_\beta)$ and $P_\beta : \mathrm{L}^1(Y) \to \mathrm{L}^1(Y)$ the Markov embeddings, conditional expectations and Markov projections corresponding to the factors $(Y_\beta, \nu_\beta, S_\beta)$. Let $0 < \varepsilon < \frac{c^2}{4(k+1)}$, and let β be such that $\|f - P_\beta f\|_{\mathrm{L}^1(Y)} < \varepsilon$. Set $A := [E_\beta f > \frac{c}{2}] \subset Y_\beta$, $B := [f > 0] \subset Y$. Since $\|E_\beta f\|_\infty \leq 1$ we have

$$c = \int_{Y_\beta} E_\beta f \,\mathrm{d}\nu = \int_{Y_\beta \setminus A} E_\beta f \,\mathrm{d}\nu_\beta + \int_A E_\beta f \,\mathrm{d}\nu_\beta \leq \frac{c}{2} + \nu_\beta(A),$$

so we conclude $\nu_\beta(A) \geq \frac{c}{2}$. We have $\frac{c}{2}\mathbf{1}_A \leq E_\beta f$, so $\frac{c}{2}\mathbf{1}_{Y\setminus B} J_\beta \mathbf{1}_A \leq |f - P_\beta f|$, and we arrive at

$$\frac{c}{2}\int_{Y_\beta} E_\beta(\mathbf{1}_{Y\setminus B} J_\beta \mathbf{1}_A) \,\mathrm{d}\nu_\beta \leq \int_{Y_\beta} E_\beta |f - P_\beta f| \,\mathrm{d}\nu_\beta = \|f - P_\beta f\|_{\mathrm{L}^1(Y)} < \varepsilon.$$

This implies

$$\nu_\beta\Big(\Big[E_\beta(\mathbf{1}_{Y\setminus B} J_\beta \mathbf{1}_A) \geq \tfrac{1}{k+1}\Big]\Big) \leq \frac{2\varepsilon(k+1)}{c},$$

and then it follows by Proposition 8.52(c) that

$$\begin{aligned} 1 - \frac{2\varepsilon(k+1)}{c} &\leq \nu_\beta\Big(\Big[E_\beta((\mathbf{1} - \mathbf{1}_B) J_\beta \mathbf{1}_A) < \tfrac{1}{k+1}\Big]\Big) \\ &= \nu_\beta\Big(\Big[\mathbf{1}_A \cdot (\mathbf{1} - E_\beta(\mathbf{1}_B)) < \tfrac{1}{k+1}\Big]\Big) \end{aligned}$$

$$\le \nu_\beta\Big(\Big[\mathbf{1} - E_\beta(\mathbf{1}_B) < \tfrac{1}{k+1}\Big]\Big) + \nu_\beta(Y_\beta \setminus A)$$
$$\le \nu_\beta\Big(\Big[\mathbf{1} - E_\beta(\mathbf{1}_B) < \tfrac{1}{k+1}\Big]\Big) + 1 - \frac{c}{2}.$$

Finally, we conclude

$$0 < \frac{c}{2} - \frac{2\varepsilon(k+1)}{c} \le \nu_\beta\Big(\Big[E_\beta(\mathbf{1}_B) > 1 - \tfrac{1}{k+1}\Big]\Big).$$

Step 2: Using (MR_k) *for* $(Y_\beta, \mu_\beta, S_\beta)$. For $m \in \mathbb{N}$ define $B_m := [f > \frac{1}{m}] \subset Y$ and $C_m := [E_\beta(\mathbf{1}_{B_m}) > 1 - \frac{1}{k+1} + \frac{1}{m}] \subset Y_\beta$. Since $E_\beta(\mathbf{1}_{B_m}) - \frac{1}{m}\mathbf{1} \to E_\beta(\mathbf{1}_B)$ monotonically as $m \to \infty$ (see, e.g., the discussion in Remark 8.56(a)), and since, as proved above, $\nu_\beta\big(\big[E_\beta(\mathbf{1}_B) > 1 - \frac{1}{k+1}\big]\big) > 0$, there is an $m \in \mathbb{N}$ such that $\nu_\beta(C_m) > 0$, i.e., $\mathbf{1}_{C_m} > 0$ in $\mathrm{L}^1(Y_\beta)$. Since $(Y_\beta, \nu_\beta, S_\beta)$ satisfies (MR_k) we have

$$d := \liminf_{N\to\infty} \frac{1}{N} \sum_{n=1}^{N} \int_{Y_\beta} \mathbf{1}_{C_m} S_\beta^n \mathbf{1}_{C_m} \cdots S_\beta^{kn} \mathbf{1}_{C_m} \,\mathrm{d}\nu_\beta > 0. \tag{13.50}$$

Since $f \ge \frac{1}{m}\mathbf{1}_{B_m}$, we have

$$f \cdot S^n f \cdots S^{kn} f \ge \frac{1}{m^{k+1}} \mathbf{1}_{B_m} \mathbf{1}_{S^{-n}B_m} \cdots \mathbf{1}_{S^{-kn}B_m}$$
$$= \frac{1}{m^{k+1}} (\mathbf{1} - \mathbf{1}_{Y\setminus B_m})(\mathbf{1} - \mathbf{1}_{S^{-n}(Y\setminus B_m)}) \cdots (\mathbf{1} - \mathbf{1}_{S^{-kn}(Y\setminus B_m)})$$
$$\ge \frac{1}{m^{k+1}} \Big(\mathbf{1} - \sum_{j=0}^{k} \mathbf{1}_{S^{-jn}(Y\setminus B_m)}\Big).$$

Take conditional expectation to arrive by (8.8) at

$$E_\beta(f \cdot S^n f \cdots S^{kn} f) \ge \frac{1}{m^{k+1}} \Big(\mathbf{1} - \sum_{j=0}^{k} E_\beta(S^{jn} \mathbf{1}_{Y\setminus B_m})\Big)$$
$$= \frac{1}{m^{k+1}} \Big(-k\mathbf{1} + \sum_{j=0}^{k} S_\beta^{jn} E_\beta(\mathbf{1}_{B_m})\Big).$$

If we multiply this by $g_n := \mathbf{1}_{C_m} S_\beta^n \mathbf{1}_{C_m} \cdots S_\beta^{kn} \mathbf{1}_{C_m}$, then we obtain by the definition of C_m and g_n

$$g_n E_\beta(f \cdot S^n f \cdots S^{kn} f) \geq \frac{1}{m^{k+1}}\Big(- k g_n + \sum_{j=0}^{k} g_n S_\beta^{jn} E_\beta(\mathbf{1}_{B_m})\Big)$$

$$\geq \frac{1}{m^{k+1}}\Big(- k g_n + \sum_{j=0}^{k} (1 - \tfrac{1}{k+1} + \tfrac{1}{m}) g_n\Big) = \frac{k+1}{m^{k+2}} g_n.$$

Integrating and averaging yield

$$\liminf_{N\to\infty} \frac{1}{N} \sum_{n=1}^{N} \int_Y f \cdot S^n f \cdots S^{kn} f \,\mathrm{d}\nu$$

$$= \liminf_{N\to\infty} \frac{1}{N} \sum_{n=1}^{N} \int_{Y_\beta} E_\beta(f \cdot S^n f \cdots S^{kn} f) \,\mathrm{d}\nu_\beta$$

$$\geq \liminf_{N\to\infty} \frac{1}{N} \sum_{n=1}^{N} \int_{Y_\beta} g_n E_\beta(f \cdot S^n f \cdots S^{kn} f) \,\mathrm{d}\nu_\beta$$

$$\geq \liminf_{N\to\infty} \frac{1}{N} \sum_{n=1}^{N} \int_{Y_\beta} \frac{k+1}{m^{k+2}} g_n \,\mathrm{d}\nu_\beta \overset{(13.50)}{=} \frac{k+1}{m^{k+2}} d > 0,$$

proving the (MR_k) property for (Y, ν, S). ■

We are now ready to prove Furstenberg's multiple recurrence theorem.

First Proof of Theorem 13.5 Let α and $(Y_\beta, \nu_\beta, S_\beta)_{\beta\leq\alpha}$ be furnished by the Furstenberg Zimmer Structure Theorem 13.50. The trivial one-point factor Y_0 clearly satisfies (MR). By Propositions 13.44, 13.51 and transfinite induction, the factor $(Y_\alpha, \nu_\alpha, S_\alpha)$ satisfies (MR) and hence so does (X, μ, T) by Proposition 13.37. ■

13.8 Multiple Recurrence via Distal Factors of Finite Order

In this section we briefly discuss another proof of multiple recurrence being a mixture of Furstenberg's original proof [445] (with an alternative proof of one step due to Frantzikinakis [404]) and the Furstenberg–Katznelson proof discussed above.

As already Furstenberg [445] observed, for (MR_k) one in fact needs a tower of only $k - 1$ factors which are compact extensions of each other.

Definition 13.52 Let (X, μ, T) be an ergodic measure-preserving system. We define the sequence of invertible factors inductively as follows. Start with the subspace $D_0 = \mathbb{C}\mathbf{1}$ induced by the trivial one-point factor $\mathcal{D}_0$. For every $k \in \mathbb{N}$ denote by $\mathcal{D}_k$ the (Markov) invertible factor inducing the subspace $D_k := \overline{A(X|\mathcal{D}_{k-1})} = \overline{E(X|\mathcal{D}_{k-1})}$ (see Lemma 13.38 and (13.26)). We call $\mathcal{D}_k$ the **k-step distal factor** of (X, μ, T).

Remark 13.53

(a) Since T acts on every finite-dimensional T-invariant subspace as a unitary matrix which is diagonalizable, we see that $\mathcal{D}_1$ is the Kronecker factor of (X, μ, T).
(b) Continuing the series $\mathcal{D}_0, \mathcal{D}_1, \ldots$ transfinitely gives the maximal distal factor from Theorem 13.50.

We start with the following deep result providing a description of conditional eigenfunctions of the conditional product for regular systems due to Furstenberg [445, Thm. 7.1]. We will denote the conditional product system $(X \times X, \mu \times_Y \mu, T \times T)$ by $X \times_Y X$ (see Definition 13.27 and Remark 13.30(b)).

Theorem 13.54 (Conditional Eigenfunctions of Conditional Product) *Let (X, μ, T) be an invertible, regular measure-preserving system with ergodic invertible factor (Y, ν, S). Then*

$$\overline{E(X \times_Y X|Y)} = \overline{\mathrm{lin}}\{f_1 \otimes f_2 : f_1, f_2 \in E(X|Y) \cap \mathrm{L}^\infty(X)\}.$$

For the proof (whose ideas inspired the proof of Theorem 13.24) we refer[15] to Furstenberg [445, Thm. 7.1] and Glasner [468, Ch. 9]. Note that we will not need the full strength of the non-trivial inclusion "$\subset$" and only need that the *fixed functions* of the conditional product system are included in the closure of all linear combinations of functions of the form $f_1 \otimes f_2$, where $f_1 \in E(X|Y)$ and $f_2 \in E(X|Y)$ are bounded. (Note however that this case is essentially as hard to prove as the general case of conditional eigenfunctions of the conditional product system.)

We now deduce from Theorem 13.54 Furstenberg's result that $\mathcal{D}_{k-1}$ is characteristic for averages (13.3) following the argument by Frantzikinakis [404]. An alternative proof for general ergodic systems will be presented in Chap. 15, see Remark 15.30.

Theorem 13.55 (Distal Factor of Step $k-1$ is Characteristic for k-Term Multiple Averages) *Let (X, μ, T) be an ergodic, invertible, regular measure-preserving system. Then its $(k-1)$-step distal factor $\mathcal{D}_{k-1}$ is characteristic for k-term multiple ergodic averages* (13.3), *i.e., satisfies* (13.41).

[15] Although Furstenberg's definition of generalized eigenfunctions is slightly different than the definition of conditional eigenfunctions we use, his argument also works for our definition.

Proof Let (X, μ, T) be an ergodic, invertible, regular measure-preserving system. The proof goes by induction, where the case $k = 1$ corresponds to the mean ergodic theorem. Let $k \geq 2$ and assume that the assertion holds for $k - 1$.

Let $f_1, \dots, f_k \in \mathrm{L}^\infty(X)$ and assume without loss of generality that $\|f_j\|_\infty \leq 1$ holds for every $j \in \{1, \dots, k\}$. Moreover, by (13.43) we can assume that $E_{\mathcal{D}_{k-1}} f_l = 0$ a.e. for some $l \in \{1, \dots, k\}$. Define for $n \in \mathbb{N}$

$$u_n := T^n f_1 \cdot T^{2n} f_2 \cdots T^{kn} f_k$$

in order to use the van der Corput trick and denote $Y := \mathcal{D}_{k-2}$. As in the proof of Proposition 13.36 using the induction hypothesis one has

$$(u_n | u_{n+h}) = \int_X \big(f_1 T^h \overline{f_1}\big) \cdot T^n \big(f_2 T^{2h} \overline{f_2}\big) \cdots T^{(k-1)n} \big(f_k T^{kh} \overline{f_k}\big) \, \mathrm{d}\mu$$

and

$$\gamma_h := \limsup_{N \to \infty} \left| \frac{1}{N} \sum_{n=0}^{N-1} (u_n | u_{n+h}) \right| \leq \|\langle T^{lh} f_l, f_l \rangle_{\mathrm{L}^2(X|Y)}\|_{\mathrm{L}^1(Y)}.$$

Thus we obtain

$$\frac{1}{H} \sum_{h=1}^{H} \gamma_h \leq \frac{1}{H} \sum_{h=1}^{H} \|\langle T^{lh} f_l, f_l \rangle_{\mathrm{L}^2(X|Y)}\|_{\mathrm{L}^1(Y)}. \tag{13.51}$$

We now rewrite the expression on the right-hand side as follows. Consider the fibered product system $(X \times X, \mu \times_Y \mu, T \times T)$ (again shortened as $X \times_Y X$) and the function $F := f_l \otimes \overline{f_l}$. For $n \in \mathbb{N}$

$$\begin{aligned} \|\langle T^n f_l, f_l \rangle_{\mathrm{L}^2(X|Y)}\|^2_{\mathrm{L}^2(Y)} &- \int_Y |E_Y(T^n f_l \cdot \overline{f_l})|^2 \, \mathrm{d}\nu \\ &= \int_{X \times X} (T^n f_l \cdot \overline{f_l}) \otimes (T^n \overline{f_l} \cdot f_l) \, \mathrm{d}(\mu \times_Y \mu) \\ &= \int_{X \times X} (T \times T)^n F \cdot \overline{F} \, \mathrm{d}(\mu \times_Y \mu). \end{aligned}$$

By the mean ergodic theorem the Cesàro limit of the right-hand side equals

$$\|P_{\mathrm{fix}} F\|^2_{\mathrm{L}^2(X \times_Y X)} = \|P_{\mathrm{fix}}(f_l \otimes \overline{f_l})\|^2_{\mathrm{L}^2(X \times_Y X)}$$

which we now show to be zero, where P_{fix} denotes the orthogonal projection onto the fixed subspace of the fibered product system $X \times_Y X$. Indeed, by Theorem 13.54 (recall that $Y = \mathcal{D}_{k-2}$ is invertible by Lemma 13.38) the fixed functions of $X \times_Y X$ are limits of linear combinations of functions of the form $g_1 \otimes g_2$ for g_1, g_2 being bounded conditional eigenfunctions of (X, μ, T) w.r.t. Y. Take two such functions g_1, g_2. They belong to D_{k-1} and are hence (images of) elements of $\mathrm{L}^\infty(\mathcal{D}_{k-1})$ by the inductive definition of the finite step distal factors. It suffices to show that $f_l \otimes \overline{f_l}$ is orthogonal to $g_1 \otimes g_2$ in $\mathrm{L}^2(X \times_Y X)$. By the definition of the conditional product

$$\int_{X\times X} (f_l \otimes \overline{f_l})(\overline{g_1 \otimes g_2})\,\mathrm{d}(\mu \times_Y \mu) = \int_Y E_Y(f_l \cdot \overline{g_1}) E_Y(\overline{f_l} \cdot \overline{g_2})\,\mathrm{d}\nu.$$

Observe that by the assumption on f_l and Proposition 8.52(c)

$$E_{\mathcal{D}_{k-1}}(f_l \cdot \overline{g_1}) = \overline{g_1} E_{\mathcal{D}_{k-1}}(f_l) = 0 \text{ a.e.}$$

implying $E_Y(f_l \cdot \overline{g_1}) = 0$ a.e. since Y is a factor of $\mathcal{D}_{k-1}$. This shows

$$\lim_{N\to\infty} \frac{1}{N} \sum_{n=1}^{N} \|\langle T^n f_l, f_l\rangle_{\mathrm{L}^2(X|Y)}\|^2_{\mathrm{L}^2(Y)} = 0$$

implying by (13.51), the Cauchy–Schwarz inequality and Exercise 4.7(b)

$$\lim_{H\to\infty} \frac{1}{H} \sum_{h=1}^{H} \gamma_h = 0.$$

Thus the van der Corput Lemma 7.31 shows $\lim_{N\to\infty} \frac{1}{N} \sum_{n=1}^{N} u_n = 0$ in $\mathrm{L}^2(X)$ as desired. ■

Remark 13.56 As the above proof indicates, the factor $\mathcal{D}_{k-1}$ is not the smallest characteristic factor for k-term multiple recurrence, see Chap. 15 for a characteristic factor which is smaller in general.

We are now ready to sketch an alternative proof of multiple recurrence in the spirit of the original argument by Furstenberg [445] combined with one step from the Furstenberg–Katznelson proof discussed earlier in this chapter.

Second Proof of Theorem 13.5 for Invertible Systems[16] *Step 1: Reduction to regular systems.* We first show that we can assume without loss of generality that (X, μ, T) is regular. Indeed, let $f \in \mathrm{L}^\infty(X, \mu)$ satisfy $f > 0$. Let V be the smallest T-invariant unital, conjugation invariant subalgebra of $\mathrm{L}^\infty(X, \mu)$ containing f, which is clearly separable and gives rise to a factor (Y, ν, S) with $\mathrm{L}^1(Y, \nu)$ being separable. Since this factor is Markov isomorphic to a regular measure-preserving system (see Remark 8.50(b)) and the integrals

$$\int_X f \cdot T^n f \cdots T^{kn} f \,\mathrm{d}\mu$$

[16] Recall that by (i)⇔(i') in the full version of the correspondence principle discussed in Sect. 13.1 it is enough to show Theorem 13.5 for invertible systems. For a direct easy reduction of the general case to invertible systems see, e.g., Einsiedler and Ward [336, p. 177].

do not change by changing to this factor or to an isomorphic system, we can indeed assume without loss of generality that (X, μ, T) is regular.
Step 2: Final argument. Denote by $\mathcal{D}_j$ its j-step distal factor. By Theorem 13.55 and Remark 8.54 we can assume without loss of generality that our system (X, μ, T) coincides with $\mathcal{D}_{k-1}$. Thus we have a finite chain of factors

$$X = \mathcal{D}_{k-1} \to \mathcal{D}_{k-2} \to \cdots \to \mathcal{D}_0$$

where $\mathcal{D}_0$ is a one-point factor (corresponding to fix $T = \mathbb{C}\mathbf{1}$) and each extension in this chain is compact. Since $\mathcal{D}_0$ clearly satisfies (MR) and every system in this chain is invertible by Lemma 13.38, (X, μ, T) satisfies (MR) by Proposition 13.44. ∎

Remark 13.57 Furstenberg [445] showed multiple recurrence for the characteristic factor $\mathcal{D}_{k-1}$ using a different argument based on analysis of the properties of diagonal measures.

Remark 13.58 Let (X, μ, T) be an ergodic, invertible, regular measure-preserving system and $k \in \mathbb{N}$ with $k \geq 2$. Considering $Y := \mathcal{D}_{k-2}$ (which is invertible by Lemma 13.38) in the conditional Jacobs–de Leeuw–Glicksberg decomposition given in Theorem 13.24 and using $D_{k-1} = \overline{E(X|\mathcal{D}_{k-2})}$ we obtain the orthogonal decomposition

$$\mathrm{L}^2(X) = D_{k-1} \oplus W(X|\mathcal{D}_{k-2}) \tag{13.52}$$

which one could call a Jacobs–de Leeuw–Glicksberg decomposition of order $k-1$. For $k = 2$ the decomposition (13.52) coincides with the classical Jacobs–de Leeuw–Glicksberg decomposition (13.6) by Remark 13.53(a). By Theorem 13.55, the averages (13.3) converge to zero in L^2 whenever at least one of the functions belongs to the second part of the sum in (13.52). Thus the first part in it is characteristic for both convergence and recurrence of k-term ergodic averages and is structured enough to show (MR_k) (and even (MR)) by applying Proposition 13.44 several times. However, it is still too large to show k-term multiple convergence on it. The right decomposition for both convergence and recurrence of k-term ergodic averages will be discussed in Chap. 15 below, see (15.14) for $l = k-1$ there.

13.9 Further Developments

Furstenberg's multiple recurrence theorem initiated enormous progress in ergodic theory and its applications and connections to combinatorial number theory and various other areas. We briefly mention here some of further developments. For multiple convergence and further results see Chap. 15 and Sect. 16.6.

Szemerédi's Theorem

Szemerédi's Theorem enriches and connects several areas of mathematics due to the many proofs it found and still finds.

More than twenty years after the original, highly complicated combinatorial proof of Szemerédi's theorem and the more accessible and flexible (but completely non-quantitative) ergodic theoretic proof(s) by Furstenberg [445] and Furstenberg and Katznelson [450] further proofs appeared using different areas of mathematics. Using some ideas from the ergodic theoretic proof (as presented in Furstenberg et al. [455]), Gowers [481] presented in 2001 a Fourier analytic proof by introducing higher order Fourier analysis and Gowers uniformity norms which highly influenced further developments in combinatorics and ergodic theory, see Chap. 15. Another proof using graph theory, in particular a hypergraph generalization of Szemerédi's regularity lemma from Szemerédi's proof, appeared in 2006–2007 by Gowers [482] and independently by Nagle et al. [816] and Rödl and Skokan [875, 876], see also Tao [974]. For a proof using model theory see Towsner [1006].

Using Gowers uniformity norms, Tao [973] presented in 2006 a quantitative finitary version of the ergodic proof by Furstenberg and Furstenberg, Katznelson. In 2010 Green and Tao [487] gave a proof of Szemerédi's theorem using the inverse theorem for Gowers norms which they developed with Ziegler in [492] for finding solutions of linear equations in primes [486]. There are several ergodic theoretic proofs using the Host–Kra structure theory discussed in Chap. 15, see, e.g., Bergelson et al. [135] (the argument was implicit in Bergelson et al. [123]) and the proofs discussed in Sect. 15.8. For a recent simplification of the original proof by Szemerédi see Tao [985].

This list is far from being complete. As Tao mentioned in [978], there were at least sixteen genuinely different proofs of Szemerédi's theorem as of 2010, and this number is constantly growing. Most of these proofs (if not all) share similar philosophy, namely crystallization of structure by eliminating randomness, see Tao [973, 975] for a discussion and for calling this philosophy a "Rosetta stone" connecting various fields.

Furstenberg–Zimmer Structure Theorem

Furstenberg and Katznelson [450] extended the Furstenberg–Zimmer Structure Theorem 13.50 to $\mathbb{Z}^d$-actions on separable probability spaces, see also Furstenberg [446, Ch. 6]. Combining Furstenberg's and Zimmer's approaches, Glasner [468, Ch. 10] extended it to actions of countable discrete groups on separable spaces, whereas Kerr and Li [640, Ch. 3] dropped the separability assumption using Tao's abstract framework. For two recent different approaches to the Furstenberg–Zimmer Structure Theorem for general groups on general probability spaces see Jamneshan [576] and Edeko et al. [331].

Multiple Recurrence for Commuting Transformations

Furstenberg and Katznelson [450] showed multiple recurrence for commuting (invertible) transformations, i.e., that for every $k \in \mathbb{N}$, commuting measure-preserving transformations $T_1, \ldots, T_k$ on a probability space (X, μ) and $f \in \mathrm{L}^\infty(X, \mu)$ with $f > 0$

$$\liminf_{N\to\infty} \frac{1}{N} \sum_{n=1}^{N} \int_X T_1^n f \cdots T_k^n f \,\mathrm{d}\mu > 0.$$

This lead to the following generalization of Theorem 13.2, called **multidimensional Szemerédi theorem**, see Exercise 13.13: Let $d \in \mathbb{N}$ and F be a finite subset of $\mathbb{N}_0^d$. Then every set $E \subset \mathbb{N}^d$ with positive upper density[17] contains a subset of the form $a + nF$ for some $a \in \mathbb{N}^d$ and $n \in \mathbb{N}$. (One recovers Szemerédi's theorem by taking $d = 1$ and $F := \{0, \ldots, k\}$.) Note that the first combinatorial proofs of this result appeared only in 2006–2007, see Gowers [482], Nagle et al. [816], Rödl and Skokan [875, 876] and Tao [974].

Polynomial Multiple Recurrence

Developing the so-called PET[18]-induction based on the van der Corput lemma, Bergelson [103] showed that every invertible weakly mixing system (X, μ, T) is polynomially weakly mixing of all orders, i.e., for every $k \in \mathbb{N}$ and every polynomials $P_1, \ldots, P_k : \mathbb{Z} \to \mathbb{Z}$ which are essentially distinct, the polynomial multiple averages

$$\frac{1}{N} \sum_{n=1}^{N} T_1^{P_1(n)} f_1 \cdots T_k^{P_k(n)} f_k$$

converge in $\mathrm{L}^2(X)$ to $\int_X f_1 \,\mathrm{d}\mu \cdots \int_X f_k \,\mathrm{d}\mu$ for all $f_1, \ldots, f_k \in \mathrm{L}^\infty(X)$. This in particular shows multiple recurrence for such polynomial averages for invertible weakly mixing systems.

Bergelson and Leibman [129] extended this to the following polynomial version of multiple recurrence for commuting transformations: For every $k \in \mathbb{N}$, every commuting invertible measure-preserving transformations $T_1, \ldots, T_k$ on a

[17] The upper density of $E \subset \mathbb{N}^d$ is defined by $\overline{\mathrm{d}}(E) := \limsup_{N\to\infty} \frac{|E\cap\{1,\ldots,N\}^d|}{N^d}$. The result was actually proved for upper Banach density defined as $\overline{\mathrm{Bd}}(E) := \limsup_{N_1-M_1\to\infty,\ldots,N_d-M_d\to\infty} \frac{|E\cap(\{M_1,\ldots,N_1\}\times\cdots\times\{M_d,\ldots,N_d\})|}{(N_1-M_1+1)\cdots(N_d-M_d+1)}$.

[18] PET is an abbreviation for both "polynomial ergodic theorem" and "polynomial exhaustion technique". Here the second meaning is used.

probability space (X, μ), every polynomials $P_1, \ldots, P_k : \mathbb{Z} \to \mathbb{Z}$ satisfying $P_1(0) = 0, \ldots, P_k(0) = 0$ and every $f \in L^\infty(X, \mu)$ with $f > 0$

$$\liminf_{N\to\infty} \frac{1}{N} \sum_{n=1}^{N} \int_X T_1^{P_1(n)} f \cdots T_k^{P_k(n)} f \, d\mu > 0,$$

see also Bergelson and McCutcheon [138] and Bergelson et al. [127, Appendix A]. By the multidimensional correspondence principle, this implies a multidimensional version of the following **polynomial Szemerédi theorem**: Every set of natural numbers with positive upper density contains polynomial progressions of the form $a + P_1(n), \ldots, a + P_k(n)$ for every $k \in \mathbb{N}$ and every polynomials $P_1, \ldots, P_k : \mathbb{Z} \to \mathbb{Z}$ satisfying $P_1(0) = 0, \ldots, P_k(0) = 0$, see Exercise 13.14(a) for the precise statement. Note that a combinatorial proof of the polynomial Szemerédi theorem is not yet found, see Prendiville [857] and Peluse [847] for some partial quantitative results.

Although the assumption $P_1(0) = 0, \ldots, P_k(0) = 0$ on the polynomials cannot be dropped, see Exercise 13.14(b), it is not the optimal one. Bergelson et al. [135] showed that polynomials $P_1, \ldots, P_k$ are good for the polynomial Szemerédi theorem if and only if they are **jointly intersective**, i.e., such that for every $m \in \mathbb{N}$ there exists $n \in \mathbb{N}$ such that $P_1(n), \ldots, P_k(n) \in m\mathbb{Z}$.

Leibman [714] proved a general multiple recurrence result including the case when $T_1, \ldots, T_k$ generate a nilpotent group and the powers are polynomial, i.e., of the form $P_1(n), \ldots, P_k(n)$ with polynomials $P_j : \mathbb{Z} \to \mathbb{Z}$ satisfying $P_j(0) = 0$, $j \in \{1, \ldots, k\}$.

For a refinement of the polynomial Szemerédi theorem for level sets of multiplicative functions see Bergelson et al. [128].

IP-recurrence

By the uniform version of Furstenberg's recurrence theorem (see Remark 13.4), for a given (invertible) measure-preserving system (X, μ, T) and a measurable set $A \subset X$ with $\mu(A) > 0$ for each k the set

$$\{n \in \mathbb{N} : \mu(A \cap T^{-n} A \cap \cdots \cap T^{-kn} A) > 0\}$$

is syndetic. In fact, one can prove, see Furstenberg and Katznelson [451], that it is an **IP* set**, i.e., intersects each IP-subset of $\mathbb{N}$ (see Sect. 11.5 for the definitions, and recall that every IP* is syndetic). Moreover, such recurrence results can be generalized to so-called IP-systems as initiated by Furstenberg and Katznelson in [447] and [451].

Let $T_1, \ldots, T_n, \ldots,$ be commuting invertible measure-preserving transformations on a probability space (X, μ). Given a multi-index $\alpha = (i_1, \ldots, i_m) \in \mathbb{N}^m$ with $i_1 < i_2 < \cdots < i_m$ we set $T_\alpha := T_{i_1} T_{i_2} \ldots T_{i_m}$. We call $(T_\alpha)_\alpha$ an **IP-**

system of measure-preserving transformations. Furstenberg and Katznelson proved the following recurrence theorem for these systems, see Theorem A in [451].

Theorem *For $k \in \mathbb{N}$ let $(T_\alpha^{(1)})_\alpha, \dots, (T_\alpha^{(k)})_\alpha$ be IP-systems of measure-preserving transformations which commute with each other. For each measurable set A with $\mu(A) > 0$ there is a multi-index α with*

$$\mu(A \cap (T_\alpha^{(1)})^{-1} A \cap \cdots \cap (T_\alpha^{(k)})^{-1} A) > 0.$$

Further extensions (to not necessarily commuting transformations, to (generalized) polynomials, to $\mathbb{Z}^d$-actions etc.) and applications in combinatorics or number theory can be found in Furstenberg and Katznelson [451, 452], Bergelson et al. [117], Bergelson et al. [119], Bergelson and McCutcheon [139], Zorin-Kranich [1069], McCutcheon [784], Bergelson and Ruzsa [146], Bergelson and Leibman [133]. The surveys of Bergelson [112] and Furstenberg [449] also provide a good start for the study of the relevance of IP-systems in dynamics.

Sets of Multiple Recurrence

A set $S \subset \mathbb{N}$ is called a **set of k-recurrence** if for every measure-preserving system (X, μ, T) and every measurable set $A \subset X$ with $\mu(A) > 0$ there exists $n \in S$ with $\mu(A \cap T^{-n} A \cdots T^{-kn} A) > 0$, and a **set of multiple recurrence** if it is a set of k-recurrence for every $k \in \mathbb{N}$. By Furstenberg's correspondence principle, S is a set of k-recurrence if and only if it is **k-intersective**, i.e., for every set $E \subset \mathbb{N}$ with positive upper density there exists $n \in S$ such that $a, a+n, \dots, a+kn \in E$ for some $a \in \mathbb{N}$. Analogously one defines[19] sets of k-recurrence and of multiple recurrence for subsets of $\mathbb{Z}$ using invertible measure-preserving systems.

Examples of sets of 1-recurrence, i.e., sets of recurrence, were discussed in Remark 9.26. Furstenberg [446, pp. 177–178] showed that there exists a set of 1-recurrence which is not a set of 2-recurrence which was generalized to all k by Frantzikinakis et al. [428, 429]. By Bergelson and Leibman [129], polynomial sets $p(\mathbb{N})$ for integer polynomials with $p(0) = 0$ are sets of multiple recurrence. Moreover, this is the case for sets of the form $\lfloor p(\mathbb{N}) \rfloor$ for real-valued polynomials p with $p(0) = 0$ which follows from [129] together with a trick from Bergelson and Håland [118], see Frantzikinakis and Wierdl [432, Sec. 2.1.2]. As has been said above, by Furstenberg and Katznelson [451] IP-sets are sets of multiple recurrence. For examples of sets of multiple recurrence coming from functions from Hardy fields (such as $\lfloor a(\mathbb{N}) \rfloor$ for a logarithmico-exponential function a of certain subpolynomial growth, e.g., $(\lfloor n^\delta \rfloor)$, $(\lfloor n \log n \rfloor)$) see Frantzikinakis and Wierdl

[19] As in the full version of Furstenberg's correspondence principle (or by a direct argument using invertible extensions, see Proposition 8.51) these two definitions coincide for subsets of $\mathbb{N}$.

[432], Frantzikinakis [407, 408, 413]. Random sets of k-recurrence were studied in Frantzikinakis et al. [431] and Altman [31]. For a related result concerning Betty sequences see Moreira and Richter [806, Cor. 2.6].

Wooley and Ziegler [1049] showed that the shifted primes $\mathbb{P}+1$ and $\mathbb{P}-1$ are sets of **polynomial multiple recurrence**, i.e., intersect every set of the form

$$\{n \in \mathbb{N} : \mu(A \cap T^{-P_1(n)}A \cap \ldots \cap T^{-P_k(n)}A) > 0\},$$

where $P_1, \ldots, P_k$ are integer polynomials with $P_1(0) = \ldots = P_k(0) = 0$. This leads to the existence of polynomial progressions of the form $a, a + P_1(n), \ldots, a + P_k(n)$ for some $a \in \mathbb{N}$ and $n \in \mathbb{P}+1$ or $n \in \mathbb{P}-1$, respectively, in every set of positive upper density, see Tao and Teräväinen [990] for a quantitative version. The linear case was shown before by Frantzikinakis et al. [419] for $k = 2$ and for higher k conditionally to the inverse Gowers conjecture proved later by Green et al. [492], see Sect. 15.9 for a short discussion. For a related result see Sun [962]. A multidimensional version is obtained by Bergelson et al. [136], with a polynomial generalization due to Fratzikinakis et al. [420], see also Koutsogiannis [665]. For more examples of sets of polynomial multiple recurrence see Bergelson et al. [127].

For results concerning largeness of the set of multiple returns of the form

$$\{n \in \mathbb{N} : \mu(A \cap T^{-\lfloor a(n) \rfloor}A \cap \ldots \cap T^{-\lfloor a(n+k) \rfloor}A) > 0\}$$

for a large class of non-polynomial functions a see Bergelson et al. [144].

Large Intersections

Recall that by Khinchin's generalization of Poincaré's theorem, for every measure-preserving system, every positive measure set A and every $\varepsilon > 0$ the set $\{n : \mu(A \cap T^{-n}A) > \mu(A)^2 - \varepsilon\}$ is syndetic, see Theorem 7.6. Bergelson et al. [123] showed that for ergodic systems the set

$$\{n : \mu(A \cap T^{-n}A \cap \ldots \cap T^{-kn}A) > \mu(A)^{k+1} - \varepsilon\}$$

is syndetic for $k = 2$ and $k = 3$.[20] By a version of Furstenberg's correspondence principle (see [123, Prop. 3.1]) this implies that for such k and every set $E \subset \mathbb{N}$ of **positive upper Banach density** the set

$$\{n : \overline{\mathrm{Bd}}(E \cap (E+n) \cap \ldots \cap (E+kn)) > \overline{\mathrm{Bd}}(E)^{k+1} - \varepsilon\}$$

[20] Recall that Furstenberg's recurrence for the averages $\frac{1}{N-M}\sum_{n=M+1}^{N}$ implies that for every $k \in \mathbb{N}$ there exists a constant $c = c(A) > 0$ such that the set $\{n : \mu(A \cap T^{-n}A \cap \ldots \cap T^{-kn}A) > c\}$ is syndetic, see Exercise 13.15.

is syndetic. Moreover, they presented counterexamples for all higher k and showed that, in opposite to Khinchin's result, the ergodicity assumption cannot be dropped. For related results, also regarding polynomial and non-polynomial powers, commuting transformations and group actions, see Frantzikinakis and Kra [424], Frantzikinakis [406], McCutcheon and Quas [785], Chu [245], Chu and Zorin-Kranich [248], Bergelson et al. [147], Donoso and Sun [311], Bergelson et al. [124], Donoso et al. [308], Ferré [389], Ackelsberg et al. [9], Durcik et al. [326], Hirayama et al. [537], Shalom [932], Jamneshan and Pan [578], and Frantzikinakis and Kuca [426].

Hales–Jewett Theorem

Hales–Jewett theorem [512] is a combinatorial statement on the existence of a monochromatic combinatorial line in a finite colored multidimensional cube being in the spirit of the van der Waerden theorem but considerably stronger. As van der Waerden's theorem can be shown via a recurrence property of topological dynamical systems (see Furstenberg and Weiss [457]), this is also the case for the Hales–Jewett theorem, see its polynomial version by Bergelson and Leibman [130].

In 1991, Furstenberg and Katznelson [454] used their ideas from [450] discussed in this chapter to give a (considerably more complicated) ergodic theoretic proof of the so-called density Hales–Jewett theorem being a stronger statement than Szemerédi's theorem. Another ergodic theoretic proof was given by Austin [61]. A purely finitary combinatorial proof of the density Hales–Jewett theorem was achieved in a Polymath project [856] in 2009, providing a new elementary proof of Szemerédi's theorem, see also Dodos et al. [306] for a simplified argument.

Green–Tao Theorem

The philosophy of the Furstenberg–Katznelson proof of Szemerédi's theorem as well as all other existing proofs from different areas inspired Green and Tao [485] to confirm the conjecture of Lagrange and Waring from 1770 and show that the primes (which have density zero) contain arbitrarily long arithmetic progressions. In [486] they generalized it to solutions of linear equations in primes giving an alternative proof which uses more ideas from ergodic theory, namely from the study of multiple convergence discussed in Chap. 15. See Sect. 15.9 for a brief discussion of these proofs and some references. A multidimensional Green–Tao theorem was shown by Tao and Ziegler [993] and independently by Cook et al. [259] with a simplification due to Fox and Zhao [403], where the proof in [259] is combinatorial without using ergodic theory. For a polynomial version of the Green–Tao theorem see Tao and Ziegler [991, 992, 995], see also Lê and Wolf [711] for a refinement along shifted primes and Matthiesen et al. [779] for a quantitative version.

Exercises

Exercise 13.1 Prove the implication (vi)⇒(iv) and (vii)⇒(v) in Furstenberg's correspondence principle, i.e., deduce the density/Banach density version of Szemerédi's theorem from the corresponding finitary version.

Exercise 13.2 (van der Waerden's Theorem) Show that Theorem 13.6 implies Theorem 13.7 using the following argument called by Tao *"compactness and contradiction"*.

Step 1. Assume the contrary and take $k, m \in \mathbb{N}$ such that for every $N \in \mathbb{N}$ there exists a coloring of $\{1, \ldots, N\}$ into m colors without a monochromatic arithmetic progression of length k.
Step 2. Using a diagonalization argument, choose a subsequence (N_n) of $\mathbb{N}$ such that the colorings of $\{1, \ldots, N_n\}$ from Step 1 stabilize (i.e., eventually become constant) at every element of $\mathbb{N}$.
Step 3. Apply Theorem 13.6 to the coloring of $\mathbb{N}$ which one obtains from Step 2.

Exercise 13.3 (Multiple Recurrence for Systems with Discrete Spectrum) Show multiple recurrence for ergodic systems with discrete spectrum by adapting the argument in the proof of double recurrence (Theorem 9.30) to higher k.

Exercise 13.4 (Conditional Almost Periodicity and Conditional Weak Mixing)

(a) Prove the assertions from Example 13.18(b),(c).
(b) Describe all conditional eigenfunctions, conditionally almost periodic functions and conditionally weakly mixing functions in the case when (X, μ, T) and (Y, ν, S) coincide.
(c) Show that (13.15) is equivalent to (13.17) for functions in $\mathrm{L}^2(X|Y)$.
(d) Give an alternative proof of Lemma 13.20(a) as follows. Assume first that $f, g \in \mathrm{L}^2(X|Y)$ and use the trick from (c) together with the formula $|z|^2 = z\overline{z}$ to reduce to convergence of the averages $\frac{1}{N}\sum_{n=1}^{N}\langle g, T^n f\rangle_{\mathrm{L}^2(X|Y)} T^n f$ to zero in $\mathrm{L}^2(X)$. Show this convergence using the van der Corput Lemma 7.31. For the approximation argument in the general case $f, g \in \mathrm{L}^2(X)$ use (13.16).

Exercise 13.5 (Conditional Cauchy–Schwarz Inequality) Show inequality (13.13) in two different ways, namely (a) adapting the proof of the classical Cauchy–Schwarz inequality replacing numbers by functions in $\mathrm{L}^\infty(Y)$ and (b) for regular systems by using disintegration, in particular (8.10). Prove that $\|\cdot\|_{\mathrm{L}^{2,\infty}(X|Y)}$ is a norm on $\mathrm{L}^2(X|Y)$ making it a Banach space.

Exercise 13.6 Work out the details of the proof of Lemma 13.20(b).

Exercise 13.7 (Conditional Product) Prove the properties of conditional product spaces stated in Remark 13.30.

Exercise 13.8 (Conditional Product: an Example) Let (X, μ, T) be the skew rotation system on $\mathbb{T}^2$ from Example 2.15 and (Y, ν, S) be the corresponding rotation on $\mathbb{T}$ by a, see Example 8.4. Determine the conditional product system $(X \times_Y X, \mu \times_Y \mu, T \times T)$.

Exercise 13.9 (Weakly Mixing Extensions)

(a) Prove that a measure-preserving system (X, μ, T) is a weakly mixing extension of its factor (Y, ν, S) if and only if the conditional product system $X \times_Y X$ is an ergodic extension of (Y, ν, S). (Hint: For the "only if" direction show first that

$$\lim_{N\to\infty} \frac{1}{N} \sum_{n=1}^{N} ((T \times T)^n (f_1 \otimes f_2) | g_1 \otimes g_2)_{\mathrm{L}^2(\mu \times_Y \mu)} = 0$$

for every $g_1, g_2 \in \mathrm{L}^\infty(X)$ whenever f_1 or f_2 is conditionally weakly mixing. For the "if" direction use the formula $|z|^2 = z\overline{z}$ and the mean ergodic theorem for $X \times_Y X$.)

(b) Show that an ergodic system (X, μ, T) is a weakly mixing extension of its factor (Y, ν, S) if and only if the conditional product system $X \times_Y X$ is ergodic. (Hint: Use the characterization in (a).)

Exercise 13.10 (Abramov Factors as Compact Extensions) Let $k \in \mathbb{N}$. Show the assertion of Example 13.19 that every generalized eigenfunction of order $\leq k+1$ is a conditional eigenfunction thus conditionally almost periodic with respect to the factor $A_k(X, \mu, T)$. Conclude that $(k+1)^{\mathrm{st}}$ Abramov factor of an ergodic system is a compact extension of the kth Abramov factor. Conclude that the skew rotation $(\mathbb{T}^2, \mathrm{m}^2, T_a)$ discussed in Example 8.34 is a compact extension of the torus rotation $(\mathbb{T}, \mathrm{m}, a)$ for irrational a and show that the same holds for rational a as well.

Exercise 13.11 (Inverse Limit) Show the characterization of inverse limits of factors given in Remark 13.47.

Exercise 13.12 (Inverse Limit of Invertible Factors) Prove that inverse limit of (Markov) invertible factors is (Markov) invertible.

Exercise 13.13 (Multidimensional Correspondence Principle) Show the following multidimensional analogue of Furstenberg's correspondence principle for arithmetic progressions due to Furstenberg and Katznelson [450]. Assume that for all $k \in \mathbb{N}$ and all commuting measure-preserving transformations $T_1, \ldots, T_k$ on a probability space (X, μ), for every $A \subset X$ with $\mu(A) > 0$ there is $n \in \mathbb{N}$ such that

$$\mu\Big(T_1^{-n}(A) \cap \cdots \cap T_k^{-n}(A)\Big) > 0.$$

Then for all $d \in \mathbb{N}$ and all finite sets $F \subset \mathbb{N}_0^d$, every subset $E \subset \mathbb{N}^d$ with $\overline{\mathrm{d}}(E) > 0$ contains an affine image of F, i.e., there exist $a \in \mathbb{N}^d$ and $n \in \mathbb{N}$ with $a + nF \in E$.

Exercise 13.14 (Polynomial Multiple Recurrence for Commuting Transformations)

(a) Prove the following polynomial version of the multidimensional correspondence principle from Exercise 13.13 due to Bergelson and Lebman [129]. Assume that for all $k \in \mathbb{N}$, all polynomials $p_1, \ldots, p_k \in \mathbb{Z}[\cdot]$ with $p_1(0) = \ldots = p_k(0) = 0$ and all invertible commuting measure-preserving transformations $T_1, \ldots, T_k$ on a probability space (X, μ), for every $A \subset X$ with $\mu(A) > 0$ there is $n \in \mathbb{N}$ such that

$$\mu\left(T_1^{-p_1(n)}(A) \cap \cdots \cap T_k^{-p_k(n)}(A)\right) > 0.$$

Then for all $d, k \in \mathbb{N}$, all polynomials $p_1, \ldots, p_k \in \mathbb{Z}[\cdot]$ with $p_1(0) = \ldots = p_k(0) = 0$ and all $\{v_1, \ldots, v_k\} \subset \mathbb{N}^d$, for every subset $E \subset \mathbb{N}^d$ with $\overline{\mathrm{d}}(E) > 0$ there exist $a \in \mathbb{Z}^d$ and $n \in \mathbb{N}$ with $a + p_1(n)v_1, \ldots, a + p_k(n)v_k \in E$.

(b) Show that one cannot drop the assumption $p_1(0) = 0, \ldots, p_k(0) = 0$ on the polynomials both in the multiple polynomial multiple recurrence result and in the corresponding version of Szemerédi's theorem discussed in (a).

Exercise 13.15 Let (a_n) be a sequence in $\mathbb{R}$. Prove that if

$$\liminf_{N-M\to\infty} \frac{1}{N-M} \sum_{n=M+1}^{N} a_n > 0,$$

then there exists $c > 0$ such that the set $E := \{n : a_n > c\}$ is syndetic. Deduce the implication (i*)⇒(iii*) from Remark 13.4.

Chapter 14
Nilsystems

Nilsystems are rotations on compact homogeneous spaces of a special kind. What makes them so particular is the nilpotency of the underlying group, allowing for inductive arguments, and the Lie group property, which allows the utilization of the euclidean structure. Nilsystems have remarkable algebraic and ergodic properties to be discussed here. Amazingly, as we will see in Sect. 15.6, they serve as building blocks for general (regular) measure-preserving systems. The exposition here is mainly based on Leibman [719], see also Malcev [767], Auslander et al. [59], Parry [841, 842], Lesigne [733], Tao [977, Sec. 2.16], Green and Tao [489], and Host and Kra [553, Part 3] for more information.

14.1 Definition and Examples

Nilpotent groups lie at the heart of the theory of nilsystems, therefore we start by recalling some basic facts about this class of groups.

Let G be a group. Its identity element is denoted by id_G or simply by id. The lower central series of G is defined recursively as follows:

$$G_1 := G, \quad G_2 := [G, G], \ldots, G_j := [G, G_{j-1}] \ldots.$$

Here, $[G, H]$ denotes the **commutator** (subgroup) of G and the subgroup H, i.e., the subgroup generated by elements of the form $[g, h] := g^{-1}h^{-1}gh$, $g \in G$, $h \in H$. Note that each subgroup G_j is normal and is generated by elements of the form $[[[g_1, g_2], g_3] \ldots, g_j]$ with $g_1, \ldots, g_j \in G$, see Exercise 14.1. The chain of normal subgroups

$$G = G_1 \trianglerighteq G_2 \trianglerighteq \cdots \trianglerighteq G_j \trianglerighteq \cdots$$

T. Eisner, B. Farkas, *A Journey Through Ergodic Theorems*, Birkhäuser Advanced Texts Basler Lehrbücher, https://doi.org/10.1007/978-3-031-96506-7_14

is then called the **lower central series** of G. Note further that each G_j is invariant under automorphisms of the group G.

Given $l \in \mathbb{N}_0$ the group G is called **l-step nilpotent** if $G_{l+1} = \{\mathrm{id}\}$, so that the lower central series takes the form

$$G = G_1 \trianglerighteq G_2 \trianglerighteq \cdots \trianglerighteq G_{l+1} = \{\mathrm{id}\}.$$

(Note that we do not require l to be minimal with the above property, i.e., the step is not uniquely defined and G is of step $l + 1$ if it is of step l.) Thus 1-step nilpotent groups are the commutative ones.

The following facts will be used throughout this chapter without further reference.

Remark 14.1 (Properties of the Lower Central Series)

(a) That G is l-step nilpotent means that for each $g \in G$ and $g_l \in G_l$ we have $[g, g_l] = \mathrm{id}$. So G_l is in the center $Z(G)$ of G, and the group G_l is Abelian.
(b) The quotient group G/G_2 is Abelian, and it is the maximal Abelian factor of G, in the sense that if H is a normal subgroup in G with G/H Abelian, then $G_2 \leq H$. More generally, the quotient groups G_j/G_{j+1} are Abelian.
(c) For each $j \in \mathbb{N}$ the group G_j/G_{j+1} is central in G/G_{j+1}, i.e., its elements commute with each element of G/G_{j+1}. So G_j/G_{j+1} is a normal subgroup in G/G_{j+1}, and then by the isomorphism theorem, the groups G_j/G_{j+1} and $(G/G_{j+1})/(G_j/G_{j+1})$ are isomorphic.
(d) If $H \leq G$ is a subgroup, then by the normality of G_j, the set $G_jH = HG_j$ is actually a subgroup.
(e) Finite products, subgroups and quotients of nilpotent groups are nilpotent. Being nilpotent is an isomorphism invariant.

Recall from Chap. 11 that a topological group is a topological space and a group such that the multiplication and taking the inverse are continuous mappings. If the group carries the structure of a finite-dimensional smooth (meaning C^∞) manifold such that the operations of multiplication and taking the inverse are smooth (meaning C^∞), then we speak of a **Lie group**. We list some properties of Lie groups which will be important for us, for details and more information see, e.g., San Martin [904].

Remark 14.2 (Lie Groups)

(a) A Lie group G is locally connected, so that the connected component G° of the identity is an open (and closed) normal subgroup of G.
(b) Every closed subgroup of a Lie group is a Lie group. This is known as Cartan's **closed subgroup theorem**, see [904, Sec. 6.5].
(c) The quotient of a Lie group G by a closed, normal subgroup is a Lie group. By this and the previous properties G° and G/G° are Lie groups.
(d) Every continuous homomorphism between Lie groups is smooth.

(e) Every compact, connected, Abelian Lie group is isomorphic to a finite dimensional torus, see, e.g., Hofmann and Morris [539, Prop. 2.42].

Recall from Chap. 11 that for a closed subgroup Γ of a Hausdorff topological group G the quotient mapping $q : G \to G/\Gamma$ is open (and continuous), and that Γ is called **cocompact** (or **uniform**) if the homogeneous space G/Γ is compact. Moreover, recall that a discrete subgroup of a Hausdorff topological group is closed.

Definition 14.3 Let G be an l-step nilpotent Lie group and Γ be a discrete cocompact subgroup of G. The compact quotient space G/Γ is called an **l-step nilmanifold**. For $a \in G$, consider the left rotation given by $T : h\Gamma \mapsto ah\Gamma$ which is an invertible, continuous transformation on G/Γ. Denote by m the **Haar measure** on G/Γ, i.e., the unique Borel probability measure which is invariant under all rotations (see Sect. 11.4). The triple $(G/\Gamma, \mathrm{m}, T)$ is called a **nilsystem**.

So nilsystems can be considered as both topological and measure-theoretical dynamical systems. We often write $(G/\Gamma, \mathrm{m}, a)$ instead of $(G/\Gamma, \mathrm{m}, T)$, or just $(G/\Gamma, a)$ if only the topological structure is to be emphasized. Again we note explicitly that G/Γ is only a G-homogeneous space and is in general not a group.

Example 14.4 (Torus) The simplest example of a nilmanifold is the compact Abelian group $\mathbb{T}^d$ with $d \in \mathbb{N}$. Indeed, $\mathbb{T}^d \cong \mathbb{R}^d/\mathbb{Z}^d$, so torus rotations can be represented as nilsystems. Note that one also has $\mathbb{T}^d \cong \mathbb{T}^d/\{1\}$ and for example $\mathbb{T}^d \cong G/\Gamma$ with $G := \mathbb{Z}^d \times \mathbb{R}^d$ and $\Gamma := \mathbb{Z}^{2d}$.

Generalizing this example we have the following.

Example 14.5 (Groups) Every compact, nilpotent Lie group can itself be represented as a nilmanifold, and any corresponding rotation system as a nilsystem.

A simple but important property is that the Cartesian product of finitely many nilsystems can be represented as a nilsystem.

Example 14.6 (Products) Let $n \in \mathbb{N}$ and $X_1 = H_1/\Gamma_1, \dots, X_n = H_n/\Gamma_n$ be nilmanifolds. Then $\prod_{j=1}^n X_j$ is a $H_1 \times \cdots \times H_n$-homogeneous system and it can be represented as the nilmanifold $X := (H_1 \times \cdots \times H_n)/(\Gamma_1 \times \ldots \times \Gamma_n)$. Note that the nilpotency step of X is the maximum of the nilpotency steps of the nilmanifolds $X_1, \dots, X_n$. Thus the product of finitely many nilsystems can be viewed again as a nilsystem. We leave the details as Exercise 14.2.

We leave as Exercise 14.3 to work out the details of the following two classical examples.

Example 14.7 (Heisenberg System) We recall the Heisenberg system from Example 8.35: Consider $G := \mathbb{R}^3$ endowed with the multiplication

$$[a, b, c] \cdot [a', b', c'] := [a + a', b + b', c + c' + ab'],$$

and identify an element $[a, b, c]$ with the matrix $\begin{pmatrix} 1 & a & c \\ 0 & 1 & b \\ 0 & 0 & 1 \end{pmatrix}$. Then

$$[G, G] = G_2 = \left\{ \begin{pmatrix} 1 & 0 & c \\ 0 & 1 & 0 \\ 0 & 0 & 1 \end{pmatrix} : c \in \mathbb{R} \right\} =: \begin{pmatrix} 1 & 0 & \mathbb{R} \\ 0 & 1 & 0 \\ 0 & 0 & 1 \end{pmatrix}$$

and $G_3 = \{I\}$. Take $\Gamma := \mathbb{Z}^3$ or using the above identification

$$\Gamma = \begin{pmatrix} 1 & \mathbb{Z} & \mathbb{Z} \\ 0 & 1 & \mathbb{Z} \\ 0 & 0 & 1 \end{pmatrix} := \left\{ \begin{pmatrix} 1 & n & k \\ 0 & 1 & m \\ 0 & 0 & 1 \end{pmatrix} : n, m, k \in \mathbb{Z} \right\}.$$

Then G/Γ is a 2-step nilmanifold.

Generalizing this example in a natural way to higher dimensions leads to nilmanifolds of arbitrary step.

A related example is the following.

Example 14.8 Using the matrix representation from the previous example consider the group

$$G := \begin{pmatrix} 1 & \mathbb{Z} & \mathbb{R} \\ 0 & 1 & \mathbb{R} \\ 0 & 0 & 1 \end{pmatrix} := \left\{ \begin{pmatrix} 1 & n & c \\ 0 & 1 & b \\ 0 & 0 & 1 \end{pmatrix} : n \in \mathbb{Z}, b, c \in \mathbb{R} \right\}$$

and the discrete cocompact subgroup $\Gamma := \begin{pmatrix} 1 & \mathbb{Z} & \mathbb{Z} \\ 0 & 1 & \mathbb{Z} \\ 0 & 0 & 1 \end{pmatrix}$. Then G/Γ is a 2-step nilmanifold homeomorphic to $\mathbb{T}^2$ and therefore connected, although G is disconnected. It is easy to modify this example to obtain a disconnected nilsystem.

We now show using the last example that the skew shift system defined in Example 2.15 can be represented as a nilsystem.

Example 14.9 (Skew Shift) Let $\alpha, \beta \in \mathbb{R}$. For G/Γ from Example 14.8 take the rotation by the element $g := [1, \alpha, \beta]$. From

$$\begin{pmatrix} 1 & 1 & \beta \\ 0 & 1 & \alpha \\ 0 & 0 & 1 \end{pmatrix} \begin{pmatrix} 1 & n & y \\ 0 & 1 & x \\ 0 & 0 & 1 \end{pmatrix} = \begin{pmatrix} 1 & n+1 & y+x+\beta \\ 0 & 1 & x+\alpha \\ 0 & 0 & 1 \end{pmatrix}$$

we see that the nilsystem $(G/\Gamma, g)$ is isomorphic to the skew shift given by

$$(x, y) \mapsto (x + \alpha, y + x + \beta)$$

on $(\mathbb{R}/\mathbb{Z})^2$. Analogously one shows that the higher dimensional skew shift

$$(x_1, \ldots, x_d) \mapsto (x_1 + \alpha_1, x_2 + x_1 + \alpha_2, \ldots, x_d + x_{d-1} + \alpha_d) \tag{14.1}$$

on $(\mathbb{R}/\mathbb{Z})^d$ is isomorphic to a d-step nilsystem, see Exercise 14.4.

Analogously, one can represent the Heisenberg system from Example 14.7 as a skew product system.

Example 14.10 (Heisenberg Systems as a Skew Product System) Let $\alpha, \beta, \gamma \in \mathbb{R}$. For G/Γ from Example 14.7 take the rotation by the element $g := [\alpha, \beta, \gamma]$. Observe that for $[x, y, z] \in G$, using the matrix representation,

$$\begin{pmatrix} 1 & \alpha & \gamma \\ 0 & 1 & \beta \\ 0 & 0 & 1 \end{pmatrix} \cdot \begin{pmatrix} 1 & x & z \\ 0 & 1 & y \\ 0 & 0 & 1 \end{pmatrix} = \begin{pmatrix} 1 & x+\alpha & z+\gamma+\alpha y \\ 0 & 1 & y+\beta \\ 0 & 0 & 1 \end{pmatrix}.$$

Multiplying this element with

$$\begin{pmatrix} 1 & -\lfloor x+\alpha \rfloor & 0 \\ 0 & 1 & -\lfloor y+\beta \rfloor \\ 0 & 0 & 1 \end{pmatrix} \in \Gamma$$

we obtain

$$\begin{pmatrix} 1 & \{x+\alpha\} & z+\gamma+\alpha y-(x+\alpha)\lfloor y+\beta \rfloor \\ 0 & 1 & \{y+\beta\} \\ 0 & 0 & 1 \end{pmatrix}.$$

This multiplied with

$$\begin{pmatrix} 1 & 0 & -\lfloor z+\gamma+\alpha y-(x+\alpha)\lfloor y+\beta \rfloor \rfloor \\ 0 & 1 & 0 \\ 0 & 0 & 1 \end{pmatrix} \in \Gamma$$

equals

$$\begin{pmatrix} 1 & \{x+\alpha\} & \{z+\gamma+\alpha y-(x+\alpha)\lfloor y+\beta \rfloor\} \\ 0 & 1 & \{y+\beta\} \\ 0 & 0 & 1 \end{pmatrix}$$

which can be identified with $(\{x + \alpha\}, \{y + \beta\}, \{z + \gamma + \alpha y - (x + \alpha)\lfloor y + \beta \rfloor\}) \in (\mathbb{R}/\mathbb{Z})^3$. Thus the Heisenberg nilsystem $(G/\Gamma, g)$ is isomorphic to the three-dimensional skew product system given by

$$(x, y, z) \mapsto (x + \alpha, y + \beta, z + \gamma + \alpha y - (x + \alpha)\lfloor y + \beta \rfloor)$$

on $(\mathbb{R}/\mathbb{Z})^3$.

A large class of nilsystems is given by affine systems defined next. We leave the details as Exercise 14.5.

Example 14.11 (Affine Systems) Let $A \in \mathbb{Z}^{l\times l}$ be a matrix with integer entries which is unipotent, i.e., satisfies $(A - I)^l = 0$. Let G be the group of all transformations of $(\mathbb{R}/\mathbb{Z})^l$ of the form $x \mapsto A^m x + a$ for $a \in (\mathbb{R}/\mathbb{Z})^l$ and $m \in \mathbb{Z}$. Then each G_j, $j \in \{1, \dots, l\}$, is the subgroup of translations by elements in $\operatorname{ran}(A - I)^{j-1}$. So G is s-step nilpotent where s is the nilpotency step of $A - I$. Consider now the subgroup Γ of G consisting of all transformations of $(\mathbb{R}/\mathbb{Z})^l$ by A^m for some $m \in \mathbb{Z}$, i.e., of the form $x \mapsto A^m x$. It is discrete and cocompact in G giving rise to the nilmanifold G/Γ, which is homeomorphic to $\mathbb{T}^l$.

It is easy to see that taking $A = \left(\begin{smallmatrix}1 & 1\\ 0 & 1\end{smallmatrix}\right)$ in this construction gives back the group from Example 14.8.

We will need the following notion of a nilsequence introduced by Bergelson et al. [123] under the name of a **basic nilsequence** (note that the terminology is not coherent in the literature).

Definition 14.12 For $l \in \mathbb{N}_0$, we call a complex sequence $(a_n)_{n\in\mathbb{N}}$ in $\mathbb{C}$ a **nilsequence of step** l if there exists a nilsystem $(G/\Gamma, \mathrm{m}, T)$ of step l, $x \in G/\Gamma$ and a continuous function $F \in \mathrm{C}(G/\Gamma)$ with

$$a_n = F(T^n x) \text{ for all } n \in \mathbb{N}. \tag{14.2}$$

Note that for a nilsequence the underlying groups G and Γ, the left rotation T, the function F and the point x are not determined uniquely, and the step of a nilsequence is also not unique.

Example 14.13

(a) By Example 14.6, the product of finitely many nilsequences is again a nilsequence.
(b) Clearly, constant sequences are nilsequences of step 0. Moreover, for $\lambda \in \mathbb{T}$ the linear sequence (λ^n) is a nilsequence of step 1. Indeed, take $G := \mathbb{T}$, $\Gamma := \{\mathrm{id}\}$, the rotation T by λ, $x := \Gamma$ and F given by $F(g\Gamma) := g$ or, e.g., $G := \mathbb{R}$, $\Gamma := \mathbb{Z}$, the translation T by α with[1] $e(\alpha) = \lambda$, $x = \mathbb{Z}$ and F given by $F(g + \mathbb{Z}) := e(g)$.
(c) More generally, polynomial sequences of the form $(e(P(n)))$ for a real polynomial P are nilsequences of step $\deg P$. By (a) and (b) it is enough to show this for monomials of the form $e(\alpha n^k)$ for $\alpha \in \mathbb{R}$ and $k \geq 2$ (or, equivalently, (λ^{n^k}) for $\lambda \in \mathbb{T}$). We will show it for $k = 2$, the general case (being analogous) is left as Exercise 14.7. The idea is to use the quadratic formula from Proposition 9.15 and the representation of the skew shift as a nilsystem from Example 14.9, where we can easily modify the underlying group to be connected and simply connected. Let $\alpha \in \mathbb{R}$ and consider the Heisenberg system from Example 14.7 with the left rotation T by

$$g := \begin{pmatrix} 1 & 1 & 0 \\ 0 & 1 & 2\alpha \\ 0 & 0 & 1 \end{pmatrix}.$$

It is easy to see that

$$g^n = \begin{pmatrix} 1 & n & n(n-1)\alpha \\ 0 & 1 & 2n\alpha \\ 0 & 0 & 1 \end{pmatrix}$$

[1] Recall the notation $e(\alpha) := \mathrm{e}^{2\pi i \alpha}$.

holds. This implies $e(\alpha n(n-1)) = F(g^n x)$, $n \in \mathbb{N}$, for $x := \Gamma$ and the continuous function F given by

$$F\left(\left(\begin{smallmatrix} 1 & a & c \\ 0 & 1 & b \\ 0 & 0 & 1 \end{smallmatrix}\right)\Gamma\right) := e(c).$$

Thus $(e(\alpha n^2))$ is a nilsequence of step 2 by (a) and (b).

Remark 14.14 In particular, by the construction in Example 14.13 polynomial sequences of the form $(e(P(n)))$ for a real polynomial P are nilsequences of step $\deg P$ where the underlying group G can be taken to be connected and simply connected with a Lipschitz function F.

Example 14.15 (Generalized Polynomials) A **generalized polynomial** (also called **bracket polynomial**) is a real-valued function which can be obtained from real polynomials by using the operations of addition, multiplication, and taking the integer part. As we saw in Example 14.10, bracket polynomials appear naturally in the context of rotations on Heisenberg groups.

Using the notation from Examples 14.7 and 14.10, it is easy to see that for $\alpha, \beta \in \mathbb{R}$ and $g := [-\alpha, \beta, 0] \in G$ one has

$$g^n\Gamma = \begin{pmatrix} 1 & -n\alpha & 0 \\ 0 & 1 & n\beta \\ 0 & 0 & 1 \end{pmatrix}\Gamma = \begin{pmatrix} 1 & \{-n\alpha\} & \{n\alpha\lfloor n\beta\rfloor\} \\ 0 & 1 & \{n\beta\} \\ 0 & 0 & 1 \end{pmatrix}\Gamma. \tag{14.3}$$

Thus the sequences $(\{n\alpha\lfloor n\beta\rfloor\})$ and $(e(n\alpha\lfloor n\beta\rfloor))$ are both so-called **Riemannian nilsequences** of step 2, i.e., satisfy (14.2) for some Riemann integrable[2] function F. Indeed, for the first sequence F can be taken to be the third coordinate function given by $F([x, y, z]\Gamma) := z$ for $x, y, z \in [0, 1)$ which is Riemann integrable but not continuous, and for the second one F can be taken to be of the form $F([x, y, z]\Gamma) := e(z)$ for $x, y, z \in [0, 1)$, which is continuous. Moreover, this together with Example 14.13(a), (c) shows that sequences of the form $(\{n\alpha\lfloor n\beta\rfloor\})$ and $(e(n\alpha\{n\beta\}))$ are Riemannian nilsequences as well. We leave the details as Exercise 14.10.

Bergelson and Leibman [132, Thm. A] showed that every sequence of the form $(f(n))$ for a bounded generalized polynomial f can be represented as a Riemannian nilsequence. Such sequences play an important role in additive combinatorics as indicated in Sect. 15.9.

For a description of 2-step nilsequences see Host and Kra [550].

[2] Under a Riemann integrable function we understand one whose points of discontinuity form a set of reference measure 0, in this case the Haar measure on the nilmanifold.

14.2 Basic Facts and Vertical Fourier Analysis

We next present a very useful tool called *vertical Fourier analysis*. It is an adaptation of the classical Fourier analysis to nilmanifolds.

For an l-step nilsystem G/Γ denote $\Gamma_j := G_j \cap \Gamma$ for $j \in \{1, \dots, l\}$. Using so-called *Malcev bases* one can show that Γ_j is cocompact in G_j and that each G_j is closed, see Leibman [719, §2.11].

Definition 14.16 Let $X = G/\Gamma$ be an l-step nilmanifold. A function $F \in \mathrm{L}^2(X)$ is called a **vertical character** if there exists a character χ on the compact, Abelian group G_l/Γ_l such that

$$F(g_l x) = \chi(g_l \Gamma_l) F(x)$$

holds for almost every $x \in X$ and every $g_l \in G_l$. In this case F is called a **vertical character associated** to χ.

We will see shortly that there are *many* vertical characters, but first we need to make some remarks. In the following we identify a function f on G/Γ with its canonical extension to G, i.e., we write f also for the function $f \circ q$, where $q : G \to G/\Gamma$ is the quotient mapping. Moreover, we denote by $\widehat{G_l/\Gamma_l}$ the (discrete) dual group of the compact Abelian group G_l/Γ_l, see Sect. 11.1. Note that if X is connected, then so is G_l/Γ_l, therefore it is a torus of dimension $d := \dim G_l$ with the dual group $\mathbb{Z}^d$.

Remark 14.17

(a) Let $x = g\Gamma$. Since G_l is central in G, for each $g_l \in G_l$ and $\gamma_l \in \Gamma_l$ we have $g_l \gamma_l x = g_l \gamma_l g \Gamma = g_l g \gamma_l \Gamma = g_l g \Gamma = g_l x$ showing that $g_l x$ depends only on the coset $g_l \Gamma_l$ and not on g_l.

(b) Let $F \in \mathrm{L}^2(X)$ and consider the function $(x, g_l) \mapsto F(g_l x)$. Note that by (a), by the rotation invariance of the Haar measure and by Fubini's theorem we can write

$$\begin{aligned}\int_X \int_{G_l/\Gamma_l} |F(g_l x)|^2 \,\mathrm{d}g_l\Gamma_l \,\mathrm{d}x &= \int_{G_l/\Gamma_l} \int_X |F(g_l x)|^2 \,\mathrm{d}x \,\mathrm{d}g_l\Gamma_l \\ &= \int_{G_l/\Gamma_l} \|F\|^2_{\mathrm{L}^2(X)} \,\mathrm{d}g_l\Gamma_l = \|F\|^2_{\mathrm{L}^2(X)}.\end{aligned}$$

It also follows that for almost every $x \in X$ the function $g_l\Gamma_l \mapsto F(g_l x)$ is in $\mathrm{L}^2(G_l/\Gamma_l)$, and for each such $x \in X$ and every $h_l \in G_l$ we have that $g_l \mapsto F(g_l h_l x)$ is in $\mathrm{L}^2(G_l/\Gamma_l)$ (by the invariance of the Haar measure).

Lemma 14.18 *For each $F \in L^2(X)$ and each character $\chi \in \widehat{G_l/\Gamma_l}$*

$$F_\chi(x) := \int_{G_l/\Gamma_l} F(g_l x)\overline{\chi(g_l\Gamma_l)}\,\mathrm{d}g_l\Gamma_l \quad (a.e.\ x \in G/\Gamma) \tag{14.4}$$

defines a vertical character associated to χ.

Proof Note that, since each $\chi \in \widehat{G_l/\Gamma_l}$ is bounded, $F_\chi(x)$ is indeed for almost every $x \in X$ well defined by (14.4), see Remark 14.17. Moreover, $F_\chi \in L^2(X)$ and for each $h_l \in G_l$ also $F_\chi(h_l\cdot) \in L^2(G_l/\Gamma_l)$.

Using (14.4), Remark 14.17 and the rotation invariance of the Haar measure we conclude for almost every $x \in X$ and every $h_l \in G_l$ that

$$\begin{aligned} F_\chi(h_l x) &= \int_{G_l/\Gamma_l} F(g_l h_l x)\overline{\chi(g_l\Gamma_l)}\,\mathrm{d}g_l\Gamma_l \\ &= \int_{G_l/\Gamma_l} F(\tilde{g}_l x)\overline{\chi(\tilde{g}_l h_l^{-1}\Gamma_l)}\,\mathrm{d}\tilde{g}_l\Gamma_l \\ &= \chi(h_l\Gamma_l)\int_{G_l/\Gamma_l} F(\tilde{g}_l x)\overline{\chi(\tilde{g}_l\Gamma_l)}\,\mathrm{d}\tilde{g}_l\Gamma_l = \chi(h_l\Gamma_l)F_\chi(x). \end{aligned}$$

This proves that F_χ is a vertical character associated to χ. ■

It is Exercise 14.11 to prove that in fact every vertical character arises in this manner.

Proposition 14.19 (Decomposition into Vertical Characters) *For every $F \in L^2(X)$ we have the decomposition*

$$F = \sum_{\chi\in\widehat{G_l/\Gamma_l}} F_\chi \tag{14.5}$$

into a sum of vertical characters $(F_\chi)_{\chi\in\widehat{G_l/\Gamma_l}}$, where the convergence is unconditional in $L^2(X)$ and each F_χ is given by

$$F_\chi(x) = \int_{G_l/\Gamma_l} F(g_l x)\overline{\chi(g_l\Gamma_l)}\,\mathrm{d}g_l\Gamma_l \quad (a.e.\ x \in X).$$

Proof From Lemma 14.18 we already know that for each $\chi \in \widehat{G_l/\Gamma_l}$ the function F_χ is a vertical character associated to χ. By the classical Fourier decomposition, see Proposition 11.23 and Remark 14.17(b), for almost every $x \in X$ we can write

$$F(\cdot x) = \sum_{\chi\in\widehat{G_l/\Gamma_l}} F_\chi(x)\chi,$$

with orthogonal, hence unconditionally convergent sum in $\mathrm{L}^2(G_l/\Gamma_l)$. Since $\widehat{G_l/\Gamma_l}$ is countable, see Remark 11.24, we can take finite sets $A_n \subset \widehat{G_l/\Gamma_l}$ with $A_n \subset A_{n+1}$ and $\bigcup_n A_n = \widehat{G_l/\Gamma_l}$. So

$$F(\cdot x) = \lim_{n\to\infty} \sum_{\chi\in A_n} F_\chi(x)\chi \tag{14.6}$$

holds in $\mathrm{L}^2(G_l/\Gamma_l)$. We can write, by the rotation invariance of the Haar measure and by Fubini's theorem

$$\begin{aligned}
\Big\| F - \sum_{\chi\in A_n} F_\chi \Big\|^2_{\mathrm{L}^2(X)} &= \int_{G_l/\Gamma_l} \Big\| F - \sum_{\chi\in A_n} F_\chi \Big\|^2_{\mathrm{L}^2(X)} \,\mathrm{d}g_l\Gamma_l \\
&= \int_{G_l/\Gamma_l}\int_X \Big| F(g_l x) - \sum_{\chi\in A_n} F_\chi(g_l x)\Big|^2 \,\mathrm{d}x\,\mathrm{d}g_l\Gamma_l \\
&= \int_{G_l/\Gamma_l}\int_X \Big| F(g_l x) - \sum_{\chi\in A_n} F_\chi(x)\chi(g_l\Gamma_l)\Big|^2 \,\mathrm{d}x\,\mathrm{d}g_l\Gamma_l \\
&= \int_X\int_{G_l/\Gamma_l} \Big| F(g_l x) - \sum_{\chi\in A_n} F_\chi(x)\chi(g_l\Gamma_l)\Big|^2 \,\mathrm{d}g_l\Gamma_l\,\mathrm{d}x.
\end{aligned}$$

Now, by (14.6)

$$\int_{G_l/\Gamma_l} \Big| F(g_l x) - \sum_{\chi\in A_n} F_\chi(x)\chi(g_l\Gamma_l)\Big|^2 \,\mathrm{d}g_l\Gamma_l \to 0 \quad (\text{as } n\to\infty)$$

for almost every $x \in X$, thus it suffices to find an integrable majorant to conclude, by Lebesgue's dominated convergence theorem, that (14.5) holds.

For almost every $x \in X$ we have by Pythagoras' theorem

$$\int_{G_l/\Gamma_l} \Big| F(g_l x) - \sum_{\chi\in A_n} F_\chi(x)\chi(g_l\Gamma_l)\Big|^2 \,\mathrm{d}g_l\Gamma_l \le \|F(\cdot x)\|^2_{\mathrm{L}^2(G_l/\Gamma_l)}.$$

The function here on the right-hand side is integrable by Remark 14.17, completing the proof. ■

Example 14.20 (Heisenberg System) Consider the system $(G/\Gamma, \mathrm{m}, T)$ from Example 14.7. We have $l = 2$ and

$$G_2/\Gamma_2 = \begin{pmatrix} 1 & 0 & \mathbb{R} \\ 0 & 1 & 0 \\ 0 & 0 & 1 \end{pmatrix} \Big/ \begin{pmatrix} 1 & 0 & \mathbb{Z} \\ 0 & 1 & 0 \\ 0 & 0 & 1 \end{pmatrix}$$

which is isomorphic to $\mathbb{R}/\mathbb{Z}$ with the usual addition. Let $F : G \to \mathbb{C}$ be right-Γ-invariant. Then the induced function on G/Γ is a vertical character if and only if there exists $m \in \mathbb{Z}$ such that

$$F[x, y, z + w] = e^{2\pi i m w} F[x, y, z] \quad \text{for every } w \in \mathbb{R} \text{ and a.e. } x, y, z \in \mathbb{R}.$$

Thus each such F is of the form $F[x, y, z] = e^{2\pi i m z} f(x, y)$ for some $m \in \mathbb{Z}$ and a function[3] $f : \mathbb{R}^2 \to \mathbb{C}$.

14.3 Minimality and Ergodicity Properties

We now come to remarkable minimality and ergodicity properties of nilsystems as well as to everywhere convergence of the corresponding ergodic averages.

Let $X = G/\Gamma$ be an l-step nilmanifold and let $j \in \{1, \dots, l + 1\}$. By the normality of G_j we have that $G_j\Gamma$ is a subgroup, and in fact a closed one. To see closedness notice first that $G_j\Gamma/\Gamma$ is homeomorphic to the compact space G_j/Γ_j, so it is compact itself. Now $G_j\Gamma$ is the preimage of the set $G_j\Gamma/\Gamma$ under the continuous quotient mapping $q : G \to G/\Gamma$, so it is closed.

Consider the homogeneous spaces $X_j := G/G_j\Gamma$ with the corresponding quotient map $p_j : G \to X_j$. It is easy to see that the mapping

$$\varphi_j : X_j \to G_j\backslash X, \quad g(G_j\Gamma) \mapsto G_j g\Gamma = \{g_j g\Gamma : g_j \in G_j\}$$

is well-defined and a homeomorphism, where the space $G_j\backslash X$ is endowed with the quotient topology defined by the mapping

$$X \to G_j\backslash X, \quad g\Gamma \mapsto G_j g\Gamma.$$

In other words $G_j\backslash X$ is the orbit space of the left action of G_j on X. It follows that X_j is compact (and Hausdorff).

Remark 14.21 We will often identify X_j with $G_j\backslash X$.

The following shows that one can give the structure of a nilmanifold to each X_j.

Lemma 14.22 (Representation of the Spaces X_j) *Let $X = G/\Gamma$ be a nilmanifold of step l, let $j \in \{1, \dots, l\}$ and let $q_j : G \to G/G_j$ the quotient map. Then $G/G_j/q_j(\Gamma)$ is a nilmanifold. The group G acts by left multiplication on*

---

[3] Note that F is right-Γ-invariant if and only if f satisfies the functional equation

$$e^{2\pi i m k x} f(x + n, y + k) = f(x, y)$$

for every $n, k \in \mathbb{Z}$ and a.e. $x, y \in \mathbb{R}$.

$G/G_j/q_j(\Gamma)$. The G-homogeneous spaces X_j and the nilmanifold $G/G_j/q_j(\Gamma)$ are isomorphic (i.e., homeomorphic under a mapping which commutes with the action of G). In particular, for each $a \in G$ the topological systems (X_j, a) and $(G/G_j/q_j(\Gamma), a)$ are isomorphic.

Proof First observe that $q_j(\Gamma)$ is homeomorphic to $\Gamma/\Gamma \cap G_j$ and hence discrete in G/G_j, see Exercise 14.12(c). Next, by the normality of G_j we have for every $g \in G$

$$gG_jq_j(\Gamma) = \{gG_j\gamma G_j : \gamma \in \Gamma\} = \{gG_j\gamma : \gamma \in \Gamma\}. \tag{14.7}$$

Consider now $\varphi : G/G_j\Gamma \to G/G_j/q_j(\Gamma)$ defined by

$$\varphi(gG_j\Gamma) := gG_jq_j(\Gamma).$$

It follows from (14.7) that φ is well defined. It is also easy to see that φ is bijective and continuous. Since X_j is compact, φ is a homeomorphism and $q_j(\Gamma)$ is cocompact in G/G_j. Thus, G/G_j being a nilpotent Lie group and $q_j(\Gamma)$ a uniform lattice, the G-homogeneous space $G/G_j/q_j(\Gamma)$ is indeed a nilmanifold. It is also clear that for each $a \in G$ and $x \in X_j$ we have $\varphi(ax) = a\varphi(x)$. ■

Remark 14.23 Let us note the following important property of X_2. Since G/G_2 is an Abelian group, X_2 is homeomorphic to the compact Abelian group $G/G_2/q_2(\Gamma)$, so X_2 can be turned into a compact Abelian group with the multiplication given by

$$gG_2\Gamma \cdot hG_2\Gamma := ghG_2\Gamma.$$

The mapping

$$G \to X_2, \quad g \mapsto gG_2\Gamma$$

becomes thereby a (continuous) group homomorphism. The special importance of X_2 will be clear, for example, in Theorem 14.27 below. Note that if X is connected then so is X_2, which is called in this case the **horizontal torus** associated with G/Γ.

So we have the continuous surjections $\pi_j : X \to X_j$, $j = 1, \ldots, l+1$ and $\eta_j : X_{j+1} \to X_j$, $gG_{j+1}\Gamma \mapsto gG_j\Gamma$

$$X = X_{l+1} \stackrel{\eta_l}{\to} X_l \to \cdots \to X_2 \stackrel{\eta_1}{\to} X_1 = \{\cdot\},$$

intertwining the G-actions and giving rise to the tower of factors for any $a \in G$

$$(X, a) = (X_{l+1}, a) \to (X_l, a) \to \cdots \to (X_2, a) \to (X_1, a) = (\{\cdot\}, \mathrm{id})$$

with continuous factor maps. Note that these maps are measure-preserving by the uniqueness of the G-invariant Haar measure.

Remark 14.24 If one starts with some X_j as a $(j-1)$-step nilmanifold, then the corresponding tower of factors is still the same

$$X_j \to X_{j-1} \to \cdots \to X_2 \to X_1 = \{\cdot\},$$

see Exercise 14.12(b). This allows for inductive proofs as we will see below.

We begin with the following property of nilmanifolds as topological dynamical systems.

Proposition 14.25 (Distality of Nilsystems) *Every nilsystem* (X, T) *is* distal, *i.e., every* $x, y \in X$ *with* $x \neq y$ *satisfy* $\inf_{n\in\mathbb{N}_0} d(T^n x, T^n y) > 0$.

Proof Let $X = G/\Gamma$ with G being l-step nilpotent and let T be the rotation by $a \in G$. For $x, y \in X$ with $x \neq y$ take $j \in \{1, \ldots, l\}$ such that the images of x and y in X_{j+1} under the factor map $\pi := \pi_{j+1} : X \to X_{j+1}$ are distinct and the corresponding images in X_j are equal. Let d_{j+1} be a compatible metric on X_{j+1}. By the uniform continuity of π it suffices to show that

$$\inf_{n\in\mathbb{N}_0} d_{j+1}(\pi(a^n x), \pi(a^n y)) > 0.$$

By the choice of j we have $G_j\pi(y) = G_j\pi(x)$ (see Remark 14.21), so $\pi(y) = c\pi(x)$ for some $c \in G_j$. Since G_j/G_{j+1} is central in G/G_{j+1} and since $c \in G_j$, we have $cbG_{j+1} = bcG_{j+1}$ for every $b \in G$. Since $\pi(x) \neq \pi(y) = c\pi(x)$, we obtain $c \notin G_{j+1}\Gamma$ and $cz \neq z$ for each $z \in X_{j+1}$. By compactness it follows that

$$\delta := \inf_{z\in X_{j+1}} d_{j+1}(z, cz) > 0.$$

By $\pi(a^n y) = a^n\pi(y) = a^n c\pi(x) = ca^n\pi(x) = c\pi(a^n x)$ we conclude that

$$d_{j+1}(\pi(a^n x), \pi(a^n y)) = d_{j+1}(\pi(a^n x), c\pi(a^n x)) \geq \delta$$

for all $n \in \mathbb{N}_0$, as desired. ∎

Recall from Theorem 11.48 of Ellis that distal systems decompose into the disjoint union of minimal subsystems. Since in a minimal system every (one-sided or two-sided) orbit is dense by Remark 3.1, we obtain the following equivalence as a direct corollary of Proposition 14.25.

Corollary 14.26 *A nilsystem is minimal if and only if it is transitive if and only if it is two-sided transitive.*

The following very useful characterization was shown by Leibman [719] in a more general context. In the case of connected, simply connected groups, the assertion in (a) was proved by Green [493] and [59, Ch. 5] with an alternative proof by Parry [841, 842] using the vertical Fourier transform. For arbitrary closed subgroups of a connected, simply connected group (a) was proved by Lesigne [733, Thm. 3]. As before we denote by G° the connected component of G containing id. Recall that G° is an open normal subgroup of G, see Exercise 11.6.

Theorem 14.27 (Lifting Criterion) *Let (X, m, a) be a nilsystem with $X = G/\Gamma$ and $G = \langle G^\circ, a\rangle$. Then the following assertions hold.*

(a) *(X, m, a) is ergodic if and only if its Abelian factor $(X_2, \mathrm{m}_{X_2}, a)$ is ergodic.*
(b) *(X, a) is uniquely ergodic if and only if (X_2, a) is uniquely ergodic.*
(c) *(X, a) is minimal if and only if (X_2, a) is minimal.*
(d) *(X, a) is transitive if and only if (X_2, a) is transitive.*

Proof Since any of the above properties for a system imply the same property on every factor (see Exercise 8.2), we only have to prove the "if" implication in the above assertions.
(a) We argue by induction on the nilpotency step l of G. For $l = 1$ there is nothing to prove. Assume that the assertion holds for $(l-1)$-step nilsystems and let (X, m, a) be an l-step nilsystem with $X = G/\Gamma$ and $G = \langle G^\circ, a\rangle$. By the induction hypothesis, see Remark 14.24, the factor X_l is ergodic.

Recall that L_g denotes the Koopman operator of the left rotation by g for $g \in G$. Let $f \in \mathrm{L}^2(X)$ satisfy $L_a f = f$, $f \neq 0$. We need to show that f is constant, see Proposition 3.24. Let $f = \sum_{\chi \in \widehat{G_l/\Gamma_l}} f_\chi$ be the vertical Fourier decomposition of f from Proposition 14.19 with

$$f_\chi(x) = \int_{G_l/\Gamma_l} f(g_l x)\overline{\chi(g_l\Gamma_l)}\, \mathrm{d}g_l\Gamma_l \quad \text{for a.e. } x \in G/\Gamma.$$

Since G_l belongs to the center of G, each f_χ is a-invariant. It suffices to prove that each a-invariant vertical character, in particular each f_χ, is constant. Thus we can assume without loss of generality that f itself is a vertical character associated to some character χ on the compact Abelian group G_l/Γ_l such that for $g_l \in G_l$ and almost every $x \in X$

$$f(g_l x) = \chi(g_l\Gamma_l) f(x). \tag{14.8}$$

It follows that $|f(g_l x)| = |f(x)|$ for each $g_l \in G_l$, so that the function $|f|$ is well-defined on $X_l = G_l\backslash X$ (see Remark 14.21) and is constant by the ergodicity of $(X_l, \mathrm{m}_{X_l}, a)$ (since $|f|$ is a-invariant), so we can assume that $|f|$ is constant 1. The proof will be complete if we can show that $\chi = \mathbf{1}$ (constant 1), since then f itself is well-defined on X_l and has to be constant again by the ergodicity of $(X_l, \mathrm{m}_{X_l}, a)$.

For $g_{l-1} \in G_{l-1}$ consider the function $\tilde{f} := (L_{g_{l-1}} f) f^{-1} = (L_{g_{l-1}} f)\overline{f}$. Since $[G, G_{l-1}] = G_l$ belongs to the center of G, we can write

$$(L_a(L_{g_{l-1}} f))(\cdot) = f(g_{l-1}a\cdot) = f(ag_{l-1}[g_{l-1}, a]\cdot)$$

$$= f(g_{l-1}[g_{l-1}, a]\cdot) = f([g_{l-1}, a]g_{l-1}\cdot)$$
$$= \chi([g_{l-1}, a]\Gamma_l)(L_{g_{l-1}} f)(\cdot),$$

implying that $\tilde{f}$ is an eigenfunction of the Koopman operator L_a. Moreover, $\tilde{f}$ is well-defined on $X_l = G_l \backslash X$, since

$$\tilde{f}(g_l x) = f(g_l g_{l-1} x)\overline{f}(g_l x) = \chi(g_l \Gamma_l) f(g_{l-1} x)\overline{\chi(g_l \Gamma_l) f(x)} = \tilde{f}(x)$$

for all $g_l \in G_l$ and almost every $x \in X$.

First, extend χ to G_l by defining χ to be constant $\chi(g_l \Gamma_l)$ on each coset $g_l \Gamma_l$. It is easy to see that χ is then multiplicative on G_l. Let E denote the group of eigenfunctions of L_a on $\mathrm{L}^2(X_l)$, see Proposition 3.26, and extend χ to a function $\chi : G_{l-1} \to E$ by setting $\chi(g_{l-1}) := (L_{g_{l-1}} f) f^{-1}$. Observe that for $g_{l-1} \in G_l$ these definitions are consistent since f is a vertical character (see (14.8)) and that χ is continuous on G_{l-1} (where E is supplied with the restriction of the L^2-topology). We obtain a continuous function $G_{l-1} \to E/C$, where C denotes the (closed) subgroup of constant functions in E. By ergodicity of $(X_l, \mathrm{m}_{X_l}, a)$, the group E/C is discrete (recall that eigenfunctions to different eigenvalues are orthogonal, see Proposition 6.1, and the eigenspace to each eigenvalue is one-dimensional, see again Proposition 3.26). So $\chi((G_{l-1})^\circ) \subset C$ for the connected identity component $(G_{l-1})^\circ$ of G_{l-1}. Since G_{l-1} is normal in G we have $a^{-1} G_{l-1} a \subset G_{l-1}$, and hence $a^{-1}(G_{l-1})^\circ a$ is a connected component of G_{l-1} containing id, i.e.,

$$a^{-1}(G_{l-1})^\circ a = (G_{l-1})^\circ. \tag{14.9}$$

Observe that $G_{l-1} \subset \langle (G_{l-1})^\circ, a \rangle$, see Exercise 14.6, therefore by (14.9) each element of G_{l-1} is of the form $a^m g$ for some $g \in (G_{l-1})^\circ$ and $m \in \mathbb{Z}$. The a-invariance of f implies $\chi(G_{l-1}) \subset C$, so χ can be identified with a complex-valued function. Extend now χ to a complex-valued function on G by setting

$$\chi(g) := \int_X (L_g f) f^{-1} \,\mathrm{dm} \quad (g \in G).$$

A short calculation shows that this definition indeed yields an *extension* of χ, and we immediately see that χ is continuous and that $\chi(ag) = \chi(g)$ for every $g \in G$. By rotation invariance of m we conclude $\chi(g^{-1}) = \overline{\chi(g)}$ for every $g \in G$. Recall that $L_{g_{l-1}} f = \chi(g_{l-1}) f$, so $f = \chi(g_{l-1}) L_{g_{l-1}^{-1}} f$ for every $g_{l-1} \in G_{l-1}$. By the rotation invariance of m and by $f^{-1} = \overline{f}$ we conclude for every $g \in G$ and $g_{l-1} \in G_{l-1}$ that

$$\chi(g g_{l-1}) = \int_X (L_{g g_{l-1}} f) f^{-1} \,\mathrm{dm} = \int_X (L_{g_{l-1}} L_g f) f^{-1} \,\mathrm{dm}$$
$$= \int_X (L_g f) \overline{L_{g_{l-1}^{-1}} f} \,\mathrm{dm} = \chi(g) \int_X (L_g f) f^{-1} \,\mathrm{dm} = \chi(g)\chi(g_{l-1}).$$

On the other hand, again by rotation invariance, we can write

$$
\begin{aligned}
\chi(g_{l-1}g) &= \int_X (L_{g_{l-1}g} f) f^{-1}\,\mathrm{dm} = \int_X (L_g L_{g_l} f) f^{-1}\,\mathrm{dm} \\
&= \int_X (L_{g_l} f) L_{g^{-1}} f^{-1}\,\mathrm{dm} = \chi(g_{l-1}) \int_X f L_{g^{-1}} f^{-1}\,\mathrm{dm} \\
&= \chi(g_{l-1}) \int_X L_g f f^{-1}\,\mathrm{dm} = \chi(g_{l-1})\chi(g),
\end{aligned} \tag{14.10}
$$

implying $\chi(gg_{l-1}) = \chi(g_{l-1}g)$. Since f is a vertical character associated to χ we have $\chi(gg_{l-1}) = \chi(g_{l-1}g)\chi([g, g_{l-1}])$ and therefore by the above $\chi(g_{l-1})\chi(g) = \chi(g_{l-1})\chi(g)\chi([g, g_{l-1}])$. Recall that, by definition, $\chi(g_{l-1})$ is unimodular and thus can be cancelled. We conclude $\chi(g) = \chi(g)\chi([g, g_{l-1}])$. Now, by continuity of χ there exists a symmetric open neighborhood U of the identity in G with $\chi(g) \neq 0$ for every $g \in U$. Thus we have $\chi([g, g_{l-1}]) = 1$ for every $g \in U$. We now show that this holds for every $g \in G$. Let first $g \in G^\circ$. Then there is $m \in \mathbb{N}$ with $g \in (U \cap G^\circ)^m \subset U^m$ (see Proposition 11.3), i.e., $g = h_1 h_2 \cdots h_m$ for some $h_1, \ldots, h_m \in U \cap G^\circ$. Since $[h_j, g_{l-1}] \in G_l$ and G_l is contained in the center of G we obtain by Exercise 14.1(b)

$$
\begin{aligned}
&[h_m, g_{l-1}][h_{m-1}, g_{l-1}] \cdots [h_1, g_{l-1}] \\
&\quad = [h_{m-1}h_m, g_{l-1}][h_{m-2}, g_{l-1}] \ldots [h_1, g_{l-1}] \\
&\quad = \cdots = [h_1 \ldots h_{m-1}h_m, g_{l-1}] = [g, g_{l-1}].
\end{aligned}
$$

Since χ is multiplicative on G_l and $[h_j, g_{l-1}] \in G_l$, we conclude

$$
\chi([g, g_{l-1}]) = \chi([h_m, g_{l-1}]) \cdots \chi(h_1, g_{l-1}]) = 1,
$$

showing $\chi = 1$ on $[G^\circ, G_{l-1}]$.

By the commutator identities in Exercise 14.1 (b) and (a) and the assumption $G = \langle a, G^\circ \rangle$, the group $G_l = [G, G_{l-1}]$ is generated by $[G^\circ, G_{l-1}]$ and $[a^m, G_{l-1}]$, $m \in \mathbb{Z}$. So by multiplicativity of χ on G_l it remains to show that $\chi = 1$ on $[a^m, G_{l-1}]$ for every $m \in \mathbb{Z}$. But this follows from (14.10) and the a-invariance of f:

$$
\begin{aligned}
\chi([a^m, g_{l-1}]) &= \chi(g_{l-1}^{-1} a^m g_{l-1}) = \chi(g_{l-1}^{-1})\chi(a^m g_{l-1}) \\
&= \chi(g_{l-1}^{-1})\chi(g_{l-1}) = 1.
\end{aligned}
$$

Thus $\chi|_{G_l} - \mathbf{1}$ constant 1. The proof of (a) is complete.

(b) We again argue by induction on the nilpotency step l. For $l = 1$ there is nothing to prove. Let (X, m, a) be an l-step nilsystem and assume that the assertion holds for $(l-1)$-step nilsystems. Assume that the factor $X_2 = G/G_2\Gamma$ is uniquely ergodic. By the induction hypothesis, see Remark 14.24, the factor $X_l = G/G_l\Gamma$ is uniquely ergodic. Moreover, since unique ergodicity implies ergodicity of the measure, (X, m, a) is ergodic by the already proved part (a).

Let $\pi := \pi_l : X \to X_l$ be the canonical factor map and denote by A the set of generic points for $\mathrm{C}(X)$ with respect to m, i.e., the set of $x \in X$ satisfying $\lim_{N\to\infty} \frac{1}{N}\sum_{n=1}^{N} f(a^n x) = \int_X f \,\mathrm{dm}$ for every $f \in \mathrm{C}(X)$. By ergodicity of (X, m, a) we have $\mathrm{m}(A) = 1$, see Remark 7.49. Moreover, since G_l belongs to the center of G and m is rotation invariant, $g_l x \in A$ holds for every $g_l \in G_l$ and every $x \in A$, i.e., $G_l A = A$. For $B := \pi(A)$ this means $A = \pi^{-1}(B)$. Note that A is a cylinder Borel set, and as a consequence $B = \pi(A)$ is a Borel set.

Let now ν be an ergodic invariant measure for (X, a). By unique ergodicity of X_l, the push-forward measures of m and ν on X_l coincide. So $\nu(A) = \pi_*\nu(B) = \pi_*\mathrm{m}(B) = \mathrm{m}(A) = 1$. Since by the ergodicity of ν, for ν-almost every $x \in X$ one has $\lim_{N\to\infty} \frac{1}{N}\sum_{n=1}^{N} f(a^n x) = \int_X f \,\mathrm{d}\nu$ (see Birkhoff's Theorem 7.40), it follows that $\int_X f \,\mathrm{dm} = \int_X f \,\mathrm{d}\nu$ for every $f \in \mathrm{C}(X)$. Thus $\mathrm{m} = \nu$ and (X, a) is uniquely ergodic by Remark 3.35.

(c) Assume that the factor X_2 is minimal. By a property of compact Abelian groups, see Theorem 11.26, it is uniquely ergodic implying by (b) unique ergodicity of (X, a). Since m is rotation invariant and has full support, by Corollary 3.37 it follows that (X, a) is minimal.

(d) is analogous to (c) or, alternatively, follows from (c) and Corollary 14.26. ∎

Remark 14.28

(a) None of the properties mentioned in Theorem 14.27 (unique ergodicity, ergodicity, minimality, transitivity) imply $G = \langle G^\circ, a\rangle$ as the following example shows: Consider $G := \mathbb{Z} \times \mathbb{R}$, $\Gamma := \mathbb{Z} \times \mathbb{Z}$ and $a := (2, \alpha)$ for an irrational α. Then $G^\circ = \{0\} \times \mathbb{R}$, $\langle G^\circ, a\rangle = 2\mathbb{Z} \times \mathbb{R} \neq G$ and the nilsystem is isomorphic to the unit circle with an irrational rotation, hence uniquely ergodic and minimal.

(b) A natural modification of the argument in the proof of Theorem 14.27(a) shows the following: If (X, a), $X = G/\Gamma$ is a uniquely ergodic nilsystem with $G = \langle G^\circ, a\rangle$, then every eigenfunction $f \in \mathrm{L}^2(X)$ of the Koopman operator L_a is of the form $f = g \circ \pi$, where $\pi : X \to X_2$ is the canonical factor map and g is an eigenfunction on the factor $(X_2, \mathrm{m}, \pi(a))$. (For connected, simply connected Lie groups this result is again due to Green [493] and [59, Ch. 5] with an alternative proof by Parry [841].) In particular, in this case $(X_2, \mathrm{m}, \pi(a))$ is the Kronecker factor of (X, m, a) by the Halmos–von Neumann Theorem 8.32 (a).

Example 14.29 Let $(G/\Gamma, a)$ be a three dimensional Heisenberg nilsystem from Example 14.7. Using the notation from this example we write $a = [\alpha, \beta, \gamma]$. The Abelian factor X_2 can be identified with $\begin{pmatrix} 1 & \mathbb{R}/\mathbb{Z} & 0 \\ 0 & 1 & \mathbb{R}/\mathbb{Z} \\ 0 & 0 & 1 \end{pmatrix}$, i.e., with $\mathbb{T}^2$ with the usual multiplication, with (α, β) corresponding to $\pi(a)$. Thus by Theorem 14.27 and Corollary 3.41 each of the properties ergodicity, unique ergodicity, minimality and transitivity for the Heisenberg system is equivalent to the rational independence of α and β. We leave the details as Exercise 14.3.

Next we define subnilmanifolds and recall that by Cartan's closed subgroup theorem, every closed subgroup of a Lie group is itself a Lie group.

Definition 14.30 Let $X = G/\Gamma$ be a nilmanifold, $x \in X$ and let H be a closed subgroup of G. If Hx is closed, it is called a **subnilmanifold** of X. If $a \in G$ and $Hx = aHx$, then (Hx, a) is called a **subnilsystem** of (X, a).

For a closed subgroup H of G, the set Hx is in general not closed as the example $X := \mathbb{R}/\mathbb{Z}$, $x := \mathbb{Z}$ and $H := \{na : n \in \mathbb{Z}\} \subset \mathbb{R}$ for an irrational $a \in \mathbb{R}$ shows.

As the name suggests, every subnilmanifold of a nilmanifold can be given the structure of a nilmanifold in the following sense.

Lemma 14.31 *Let $Y := Hx$ be a subnilmanifold of a nilmanifold G/Γ for $x = g\Gamma$. Then there exists a homeomorphism between Y and $H/(H \cap g\Gamma g^{-1})$ commuting with the action of H and $H/(H \cap g\Gamma g^{-1})$ is a nilmanifold.*

Proof Consider the map $\varphi : Y \to H/(H \cap g\Gamma g^{-1})$ given by

$$\varphi : hx \mapsto h(H \cap g\Gamma g^{-1}).$$

It is easy to see that φ is well defined, bijective and continuous (i.e., a homeomorphism) and commutes with the action of H. In particular, $H \cap g\Gamma g^{-1}$ is cocompact in H. Since $H \cap g\Gamma g^{-1}$ is clearly discrete in H we see that $H/(H \cap g\Gamma g^{-1})$ is a nilmanifold. ■

Closed subgroups H of G such that Hx is closed for $x := \mathrm{id}\Gamma \in X$ are called **rational** (or **Γ-rational**). By Lemma 14.31, if H is rational, then $H \cap \Gamma$ is cocompact in H and Hx is homeomorphic to $H/(H \cap \Gamma)$. We will use the following characterization of rational subgroups.

Lemma 14.32 *For a nilmanifold $X := G/\Gamma$, a closed subgroup H of G is rational if and only if $H\Gamma$ is a closed subset of G.*

Proof Let $q : G \to G/\Gamma$ be the canonical projection and denote $x := \mathrm{id}\Gamma \in X$. The assertion follows from $H\Gamma = q^{-1}(Hx)$ and the definition of the topology of G/Γ. ■

Remark 14.33 By a quick application of Baire's category theorem one can show that the closedness of $H\Gamma$ in G already implies the closedness of H. Moreover, one can show that the compactness of $H/(H \cap \Gamma)$ is not only necessary but also sufficient for H being rational, just providing one more equivalent condition in Lemma 14.32. For details see Host and Kra [553, Lem. 10.14].

Example 14.34 Denote by G° the connected component of the identity of G (and recall that it is an open normal subgroup of G by Exercise 11.6(b) and Exercise 11.7(a)). Let $X := G/\Gamma$ be a nilmanifold and S be an arbitrary subset of

G. We show that the subgroup $H := \langle G^\circ, S\rangle$ (which is closed by Exercise 11.7(c)) is rational. The normality of G° implies $H\Gamma = \cup_{g\in\langle S\rangle\Gamma} G^\circ g$. So the set

$$G \setminus (H\Gamma) = \cup_{g\notin G^\circ\langle S\rangle\Gamma} G^\circ g$$

is open, i.e., H is rational by Lemma 14.32.

We now present the so-called **orbit closure theorem** due to Leibman [719, Thm. 2.21] and Lesigne [733] (in the case of connected, simply connected groups, cf. Ratner [865] and Shah [931, Thm. 1.3] for the so-called Ratner orbit closure theorem for unipotent flows).

The proof we present is based on a modification of Leibman's proof due to Host and Kra [553, Sec. 11.2.3].

Theorem 14.35 (Orbit Closures are Subnilmanifolds) *Let (X, m, a), $X = G/\Gamma$, be a nilsystem, $x \in X$ and $Y := \overline{\mathrm{orb}}(x) = \overline{\{a^n x : n \in \mathbb{Z}\}}$. Then there exists a closed subgroup H of G containing a with $Y = Hx$. In particular, Y is a subnilmanifold and hence has the structure of a nilmanifold.*

Note that for $x = \mathrm{id}\Gamma$ the subgroup H from Theorem 14.35 is automatically rational.

Proof We first show that one can assume without loss of generality that $x = q(\mathrm{id}) = \Gamma$ for the canonical factor map $q : G \to X$. Indeed, for $x = g\Gamma$ consider $b := g^{-1}ag$. If there exists a closed subgroup F of G containing b with $\overline{\{b^n\Gamma : n \in \mathbb{Z}\}} = F\Gamma$, then for the closed subgroup $H := gFg^{-1}$ we have $a \in H$ and $g^{-1}\overline{\{a^n x : n \in \mathbb{Z}\}} = g^{-1}Hg\Gamma$, proving together with Lemma 14.31 the assertion.

So we assume $x = q(\mathrm{id}) = \mathrm{id}\Gamma$ and make the following simple observation. Assume that we have a rational subgroup E of G containing a with $Y \subset Ex$. We show that we can assume without loss of generality $G = E$. Indeed, by Lemma 14.31 there exists a homeomorphism $\varphi : Ex \to E/(E \cap \Gamma)$ commuting with the action of E. Assume that there exists a closed subgroup $H \subset E$ containing a such that $\varphi(Y) = \overline{\{a^n\varphi(x) : n \in \mathbb{Z}\}} = H\varphi(x) = \varphi(Hx)$. Then $Y = Hx$ with H being a closed subgroup of G. Thus we can, indeed, assume $G = E$.

We first apply this observation to the subgroup $\langle G^\circ, a\rangle$, which is rational by Example 14.34, and assume without loss of generality $G = \langle G^\circ, a\rangle$. Consider the compact Abelian group $Z := X_2 = G_2\backslash X$ with the canonical factor map $\pi := \pi_2 : X \to Z$. If $Y = X$ then there is nothing to prove, otherwise by $x = \Gamma$, the lifting criterion (Theorem 14.27) and Theorem 11.26 we have

$$Z' := \overline{\{a^n \mathrm{id}_Z : n \in \mathbb{Z}\}} = \overline{\{\pi(a^n x) : n \in \mathbb{Z}\}} \neq Z.$$

It is easy to see that $\pi \circ q : G \to Z$ is a homomorphism. Consider the closed subgroup $G' := (\pi \circ q)^{-1}Z'$ containing a. Since $G = \langle G^\circ, a\rangle$, we have $G = \bigcup_{n\in\mathbb{Z}} a^n G^\circ$, so from $a \in G'$ we conclude $G' = \bigcup_{n\in\mathbb{Z}} a^n(G' \cap G^\circ)$. Now G' is a proper subgroup of G, so $G' \cap G^\circ$ must be a proper subgroup of

G°. By Proposition 11.4(b) $G'^\circ \subset G^\circ$, so $G'^\circ \subset G' \cap G^\circ$. Thus, by the above, G'° is a proper subgroup of G°. We conclude that $\dim G' < \dim G$. Moreover, $G'x = q(G') = \pi^{-1}Z'$ is closed and hence G' is rational. By construction we have $Y \subset G'x$ and by the above observation we can assume without loss of generality $G = G'$. Applying the observation one more time to the new subgroup $\langle G^\circ, a\rangle$, which is rational by Example 14.34, we can further keep the property $G = \langle G^\circ, a\rangle$. By repeating the procedure we are finished after finitely many steps (at latest by arriving at $\dim G = 0$).
The last statement follows from Lemma 14.31. ■

Remark 14.36

(a) In general one cannot take $H = \overline{\{a^n : n \in \mathbb{Z}\}}$ in Theorem 14.35 as the following example shows. Consider $G := \mathbb{R}$, $\Gamma := \mathbb{Z}$, $x := \mathbb{Z}$ and let a be an irrational number. Then $\overline{\{na : n \in \mathbb{Z}\}} + x = \{na : n \in \mathbb{Z}\} + x$ is countable while $\overline{\{na + x : n \in \mathbb{Z}\}} = \mathbb{R}/\mathbb{Z}$.
(b) There are also closed subgroups E of G with $Y = Ex$ which do not contain a: Consider for example $G := \mathbb{R}$, $\Gamma := \mathbb{Z}$, $x := \mathbb{Z}$ and $a := \frac{1}{2}$. Then $E := \frac{3}{2}\mathbb{Z}$ is such a subgroup.
(c) The maximal closed subgroup E with $Y = Ex$ is given by $E = E_{\max} := \{g \in G : gY = Y\}$ and it contains a, see Exercise 14.14.

The following is a generalization of the notion of equidistribution to compact Borel probability spaces.

Definition 14.37 Let (Y, μ_Y) be a compact Borel probability space. We call a sequence $(a_n)_{n\in\mathbb{N}}$ in Y **equidistributed in** Y if for every $f \in \mathrm{C}(Y)$

$$\lim_{N\to\infty} \frac{1}{N} \sum_{n=1}^{N} f(a_n) = \int_Y f \, \mathrm{d}\mu_Y.$$

Remark 14.38 We leave as Exercise 14.15 to show that every sequence which is equidistributed in Y is dense in the support of μ_Y.

For nilmanifolds we will consider equidistribution with respect to the Haar measure. Moreover, for a subnilmanifold $Y = Hx$ of a nilmanifold we will consider equidistribution in Y with respect to the unique H-invariant measure which exists by Lemma 14.31.

The following equidistribution result due to Leibman [719] shows in particular everywhere convergence of ergodic averages for nilsystems and continuous functions, see Lesigne [733] for the case of connected, simply connected groups.

Theorem 14.39 (Equidistribution and Convergence of Ergodic Averages) *Let (X, m, a) be a nilsystem, $x \in X$ and $Y := \overline{\mathrm{orb}}(x) = \overline{\{a^n x : n \in \mathbb{Z}\}}$. Then the sequence $(a^n x)_{n\in\mathbb{N}}$ is equidistributed in Y. Moreover, for a nilsystem (X, m, a) with Koopman operator T the ergodic averages $\frac{1}{N}\sum_{n=1}^{N} T^n f$ converge* pointwise everywhere *for every* $f \in \mathrm{C}(X)$. *If the nilsystem is minimal, the convergence is* uniform *for every* $f \in \mathrm{C}(X)$.

Proof By Theorem 14.35 we can assume that Y is itself a nilmanifold, say $Y = \tilde{G}/\tilde{\Gamma}$. Then the system (Y, a) is by definition transitive, hence so is its horizontal torus. By Theorem 11.26 this compact, Abelian group factor is uniquely ergodic, therefore the nilsystem (Y, a) is uniquely ergodic by Theorem 14.27. The assertion follows from the pointwise ergodic theorem for uniquely ergodic systems, see Theorem 7.23. ■

Remark 14.40 It follows from the previous result that every nilsequence $(a_n)_{n\in\mathbb{N}}$ is Cesàro convergent, i.e.,

$$\lim_{N\to\infty} \frac{1}{N}\sum_{n=1}^{N} a_n \quad \text{exists.}$$

Now we can prove the equivalence of various irreducibility properties of nilsystems, a result whose analogue was proved for compact group rotations in Theorem 11.26.

Theorem 14.41 (Equivalence of Minimality and Ergodicity Properties) *For a nilsystem* (X, m, a) *the following assertions are equivalent.*

(i) (X, m, a) *is ergodic.*
(ii) (X, a) *is uniquely ergodic.*
(iii) (X, a) *is minimal.*
(iv) (X, a) *is transitive.*
(v) *For every* $x \in X$ *the sequence* $(a^n x)_{n\in\mathbb{N}}$ *is equidistributed in* X.

In this case, the ergodic averages of the Koopman operator T *satisfy for every* $f \in \mathrm{C}(X)$

$$\lim_{N\to\infty} \frac{1}{N}\sum_{n=1}^{N} T^n f = \int_X f \,\mathrm{dm},$$

where the convergence is uniform.

Proof (ii)⇒(i) is clear, while the equivalence of (iii) and (iv) was shown in Corollary 14.26.
(ii)⇒(iii): The Haar measure m has full support and a uniquely ergodic system with such an invariant measure is minimal by Corollary 3.37.
(iii)⇒(ii): By minimality, the closure of every orbit is X implying by Theorem 14.39

$$\lim_{N\to\infty} \frac{1}{N}\sum_{n=1}^{N} f(a^n x) = \int_X f \,\mathrm{dm} \quad \text{for all } f \in \mathrm{C}(X),\ x \in X. \tag{14.11}$$

Thus unique ergodicity follows from Proposition 7.24.

(i)⇒(iv): Assume that m is ergodic. Since X is metrizable, by Birkhoff's ergodic theorem or, more precisely, by Remark 7.49 there exists a full measure set $X' \subset X$ with

$$\lim_{N\to\infty} \frac{1}{N} \sum_{n=1}^{N} f(a^n x) = \int_X f \,\mathrm{dm} \tag{14.12}$$

for every $x \in X'$ and $f \in \mathrm{C}(X)$. Take one such $x \in X'$ and set $Y := \overline{\mathrm{orb}_+}(x) \subset X$. By Theorem 14.39 we obtain $\int_X f \,\mathrm{dm} = \int_Y f \,\mathrm{dm}_Y$ for every $f \in \mathrm{C}(X)$ implying $\mathrm{m} = \mathrm{m}_Y$ and hence $Y = X$. Thus x is a transitive point for (X, a).
(iii)⇔(v) follows from Remark 14.38, Theorem 14.39 and the fact that minimality is equivalent to the property that every two-sided orbit is dense (see Remark 3.1). The last assertion follows from the pointwise ergodic theorem for uniquely ergodic systems, see Theorem 7.23. ■

Remark 14.42

(a) If the nilsystem $(G/\Gamma, a)$ satisfies $G = \langle G^\circ, a\rangle$, then the equivalence in Theorem 14.41 follows directly from Theorem 14.27 and the corresponding equivalence for compact Abelian groups, see Theorem 11.26.
(b) For a connected nilsystem one can add one more equivalent property to the ones in Theorem 14.41, namely **total ergodicity**, i.e., the property that the system (X, m, a^k) is ergodic for each $k \in \mathbb{N}$, see Exercise 14.13.
(c) The lifting criterion (Theorem 14.27) leads to the following equivalent condition to the ones in Theorem 14.41 under the additional assumption $G = \langle G^\circ, a\rangle$, where $\pi : X \to X_2 = G/G_2\Gamma$ denotes the canonical factor map:

(vi) $\{\pi(a^n) : n \in \mathbb{N}\}$ *is dense in* X_2.

We finish this section with one more lifting property in the spirit of Theorem 14.27 known as (Leon) Green's theorem, see B. Green and T. Tao [489]. It follows directly from Theorem 14.41 and the lifting criterion (Theorem 14.27).

Corollary 14.43 *Let (X, a) be a nilsystem with $X = G/\Gamma$, $G = \langle G^\circ, a\rangle$, let $\pi : X \to X_2 = G/G_2\Gamma$ be the canonical projection and let $x \in X$. Then the sequence $(a^n x)_{n\in\mathbb{N}}$ is equidistributed in X if and only if its image $(\pi(a^n x))_{n\in\mathbb{N}}$ is equidistributed in X_2.*

14.4 Multiple and Polynomial Ergodic Theorems

Since a finite product of nilsystems has structure of a nilsystem, one immediately obtains the following multiple version of the pointwise ergodic theorem.

Proposition 14.44 (Everywhere Convergence of Multiple Ergodic Averages) *Let (X, a) be a nilsystem and let $k \in \mathbb{N}$. For every $f_1, \ldots, f_k \in \mathrm{C}(X)$ the multiple* *ergodic averages*

$$\frac{1}{N}\sum_{n=1}^{N} T^n f_1 \cdot T^{2n} f_2 \cdots T^{kn} f_k \tag{14.13}$$

converge everywhere.

Proof Let $X = G/\Gamma$ and consider the nilsystem given by the rotation $\tilde{T}$ by $(a, a^2, \ldots, a^k)$ on the nilmanifold $\tilde{X} := G^k/\Gamma^k$ and consider the function $F \in \mathrm{C}(\tilde{X})$ defined by $F(x_1, \ldots, x_k) := f_1(x_1) \cdot \ldots \cdot f_k(x_k)$. By Theorem 14.39, the averages

$$\frac{1}{N}\sum_{n=1}^{N}(\tilde{T}^n F)(x, \ldots, x) = \frac{1}{N}\sum_{n=1}^{N} f_1(a^n x) \cdot \ldots \cdot f_k(a^{kn} x)$$

converge for every $x \in X$, proving the assertion. ■

For an explicit formula for the limit of the averages (14.13) in almost every point see Ziegler [1064] as well as Lesigne [731] for the case $k = 3$ and Bergelson et al. [123] for an alternative proof.

Finally, we come to the polynomial ergodic theorem for nilsystems. An easy computation shows that for g, h in a nilpotent group G of step 2 and $n \in \mathbb{N}$ one has $g^n h^n = (gh)^n [h, g]^{\binom{n}{2}}$. The longer expression for general groups is known as *Hall–Petresco formula*

$$g^n h^n = (gh)^n g_2^{\binom{n}{2}} \cdots g_{n-1}^{\binom{n}{n-1}} g_n$$

for some appropriate $g_2 \in G_2, \ldots, g_n \in G_n$. It follows that in a nilpotent group the multiplication is of polynomial nature and it is interesting and natural to consider **polynomial sequences**, i.e., sequences $(g(n))_{n\in\mathbb{Z}}$ of the form

$$g(n) = a_1^{p_1(n)} a_2^{p_2(n)} \cdots a_k^{p_k(n)}, \quad n \in \mathbb{Z}$$

for $k \in \mathbb{N}$, $a_1, \ldots, a_k \in G$ and polynomials $p_1, \ldots, p_k : \mathbb{Z} \to \mathbb{Z}$. (For an abstract alternative definition of polynomial sequences see Leibman [715] and Green and Tao [489] for the equivalence of these definitions.) Polynomial sequences form a group with respect to multiplication, see Leibman [715]. The following equidistribution property of polynomial orbits in the spirit of Theorem 14.39 is due to Leibman, for a proof we refer to [719].

Theorem 14.45 (Equidistribution of Polynomial Sequences and Polynomial Ergodic Theorem) *Let $X = G/\Gamma$ be a nilmanifold, $(g(n))_{n\in\mathbb{Z}}$ be a polynomial sequence in G and $x \in X$. Then there exists a connected, closed subgroup E of G,* *a natural number $k \in \mathbb{N}$ and (not necessarily distinct) points $x_1, \dots, x_k \in X$ such that the following assertions hold.*

(a) *The set Ex_j is closed and hence a subnilmanifold for every $j \in \{1, \dots, k\}$.*
(b) *The polynomial orbit of x satisfies $\overline{\{g(n)x : n \in \mathbb{Z}\}} = \cup_{j=1}^k Ex_j$.*
(c) *For every $j \in \{1, \dots, k\}$ the sequence $(g(j + kn)x)_{n\in\mathbb{N}}$ is equidistributed in Ex_j.*

In particular, for every nilsystem (X, a) with Koopman operator T the polynomial ergodic averages

$$\frac{1}{N}\sum_{n=1}^{N} T^{P(n)} f$$

converge everywhere *for every $f \in \mathrm{C}(X)$ and every polynomial $P : \mathbb{Z} \to \mathbb{Z}$.*

A quantitative version of this result due to Green and Tao [489, 490] leads to remarkable applications to the number theoretic Sarnak conjecture, which is discussed in Chap. 17 below.

The following example shows that even for simplest systems, unlike the linear case, a polynomial orbit need not be equidistributed in a single subnilmanifold.

Example 14.46 Consider $G := \mathbb{R}$, $\Gamma := \mathbb{Z}$, $g(n) := n^2$ and $x := \frac{1}{3}$. Then $g(n)x$ takes two values $0, \frac{1}{3}$ with different frequency leading to $E = \{0\}$ and $x_1 = x_2 = \frac{1}{3}$ and $x_3 = 0$ in Theorem 14.45.

We close this chapter with the following polynomial generalization of Proposition 14.44.

Corollary 14.47 (Everywhere Convergence of Multiple Polynomial Ergodic Averages) *Let (X, a) be a nilsystem. Then for every $k \in \mathbb{N}$, all polynomials $p_1, \dots, p_k : \mathbb{Z} \to \mathbb{Z}$ and every $f_1, \dots, f_k \in \mathrm{C}(X)$ the multiple polynomial ergodic averages*

$$\frac{1}{N}\sum_{n=1}^{N} T^{p_1(n)} f_1 \cdot T^{p_2(n)} f_2 \cdots T^{p_k(n)} f_k \tag{14.14}$$

converge everywhere.

Proof The assertion follows immediately from Theorem 14.45 applied to the nilmanifold $\tilde{X} := G^k/\Gamma^k$ and the polynomial sequence

$$(a^{p_1(n)}, \dots, a^{p_k(n)})_{n\in\mathbb{Z}} = (b_1^{p_1(n)} \cdots b_k^{p_k(n)} y)_{n\in\mathbb{Z}},$$

where $b_1 := (a, \mathrm{id}, \dots, \mathrm{id}), \dots, b_k := (\mathrm{id}, \dots, \mathrm{id}, a)$ and $y := (x, \dots, x)$. ■

It is much more difficult to find an explicit formula for the limit of the averages (14.14) than in the linear case, see, e.g., Frantzikinakis and Kra [423] and Leibman [721].

As will be discussed in Chap. 15, nilsystems serve as building blocks for general (regular) measure-preserving systems and the above results lead to mean convergence of the corresponding averages for general systems.

Remark 14.48 (Everywhere Convergence along Primes and Polynomials of Primes) For everywhere convergence of ergodic averages along primes and polynomials of primes for nilsystems and continuous functions see Green and Tao [488, Thm. 7.1], [340], and Le [709].

Exercises

Exercise 14.1 (Commutator Identities) Let G be a group, $a, b, c \in G$ and denote $a^b := b^{-1}ab$. Show the following identities:

(a) $[a,b]^{-1} = [b,a], \quad c^{-1}[a,b]c = [a^c, b^c]$.
(b) $[ab,c] = [a,c][[a,c],b][b,c], \quad [a,bc] = [a,c][a,b][[a,b],c]$.
(c) $[ab,c] = [a^b, c^b][b,c]$.
(d) $[[a,b],c^a][[c,a],b^c][[b,c],a^b] = \mathrm{id}$.
(e) For normal subgroups A, B, C of G

$$[[A,B],C] \subset [[C,A],B][[B,C],A].$$

Deduce that for a group G, each subgroup G_j from the lower central series is normal, satisfies $G_{j-1} \trianglerighteq G_j$ and is generated by elements of the form $[[[g_1, g_2], g_3] \dots, g_j]$ with $g_1, \dots, g_j \in G$.

Exercise 14.2 (Products of Nilsystems) Prove the assertions in Example 14.6.

Exercise 14.3 (Heisenberg System) Work out the details in Examples 14.7, 14.8 and 14.29.

Exercise 14.4 (Skew Shifts) Let $d \in \mathbb{N}$ with $d \geq 2$. Show that the skew shift on $\mathbb{T}^d$ given by formula (14.1) is isomorphic to the d-step Heisenberg system $(G/\Gamma, a)$ with

$$a = \begin{pmatrix} 1\,1\,0 \dots 0\ \alpha_1 \\ 0\,1\,1 \dots 0\ \alpha_2 \\ \dots \\ 0\,0\,0 \dots 1\ \alpha_d \\ 0\,0\,0 \dots 0\ \ 1 \end{pmatrix}.$$

Exercise 14.5 (Affine Systems) Work out the details in Example 14.11.

Exercise 14.6 Let G be a topological group and denote by G° the connected component containing id. Prove that if $G = \langle G^\circ, S\rangle$ for a set $S \subset G$, then every element of the lower central series satisfies $G_j = \langle (G_j)^\circ, G_j \cap \langle S\rangle\rangle$. (Hint: As in Exercise 14.1 show that G_j is generated by elements of the form $[[[g_1, g_2], g_3] \ldots, g_j]$ with $g_1, \ldots, g_j \in G^\circ \cup S$ and that such an element is in $(G_j)^\circ$ as soon as at least one g_k is in G°.)

Exercise 14.7 (Polynomial Sequences) Work out the case $k \geq 3$ in Example 14.13(c).

Exercise 14.8 (Connected Nilmanifolds) Let $X = G/\Gamma$ be a connected nilmanifold, $q : G \to \Gamma$ be the canonical projection and recall that G° denotes the connected component of G containing the identity. Prove the following assertions.

(a) One has $G = G^\circ\Gamma$. (Hint: Use the result in Exercise 11.7 (c) and the fact that q is an open map). Can one drop the assumption of connectedness of the nilmanifold?
(b) $X = G^\circ\Gamma/\Gamma$ is isomorphic to $G^\circ/(\Gamma \cap G^\circ)$. In particular, X can be represented as a nilmanifold coming from a connected group, cf. Exercise 14.6 (b).

Exercise 14.9 (Connected Components) Let $X := G/\Gamma$ be a nilmanifold with canonical projection $q : G \to X$ and denote by X° the connected component of X containing $q(\mathrm{id})$. Show the following assertions.

(a) $X^\circ = q(G^\circ) = G^\circ x$ and hence X° is a subnilmanifold of X which can be represented as $G^\circ/(G^\circ \cap \Gamma)$ in the sense of Lemma 14.31 (cf. Exercise 14.8 for the case of connected G).
(b) X has finitely many connected components and every connected component of X is of the form gX° for some $g \in G$. (Hint: Use (a) and Exercise 11.7(a).)

Exercise 14.10 (Generalized Polynomials) Provide the details for Example 14.15. (Hint: For the Riemann integrability property of the respective function F show that the set of discontinuities of F has measure zero.)

Exercise 14.11 (Vertical Characters) Let X be an l-step nilmanifold. Show that every vertical character is of the form (14.4).

Exercise 14.12 (The Spaces X_j) Let $X := G/\Gamma$ be a nilmanifold of step l, $j \in \{1, \ldots, l\}$ and $X_j = \tilde{G}/\tilde{\Gamma}$ be the representation from Lemma 14.22. For $k \in \{1, \ldots, j\}$ denote by $\tilde{X}_k$ the space which relates to X_j in the same manner as X_k relates to X.

(a) Show that one has $\tilde{X}_k = G/G_j/q_j(G_k\Gamma)$ for the quotient map $q_j : G \to G/G_j$.
(b) Deduce that there exists a homeomorphism between $\tilde{X}_k$ and X_k intertwining the action of G.

(c) Prove that $q_j(\Gamma)$ is homeomorphic to $\Gamma/(\Gamma \cap G_j)$ and therefore discrete in G/G_j.

Exercise 14.13 (Total Ergodicity) Show that a *connected* nilsystem is ergodic if and only if it is totally ergodic[4] in two different ways:

(a) By using Fourier analysis.
(b) By first proving that a topological system (K, T) with K connected is minimal if and only if (K, T^k) is minimal for each $k \in \mathbb{N}$ (a system with this latter property is called **totally minimal**).

Exercise 14.14 (Orbit Closures)

(a) Using the notation introduced in the proof of Theorem 14.35 show that $Z' = \pi(q(G'))$ is isomorphic to $G'/([G', G']\Gamma)$.
(b) Let (X, m, a) be a nilsystem, $x \in X$ and $Y := \overline{\mathrm{orb}}(x) = \overline{\{a^n x : n \in \mathbb{Z}\}}$. Show that $E = E_{\mathrm{max}} := \{g \in G : gY = Y\}$ is the maximal closed subgroup of G with $Ex = Y$ and it contains a. Can this part of the exercise be used to give an alternative proof of Theorem 14.35?

Exercise 14.15 (Equidistribution) Prove the assertion of Remark 14.38.

[4] See Exercise 9.6.

Chapter 15
Gowers–Host–Kra Seminorms and Multiple Convergence

After Furstenberg showed multiple recurrence (see Theorem 13.5), multiple convergence remained open for over twenty years and inspired a lot of research. In his proof of multiple recurrence, Furstenberg [445] showed L^2-convergence of multiple ergodic averages

$$\frac{1}{N}\sum_{n=1}^{N} T^n f_1 \cdot T^{2n} f_2 \cdots T^{kn} f_k \tag{15.1}$$

for bounded functions in the weakly mixing case, the limit being $\prod_{j=1}^{k}\int_X f_j \,\mathrm{d}\mu$. Moreover, he showed L^2-convergence for general systems for $k = 2$ (see Theorem 9.29). The case $k = 3$ was treated by Conze and Lesigne [256–258] under the assumption of total ergodicity. Finally, Host and Kra [549] solved the question in the general case, with an independent proof given subsequently by Ziegler [1065], as follows.

Theorem 15.1 (Multiple Convergence) *Let* (X, μ, T) *be a measure-preserving system,* $k \in \mathbb{N}$ *and* $f_1, \ldots, f_k \in \mathrm{L}^\infty(X, \mu)$. *Then the multiple averages* (15.1) *converge in* L^2.

Remark 15.2 Since multiple averages (15.1) involve only the functions $f_1, \ldots, f_k$ (and their orbits), we can reduce to the separable factor generated by these functions. Moreover, since the L^2-convergence of (15.1) is clearly preserved by Markov isomorphisms, by Remark 8.50(b) we can further assume that our system is regular.[1]

Thus it will be convenient, and no loss of generality, to have regularity of the measure-preserving system as a common assumption in this chapter.

[1] Recall that a system (X, μ, T) is called regular if X is a compact metric space and μ is a Borel probability measure on it.

T. Eisner, B. Farkas, *A Journey Through Ergodic Theorems*, Birkhäuser Advanced Texts Basler Lehrbücher, https://doi.org/10.1007/978-3-031-96506-7_15

Note that both Host, Kra and Ziegler constructed a characteristic factor for the averages (15.1) (see the end of Sect. 15.1 for a precise definition in this context) to show convergence, where Ziegler's factor is the minimal one which is characteristic for more general averages

$$\frac{1}{N}\sum_{n=1}^{N} T^{a_1 n} f_1 \cdot T^{a_2 n} f_2 \cdots T^{a_k n} f_k$$

for arbitrary positive integers $a_1, \ldots, a_k$. Leibman [718] showed that for each $k \geq 2$ the corresponding Host–Kra factor and Ziegler's factor coincide. See also Host and Kra [553, Sec. 21.2] for their treatment of these more general averages.

We present some ideas behind the proof of Theorem 15.1 by Host and Kra who introduced and used the so-called Gowers–Host–Kra (or uniformity) seminorms to reduce the problem to multiple convergence *for nilsystems* discussed in Proposition 14.44. We use Host and Kra [549, 553], Kra [669], Zorin-Kranich [1073] as well as Gowers [481] who originally introduced uniformity seminorms in his proof of Szemerédi's theorem and whose ideas inspired our presentation. While we present the construction up to the definition of the factors in detail, we spend only a very few words on the long and technical proof of the structure theorem (Theorem 15.45). For full details on it we refer to the works of Host and Kra [549, 553]. We finish this chapter with a brief review of some further developments and research directions.

15.1 Definition of Gowers–Host–Kra Seminorms

The uniformity seminorms were introduced by Gowers [481] in his Fourier analytic proof of Szemerédi's theorem for the shift on finite systems $\mathbb{Z}/N\mathbb{Z}$ and then extended by Host and Kra [549] to general ergodic measure-preserving systems. The definition in the non-ergodic case is due to Chu et al. [247], see also Host and Kra [553, Ch. 8].

Definition 15.3 (Gowers–Host–Kra (Uniformity) Seminorms) Suppose that (X, μ, T) is an ergodic measure-preserving system and let $f \in \mathrm{L}^\infty(X, \mu)$. The **Gowers–Host–Kra** (or **uniformity**) **seminorms** are defined recursively by

$$\|f\|_{U^1} := \left|\int_X f \,\mathrm{d}\mu\right|,$$

$$\|f\|_{U^{l+1}}^{2^{l+1}} := \limsup_{N\to\infty} \frac{1}{N}\sum_{n=1}^{N} \|T^n f \cdot \overline{f}\|_{U^l}^{2^l}, \quad l \in \mathbb{N}. \tag{15.2}$$

It is not yet clear why these expressions indeed define seminorms, we shall establish this property soon. For general measure-preserving systems the Gowers–Host–Kra seminorms are defined in the same manner starting with

$$\|f\|_{U^1}^2 := \lim_{N\to\infty} \frac{1}{N}\sum_{n=1}^{N} \int_X T^n f \cdot \overline{f}\, d\mu \tag{15.3}$$

$$= \int_X P_{\text{fix}\, T} f \cdot \overline{f}\, d\mu = \|P_{\text{fix}\, T} f\|_2^2 \tag{15.4}$$

where the equalities (and the existence of the limit) follow from the mean ergodic theorem and orthogonality, respectively.

Remark 15.4 The two definitions of the first seminorm coincide for ergodic systems by (15.4) since for ergodic systems $P_{\text{fix}\, T} f = \int_X f\, d\mu \cdot \mathbf{1}$. Note that the representation

$$\|f\|_{U^1} = \|P_{\text{fix}\, T} f\|_2 \tag{15.5}$$

implies in particular $\|f\|_{U^1} \geq 0$ for every $f \in L^\infty(X, \mu)$.

Remark 15.5 For a measure-preserving system (X, μ, T) and $f \in L^\infty(X, \mu)$ we see that

$$\|f\|_{U^1}^2 = \int_X P_{\text{fix}\, T} f \cdot \overline{f}\, d\mu = \int_X P_{\text{fix}\, T} f \cdot P_{\text{fix}\, T} \overline{f}\, d\mu = \|\overline{f}\|_{U^1}^2.$$

Thus, by induction

$$\|f\|_{U^l} = \|\overline{f}\|_{U^l} \text{ for all } l \in \mathbb{N}. \tag{15.6}$$

Moreover, one easily shows by induction the inequality

$$\|f\|_{U^l} \leq \|f\|_\infty \text{ for every } l \in \mathbb{N}. \tag{15.7}$$

Up to the triangle inequality, the properties of a seminorm are easily established. We will show the triangle inequality (which is clear for $\|\cdot\|_{U^1}$ anyway) later in Corollary 15.26. We will also see in Proposition 15.19 below that the limit superior in the above definition in (15.2) is actually a limit.

Remark 15.6

(a) For ergodic systems, the second seminorm satisfies

$$\|f\|_{U^2}^4 = \lim_{N\to\infty} \frac{1}{N}\sum_{n=1}^{N} |(T^n f | f)|^2. \tag{15.8}$$

Recall that by Proposition 6.22 this limit exists. In particular, for ergodic systems $\|f\|_{U^2} = 0$ if and only if f is weakly mixing.[2]

(b) For general systems one only has the inequality

$$\|f\|_{U^2}^4 \geq \lim_{N\to\infty} \frac{1}{N} \sum_{n=1}^{N} |(T^n f | f)|^2 \tag{15.9}$$

by

$$\begin{aligned}\|f\|_{U^2}^4 &= \limsup_{N\to\infty} \frac{1}{N} \sum_{n=1}^{N} \|P_{\mathrm{fix}\,T}(T^n f \cdot \overline{f})\|_2^2 \\ &\geq \limsup_{N\to\infty} \frac{1}{N} \sum_{n=1}^{N} \|P_{\mathbb{C}\mathbf{1}}(T^n f \cdot \overline{f})\|_2^2 = \lim_{N\to\infty} \frac{1}{N} \sum_{n=1}^{N} |(T^n f | f)|^2.\end{aligned}$$

Remark 15.7 By (15.8) and Proposition 6.22 one has the following representation for ergodic systems:

$$\|f\|_{U^2}^4 = \sum_{a\in\mathbb{T}} \|P_a f\|^4 \leq \|P_{\mathrm{kr}} f\|^4, \tag{15.10}$$

where P_a again denotes the orthogonal projection of T onto the eigenfunctions of T corresponding to the eigenvalue $a \in \mathbb{T}$. Moreover, the inequality in (15.10) is strict if and only if f correlates with at least two eigenfunctions of T corresponding to different eigenvalues, see Remark 6.15. Note that for general systems both the equality in (15.10) and the inequality $\|f\|_{U^2} \leq \|P_{\mathrm{kr}} f\|$ fail in general, see the non-ergodic skew shift example in Exercise 15.1.

Proposition 15.8 (Monotonicity of the Uniformity Seminorms) *For each measure-preserving system (X, μ, T) and each $f \in \mathrm{L}^\infty(X, \mu)$ one has*

$$\|f\|_{U^l} \leq \|f\|_{U^{l+1}} \text{ for every } l \in \mathbb{N}.$$

We give here the proof for the ergodic case. The proof for the general case (which we will not need further) is based on one of the alternative definitions of the seminorms discussed in Sect. 15.2 and is left as Exercise 15.6.

Proof for Ergodic Systems Assume that (X, μ, T) is ergodic (for the general case see Exercise 15.6) and let $f \in \mathrm{L}^\infty(X, \mu)$. We show the assertion by induction on l.

[2] i.e., belongs to the almost weakly stable part, see Remark 6.27.

For $l = 1$ we have by (15.10)

$$\|f\|_{U^2}^4 = \sum_{a\in\mathbb{T}} \|P_a f\|^4 \geq \|P_{\mathrm{fix}\,T} f\|^4 = \|f\|_{U^1}^4,$$

where P_a again denotes the orthogonal projection onto the eigenfunctions of T corresponding to the eigenvalue $a \in \mathbb{T}$ (so $P_1 = P_{\mathrm{fix}\,T}$).

Let now $l \geq 2$ and assume that $\|f\|_{U^{l-1}} \leq \|f\|_{U^l}$ holds. Then we have by the Cauchy–Schwarz inequality

$$\begin{aligned}\|f\|_{U^l}^{2^{l+1}} &= \limsup_{N\to\infty} \Big(\frac{1}{N}\sum_{n=1}^{N} \|T^n f\cdot \overline{f}\|_{U^{l-1}}^{2^{l-1}}\Big)^2 \leq \limsup_{N\to\infty} \frac{1}{N}\sum_{n=1}^{N} \|T^n f\cdot \overline{f}\|_{U^{l-1}}^{2^l}\\ &\leq \limsup_{N\to\infty} \frac{1}{N}\sum_{n=1}^{N} \|T^n f\cdot \overline{f}\|_{U^l}^{2^l} = \|f\|_{U^{l+1}}^{2^{l+1}},\end{aligned}$$

proving the assertion. ∎

Remark 15.9 As we will see in Lemma 15.28 below, for ergodic systems for all $l \in \mathbb{N}$ and $p_l := \frac{2^l}{l+1}$ the estimate $\|f\|_{U^l} \leq \|f\|_{p_l}$ holds, and therefore one can extend the definition of the lth uniformity seminorm to $\mathrm{L}^{p_l}(X, \mu)$ by continuity.

An important class of examples are rotations on compact Abelian groups.

Example 15.10 (Cyclic Systems and Ergodic Group Rotations) Consider the additive cyclic group $G := \mathbb{Z}/N\mathbb{Z}$ with the normalized counting measure m and define $T^h : \mathrm{L}^\infty(G, \mathrm{m}) \to \mathrm{L}^\infty(G, \mathrm{m})$ by $(T^h f)(x) := f(x + h)$. It is easy to see that the seminorms with respect to the ergodic shift transformation $x \mapsto x + 1$ satisfy

$$\begin{aligned}\|f\|_{U^1} &= \Big|\int_G f \,\mathrm{dm}\Big|,\\ \|f\|_{U^{l+1}}^{2^{l+1}} &= \int_G \|T^h f\cdot \overline{f}\|_{U^l}^{2^l} \,\mathrm{dm}(h) \text{ for every } l \in \mathbb{N}.\end{aligned}$$

These formulas provided the original uniformity (semi)norms as defined by Gowers [481] in his proof of Szemerédi's theorem. Note however that he used an equivalent definition discussed in Example 15.20 below, cf. also Exercise 15.5.

Moreover, the analogous representation holds for every compact Abelian group G, the normalized Haar measure m and an ergodic transformation $x \mapsto x + \alpha$, see Exercise 15.5. In addition, for such systems $\|\cdot\|_{U^l}$ is a norm for $l \geq 2$ and $\|\cdot\|_{U^2}$ has the simple representation

$$\|f\|_{U^2} = \Big(\sum_{\chi\in\hat{G}} |\hat{f}(\chi)|^4\Big)^{1/4}, \tag{15.11}$$

where $\hat{f}(\chi)$ denotes the Fourier coefficient of f corresponding to the character χ, see Exercise 15.2. In particular, since for ergodic systems the Kronecker factor is isomorphic to an ergodic rotation on a compact Abelian group by the Halmos–von Neumann theorem (Theorem 8.32), we see that for general ergodic measure-preserving systems the second seminorm $\|g\|_{U^2}$ of a function g can be computed, by (15.10), as in (15.11) using Fourier coefficients of the function f on G corresponding to $P_{\mathrm{kr}}g$ under the isomorphism.

The following property characterizes weak mixing in terms of the uniformity seminorms.

Proposition 15.11 (Weakly Mixing Systems) *For a measure-preserving system (X, μ, T) the following assertions hold.*

(a) *If a function $f \in \mathrm{L}^\infty(X, \mu)$ satisfies $\|f\|_{U^2} = 0$, then f is weakly mixing. For ergodic systems also the converse implication holds.*
(b) *The system (X, μ, T) is weakly mixing if and only if $\|f\|_{U^2} = \|f\|_{U^1}$ holds for every $f \in \mathrm{L}^\infty(X, \mu)$. In this case one has $\|f\|_{U^l} = \|f\|_{U^1}$ for every $l \in \mathbb{N}$ and every $f \in \mathrm{L}^\infty(X, \mu)$.*

Note that the converse implication in (a) fails in general, see Exercise 15.1.

Proof (a) The assertion for ergodic systems was shown in Remark 15.6(a). The assertion for general systems follows from (15.9).
(b) Let (X, μ, T) be weakly mixing (and in particular ergodic). By (15.10), Proposition 6.28 and (15.5) we obtain $\|f\|_{U^2} \leq \|f\|_{U^1}$ which, together with Proposition 15.8, leads to $\|f\|_{U^2} = \|f\|_{U^1}$ for every $f \in \mathrm{L}^\infty(X, \mu)$. For $l \geq 2$, the assertion is proved by induction. Suppose we have proved the equality $\|\cdot\|_{U^1} = \|\cdot\|_{U^l}$ for some $l \geq 2$. This induction hypothesis and ergodicity yield

$$\begin{aligned}\|f\|_{U^{l+1}}^{2^{l+1}} &= \limsup_{N\to\infty} \frac{1}{N}\sum_{n=1}^{N} \|T^n f \cdot \overline{f}\|_{U^l}^{2^l} = \limsup_{N\to\infty} \frac{1}{N}\sum_{n=1}^{N} \|T^n f \cdot \overline{f}\|_{U^1}^{2^l} \\ &= \limsup_{N\to\infty} \frac{1}{N}\sum_{n=1}^{N} |(T^n f | f)|^{2^l}.\end{aligned}$$

By decomposing f as a sum of the constant function $\int_X f\,\mathrm{d}\mu\cdot\mathbf{1}$ and a weakly mixing function and by using the Koopman–von Neumann lemma (Proposition 4.19), we see that $|(T^n f | f)|$ converge to $|\int_X f\,\mathrm{d}\mu|^2$ along a subsequence of density one. We thus obtain

$$\|f\|_{U^{l+1}}^{2^{l+1}} = \left|\int_X f\,\mathrm{d}\mu\right|^{2^{l+1}} = \|f\|_{U^1}^{2^{l+1}}.$$

This finishes the proof of the "only if" direction and the additional assertion.

For the "if" direction assume now that $\|f\|_{U^2} = \|f\|_{U^1}$ for every $f \in \mathrm{L}^\infty(X, \mu)$. Then by (15.9) and Proposition 6.14

$$\|f\|_{U^2}^4 \geq \lim_{N\to\infty} \frac{1}{N}\sum_{n=1}^{N} |(T^n f\,|\,f)|^2 = \sum_{a\in\mathbb{T}} \|P_a f\|^4 \geq \|P_{\mathrm{fix}\,T} f\|^4 = \|f\|_{U^1}^4$$

and, by assumption, we have here equalities. Hence $P_a f = 0$ for every $a \neq 1$ and by Proposition 6.28 it remains to show $\mathrm{fix}\, T = \mathbb{C}\cdot\mathbf{1}$. Let $f \in \mathrm{fix}\, T$. Then by the definition of the seminorms we have $\|f\|_{U^1} = \|f\|_2$ and

$$\|f\|_{U^2}^4 = \||f|^2\|_2^2 = \|f\|_4^4.$$

Thus $\|f\|_{U^2} = \|f\|_{U^1}$ means equality in the Cauchy–Schwarz inequality $\|f\|_2^2 = (|f|^2|\mathbf{1}) \leq \||f|^2\|_2$ implying that $|f|$ is constant, say $|f| = d\cdot\mathbf{1}$. Since $f + d\cdot\mathbf{1} \in \mathrm{fix}\, T$ and hence has constant absolute value by the above, $\mathrm{Re} f$ is constant. By the same argument applied to $f + id\cdot\mathbf{1} \in \mathrm{fix}\, T$ also $\mathrm{Im} f$ is constant, i.e., f is constant. Thus (X, μ, T) is weakly mixing. ∎

Remark 15.12 (Maximality of the Gowers–Host–Kra Seminorms) For ergodic systems, in contrast to Proposition 15.11(a), the lth seminorm is maximal (in the sense $\|f\|_{U^l} = \|f\|_\infty$) on generalized eigenfunctions of order $l-1$, see Exercise 15.3 and Example 8.33 for the definition of generalized eigenfunctions. In fact, this characterizes generalized eigenfunctions, see Eisner and Tao [348]. In particular, for ergodic systems $\|f\|_{U^2} = \|f\|_\infty$ holds if and only if $Tf = \lambda f$ for some $\lambda \in \mathbb{T}$. Note that for general systems this is not true as the example of $T = I$ shows (in this case $\|f\|_{U^2} = \|f\|_\infty$ holds if and only if $|f|$ is constant by the same argument as at the end of the proof of Proposition 15.11 using the monotonicity of the seminorms).

Remark 15.13 (U^{1+}-seminorm) Motivated by applications in combinatorics, Tao [971] introduced an intermediate seminorm $\|\cdot\|_{U^{1+}}$ between $\|\cdot\|_{U^1}$ and $\|\cdot\|_{U^2}$ having the property that $\|f\|_{U^{1+}} = 0$ if and only if f is orthogonal to the rational spectrum factor (recall Example 8.31(b)) together with a combinatorial counterpart.

The following estimate shows that multiple ergodic averages are controlled by the Gowers–Host–Kra seminorms explaining the relevance and the definition of the seminorms.

Proposition 15.14 (Generalized von Neumann Inequality) *Let (X, μ, T) be a measure-preserving system, $k \in \mathbb{N}$ and $f_1, \ldots, f_k \in \mathrm{L}^\infty(X, \mu)$. Then*

$$\limsup_{N\to\infty} \Big\| \frac{1}{N}\sum_{n=1}^{N} T^n f_1 \cdot T^{2n} f_2 \cdots T^{kn} f_k \Big\|_2 \leq \min_{j=1,\ldots,k} j\|f_j\|_{U^k}.$$

For the proof we recall[3] here van der Corput's inequality from Chap. 7, see Remark 7.32: For each Hilbert space H and each bounded sequence $(u_n)_{n\in\mathbb{N}}$ in H

$$\limsup_{N\to\infty}\left\|\frac{1}{N}\sum_{n=0}^{N-1}u_n\right\|^2 \le \limsup_{J\to\infty}\frac{1}{J}\sum_{h=0}^{J-1}\limsup_{N\to\infty}\frac{1}{N}\left|\sum_{n=0}^{N-1}(u_n|u_{n+h})\right|. \tag{15.12}$$

Proof The proof goes by induction on k. For $k=1$ we have by the mean ergodic theorem and (15.5)

$$\lim_{N\to\infty}\left\|\frac{1}{N}\sum_{n=1}^{N}T^n f_1\right\|_2 = \|P_{\mathrm{fix}\,T} f_1\|_2 = \|f_1\|_{U^1}.$$

Let now $k\ge 2$ and suppose that the assertion holds for $k-1$. Without loss of generality we can assume that $\|f_i\|_\infty \le 1$ holds for all $i\in\{1,\dots,k\}$. We apply the van der Corput inequality (15.12) to the functions

$$u_n := T^n f_1\cdot T^{2n} f_2\cdots T^{kn} f_k.$$

Take first $j\in\{2,\dots,k\}$ and observe

$$\begin{aligned}(u_{n+h}|u_n) &= \int_X T^n(T^h f_1\overline{f_1})\cdot T^{2n}(T^{2h} f_2\overline{f_2})\cdots T^{kn}(T^{kh} f_k\overline{f_k})\,\mathrm{d}\mu\\ &= \int_X (T^h f_1\overline{f_1})\cdot T^{n}(T^{2h} f_2\overline{f_2})\cdots T^{(k-1)n}(T^{kh} f_k\overline{f_k})\,\mathrm{d}\mu.\end{aligned}$$

This implies using the Cauchy–Schwarz inequality and $\|f_1\|_\infty\le 1$

$$\left|\frac{1}{N}\sum_{n=1}^{N}(u_{n+h}|u_n)\right| \le \left\|\frac{1}{N}\sum_{n=1}^{N}T^n(T^{2h} f_2\overline{f_2})\cdots T^{(k-1)n}(T^{kh} f_k\overline{f_k})\right\|_2$$

leading by the induction hypothesis to

$$\gamma_h := \limsup_{N\to\infty}\left|\frac{1}{N}\sum_{n=1}^{N}(u_{n+h}|u_n)\right| \le (j-1)\|T^{jh} f_j\overline{f_j}\|_{U^{k-1}} \le j\|T^{jh} f_j\overline{f_j}\|_{U^{k-1}}.$$

[3] We proved only the version of van der Corput's inequality with the factor 2 on the right hand side, see Lemma 7.31, which shows the generalized von Neumann inequality with a factor on the right hand side as well. This weaker version completely suffices for our purposes—we will only need that the multiple ergodic averages converge to zero as soon as at least one of the functions has zero k-th uniformity seminorm.

By the van der Corput inequality (15.12) we have

$$\limsup_{N\to\infty}\left\|\frac{1}{N}\sum_{n=1}^{N}u_n\right\|_2^2 \le \limsup_{N\to\infty}\frac{1}{N}\sum_{h=1}^{N}\gamma_h \le j\limsup_{N\to\infty}\frac{1}{N}\sum_{h=1}^{N}\|T^{jh}f_j\overline{f_j}\|_{U^{k-1}}$$

$$\le j^2\limsup_{N\to\infty}\frac{1}{jN}\sum_{h=1}^{jN}\|T^{n}f_j\overline{f_j}\|_{U^{k-1}} \le j^2\limsup_{N\to\infty}\frac{1}{N}\sum_{n=1}^{N}\|T^{n}f_j\overline{f_j}\|_{U^{k-1}},$$

which is, by an iterated use of the Cauchy–Schwarz inequality (or by Jensen's inequality for the convex function $t\mapsto t^{2^{k-1}}$), bounded from above by

$$j^2\limsup_{N\to\infty}\left(\frac{1}{N}\sum_{n=1}^{N}\|T^{n}f_j\overline{f_j}\|_{U^{k-1}}^{2^{k-1}}\right)^{1/2^{k-1}} = j^2\|f_j\|_{U^k}^2.$$

The claim follows. The case $j = 1$ is treated analogously (by using $\|f_k\|_\infty \le 1$) and is left to the reader as Exercise 15.4. ■

Remark 15.15 Proposition 15.14 together with Proposition 15.11 already shows multiple convergence for weakly mixing systems.

We now define for $l \in \mathbb{N}_0$ the closed linear subspace

$$Z_l := \{f \in \mathrm{L}^\infty(X,\mu) : \|f\|_{U^{l+1}} = 0\}^\perp \tag{15.13}$$

of $\mathrm{L}^2(X,\mu)$ and call it **Host–Kra subspace of order** l. Note that $Z_0 = \operatorname{fix} T$ by (15.5) corresponding to the trivial orthogonal decomposition[4]

$$\mathrm{L}^2(X,\mu) = \operatorname{fix} T \oplus \{f \in \mathrm{L}^2(X,\mu) : P_{\operatorname{fix} T} f = 0\}.$$

For ergodic systems Z_1 is the closed linear hull of the eigenfunctions of T by the Jacobs–de Leeuw–Glicksberg decomposition

$$\mathrm{L}^2(X,\mu) = \overline{\operatorname{lin}}\{f : Tf = \lambda f \text{ for some } \lambda \in \mathbb{T}\} \oplus \{f \text{ weakly mixing}\},$$

(see (6.10)), Proposition 15.11(a), Remark 15.6(a) and Exercise 8.12. In general one has the orthogonal decompositions

$$\mathrm{L}^2(X,\mu) = Z_l \oplus \overline{\left\{f \in \mathrm{L}^\infty(X,\mu) : \|f\|_{U^{l+1}} = 0\right\}}, \quad l \in \mathbb{N}_0, \tag{15.14}$$

[4] It is easy to see that if a sequence $(f_n)\subset \mathrm{L}^\infty(X,\mu)$ approximates f from the second subspace in L^2 then the sequence $(P_{(\operatorname{fix} T)^\perp} f_n)$ again approximates f in L^2 and is a subset of L^∞ since fix T is induced by a factor, see (8.2).

and Z_l increase by Proposition 15.8. Note that the linear subspace property of $\ker \|\cdot\|_{U^{l+1}}$ here follows from the not yet proven triangle inequality for $\|\cdot\|_{U^{l+1}}$ (see Corollary 15.26 below).

Thus the following is a direct consequence of Proposition 15.14 and the triangle inequality for the Gowers-Host-Kra seminorms (see Corollary 15.26 below).

Corollary 15.16 *Let (X, μ, T) be a measure-preserving system, $k \in \mathbb{N}$ and $f_1, \dots, f_k \in \mathrm{L}^\infty(X, \mu)$. Then*

$$\lim_{N\to\infty} \frac{1}{N} \sum_{n=1}^{N} T^n f_1 \cdot T^{2n} f_2 \cdots T^{kn} f_k = 0$$

whenever $f_j \perp Z_{k-1}$ for some $j \in \{1, \dots, k\}$.

Proof *(Assuming the Triangle Inequality for $\|\cdot\|_{U^k}$)* Let $f_j \perp Z_{k-1}$ for some $j \in \{1, \dots, k\}$ and assume without loss of generality that $\|f_1\|_\infty \leq 1, \dots, \|f_k\|_\infty \leq 1$. By (15.14) for $l := k-1$ there exists a sequence $(g_m) \subset \mathrm{L}^\infty(X)$ satisfying $\|g_m\|_{U^k} = 0$ for all $m \in \mathbb{N}$ with $\lim_{m\to\infty} \|f_j - g_m\|_2 = 0$. By decomposing $f_j = (f_j - g_m) + g_m$, we have for every $m \in \mathbb{N}$ by Proposition 15.14 and the triangle inequality for the L^2-norm

$$\limsup_{N\to\infty} \left\| \frac{1}{N} \sum_{n=1}^{N} T^n f_1 \cdot T^{2n} f_2 \cdots T^{kn} f_k \right\|_2 \leq j \|f_j - g_m\|_{\mathrm{L}^2(X)}.$$

By letting $m \to \infty$, the assertion follows. ∎

This motivates the following strategy to show multiple convergence, which will be realized in Sect. 15.7 below with the preparation done in Sects. 15.2–15.6.

Strategy First we show that the Gowers–Host–Kra seminorms are indeed seminorms, i.e., that the triangle inequality holds (see Corollary 15.26 below).

Let $k \in \mathbb{N}$. Decompose now every f_j into its projection $P_{Z_{k-1}} f_j$ onto Z_{k-1} and the projection onto its orthogonal complement

$$Z_{k-1}^\perp = \overline{\{f \in \mathrm{L}^\infty(X, \mu) : \|f\|_{U^k} = 0\}}.$$

If we know that the first projection (and then also the second one) belongs to $\mathrm{L}^\infty(X, \mu)$, then by the multilinearity of multiple ergodic averages and Corollary 15.16

$$\lim_{N\to\infty} \left\| \frac{1}{N} \sum_{n=1}^{N} T^n f_1 \cdots T^{kn} f_k - \frac{1}{N} \sum_{n=1}^{N} T^n P_{Z_{k-1}} f_1 \cdots T^{kn} P_{Z_{k-1}} f_k \right\|_2 = 0.$$

This means that we can assume without loss of generality that $f_1, \ldots, f_k \in Z_{k-1}$. One says that the subspace Z_{k-1} is **characteristic for k-term multiple ergodic averages** (15.1), see also Definition 9.23.

The missing property of the boundedness of both projections will follow from the fact that for regular systems Z_{k-1} comes from a factor, called *the Host–Kra factor* $\mathcal{Z}_{k-1}$, see Definition 15.39 below (recall that we assumed our system to be regular). This factor is also said to be **characteristic** for k-term multiple ergodic averages and we call the decomposition (15.14) for $l := k-1$, i.e.,

$$\mathrm{L}^2(X, \mu) = Z_{k-1} \oplus \overline{\{f \in \mathrm{L}^\infty(X, \mu) : \|f\|_{U^k} = 0\}} \tag{15.15}$$

the **Host–Kra decomposition of order** $k-1$. Note that by this, $\mathcal{Z}_{k-1}$ is the maximal factor of (X, μ, T) such that $\|\cdot\|_{U^k}$ is a norm on it. It is the most difficult part to give structure to $\mathcal{Z}_{k-1}$ so that convergence of multiple ergodic averages on it follows by using this structure. It occurs that for ergodic systems $\mathcal{Z}_{k-1}$ is an inverse limit of nilsystems of step $k-1$ (Theorem 15.45 below), so everywhere convergence of multiple ergodic averages for nilsystems (Proposition 14.44) and ergodic decomposition (see Sect. 8.6) together with an approximation argument complete the proof.

Following the terminology of Tao [975], in the Host–Kra decomposition (15.15) one calls Z_{k-1} the **structured subspace** and its orthogonal complement $\overline{\{f \in \mathrm{L}^\infty(X, \mu) : \|f\|_{U^k} = 0\}}$ the **pseudorandom subspace** of order $k-1$.

15.2 Alternative Definitions and the Triangle Inequality

The goal of this subsection is to show the triangle inequality for the Gowers–Host–Kra seminorms. Since the triangle inequality involves only two functions (and their sum), we can always assume that our system is separable. Moreover, it is easy to see that the seminorms do not change under Markov isomorphisms, so we can assume by Remark 8.50(b) that our system is regular, i.e., X is a compact metric space and μ is a (regular) Borel measure. For such systems we start with an alternative definition of the Gowers–Host–Kra seminorms.

Thus let (X, μ, T) be a regular measure-preserving system. Observe that by definition we have for $f \in \mathrm{L}^\infty(X, \mu)$ by (15.6)

$$\begin{aligned}
\|f\|_{U^2}^4 &= \limsup_{M\to\infty} \frac{1}{M} \sum_{m=1}^{M} \lim_{N\to\infty} \frac{1}{N} \sum_{n=1}^{N} \int_X T^n(T^m f \cdot \overline{f}) \cdot \overline{T^m f \cdot \overline{f}} \,\mathrm{d}\mu \\
&= \limsup_{M\to\infty} \frac{1}{M} \sum_{m=1}^{M} \lim_{N\to\infty} \frac{1}{N} \sum_{n=1}^{N} \int_X T^{m+n} f \cdot T^n \overline{f} \cdot T^m \overline{f} \cdot f \,\mathrm{d}\mu.
\end{aligned}$$

To see that the limit superior above is actually a limit we need the following reformulation. For $f_1, f_2 \in \mathrm{L}^2(X, \mu)$, by using the mean ergodic theorem, we can write

$$\lim_{N\to\infty} \frac{1}{N}\sum_{n=1}^{N} \int_X T^n f_1 \cdot f_2 \,\mathrm{d}\mu = \int_X P_{\mathrm{fix}\, T} f_1 \cdot P_{\mathrm{fix}\, T} f_2 \,\mathrm{d}\mu =: \int_{X^2} f_1 \otimes f_2 \,\mathrm{d}\mu^{[1]} \tag{15.16}$$

which we use as definition for the measure $\mu^{[1]}$ on the product σ-algebra over $X^2 = X \times X$. The existence and the uniqueness of this measure is provided by Proposition 13.28 applied[5] to the fixed factor, leading to a regular probability space $(X^2, \mu^{[1]})$. This measure is clearly invariant under $T^{[1]} := T \times T$ since T is μ-preserving. (Moreover, $\mu^{[1]}$ is invariant under $T \times \mathrm{id}$ and $\mathrm{id} \times T$ by shift invariance of the Cesàro limit.) If we put $f_1 := T^m f \cdot \overline{f}$ and $f_2 := T^m \overline{f} \cdot f$ in the above formula, we arrive at

$$\begin{aligned}\lim_{N\to\infty} \frac{1}{N}\sum_{n=1}^{N} \int_X T^{m+n} f \cdot T^n \overline{f} \cdot T^m \overline{f} \cdot f \,\mathrm{d}\mu &= \int_{X^2} (T^m f \cdot \overline{f}) \otimes (T^m \overline{f} \cdot f) \,\mathrm{d}\mu^{[1]} \\ &= \int_{X^2} (T \times T)^m (f \otimes \overline{f}) \cdot (\overline{f} \otimes f) \,\mathrm{d}\mu^{[1]}.\end{aligned}$$

Thus, by the mean ergodic theorem, we conclude that the Cesàro limit of this expression in m exists and

$$\begin{aligned}\|f\|_{U^2}^4 &= \lim_{M\to\infty} \frac{1}{M}\sum_{m=1}^{M} \int_{X^2} (T \times T)^m (f \otimes \overline{f}) \cdot (\overline{f} \otimes f) \,\mathrm{d}\mu^{[1]} \\ &= \int_{X^2} P_{\mathrm{fix}\, T^{[1]}} (f \otimes \overline{f}) \cdot P_{\mathrm{fix}\, T^{[1]}} (\overline{f} \otimes f) \,\mathrm{d}\mu^{[1]} \\ &=: \int_{X^4} f \otimes \overline{f} \otimes \overline{f} \otimes f \,\mathrm{d}\mu^{[2]},\end{aligned}$$

where the measure $\mu^{[2]}$ is the measure on the product σ-algebra of $X^4 = (X \times X)^2$ defined via

$$\begin{aligned}\int_{X^4} F_1 \otimes F_2 \,\mathrm{d}\mu^{[2]} &:= \lim_{N\to\infty} \frac{1}{N}\sum_{n=1}^{N} \int_{X^2} (T \times T)^n F_1 \cdot F_2 \,\mathrm{d}\mu^{[1]} \\ &= \int_{X^2} P_{\mathrm{fix}\, T^{[1]}} F_1 \cdot P_{\mathrm{fix}\, T^{[1]}} F_2 \,\mathrm{d}\mu^{[1]}\end{aligned}$$

[5] Recall that for a factor (Y, ν, S) of (X, μ, T) with Markov factor map J one has $P_Y = J E_Y$.

for $F_1, F_2 \in \mathrm{L}^\infty(X^{[1]}, \mu^{[1]})$. (For the existence and the uniqueness of $\mu^{[2]}$ see the discussion above for $\mu^{[1]}$.) Again, $\mu^{[2]}$ is invariant under $T^{[2]} := T \times T \times T \times T$ since $\mu^{[1]}$ is invariant under $T^{[1]} = T \times T$.

By recursion we obtain the case of general l as follows.

Definition 15.17 (Cube Spaces and Cube Measures) Let (X, μ, T) be a regular measure-preserving system and let $l \in \mathbb{N}$. Set $X^{[0]} := X$, $T^{[0]} := T$, define $X^{[l]} := X^{2^l} = (X^{[l-1]})^2$ and consider elements of it as $x = (x_\varepsilon)_{\varepsilon\in\{0,1\}^l}$, i.e., indexed by the **Euclidean cube** $\{0, 1\}^l$. We also use the notation $x = (x', x'')$ with $x', x'' \in X^{[l-1]}$ given by $x' := (x_{\omega 0})_{\omega\in\{0,1\}^{l-1}}$ and $x'' := (x_{\omega 1})_{\omega\in\{0,1\}^{l-1}}$. On $X^{[l]}$ consider the product transformation $T^{[l]} := T \times T \times \ldots \times T$ (taken 2^l times), i.e., $T^{[l]} = T^{[l-1]} \times T^{[l-1]}$. We define $T^{[l]}$-invariant measures $\mu^{[l]}$ on the product σ-algebra of $X^{[l]}$ inductively as follows. Let $\mu^{[0]} := \mu$, and assume that $\mu^{[l]}$ is already defined. Define $\mu^{[l+1]}$ by the law

$$\int_{X^{[l+1]}} F(x')G(x'')\,\mathrm{d}\mu^{[l+1]}(x', x'') := \lim_{N\to\infty} \frac{1}{N} \sum_{n=1}^{N} \int_{X^{[l]}} (T^{[l]})^n F \cdot G\,\mathrm{d}\mu^{[l]} \tag{15.17}$$

$$= \int_{X^{[l]}} P_{\mathrm{fix}\,T^{[l]}} F \cdot P_{\mathrm{fix}\,T^{[l]}} G\,\mathrm{d}\mu^{[l]} \tag{15.18}$$

for $F, G \in \mathrm{L}^\infty(X^{[l]}, \mu^{[l]})$, where we used $T^{[l]}$-invariance of $\mu^{[l]}$ and the mean ergodic theorem. (The existence and the uniqueness of the (regular Borel) measure $\mu^{[l+1]}$ follow recursively, by using Proposition 13.28.) Again by $T^{[l]}$-invariance of $\mu^{[l]}$ we see that $\mu^{[l+1]}$ is $T^{[l+1]}$-invariant and the inductive definition is consistent. The spaces $X^{[l]}$ together with the measures $\mu^{[l]}$ and the transformations $T^{[l]}$ are called **Host–Kra cube spaces** and the measures $\mu^{[l]}$ are called **Host–Kra cube measures**.

Remark 15.18

(a) If (X, μ, T) is ergodic, one has $P_{\mathrm{fix}\,T^{[0]}} F = \int_X F\,\mathrm{d}\mu \cdot \mathbf{1}$ and therefore

$$\mu^{[1]} = \mu \times \mu. \tag{15.19}$$

Moreover, if we start with a weakly mixing system, then by Proposition 8.42 we have $\mu^{[l]} = \mu^{2^l}$.

(b) By (15.17), orthogonality and the mean ergodic theorem we also have the representation

$$\int_{X^{[l+1]}} F(x')G(x'')\,\mathrm{d}\mu^{[l+1]}(x', x'') = \lim_{N\to\infty} \frac{1}{N} \sum_{n=1}^{N} \int_{X^{[l]}} F \cdot (T^{[l]})^n G\,\mathrm{d}\mu^{[l]}, \tag{15.20}$$

so the measure $\mu^{[l+1]}$ is invariant under the transformation $(x', x'') \mapsto (x'', x')$ corresponding to interchanging F and G.

(c) In view of representation (15.18) the system $(X^{[l+1]}, \mu^{[l+1]}, T^{[l+1]})$ is called the **relatively independent joining** of two copies of $(X^{[l]}, \mu^{[l]}, T^{[l]})$ over the invariant factor fix $T^{[l]}$, with the notation $\mu^{[l+1]} = \mu^{[l]} \times_{\mathrm{fix}\, T^{[l]}} \mu^{[l]}$, recall Definition 13.27.

(d) If the original system (X, μ, T) is (essentially) invertible, then so is $(X^{[l]}, \mu^{[l]}, T^{[l]})$ for every l with the inverse transformation given by $T^{-[l]} := T^{-1} \times \ldots \times T^{-1}$.

(e) If the original system (X, μ, T) is regular then so is the cube space $(X^{[l]}, \mu^{[l]}, T^{[l]})$ for every $l \in \mathbb{N}$.

More properties of these measures will be shown in the following sections.

We can now represent the Host–Kra measures in two different ways, one staying in the original system and one using the cube spaces. Here and later we denote by C the operation of complex conjugation, i.e., $Cf := \overline{f}$. Moreover, for $\omega \in \{0, 1\}^l$ denote $|\omega| := \omega_1 + \cdots + \omega_l$, write $\omega \cdot n := \omega_1 n_1 + \cdots + \omega_l n_l$ for $n = (n_1, \ldots, n_l)$ and use the notation

$$\operatorname*{C-lim}_{j\to\infty} := \lim_{N\to\infty} \frac{1}{N} \sum_{j=1}^{N}.$$

Proposition 15.19 (Two Representations of the Seminorms) *For each measure-preserving system (X, μ, T), $f \in \mathrm{L}^\infty(X, \mu)$ and $l \in \mathbb{N}$ one has*

$$\|f\|_{U^l}^{2^l} = \operatorname*{C-lim}_{n_l\to\infty} \cdots \operatorname*{C-lim}_{n_1\to\infty} \int_X \prod_{\omega\in\{0,1\}^l} T^{\omega\cdot n} C^{|\omega|} f \,\mathrm{d}\mu. \tag{15.21}$$

Moreover, the limit superior in the definition of the Gowers–Host–Kra seminorms is actually a limit. For regular systems one also has the representation

$$\|f\|_{U^l}^{2^l} = \int_{X^{[l]}} \bigotimes_{\omega\in\{0,1\}^l} C^{|\omega|} f \,\mathrm{d}\mu^{[l]}. \tag{15.22}$$

We will see in Proposition 15.32 below that one can arbitrarily interchange the order of the Cesàro limits in (15.21), see also Remark 15.33.

Proof Since the expressions in (15.21) and in the definition of the Gowers–Host–Kra seminorms deal with only one function (and its orbit) and the corresponding quantities are preserved by Markov isomorphisms, by Remark 8.50(b) we can assume without loss of generality that the system is regular. The proof goes now by induction on l. The case $l = 1$ follows directly from (15.3) and (15.16). Let now $l \geq 2$ and assume that the claim holds for $l - 1$. Then the induction hypothesis and (15.6) imply

$$\|T^{n_l} f \cdot \overline{f}\|_{U^{l-1}}^{2^{l-1}} = \|T^{n_l} \overline{f} \cdot f\|_{U^{l-1}}^{2^{l-1}}$$
$$= \underset{n_{l-1}\to\infty}{\text{C-}\lim} \cdots \underset{n_1\to\infty}{\text{C-}\lim} \int_X \prod_{\omega\in\{0,1\}^{l-1}} T^{\omega_1 n_1+\cdots+\omega_{l-1}n_{l-1}} C^{|\omega|}(T^{n_l}\overline{f} \cdot f)\,\mathrm{d}\mu$$

showing (15.21) with $\limsup_{N_l\to\infty} \frac{1}{N_l}\sum_{n_l=1}^{N_l}$ instead of $\underset{n_l\to\infty}{\text{C-}\lim}$. To show that one can replace $\limsup$ by $\lim$ here and in (15.2) and to prove (15.22) observe that by the induction hypothesis and (15.6) again

$$\|T^{n_l} f \cdot \overline{f}\|_{U^{l-1}}^{2^{l-1}} = \int_{X^{[l-1]}} \bigotimes_{\omega\in\{0,1\}^{l-1}} (T^{n_l} C^{|\omega|+1} f \cdot C^{|\omega|} f)\,\mathrm{d}\mu^{[l-1]}$$
$$= \int_{X^{[l-1]}} (T^{[l-1]})^{n_l}\Big(\bigotimes_{\omega\in\{0,1\}^{l-1}} C^{|\omega|+1} f\Big) \cdot \Big(\bigotimes_{\omega\in\{0,1\}^{l-1}} C^{|\omega|} f\Big)\,\mathrm{d}\mu^{[l-1]}.$$

The mean ergodic theorem applied to the cube system $(X^{[l-1]}, \mu^{[l-1]}, T^{[l-1]})$ implies that the Cesàro limit of $\|T^{n_l}\overline{f} \cdot f\|_{U^{l-1}}^{2^{l-1}}$ exists and equals

$$\int_{X^{[l]}} \Big(\bigotimes_{\omega\in\{0,1\}^{l-1}} C^{|\omega|+1} f\Big) \otimes \Big(\bigotimes_{\omega\in\{0,1\}^{l-1}} C^{|\omega|} f\Big)\,\mathrm{d}\mu^{[l]} = \int_{X^{[l]}} \bigotimes_{\omega\in\{0,1\}^{l}} C^{|\omega|} f\,\mathrm{d}\mu^{[l]}$$

completing the proof. ∎

Example 15.20 Return now to the finite system $\mathbb{Z}/N\mathbb{Z}$ with the rescaled counting measure μ and the transformation T given by $Tx = x + 1$ from Example 15.10. Then the first seminorm satisfies

$$\|f\|_{U^1}^2 = \left|\int_{\mathbb{Z}/N\mathbb{Z}} f\,\mathrm{d}\mu\right|^2 = \int_{(\mathbb{Z}/N\mathbb{Z})^2} f(x)\overline{f(x+t)}\,\mathrm{d}\mu(x)\,\mathrm{d}\mu(t)$$

and the representation (15.21) takes the form

$$\|f\|_{U^l}^{2^l} = \int_{(\mathbb{Z}/N\mathbb{Z})^{l+1}} \prod_{\omega\in\{0,1\}^l} C^{|\omega|} f(x+\omega_1 t_1+\cdots+\omega_l t_l)\,\mathrm{d}\mu(x)\,\mathrm{d}\mu(t_1)\cdots\,\mathrm{d}\mu(t_l).$$

Note that this formula (up to minor variation) was used by Gowers for the original definition of the uniformity (semi)norms in his proof of Szemerédi's theorem in [481]. For $l \geq 2$ these seminorms are norms, called **Gowers box norms**. The name comes from the fact that the sets $\{\omega_1 n_1 + \cdots + \omega_l n_l : \omega \in \{0,1\}^l\}$ (also called parallelepipeds, see Tao [980, Sec. 1.3]) correspond to images of boxes $\{0,1\}^l$ in $\mathbb{Z}^l$. (Although we called the sets $\{0,1\}^l$ (Euclidean) cubes before, we keep both names to distinguish the two approaches.) Moreover, analogous representation is valid for every ergodic rotation on a compact Abelian group. The details are left as Exercise 15.5.

Using the idea of Gowers [481], see also Tao [980], and its variation by Host and Kra, we introduce a multidimensional scalar product in order to show the triangle inequality for the Gowers–Host–Kra seminorms. We again present it in two forms: as a box scalar product staying in the original space and as a cube scalar product going to the cube spaces.

Proposition 15.21 (Box/Cube Scalar Product) *Let (X, μ, T) be a measure-preserving system, $l \in \mathbb{N}$ and let $(f_\omega)_{\omega\in\{0,1\}^l}$ be a family of functions in $\mathrm{L}^\infty(X, \mu)$. Then the multiple Cesàro limit*

$$\left\langle (f_\omega)_{\omega\in\{0,1\}^l} \right\rangle := \underset{n_l\to\infty}{\text{C-lim}} \cdots \underset{n_1\to\infty}{\text{C-lim}} \int_X \prod_{\omega\in\{0,1\}^l} T^{\omega\cdot n} C^{|\omega|} f_\omega \,\mathrm{d}\mu \tag{15.23}$$

exists. Moreover, for regular systems one has the representation

$$\left\langle (f_\omega)_{\omega\in\{0,1\}^l} \right\rangle = \int_{X^{[l]}} \bigotimes_{\omega\in\{0,1\}^l} C^{|\omega|} f_\omega \,\mathrm{d}\mu^{[l]}. \tag{15.24}$$

We call $\left\langle (f_\omega)_{\omega\in\{0,1\}^l} \right\rangle$ the **box** (or **cube**) **scalar product** of the family $(f_\omega)_{\omega\in\{0,1\}^l}$. More precisely, the representation in (15.23) yields the name *box scalar product*, while the representation in (15.24) amounts to the term *cube scalar product*. This function is $\mathbb{R}$-multilinear, homogeneous in the arguments f_ω with $|\omega|$ even and conjugate homogeneous in the other arguments.

For example,

$$\langle (f_0, f_1) \rangle = \underset{n\to\infty}{\text{C-lim}} \int_X T^n \overline{f_1} \cdot f_0 \,\mathrm{d}\mu = \int_{X^{[1]}} \overline{f_1} \otimes f_0 \,\mathrm{d}\mu^{[1]}$$

and

$$\begin{aligned} \langle (f_{00}, f_{01}, f_{10}, f_{11}) \rangle &= \underset{n_2\to\infty}{\text{C-lim}}\, \underset{n_1\to\infty}{\text{C-lim}} \int_X T^{n_1+n_2} f_{11} \cdot T^{n_1} \overline{f_{10}} \cdot T^{n_2} \overline{f_{01}} \cdot f_{00} \,\mathrm{d}\mu \\ &= \int_{X^{[2]}} f_{11} \cdot \overline{f_{10}} \cdot \overline{f_{01}} \cdot f_{00} \,\mathrm{d}\mu^{[2]}, \end{aligned}$$

where the representations on the right-hand side hold under the regularity assumption.

Proof As in the proof of Proposition 15.19 we can assume without loss of generality that the system is regular. We have to show that the Cesàro limits on the right-hand side of (15.23) exist and the equality between the two representations. We do this by induction on l. The mean ergodic theorem and (15.16) imply

$$\int_{X^1} \overline{f_1} \otimes f_0 \, \mathrm{d}\mu^{[1]}(x) = \int_X P_{\mathrm{fix}\,T} \overline{f_1} \cdot P_{\mathrm{fix}\,T} f_0 \, \mathrm{d}\mu = \underset{n\to\infty}{\text{C-}\lim} \int_X T^n \overline{f_1} \cdot f_0 \, \mathrm{d}\mu,$$

showing the assertion for $l = 1$.

Assume now that the assertion holds for l and we show that it holds for $l + 1$. For $(f_\omega)_{\omega\in\{0,1\}^{l+1}} \subset \mathrm{L}^\infty(X, \mu)$ the representation (15.20) implies

$$\begin{aligned}
&\int_{X^{[l+1]}} \bigotimes_{\omega\in\{0,1\}^{l+1}} C^{|\omega|} f_\omega \, \mathrm{d}\mu^{[l+1]} \\
&= \underset{n_{l+1}\to\infty}{\text{C-}\lim} \int_{X^{[l]}} \Big(\bigotimes_{\omega\in\{0,1\}^l} C^{|\omega|} f_{\omega 0} \Big) \cdot \Big(\bigotimes_{\omega\in\{0,1\}^l} T^{n_{l+1}} C^{|\omega|+1} f_{\omega 1} \Big) \, \mathrm{d}\mu^{[l]} \\
&= \underset{n_{l+1}\to\infty}{\text{C-}\lim} \int_{X^{[l]}} \bigotimes_{\omega\in\{0,1\}^l} C^{|\omega|} \Big(f_{\omega 0} T^{n_{l+1}} \overline{f_{\omega 1}} \Big) \, \mathrm{d}\mu^{[l]}.
\end{aligned}$$

By the induction hypothesis, this equals

$$\underset{n_{l+1}\to\infty}{\text{C-}\lim} \ldots \underset{n_1\to\infty}{\text{C-}\lim} \int_X \prod_{\omega\in\{0,1\}^l} T^{n\cdot\omega} C^{|\omega|} f_{\omega 0} \cdot T^{n\cdot\omega + n_{l+1}} C^{|\omega|} \overline{f_{\omega 1}} \, \mathrm{d}\mu,$$

which can be rewritten as

$$\underset{n_{l+1}\to\infty}{\text{C-}\lim} \ldots \underset{n_1\to\infty}{\text{C-}\lim} \int_X \prod_{\omega\in\{0,1\}^{l+1}} T^{n\cdot\omega} C^{|\omega|} f_\omega \, \mathrm{d}\mu.$$

The proof is complete. ■

Remark 15.22 By construction we have the relation

$$\|f\|_{U^l}^{2^l} = \big\langle (f)_{\omega\in\{0,l\}^l} \big\rangle. \tag{15.25}$$

Moreover, it will be shown in Proposition 15.32 below that in (15.23) one can arbitrarily change the order in $\text{C-}\lim_{n_l\to\infty} \ldots \text{C-}\lim_{n_1\to\infty}$, see also Remark 15.33 below regarding the joint limit.

Example 15.23 For ergodic rotations on a compact (Abelian) group one has the following representation analogous to the representation of the Gowers–Host–Kra seminorms in Example 15.20. Let G be a compact Abelian group with Haar measure m and an ergodic rotation, let $l \in \mathbb{N}$ and $(f_\omega)_{\omega\in\{0,1\}^l} \subset \mathrm{L}^\infty(G, \mathrm{m})$. Then

$$\big\langle (f_\omega)_{\omega\in\{0,1\}^l} \big\rangle = \int_{G^{l+1}} \prod_{\omega\in\{0,1\}^l} C^{|\omega|} f(x + \omega_1 h_1 + \cdots + \omega_l h_l) \, \mathrm{d}x \, \mathrm{d}h_1 \ldots \mathrm{d}h_l.$$

The proof is left as Exercise 15.5. For the group $\mathbb{Z}/N\mathbb{Z}$ with $Tx = x + 1 \bmod 1$ this is (up to rescaling) the original definition used by Gowers in [481].

The following is an analogue of the Cauchy–Schwarz inequality for the box/cube scalar product due to Gowers.

Proposition 15.24 (Gowers–Cauchy–Schwarz Inequality) *Let (X, μ, T) be a measure-preserving system, $l \in \mathbb{N}$ and let $(f_\omega)_{\omega\in\{0,1\}^l} \subset \mathrm{L}^\infty(X, \mu)$. Then*

$$\left|\left\langle (f_\omega)_{\omega\in\{0,1\}^l}\right\rangle\right| \le \prod_{\omega\in\{0,1\}^l} \|f_\omega\|_{U^l}.$$

Proof We will use the representation (15.23) and begin with a simple observation. Let $f, g \in \mathrm{L}^\infty(X, \mu)$. Then by the mean ergodic theorem, the Cauchy–Schwarz inequality and (15.5)

$$\begin{aligned}\left|\operatorname*{C-lim}_{n\to\infty}\int_X T^n\overline{f}\cdot g\,\mathrm{d}\mu\right|^2 &= \left|\int_X P_{\mathrm{fix}\,T}\overline{f}\cdot g\,\mathrm{d}\mu\right|^2 = \left|\int_X P_{\mathrm{fix}\,T}\overline{f}\cdot P_{\mathrm{fix}\,T}g\,\mathrm{d}\mu\right|^2\\ &\le \|P_{\mathrm{fix}\,T}\overline{f}\|_2^2\|P_{\mathrm{fix}\,T}g\|_2^2 = \|\overline{f}\|_{U^1}^2\|g\|_{U^1}^2\\ &= \operatorname*{C-lim}_{n\to\infty}\int_X T^n\overline{f}\cdot f\,\mathrm{d}\mu\cdot\operatorname*{C-lim}_{n\to\infty}\int_X T^n\overline{g}\cdot g\,\mathrm{d}\mu.\end{aligned}$$

This, in particular, establishes the case $l = 1$ using (15.6).

Consider now $l \ge 2$. The idea is to split the box $\{0,1\}^l$ into two "sides" with fixed first coordinate $\omega_1 = 0$ and $\omega_1 = 1$, respectively, and then repeat the procedure with every side. To be more specific, let $(f_\omega)_{\omega\in\{0,1\}^l}$ be a family of functions in $\mathrm{L}^\infty(X, \mu)$. By the Cauchy–Schwarz inequality applied $l - 1$ times we have

$$\left|\left\langle (f_\omega)_{\omega\in\{0,1\}^l}\right\rangle\right|^2 \le \operatorname*{C-lim}_{n_l\to\infty}\dots\operatorname*{C-lim}_{n_2\to\infty}\left|\operatorname*{C-lim}_{n_1\to\infty}\int_X\prod_{\omega\in\{0,1\}^l}T^{\omega\cdot n}C^{|\omega|}f_\omega\,\mathrm{d}\mu\right|^2.$$

Writing the integral on the right-hand side as $\int_X T^{n_1}\overline{f}\cdot g\,\mathrm{d}\mu$ for

$$f := \prod_{\omega\in\{0,1\}^{l-1}} T^{\omega\cdot(n_2,\dots,n_l)}C^{|\omega|}f_{1\omega},\quad g := \prod_{\omega\in\{0,1\}^{l-1}} T^{\omega\cdot(n_2,\dots,n_l)}C^{|\omega|}f_{0\omega}$$

and using the observation above yield that

$$\left|\operatorname*{C-lim}_{n_1\to\infty}\int_X\prod_{\omega\in\{0,1\}^l}T^{\omega\cdot n}C^{|\omega|}f_\omega\,\mathrm{d}\mu\right|^2$$

is bounded by the product of

$$\operatorname*{C-lim}_{n_1\to\infty} \int_X \prod_{\omega\in\{0,1\}^{l-1}} T^{n_1+\omega\cdot(n_2,\dots,n_l)} C^{|\omega|+1} f_{1\omega} \cdot T^{\omega\cdot(n_2,\dots,n_l)} C^{|\omega|} f_{1\omega} \, \mathrm{d}\mu$$

and

$$\operatorname*{C-lim}_{n_1\to\infty} \int_X \prod_{\omega\in\{0,1\}^{l-1}} T^{n_1+\omega\cdot(n_2,\dots,n_l)} C^{|\omega|+1} f_{0\omega} \cdot T^{\omega\cdot(n_2,\dots,n_l)} C^{|\omega|} f_{0\omega} \, \mathrm{d}\mu.$$

Note that these are the same expressions as appearing (after several Cesàro limits) in the box scalar product in (15.23), but with a fixed first coordinate, so using only the half of the functions, each of which appearing twice.

We now repeat the procedure using $l-1$ times the inequality

$$\left|\limsup_{N\to\infty} \frac{1}{N}\sum_{n=1}^{N} a_n b_n\right|^2 \le \limsup_{N\to\infty} \frac{1}{N}\sum_{n=1}^{N} |a_n|^2 \cdot \limsup_{N\to\infty} \frac{1}{N}\sum_{n=1}^{N} |b_n|^2$$

expanding to more and more terms by fixing more and more coordinates. Finally we arrive at the product in ω of the same expressions as the original scalar product but for the fixed function f_ω appearing 2^l times, i.e., by (15.25) we arrive at

$$\left|\left\langle (f_\omega)_{\omega\in\{0,1\}^l}\right\rangle\right|^{2^l} \le \prod_{\omega\in\{0,1\}^l} \langle (f_\omega)\rangle = \prod_{\omega\in\{0,1\}^l} \|f_\omega\|_{U^l}^{2^l}.$$

The proof is thus complete. ■

Remark 15.25 One can also show the Gowers–Cauchy–Schwarz inequality for regular systems using the representation as cube scalar product instead of box scalar product. The details are left as Exercise 15.7 (b).

The triangle inequality for the Gowers–Host–Kra seminorms is now a simple corollary, following from Proposition 15.24 in the same way as the L^2-triangle inequality follows from the classical Cauchy–Schwarz inequality. The proof is left as Exercise 15.8.

Corollary 15.26 (Triangle Inequality) *Let (X,μ,T) be a measure-preserving system, $l\in\mathbb{N}$ and $f,g\in \mathrm{L}^\infty(X,\mu)$. Then*

$$\|f+g\|_{U^l} \le \|f\|_{U^l} + \|g\|_{U^l}.$$

This justifies the name **Gowers–Host–Kra seminorm** and shows the promised orthogonal decomposition

$$\mathrm{L}^2(X,\mu) = Z_{k-1} \oplus \overline{\{f \in \mathrm{L}^\infty(X,\mu) : \|f\|_{U^k} = 0\}}$$

for each $k \in \mathbb{N}$ from the end of Sect. 15.1.

15.3 Multiple Recurrence via Gowers–Host–Kra Seminorms

We interrupt, for this section, the discussion of multiple convergence and present a proof of multiple recurrence (Theorem 13.5) using the Gowers–Host–Kra seminorms following [341]. A second such proof will follow in Sect. 15.8. The idea is to compare the Host–Kra subspace Z_l defined by (15.13) to the subspace D_l induced by Furstenberg's distal factor of order l defined in Sect. 13.8, see Proposition 15.29 below.

We first need some preparation. The following is a relative version of the statement in Exercise 8.12. (Recall that the subspace $W(X|Y)$ of conditionally weakly mixing functions was defined in Sect. 13.3.)

Lemma 15.27 *Let (X,μ,T) be a measure-preserving system with (Markov) invertible factor Y. Then $W(X|Y) \cap \mathrm{L}^\infty(X)$ is dense in $W(X|Y)$ w.r.t. the* L^2*-norm.*

Proof Take $f \in W(X|Y)$. For a sequence (f_n) in $\mathrm{L}^\infty(X)$ converging to f consider the sequence $(P_{W(X|Y)}f_n)$ in $W(X|Y)$ where $P_{W(X|Y)}$ denotes the orthogonal projection onto $W(X|Y)$. By continuity this sequence converges to f, and it remains to show $P_{W(X|Y)}f_n \in \mathrm{L}^\infty(X)$ for every n. This follows from

$$P_{W(X|Y)}f_n = f_n - P_{\overline{E(X|Y)}}f_n \in \mathrm{L}^\infty(X)$$

by the relative Jacobs–de Leeuw–Glicksberg decomposition in Theorem 13.24 and Lemma 13.38. (Recall that for a factor Z with corresponding Markov embedding J one has $P_Z = JE_Z$ and both J and E_Z are Markov operators, so contractive w.r.t. the L^∞-norm.) ∎

We will also need the following property of the Gowers–Host–Kra seminorms for ergodic systems, see, e.g., Eisner and Tao [348, Eq. (12)].

Lemma 15.28 *Let (X,μ,T) be an ergodic measure-preserving system and $k \in \mathbb{N}$. Then for $p_k := \frac{2^k}{k+1}$ we have*

$$\|f\|_{U^k} \le \|f\|_{\mathrm{L}^{p_k}(X)} \textit{ for every } f \in \mathrm{L}^\infty(X).$$

Proof We first show the so-called **Young's inequality** for ergodic systems

$$\limsup_{N\to\infty}\left(\frac{1}{N}\sum_{n=1}^{N}\|T^n f\cdot g\|_{\mathrm{L}^r(X)}^s\right)^{1/s}\le \|f\|_{\mathrm{L}^p(X)}\|g\|_{\mathrm{L}^q(X)} \tag{15.26}$$

for every $f\in\mathrm{L}^p(X)$ and $g\in\mathrm{L}^q(X)$, where $0<r<p$ and $p,q<s$ satisfy $\frac{1}{r}+\frac{1}{s}=\frac{1}{p}+\frac{1}{q}$, see [348, Prop. 7.1]. For $n\in\mathbb{N}$ write

$$|T^n f\cdot g|=T^n|f|^{\frac{s-p}{s}}|g|^{\frac{s-q}{s}}(T^n|f|^p\cdot|g|^q)^{\frac{1}{s}}.$$

This, the identity

$$\frac{1}{r}=\frac{s-p}{sp}+\frac{s-q}{sq}+\frac{1}{s}$$

and the generalized Hölder inequality (1.1) imply

$$\begin{aligned}\|T^n f\cdot g\|_{\mathrm{L}^r(X)}^s&\le \|T^n|f|^{\frac{s-p}{s}}\|^s_{\mathrm{L}^{\frac{sp}{s-p}}(X)}\||g|^{\frac{s-q}{s}}\|^s_{\mathrm{L}^{\frac{sq}{s-q}}(X)}\|(T^n|f|^p\cdot|g|^q)^{\frac{1}{s}}\|^s_{\mathrm{L}^s(X)}\\&=\|f\|^{s-p}_{\mathrm{L}^p(X)}\|g\|^{s-q}_{\mathrm{L}^q(X)}\int_X T^n|f|^p\cdot|g|^q\,\mathrm{d}\mu.\end{aligned}$$

The mean ergodic theorem together with the ergodicity assumption finishes the proof of (15.26).

We now show the statement of the theorem by induction. The inequality $\|f\|_{U^1}\le\|f\|_{\mathrm{L}^1(X)}$ follows directly from the definition of the first seminorm for ergodic systems. Assume now that the assertion holds for $k\in\mathbb{N}$ and we show it for $k+1$. The induction hypothesis and (15.26) for $g=\overline{f}\in\mathrm{L}^\infty(X)$, $r:=p_k$, $p:=q:=p_{k+1}$ and $s:=2^k$ lead to

$$\begin{aligned}\|f\|_{U^{k+1}}^{2^{k+1}}&=\limsup_{N\to\infty}\frac{1}{N}\sum_{n=1}^{N}\|T^n f\cdot\overline{f}\|_{U^k}^{2^k}\\&\le\limsup_{N\to\infty}\frac{1}{N}\sum_{n=1}^{N}\|T^n f\cdot\overline{f}\|_{\mathrm{L}^{p_k}(X)}^{2^k}\le\|f\|_{\mathrm{L}^{p_{k+1}}(X)}^{2^{k+1}},\end{aligned}$$

finishing the proof. ■

We now compare the Host–Kra subspaces with distal factors of the corresponding step defined in Sect. 13.8. (The stronger property due to Host and Kra [549] and Ziegler [1065] will be discussed in Proposition 15.49 below.)

Proposition 15.29 (Host–Kra–Ziegler Factors vs. Distal Factors of Finite Step) *Let (X, μ, T) be an ergodic system and let $k \in \mathbb{N}$. Then $Z_k \subset D_k$.*

Remark 15.30 Proposition 15.29 together with the generalized von Neumann inequality (Proposition 15.14) provide an alternative proof of the fact that the factor $\mathcal{D}_{k-1}$ is characteristic for multiple averages (13.3) (Theorem 13.55).

Proof Let $k \in \mathbb{N}_0$. By the orthogonal decomposition

$$\mathrm{L}^2(X) = D_{k+1} \oplus W(X|\mathcal{D}_k) \tag{15.27}$$

from Remark 13.58 and the orthogonal Host–Kra decomposition (15.14) for $l = k+1$ we have to show

$$W(X|\mathcal{D}_k) \subset \overline{\{f \in \mathrm{L}^\infty(X) : \|f\|_{U^{k+2}} = 0\}}.$$

By Lemma 15.27 it suffices to show

$$\|f\|_{U^{k+2}} = 0 \text{ for every } f \in W(X|\mathcal{D}_k) \cap \mathrm{L}^\infty(X). \tag{15.28}$$

This we do by induction on k. For $k = 0$ let $f \in \mathrm{L}^\infty(X)$ be weakly mixing (see Example 13.18(b)). Then $\|f\|_{U^2} = 0$ by Remark 15.6(a).

Let now $k \in \mathbb{N}$ and assume that (15.28) holds for $k-1$. Let $f \in W(X|\mathcal{D}_k)$ be bounded and assume without loss of generality $\|f\|_\infty \le 1$. Denote by P_{D_k} and $P_{W(X|\mathcal{D}_{k-1})}$ the orthogonal projections onto D_k and $W(X|\mathcal{D}_{k-1})$, respectively. (Recall that since D_k is induced by a factor, by (8.2) both P_{D_k} and $P_{W(X|\mathcal{D}_{k-1})} = I - P_{D_k}$ map bounded functions to bounded functions.) Take $n \in \mathbb{N}$. By the decomposition (13.52) for $k+1$, the triangle inequality and the induction hypothesis

$$\begin{aligned}\|T^n f \cdot \overline{f}\|_{U^{k+1}} &\le \|P_{D_k}(T^n f \cdot \overline{f})\|_{U^{k+1}} + \|P_{W(X|\mathcal{D}_{k-1})}(T^n f \cdot \overline{f})\|_{U^{k+1}} \\ &= \|P_{D_k}(T^n f \cdot \overline{f})\|_{U^{k+1}}.\end{aligned} \tag{15.29}$$

Recall that $P_{D_k} = JE_{\mathcal{D}_k}$ for the corresponding Markov embedding map $J : \mathrm{L}^1(X) \to \mathrm{L}^1(\mathcal{D}_k)$ and J acts as a contraction with respect to every L^p-norm, see Proposition 8.11(g). Denoting $p_{k+1} := \frac{2^{k+1}}{k+2}$ we obtain by (15.29), Lemma 15.28 and $\|f\|_\infty \le 1$ together with (8.6)

$$\begin{aligned}\|T^n f \cdot \overline{f}\|_{U^{k+1}} &\le \|P_{D_k}(T^n f \cdot \overline{f})\|_{U^{k+1}} \le \|P_{D_k}(T^n f \cdot \overline{f})\|_{\mathrm{L}^{p_{k+1}}(X)} \\ &\le \|E_{\mathcal{D}_k}(T^n f \cdot \overline{f})\|_{\mathrm{L}^{p_{k+1}}(\mathcal{D}_k)} \le \|E_{\mathcal{D}_k}(T^n f \cdot \overline{f})\|_{\mathrm{L}^1(\mathcal{D}_k)}^{1/p_{k+1}}.\end{aligned}$$

Observe that $\|T^n f \cdot \overline{f}\|_{U^{k+1}} \le 1$ by $\|f\|_\infty \le 1$ and (15.7). Thus the above implies

$$\|T^n f \cdot \overline{f}\|_{U^{k+1}}^{2^{k+1}} \le \|T^n f \cdot \overline{f}\|_{U^{k+1}}^{p_{k+1}} \le \|E_{\mathcal{D}_k}(T^n f \cdot \overline{f})\|_{\mathrm{L}^1(\mathcal{D}_k)}$$

and, by $f \in W(X|\mathcal{D}_k)$, we have

$$\|f\|_{U^{k+2}}^{2^{k+2}} = \limsup_{N\to\infty} \frac{1}{N}\sum_{n=1}^{N} \|T^n f \cdot \overline{f}\|_{U^{k+1}}^{2^{k+1}} = 0$$

proving (15.28). The proof is complete. ■

We now show how one derives from Proposition 15.29 multiple recurrence using the same finite tower of extensions as in the second proof of multiple recurrence presented in Sect. 13.8.

Third Proof of Theorem 13.5 Using Proposition 15.29 Let (X, μ, T) be an ergodic measure-preserving system and $k \in \mathbb{N}$. For $f \in L^\infty(X, \mu)$ with $f > 0$, decompose f into the sum of $P_{D_{k-1}} f$ and $f - P_{D_{k-1}} f$, where $P_{D_{k-1}}$ denotes the orthogonal projection onto the subspace D_{k-1}. Recall that both functions are bounded and $P_{D_{k-1}} f > 0$ by the fact that D_{k-1} is induced by a factor, or, equivalently, by the fact that $P_{D_{k-1}}$ is the restriction of a Markov projection, see Lemma 13.38, (8.2) and (8.3). Thus by linearity, Proposition 15.29 and Corollary 15.16 we obtain

$$\lim_{N\to\infty} \left\| \frac{1}{N}\sum_{n=1}^{N} T^n f \cdots T^{kn} f - \frac{1}{N}\sum_{n=1}^{N} T^n P_{D_{k-1}} f \cdots T^{kn} P_{D_{k-1}} f \right\|_{L^2(X)} = 0 \tag{15.30}$$

and therefore we can assume without loss of generality that $\mathcal{D}_{k-1}$ is the entire system.

We can now proceed as in the second proof of Theorem 13.5 in Sect. 13.8. Consider a finite sequence of invertible factors

$$X = \mathcal{D}_{k-1} \to \mathcal{D}_{k-2} \to \ldots \to \mathcal{D}_1 \to \mathcal{D}_0 = \{\cdot\}$$

which are compact extensions of each other by definition. Since the trivial one-point factor clearly satisfies (MR) and compact extensions preserve (MR) by Proposition 13.44, $X = \mathcal{D}_{k-1}$ satisfies (MR). The proof is complete. ■

15.4 Symmetry Properties of Cube Measures

In this section we take a closer look at the cube measures $\mu^{[l]}$ defined above for regular systems and study their symmetry properties which will be important later.

We start with a disintegration type statement for the measures $\mu^{[l]}$. Note that even if starting with an ergodic system (X, μ, T), already the measure $\mu^{[1]} = \mu \times \mu$ is ergodic if only if (X, μ, T) is weakly mixing, see Proposition 8.42. (Recall that in this case every measure $\mu^{[l]}$ is just the product of 2^l copies of μ.)

Lemma 15.31 (Disintegration of Measures $\mu^{[l]}$) *Let X be a compact metric space, $T : X \to X$continuous, let (Y, ν) be a probability space and let $y \mapsto \mu_y$ be a measurable map from Y to the space $\mathrm{M}_1(X, T)$ of T-invariant (regular) Borel probability measures on X. Denote $\mu := \int_Y \mu_y \, \mathrm{d}y$, i.e., for every $f \in \mathrm{C}(X)$ we have*

$$\int_X f \, \mathrm{d}\mu = \int_Y \int_X f \, \mathrm{d}\mu_y \, \mathrm{d}\nu(y).$$

Then $\mu \in \mathrm{M}_1(X, T)$ and for every $l \in \mathbb{N}$ one has[6]

$$\mu^{[l]} = \int_Y \mu_y^{[l]} \, \mathrm{d}\nu(y).$$

Proof The first assertion is clear. To show the second one, let $l \in \mathbb{N}$ and take a family $(f_\omega)_{\omega \in \{0,1\}^l}$ in $\mathrm{C}(X)$. Then for every $y \in Y$ we have by Proposition 15.21

$$\int_{X^{[l]}} \bigotimes_{\omega \in \{0,1\}^l} f_\omega \, \mathrm{d}\mu_y^{[l]} = \operatorname*{C-lim}_{n_l \to \infty} \ldots \operatorname*{C-lim}_{n_1 \to \infty} \int_X \prod_{\omega \in \{0,1\}^l} T^{n \cdot \omega} f_\omega \, \mathrm{d}\mu_y.$$

Integrating over Y and using the Lebesgue dominated convergence theorem l times, we obtain again by Proposition 15.21

$$\begin{aligned} &\int_Y \int_{X^{[l]}} \bigotimes_{\omega \in \{0,1\}^l} f_\omega \, \mathrm{d}\mu_y^{[l]} \, \mathrm{d}\nu(y) \\ &\quad = \operatorname*{C-lim}_{n_l \to \infty} \ldots \operatorname*{C-lim}_{n_1 \to \infty} \int_Y \int_X \prod_{\omega \in \{0,1\}^l} T^{n \cdot \omega} f_\omega \, \mathrm{d}\mu_y \, \mathrm{d}\nu(y) \\ &\quad = \operatorname*{C-lim}_{n_l \to \infty} \ldots \operatorname*{C-lim}_{n_1 \to \infty} \int_X \prod_{\omega \in \{0,1\}^l} T^{n \cdot \omega} f_\omega \, \mathrm{d}\mu = \int_{X^{[l]}} \bigotimes_{\omega \in \{0,1\}^l} f_\omega \, \mathrm{d}\mu, \end{aligned}$$

whence the assertion follows. ■

As a simple corollary of the representation given in Proposition 15.21 we obtain the following symmetry property of the measures $\mu^{[l]}$ which will be important later. A map $\tau : \{0, 1\}^l \to \{0, 1\}^l$ is thereby called **digit permutation** if it is of the form $\tau(\omega_1, \ldots, \omega_l) = (\omega_{\sigma(1)}, \ldots, \omega_{\sigma(l)})$ for a permutation σ on $\{1, \ldots, k\}$ and **reflection** if it is of the form $\tau(\omega_1, \ldots, \omega_j, \ldots, \omega_l) = (\omega_1, \ldots, 1 - \omega_j, \ldots, \omega_l)$ for some $j \in \{1, \ldots, k\}$. Note that these mappings induce mappings on $X^{[l]}$. The following describes symmetry properties of the measures $\mu^{[l]}$ with respect to these mappings.

[6] We use here the notation $\mu_y^{[l]} := (\mu_y)^{[l]}$.

Proposition 15.32 (Symmetry under Digit Permutations and Reflections) *Let (X, μ, T) be a measure-preserving system and $l \in \mathbb{N}$. Then in the box scalar product (15.23) one can arbitrarily change the order in $\operatorname{C-lim}_{n_l\to\infty}\dots\operatorname{C-lim}_{n_1\to\infty}$. For regular systems, equivalently, each measure $\mu^{[l]}$ is invariant under digit permutations. Moreover, each measure $\mu^{[l]}$ is reflection invariant.*

Proof As in the proof of Proposition 15.19 we can again assume that our system is regular.

For the case $l = 1$ one needs to check only the reflection invariance which follows from (15.16).

We first show the assertion about interchanging the order of the limits. For $l = 2$ assume first that (X, μ, T) is ergodic. Then by the Gowers–Cauchy–Schwarz inequality (Proposition 15.24) and (15.8) together with (6.10) we can replace every function f_{ij} $(i, j \in \{0, 1\})$ by its projection onto the Kronecker factor. Thus we can assume by the Halmos–von Neumann theorem (Theorem 8.32) that our system is an ergodic rotation on compact Abelian group G. By Example 15.23 we thus have the representation

$$\begin{aligned}&\operatorname*{C-lim}_{n_2\to\infty}\operatorname*{C-lim}_{n_1\to\infty}\int_X T^{n_1+n_2}f_{11}\cdot T^{n_1}\overline{f_{10}}\cdot T^{n_2}\overline{f_{01}}\,f_{11}\,\mathrm{d}\mu\\&\quad=\int_{G^3} f_{11}(x+h_1+h_2)\overline{f_{10}}(x+h_1)\overline{f_{01}}(x+h_2)\cdot f_{00}(x)\,\mathrm{d}\mu(x)\,\mathrm{d}\mu(h_1)\,\mathrm{d}\mu(h_2)\end{aligned}$$

and one can interchange f_{01} with f_{10} by Fubini's theorem. This corresponds to interchanging n_1 and n_2 under the integral as desired. For regular, not necessarily ergodic systems the assertion follows from the disintegration, see Proposition 8.61 (ergodic decomposition) and Lemma 15.31.

Let now $l \geq 3$. By induction one can arbitrarily interchange the order of taking the limit of the averages for $n_{l-1}, \dots, n_1$. Since every permutation of $\{1, \dots, l\}$ is a composition of a permutation of $\{1, \dots, l-1\}$ and $(l-1, l)^m$ $(m \in \{0, 1\})$, it remains to show that one can interchange n_{l-1} and n_l. Applying (15.17) twice leads to

$$\int_{X^{[l]}} \bigotimes_{\omega\in\{0,1\}^l} C^{|\omega|} f_\omega\,\mathrm{d}\mu^{[l]} =$$

$$\operatorname*{C-lim}_{n_l\to\infty}\operatorname*{C-lim}_{n_{l-1}\to\infty}\int_{X^{[l-2]}}(T^{[l-2]})^{n_l+n_{l-1}}g_{11}\cdot(T^{[l-2]})^{n_{l-1}}\overline{g_{10}}\cdot(T^{[l-2]})^{n_l}\overline{g_{01}}\cdot g_{00}\,\mathrm{d}\mu^{[l-2]}$$

for

$$g_{ij} := \bigotimes_{\omega\in\{0,1\}^{l-2}} C^{|\omega|} f_{\omega ij},\quad i, j \in \{0, 1\}.$$

Thus we can interchange n_l and n_{l-1} by the already established case $l = 2$ above. This proves the first part of the proposition.

Invariance of $\mu^{[l]}$ under digit permutations follows from the first part and the fact that an arbitrary permutation of the digits $1, \ldots, l$ corresponds to a permutation of $\omega_1, \ldots, \omega_l$ which in turn corresponds to a permutation of $n_1, \ldots, n_l$.

Finally, invariance of $\mu^{[l]}$ under the reflection corresponding to the last coordinate follows directly from (15.18). The case of other reflections follows now from this and the invariance under digit permutations showed above. ∎

Remark 15.33 In fact, for every $l \in \mathbb{N}$ the joint limit

$$\lim_{N_1,\ldots,N_l\to\infty} \frac{1}{N_1}\sum_{n_1=1}^{N_1} \cdots \frac{1}{N_l}\sum_{n_l=1}^{N_l} \int_X \prod_{\omega\in\{0,1\}^l} T^{\omega\cdot n} C^{|\omega|} f_\omega \, \mathrm{d}\mu$$

exists, see, e.g., Tao and Ziegler [994, Thm. 2.1] for a much more general situation. In particular, the box scalar products and the Gowers–Host–Kra seminorms can be described via such joint limit, see Remark 15.22.

To formulate further symmetry properties we first need some more notions of geometric flavor.

Let (X, μ, T) be a regular measure-preserving system and $l \geq 2$. We call each element of the Euclidean cube $\{0, 1\}^l$ a **vertex**. A set $\alpha \subset \{0, 1\}^l$ is called a **side** if it is a cylinder set with one fixed coordinate with other coordinates being arbitrary, e.g., $\alpha = \{0\} \times \{0, 1\}^{l-1}$.

For $x \in X^{[l]}$ we recall the notation $x = (x', x'')$ with $x', x'' \in X^{[l-1]}$ given by

$$x'_\omega := x_{\omega 0} \text{ and } x''_\omega := x_{\omega 1}, \quad \omega \in \{0, 1\}^{l-1}.$$

This corresponds to diving the cube $\{0, 1\}^l$ into two sides with the fixed last coordinate.

The following are direct consequences of the definition of the cube measures.

Proposition 15.34 *Let (X, μ, T) be a regular measure-preserving system and let $l \in \mathbb{N}$.*

(a) *For $\omega \in \{0, 1\}^l$ let $\pi_\omega : X^{[l]} \to X$ be the projection onto the ωth coordinate. Then $\pi_{\omega *}\mu^{[l]} = \mu$, i.e., each marginal[7] of $\mu^{[l]}$ is the measure μ.*
(b) *For $j \in \{0, 1\}$ let $\pi_j : X^{[l+1]} = X^{[l]} \times X^{[l]} \to X^{[l]}$ be the projection onto the first ($j = 0$) and the second ($j = 1$) component, respectively, i.e., $\pi_0(x) = x'$ and $\pi_1(x) = x''$ for $x = (x', x'')$ as above. Then $\pi_{j*}\mu^{[l+1]} = \mu^{[l]}$.*

We leave the proof of this proposition as Exercise 15.9.

Functions on $X^{[l]}$ invariant under $T^{[l]}$ have remarkable symmetry properties.

[7] This is usually expressed by saying that $\mu^{[l]}$ is a 2^l-fold **coupling** of μ with itself.

Lemma 15.35 *Let (X, μ, T) be a regular measure-preserving system and $l \in \mathbb{N}_0$. Then every $T^{[l]}$-invariant function $F \in \mathrm{L}^1(X^{[l]}, \mu^{[l]})$ satisfies $F(x') = F(x'')$ for $\mu^{[l+1]}$-a.e. $x = (x', x'') \in X^{[l+1]}$.*

Proof By the definition of $\mu^{[l+1]}$ and $T^{[l]}$-invariance of F (and hence of $|F|^2 = F \cdot \overline{F}$) one has

$$\begin{aligned}
&\int_{X^{[l+1]}} |F(x') - F(x'')|^2 \,\mathrm{d}\mu^{[l+1]} \\
&\quad = \int_{X^{[l+1]}} |F|^2(x') \,\mathrm{d}\mu^{[l+1]} + \int_{X^{[l+1]}} |F|^2(x'') \,\mathrm{d}\mu^{[l+1]} \\
&\qquad - 2\mathrm{Re} \int_{X^{[l+1]}} F(x')\overline{F}(x'') \,\mathrm{d}\mu^{[l+1]} \\
&\quad = 2 \int_{X^{[l]}} P_{\mathrm{fix}\, T^{[l]}}(|F|^2) \,\mathrm{d}\mu^{[l]} - 2 \int_{X^{[l]}} |P_{\mathrm{fix}\, T^{[l]}} F|^2 \,\mathrm{d}\mu^{[l]} \\
&\quad = 2 \int_{X^{[l]}} |F|^2 \,\mathrm{d}\mu^{[l]} - 2 \int_{X^{[l]}} |F|^2 \,\mathrm{d}\mu^{[l]} = 0,
\end{aligned}$$

proving the assertion. ∎

Every side $\alpha \subset \{0, 1\}^l$ induces the **side transformation** T_α on $X^{[l]}$ defined by

$$(T_\alpha x)_\omega := \begin{cases} Tx_\omega & \text{if } \omega \in \alpha, \\ x_\omega & \text{otherwise,} \end{cases}$$

i.e., T is applied only to the coordinates of x which belong to the side α (Fig. 15.1).

Lemma 15.36 (Invariance under Side Transformations) *Let (X, μ, T) be a regular measure-preserving system and $l \in \mathbb{N}$. Then the measure $\mu^{[l]}$ is invariant under all side transformations.*

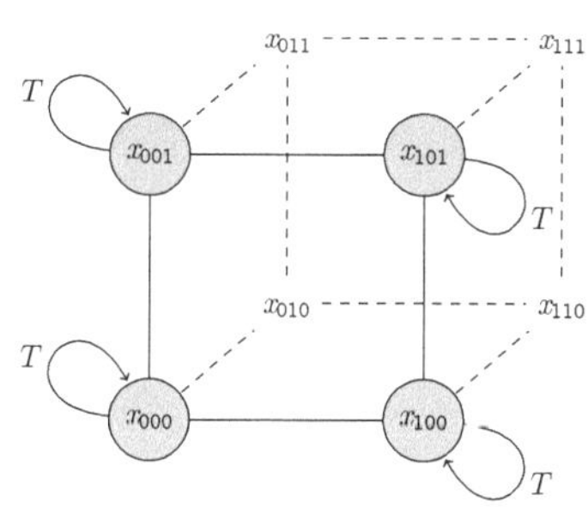

Fig. 15.1 Side transformation T_α corresponding to the side $\alpha = \{0, 1\} \times \{0\} \times \{0, 1\}$

Proof By Proposition 15.32 it suffices to show that $\mu^{[l]}$ is invariant under the side transformation for the side $\alpha = \{\omega \in \{0,1\}^l : \omega_l = 0\}$. By (15.17) one has

$$\begin{aligned}
&\int_{X^{[l]}} \Big(\bigotimes_{\omega \in \{0,1\}^{l-1}} T f_{\omega 0} \Big) \otimes \Big(\bigotimes_{\omega \in \{0,1\}^{l-1}} f_{\omega 1} \Big) \, \mathrm{d}\mu^{[l]} \\
&\quad = \lim_{N\to\infty} \frac{1}{N} \sum_{n=1}^{N} \int_{X^{[l-1]}} \Big(\bigotimes_{\omega \in \{0,1\}^{l-1}} T^{n+1} f_{\omega 0} \Big) \cdot \Big(\bigotimes_{\omega \in \{0,1\}^{l-1}} f_{\omega 1} \Big) \, \mathrm{d}\mu^{[l-1]},
\end{aligned}$$

and the assertion follows from the shift invariance of the Cesàro limit and (15.17) again. ■

15.5 Construction of the Factors

To define the Host–Kra–Ziegler factors, the zero vertex of the Euclidean cube and the corresponding zero coordinate will play a distinguished role. (Note that by symmetry properties presented in Proposition 15.32 any other vertex of the Euclidean cube can be taken as a distinguished vertex.) We say that a side α **does not contain zero** if it does not contain the point $\vec{0} := (0, \ldots, 0)$ and use the notation $\{0,1\}^{l*} := \{0,1\}^l \setminus \{\vec{0}\}$. We further decompose $x \in X^{[l]}$ into $x = (x_0, x^*)$ with $x_0 := x_{\vec{0}} \in X$ and the rest $x^* \in X^{[l]*} := X^{2^l-1}$.

The following result provides the key property for defining the factor.

Proposition 15.37 *Let (X, μ, T) be a regular measure-preserving system. A function $F \in \mathrm{L}^\infty(X^{[l]}, \mu^{[l]})$ is invariant under all side transformations T_α with α not containing zero if and only if there exists $f \in \mathrm{L}^\infty(X, \mu)$ with*

$$F(x) = f(x_0) \text{ for } \mu^{[l]}\text{-a.e. } x = (x_0, x^*) \in X^{[l]}. \tag{15.31}$$

Proof Since every side transformation T_α with α not containing zero leaves the zero coordinate invariant, the "if" direction of the equivalence follows immediately.

To show the "only if" direction, consider the side transformation $T_\alpha = \mathrm{id}^{[l-1]} \times T^{[l-1]}$ corresponding to the side $\alpha = \{0,1\}^{l-1} \times \{1\}$. By Birkhoff's ergodic theorem and Lemma 15.35, for every $F', F'' \in \mathrm{L}^\infty(X^{[l-1]}, \mu^{[l-1]})$ and $\mu^{[l]}$-a.e. $x = (x', x'')$

$$\begin{aligned}
P_{\mathrm{fix}\, T_\alpha}(F' \otimes F'')(x) &= \lim_{N\to\infty} \frac{1}{N} \sum_{n=1}^{N} (\mathrm{id}^{[l-1]} \times T^{[l-1]})^n (F' \otimes F'')(x) \\
&= F'(x') \lim_{N\to\infty} \frac{1}{N} \sum_{n=1}^{N} (T^{[l-1]})^n F''(x'')
\end{aligned}$$

$$= F'(x')P_{\text{fix } T^{[l-1]}}F''(x'')$$
$$= (F' \cdot P_{\text{fix } T^{[l-1]}}F'')(x').$$

Since the linear hull of functions of the form $F' \otimes F''$ with $F', F'' \in \mathrm{L}^\infty(X^{[l-1]}, \mu^{[l-1]})$ is dense in $\mathrm{L}^2(X^{[l]}, \mu^{[l]})$, we see that the orthogonal projection of any L^2-function to fix T_α is independent of x'' $\mu^{[l]}$-a.e. In particular, the given T_α-invariant function $F = P_{\text{fix } T_\alpha}F$ does not depend on the coordinate x'' for $\mu^{[l]}$-a.e. x.

Repeating the procedure for sides involving other fixed coordinates (except the zero coordinate) we obtain inductively after $l-1$ steps that $\mu^{[l]}$-a.e. F does not depend on any coordinate other than 0 finishing the proof. ■

Remark 15.38 As mentioned above, even if μ is ergodic then already $\mu^{[1]} = \mu \times \mu$ is not necessarily ergodic (under $T^{[1]}$). However, by Proposition 15.37, for ergodic μ and every $l \in \mathbb{N}$ the measure $\mu^{[l]}$ is ergodic under the action of the group generated by $T^{[l]}$ and all side transformations T_α for α not containing zero, meaning that every bounded measurable function which is invariant under all of these maps is constant almost everywhere.

The Host–Kra–Ziegler factors can now be defined as follows. We leave it as Exercise 15.10 to show that the corresponding set of functions indeed defines a factor of (X, μ, T).

Definition 15.39 (Host–Kra–Ziegler Factors) Let (X, μ, T) be a regular measure-preserving system and $l \in \mathbb{N}$. We define the σ-algebra $\mathcal{Z}_l$ by requiring a function $f \in \mathrm{L}^\infty(X, \mu)$ to be **measurable w.r.t.** $\mathcal{Z}_l$ if there exists a function $F \in \mathrm{L}^\infty(X^{[l+1]}, \mu^{[l+1]})$ *not* depending on the zero coordinate which satisfies (15.31), i.e.,

$$F(x) = f(x_0) \text{ for } \mu^{[l+1]}\text{-a.e. } x = (x_0, x^*) \in X^{[l+1]}.$$

The factor induced by this σ-algebra (see Exercise 15.10) is called[8] the **Host–Kra–Ziegler factor** (or **Host–Kra factor**) **of order** l and is denoted by Z_l.

In other words, denoting by $\pi_{\vec{0}} : X^{[l+1]} \to X$ the projection onto the zero coordinate, the factor Z_l is generated by all functions $f \in \mathrm{L}^\infty(X, \mu)$ such that the function $f \circ \pi_{\vec{0}}$ (which depends only on the zero coordinate) is $\mu^{[l+1]}$-a.e. equal to a function which does not depend on the zero coordinate.

Remark 15.40 (Description via Sub-σ-Algebras) By restricting to characteristic functions one can explicitly specify which sets belong to the corresponding σ-algebra, see the characterization of factors via sub-σ-algebras (Theorem 8.18). More specifically, if the function F is of the form $F = \mathbf{1}_{X \times B}$ for some measurable

---

[8] Calling these factors Host–Kra–Ziegler factors we follow the terminology used by, e.g., Jamneshan, Shalom, Tao in [579].

$X \times B \subset X^{[l+1]}$ and satisfies the assertion of Proposition 15.37, then the function f in (15.31) is of the form $\mathbf{1}_A$ for some measurable $A \subset X$. Thus we see that a measurable set $A \subset X$ belongs to the σ-algebra $\mathcal{Z}_l$ if and only if there exists a set $B \subset X^{[l+1]*}$ with

$$\mathbf{1}_A(x_0) = \mathbf{1}_B(x^*) \text{ for } \mu^{[l+1]}\text{-a.e. } x = (x_0, x^*) \in X^{[l+1]},$$

i.e., with

$$X \times B = A \times X^{[l+1]*} \text{ up to } \mu^{[l+1]}\text{-null sets.} \tag{15.32}$$

Remark 15.41 This somewhat counterintuitive definition shows that the measures $\mu^{[l]}$ are in some sense between measures with full support and measures supported on the diagonal. Indeed, for a measure with full support the only sets A satisfying the a.e.-equality $X \times B = A \times X^{[l+1]*}$ in (15.32) for some $B \subset X^{[l+1]*}$ are X and $\emptyset$ up to null sets, whereas for a diagonal measure every set has the corresponding property (take $B := A \times \ldots \times A$).

Remark 15.42 It is easy to see that the fixed factor fix T is contained in the lth Host–Kra–Ziegler factor Z_l for each $l \in \mathbb{N}$. We leave the proof as Exercise 15.11.

We now show that for regular systems the Host–Kra–Ziegler factors defined above satisfy (15.14), i.e.,

$$\mathrm{L}^2(X, \mu) = Z_l \oplus \overline{\left\{f \in \mathrm{L}^\infty(X, \mu) : \|f\|_{U^{l+1}} = 0\right\}}, \quad l \in \mathbb{N}_0.$$

Here Z_l is, as usual with a slight abuse of notation, the L^2-closure of the subspace of bounded functions generating the factor $\mathcal{Z}_l$. In particular, we obtain that Z_1 is the Kronecker subspace.

For this decomposition, we first need an analogue of (15.18) for the decomposition $x = (x_0, x^*)$ instead of $x = (x', x'')$. Here and below we identify a function on $X^{[l]}$ which is independent of the zero coordinate with a function on $X^{[l]*}$.

Lemma 15.43 *Let (X, μ, T) be a regular measure-preserving system, $l \in \mathbb{N}$, $f \in \mathrm{L}^\infty(X, \mu)$ and let $G \in \mathrm{L}^\infty(X^{[l]}, \mu^{[l]})$ be independent of the zero coordinate. Then*

$$\int_{X^{[l]}} f(x_0) G(x) \,\mathrm{d}\mu^{[l]}(x) = \int_{X^{[l]}} (P_{Z_{l-1}} f)(x_0) \tilde{G}(x) \,\mathrm{d}\mu^{[l]}(x)$$

for some function $\tilde{G} \in \mathrm{L}^\infty(X^{[l]}, \mu^{[l]})$ invariant under all side transformations T_α with α not containing zero.

Proof Denote by $T_1, \ldots, T_l$ the side transformations with sides not containing zero (such sides are of the form $\{\omega : \omega_j = 1\}$ for $j \in \{1, \ldots, l\}$). Since $\mu^{[l]}$ is invariant under T_1 by Lemma 15.36, we obtain by the mean ergodic theorem

$$\int_{X^{[l]}} f(x_0)G(x)\,\mathrm{d}\mu^{[l]}(x) = \int_{X^{[l]}} f(x_0)\frac{1}{N}\sum_{n=1}^{N} T_1^n G(x)\,\mathrm{d}\mu^{[l]}(x)$$
$$\to \int_{X^{[l]}} f(x_0)G_1(x)\,\mathrm{d}\mu^{[l]}(x) \text{ as } N \to \infty$$

for some function $G_1 \in \operatorname{fix} T_1$ (one can also use (8.4)). Note that G_1 does not depend on the zero coordinate since the corresponding side does not contain zero. Repeating the procedure with G_1 and T_2 we obtain the needed equality with a function $G_2 \in \operatorname{fix} T_1 \cap \operatorname{fix} T_2$ which is independent of the zero coordinate, etc. In l steps we get the desired function $\tilde{G}$ which satisfies

$$\int_{X^{[l]}} f(x_0)G(x)\,\mathrm{d}\mu^{[l]}(x) = \int_{X^{[l]}} f(x_0)\tilde{G}(x)\,\mathrm{d}\mu^{[l]}(x).$$

Now, by Proposition 15.37 there is $\tilde{g} \in \mathrm{L}^\infty(X,\mu)$ with $\tilde{g}(x_0) = \tilde{G}(x)$ for $\mu^{[l]}$-almost every $x = (x_0, x^*)$. Note that, since by construction $\tilde{G}$ does not depend on the zero coordinate, $\tilde{g}$ belongs to the factor Z_{l-1}, i.e., $P_{Z_{l-1}}\tilde{g} = \tilde{g}$. This yields as above (or by (8.4))

$$\int_{X^{[l]}} f(x_0)G(x)\,\mathrm{d}\mu^{[l]}(x) = \int_{X^{[l]}} f(x_0)\tilde{G}(x)\,\mathrm{d}\mu^{[l]}(x)$$
$$= \int_{X^{[l]}} f(x_0)\tilde{g}(x_0)\,\mathrm{d}\mu^{[l]}(x) = \int_X f(x_0)\tilde{g}(x_0)\,\mathrm{d}\mu(x_0)$$
$$= \int_X P_{Z_{l-1}} f(x_0)\tilde{g}(x_0)\,\mathrm{d}\mu(x_0) = \int_{X^{[l]}} P_{Z_{l-1}} f(x_0)\tilde{G}(x)\,\mathrm{d}\mu^{[l]}(x),$$

where we used also that the marginal of $\mu^{[l]}$ is μ. The proof is complete. ∎

Next we can relate the kernels of the projections onto the factors Z_l and the kernels of Gower–Host–Kra seminorms.

Proposition 15.44 *Let (X,μ,T) be a regular measure-preserving system, $l \in \mathbb{N}$ and $f \in \mathrm{L}^\infty(X,\mu)$. Then the equivalence*

$$P_{Z_l} f = 0 \quad \Longleftrightarrow \quad \|f\|_{U^{l+1}} = 0.$$

holds.

Proof Assume first that $\|f\|_{U^{l+1}} = 0$. By the Gowers–Cauchy–Schwarz inequality (Proposition 15.24) and (15.24) we see that

$$\int_{X^{[l+1]}} f(x_0)G(x)\,\mathrm{d}\mu^{[l+1]}(x) = 0$$

for every function $G \in \mathrm{L}^\infty(X^{[l+1]}, \mu^{[l+1]})$ of the form

$$G(x) = \prod_{\omega \in \{0,1\}^{l*}} f_\omega(x_\omega)$$

for $x = (x_\omega)_{\omega \in \{0,1\}^l}$ and $f_\omega \in \mathrm{L}^\infty(X, \mu)$ for every $\omega \in \{0, 1\}^{l*}$. Note that one can approximate in $\mathrm{L}^2(X^{[l+1]}, \mu^{[l+1]})$ every $\mathrm{L}^\infty(X^{[l+1]}, \mu^{[l+1]})$-function which does not depend on the zero coordinate by linear combinations of such functions arbitrarily well. So, by the definition of Z_l, we obtain that f is orthogonal to every function in Z_l, i.e., $P_{Z_l} f = 0$.

To show the converse implication suppose that $P_{Z_l} f = 0$. By Proposition 15.19 and Lemma 15.43, the function $G \in \mathrm{L}^\infty(X^{[l+1]}, \mu^{[l+1]})$ defined by $G(x) := \prod_{\omega \in \{0,1\}^{(l+1)*}} C^{|\omega|} f(x_\omega)$ satisfies for some function $\tilde{G} \in \mathrm{L}^\infty(X^{[l+1]}, \mu^{[l+1]})$

$$\begin{aligned} \|f\|_{U^{l+1}} &= \int_{X^{[l+1]}} f(x_0)G(x)\,\mathrm{d}\mu^{[l+1]}(x) \\ &= \int_{X^{[l+1]}} (P_{Z_l} f)(x_0)\tilde{G}(x)\,\mathrm{d}\mu^{[l+1]}(x) = 0. \end{aligned}$$

The proof is complete. ∎

15.6 Structure Theorem of Host and Kra

By the decomposition (15.14), to show convergence of multiple averages (15.1) it suffices by Remark 15.2 to assume that all functions $f_1, \ldots, f_l$ belong to the Host–Kra–Ziegler factor $\mathcal{Z}_{l-1}$. One of the main ingredients in the proof of multiple convergence is the following structure theorem which describes the nilpotent nature of the factors $\mathcal{Z}_l$, thereby reducing the problem to nilsystems. For the definition of inverse limit see the beginning of Sect. 13.7.

Theorem 15.45 (Host–Kra Structure Theorem) *Let (X, μ, T) be a regular ergodic measure-preserving system and $l \in \mathbb{N}$. Then the Host–Kra–Ziegler factor $\mathcal{Z}_l$ can be represented as an inverse limit of its factors, each of which is isomorphic to a nilsystem of step l.*

As mentioned in [553, p. 268], in general, one cannot drop the assumption of ergodicity here. To shorten the statement of Theorem 15.45 we say that $\mathcal{Z}_l$ is *an inverse limit of nilsystems of step l*.

For the lengthy proof of this theorem we refer to [549, 553]. Here we just spend a few words on some of its main ingredients and outline a very rough sketch following [553, Ch. 18–20].

Let (X, μ, T) be a regular ergodic measure-preserving system. By the regularity assumption all factors of the system are separable. By the Halmos–von Neumann theorem (Theorem 8.32) the Kronecker factor $\mathcal{Z}_1$ is isomorphic to an ergodic rotation on a compact Abelian group. This group is however not necessarily a Lie group. (Note that a compact Abelian group is a Lie group if and only if its dual group is finitely generated.) Writing the dual group as an increasing countable union of finitely generated Abelian groups, one can easily show that $(\mathcal{Z}_1, \mathrm{m}, T)$ is an inverse limit of rotations on compact Abelian *Lie* groups, i.e., of nilsystems of step one. For details see Host and Kra [553, Sec. 18.1].

To handle the case $l \geq 2$, assume without loss of generality that $X = \mathcal{Z}_l$ (such systems Host and Kra call **systems of order** l). An equivalent description is by the decomposition (15.14) that the seminorm $\|\cdot\|_{U^{l+1}}$ is a norm on $\mathrm{L}^\infty(X, \mu)$.

The rest of the proof is then divided into three parts.

Part 1. Recall Rokhlin's skew product theorem (Theorem 8.10) giving a representation of an ergodic extension as a skew rotation system. Consider the finite sequence of factors

$$X = \mathcal{Z}_l \to \mathcal{Z}_{l-1} \to \cdots \to \mathcal{Z}_2 \to \mathcal{Z}_1.$$

One shows first that for every $j \in \{2, \ldots, l\}$, $\mathcal{Z}_j$ is an extension of $\mathcal{Z}_{j-1}$ by a connected compact Abelian group being as such the inverse limit of tori. Moreover, the corresponding cocycle ρ satisfies a certain functional equation (such cocycles are called **cocycles of type** j). Note that in this section we use the terminology of Host and Kra [553] and call the mapping $\rho = \widehat{\rho}(1, \cdot)$ the cocycle of the skew product (where $\widehat{\rho}$ is the cocycle of the skew product as defined in Example 8.2). This functional equation for $j = 2$ corresponds to the so-called **Conze–Lesigne equation**[9] introduced by Conze and Lesigne [256–258] in their proof of convergence for triple ergodic averages for totally ergodic systems, see [553, Sec. 18.3.3]. In the converse direction, if ρ is a cocycle of type j defined on a system of order j with values in a compact Abelian group such that the corresponding extension is ergodic, then this extension is a system of order j, see [553, Sec. 18.4.2].

Part 2. One proves a structure result in a special case being the heart of the argument. Assume namely that (X, μ, T) is a torus extension of $\mathcal{Z}_{l-1}$ by a cocycle of type l and assume further that $\mathcal{Z}_{l-1}$ is isomorphic to a nilsystem of step $l - 1$. Then (X, μ, T) is isomorphic to a nilsystem of step l. Thereby the corresponding group G consists of certain transformations S_g on X preserving the cube structure of

---

[9] One says that a cocycle $\rho : Z \to K$ for a rotation system (Z, μ, a) and a compact Abelian group K obeys the **Conze–Lesigne equation** if for every $s \in Z$ the *discrete derivative* $R_s\rho \cdot \overline{\rho}$ is a *quasi-coboundary*, i.e., for every $s \in Z$ one has

$$\rho(sz)\overline{\rho}(z) = cF(az)\overline{F}(z) \text{ for a.e. } z \in Z$$

for some $F : Z \to K$ and $c \in K$. The additive version of this equation is

$$\rho(s + z) - \rho(z) = F(a + z) - F(z) + c.$$

X, see [553, Sec. 19.2]. Moreover, by identifying (X, μ, G) with an appropriate topological model one can assume that $\mathcal{Z}_{l-1}$ is a nilsystem, that the action of G on X is continuous (meaning that $g \mapsto S_g f$ is continuous for all continuous functions f on X) and the factor projection $\pi : X \to \mathcal{Z}_{l-1}$ is continuous as well. This allows to show transitivity of G and construct the corresponding cocompact subgroup as the stabilizer[10] of a fixed preimage of identity in the nilsystem $\mathcal{Z}_{l-1}$ with respect to π , see [553, Sec. 19.3].

Part 3. Let now (X, μ, T) be an arbitrary regular ergodic system with $X = \mathcal{Z}_l$, denote $Y := \mathcal{Z}_{l-1}$ and denote by K the connected compact Abelian group corresponding to the extension $X = \mathcal{Z}_l$ of Y given by Part 1. Passing to the dual group of K one sees that X is an inverse limit of systems $(\tilde{X}_n)$ each of which is an extension of Y by a torus. Thus it suffices to show that each $\tilde{X}_n$ is an inverse limit of nilsystems, or, in other words, one can assume that K is a torus.

The proof goes by induction on l, where the case $l = 1$ was discussed above. By the induction hypothesis, Y is an inverse limit of ergodic nilsystems (Y_n) of step $l - 1$. For each n one constructs an appropriate cocycle on $X_n := Y_n \times K$ of type l leading to factors X_n of X such that X is an inverse limit of (X_n) and $\mathcal{Z}_{k-1}(X_n) = Y_n$ for every n. Thus by Part 2 each X_n is isomorphic to a nilsystem of step l finishing the proof, see [553, pp. 325–326].

For a non-ergodic version of the structure theorem recall the notion of a nilsequence from Sect. 14.1.

Remark 15.46 (Non-ergodic Case) It is essential for Theorem 15.45 to assume ergodicity of the system. The following non-ergodic function version of the structure theorem is due to Chu et al. [247], see also Host and Kra [553, Sec. 16.2]: Let (X, μ, T) be a regular system, $l \in \mathbb{N}$, f be a bounded measurable function on the factor $\mathcal{Z}_l$ and $\varepsilon > 0$. Then f admits the decomposition

$$f = f_{\text{nil}} + f_{\text{small}},$$

where $\|f_{\text{small}}\|_1 < \varepsilon$, $\|f_{\text{small}}\|_\infty \le \|f\|_\infty$ and the sequence $(f_{\text{nil}}(T^n x))_{n=1}^\infty$ is for a.e. $x \in X$ a nilsequence of step l. Note that for $f \in \mathrm{L}^\infty(X, \mu)$ one has by (15.14) the decomposition

$$f = f_{\text{unif}} + f_{\text{nil}} + f_{\text{small}},$$

where the additional term f_{unif} satisfies $\|f_{\text{unif}}\|_{U^{l+1}} = 0$.

Remark 15.47 (More General Setting) The Host–Kra structure theory was extended to $\mathbb{Z}^d$-actions by Griesmer [497] and to finitely generated Abelian or nilpotent group actions by Gutman and Lian [504, Sec. 5] and Candela and Szegedy

[10] cf. Proposition 11.28.

[221, Thm. 5.12], respectively. For some steps in the direction of an extension of the Host–Kra structure theory to general (not necessarily separable) systems and general (ergodic) group actions see Jamneshan et al. [579], Jamneshan and Tao [580], and also Bergelson et al. [147], Shalom [932–934].

Remark 15.48 (Topological Structure Theorem) A topological version of the Host–Kra structure theorem for minimal distal systems was presented by Host et al. [554] (with the case of $l = 2$ previously done by Host and Maass [556]). It was generalized to minimal systems (and Abelian group actions) by Shao and Ye [935] and to general minimal group actions by Glasner et al. [469] and Gutman et al. [507] who presented a connection to nilspaces introduced by Antolín Camarena and Szegedy in [33] and briefly discussed in Sect. 15.9 below.

15.7 Finishing the Proof of Multiple Convergence

We now summarize the proof of multiple convergence putting all ingredients together.

Proof of Theorem 15.1 By Remark 15.2 we can assume that the measure-preserving system is regular. We show that one can further assume without loss of generality that the system (X, μ, T) is ergodic. Let Y be its fixed factor and $\mu = \int_Y \mu_y \, \mathrm{d}\mu(y)$ be the ergodic decomposition, see Sect. 8.6. Assume that the statement of Theorem 15.1 holds for ergodic systems and denote $S_N := \frac{1}{N}\sum_{n=1}^{N} T^n f_1 \cdots T^{kn} f_k$ for $N \in \mathbb{N}$. Then Lebesgue's dominated convergence theorem yields[11]

$$\int_X |S_N - S_M|^2 \, \mathrm{d}\mu = \int_Y \int_X |S_N - S_M|^2 \, \mathrm{d}\mu_y \, \mathrm{d}\mu(y) \xrightarrow[N,M\to\infty]{} 0.$$

Thus we can and will assume that our system (X, μ, T) is ergodic.
Using the decomposition (15.14), write each function f_j, $j \in \{1, \dots, k\}$ as $f_j = P_{Z_{k-1}} f_j + g_j$, where $\|g_j\|_{U^k} = 0$. Note that since $\mathcal{Z}_{k-1}$ is a factor, each function $P_{Z_{k-1}} f_j$ and hence g_j is bounded and the uniformity seminorm of g_j is indeed defined. By linearity and the generalized von Neumann inequality (Proposition 15.14), it is no loss of generality to assume that $f_j = P_{Z_{k-1}} f_j$ for every $j \in \{1, \dots, k\}$ or, in other words, to assume that the system equals $\mathcal{Z}_{k-1}$. By the Host–Kra structure theorem (Theorem 15.45), the system is an inverse limit of nilsystems X_n, $n \in \mathbb{N}$, of step $k-1$. By Remark 13.47 and the standard approximation argument (recall that T is a contraction on each L^p) we can assume without loss of generality that each f_j belongs to the subspace generated by

[11] Recall that every μ-null set is also a μ_y-set for a.e. $y \in Y$ by Theorem 8.57(a), so one has $\|\cdot\|_{\mathrm{L}^\infty(X,\mu_y)} \le \|\cdot\|_{\mathrm{L}^\infty(X,\mu)}$ for such y.

some X_n, where n is independent of j by the increasing property of factors. By another approximation argument we can assume that each f_j is continuous on the corresponding space.

Thus by multiple convergence for nilsystems and continuous functions, see Proposition 14.44, the proof is complete. ■

15.8 Multiple Recurrence via Host–Kra–Ziegler Factors

In this section we present one more proof of multiple recurrence (Theorem 13.5) using the conditional Jacobs–de Leeuw–Glicksberg decomposition and more detailed analysis of the Host–Kra–Ziegler factors. It is based on the following property of the Host–Kra–Ziegler factors due to Host and Kra [549, Lem. 6.2], [553, Lem. 18.2][12] and Ziegler [1065, Thm. 6.1], namely that they are compact extensions of each other. This property is clearly stronger than the one in Proposition 15.29 by the definition of the distal factors of finite step. We follow the proof by Zorin-Kranich [1073, Sec. 10] and will need the description of the Host–Kra–Ziegler factors via the cube spaces from Sect. 15.5.

Proposition 15.49 (Host–Kra–Ziegler Factors are Compact Extensions of Each Other) *Let (X, μ, T) be a regular ergodic system and let $k \in \mathbb{N}$. Then $Z_{k+1} \subset \overline{A(X|\mathcal{Z}_k)}$, i.e., $\mathcal{Z}_{k+1}$ is a compact extension of $\mathcal{Z}_k$.*

Proof We may suppose that the whole system is $\mathcal{Z}_{k+1}$, i.e., $\mathrm{L}^2(X) = Z_{k+1}$, and apply the conditional Jacobs–de Leeuw–Glicksberg decomposition presented in Theorem 13.24

$$\mathrm{L}^2(\mathcal{Z}_{k+1}) = Z_{k+1} = W(X|\mathcal{Z}_k) \oplus \overline{A(X|\mathcal{Z}_k)}.$$

By the decomposition in (15.14) it suffices to prove that for each bounded $f \in W(X|\mathcal{Z}_k)$ one has $\|f\|_{U^{k+1}} = 0$. Take a bounded $f \in W(X|\mathcal{Z}_k)$ and denote $f_\omega := C^{|\omega|} f$ for $\omega \in \{0,1\}^k$ or $\omega \in \{0,1\}^{k+1}$, respectively. Recall that by (15.22) and (15.18)

$$\|f\|_{U^{k+1}}^{2^{k+1}} = \int_{X^{[k+1]}} \bigotimes_{\omega\in\{0,1\}^{k+1}} f_\omega \, \mathrm{d}\mu^{[k+1]} = \Big\| P_{\mathrm{fix}\, T^{[k]}} \bigotimes_{\omega\in\{0,1\}^k} f_\omega \Big\|^2_{\mathrm{L}^2(X^{[k]},\mu^{[k]})},$$

so we have to show that $\bigotimes_{\omega\in\{0,1\}^k} f_\omega \perp \mathrm{fix}\, T^{[k]}$.

Since $\mathrm{fix}\, T^{[k]}$ is spanned by one-dimensional $T^{[k]}$-invariant subspaces, we have $\mathrm{fix}\, T^{[k]} \subset \overline{A(X^{[k]}|\mathcal{Z})}$ for every factor $\mathcal{Z}$ of $X^{[k]}$. Thus by Theorem 13.24, applied to

[12] See Glasner [468, Thm. 9.14] for the equivalence of the property of being an isometric extension in the sense of Furstenberg [445] proved by Host and Kra and the definition of a compact extension here (Recall that one has $\overline{E(X|Y)} = \overline{A(X|Y)}$ by Theorem 13.24.).

the system $(X^{[k]}, \mu^{[k]}, T^{[k]})$, it suffices to show that there exists a factor $\mathcal{Z}$ of this system with

$$\bigotimes_{\omega\in\{0,1\}^k} f_\omega \in W(X^{[k]}|\mathcal{Z}). \tag{15.33}$$

Consider now the factor $\mathcal{Z}$ generated by functions of the form $\otimes_{\omega\in\{0,1\}^k} g_\omega$ with $g_{\vec{0}} \in Z_k$ and arbitrary $g_\omega \in \mathrm{L}^\infty(X)$ for $\omega \neq \vec{0}$. We first show that

$$E_{\mathcal{Z}}(g \circ \pi_{\vec{0}}) = (P_{Z_k} g) \circ \pi_{\vec{0}} \tag{15.34}$$

holds for every $g \in \mathrm{L}^\infty(X)$. By approximation we can assume that $g \in \mathrm{L}^\infty(X)$, and hence by (15.14) we can assume without loss of generality that either $g \in Z_k$ or $\|g\|_{U^{k+1}} = 0$. In the first case, both sides of (15.34) equal $g \circ \pi_{\vec{0}}$ by the construction of $\mathcal{Z}$. Assume now that $\|g\|_{U^{k+1}} = 0$. Then the right-hand side of (15.34) vanishes. To prove that the left-hand side vanishes as well, it suffices to show that $g \circ \pi_{\vec{0}} \perp \otimes_{\omega\in\{0,1\}^k} g_\omega$ in $\mathrm{L}^2(X^{[k]}, \mu^{[k]})$ for every bounded $g_{\vec{0}} \in Z_k$ and $g_\omega \in \mathrm{L}^\infty(X)$, $\vec{0} \neq \omega \in \{0,1\}^k$. By the properties of Markov projections, see Proposition 8.25, $P_{Z_k}(g\overline{g_{\vec{0}}}) = \overline{g_{\vec{0}}} P_{Z_k} g = 0$ implying $\|g\overline{g_{\vec{0}}}\|_{U^{k+1}} = 0$ by (15.14) (and by the boundedness of both g and $g_{\vec{0}}$). By the Gowers–Cauchy–Schwarz inequality, see Proposition 15.24, we obtain

$$\int_{X^{[k]}} (g\overline{g_{\vec{0}}}) \otimes \bigotimes_{\omega\in\{0,1\}^k\setminus\{\vec{0}\}} \overline{g_\omega}\, \mathrm{d}\mu^{[k]} = 0,$$

and so finally (15.34) is proved.

By (15.34), Proposition 8.52(c) and the definition of $\mathcal{Z}$ we have for every $(g_\omega) \subset \mathrm{L}^\infty(X)$

$$E_{\mathcal{Z}}\Biggl(\bigotimes_{\omega\in\{0,1\}^k} g_\omega\Biggr) = P_{Z_k} g_{\vec{0}} \otimes \bigotimes_{\omega\in\{0,1\}^{k*}} g_\omega$$

and thus

$$\begin{aligned}
&\Bigl\| E_{\mathcal{Z}}\Bigl(T^{[k]^n}\Bigl(\bigotimes_{\omega\in\{0,1\}^k} f_\omega\Bigr) \cdot \bigotimes_{\omega\in\{0,1\}^k} \overline{f_\omega}\Bigr)\Bigr\|_{\mathrm{L}^1(\mathcal{Z})} \\
&= \Bigl\| E_{\mathcal{Z}}\Bigl(\bigotimes_{\omega\in\{0,1\}^k} (T^n f_\omega \cdot \overline{f_\omega})\Bigr)\Bigr\|_{\mathrm{L}^1(\mathcal{Z})} \\
&= \Bigl\| P_{Z_k}(T^n f \cdot \overline{f}) \otimes \bigotimes_{\omega\in\{0,1\}^k\setminus\{\vec{0}\}} (T^n f_\omega \cdot \overline{f_\omega})\Bigr\|_{\mathrm{L}^1(X^{[k]})} \\
&\le \|(P_{Z_k}(T^n f \cdot \overline{f})) \circ \pi_{\vec{0}}\|_{\mathrm{L}^1(X^{[k]})} \\
&= \|P_{Z_k}(T^n f \cdot \overline{f})\|_{\mathrm{L}^1(X)},
\end{aligned}$$

where for the last equality we used Proposition 15.34. Since $f \in W(X|\mathcal{Z}_k)$ the last term above converges to zero in density as $n \to \infty$, yielding (15.33). The proof is complete. ■

The fourth proof of Theorem 13.5 via Proposition 15.49 is analogous to the third proof via Proposition 15.29 using the finite sequence of factors

$$X = Z_k \to \ldots \to Z_1 \to Z_0 = \{\cdot\}.$$

We leave the details as Exercise 15.12.

15.9 Further Developments

Here we mention some further developments on convergence of multiple ergodic averages, for more see Sect. 16.6.

Green–Tao Theorem Revisited, Inverse Gowers Conjecture and Nilspaces

Gowers norms play an important role in combinatorics. As mentioned above, Gowers introduced them in his proof of Szemerédi's theorem. In their seminal proof of existence of arithmetic progressions in the primes [485], Green and Tao used ideas with ergodic theoretic flavor such as characteristic factor and a decomposition of a function into a structured part (with large Gowers uniformity norm) and a pseudorandom part (with small Gowers uniformity norm), where the pseudorandom part does not contribute to the existence of arithmetic progressions by a version of Proposition 15.14. See Kra [668] for an exposition and parallels to ergodic theory and Tao [975] for a discussion of the dichotomy between structure and randomness.

In a subsequent proof generalizing the result to linear equations in primes, Green and Tao [486] used a much simpler decomposition of the modified Λ'-function (being a variant of the von Mangoldt function, see (10.9)) into a structured and a pseudorandom part, where the structured part is just the constant one function. Thus the work was to show that the difference is pseudorandom, i.e., has sufficiently small Gowers uniformity norm. To do this they explicitly used ergodic theory, namely nilsequences (coming from Riemann integrable functions rather then continuous ones). In particular, they use the following fact, known as *inverse conjecture/theorem for the Gowers norms* and proved by Green and Tao [484] and Green et al. [491, 492]: If the (rescaled) Gowers norm of order

s of a function on $\{1, \ldots, N\}$ is greater than δ, then there exists a bounded[13] complexity Riemannian[14] nilsequence (a_n) (depending on N) of step $s - 1$ and bounded by one with $\left|\frac{1}{N}\sum_{n=1}^{N} f(n)a_n\right| \geq C_{s,\delta}$, i.e., largeness of the Gowers norm of order s implies non-trivial correlation with a nilsequence of step $s - 1$. Thus in order to show that the Gowers norm is small it suffices to show small correlation with nilsequences of the corresponding step. In [487] Green and Tao showed how to derive Szemerédi's theorem from the inverse theorem for the Gowers norms. Moreover, using Manners' quantitative version of the inverse theorem for the Gowers norms [769], Tao and Teräväinen [990] provided quantitative bounds for solutions of linear equations in primes, see Leng [726] for improved bounds and Matomäki et al. [778] for an estimate of solutions of linear equations in primes in short intervals. Moreover, Leng et al. [730] presented quasi-polynomial bounds in the inverse theorem for the Gowers norms (with the case $s = 4$ done previously by Leng [727]) and used them in [729] to improve bounds in Szemerédi's theorem.

Szegedy [968] presented an algebraic view on higher order Fourier analysis and inverse theorems for the Gowers norms, see also [969]. Together with Antolín Camarena [33] he presented an abstract approach to Host–Kra–Ziegler factors and Gowers uniformity norms unifying combinatorics and ergodic theory by introducing so-called **nilspaces**. Nilspaces are spaces endowed with a collection of cubes satisfying some abstract axioms being a variant of the axioms introduced in Host and Kra [551] for parallelepiped structures. Using nilspaces, Candela and Szegedy [221] presented a structure theorem for characteristic factors for Gowers–Host–Kra type seminorms for actions of countable nilpotent groups providing in particular an alternative proof of Theorem 15.45. Another approach to the Host–Kra–Ziegler factors $\mathcal{Z}_l$ using nilspaces is due to Gutman and Lian [504] who found a topological model of the factor $\mathcal{Z}_l$ which is a nilspace. This together with the structure theory due to Gutman et al. [505–507] representing every nilspace as an inverse limit of nilmanifolds gives one more alternative proof of Theorem 15.45. For related resuts see Candela and Szegedy [220], Candela et al. [219] and Glasner et al. [469].

Polynomial Multiple Averages and Hardy Fields

Using the Host–Kra theory, Leibman [717] extended in 2005 convergence of polynomial multiple ergodic averages

$$\frac{1}{N}\sum_{n=1}^{N} T^{P_1(n)} f_1 \cdot \ldots \cdot T^{P_k(n)} f_k, \tag{15.35}$$

[13] The complexity bound here depends on s and δ. For the precise formulation of the inverse Gowers conjecture/theorem see Green et al. [492, Conjecture 1.2].

[14] such as a sequence of the form $(e(\alpha n\{\beta n\}))$ or, more generally, $(e(f(n)))$ for a bracket polynomial f, see Example 14.15.

$P_1, \ldots, P_k : \mathbb{Z} \to \mathbb{Z}$ being polynomials, for nilsystems (Corollary 14.44) to norm convergence of such averages for general invertible measure-preserving systems. To do this he showed the following variation of the generalized von Neumann inequality (Proposition 15.14): If $P_1, \ldots, P_k$ are essentially distinct (i.e., the difference of any two of them is non-constant) and have degree at most d, then there exists r depending only on k and d such that the factor $\mathcal{Z}_r$ is characteristic for the averages (15.35), i.e., (15.35) converge in L^2 to zero whenever at least one of the functions satisfies $\|f_j\|_{U^{r+1}} = 0$. By ergodic decomposition, linearity and the structure theorem (Theorem 15.45) one may then assume that the system is a nilsystem and use Corollary 14.47. This generalized a partial result on norm convergence of (15.35) done by Host and Kra [548] as well as the weakly mixing case proved by Bergelson [103]. For some results for finding the minimal r such that the factor $\mathcal{Z}_r$ is characteristic for the averages (15.35) see Kuca [689]. Moreover, Frantzikinakis [408] showed that the nilfactor (being the inverse limit of $\mathcal{Z}_r$) is characteristic for multiple averages

$$\frac{1}{N}\sum_{n=1}^{N} T^{[a(n)]} f_1 \cdot \ldots \cdot T^{k[a(n)]} f_k, \tag{15.36}$$

where a is a real-valued polynomial or, more generally, a function from a Hardy field with certain (subpolynomial) growth, showing convergence of these averages for such functions a. See also Frantzikinakis [409, 413] for a multidimensional version for functions of different growth and a similar result along primes for linearly independent fractional polynomials, respectively.

Convergence of polynomial multiple averages along primes of the form

$$\frac{1}{\pi(N)} \sum_{p<N, p \text{ prime}} T^{P_1(p)} f_1 \cdot \ldots \cdot T^{P_k(p)} f_k$$

for the prime counting function π was shown by Wooley and Ziegler [1049] by comparing them to polynomial multiple averages (15.35) along $\mathbb{N}$ using the uniformity result of the von Mangoldt function by Green and Tao [486] (conditional to [488] and [492]). The linear case for $k = 2$ and for higher k conditional to [488] and [492] was done by Frantzikinakis et al. [419], see also Moreira and Richter [806]. For more results in this direction, also including more general sequences of powers, we refer to Fratzikinakis et al. [420], Sun [962], Koutsogiannis [665], Frantzikinakis [413], Frantzikinakis and Kuca [426], Koutsogiannis and Tsinas [667]. For convergence of multiple ergodic averages along primes in short intervals see Matomäki et al. [778].

Multiple Averages for Commuting and Non-commuting Transformations

Convergence of multiple ergodic averages

$$\frac{1}{N}\sum_{n=1}^{N} T_1^n f_1 \cdot \ldots \cdot T_k^n f_k \tag{15.37}$$

for commuting transformations $T_1, \ldots, T_k$ was shown by Tao [976] in 2008 neither using nilsystems nor characteristic factors. He deduced convergence of averages (15.37) from a combinatorial result using a reverse version of Furstenberg's correspondence principle, cf. Theorem 9.21. The case $k = 2$ was done before by Conze and Lesigne [256] (see Durcik et al. [327] for a quantitative version), $k = 3$ by Zhang [1062] and general k with an additional assumption by Frantzikinakis and Kra [422]. Note that Tao's proof, unlike the earlier ones which used characteristic factors, does not give any information about the limit. Alternative proofs of Tao's result were given by Towsner [1006] using non-standard analysis and Austin [60] and Host [547] using more ergodic theory. In particular, Host introduced a version of the Gowers–Host–Kra seminorms for commuting transformations. Convergence of polynomial multiple averages for commuting transformations along primes, again by comparing them to polynomial multiple averages along $\mathbb{N}$ using the uniformity result of the von Mangoldt function by Green and Tao [486] (conditional to [488] and [492]), was shown by Frantzikinakis et al. [420].

Confirming a conjecture of Bergelson and Leibman, see [104, 131], and again using a translation into a combinatorial problem, Walsh [1031] proved in 2012 convergence of multiple polynomial ergodic averages

$$\frac{1}{N}\sum_{n=1}^{N}\prod_{j=1}^{k}(T_1^{P_{1,j}}(n)\cdots T_l^{P_{l,j}(n)})f_j \tag{15.38}$$

where $T_1, \ldots, T_k$ generate a nilpotent group and $P_{m,j} : \mathbb{Z} \to \mathbb{Z}$, $m \in \{1, \ldots, l\}$, $j \in \{1, \ldots, k\}$, are polynomials, see also Zorin-Kranich [1074] and Austin [64, 65]. (Note that Bergelson and Leibman showed in [131] that the limit of $\frac{1}{N}\sum_{n=1}^{N} T^n f \cdot S^n g$ does not have to exist if T and S generate a solvable group.) Also here there is no information about the limit. The special case of commuting transformations and polynomials of distinct degrees was done before by Chu et al. [247] where a characteristic factor was identified as a mixture of inverse limits of nilsystems.

There has been an effort to extend the structure theory of Host and Kra to commuting transformations and multiple polynomial averages

$$\frac{1}{N}\sum_{n=1}^{N} T_1^{P_1(n)} f_1 \cdot \ldots \cdot T_k^{P_k(n)} f_k, \tag{15.39}$$

in particular to averages (15.37) in order to obtain a formula or at least some information about the limit of these averages. For partial results in this directions see Frantzikinakis and Kra [422], Johnson [582], Austin [60, 62, 63], Host [547], Chu et al. [247], Tao and Ziegler [994], Donoso et al. [307], Frantzikinakis and Kuca [425, 426]. In particular, Frantzikinakis and Kuca [426] showed that for pairwise independent[15] polynomials (with zero constant term), the Host–Kra–Ziegler factor of sufficiently large step is characteristic for the averages (15.39). They also showed that the rational Kronecker factor is characteristic for these averages whenever the polynomials are linearly independent. The general case (as well as the case of linear averages (15.37)) is still a major open problem in the area.

For convergence of averages (15.37) for ergodic disjoint (not necessarily commuting) transformations see Hirayama et al. [537]. Moreover, Frantzikinakis and Host [418] showed convergence and recurrence for double polynomial averages

$$\frac{1}{N}\sum_{n=1}^{N} T^{P(n)}f\cdot S^{Q(n)}g \tag{15.40}$$

for invertible, not necessarily commuting transformations T, S with T being of zero (measure-theoretic) entropy and $P, Q\in\mathbb{Z}[\cdot]$ whenever $P(n)=n$ and $\deg Q\geq 2$. (They also showed that the entropy assumption cannot be dropped.) For a multiple version of this result see Huang et al. [559]. Counterexamples for convergence in the linear case $P(n)=Q(n)=n$ were constructed by Austin [66], Huang et al. [560] and Ryzhikov [902]. See also Frantzikinakis [412] for convergence of the averages of the form $\frac{1}{N}\sum_{n=1}^{N}T^nf\cdot S^{\lfloor n^\alpha\rfloor}g$ for such transformations and $\alpha\in\mathbb{R}_+\setminus\mathbb{Z}$ and related results. Huang et al. [560] showed that the double polynomial averages (15.40) diverge in general for ergodic, invertible zero entropy systems whenever $\deg P, \deg Q\geq 5$. Their ideas were used by Kosloff and Sanadhya [663] to treat polynomials satisfying $\deg P, \deg Q\geq 2$ and to show failure of double recurrence for polynomials of the form $P(n)=Q(n)=Cn^d$ with $d\geq 3$ and large C. A counterexample for double recurrence in the linear case $P(n)=Q(n)=n$ was presented by Ryzhikov [903] who also showed failure of double convergence for general integer sequences $(P(n))$, $(Q(n))$ satisfying $\lim_{n\to\infty}(P(n+1)-P(n))=\lim_{n\to\infty}(Q(n+1)-Q(n))=0$. (Note that failure of double convergence and double recurrence for general sequences without the entropy zero assumption was already proved by Frantzikinakis et al. [430, Theorem 1.7].) For some convergence results on multiple polynomial averages involving a transformation with singular spectrum and a totally ergodic transformation (again not necessarily commuting) see Farhangi [378].

[15] i.e., P_j/P_k is non-constant whenever $j\neq k$.

Multiple Correlation Sequences

Using the Host–Kra theory, Bergelson et al. [123] showed that *multiple correlation sequences* being sequences of the form

$$\int_X f_0 \cdot T^n f_1 \cdots T^{kn} f_k \,\mathrm{d}\mu, \quad n \in \mathbb{N},$$

decompose as a sum of a sequence being a uniform limit of k-step nilsequences and a sequence converging to zero in uniform density, i.e., a sequence $(b_n)_{n\in\mathbb{N}}$ satisfying

$$\lim_{N\to\infty} \sup_{M\in\mathbb{N}} \sum_{n=M}^{M+N-1} |b_n| = 0.$$

Taking characteristic functions leads to a description of the multiple return times sequence defined by

$$a_n := \mu(A_0 \cap T^{-n}A_1 \cap \ldots \cap T^{-kn}A_k).$$

In particular, they showed that for ergodic systems, $A = A_0 = \cdots = A_k$ and every $\varepsilon > 0$ the set $\{n \in \mathbb{N} : a_n > \mu(A)^{k+1} - \varepsilon\}$ is syndetic for $k = 2$ and $k = 3$ (the case $k = 1$ is Khinchin's recurrence, see Theorem 7.6) while for $k \geq 4$ this is no longer true.

Leibman [720] extended the decomposition result of Bergelson, Host and Kra mentioned above to polynomial multicorrelation sequences (and polynomial nilsequences, respectively) with a multiparameter non-ergodic version done in [722]. For further results see Griesmer [497], Frantzikinakis [410], Frantzikinakis and Host [416], Moreira and Richter [806], Le [709], Ferré Moragues [389], Le et al. [710], Koutsogiannis et al. [666], Briët and Green [193], Frantzikinakis and Kuca [426], Leng [728].

Cubic Averages

As a byproduct of their theory, Host and Kra [549] showed that cubic averages of the form

$$\frac{1}{N^k} \sum_{n_1,\ldots,n_k=1}^{N} \prod_{\omega\in\{0,1\}^{l*}} T^{n\cdot\omega} f_\omega \tag{15.41}$$

converge in L^2 generalizing the case $k = 2$ due to Bergelson [106], with the corresponding characteristic factor being an inverse limit of nilsystems. An analogous result for $\mathbb{Z}^l$-actions, i.e., for commuting transformations, was shown by Griesmer [497], see also Gutman and Lian [504] for actions of finitely generated Abelian groups. For results for non-finitely generated groups see Bergelson et al. [147] and Shalom [934].

Assani [45] proved almost everywhere convergence of cubic averages (15.41). For pointwise convergence for different, not necessarily commuting transformations (i.e., with $T^{n \cdot \omega} f_\omega$ replaced by $T_\omega^{n \cdot \omega} f_\omega$) see Assani [45], Chu and Frantzikinakis [246] and Host and Kra [553, Sec. 22.3].

Pointwise Convergence of Multiple (Subsequential) Averages

In contrast to the above cubic averages, pointwise convergence of the original multiple averages (15.1) is for $k \geq 3$ a major open question in the area. The case $k = 2$ was proved by Bourgain [184] in his famous double convergence theorem showing pointwise convergence of the averages

$$\frac{1}{N} \sum_{n=1}^{N} T^{an} f \cdot T^{bn} g$$

for $a, b \in \mathbb{N}$, see also Demeter [285] for an alternative proof, Do et al. [305] for a quantitative version and Krause [674] for convergence of the averages $\frac{1}{N} \sum_{n=1}^{N} T^{\lfloor \alpha n \rfloor} f \cdot T^{\lfloor \gamma n \rfloor} g$ for $\alpha \in \mathbb{R}$ and $\gamma \in \mathbb{Q}$. Krause et al. [675] extended Bourgain's theorem and showed pointwise convergence of double averages of the form

$$\frac{1}{N} \sum_{n=1}^{N} T^{n} f \cdot T^{P(n)} g$$

for a polynomial $P \in \mathbb{Z}[\cdot]$ with $\deg P \geq 2$, see also Krause [671, Ch. 15, 16]. (This result was first announced by El Abdalaoui [350].) Pointwise convergence of such averages along primes was shown by Krause et al. [676]. For a double convergence result for commuting transformations with random powers see Frantzikinakis et al. [430].

For some partial results for general k of multiple (linear and polynomial) averages see Derrien and Lesigne [291], Assani [40], Gutman et al. [502], Huang et al. [558] and Xiao [1055], see also Donoso and Sun [311]. In particular, Huang et al. [558] showed pointwise convergence of multiple averages (15.1) on the maximal distal factor defined in Sect. 13.7 extending the result of Leibman [719] on pointwise convergence of multiple polynomial averages (15.35) for nilsystems (see Corollary 14.47).

Finally, Kosz et al. [664] proved pointwise convergence of multiple polynomial averages (15.39) for commuting invertible transformations whenever the polynomials $P_1, \ldots, P_k \in \mathbb{Z}[\cdot]$ have distinct degrees.

Exercises

Exercise 15.1 (Skew Shift) Consider the non-ergodic skew shift T on $X = [0,1)^2$ endowed with the two-dimensional Lebesgue measure given by $T(x,y) := (x + \frac{1}{2}, x + y)$.

(a) Determine all eigenfunctions of T with the corresponding eigenvalues. Deduce the fixed factor and the Kronecker factor of (X, μ, T). (Hint: See Example 8.34.)
(b) Show that the function $g \in \mathrm{L}^\infty(X,\mu)$ given by $g(x,y) := \mathrm{e}^{2\pi i y}$ satisfies $P_{\mathrm{kr}} g = 0$ while $\|g\|_{U^2} = \frac{1}{\sqrt[4]{2}}$.

Exercise 15.2 (Ergodic Rotations on Compact Abelian Groups I) Let (G, m, a) be an ergodic rotation system for compact Abelian group G. Show the representation (15.11) of $\|\cdot\|_{U^2}$. Show further that $\|\cdot\|_{U^l}$ are norms for all $l \geq 2$. (Hint: Use Proposition 6.14 to show (15.11).)

Exercise 15.3 (Maximality of Gowers–Host–Kra Seminorms) Let (X, μ, T) be an ergodic measure-preserving system and let $f \in \mathrm{L}^\infty(X,\mu)$ be a generalized eigenfunction of order $l-1$ for some $l \geq 2$. Prove $\|f\|_{U^l} = \|f\|_\infty$.

Exercise 15.4 (Gowers–Host–Kra Seminorms Control Multiple Ergodic Averages) Complete the proof of Proposition 15.14 by showing

$$\lim_{N\to\infty} \left\| \frac{1}{N} \sum_{n=1}^{N} T^n f_1 \cdot T^{2n} f_2 \cdots T^{kn} f_k \right\|_2 \leq \|f_1\|_{U^k}.$$

Exercise 15.5 (Ergodic Rotations on Compact Abelian Groups II) Show by induction the representation of the box scalar product for ergodic rotations given in Example 15.23 and the corresponding representation of the Gowers–Host–Kra seminorms discussed in Example 15.20.

Exercise 15.6 (Monotonicity of the Gowers–Host–Kra Seminorms) Prove the assertion of Proposition 15.8 for general (not necessarily ergodic) systems using the representation (15.22), the inductive definition of the measures $\mu^{[l]}$ and the classical Cauchy–Schwarz inequality.

Exercise 15.7 (Gowers–Cauchy–Schwarz Inequality)

(a) Complete the proof of Proposition 15.24.
(b) Give an alternative proof of Proposition 15.24 using the description as cube scalar product given by (15.24).

Exercise 15.8 (Triangle Inequality for Gowers–Host–Kra Seminorms) Prove Corollary 15.26.

Exercise 15.9 (Cube Measures as Couplings) Prove Proposition 15.34.

Exercise 15.10 (Host–Kra–Ziegler Factors) Let (X, μ, T) be ergodic. Consider the set $\mathcal{A}$ of all functions $f \in L^\infty(X, \mu)$ such that there exists $F \in L^\infty(X^{[l]}, \mu^{[l]})$ satisfying (15.31). Show that $\mathcal{A}$ is a conjugation invariant subalgebra of $L^\infty(X, \mu)$ which contains $\mathbf{1}$ and is T-invariant. Conclude, by using Proposition 8.19, that this set defines a factor of (X, μ, T).

Exercise 15.11 (The Fixed Factor) Prove Remark 15.42.

Exercise 15.12 (Fourth Proof of Multiple Recurrence) Derive a proof of Theorem 13.5 using Proposition 15.49. (Hint: Change the third proof of Theorem 13.5 via Proposition 15.29 accordingly using the finite sequence of factors $X = \mathcal{Z}_k \mapsto \ldots \mapsto \mathcal{Z}_1 \mapsto \mathcal{Z}_0 = \{\cdot\}$.)

Chapter 16
Weighted Ergodic Theorems

We discuss yet another variant of the classical ergodic theorems. Given a sequence $(a_n)_{n\in\mathbb{N}}$ of complex numbers and a measure-preserving system (X, μ, T) with Koopman operator also denoted by T we are interested in the convergence of the **weighted ergodic averages**

$$\frac{1}{N}\sum_{n=1}^{N} a_n T^n f. \tag{16.1}$$

For the sake of simplicity, we shall consider invertible measure-preserving systems only.

Remark 16.1 The terminology is not consistent in the literature. The term *weighted ergodic averages* is used by some authors for averages[1] of the form $\frac{1}{W_N}\sum_{n=1}^{N} w_n T^n f$, where $(w_n) \subset [0, \infty)$ and $W_N := \sum_{n=1}^{N} w_n \to \infty$ as $N \to \infty$, see, e.g., Jaminon et al. [575], Lin and Weber [750], Cuny and Weber [269] and Buczolich [204]. Thereby averages (16.1) are often called **modulated ergodic averages** with **modulating sequence** (a_n), see, e.g., Lin et al. [747] and Berend et al. [99]. By choosing our terminology we follow Baxter and Olsen [84], Bellow and Losert [97], Çömez et al. [231], Assani [41] and others.

[1] Clearly, convergence of the averages $\frac{1}{A_n}\sum_{n=1}^{N} a_n T^n f$ with $A_N := \sum_{n=1}^{N} a_n$ is equivalent to convergence of averages (16.1) whenever $\lim_{N\to\infty} \frac{A_N}{N} > 0$ (and in particular finite).

T. Eisner, B. Farkas, *A Journey Through Ergodic Theorems*, Birkhäuser Advanced Texts Basler Lehrbücher, https://doi.org/10.1007/978-3-031-96506-7_16

16.1 Mean Good Weights

We begin with the study of weights that are good for L^2-mean convergence. The assumption of invertibility makes the Koopman operator unitary on L^2, so we can directly apply the spectral theorem. As we will see at the end of Remark 16.4 below, this is, however, just a technical restriction, and the analogous results are true for general, not necessarily invertible, measure-preserving systems.

Definition 16.2 A sequence $(a_n)_{n\in\mathbb{N}}$ in $\mathbb{C}$ is called a **good weight for the mean ergodic theorem** (or a **mean good weight**) if for every invertible measure-preserving system (X, μ, T) and every $f \in \mathrm{L}^2(X, \mu)$ the weighted averages in (16.1) converge in L^2.

Thus, the sequence $(1, 1, 1, \ldots)$ is a mean good weight by von Neumann's ergodic theorem (Theorem 7.2).

The first result is a characterization of mean good weights, see Berend et al. [99] for a more general result.

Proposition 16.3 (Characterization of Mean Good Weights) *Suppose that the sequence $(a_n)_{n\in\mathbb{N}}$ in $\mathbb{C}$ satisfies $\sup_{N\in\mathbb{N}} \frac{1}{N}\sum_{n=1}^{N} |a_n| < \infty$. Then the following assertions are equivalent.*[2]

(i) *$(a_n)_{n\in\mathbb{N}}$ is a good weight for the mean ergodic theorem.*
(ii) *For every $\lambda \in \mathbb{T}$ the limit*

$$c(\lambda) := \lim_{N\to\infty} \frac{1}{N}\sum_{n=1}^{N} a_n\lambda^n \in \mathbb{C} \textit{ exists.}$$

(iii) *For every unitary operator $S \in \mathscr{L}(H)$ on a Hilbert space H and every $f \in H$ the averages*

$$\frac{1}{N}\sum_{n=1}^{N} a_n S^n f \tag{16.2}$$

converge.

When either of these conditions holds, the set $\{\lambda \in \mathbb{T} : c(\lambda) \neq 0\}$ is at most countable and for every Hilbert space H, every unitary operator $S \in \mathscr{L}(H)$ and

[2] Complex sequences (a_n) satisfying the condition in Proposition 16.3(ii) are called **sequences having Fourier coefficients**. Some authors also call such sequences **Hartman almost periodic** (or just **Hartman**), see Kahane [608, p. 72], Rosenblatt [880] and Lin et al. [747].

every $f \in H$ *we have*

$$\lim_{N\to\infty} \frac{1}{N} \sum_{n=1}^{N} a_n S^n f = \sum_{\lambda\in\mathbb{T}} c(\lambda) P_\lambda f. \tag{16.3}$$

Here, $P_\lambda = P_{\ker(\lambda - S)}$ *denotes the orthogonal projection onto* $\ker(\lambda - S)$.

The proof is analogous to the one of Proposition 12.9 and is left as Exercise 16.3 up to at most countability of the set $\{\lambda \in \mathbb{T} : c(\lambda) \neq 0\}$. This statement is due to Kahane [607], we follow here a simple and elegant proof for bounded sequences due to Boshernitzan from Rosenblatt [880, Thm. 41], which does not use the existence of the limit in (ii).

Proof Assume that (a_n) satisfies $C := \limsup_{N\to\infty} \frac{1}{N} \sum_{n=1}^{N} |a_n|^2 < \infty$ and let $\delta > 0$. It suffices to show that there are at most finitely many $\lambda \in \mathbb{T}$ with

$$\liminf_{N\to\infty} \left| \frac{1}{N} \sum_{n=1}^{N} a_n \overline{\lambda}^n \right| \geq \delta. \tag{16.4}$$

Let $\lambda_1, \dots, \lambda_M \in \mathbb{T}$ be distinct and satisfy (16.4). Observe that for distinct $j, l \in \{1, \dots, M\}$ one has $\lim_{N\to\infty} \frac{1}{N} \sum_{n=1}^{N} \lambda_j^n \overline{\lambda_l}^n = 0$.

Choose a subsequence (N_k) of $\mathbb{N}$ such that the limits $\lim_{k\to\infty} \frac{1}{N_k} \sum_{n=1}^{N_k} |a_n|^2$ and $\lim_{k\to\infty} \frac{1}{N_k} \sum_{n=1}^{N_k} a_n \overline{\lambda_j}^n$, $i \in \{1, \dots, M\}$, simultaneously exist. On the finite-dimensional subspace H of ℓ^∞ spanned by $a := (a_n)$ and $b_j := (\lambda_j^n)$, $j \in \{1, \dots, M\}$, define the semi-scalar product given by

$$(t|s) := \lim_{k\to\infty} \frac{1}{N_k} \sum_{n=1}^{N_k} t_n \overline{s_n}.$$

Since $b_1, \dots, b_M$ build an orthonormal family with respect to this semi-scalar product, Bessel's inequality[3] and the assumption on $\lambda_1, \dots, \lambda_M$ imply

$$\begin{aligned} M\delta^2 &\leq \sum_{j=1}^{M} \lim_{k\to\infty} \left| \frac{1}{N_k} \sum_{n=1}^{N_k} a_n \overline{\lambda_j}^n \right|^2 = \sum_{j=1}^{M} |(a|b_j)|^2 \\ &\leq |(a|a)| = \lim_{k\to\infty} \frac{1}{N_k} \sum_{n=1}^{N_k} |a_n|^2 \leq C. \end{aligned}$$

Thus $M \leq \frac{C}{\delta^2}$ proving the assertion. ∎

[3] Recall that Bessel's inequality holds for spaces with a semi-scalar product.

Remark 16.4

(a) Note that if $c(\lambda)$ exists, then

$$|c(\lambda)| \leq \sup_{N\in\mathbb{N}} \frac{1}{N}\sum_{n=1}^{N} |a_n|.$$

So for bounded mean good weights (or, more generally, mean good weights satisfying $\sup_{N\in\mathbb{N}} \frac{1}{N}\sum_{n=1}^{N} |a_n| < \infty$) the function $c : \mathbb{T} \to \mathbb{C}$ is a bounded (and measurable) function. Similarly to Remark 12.10 the series on the right-hand side of (16.3) is unconditionally convergent.

(b) Moreover, as in Remark 12.10 one can add the following assertion to the list of equivalent conditions in Proposition 16.3, see Lin et al. [747]:

(iii') *For every contraction $S \in \mathscr{L}(H)$ on a Hilbert space H and every $f \in H$ the averages* (16.2) *converge.*

Also in this case, the identity in (16.3) holds, see Exercise 16.3. In particular, the assumption of invertibility of the measure-preserving systems in the definition of mean good sequences can be dropped.

Now, Proposition 16.3 or the mean ergodic theorem (Theorem 7.2) applied to the operator λT yields that for given $\lambda \in \mathbb{T}$ the sequence $(\lambda^n)_{n\in\mathbb{N}}$ is a mean good weight. Moreover, for a given a unitary operator $S \in \mathscr{L}(H)$ and $f \in H$ we have

$$\lim_{N\to\infty} \frac{1}{N}\sum_{n=1}^{N} \lambda^n S^n f = P_{\mathrm{fix}(\lambda S)} f = P_{\ker(\overline{\lambda}-S)} f.$$

Example 16.5 (Besicovitch Sequences) A sequence $(a_n) \subset \mathbb{C}$ is called of **Besicovitch class** 1 if it belongs to the closure of the set $\mathrm{lin}\{(\lambda^n) : \lambda \in \mathbb{T}\} \subset \mathbb{C}^{\mathbb{N}}$ with respect to the seminorm given by

$$\|a - b\|_{B(1)} := \limsup_{N\in\mathbb{N}} \frac{1}{N}\sum_{n=1}^{N} |a_n - b_n|,$$

and a **bounded Besicovitch sequence** is a bounded sequence in $\mathbb{C}$ which is of Besicovitch class 1. By using Proposition 16.3 it is easy to see that Besicovitch sequences of class 1 are mean good weights, see Exercise 16.11(a).

The following is a polynomial generalization of the linear weights (λ^n), $\lambda \in \mathbb{T}$. Recall the notation $e(s) := \mathrm{e}^{2\pi i s}$ for $s \in \mathbb{R}$.

Example 16.6 (Polynomial Weights) We show that for every polynomial $P \in \mathbb{R}[\cdot]$ the sequence $(e(P(n)))_{n\in\mathbb{N}}$ is a good weight for the mean ergodic theorem. Let $\lambda \in \mathbb{T}$ and take $s \in \mathbb{R}$ such that $\lambda = e(s)$. For the polynomial Q given by

$Q(x) := P(x) + sx$ we have

$$\frac{1}{N}\sum_{n=1}^{N} e(P(n))\lambda^n = \frac{1}{N}\sum_{n=1}^{N} e(P(n) + sn) = \frac{1}{N}\sum_{n=1}^{N} e(Q(n)).$$

If Q has at least one irrational non-constant coefficient, then by the equidistribution of polynomials (Theorem 10.17) and Weyl's criterion (Theorem 10.3) one has $\lim_{N\to\infty} \frac{1}{N}\sum_{n=1}^{N} e(Q(n)) = 0$. If all non-constant coefficients of Q are rational, then for $d \in \mathbb{N}$ being a common multiple of their denominators the sequence $(e(Q(n)))$ is d-periodic, so its Cesàro limits exist and the sequence $(e(P(n))_{n\in\mathbb{N}}$ is indeed a mean good weight by Proposition 16.3. (Note that for P of the form $P(n) = s\tilde{P}(n)$ for some $s \in \mathbb{R}$ and $\tilde{P} \in \mathbb{Z}[\cdot]$ we obtain weights of the form $(\lambda^{\tilde{P}(n)})_{n\in\mathbb{N}}$ for $\lambda \in \mathbb{T}$.)

Example 16.7 (Nilsequences) Nilsequences are good weights for the mean ergodic theorem. This follows from the characterization in Proposition 16.3, since the sequences (λ^n), $\lambda \in \mathbb{T}$, are nilsequences and the product of nilsequences is again a nilsequence (see Example 14.13), so Remark 14.40 yields the required convergence in Proposition 16.3(ii). This generalizes Example 16.6, see Example 14.13(c).

Example 16.8 (Primes and Polynomials of Primes) By Example 12.14 and Proposition 16.3, every sequence of the form $(e(sP(p_n)))_{n\in\mathbb{N}}$ for $s \in \mathbb{R}$ and an integer polynomial P (or, equivalently, $(\lambda^{P(p_n)})_{n\in\mathbb{N}}$ for some $\lambda \in \mathbb{T}$) is a good weight for the mean ergodic theorem. This is in fact true for sequences of the form $(e(P(p_n)))_{n\in\mathbb{N}}$ for any real polynomial P, which can be derived in a similar way from the equidistribution result of Rhin, see [869], for polynomials with at least one non-constant irrational coefficient.

Example 16.9 (Modified von Mangoldt Function) We show following Eisner and Lin [346] that the modified von Mangoldt function Λ' defined as $\Lambda'(n) := \log(n)\mathbf{1}_{\mathbb{P}}(n)$ for $n \in \mathbb{N}$, see Sect. 10.4, is a good weight for the mean ergodic theorem. By (10.7) and Proposition 16.3 it suffices to show that $\frac{1}{N}\sum_{n=1}^{N} \Lambda'(n)\lambda^n$ converges for every $\lambda \in \mathbb{T}$. By Lemma 10.23 this is equivalent to convergence of the averages along primes $\frac{1}{N}\sum_{n=1}^{N} \lambda^{p_n}$ for every $\lambda \in \mathbb{T}$ which was shown in Example 12.14.

The following class of slowly oscillating sequences has a stronger property than being a good weight, cf. Cohen [253].

Example 16.10 (Good Weights for Mean Ergodic Operators) Let $c \in \mathbb{C}$ and $(a_n) \subset \mathbb{C}$ satisfy

(1) $\lim_{N\to\infty} \frac{1}{N}\sum_{n=1}^{N} a_n = c$;
(2) $\sup_{N\in\mathbb{N}} \frac{1}{N}\sum_{n=1}^{N} |a_n| < \infty$;
(3) $\lim_{N\to\infty} \frac{1}{N}\sum_{n=1}^{N} |a_{n+1} - a_n| = 0$.

It is easy to show using von Neumann's decomposition that for every mean ergodic operator S on a Banach space E and every $x \in E$ one has

$$\lim_{N\to\infty} \frac{1}{N}\sum_{n=1}^{N} a_n S^n x = c P_{\mathrm{fix}\, S} x.$$

In particular, (a_n) is a mean good weight and for every uniquely ergodic topological dynamical system (K, T) with unique invariant measure μ, and for every $f \in \mathrm{C}(K)$

$$\lim_{N\to\infty} \frac{1}{N}\sum_{n=1}^{N} a_n T^n f = c \int_X f \,\mathrm{d}\mu$$

holds uniformly on K. We leave the proof as Exercise 16.4.

We finish this section with a weighted analogue of the Blum–Hanson theorem and its operator theoretic version (Theorems 12.2 and 12.1), see Çömez et al. [231] and Lin et al. [747] and Eisner and Lin [346] for more general and related results.

Proposition 16.11 *Let S be a contraction on a Hilbert space H and let $x, x_0 \in H$. Then the following assertions are equivalent.*

(i) $\mathrm{D}\text{-}\lim_{n\to\infty} S^n x = x_0$ *weakly, i.e.,* $x \in H_{\mathrm{aws}}$.
(ii) $\lim_{N\to\infty} \frac{1}{N}\sum_{n=1}^{N} a_n S^n x = x_0$ *for every bounded sequence* $(a_n) \subset \mathbb{C}$.

In particular, a measure-preserving system (X, μ, T) is weakly mixing if and only if the averages (16.1) converge to zero for every $f \in \mathrm{L}^2(X, \mu)$ with $\int_X f \,\mathrm{d}\mu = 0$ and every bounded sequence $(a_n) \subset \mathbb{C}$.

We leave the proof as Exercise 16.2.

16.2 Pointwise Good Weights

The study of pointwise convergence of weighted ergodic averages is more delicate. Also here the assumption of invertibility is technical and can be removed, see Exercise 16.6.

Definition 16.12 For $p \in [1, \infty]$, a sequence $(a_n)_{n\in\mathbb{N}}$ in $\mathbb{C}$ is called a **good weight for the pointwise ergodic theorem in** L^p (or L^p**-pointwise good** for short) if for every ergodic, invertible measure-preserving system (X, μ, T) and every $f \in \mathrm{L}^p(X, \mu)$ (more precisely, for every $f \in \mathscr{L}^p(X, \mu)$) the weighted ergodic averages (16.1) converge almost everywhere.

Note that every L^∞-pointwise good weight (a_n) satisfying $\sup_{N\in\mathbb{N}} \frac{1}{N}\sum_{n=1}^{N} |a_n| < \infty$ is mean good by Proposition 16.3 and by considering ergodic rotations on the unit circle and on the finite sets of roots of unity, respectively.

Clearly, the sequence $(1, 1, 1, \dots)$ is a pointwise good weight in L^1 by Birkhoff's ergodic theorem (Theorem 7.40). The following example contains this weight as a special case.

Example 16.13 We show that the sequence $(\lambda^n)_{n\in\mathbb{N}}$ is a pointwise good weight in L^1 for every $\lambda \in \mathbb{T}$. Let (X, μ, T) be a measure-preserving system and $f \in \mathrm{L}^1(X, \mu)$. Consider the product $Y := X \times \mathbb{T}$ with the product measure $\nu := \mu \times \mathrm{m}$, the transformation $S : Y \to Y$ given by $S(x, z) := (Tx, \lambda z)$, and let $g \in \mathrm{L}^1(Y, \nu)$ be defined by $g(x, z) := zf(x)$. Then we have

$$g(S^n(x, z)) = g(T^n x, \lambda^n z) = \lambda^n f(T^n x)z,$$

which implies

$$\frac{1}{N}\sum_{n=1}^{N}\lambda^n f(T^n x) = \frac{1}{N}\sum_{n=1}^{N} g(S^n(x, z))\overline{z}.$$

Thus almost everywhere convergence of the left-hand side follows from Birkhoff's ergodic theorem applied to the product system $(X \times Y, \mu \times \mathrm{m}, S)$ and the function g. For a different proof, not using product systems, see Exercise 16.8. Note moreover that the assertion is an immediate consequence of Theorem 7.43 since λT is Dunford–Schwartz for every Dunford–Schwartz operator T.

The following strengthening of Example 16.5 is due to Tempelman [998] for $p > 1$ and Ryll-Nardzewski [897] for $p = 1$. For a closely related class of pointwise good weights see Baxter and Olsen [84].

Example 16.14 (Besicovitch Sequences) A sequence (a_n) is called **Besicovitch of class** p for $p \geq 1$ if it belongs to the closure of the set $\mathrm{lin}\{(\lambda^n) : \lambda \in \mathbb{T}\} \subset \mathbb{C}^{\mathbb{N}}$ with respect to the seminorm given by

$$\|a - b\|_{B(p)} := \left(\limsup_{N\in\mathbb{N}} \frac{1}{N}\sum_{n=1}^{N} |a_n - b_n|^p\right)^{\frac{1}{p}}.$$

Besicovitch sequences of class p are q-pointwise good weights for q satisfying $\frac{1}{p} + \frac{1}{q} = 1$ (with $q := \infty$ for $p = 1$), see Exercise 16.11(b). In particular, by Proposition 16.18 we see that bounded Besicovitch sequences are L^1-pointwise good weights.

Remark 16.15 It follows from Remark 12.26 that every subsequence (k_n) of $\mathbb{N}$ with positive density which is a pointwise good subsequence in L^p provides an example of a pointwise good weight in L^p by defining $a_n := 1$ if $n \in \{k_1, k_2, \dots\}$ and $a_n := 0$ otherwise, i.e., (a_n) is the characteristic sequence of this latter set.

The following sufficient condition for a sequence to be a pointwise good weight in L^1 was announced by Sarnak [909]. We follow the proof given by El Abdalaoui, Kułaga-Przymus, Lemańczyk, de la Rue who formulated it for the Möbius function, see [357, Prop. 3.1 and its proof]. It has the important corollary that the Möbius function is a pointwise good weight in L^1, see Theorem 17.3 below.

Theorem 16.16 (A Sufficient Condition for Pointwise Good Weights in L^1) *Let $(a_n)_{n\in\mathbb{N}}$ be a bounded sequence such that for some*[4] *$A > 1$ and $C \geq 0$*

$$\sup_{\lambda\in\mathbb{T}} \Bigl| \frac{1}{N} \sum_{n=1}^{N} a_n \lambda^n \Bigr| \leq \frac{C}{\log^A(N)} \text{ for all } N \geq 2. \tag{16.5}$$

Then for every measure-preserving system (X, μ, T) and every $f \in \mathrm{L}^1(X, \mu)$

$$\lim_{N\to\infty} \frac{1}{N} \sum_{n=1}^{N} a_n f(T^n x) = 0 \text{ for almost every } x \in X. \tag{16.6}$$

In particular, $(a_n)_{n\in\mathbb{N}}$ is a good weight for the pointwise ergodic theorem in L^1.

For the proof we use the strategy that we employed in the proof of Birkhoff's ergodic theorem. We prove a maximal inequality and the convergence on a dense subset, then we deduce the statement from Banach's principle. The first ingredient is hence the corresponding maximal inequality, this is due to Bellow and Losert [97, Prop. 1.3] who formulated it for Dunford–Schwartz operators.

Proposition 16.17 (Maximal Inequality for Weighted Averages) *Let (X, ν, T) be a measure-preserving system with Koopman operator T, and let $(a_n)_{n\in\mathbb{N}}$ be a bounded sequence in $\mathbb{C}$. For $N \in \mathbb{N}$ and $f \in \mathrm{L}^1(X, \mu)$ define*

$$S_N f := \frac{1}{N} \sum_{n=1}^{N} a_n T^n f \quad \text{and} \quad S^* f := \sup_{N\in\mathbb{N}} |S_N f|.$$

Then for every $f \in \mathrm{L}^1(X, \mu)$

$$\mu[S^* f > \lambda] \leq \sup_{n\in\mathbb{N}} |a_n| \cdot \frac{\|f\|_1}{\lambda} \text{ for all } \lambda > 0.$$

We leave the proof of this proposition as Exercise 16.9.

[4] For a refinement of this result with $A > \frac{1}{2}$ see Fan [376].

Proof of Theorem 16.16 First we show that for every $f \in \mathrm{L}^2(X,\mu)$ with $\|f\|_2 = 1$ the inequality

$$\Big\|\frac{1}{N}\sum_{n=1}^{N} a_n T^n f\Big\|_2 \le \frac{C}{\log^A(N)}$$

holds. For this, we can assume, by passing to a unitary dilation of the Koopman operator, that T is unitary on the Hilbert space H. By restricting to the cyclic subspace $Z(f) := \overline{\mathrm{lin}}\{T^n f : n \in \mathbb{Z}\}$ of H we may assume that f is a cyclic vector, i.e., $Z(f) = H$. By the spectral theorem (Theorem 5.13) we conclude by assumption (16.5)

$$\Big\|\frac{1}{N}\sum_{n=1}^{N} a_n T^n f\Big\|_2 = \Big\|\frac{1}{N}\sum_{n=1}^{N} a_n \mathbf{z}^n\Big\|_{\mathrm{L}^2(\mathbb{T},\sigma_f)} \le \Big\|\frac{1}{N}\sum_{n=1}^{N} a_n \mathbf{z}^n\Big\|_\infty \le \frac{C}{\log^A(N)},$$

where σ_f is the corresponding spectral measure.

Now we turn the proof of the assertion in (16.6) for bounded functions. Let $f \in \mathrm{L}^\infty(X,\mu)$ satisfy $\|f\|_2 = 1$. Take $\rho > 1$ and consider $N \in \mathbb{N}$ of the form $N := \lfloor \rho^m \rfloor$ with $m \in \mathbb{N}$. The previous considerations imply

$$\Big\|\frac{1}{\lfloor \rho^m \rfloor}\sum_{n=1}^{\lfloor \rho^m \rfloor} a_n T^n f\Big\|_2 \le \frac{C'}{(m\log(\rho))^A}$$

for some constant C'. From this we conclude

$$\sum_{m=1}^{\infty}\Big\|\frac{1}{\lfloor \rho^m \rfloor}\sum_{n=1}^{\lfloor \rho^m \rfloor} a_n T^n f\Big\|_2 < \infty,$$

which implies that the function

$$\sum_{m=1}^{\infty}\Big|\frac{1}{\lfloor \rho^m \rfloor}\sum_{n=1}^{\lfloor \rho^m \rfloor} a_n T^n f\Big|$$

belongs to $\mathrm{L}^2(X,\mu)$. So in particular

$$\lim_{m\to\infty}\frac{1}{\lfloor \rho^m \rfloor}\sum_{n=1}^{\lfloor \rho^m \rfloor} a_n T^n f = 0 \text{ almost everywhere.} \tag{16.7}$$

Let now $N \in \mathbb{N}$ be arbitrary and take $m \in \mathbb{N}$ with $\lfloor \rho^m \rfloor \le N < \lfloor \rho^{m+1} \rfloor$. We have

$$\left|\frac{1}{N}\sum_{n=1}^{N} a_n T^n f\right| \le \left|\frac{1}{N}\sum_{n=1}^{\lfloor \rho^m \rfloor} a_n T^n f\right| + \left|\frac{1}{N}\sum_{n=\lfloor \rho^m \rfloor+1}^{N} a_n T^n f\right|$$

$$\le \left|\frac{1}{\lfloor \rho^m \rfloor}\sum_{n=1}^{\lfloor \rho^m \rfloor} a_n T^n f\right| + \sup_{n\in\mathbb{N}} |a_n| \frac{\lfloor \rho^{m+1} \rfloor - \lfloor \rho^m \rfloor}{\lfloor \rho^m \rfloor} \|f\|_\infty.$$

Since $\lim_{m\to\infty} \frac{\lfloor \rho^{m+1} \rfloor - \lfloor \rho^m \rfloor}{\lfloor \rho^m \rfloor} = \rho - 1$, using (16.7) and taking ρ arbitrarily close to 1 we obtain

$$\lim_{N\to\infty} \left|\frac{1}{N}\sum_{n=1}^{N} a_n T^n f\right| = 0 \text{ almost everywhere.}$$

We have proved (16.6) for every $f \in \mathrm{L}^\infty(X, \mu)$. The assertion follows now from Banach's principle (Proposition 7.45) and the maximal inequality for weighted averages (Proposition 16.17). ■

At the end of the above proof we used the following direct corollary of the maximal inequality for weighted averages (Proposition 16.17) and Banach's principle (Proposition 7.45).

Proposition 16.18 *Let $(a_n)_{n\in\mathbb{N}}$ be a bounded sequence in $\mathbb{C}$. If $(a_n)_{n\in\mathbb{N}}$ is a good weight for the pointwise ergodic theorem in L^∞, then it is a good weight for the pointwise ergodic theorem in L^1.*

Note that one can show Proposition 16.18 by elementary methods without using the maximal inequality, we leave this task as Exercise 16.10.

Remark 16.19 The assertion of Proposition 16.18 is false in general for unbounded weights. For example, by Exercise 12.8, Theorems 12.30 and 12.38, the sequence $(\Lambda'(n))_{n\in\mathbb{N}}$ for the modified von Mangoldt function Λ' is a pointwise good weight in L^p for every $p > 1$ but not for $p = 1$.

We will use Theorem 16.16 in Chap. 17 to conclude that the Möbius function $(\mu(n))_{n\in\mathbb{N}}$ is an L^1-pointwise good weight, see Theorem 17.3.

16.3 The Wiener–Wintner Theorem

We now strengthen the assertion in Example 16.13 and show that the weights $(\lambda^n)_{n\in\mathbb{N}}$, as λ varies in $\mathbb{T}$, are *simultaneously good* for the pointwise ergodic theorem in L^1 with the set of joint pointwise convergence having full measure. This is the Wiener–Wintner theorem [1046].

Theorem 16.20 (Wiener–Wintner) *Let (X, μ, T) be an ergodic measure-preserving system. Then for every $f \in \mathrm{L}^1(X, \mu)$ there exists $X' \subset X$ with $\mu(X') = 1$ such that the averages*

$$\frac{1}{N}\sum_{n=1}^{N} \lambda^n f(T^n x) \tag{16.8}$$

converge for every $x \in X'$ and every $\lambda \in \mathbb{T}$. For $f \in \mathrm{L}^2(X, \mu)$ the limit is almost everywhere equal to $P_{\overline{\lambda}} f$, where $P_{\overline{\lambda}}$ is the orthogonal projection onto $\ker(\overline{\lambda} - T)$ *in* $\mathrm{L}^2(X, \mu)$.

There are various proofs of this theorem, see, e.g., Assani [44]. The idea of the proof of the Wiener–Wintner theorem we present here is to establish the statement for f from a dense subset of $\mathrm{L}^1(X, \mu)$, in our case $\mathrm{L}^2(X, \mu)$, and then to make an approximation argument. This last step is taken care of by the following lemma. Recall that for an *ergodic* measure-preserving system and for $f \in \mathscr{L}^1(X, \mu)$ a point $x \in X$ is called **generic** for f with respect to T if

$$\lim_{N\to\infty} \frac{1}{N}\sum_{n=1}^{N} f(T^n x) = \int_X f \, \mathrm{d}\mu.$$

Note that Definition 7.48 involved averages of the form $\frac{1}{N}\sum_{n=0}^{N-1}$, but of course this is immaterial by the shift invariance of the Cesàro limit.

Lemma 16.21 *Let (X, μ, T) be an ergodic measure-preserving system, let $(a_n)_{n\in\mathbb{N}}$ be a bounded sequence in $\mathbb{C}$, and let $f, f_j \in \mathscr{L}^1(X, \mu)$, $j \in \mathbb{N}$, be such that $\lim_{j\to\infty} \|f - f_j\|_1 = 0$. Suppose that $x \in X$ is a generic point for $|f_j - f_k|$ and for $|f - f_j|$ for every $j, k \in \mathbb{N}$, and also that the limit*

$$\lim_{N\to\infty} \frac{1}{N}\sum_{n=1}^{N} a_n f_j(T^n x) =: b_j$$

exists for every $j \in \mathbb{N}$. Then the limit $\lim_{j\to\infty} b_j$ exists and

$$\lim_{N\to\infty} \frac{1}{N}\sum_{n=1}^{N} a_n f(T^n x) = \lim_{j\to\infty} b_j.$$

Proof We can assume without loss of generality $|a_n| \le 1$ for every $n \in \mathbb{N}$. The first assertion follows from the genericity assumption by

$$|b_j - b_k| \le \lim_{N\to\infty} \frac{1}{N} \sum_{n=1}^{N} |f_j - f_k|(T^n x) = \|f_j - f_k\|_1 \to 0 \text{ as } j, k \to \infty.$$

Denote $b := \lim_{j\to\infty} b_j$. For the second assertion observe for every $N, j \in \mathbb{N}$

$$\left|\frac{1}{N}\sum_{n=1}^{N} a_n f(T^n x) - b\right| \le \frac{1}{N}\sum_{n=1}^{N} |f - f_j|(T^n x) + \left|\frac{1}{N}\sum_{n=1}^{N} a_n f_j(T^n x) - b_j\right| + |b - b_j|$$

implying by assumption

$$\limsup_{N\to\infty} \left|\frac{1}{N}\sum_{n=1}^{N} a_n f(T^n x) - b\right| \le \|f - f_j\|_1 + |b - b_j|.$$

Letting $j \to \infty$ finishes the proof. ∎

Before we prove the Wiener–Wintner theorem some further remarks are in order. In what follows we will work with fixed representatives of functions from $\mathrm{L}^1(X, \mu)$ and the generic points are always to be understood with respect to these. Moreover, in the proof, for a given $f \in \mathrm{L}^1(X, \mu)$, we will work with only countably many functions from L^1 (e.g., in the approximation argument), so we can take a subset $X' \subset X$ with $\mu(X') = 1$ such that every $x \in X'$ is generic for *each* occurring function (this is so by Birkhoff's ergodic theorem and by taking intersection of countably many sets of full measure).

Proof of Theorem 16.20 By Lemma 16.21, it is enough to prove the assertion for $f \in \mathrm{L}^2(X, \mu)$. By the Jacobs–de Leeuw–Glicksberg decomposition (Theorem 6.20) it suffices to consider the cases when $f \in H_{\mathrm{kr}}$ and $f \in H_{\mathrm{aws}}$ separately.
To handle the case $f \in H_{\mathrm{kr}}$ we can suppose by linearity and by Lemma 16.21 that f is an eigenfunction of T with $Tf = \lambda_0 f$ for some $\lambda_0 \in \mathbb{T}$. Then $T^n f = \lambda_0^n f$ and

$$\lim_{N\to\infty} \frac{1}{N}\sum_{n=1}^{N} \lambda^n T^n f = \lim_{N\to\infty} \frac{1}{N}\sum_{n=1}^{N} (\lambda_0\lambda)^n f = \mathbf{1}_{\{\overline{\lambda}\}}(\lambda_0) f = P_{\overline{\lambda}} f,$$

see also Proposition 5.17. The convergence here is almost everywhere, and the additional statement about the limit follows at once.
Suppose now that $f \in H_{\mathrm{aws}}$ and define for $x \in X$ and for $n \in \mathbb{N}$

$$u_n(x) := \lambda^n (T^n f)(x) \in \mathbb{C}.$$

For $h \in \mathbb{N}$ we have by Birkhoff's ergodic theorem (Theorem 7.40) that there is a set $X_h \subset X$ with $\mu(X_h) = 1$ such that

$$\begin{aligned}\frac{1}{N}\sum_{n=1}^{N} u_{n+h}(x)\overline{u_n}(x) &= \frac{1}{N}\sum_{n=1}^{N} \lambda^{n+h}(T^{n+h} f)(x) \cdot \lambda^{-n}(T^n \overline{f})(x)\\ &= \frac{1}{N}\sum_{n=1}^{N} \lambda^h (T^n(T^h f \cdot \overline{f}))(x)\end{aligned}$$

converges for every $x \in X_h$ to

$$\lambda^h \int_X T^h f \cdot \overline{f}\, \mathrm{d}\mu = \lambda^h (T^h f | f).$$

This implies

$$\gamma_h(x) := \lim_{N\to\infty} \left| \frac{1}{N}\sum_{n=1}^{N} u_{n+h}(x)\overline{u_n}(x)\right| = |\lambda^h (T^h f | f)| = |(T^h f | f)|.$$

Define $X' := \bigcap_{h\in\mathbb{N}} X_h$. Then $\mu(X') = 1$ and, since $f \in H_{\mathrm{aws}}$ by assumption, for every $x \in \bigcap_{h\in\mathbb{N}} X_h$ we have

$$\lim_{N\to\infty} \frac{1}{N}\sum_{h=1}^{N} \gamma_h(x) = \lim_{N\to\infty} \frac{1}{N}\sum_{h=1}^{N} |(T^h f | f)| = 0.$$

Van der Corput's Lemma 7.31 implies

$$\lim_{N\to\infty} \frac{1}{N}\sum_{n=1}^{N} u_n(x) - \lim_{N\to\infty} \frac{1}{N}\sum_{n=1}^{N} \lambda^n (T^n f)(x) = 0$$

for every $x \in X'$. The proof is complete. ∎

Remark 16.22 If we use the van der Corput *inequality*, see Remark 7.32, we obtain the upper bound of the Wiener–Wintner averages for all functions $f \in \mathrm{L}^\infty(X, \mu)$. In fact, for almost every $x \in X$, we have the estimate

$$\sup_{\lambda\in\mathbb{T}} \limsup_{N\to\infty} \left| \frac{1}{N}\sum_{n=1}^{N} \lambda^n f(T^n x)\right|^2 \le \limsup_{N\to\infty} \frac{1}{N}\sum_{n=1}^{N} |(T^n f | f)|. \tag{16.9}$$

Bourgain observed that if one uses the finitary version of the van der Corput inequality (Lemma 7.30) one obtains an even stronger assertion for weakly mixing functions: For almost every $x \in X$ we have

$$\limsup_{N\to\infty} \sup_{\lambda\in\mathbb{T}} \Bigl|\frac{1}{N}\sum_{n=1}^{N} \lambda^n f(T^n x)\Bigr|^2 \le \limsup_{N\to\infty} \frac{1}{N}\sum_{n=1}^{N} |(T^n f\,|\,f)|.$$

The interested reader can check this (or see [44, Sec. 2.4.2]). This leads to *uniform* convergence (w.r.t. λ) to zero in (16.8) for each weakly mixing function f.

Remark 16.23 Another careful inspection of the proof and the uniform convergence of ergodic averages for uniquely ergodic systems (Theorem 7.23) yield the following: If (X, μ, T) is ergodic with (X, T) a uniquely ergodic topological system, then one can take $X' = X$ in the Wiener–Wintner Theorem 16.20 and the convergence is uniform for $x \in X$ and $f \in \mathrm{C}(X)$. Combining this with the previous remark, one easily shows

$$\lim_{N\to\infty} \sup_{\lambda\in\mathbb{T}} \Bigl\|\frac{1}{N}\sum_{n=1}^{N} \lambda^n T^n f\Bigr\|_\infty = 0$$

whenever the system is uniquely ergodic and $f \in \mathrm{C}(X)$ is a weakly mixing function w.r.t. μ (Exercise 16.13). This observation is due to Assani [44, Thm. 2.10].

More information on Wiener–Wintner ergodic theorems can be found in the book [44] of Assani. For a version of the Wiener–Wintner theorem for flows see also El Abdalaoui [353] and for amenable group actions see Ornstein and Weiss [838], Zorin-Kranich [1070, Cor. 4.1] and El Abdalaoui [353]. We also refer to Assani [43, 44], Assani and Nicolaou [54] for properties of the so-called Wiener–Wintner dynamical systems (ergodic systems with certain decay rate of the Wiener–Wintner averages for weakly mixing functions) and a simple proof of Bourgain's double convergence result for such systems. Higher order Wiener–Wintner dynamical systems and pointwise convergence of multiple ergodic averages as well as a multiple version of the return times theorem[5] for them are studied in Assani et al. [50].

The following result of Lesigne [732, 734], which we state without proof, generalizes the Wiener–Wintner theorem to polynomial weights. (Recall the notation $e(s) := \mathrm{e}^{2\pi i s}$.)

Theorem 16.24 (Lesigne) *Let (X, μ, T) be an ergodic measure-preserving system. Then for every $f \in \mathrm{L}^1(X, \mu)$ there exists a set $X' \subset X$ with $\mu(X') = 1$ such that the averages*

$$\frac{1}{N}\sum_{n=1}^{N} e(P(n))(T^n f)(x)$$

converge as $N \to \infty$ for every $x \in X'$ and every polynomial $P \in \mathbb{R}[\cdot]$.

[5] The return times theorem is discussed in the next section.

Remark 16.25 The corresponding decomposition used by Lesigne [734] is

$$L^2(X, \mu) = \overline{\text{lin}}\{\text{generalized eigenfunctions of } T\} \oplus \text{Rest},$$

see Example 8.33 for the definition of generalized eigenfunctions which generate the Abramov factor. In particular, Lesigne showed that for totally ergodic systems

$$\lim_{N\to\infty} \frac{1}{N} \sum_{n=1}^{N} e(P(n))T^n f = 0 \quad \text{a.e.} \tag{16.10}$$

whenever f is orthogonal to all generalized eigenfunctions (i.e., belongs to Rest), with a uniform version given by Frantzikinakis [405]. Moreover, Frantzikinakis showed by a counterexample that the assumption of total ergodicity cannot be dropped for (16.10) to hold for all functions from Rest. Note that, although based on a decomposition, the proof of Lesigne (and Frantzikinakis) did not use any upper bound estimate analogous to (16.9).

Definition 16.26 We call a family $\mathcal{A}$ of sequences $(a_n)_{n\in\mathbb{N}}$ in $\mathbb{C}$ **a Wiener–Wintner family of good weights** (or, shortly, a **WW-family**) if for every ergodic system (X, μ, T) and every $f \in L^1(X, \mu)$ there exists a set $X' \subset X$ with $\mu(X') = 1$ such that the averages

$$\frac{1}{N} \sum_{n=1}^{N} a_n (T^n f)(x)$$

converge for every $x \in X'$ and every $(a_n)_{n\in\mathbb{N}} \in \mathcal{A}$.

Thus, a short formulation of the Wiener–Wintner theorem is that $\{(\lambda^n)_{n\in\mathbb{N}} : \lambda \in \mathbb{T}\}$ is a WW-family. Lesigne's Theorem 16.24 shows that $\{(e(P(n)))_{n\in\mathbb{N}} : P \in \mathbb{R}[\cdot]\}$ is a WW-family. Host and Kra [552, Thm. 2.22] showed that nilsequences build a WW-family, with an alternative proof and a uniform version due to Eisner and Zorin-Kranich [349], generalizing Lesigne's Theorem 16.24 by Example 14.13(c).

The Wiener–Wintner theorem in combination with Proposition 16.3 can be used to produce good weights for the mean ergodic theorem as follows, see Exercise 16.12

Proposition 16.27 *Let (X, μ, T) be an ergodic measure-preserving system and let $f \in L^1(X, \mu)$. Then for almost every $x \in X$ the sequence $(f(T^n x))_{n\in\mathbb{N}}$ is a good weight for the mean ergodic theorem.*

Remark 16.28 Analogously, Theorem 16.24 implies that for an ergodic measure-preserving system (X, μ, T), a function $f \in L^\infty(X, \mu)$ and almost every $x \in X$, the sequence $(f(T^n x))_{n\in\mathbb{N}}$ is a good weight for the polynomial mean ergodic theorem (cf. Theorem 7.28), see Exercise 16.14.

The next example connects the previous proposition to the mean good subsequences from Sect. 12.2.

Example 16.29 (Most Return Time Subsequences are Mean Good) Let (X, μ, T) be an ergodic measure-preserving system. For $A \subset X$ with $\mu(A) > 0$ and $x \in X$ consider the set

$$R_A(x) := \{n \in \mathbb{N} : T^n x \in A\}$$

of return times and the corresponding subsequence $(r_{A,x}(n))_{n\in\mathbb{N}}$ of $\mathbb{N}$, where $r_{A,x}(n)$ is the time of the nth visit of x in A under the dynamics. By Proposition 16.27 the sequence $(T^n \mathbf{1}_A)_{n\in\mathbb{N}}$ is a mean good weight for almost every $x \in X$, so by Remark 12.26 and since the density of $R_A(x)$ is $\mu(A) > 0$ for almost every $x \in X$, see Exercise 12.7(a), we obtain that $(r_{A,x}(n))_{n\in\mathbb{N}}$ is a mean good subsequence for almost every $x \in X$.

16.4 Bourgain's Return Times Theorem

The following result of Bourgain is much harder to prove and yields good subsequences and good weights for the pointwise ergodic theorem, see Bourgain [182] and Bourgain et al. [186].

Theorem 16.30 (Bourgain, Return Times Theorem) *Let (X, μ, T) be an ergodic measure-preserving system and let $f \in \mathrm{L}^\infty(X, \mu)$. Then for almost every $x \in X$ the sequence $(f(T^n x))_{n\in\mathbb{N}}$ is a good weight for the pointwise ergodic theorem in L^1.*

From this it is easy to deduce that for every ergodic measure-preserving system (X, μ, T), for every $A \subset X$ with $\mu(A) > 0$ and for almost every $x \in X$ the set of return times $R_A(x)$ yields (as in Example 16.29) a pointwise good subsequence, see Example 12.27.

Remark 16.31 By considering rotations on the unit circle it is easy to deduce the Wiener–Wintner theorem from the return times theorem.

The proof in [186] of the return times theorem is based on the following "instrinsic" sufficient condition for the sequence $(f(T^n x))_{n\in\mathbb{N}}$ to be a pointwise good weight with limit zero. (We refer to Definition 7.48 for the notion of generic points.)

Theorem 16.32 (Bourgain–Furstenberg–Katznelson–Ornstein Orthogonality Criterion) *Let (X, μ, T) be an ergodic, invertible, regular measure-preserving system, and let $f \in \mathscr{L}^\infty(X, \mu)$. Assume that $x \in X$ is generic for f and satisfies*

$$\lim_{N\to\infty} \frac{1}{N} \sum_{n=1}^{N} f(T^n x)\overline{f(T^n y)} = 0$$

for a.e. $y \in X$. Then the sequence $(f(T^n x))_{n\in\mathbb{N}}$ is a good weight for the pointwise ergodic theorem in L^1 with limit zero, i.e., for every measure-preserving system (Y, ν, S) and every $g \in \mathrm{L}^1(Y, \nu)$

$$\lim_{N\to\infty} \frac{1}{N} \sum_{n=1}^{N} f(T^n x) g(S^n y) = 0$$

holds for ν-a.e. $y \in Y$.

For the proof of Theorem 16.32 we refer to Bourgain et al. [186] and to Assani [44, Ch. 4] and Fritzsch [438] for a more detailed version. Here we only show how to derive Theorem 16.30 from it.

Proof of Theorem 16.30 assuming Theorem 16.32 Let (X, μ, T) be ergodic and $f \in \mathrm{L}^\infty(X, \mu)$. By taking an invertible extension and a separable factor containing f we can assume that the system is ergodic, regular and invertible. Decompose $f = f_1 + f_2$ with $f_1 \in H_{\mathrm{kr}}$ and f_2 being weakly mixing according to the Jacobs–de Leeuw–Glicksberg decomposition (13.6). Since H_{kr} is induced by the Kronecker *factor*, f_1 and hence also f_2 both belong to $\mathrm{L}^\infty(X, \mu)$, see (8.2).

We first show that the sequence $(f_1(T^n x))_{n\in\mathbb{N}}$ is a pointwise good weight in L^1 for a.e. $x \in X$. Since eigenfunctions of T to different eigenvalues are orthogonal, by Bessel's inequality there exist at most countably many eigenfunctions (h_j) with $f \in \overline{\mathrm{lin}}\{h_j\}$. We denote the corresponding eigenvalues by (λ_j). Let x be such that $(T^n h_j)(x) = \lambda_j^n h_j(x)$ holds for all $n \in \mathbb{N}$ and all j. By Example 16.13, each sequence $(h_j(T^n x))_{n\in\mathbb{N}} = (\lambda_j^n h_j(x))_{n\in\mathbb{N}}$ is a pointwise good weight in L^1. Let (Y, ν, S) be a measure-preserving system, $g \in \mathrm{L}^\infty(Y, \nu)$ and let $y \in Y$ be such that the averages $\frac{1}{N}\sum_{n=1}^{N} h_j(T^n x) g(S^n y)$ converge for every j. By Lemma 16.21 applied to $a_n := g(S^n y)$ we see that the averages $\frac{1}{N}\sum_{n=1}^{N} f_1(T^n x) g(S^n y)$ converge. Thus $(f_1(T^n x))_{n\in\mathbb{N}}$ is a pointwise good weight in L^∞ and hence in L^1 by Proposition 16.18.

Again by Proposition 16.18 it remains to show that the sequence $(f_2(T^n x))_{n\in\mathbb{N}}$ is a pointwise good weight in L^∞ for a.e. $x \in X$. We will show that for f_2 the assumption of the orthogonality criterion (Theorem 16.32) is satisfied for a.e. $x \in X$. Denote by F the (bounded measurable) function on $X \times X$ defined by

$$F(x, y) := \lim_{N\to\infty} \frac{1}{N} \sum_{n=1}^{N} f_2(T^n x)\overline{f_2(T^n y)}$$

which exists a.e. by the pointwise ergodic theorem applied to the product system $(X \times X, \mu \times \mu, T \times T)$ and the (bounded) function $f_2 \otimes \overline{f_2}$. By Fubini's theorem and the T-invariance of μ observe

$$
\begin{aligned}
\|F\|^2_{\mathrm{L}^2(X\times X,\mu\times\mu)} &= \lim_{N\to\infty}\frac{1}{N^2}\sum_{n,k=1}^{N}\left|\int_X f_2(T^n x)\overline{f_2(T^k x)}\,\mathrm{d}\mu(x)\right|^2\\
&= \lim_{N\to\infty}\frac{1}{N^2}\sum_{n,k=1}^{N}|(T^n f_2|T^k f_2)|^2\\
&\le 2\lim_{N\to\infty}\frac{1}{N^2}\sum_{n=1}^{N}\sum_{k=1}^{n}|(T^{n-k} f_2|f_2)|^2\\
&\le 2\lim_{N\to\infty}\frac{1}{N^2}\sum_{n=1}^{N}\sum_{m=0}^{N-1}|(T^m f_2|f_2)|^2\\
&= 2\lim_{N\to\infty}\frac{1}{N}\sum_{m=0}^{N-1}|(T^m f_2|f_2)|^2 = 0,
\end{aligned}
$$

the last equality being true since f_2 is weakly mixing. This we can rewrite again by Fubini's theorem as

$$
\int_X\left(\int_X |F(x,y)|^2\,\mathrm{d}\mu(y)\right)\mathrm{d}\mu(x) = 0
$$

implying that for a.e. $x \in X$ there exists a full measure set of $y \in X$ such that $F(x, y) = 0$. Thus by the orthogonality criterion (Theorem 16.32) the sequence $(f_2(T^n x))_{n\in\mathbb{N}}$ is a pointwise good weight in L^∞ for a.e. $x \in X$. The proof is complete. ∎

Remark 16.33 By inspecting the above argument one sees that it does not give any information on the points $x \in X$ for which the orthogonality assumption in Theorem 16.32 is satisfied. Hence one does not have an explicit description of the "good" points x for which the sequence $(f_2(T^n x))_{n\in\mathbb{N}}$, and hence also $(f(T^n x))_{n\in\mathbb{N}}$, is a good weight for the pointwise ergodic theorem in L^1. Analogously to the pointwise ergodic theorem, one expects that for "good" (topological) systems and for "good" (say continuous) functions this is the case for *every* x. This is true for example for nilsystems by Host and Kra [552, Thm. 2.22], see also Lesigne et al. [736, Thm. C] and Brunel and Keane [199].

Remark 16.34 By a different argument using joinings, Rudolph [893] showed that for every $p \in [1, \infty]$, every ergodic system (X, μ, T) and every $f \in \mathrm{L}^p(X, \mu)$, the sequence $(f(T^n x))$ is L^q-pointwise good for almost all $x \in X$, where $q \in [1, \infty]$ satisfies $\frac{1}{p} + \frac{1}{q} = 1$. See Sect. 16.6 for further results which, in particular, imply that in Theorem 16.30, instead of $f \in \mathrm{L}^\infty(X, \mu)$ one can take $f \in \mathrm{L}^p(X, \mu)$, $p > 2$.

16.5 Wiener–Wintner Theorem for Uniquely Ergodic Systems

The classical (pointwise) ergodic theorem has a very strong form for uniquely ergodic systems, see Sect. 7.3. We now show, following Robinson [874], the uniform convergence of Wiener–Wintner ergodic averages for such systems.

For a topological system (K, T) we call $\lambda \in \mathbb{T}$ a **continuous eigenvalue** of T if there exists $g \in \mathrm{C}(K)$ with $g \neq 0$ such that

$$g(Tx) = \lambda g(x) \tag{16.11}$$

holds for every $x \in K$, i.e., λ is an eigenvalue of the Koopman operator T on $\mathrm{C}(K)$. We denote the set of all continuous eigenvalues of T by C_T.

Suppose now that (K, T) is uniquely ergodic with the unique, and hence ergodic, invariant measure μ (see Remark 3.35). We call $\lambda \in \mathbb{T}$ a **measurable eigenvalue** of T if there exists $g \in \mathrm{L}^\infty(K, \mu)$ with $g \neq 0$ such that (16.11) holds for μ-almost every $x \in K$. The set of all measurable eigenvalues of T will be denoted by M_T. The inclusion $C_T \subset M_T$ is clear. Examples show that one can have strict inclusion here, see, e.g., Robinson [874, Sec. 3].

The following lemma is a topological analogue of the upper bound of the Wiener–Wintner ergodic averages (16.9), but with a varying point x.

Lemma 16.35 *Let (K, T) be a uniquely ergodic system with invariant measure μ and let $(x_N)_{N\in\mathbb{N}}$ be a sequence in K. Then for every $\lambda \in \mathbb{T}$ and every $f \in \mathrm{C}(K)$ we have*

$$\limsup_{N\to\infty} \left| \frac{1}{N} \sum_{n=1}^{N} \lambda^n (T^n f)(x_N) \right| \leq \| P_{\overline{\lambda}} f \|_{\mathrm{L}^2(K,\mu)},$$

where $P_{\overline{\lambda}}$ is the orthogonal projection onto $\ker(\overline{\lambda} - T)$ in $\mathrm{L}^2(K, \mu)$.

Proof For $\lambda \in \mathbb{T}$ consider the rotation system $(\mathbb{T}, \mathrm{m}, \lambda)$ and the product system $(K \times \mathbb{T}, \mu \times \mathrm{m}, S)$, i.e., $S(x, z) = (Tx, \lambda z)$. For the function $g \in \mathrm{C}(K \times \mathbb{T})$ defined by $g(x, z) := zf(x)$ as in Example 16.13 we have

$$(S^n g)(x, z) = \lambda^n (T^n f)(x) z.$$

For $N \in \mathbb{N}$ define

$$\nu_N := \frac{1}{N} \sum_{n=1}^{N} \delta_{S^n(x_N, 1)} \in \mathrm{M}(K \times \mathbb{T}).$$

By Lemma 7.22 the sequence $(\nu_N)_{N\in\mathbb{N}}$ of probability measures is asymptotically invariant. Let $(N_j)_{j\in\mathbb{N}}$ be a subsequence of $\mathbb{N}$ such that

$$\lim_{j\to\infty}\left|\frac{1}{N_j}\sum_{n=1}^{N_j}\lambda^n(T^n f)(x_{N_j})\right| = \limsup_{N\to\infty}\left|\frac{1}{N}\sum_{n=1}^{N}\lambda^n(T^n f)(x_N)\right|.$$

By the Banach–Alaoglu theorem (Theorem 1.8) there exists an accumulation point ν of $(\nu_{N_j})_{j\in\mathbb{N}}$ with respect to the weak*-topology. By Lemma 7.21 this probability measure ν is S-invariant.

Next, observe

$$\int_{K\times\mathbb{T}} g\,\mathrm{d}\nu_{N_j} = \frac{1}{N_j}\sum_{n=1}^{N_j} g(S^n(x_{N_j},1)) = \frac{1}{N_j}\sum_{n=1}^{N_j}\lambda^n(T^n f)(x_{N_j}).$$

Letting $j\to\infty$ and by the definition of ν and $(N_j)_{j\in\mathbb{N}}$ we obtain

$$\limsup_{N\to\infty}\left|\frac{1}{N}\sum_{n=1}^{N}\lambda^n(T^n f)(x_N)\right| = \lim_{j\to\infty}\left|\frac{1}{N_j}\sum_{n=1}^{N_j}\lambda^n(T^n f)(x_{N_j})\right| = \left|\int_{K\times\mathbb{T}} g\,\mathrm{d}\nu\right|.$$

Thus, it remains to show

$$\left|\int_{K\times\mathbb{T}} g\,\mathrm{d}\nu\right| \le \|P_{\overline{\lambda}} f\|_{\mathrm{L}^2(K,\mu)}. \tag{16.12}$$

Consider the measure $\tilde{\mu}$ on K defined by

$$\tilde{\mu}(E) := \nu(E\times\mathbb{T}).$$

By the S-invariance of ν we have

$$\tilde{\mu}(T^{-1}E) = \nu(S^{-1}(E\times\mathbb{T})) = \nu(E\times\mathbb{T}) = \tilde{\mu}(E),$$

i.e., $\tilde{\mu}$ is a T-invariant probability measure. By the unique ergodicity this implies $\tilde{\mu}=\mu$. Hence we conclude

$$\begin{aligned}(S^n g|g)_{\mathrm{L}^2(K\times\mathbb{T},\nu)} &= \int_{K\times\mathbb{T}} S^n g\cdot\overline{g}\,\mathrm{d}\nu = \int_{K\times\mathbb{T}}\lambda^n T^n f\cdot\overline{f}\cdot\mathbf{z}\overline{\mathbf{z}}\,\mathrm{d}\nu \\ &= \int_{K\times\mathbb{T}}\lambda^n T^n f\cdot\overline{f}\,\mathrm{d}\nu = \lambda^n(T^n f|f)_{\mathrm{L}^2(K,\mu)}.\end{aligned} \tag{16.13}$$

Now, the mean ergodic theorem yields

$$\lim_{N\to\infty}\frac{1}{N}\sum_{n=1}^{N}(S^n g|g) = (P_{\mathrm{fix}(S)}g|g) = \|P_{\mathrm{fix}(S)}g\|^2_{\mathrm{L}^2(K\times\mathbb{T},\nu)}.$$

On the other hand, we have $P_{\mathrm{fix}(\lambda T)} = P_{\overline{\lambda}}$, so by the mean ergodic theorem

$$\lim_{N\to\infty} \frac{1}{N}\sum_{n=1}^{N} \lambda^n (T^n f \,|\, f)_{\mathrm{L}^2(K,\mu)} = (P_{\overline{\lambda}} f \,|\, f) = \|P_{\overline{\lambda}} f\|^2_{\mathrm{L}^2(K,\mu)}.$$

This and (16.13) imply $\|P_{\mathrm{fix}(S)} g\| = \|P_{\overline{\lambda}} f\|$. Let P be the orthogonal projection onto $\mathbb{C}\mathbf{1}_{K\times\mathbb{T}} \subset \mathrm{L}^2(K\times\mathbb{T},\nu)$, i.e., $Pg = \int_{K\times\mathbb{T}} g\,\mathrm{d}\nu \cdot \mathbf{1}_{K\times\mathbb{T}}$. Since $\mathrm{ran}(P) \subset \mathrm{ran}(P_{\mathrm{fix}(S)})$, we conclude

$$\begin{aligned}\|Pg\|_{\mathrm{L}^2(K\times\mathbb{T},\nu)} &= \|P P_{\mathrm{fix}(S)} g\|_{\mathrm{L}^2(K\times\mathbb{T},\nu)} \le \|P_{\mathrm{fix}(S)} g\|_{\mathrm{L}^2(K\times\mathbb{T},\nu)} \\ &= \|P_{\overline{\lambda}} f\|_{\mathrm{L}^2(K,\mu)}.\end{aligned}$$

Thus (16.12) is proved and the argument is complete. ∎

Theorem 16.36 (Robinson) *Let (K,T) be a uniquely ergodic topological system. Then for every $\lambda \notin M_T \setminus C_T$ and every $f \in \mathrm{C}(K)$ the averages*

$$\frac{1}{N}\sum_{n=1}^{N} \lambda^n T^n f$$

converge uniformly to $P_{\overline{\lambda}} f$ as $N \to \infty$.

Proof We first prove uniform convergence and distinguish two cases.
Case 1. Assume that $\lambda \in C_T$ and take $g \in \mathrm{C}(K)$ with $g \neq 0$ such that $Tg = \lambda g$. We can assume without loss of generality that $|g(x)| = 1$ for every $x \in K$ (Exercise 16.16). For $n \in \mathbb{N}$ we thus have $T^n g = \lambda^n g$ or, equivalently, $\lambda^n = T^n g \cdot \overline{g}$. This implies that for every $f \in \mathrm{C}(K)$ the averages

$$\frac{1}{N}\sum_{n=1}^{N} \lambda^n T^n f = \frac{1}{N}\sum_{n=1}^{N} T^n (fg)\cdot \overline{g}$$

converge uniformly by Theorem 7.23 since the system is uniquely ergodic.
Case 2: Suppose that $\lambda \notin M_T$. We need to show

$$\lim_{N\to\infty} \Big\| \frac{1}{N}\sum_{n=1}^{N} \lambda^n T^n f \Big\|_\infty = 0.$$

Assume the contrary, i.e., that there exist $\varepsilon > 0$, a subsequence $(N_j)_{j\in\mathbb{N}}$ of $\mathbb{N}$ and a sequence $(y_j)_{j\in\mathbb{N}}$ in K such that

$$\Big| \frac{1}{N_j}\sum_{n=1}^{N_j} \lambda^n (T^n f)(y_j) \Big| \ge \varepsilon \text{ for all } j \in \mathbb{N}.$$

For $j \in \mathbb{N}$ set $x_{N_j} := y_j$ and for $N \notin \{N_j : j \in \mathbb{N}\}$ let $x_N \in K$ be arbitrary. Then

$$\limsup_{N\to\infty} \left| \frac{1}{N} \sum_{n=1}^{N} \lambda^n (T^n f)(x_N) \right| \geq \varepsilon.$$

This and Lemma 16.35 imply $P_{\overline{\lambda}} f \neq 0$ meaning that $\overline{\lambda} \in M_T$. By Proposition 3.26 $M_T \subset \mathbb{T}$ is a subgroup of $\mathbb{T}$, so $\lambda \in M_T$, a contradiction.
We have seen the uniform convergence as stated. The mean ergodic theorem applied to the operator λT shows that the limit is as asserted. ■

Remark 16.37 Note that for $\lambda \in M_T \setminus C_T$ the divergence can happen, see Robinson [874]. A polynomial version of Robinson's theorem being a topological analogue of Lesigne's Theorem 16.24 was recently obtained by Fan [375]. For a nilsequence version of Robinson's theorem for minimal systems see Gutman and Lian [503, 504] and for an extension of Robinson's theorem to amenable semigroups of Markov operator on C(K) see Schreiber [924] and Bartozhek and Śpiewak [80]. For more topological Wiener–Wintner type theorems with an application to random ergodic theorems see Walters [1034] and Santos and Walkden [905].

16.6 Further Directions

We mention here some further research directions, for more see Chap. 17.

More Pointwise Good Weights

Bellow and Losert [97] presented a Wiener–Wintner family of good weights which include bounded Besicovitch sequences. For the necessity of the duality of p and q in the weighted ergodic theorem in L^q with Besicovitch weights of class p see Baxter et al. [83], Demeter [284] and Demeter and Jones [287].

Among others the following classes of weights were shown to be pointwise good: certain sequences having mean and correlation (see Bellow and Losert [97]), certain q-multiplicative sequences (see Lesigne et al. [736], Lesigne and Mauduit [735] and Fan [376]), certain automatic sequences (see Eisner and Konieczny [344]) and certain sequences coming from Hardy fields functions (see Eisner and Krause [345] for a Wiener–Wintner type result and also Krause and Zorin-Kranich [677, 678] for random versions). Note also that the product of a bounded L^1-pointwise good weight with a sequence of Besicovitch class 1 is a good weight, see Lesigne and Mauduit [735, Prop. C2]. Moreover, double return times sequences of the form $(f_1(T^{an}x) f_2(T^{bn}x))$ are (for bounded functions f_1, f_2 and a.e. x) pointwise good weights, see Assani [42, Thm. 2] for the weakly mixing and Zorin-Kranich [1075] for the general case.

In his extension of Bourgain's result on the pointwise ergodic theorem along primes, Wierdl [1047] showed that the modified von Mangoldt function Λ' (see Sect. 10.4) is a good weight for the pointwise ergodic theorem in L^p for every $p > 1$. This provides a natural example of a good *unbounded* weight.

Rosenblatt and Wierdl [881, Ex. 6 in Ch. III] (combined with Remark 12.26) showed that the characteristic function of the square-free integers $(|\mu(n)|)$ is L^1-pointwise good, see Cuny and Weber [269] for an alternative proof. For results on other arithmetic functions as good weights in the related setting mentioned in Remark 16.1 see Cuny and Weber [269], Buczolich [204] and Kim [648].

Note that the argument in the proof of Theorem 16.16 from [357] combined with a result of Green and Tao [488] shows that the product of the Möbius function and a nilsequence is a pointwise good weight in L^1 with corresponding limit zero. Frantzikinakis and Host [414, §1.4] showed using the Kátai–Bourgain–Sarnak–Ziegler criterion (Theorem 17.17 below) that every multiplicative function bounded by 1 is a pointwise good weight as soon as it is Cesàro convergent along arithmetic progressions.

For more examples of mean and pointwise good (and bad) weights see Ryll-Nardzewski [897], Baxter and Olsen [84], Bellow and Losert [97], Ornstein and Weiss [838], Lesigne et al. [736], Assani [39, 41, 42], Çömez et al. [231], Demeter [284], Çömez and Litvinov [232], Fan [376], and also Eisner [338].

Return Times Theorem

Ornstein and Weiss [838, Thm. 1] showed (in discussions with Furstenberg, Keane and Thouvenot) that for systems with completely positive entropy every generic point satisfies the assertion of the return times theorem and presented an example showing that this is not true for general systems. Moreover, they also extended Bourgain's return times theorem to actions of certain amenable groups, see Zorin-Kranich [1070] and El Abdalaoui [353] for further extensions to all (locally compact second countable) amenable groups.

Assani et al. [51] presented a Wiener–Wintner version of the return times theorem studying convergence of the averages of the form

$$\frac{1}{N}\sum_{n=1}^{N}\lambda^n f(T^n x)g(S^n y),$$

and showing that for a.e. x there exists a full measure set of y such that the above averages converge for every $\lambda \in \mathbb{T}$. Answering a question of Assani, Rudolph [894] presented a multiple return times theorem concerning averages

$$\frac{1}{N}\sum_{n=1}^{N} f_1(T_1^n x_1) f_2(T_2^n x_2)\dots f_k(T_k^n x_k) \tag{16.14}$$

and showed that given a measure-preserving system (X_1, μ_1, T_1) and $f_1 \in L^\infty(X_1, \mu_1)$, there exists a full measure set of x_1 such that for every measure-preserving system (X_2, μ_2, T_2) and $f_2 \in L^\infty(X_2, \mu_2)$, there exists a full measure set of $x_2, \ldots$ for every measure-preserving system (X_k, μ_k, T_k) and $f_k \in L^\infty(X_k, \mu_k)$ there exists a full measure set of x_k such that the multiple averages (16.14) converge. Note that the result of Assani et al. [51] mentioned above gives norm convergence of three-term averages.

An alternative proof of Rudolph's multiple return times theorem was given by Zorin-Kranich [1068] who used the Host–Kra structure theory discussed in Chap. 15 and showed that the Host–Kra factors of the corresponding orders are characteristic for averages (16.14). (The last assertion was proved before for ergodic systems by Assani and Presser [55].) He also proved a Wiener–Wintner return times theorem for nilsequences showing that, with the same quantifiers as in the multiple return times theorem, for every nilsequence (a_n) the averages

$$\frac{1}{N}\sum_{n=1}^{N} a_n f_1(T_1^n x_1) f_2(T_2^n x_2) \ldots f_k(T_k^n x_k)$$

converge, generalizing the result of Assani et al. [51] mentioned above to nilsequences and multiple averages.

Generalizing a result of Assani [40] for weakly mixing systems to general systems, Zorin-Kranich [1075] showed that for every invertible measure-preserving system (X, μ, T), $a_1 \neq a_2 \in \mathbb{Z}$ and $f_1, f_2 \in L^\infty(X, \mu)$, the sequence $(f_1(T^{a_1 n}x) f_2(T^{a_2 n}x))$ is for a.e. x a pointwise good weight in L^∞ (and hence in L^1 by Proposition 16.18), combining Bourgain's return times theorem with his double convergence theorem discussed in Sect. 15.9. For a corresponding Wiener–Wintner type result see Assani et al. [49], with a polynomial version due to Assani and Moore [53] and nilsequence versions due to Assani and Moore [52] (combined with Host and Kra [552, Thm. 2.24]), Zorin-Kranich [1071] and Assani [46], see also Xiao [1054]. For another connection between and unified approach to Bourgain's return times and double convergence theorems see Krause [674], who extended Bourgain's return times theorem to pointwise convergence of averages of the form $\frac{1}{N}\sum_{n=1}^{N} f(T^{\lfloor \gamma n \rfloor}x) \cdot S^{\lfloor \alpha n \rfloor} g$ for $\gamma \in \mathbb{Q}$ and $\alpha \in \mathbb{R}$.

Substantial effort has been made to understand for which pairs (p, q) the assertion of Bourgain's return times theorem is valid for $f \in L^p$ and $g \in L^q$. By standard arguments, this is the case as soon as $\frac{1}{p} + \frac{1}{q} \leq 1$, see, e.g., [893] or [288, Sec. 4]. Assani et al. [48] showed that the assertion fails for $p = q = 1$ while Demeter et al. [288] proved that it is valid for every $p > 1$ and $q \geq 2$, with an extension by Demeter [286] to pairs satisfying $\frac{1}{p} + \frac{1}{q} < \frac{3}{2}$, yielding an extension of Bourgain's return times theorem for $f \in L^p(X, \mu)$, $p > 2$, see Theorem 16.30.

For more information and results see the survey papers by Assani and Presser [56], Assani [47] and the references therein.

More General Operators

Olsen [831] showed that bounded Besicovitch sequences are L^p-pointwise good for every $p \in (1, \infty)$ for Dunford–Schwartz operators. Baxter and Olsen [84], see also Bellow and Losert [97, Thm. 1.4], proved the general fact that every bounded weight which is L^1-pointwise good is pointwise good for Dunford–Schwartz operators (see also Çömez et al. [231] who dropped the implicit restriction to Lebesgue spaces). This was extended to positive mean bounded operators on L^p-spaces for $p \in (1, \infty)$ by Assani [38, 39] and to contractions with mean ergodic modulus on such spaces by Çömez et al. [231].

Baxter and Olsen [84] introduced several classes of Besicovitch sequences as pointwise good weights for several classes of operators including operators on L^p for $p \in (1, \infty)$ dominated by a positive contraction and operators being power bounded with respect to the L^∞-norm and one other L^p-norm.

Lin et al. [747] studied Besicovitch sequences further as mean and pointwise good weights for Dunford–Schwartz operators. They completely characterized mean good weights for operators with compact orbits and for mean ergodic power bounded operators on Banach spaces, respectively. Their study of mean good weights for operators with relatively weakly compact orbits was continued by Berend et al. [99].

Çömez et al. [231] extended Bourgain's return times theorem (Theorem 16.30) to Dunford–Schwartz operators on Lebesgue spaces and showed that for every such operator T and every $f \in L^\infty$, the sequence $(f(T^n x))$ is L^1-pointwise good for a.e. x, see also Lin et al. [747]. Assani [39] proved before that for every mean bounded positive operator T on an L^p-space with $p \in (1, \infty)$ and every L^1-function f, the sequence $(f(T^n x))$ is L^∞-pointwise good for a.e. x.

It follows from the subsequential result along primes by Jones et al. [594][6] and Exercise 12.8(b) based on Wierdl [1047, pp. 316–317] that the modified von Mangoldt function is an L^p-pointwise good weight for every positive contraction and every quasi-isometric Lamperti operator on L^p for $p > 1$ (see Sect. 7.8). See also Eisner and Lin [346, Thm. 5.6] (with the idea of the proof belonging to Cuny) for norm convergence of weighted averages with the modified von Mangoldt function as weight for invertible doubly power bounded operators on L^p for $p > 1$.

For more results see Baxter and Olsen [84], Assani [39], Rosenblatt [880], Çömez et al. [231], Lin et al. [747], Eisner and Lin [346], Barthmann and Farhangi [79]. For results concerning convergence of averages mentioned in Remark 16.1 see Lin and Weber [750], Cuny and Weber [269], Buczolich [204] and Kim [648].

[6] See also the text after the proof of Theorem 5.2 and Appendix A by Wierdl in Eisner and Lin [346].

Weighted Ergodic Theorems Along Subsequences

One can also study convergence of weighted subsequential averages of the form

$$\frac{1}{N}\sum_{n=1}^{N} a_n T^{k_n} f \tag{16.15}$$

for a subsequence (k_n) of $\mathbb{N}$ and complex weights (a_n).

It follows from Bourgain [180, Thm. 6] and his maximal inequality [183, (1.7)] that the linear weights (λ^n) for $\lambda \in \mathbb{T}$ are L^p-good weights for ergodic averages along polynomials for $p > 1$, see also Mirek and Trojan [798], Eisner and Krause [345] and Krause et al. [675]. In the second paper a partial Wiener–Wintner type result for badly approximable θ is presented. (Note that the question whether a Wiener–Wintner version holds for polynomial subsequences is open even for squares $P(n) = n^2$.)

Jones et al. [592] studied pointwise convergence of averages (16.15) for Besicovitch-type weights along L^p-pointwise good subsequences (such as polynomials and primes) for Dunford–Schwartz operators and positive (or positively dominated) contractions on L^p for $p \in (1, \infty)$. Their weights include uniform sequences coming from connected spaces.

Durand and Schneider [325] studied q-multiplicative sequences as mean and L^2-pointwise good weights for subsequences including squares. Eisner and Konieczny [344] presented automatic sequences as mean and L^p-pointwise good weights for polynomial subsequences and $p > 1$.

Fan [376] showed a.e. convergence for Davenport-type weights for subsequences satisfying a strong (unweighted) maximal inequality in L^p for $p \in (1, \infty)$ and presented a class of such weights for polynomial subsequences. (Note that the case of the Möbius function and polynomials follows by a modification of the proof of Theorem 16.16 using a nilsequence version of Davenport's estimate by Green and Tao [488] as observed in [339].) He also studied the rates of convergence of weighted averages along subsequences.

Eisner and Müller [347] proved that polynomial weights of the form $(e(P(n)))$ with $P \in \mathbb{R}[\cdot]$ are mean good for subsequences of the form $(\lfloor Q(n) \rfloor)$ for $Q \in \mathbb{R}[\cdot]$ for every power bounded Hilbert space operator.[7] Moreover, they presented weighted subsequential ergodic theorems for power bounded Hilbert space operators without unimodular eigenvalues for a large class of weights (including polynomial weights) along a large class of subsequences (including the integer part of most of subpolynomial Hardy functions, e.g., logarithmico-exponential functions).

Eisner and Buczolich [205] showed divergence in L^1 of averages along squares with polynomial weights of the form $(e(P(n)\theta))$ for $P \in \mathbb{Z}[\cdot]$ for many θ. The

[7] For $Q \in \mathbb{Z}[\cdot]$ this was proved without weights before by ter Elst and Müller [999].

question whether divergence holds for every θ or, more generally, for every weight of the form $(e(P(n)))$ with $P \in \mathbb{R}[\cdot]$ remains open.

Weighted Multiple Ergodic Theorems

There are many results concerning mean convergence of weighted multiple averages of the form

$$\frac{1}{N}\sum_{n=1}^{N} a_n T^{P_1(n)} f \cdots T^{P_k(n)} f \tag{16.16}$$

for (usually bounded) $(a_n) \in \mathbb{C}$, $k \geq 2$ and integer polynomials (or other functions) $P_1, \ldots, P_k$.

Using their structure theorem (Theorem 15.45), Host and Kra [552] showed that a bounded sequence (a_n) in $\mathbb{C}$ is a mean good weight for the classical multiple averages with $P_1(n) := n, \ldots, P_k(n) := kn$ if $\frac{1}{N}\sum_{n=1}^{N} a_n b_n$ converges for every nilsequence (b_n) of step k. (Moreover, the limit of the corresponding weighted multiple averages is zero if it is zero for nilsequences.) This class of weights includes nilsequences, the Möbius function and the Liouville function (see Green and Tao [488]), as well as sequences of the form $e(P(n))$ with P being a generalized polynomial and return times sequences of the form $a_n := g(S^n y)$ for every ergodic measure-preserving system (Y, ν, S), every $g \in \mathrm{L}^\infty(Y, \nu)$ and a.e. $y \in Y$ as shown in the same paper [552]. Chu [244] showed that the same class of weights is mean good for polynomial multiple averages, see Frantzikinakis and Host [416] for the case of commuting transformations.

In their study of multiple polynomial convergence along primes, Wooley and Ziegler [1049] showed[8] that the modified von Mangoldt function is a mean good weight for multiple polynomial averages (16.16), see also Frantzikinakis et al. [419, 420]. For related results see Sun [962] and Koutsogiannis [665].

Frantzikinakis and Host [414], see also [416], showed that multiplicative functions bounded by one are mean good weights for multiple polynomial averages for commuting transformations as soon as they are Cesàro convergent along arithmetic progressions. Note that thereby the limit of the corresponding weighted averages is always zero whenever such weight is aperiodic, i.e., is Cesàro convergent to zero along all arithmetic progressions.

Assani and Moore [52] showed that sequences of the form $(f_1(T^{an}x) f_2(T^{bn}x))$ appearing in Bourgain's double recurrence theorem are (for ergodic systems and

[8] Recall Lemma 10.23.

a.e. x) mean good weights for the classical multiple ergodic averages for ergodic systems. Moreover, the limit of the corresponding weighted averages

$$\frac{1}{N}\sum_{n=1}^{N} f(T^{an}x)g(T^{bn}x)S^{n}g_1\cdots S^{kn}g_k$$

equals zero for every ergodic measure-preserving system (Y, ν, S) and every $g_1, \ldots, g_k \in \mathrm{L}^\infty(Y, \nu)$ whenever either f or g is orthogonal to the Host–Kra factor $\mathcal{Z}_{k+1}$. El Abdalaoui [351, 352] showed that the Möbius and the Liouville function are pointwise good weights for multiple averages for weakly mixing systems with a certain property and conjectured that they are pointwise good weights for double averages for general systems.

Assani et al. [49] studied pointwise convergence of weighted bilinear averages

$$\frac{1}{N}\sum_{n=1}^{N} a_n T^n f\cdot T^{2n}g$$

and showed that linear weights $(e(nt))_{n\in\mathbb{N}_0}$, $t \in \mathbb{R}$ (or, equivalently, λ^n, $\lambda \in \mathbb{T}$) form a Wiener–Wintner family for such averages for ergodic systems, with uniform convergence to zero whenever either f or g is orthogonal to the Kronecker factor. Their results were generalized to polynomial sequences of the form $e(p(n))$, $p \in \mathbb{R}[\cdot]$ by Assani and Moore [53] (with the Host–Kra factors with order depending on the degree of the polynomials instead of the Kronecker factor in the uniform version) and to nilsequences by Zorin-Kranich [1071] and Assani [46].

Very recently, Teräväinen [1000] showed

$$\lim_{N\to\infty}\frac{1}{N}\sum_{n=1}^{N}\mu(n)T^{p_1(n)}f_1\cdots T^{p_k(n)}f_k = 0 \text{ a.e.} \tag{16.17}$$

for polynomials $p_1, \ldots, p_k \in \mathbb{Z}[\cdot]$ whenever $k \leq 3$ and the polynomials are linear or k is arbitrary and the polynomials have distinct degrees. He also showed that (16.17) holds along a subsequence for arbitrary k and $p_1, \ldots, p_k$. Recall that the unweighted averages are known to converge a.e. only for $k \leq 2$ and either the polynomials being linear or having distinct degree, see Bourgain [184], Krause et. al [675] and Kosz et al. [664].

Exercises

Exercise 16.1 (Mean Good Weights and L^p-Convergence) Let (a_n) be a mean good weight satisfying $\sup_{N\in\mathbb{N}} \frac{1}{N}\sum_{n=1}^{N}|a_n| < \infty$. Prove that for every measure-preserving system (X, μ, T) and every $f \in \mathrm{L}^p(X, \mu)$, $1 \le p < \infty$, $\frac{1}{N}\sum_{n=1}^{N} a_n T^n f$ converges in the L^p-norm.

Exercise 16.2 (Weighted Averages for Almost Weakly Stable Vectors) Prove that for a contraction $S \in \mathscr{L}(H)$ on a Hilbert space and for $x \in H$ the following assertions are equivalent.

(i) x is almost weakly stable, i.e., $x \in H_{\mathrm{aws}}$.
(ii) For every bounded sequence $(a_n)_{n\in\mathbb{N}}$ in $\mathbb{C}$ one has

$$\lim_{N\to\infty} \frac{1}{N}\sum_{n=1}^{N} a_n S^n x = 0.$$

(iii) For every sequence $(a_n)_{n\in\mathbb{N}}$ in $\mathbb{C}$ with $\sup_N \frac{1}{N}\sum_{n=1}^{N} |a_n|^2 < \infty$ one has

$$\lim_{N\to\infty} \frac{1}{N}\sum_{n=1}^{N} a_n S^n x = 0.$$

This proves Proposition 16.11. (Hint: adapt the proof of Theorem 12.2.)

Exercise 16.3 (Mean Good Weights) Prove Proposition 16.3 up to the at most countability of the set $\{\lambda \in \mathbb{T} : c(\lambda) \neq 0\}$ and also the statement in Remark 16.4(b).

Exercise 16.4 (Good Weights for Mean Ergodic Operators) Show the assertion of Example 16.10.

Exercise 16.5 (Bad Weights) Give examples of **bad**, i.e., not good, bounded weights for the mean ergodic theorem.

Exercise 16.6 (Pointwise Good Weights) Show that the assumption of invertibility can be removed from the definition of pointwise good weights. (See the hint to Exercise 12.6.)

Exercise 16.7 (Periodic Sequences are Good Weights) Show using the definition only that every periodic sequence is a pointwise good weight.

Exercise 16.8 (Weights (λ^n) are Pointwise Good) For $\lambda \in \mathbb{T}$ it was shown in Remark 16.13 that the sequence (λ^n) is a pointwise good weight. Give an alternative proof to this along the lines of the proof of Birkhoff's theorem using the maximal inequality (Corollary 7.47) as a black box and Banach's principle (Proposition 7.45) for an appropriate dense set of functions.

Exercise 16.9 (Maximal Inequality for Weighted Averages) Prove Proposition 16.17.

Exercise 16.10 (Bounded Weights) Give an alternative proof of Proposition 16.18 using the following steps. Let (Y, ν, S) be a measure-preserving system and $g \in \mathrm{L}^1(Y, \nu)$.

(a) Show that the functions $g_j \in \mathrm{L}^\infty(Y, \nu)$ defined by $g_j := g\mathbf{1}_{[|g| \le j]}$ for every $j \in \mathbb{N}$ satisfy

$$\lim_{j\to\infty} \lim_{N\to\infty} \frac{1}{N} \sum_{n=1}^{N} |g - g_j|(S^n y) = 0 \quad \text{for } \nu\text{-a.e. } y \in Y.$$

(Hint:[9] Use the pointwise ergodic theorem and the fact that the invariant functions h_j defined by $h_j(y) := \lim_{N\to\infty} \frac{1}{N} \sum_{n=1}^{N} |g - g_j|(S^n y)$ are monotonously decreasing to show that the pointwise limit function $h := \lim_{k\to\infty} h_j$ satisfies $\int_Y h \, \mathrm{d}\mu = 0$.)

(b) Assume that (a_n) is a pointwise good weight in L^∞. Prove that the sequence $\left(\frac{1}{N} \sum_{n=1}^{N} a_n g(S^n y)\right)_{N\in\mathbb{N}}$ is Cauchy for ν-a.e. $y \in Y$ using (a).

Exercise 16.11 (Besicovitch Sequences as Good Weights)

(a) Show the assertion on mean convergence of Example 16.5.
(b) Prove the assertion on pointwise convergence of Example 16.14.
(c) Let (X, μ, T) be a measure-preserving system. Show that the limit of the weighted ergodic averages (16.1) is a constant function for every bounded Besicovitch sequence as weight if and only if (X, μ, T) is weakly mixing. (This observation is due to Ryll-Nardzewski [897].)

Exercise 16.12 (Mean Good Weights via Wiener–Wintner Theorem) Prove Proposition 16.27.

Exercise 16.13 (Uniform Convergence in the Wiener–Wintner Theorem for Uniquely Ergodic systems) Prove the assertion of Remark 16.23.

Exercise 16.14 (Good Weights for the Polynomial Mean Ergodic Theorem) A sequence $(a_n)_{n\in\mathbb{N}}$ in $\mathbb{C}$ is called a **good weight for the polynomial mean ergodic theorem** if for every invertible measure-preserving system (X, μ, T) and every $f \in \mathrm{L}^2(X, \mu)$ the weighted polynomial averages

$$\frac{1}{N} \sum_{n=1}^{N} a_n T^{P(n)} f$$

converge in $\mathrm{L}^2(X, \mu)$ for every integer polynomial P.

(a) Formulate and prove a characterization of good weights for the polynomial mean ergodic theorem analogous to Proposition 16.3 (without information about the limit).
(b) Use the characterization from (a) and Theorem 16.24 to prove the assertion of Remark 16.28.

[9] See [438, Behauptung 2.39].

Exercise 16.15 (Weighted Averages for Generalized Eigenfunctions) Let (X, μ, T) be an ergodic measure-preserving system, $k \in \mathbb{N}$ and let f be a generalized eigenfunction of T of order k (see Example 8.33). Compute $T^n f$ for every $n \in \mathbb{N}$ and show that the polynomial weighted averages

$$\frac{1}{N}\sum_{n=1}^{N} e(P(n))T^n f$$

converge almost everywhere for every $P \in \mathbb{R}[\cdot]$.

Exercise 16.16 (Eigenfunctions of Uniquely Ergodic Systems) Let (K, T) be a uniquely ergodic topological system and let $g \in \mathrm{C}(K)$ satisfy $Tg = \lambda g$ for some $\lambda \in \mathbb{T}$. Prove that $|g|$ is constant.

Exercise 16.17 (Return Times Theorem and Weakly Mixing Systems) Let (X, μ, T) be a weakly mixing system, let $f \in \mathrm{L}^\infty(X, \mu)$ and for $x \in X$ denote $a_n(x) = f(T^n x)$. Prove that there exist a set $X_f \subset X$ of measure 1 and a constant $C = C(f)$ such that when $x \in X_f$, for every measure-preserving system (Y, ν, S) with Koopman operator S and $g \in \mathrm{L}^1(Y, \nu)$ we have

$$\frac{1}{N}\sum_{n=1}^{N} a_n(x) S^n g \to C P_{\mathrm{fix}(S)} g \quad \nu\text{-almost everywhere.}$$

Show furthermore that $C(f) = \int_X f \,\mathrm{d}\mu$.

Chapter 17
Sarnak's Conjecture

With his conjecture, P. Sarnak[1] discovered in 2010 a deep connection between ergodic theory and number theory concerning the behavior of the prime numbers. This conjecture has initiated extensive research in both fields and has become a very active and dynamic area. We give here a flavor of Sarnak's conjecture and briefly mention some of the recent developments.

17.1 Motivation

Sarnak's conjecture connects dynamical systems to the following important number theoretic function introduced by A.F. Möbius in 1832. As before, we write p (sometimes with a subscript) to denote a prime number without mentioning this explicitly.

Definition 17.1 The **Möbius function** $\mu : \mathbb{N} \to \{-1, 0, 1\}$ is defined as $\mu(1) := 1$ and

$$\mu(n) := \begin{cases} 1, & \text{if } n = p_1 \cdots p_{2k} \text{ for some } k \in \mathbb{N} \text{ and distinct } p_1, \dots p_{2k}, \\ -1, & \text{if } n = p_1 \cdots p_{2k-1} \text{ for some } k \in \mathbb{N} \text{ and distinct } p_1, \dots p_{2k-1}, \\ 0, & \text{if } n \text{ is not square free.} \end{cases}$$

Clearly, the function μ is bounded by 1 and **multiplicative**, i.e., satisfies $\mu(nm) = \mu(n)\mu(m)$ whenever n and m are coprime. Here are its first values:

$$1, -1, -1, 0, -1, 1, -1, 0, 0, 1, -1, 0, -1, 1, 1, 0, \dots$$

[1] He formulated his conjecture in "Three Lectures on the Möbius Function, Randomness and Dynamics" [909].

T. Eisner, B. Farkas, *A Journey Through Ergodic Theorems*, Birkhäuser Advanced Texts Basler Lehrbücher, https://doi.org/10.1007/978-3-031-96506-7_17

The function μ reflects many properties of the set of primes. In particular, the generally believed irregularity of the primes corresponds heuristically to the random behavior of the Möbius function. The weakest variant of random behavior is

$$\lim_{N\to\infty} \frac{1}{N} \sum_{n=1}^{N} \mu(n) = 0 \tag{17.1}$$

(meaning that there are asymptotically as many values of μ equal to 1 as to -1), which is already a deep fact, known to be equivalent to the prime number theorem due to the works of Landau [704, 705]. We remark that the convergence rate

$$\frac{1}{N} \sum_{n=1}^{N} \mu(n) \leq \frac{C}{N^{1/2-o(1)}}$$

is equivalent to the Riemann hypothesis, see Titchmarsh [1002, Thm. 14.25(C)].

The following classical result of Davenport [272], building on the methods of Vinogradov developed in the 1930s, will be used as a black box. It is a strengthening of the asymptotics in (17.1).

Theorem 17.2 (Davenport's Estimate) *For every $A > 0$ there is $C_A \geq 0$ such that*

$$\sup_{\lambda\in\mathbb{T}} \left| \frac{1}{N} \sum_{n=1}^{N} \mu(n)\lambda^n \right| \leq \frac{C_A}{\log^A(N)} \qquad \textit{for } N \geq 2.$$

It follows that

$$\lim_{N\to\infty} \frac{1}{N} \sum_{n=1}^{N} \mu(n)\lambda^n = 0 \quad \text{for all } \lambda \in \mathbb{T}. \tag{17.2}$$

Theorem 17.2 means in particular that the Möbius function satisfies the assumptions of Theorem 16.16. Whence we immediately obtain the following result announced by Sarnak in 2010 (the argument presented here is due to El Abdalaoui et al. [357, Sec. 3]).

Theorem 17.3 (The Möbius Function is a Pointwise Good Weight in L^1 with Limit Zero) *For every measure-preserving system (X, μ, T) and every $f \in \mathrm{L}^1(X, \mu)$*

$$\lim_{N\to\infty} \frac{1}{N} \sum_{n=1}^{N} \mu(n) f(T^n x) = 0 \textit{ for almost every } x \in X. \tag{17.3}$$

In particular, $(\mu(n))_{n\in\mathbb{N}}$ is a good weight for the pointwise ergodic theorem in L^1.

The reader might have noticed the general philosophy that a statement holding almost everywhere for general systems (and functions) holds *everywhere* for sufficiently good systems (and functions), see, e.g., the pointwise ergodic Theorem 7.40 vs. Theorem 7.23 or the Wiener–Wintner Theorem 16.20 vs. Remark 16.23 and Theorem 16.36. Thus, the following question is natural.

Question For which systems and functions does (17.3) hold everywhere?

Sarnak's conjecture is an attempt to answer this question. (Note that the Möbius function does not satisfy the conditions of Example 16.10.)

Since the Möbius function is believed to have random behavior, it is believed to have zero correlation with deterministic sequences, i.e., to satisfy

$$\lim_{N\to\infty} \frac{1}{N} \sum_{n=1}^{N} \mu(n) b_n = 0$$

whenever $(b_n)_{n\in\mathbb{N}}$ is deterministic (in some sense to be defined). Sarnak's idea was to define deterministic sequences as sequences coming from deterministic dynamical systems and continuous functions. Thus, before presenting Sarnak's conjecture, we have to spend some words on deterministic dynamical systems.

17.2 Topological Entropy and Sarnak's Conjecture

We start with a definition of topological entropy for metrizable topological systems due to Bowen [190] and Dinaburg [303]. Let K be a compact[2] metrizable space with d being any metric defining the topology on K. We call a set of pairwise disjoint open balls of radius $\varepsilon > 0$ an ε**-packing**. By compactness, each ε-packing in K is finite. Denote by $s_d(\varepsilon)$ the largest[3] number of balls in ε-packings with respect to the metric d, called the ε**-packing number**, or ε-capacity.

Now, let (K, T) be a metrizable topological system. For $n \in \mathbb{N}$ and $x, y \in K$ we define

$$d_n(x, y) := \max\{d(T^j x, T^j y) : j = 0, \ldots, n\}.$$

Then d_n is a metric on K which is (uniformly) equivalent to d, i.e., the identity mapping $\mathrm{id} : (K, d) \to (K, d_n)$ is (uniformly) continuous with (uniformly) continuous inverse. We call $(x, Tx, \ldots, T^n x)$ the n-orbit of x. Observe that two points $x, y \in K$ are at least ε-apart with respect to d_n if and only if their n-orbits are at least ε-apart, meaning that $T^j x$ and $T^j y$ are at least ε-apart for some

[2] Bowen did not assume this, compactness is only needed to shorten things a bit.

[3] It is Exercise 17.1 to show that this largest number exists.

$j = 0, \dots, n$. Therefore the packing number $s_{d_n}(\varepsilon)$ is the maximal number of n-orbits which are at least 2ε-apart.

The **topological entropy** of (K, T) with respect to the metric d is defined as

$$h_d(K, T) := \lim_{\varepsilon \to 0} \limsup_{n \to \infty} \frac{1}{n} \log(s_{d_n}(\varepsilon)).$$

The existence of this, possibly infinite, limit is clear since, for given $n \in \mathbb{N}$, the mapping $\varepsilon \mapsto s_{d_n}(\varepsilon)$ is decreasing. Topological entropy quantifies the asymptotic growth of the number of "distinct n-orbits", and can be considered as a measure of complexity (or randomness) of the system (K, T). We finally call a compact metrizable system (K, T) **deterministic** if it has topological entropy 0. In particular, by monotonicity for every $\varepsilon > 0$ such systems satisfy

$$\lim_{n \to \infty} \frac{1}{n} \log(s_{d_n}(\varepsilon)) = 0,$$

i.e., the maximal number of ε-distinct n-orbits grows subexponentially, and such systems are of lowest (exponential) complexity.

Proposition 17.4 *The entropy $h_d(K, T)$ is independent of the metric d. More precisely, any two metrics defining the same topology yield the same topological entropy.*

We leave the proof as Exercise 17.2, and write $h(K, T)$ instead of $h_d(K, T)$.

Proposition 17.5 *If K is a compact metric space and $T : K \to K$ is* ***non-expansive****, i.e., $d(Tx, Ty) \le d(x, y)$ for each pair $x, y \in K$, then (K, T) is deterministic.*

Proof If T is non-expansive, then $d_n = d$ for each $n \in \mathbb{N}$. Therefore $h(K, T) = 0$. ■

As a consequence, each torus rotation $(\mathbb{T}, a)$, $a \in \mathbb{T}$, has topological entropy 0.

Recall from Example 11.43 the notion of an equicontinuous system: A topological system (K, T) is **equicontinuous** if for each continuous semimetric d on K, each $x \in K$ and each $\varepsilon > 0$ there is an open set U with $x \in U$ and such that

$$y \in U \quad \Longrightarrow \quad d(T^n x, T^n y) < \varepsilon \qquad \text{for every } n \in \mathbb{N}_0.$$

It is Exercise 17.3 to show that for metrizable systems this is equivalent to the following property: Given a compatible metric d on K, for every $\varepsilon > 0$ there is $\delta > 0$ such that for every $x, y \in K$ and $n \in \mathbb{N}_0$ one has

$$d(x, y) < \delta \quad \Longrightarrow \quad d(T^n x, T^n y) < \varepsilon.$$

Evidently, if (K, T) is a **non-expansive** topological system, i.e., T is non-expansive with respect to some compatible metric d on the compact space K, then (K, T) is an equicontinuous system. Conversely, an equicontinuous system can be turned into a non-expansive one by considering an equivalent metric, see Exercise 17.3 (cf. also Exercise 11.22 for the invertible case). As a consequence, equicontinuous systems are deterministic.

Proposition 17.6 *Let G be a compact, metric group. Then each group rotation (G, a), $a \in G$, is deterministic.*

Proof The mapping $G \times G \to G$, $(x, y) \mapsto xy$ is uniformly continuous, which implies that the set $\{\tau_x : x \in G\}$ of left rotations is equicontinuous. As a consequence (G, a) is an equicontinuous systems for every $a \in G$, and therefore has zero topological entropy by the above. ∎

We present here some more examples without proof and refer the reader, e.g., to the book by Walters [1033, Ch. 8].

Example 17.7

(a) Even though the skew shift $(\mathbb{T}^2, T_a)$ is not equicontinuous, see Exercise 11.24, it has zero topological entropy, i.e., the system is deterministic.
(b) The entropy of the topological shift (one or two-sided, see Example 2.31) on k letters is $\log(k)$.
(c) Let $A \subset \mathbb{N}$ be a subset and let $a = (a_n)_{n\in\mathbb{N}}$ be its characteristic sequence, i.e., $a = \mathbf{1}_A$. Consider the orbit closure $K = \overline{\mathrm{orb}_+(a)}$ of a in the shift system $(\{0, 1\}^{\mathbb{N}}, T)$. The topological entropy of the subsystem (K, T) equals

$$\lim_{n\to\infty} \frac{1}{n}\theta_n,$$

where θ_n denotes the number of n-tuples $(b_1, b_2, \cdots, b_n)$ such that $K \cap \{x : \exists m \in \mathbb{N} \text{ with } x_m = b_1, \dots, x_{m+n-1} = b_n\} \neq \emptyset$, i.e., the number of finite patterns of length n occurring in A. (One calls $h(K, T)$ the entropy of the sequence a or of the set A.)
(d) Host et al. [555, Thm. 4.2] showed that for transitive[4] nilsystems the quantity s_{d_n} has polynomial growth implying that such systems are deterministic.

Remark 17.8 In Sect. 2.2 we briefly mentioned *measure-theoretic entropy* as an important isomorphism invariant of measure-preserving systems. Now, the **variational principle**, due to Dinaburg [303] and Goodwyn [477], states the following: If (K, T) is a topological system, then the topological entropy is the supremum of the measure-theoretic entropies of (K, μ, T) where μ ranges over the set of T-invariant probability measures, see also [1033, Ch. 8].

[4] Host et al. [555, Thm. 4.2] showed the result for minimal nilsystems. Recall that by Corollary 14.26, for nilsystems minimality is equivalent to transitivity.

Thus, if $h(K,T) = 0$, then no matter which invariant measure we choose, the measure-preserving system (K,μ,T) has zero measure-theoretic entropy. Furthermore, for uniquely ergodic systems the measure-theoretic entropy equals the topological entropy $h(K,T)$.

We can now formulate Sarnak's conjecture from 2010. A sequence $(a_n)_{n\in\mathbb{N}}$ of complex numbers is called **deterministic** if there is a deterministic topological system (K,T), $x \in K$ and a continuous function $f \in \mathrm{C}(K)$ such that $a_n = f(T^n x)$ for every $n \in \mathbb{N}$. We moreover call two sequences $(a_n)_{n\in\mathbb{N}}$ and $(b_n)_{n\in\mathbb{N}}$ in $\mathbb{C}$ **asymptotically orthogonal** if

$$\lim_{N\to\infty} \frac{1}{N}\sum_{n=1}^{N} a_n b_n = 0.$$

Conjecture 17.9 (Sarnak [909]) Let (K,T) be a deterministic topological dynamical system. Then

$$\lim_{N\to\infty} \frac{1}{N}\sum_{n=1}^{N} \mu(n) f(T^n x) = 0 \tag{17.4}$$

holds for every $f \in \mathrm{C}(K)$ and *every* $x \in K$. In other words, μ is asymptotically orthogonal to every deterministic sequence $(a_n)_{n\in\mathbb{N}}$.

Remark 17.10 (Reduction to Transitive Systems) The orthogonality condition (17.4) involves the orbit of a (fixed arbitrary) point x. Hence, by considering the subsystem generated by the orbit of x and since subsystems of deterministic systems are clearly again deterministic, we see that it suffices to check Sarnak's conjecture for transitive systems.

Some first examples are the following.

Example 17.11 (One-point System) It is a direct consequence of (17.1), i.e., of the prime number theorem that Sarnak's conjecture holds for the one-point system.

Example 17.12 (Periodic Systems) One can show that (17.2) for rational λ implies

$$\lim_{N\to\infty} \frac{1}{N}\sum_{n=1}^{N} \mu(n) b_n = 0 \quad \text{for every periodic sequence } (b_n)_{n\in\mathbb{N}} \text{ in } \mathbb{C}. \tag{17.5}$$

Thus Sarnak's conjecture holds for periodic systems (i.e., topological systems (K,T) with $T^q = \mathrm{id}$ for some $q \in \mathbb{N}$). We leave the details as Exercise 17.7.

Example 17.13 (Rotations) Also rotations on $\mathbb{T}$ satisfy Sarnak's conjecture. This is an easy consequence of (17.2) and the approximation argument presented in Exercise 17.6. We leave it as Exercise 17.8 to work out the details. Analogously, one can show that rotations on compact, metric, Abelian groups satisfy Sarnak's conjecture.

For more examples see Example 17.21 and Sect. 17.4 below.

Connection to Chowla's Conjecture Another motivation for Sarnak's conjecture is its close relation to the famous Chowla conjecture from number theory going back to [243], which is widely believed to be true.

Conjecture 17.14 (Chowla [243]) For every $m \in \mathbb{N}$, every $a_1, \dots, a_m \in \mathbb{N}_0$ with $a_1 < \dots < a_m$, and every $k_1, \dots, k_m \in \{1, 2\}$ which are not all even

$$\lim_{N\to\infty} \frac{1}{N} \sum_{n=1}^{N} \mu^{k_1}(n+a_1)\mu^{k_2}(n+a_2)\cdots\mu^{k_m}(n+a_m) = 0.$$

Chowla's conjecture states that the Möbius function has zero multiple correlation with itself and would confirm (a strong version of) randomness of the Möbius function or, equivalently, of primes. Even the special case

$$\lim_{N\to\infty} \frac{1}{N} \sum_{n=1}^{N} \mu(n)\mu(n+2) = 0$$

is open. A certain quantitative version of this would imply the **twin primes conjecture**, i.e., that

$$\liminf_{n\to\infty}(p_{n+1} - p_n) = 2,$$

where p_n is the nth prime, see, e.g., Tao [979].

Remark 17.15 Chowla formulated his conjecture for the **Liouville function** λ defined by

$$\lambda(n) := (-1)^{\Omega(n)},$$

where $\Omega(n)$ denotes the number of all prime factors of n counted with multiplicity (in particular, $\lambda(1) = 1$), see Ramaré [864] for a discussion. Note that $\mu(n) = \lambda(n)$ holds for every square-free n. Almost everything said in this chapter remains true when replacing μ by λ (with only one exception which will be mentioned explicitly). In fact, many papers treat both functions simultaneously, whereas treating λ is often somewhat easier due to its complete multiplicativity.[5]

[5] meaning that $\lambda(nm) = \lambda(n)\lambda(m)$ for every $n, m \in \mathbb{N}$.

The following gives a relation between these fundamental conjectures of Sarnak and Chowla, for a proof we refer the interested reader to Sarnak [909], see also El Abdalaoui et al. [357, Sec. 4].

Theorem 17.16 (Sarnak) *Chowla's conjecture implies Sarnak's conjecture.*

Thus, the property of μ having zero correlation with deterministic systems (in Sarnak's sense) is formally weaker than to have zero multiple correlations with itself.

17.3 The Kátai–Bourgain–Sarnak–Ziegler Criterion

We present here an extremely useful sufficient condition for Möbius orthogonality due to Bourgain et al. [188], see also Kátai [632]. We present a slightly simplified qualitative version of the statement from [188].

Theorem 17.17 (Kátai–Bourgain–Sarnak–Ziegler Criterion) *Let $a = (a_n)_{n\in\mathbb{N}}$ be a bounded sequence in $\mathbb{C}$ such that*

$$\lim_{N\to\infty} \frac{1}{N} \sum_{n=1}^{N} a_{pn}\overline{a_{qn}} = 0 \tag{17.6}$$

for all distinct primes p, q. Then $(a_n)_{n\in\mathbb{N}}$ is asymptotically orthogonal to every bounded, multiplicative function $c : \mathbb{N} \to \mathbb{C}$, i.e.,

$$\lim_{N\to\infty} \frac{1}{N} \sum_{n=1}^{N} a_n c(n) = 0.$$

In particular, it is asymptotically orthogonal to the Möbius function.

Remark 17.18

(a) For the Möbius function, the above sufficient condition is not necessary (take $a_n := 1$ for every $n \in \mathbb{N}$ and use (17.1)).
(b) Theorem 17.17 shows that if a bounded sequence does not correlate with itself in the sense of (17.6), it does not correlate with any multiplicative function. This gives an *internal* sufficient condition for systems to satisfy Sarnak's conjecture without using the Möbius function at all.

Proof of Theorem 17.17 For the sake of convenience we assume that both $|a|$ and $|c|$ are bounded by 1. The idea is to decompose $[1, N] := \{1, \ldots, N\}$ into well-factorized sets (up to a small error) to be able to exploit the multiplicativity of c when computing the average over $[1, N]$.

Step 1: A decomposition of the interval $[1, N]$. Let $\varepsilon \in (0, \frac{1}{2})$ and let $\alpha > 0$ be small (depending on ε) to be chosen later.[6] Consider

$$j_0 := \tfrac{1}{\alpha}\big(\log \tfrac{1}{\alpha}\big)^3, \quad D_0 := (1+\alpha)^{j_0}, \quad D_1 := (1+\alpha)^{j_0^2}.$$

Take $N \in \mathbb{N}$ (in what follows, N will be chosen after fixing $\alpha > 0$) and define

$$S := S(N) := \{n \in [1, N] : n \text{ has a prime divisor in } [D_0, D_1)\}.$$

It is Exercise 17.9(a) to show that

$$\lim_{N\to\infty} \frac{|[1, N] \setminus S(N)|}{N} \leq \alpha, \tag{17.7}$$

i.e., $[1, N] = S(N)$ up to a small error. Thus for sufficiently large N we have $|[1, N] \setminus S(N)| \leq 2\alpha N$.

In what follows j and m will always denote elements of $\mathbb{N}$. The sets we construct below will depend on N, ε and α. However, to simplify notation we will not denote this dependence. For $j \in [j_0, j_0^2)$ let

$$P_j := [(1+\alpha)^j, (1+\alpha)^{j+1}) \cap \mathbb{P},$$

and define the sets

$$S_j := \Big\{n \in S : n \text{ has exactly one divisor in } P_j \text{ and no divisor in } \bigcup_{i=1}^{j-1} P_i\Big\},$$

which are by definition pairwise disjoint. We will show $[1, N] = \bigcup_{j_0 \leq j < j_0^2} S_j$ up to a small error. By definition we have

$$S \setminus \bigcup_{j_0 \leq j < j_0^2} S_j = \bigcup_{j_0 \leq j < j_0^2} \{n \in [1, N] : n \text{ has at least two prime divisors in } P_j\}.$$

Therefore, we obtain

$$\Big|S \setminus \bigcup_{j_0 \leq j < j_0^2} S_j\Big| \leq \sum_{j_0 \leq j < j_0^2} \sum_{l,k \in P_j} \frac{N}{lk} \leq N \sum_{j_0 \leq j < j_0^2} \frac{|P_j|^2}{(1+\alpha)^{2j}}. \tag{17.8}$$

We now estimate $|P_j|$ as follows. By the prime number theorem (Theorem 10.21) using the usual notation $\pi(x) := |\{p \in \mathbb{P} : p \leq x\}|$, we have for small enough α

[6] We will take $\alpha > 0$ in any case so small that $\alpha + 38\alpha \log\big(\frac{1}{\alpha}\big) < \varepsilon$, but in many places in the proof $\alpha > 0$ has to be taken potentially smaller.

(and therefore large enough j_0 and hence j)

$$
\begin{aligned}
|P_j| &\le \pi((1+\alpha)^{j+1}) - \pi((1+\alpha)^j) + 1 \\
&\le \frac{(1+\alpha)^{j+1}}{(j+1)\log(1+\alpha)} - \frac{(1+\alpha)^j}{j\log(1+\alpha)} + \varepsilon\frac{(1+\alpha)^j}{j} \\
&= (1+\alpha)^j\left(\frac{1}{\log(1+\alpha)}\left(\frac{1+\alpha}{j+1} - \frac{1}{j}\right) + \frac{\varepsilon}{j}\right) \\
&\le (1+\alpha)^j\left(\frac{\alpha}{j\log(1+\alpha)} + \frac{\varepsilon}{j}\right) \le (1+\alpha)^j\frac{1+2\varepsilon}{j} \le \frac{2(1+\alpha)^j}{j}.
\end{aligned}
\tag{17.9}
$$

Inserting this into (17.8) yields

$$\left|S \setminus \bigcup_{j_0 \le j < j_0^2} S_j\right| \le 4N \sum_{j_0 \le j < j_0^2} \frac{1}{j^2} \le 4N \int_{j_0-1}^{j_0^2} \frac{\mathrm{d}x}{x^2} \le \frac{4N}{j_0-1},$$

which implies for small enough α by the definition of j_0

$$\left|S \setminus \bigcup_{j_0 \le j < j_0^2} S_j\right| \le 4\alpha N.$$

We thus obtain

$$\left|[1, N] \setminus \bigcup_{j_0 \le j < j_0^2} S_j\right| \le 6\alpha N \qquad \text{for } N \text{ large enough} \tag{17.10}$$

meaning that $[1, N]$ is covered by the sets S_j, $j_0 \le j < j_0^2$ up to a small error.
Step 2: Refined decomposition into well-factorized sets. We now decompose each set S_j as a product set of the form $P_j Q_j$ up to a small error, where each element of Q_j is coprime to each element of P_j.
Define

$$Q_j := \left\{m \le \frac{N}{(1+\alpha)^{j+1}} : m \text{ has no prime divisors in } \bigcup_{k \le j} P_k\right\}$$

and observe that by definition

$$P_j Q_j \subset S_j.$$

Moreover, we have the inclusion

$$S_j \setminus P_j Q_j \subset P_j \cdot \left[\tfrac{N}{(1+\alpha)^{j+1}}, \tfrac{N}{(1+\alpha)^j}\right],$$

which implies, by the definition of j_0, by the estimate (17.9), and by the inequality $\log(x) \le x$ that

$$\begin{aligned}\sum_{j_0\le j<j_0^2} |S_j \setminus P_j Q_j| &\le \sum_{j_0\le j<j_0^2} \frac{\alpha N |P_j|}{(1+\alpha)^{j+1}}\\ &\le 2\alpha N \sum_{j_0\le j<j_0^2} \frac{1}{j} \le 2\alpha \log\left(j_0^2\right) N\\ &= 4\alpha \log\left(\tfrac{1}{\alpha}\left(\log\tfrac{1}{\alpha}\right)^3\right) N \le 16\alpha \log\left(\tfrac{1}{\alpha}\right) N.\end{aligned}$$

Combining this with (17.10) yields

$$\Big|[1,N] \setminus \bigcup_{j_0\le j<j_0^2} P_j Q_j\Big| \le 22\alpha \log\left(\tfrac{1}{\alpha}\right) N.$$

Thus we have decomposed $[1, N]$ into disjoint sets of the form $P_j Q_j$ up to a small error. Moreover, from the construction it follows that the mapping $P_j \times Q_j \to P_j Q_j$ $(p, q) \mapsto pq$ is injective and that every $p \in P_j$ is coprime to every $q \in Q_j$ for every j between j_0 and j_0^2.

Step 3: The correlation estimate. We now use the previous decomposition to estimate the correlation between a and c. The above and the multiplicativity of c imply (recall that $|a|$ and $|c|$ are bounded by 1)

$$\begin{aligned}\left|\sum_{n=1}^{N} a_n c(n)\right| &\le \sum_{j_0\le j<j_0^2} \Big|\sum_{p\in P_j, m\in Q_j} a_{pm} c(p) c(m)\Big| + 22\alpha \log\left(\tfrac{1}{\alpha}\right) N\\ &\le \sum_{j_0\le j<j_0^2} \sum_{m\in Q_j} \Big|\sum_{p\in P_j} a_{pm} c(p)\Big| + 22\alpha \log\left(\tfrac{1}{\alpha}\right) N.\end{aligned}$$

The inner sum can be estimated by the Cauchy–Schwarz inequality as follows:

$$\begin{aligned}\sum_{m\in Q_j} \Big|\sum_{p\in P_j} a_{pm} c(p)\Big| &\le |Q_j|^{1/2} \Big(\sum_{m\in Q_j} \Big|\sum_{p\in P_j} a_{pm} c(p)\Big|^2\Big)^{1/2}\\ &\le |Q_j|^{1/2} \Big(\sum_{m\le \frac{N}{(1+\alpha)^{j+1}}} \Big|\sum_{p\in P_j} a_{pm} c(p)\Big|^2\Big)^{1/2}\\ &= |Q_j|^{1/2} \Big(\sum_{m\le \frac{N}{(1+\alpha)^{j+1}}} \sum_{p_1,p_2\in P_j} a_{p_1 m} c(p_1) \overline{a_{p_2 m} c(p_2)}\Big)^{1/2}\end{aligned}$$

$$
\begin{aligned}
&\le |Q_j|^{1/2}\Big(\sum_{p_1,p_2\in P_j}\Big|\sum_{m\le\frac{N}{(1+\alpha)^{j+1}}} a_{p_1 m}\overline{a_{p_2 m}}\Big|\Big)^{1/2} \qquad (17.11)\\
&\le I_j + II_j,
\end{aligned}
$$

where I_j corresponds to the diagonal contribution (if $p_1 = p_2$) of the outer sum in (17.11) and II_j to the non-diagonal one (if $p_1 \neq p_2$). (For the last estimate we have used the inequality $\sqrt{a+b} \le \sqrt{a}+\sqrt{b}$.) Note that the multiplicative function c has already disappeared at this point.
Since $|a|$ is bounded by 1, we have

$$
I_j = |Q_j|^{1/2}\Big(\sum_{p\in P_j}\Big|\sum_{m\le\frac{N}{(1+\alpha)^{j+1}}} a_{pm}\overline{a_{pm}}\Big|\Big)^{1/2} \le |Q_j|^{1/2}|P_j|^{1/2}\frac{\sqrt{N}}{(1+\alpha)^{(j+1)/2}}.
$$

Thus, by $\sum_{j_0\le j<j_0^2}|P_j||Q_j| \le \sum_{j_0\le j<j_0^2}|S_j| \le N$ and by the Cauchy–Schwarz inequality we obtain

$$
\begin{aligned}
\sum_{j_0\le j<j_0^2} I_j &\le \sqrt{N}\Big(\sum_{j_0\le j<j_0^2}|Q_j||P_j|\Big)^{1/2}\Big(\sum_{j_0\le j<j_0^2}\frac{1}{(1+\alpha)^{j+1}}\Big)^{1/2}\\
&\le N\Big(\sum_{j_0\le j<j_0^2}\frac{1}{(1+\alpha)^{j+1}}\Big)^{1/2} \le N\Big(\frac{1}{\alpha(1+\alpha)^{j_0}}\Big)^{1/2} \le \alpha N,
\end{aligned}
$$

the last inequality being true for α small enough, see Exercise 17.9(b).

The hypothesis of the theorem (which we have not used until now) implies for large enough N (independently of the finitely many $p_1 \neq p_2$ occurring in the definition of II_j) that

$$
\Big|\sum_{m\le\frac{N}{(1+\alpha)^{j+1}}} a_{p_1 m}\overline{a_{p_2 m}}\Big| \le \frac{\alpha^2 N}{(1+\alpha)^{j+1}}.
$$

Therefore, by (17.9) and again by the Cauchy–Schwarz inequality we conclude

$$
\begin{aligned}
\sum_{j_0\le j<j_0^2} II_j &\le \alpha\sqrt{N}\sum_{j_0\le j<j_0^2}|Q_j|^{1/2}|P_j|(1+\alpha)^{-(j+1)/2}\\
&\le \alpha\sqrt{N}\Big(\sum_{j_0\le j<j_0^2}|P_j||Q_j|\Big)^{1/2}\Big(\sum_{j_0\le j<j_0^2}|P_j|(1+\alpha)^{-(j+1)}\Big)^{1/2}
\end{aligned}
$$

$$\leq \frac{2\alpha N}{1+\alpha} \sum_{j_0 \leq j < j_0^2} \frac{1}{j} \leq 2\alpha N \log(j_0^2) \leq 16\alpha N \log\left(\tfrac{1}{\alpha}\right)$$

also by the definition of j_0 and by the inequality $\log(x) \leq x$. Putting everything together gives

$$\left| \frac{1}{N} \sum_{n=1}^{N} a_n c(n) \right| \leq \alpha + 16\alpha \log\left(\tfrac{1}{\alpha}\right) + 22\alpha \log\left(\tfrac{1}{\alpha}\right) = \alpha + 38\alpha \log\left(\tfrac{1}{\alpha}\right) < \varepsilon$$

if α sufficiently small (and then N is sufficiently large). The proof is complete. ∎

Remark 17.19 In particular, using

$$\lim_{N\to\infty} \frac{1}{N} \sum_{n=1}^{N} \lambda^{pn}\overline{\lambda^{qn}} = \lim_{N\to\infty} \frac{1}{N} \sum_{n=1}^{N} (\lambda^{p-q})^n = 0$$

for every two distinct primes p, q and every irrational $\lambda \in \mathbb{T}$, the Kátai–Bourgain–Sarnak–Ziegler criterion provides an alternative way, without using Davenport's estimate (Theorem 17.2), to show (17.2) for such λ.

A polynomial generalization of this is the following.

Corollary 17.20 (Möbius Function is Orthogonal to Polynomial Sequences) *Let $P \in \mathbb{R}[\cdot]$ be a real polynomial. Then*[7]

$$\lim_{N\to\infty} \frac{1}{N} \sum_{n=1}^{N} \mu(n) e(P(n)) = 0.$$

We remark that polynomial sequences as above come from rotations on Heisenberg groups, see Example 14.13(c).

Proof We can assume $P(0) = 0$. If all coefficients of P are rational, then the sequence $(P(n))_{n\in\mathbb{N}}$ is periodic. Thus the assertion follows from Example 17.12.

Let now P given by $P(x) = a_1 x + \cdots + a_d x^d$ have at least one irrational coefficient and let p, q be two distinct primes. Then

$$\begin{aligned} \frac{1}{N} \sum_{n=1}^{N} e(P(pn))\overline{e(P(qn))} &= \frac{1}{N} \sum_{n=1}^{N} e(P(pn) - P(qn)) \\ &= \frac{1}{N} \sum_{n=1}^{N} e(a_1(p-q)n + \cdots + a_d(p^d - q^d)n^d) \end{aligned}$$

[7] Recall the notation $e(\alpha) := \mathrm{e}^{2\pi i\alpha}$.

converges to zero as $N \to \infty$ by Weyl's equidistribution theorem for polynomials (Theorem 10.17) and Weyl's criterion (Theorem 10.3). The Kátai–Bourgain–Sarnak–Ziegler criterion finishes the proof. ■

Example 17.21 (Skew Shifts) It is now easy to deduce from Corollary 17.20 and the approximation argument given in Exercise 17.6 that the skew rotation system $(\mathbb{T}^2, T_a)$ (which is deterministic as we mentioned in Example 17.7) given by

$$T_a : (x, y) \mapsto (ax, xy)$$

for a given $a \in \mathbb{T}$ satisfies the assertion of Sarnak's conjecture, see Exercise 17.10.

Example 17.22 (Nilsequences and Nilsystems) Green and Tao [488] showed quantitatively that the Möbius function is orthogonal to nilsequences, i.e.,

$$\lim_{N\to\infty} \frac{1}{N} \sum_{n=1}^{N} \mu(n) a_n = 0$$

holds for every[8] nilsequence (a_n) in $\mathbb{C}$. This generalizes Corollary 17.20 by Example 14.13(c) (see also Remark 14.14) and Example 17.21 by Example 14.9 and shows that Sarnak's conjecture holds for nilsystems. (Recall that by Example 17.7(d), transitive nilsystems are deterministic, and we can reduce to the transitive case by Theorem 14.35, see the discussion in Remark 17.10.)

For a quantitative estimate of convergence to zero for polynomials and nilsequences in short intervals see Matomäki et al. [778].

17.4 What Else Is Known?

In this last section we briefly indicate some of the fascinating recent developments in the direction of Sarnak's conjecture.

More Examples

Sarnak's conjecture has been verified for many (classes of) dynamical systems, often using the Kátai–Bourgain–Sarnak–Ziegler criterion, see, e.g.,[9] Green and Tao [488], Forni and Kanigowski [401] (nilsystems), Bourgain et al. [188] (horocycle flows), Bourgain [185], El Abdalaoui et al. [359], Etedadialiabadi and Gao [372] (certain

[8] They showed it for Lipschitz nilsequences coming from connected, simply connected groups, the general case for such groups easily following by approximation. Since by Wooley and Ziegler [1049, text between Cor. 3.14 and Prop. 3.15] every nilsequence is a finite sum of nilsequences coming from connected, simply connected groups, the general case follows.

[9] This list is far from being complete.

rank-one systems), Liu and Sarnak [752, 753] (certain distal flows, see also Fan [373]), Kułaga-Przymus and Lemańczyk [691, 692], Wang [1035], Huang et al. [561], Huang et al. [557], He and Wang [524] (certain extensions of rotations), Bourgain [185], Ferenczi and Mauduit [388], Chaika and Eskin [240], Karagulyan [628], Kanigowski et al. [623] (certain interval exchange transformations), Karagulyan [626] (continuous interval maps, orientation-preserving circle homeomorphisms, see also Fan and Jiang [377]), Ferenczi et al. [387] (bijective substitutions, certain regular Morse sequences and Rudin-Shapiro type systems, see also Tao [981] and Mauduit and Rivat [780]), El Abdalaoui et al. [356], Downarowicz and Kasjan [317] (regular Toeplitz systems), el Abdalaoui et al. [360], Huang et al. [562] (uniquely ergodic systems with discrete or irrational quasi-discrete spectrum), Veech [1019] (certain generalized Morse–Kakutani flows), Fan and Jiang [377] (equicontinuous flows among others), Huang et al. [561] (systems with invariant measures having discrete spectrum), Kanigowski et al. [625] (some parabolic flows), Kanigowski et al. [623] (systems with certain rigidity properties), el Abdalaoui and Devianne [355] (certain dendrite maps). Huang et al. [561] also showed that Sarnak's conjecture holds for topological systems all whose invariant measures have subpolynomial measure complexity. See also and the surveys by Ferenczi et al. [386] and Kułaga-Przymus and Lemańczyk [693] for further references.

For orthogonality of the Möbius function to automatic and q-semimultiplicative sequences see Müllner [811], Lemańczyk and Müllner [723] and Konieczny [657], respectively.

There are examples of topological systems with *arbitrarily small positive entropy* for which (17.4) fails, see Karagulyan [627]. On the other hand, there are also examples of topological systems with *arbitrarily large entropy* for which (17.4) holds (even uniformly), see Downarowicz and Serafin [318, 319]. Moreover, for systems of zero entropy convergence to zero in (17.4) can be arbitrarily slow, see Algom and Wang [30].

Polynomial Version of Sarnak Conjecture

As observed in [339], analogously to the presented proof of Theorem 17.3 (using a nilsequence version of Davenport's estimate for the Möbius function due to Green and Tao [488], Remark 12.19 and Bourgain's maximal inequality for polynomial averages [183]) one shows that μ is a pointwise good weight for polynomial ergodic averages in L^p with limit zero for every $p > 1$, i.e., for every invertible measure-preserving system (X, μ, T), every $f \in \mathrm{L}^p(X, \mu)$ and every $P \in \mathbb{Z}[\cdot]$

$$\lim_{N\to\infty} \frac{1}{N} \sum_{n=1}^{N} \mu(n) f(T^{P(n)}x) = 0 \quad \text{for almost every } x \in X.$$

However, a polynomial version of Sarnak's conjecture for minimal systems (as formulated in [339, Conjecture 2.3]) is false, see Lian and Shi [738] and Kanigowski et al. [624]. (Note that the minimal system in the counterexample in the last paper is uniquely ergodic.)

Logarithmic Sarnak's and Chowla's Conjecture

There has been much progress on a (formally) weaker, namely logarithmic, version of Sarnak's conjecture. The Nth **logarithmic average** of a sequence $(a_n)_{n\in\mathbb{N}}$ in $\mathbb{C}$ is defined as

$$\frac{1}{\log N}\sum_{n=1}^{N}\frac{a_n}{n}.$$

Note that convergence of logarithmic averages is equivalent to convergence of

$$\frac{1}{\sum_{n=1}^{N}\frac{1}{n}}\sum_{n=1}^{N}\frac{a_n}{n},$$

being just a modification of the Cesàro averages $\frac{1}{\sum_{n=1}^{N}1}\sum_{n=1}^{N}a_n$, giving a_n with small n more weight than with large n. As Exercise 17.11 shows, convergence of logarithmic averages is in general weaker than Cesàro convergence.

The following surprising results were shown for the logarithmic version of Sarnak's conjecture (where the Cesàro averages are replaced by the logarithmic ones).

- Logarithmic versions of Sarnak's and Chowla's conjectures are equivalent, see Tao [983]. Based on this result, Gomilko et al. [474] showed that the logarithmic Sarnak conjecture implies the (classical) Chowla conjecture along a subsequence. Tao [984] showed that one can even choose such a subsequence to have logarithmic density[10] 1.
- The logarithmic Chowla conjecture is true for $m = 2$ and for all odd m, see Tao [982] and Tao and Teräväinen [987, 989], respectively. For more results on the speed of convergence in the logarithmic Chowla conjecture for $m = 2$ see Helfgott and Radziwiłł [526], Helfgott [525] and Pilatte [849]. Moreover, Tao, Teräväinen showed in [988, Cor. 1.14] that the classical Chowla conjecture holds along a sequence with logarithmic density 1 for $m = 2$ and all odd m.

[10] The **logarithmic density** of a set $E\subset\mathbb{N}$ is defined by $d^{\log}(E) := \lim_{N\to\infty}\frac{1}{\log N}\sum_{n=1}^{N}\frac{\mathbb{1}_E(n)}{n}$ if it exists, with the natural counterpart for subsequences of $\mathbb{N}$.

The logarithmic Sarnak conjecture holds for deterministic invertible systems with at most countably many ergodic invariant measures (in particular, for uniquely ergodic deterministic invertible systems), see Frantzikinakis and Host [415] as well as [417] for uniform convergence. Moreover, Gomilko et al. [475] showed that the classical Sarnak conjecture holds for such systems along a sequence of logarithmic density 1 uniformly in x.

In his work on the equivalence of the logarithmically averaged Chowla's and Sarnak's conjectures, Tao presents a certain (logarithmically averaged uniform) version of Sarnak's conjecture for nilsystems which is equivalent to the logarithmic Sarnak's conjecture, see Tao [983, Rem. 1.9], cf. Huang et al. [563, Conj. 2.1]. Further, Huang et al. [563] proved the logarithmic Sarnak conjecture for systems with sublinear mean complexity and showed that the logarithmic Sarnak conjecture holds if and only if it holds for all systems with (sub)polynomial mean complexity.

McNamara [786] verified the logarithmic Sarnak conjecture for the Liouville function for subshifts with subquadratic block growth, implying that the Liouville function does not have subquadratically many sign patterns. The linear case and the fact that the Liouville function has super-linear word growth were done in Frantzikinakis and Host [415].

Uniformity in Sarnak's Conjecture and the Strong MOMO Property

It is natural to ask for uniform convergence in Sarnak's conjecture (which is a seemingly stronger property). It occurs that uniform convergence holds automatically if Sarnak's conjecture is true, see el Abdalaoui et al. [358, Cor. 10]. To show this, the authors introduced the so-called **strong MOMO**[11] **property** for topological dynamical systems (K, T) being a variation of Sarnak's Möbius orthogonality involving averages of different orbits over long blocks of time, namely for every $0 = b_0 < b_1 < b_2 < \ldots$ with $\lim_{k\to\infty}(b_{k+1} - b_k) = \infty$, every $f \in \mathrm{C}(K)$ and every sequence (x_k) in K

$$\lim_{M\to\infty} \frac{1}{b_M} \sum_{k=0}^{M-1} \left| \sum_{n=b_k}^{b_{k+1}-1} \mu(n) f(T^{n-b_k} x_k) \right| = 0.$$

They showed in particular that the strong MOMO property for all entropy zero systems is equivalent to Sarnak's conjecture. They moreover proved that the strong MOMO property for one uniquely ergodic model of an ergodic measure-preserving system implies this property as well as Möbius orthogonality for all uniquely ergodic models of the system. Gomilko et al. [475] gave an equivalent uniform

[11] MOMO stands for Möbius orthogonality on moving orbits.

representation of the strong MOMO property and presented a weaker, namely logarithmic strong MOMO property which implies uniform Möbius orthogonality on a sequence of full logarithmic density.

Ergodic Theoretic Reformulation of Chowla's and Sarnak's Conjectures

Inspired by Furstenberg's construction in his correspondence principle (see the proof of Theorem 9.21), one associates to μ a measure-preserving system (K, ν, T), where $K = \{-1, 0, 1\}^{\mathbb{Z}}$, T is the left shift on K and ν is a weak*-accumulation point of the measures $\nu_N := \frac{1}{N}\sum_{n=1}^{N} \delta_{T^n\mu}$. Here, we consider μ as an element of X by extending it to $\mathbb{Z}$ by $\mu(n) := 0$ for $n \leq 0$. Note that the measure ν might not be unique, and each such system is called a **Furstenberg system associated to** μ.

Consider now the function $F \in \mathrm{C}(K)$ being the projection onto the 0th coordinate, i.e., defined by $F((x_n)_{n\in\mathbb{Z}}) := x_0$. Note that F is T-generating in the sense that the T-invariant subalgebra generated by F and the constant one function is dense in $\mathrm{C}(K)$. Then we have

$$\mu^{k_1}(n+a_1)\cdot \ldots \cdot \mu^{k_m}(n+a_m) = (T^{a_1}F^{k_1}\cdot \ldots \cdot T^{a_m}F^{k_m})(T^n\mu)$$

implying

$$\lim_{N\to\infty}\frac{1}{N}\sum_{n=1}^{N}\mu^{k_1}(n+a_1)\cdot \ldots \cdot \mu^{k_m}(n+a_m) = \int_X T^{a_1}F^{k_1}\cdot \ldots \cdot T^{a_m}F^{k_m}\,\mathrm{d}\nu,$$

whenever the limit on the left-hand side exists. Moreover, since F is T-generating, Chowla's conjecture implies that ν (and hence the Furstenberg system) is unique.

Frantzikinakis [411] showed that the uniqueness and ergodicity of ν is equivalent to Chowla's conjecture, and hence implies Sarnak's conjecture. Moreover, Frantzikinakis and Host [415] proved that every Furstenberg system associated to μ is a factor of a system which has no irrational spectrum and whose ergodic components are isomorphic to direct products of Bernoulli systems and infinite-step nilsystems. Note that El Abdalaoui et al. [357, Lem. 4.6, Rem. 4.8], following the investigations of Sarnak [909], showed that Chowla's conjecture[12] is equivalent to the fact that the (unique) measure ν is the relatively independent extension of the Mirsky measure on the square-free system induced by μ^2. For multiplicative functions with such property of the Furstenberg system see Klurman et al. [653]. For a study of Furstenberg's systems of general bounded multiplicative functions see Frantzikinakis et al. [427].

In his unpublished lecture notes, Veech [1018] presented an internal condition on the Möbius function involving only factors of zero entropy of its Furstenberg's

[12] for the Möbius function (the case of the Liouville function is treated in [357, Prop. 4.9]).

systems. He showed that this condition (also called[13] Veech property) implies Sarnak's conjecture and conjectured the equivalence. This equivalence was shown by Kanigowski et al. [622]. They also showed that the logarithmic Sarnak's conjecture holds if and only if for each $s \in \mathbb{N}$ the Möbius function is orthogonal (in the sense of (17.4)) to systems whose all ergodic measures yield (systems with almost all ergodic components being) nilsystems of step s.

For details and more on the ergodic theoretic approach to Chowla's and Sarnak's conjecture see the above references as well as, e.g., El Abdalaoui et al. [358], Kanigowski et al. [622] and the recent chapter by Kułaga-Przymus and Lemańczyk [693].

Averaged Chowla's Conjecture and More

Using the Matomäki–Radziwiłł theory [776] on the behavior of multiplicative functions on short intervals and ideas from Bourgain et al. [188] discussed above, Matomäki et al. [777] proved a quantified version of the following result for every $m \in \mathbb{N}$ known as the **averaged version of Chowla's conjecture**:

$$\frac{1}{H^m} \sum_{a_1,\dots,a_m=1}^{H} \left| \frac{1}{N} \sum_{n=1}^{N} \mu(n+a_1)\mu(n+a_2)\cdots\mu(n+a_m) \right| \to 0$$

as $N \to \infty$ whenever $H = H(N) \leq N$ goes to infinity as $N \to \infty$. They also presented a class of aperiodic[14] multiplicative functions $u : \mathbb{N} \to \mathbb{T} \cup \{0\}$ for which the averages

$$\frac{1}{N} \sum_{n=1}^{N} u(n)\overline{u(n+1)}$$

do not converge to zero. Furstenberg's systems of such multiplicative functions were studied in Gomilko et al. [476] and Frantzikinakis et al. [427]. In the first paper, the authors also show that functions from this class satisfy Chowla's conjecture along a subsequence.

Exercises

Exercise 17.1 (ε-Packing Number) Let (K, d) be a compact metric space and let $\varepsilon > 0$. Show that the number of balls in ε-packings of K is bounded by above, i.e., that the ε-packing number of K is well defined.

---

[13] see, e.g., Kanigowski et al. [622].

[14] see Frantzikinakis et al. [427].

Exercise 17.2 (Topological Entropy) Prove Proposition 17.4.

Exercise 17.3 (Equicontinuous Systems) Prove the equivalence of the formulations of equicontinuity of topological systems given after Proposition 17.5. Show moreover the following property. Let (K, T) be an equicontinuous topological system and let d be a metric inducing the topology on K. Prove that

$$\rho(x, y) := \sup_{n\in\mathbb{N}_0} d(T^n x, T^n y) \quad (x, y \in K)$$

defines a metric on K which is (uniformly) equivalent to d, and which makes T a non-expansive transformation. (For invertible transformations this was Exercise 11.22.)

Exercise 17.4 (Contractive Systems) Let $T : K \to K$ be a **contraction** on a compact metric space (K, d), i.e., $d(Tx, Ty) < d(x, y)$ for each pair $x, y \in K$ with $x \neq y$. In this case the topological system (K, T) is called **contractive**. Prove the following assertions.

(a) The mapping $D : K \to \mathbb{R}$ defined as $D(x) := d(x, Tx)$ is continuous.
(b) Any z minimizing D is a fixed point of T. There is a unique fixed point.
(c) $(T^n x)$ converges to the unique fixed point of T for any $x \in K$.
(d) Equation (17.4) holds for every $f \in \mathrm{C}(K)$ and $x \in K$.

Exercise 17.5 (Deterministic Sequences) Give an example of a non-periodic, deterministic 0-1-sequence.

Exercise 17.6 (Approximation Argument for Topological Systems) Let $(a_n)_{n\in\mathbb{N}}$ be a bounded sequence in $\mathbb{C}$. Let further (K, T) be a topological system with K metrizable, and take $M \subset \mathrm{C}(T)$ with $\overline{\mathrm{lin}}(M) = \mathrm{C}(K)$. Show the following assertion: If for every $f \in M$

$$\lim_{N\to\infty} \frac{1}{N} \sum_{n=1}^{N} a(n) f(T^n x) = 0 \quad \text{for every } x \in K,$$

then this holds for every $f \in \mathrm{C}(K)$.

Exercise 17.7 (Periodic Systems) Show that every periodic sequence is a finite linear combination of sequences of the form $(\lambda^n)_{n\in\mathbb{N}}$, $\lambda \in \mathbb{T}$ rational. Deduce that periodic systems satisfy Sarnak's conjecture. (Why are they deterministic?)

Exercise 17.8 (Rotations on $\mathbb{T}$) Show that Sarnak's conjecture holds for rotations on the circle. (Hint: Use (17.1) and Exercise 17.6.)

Exercise 17.9 (Proof of the Kátai–Bourgain–Sarnak–Ziegler Criterion)

(a) Prove the inequality (17.7). To do this, first use the Chinese remainder theorem to estimate the left-hand side of (17.7) from above by

$$\prod_{p\in[D_0,D_1)}\left(1-\frac{1}{p}\right)$$

and then use the following consequence of the prime number theorem, a result obtained by Mertens in 1874, see, e.g., Jameson [574, Thm. 2.6.4]: There are constants A and C such that

$$\left|\sum_{p\le x}\frac{1}{p}-\log\log(x)-A\right|\le\frac{C}{\log(x)}\quad\text{for every } x\ge 2.$$

(b) Show that for $j_0:=\frac{1}{\alpha}\left(\log\frac{1}{\alpha}\right)^3$ one has $\frac{1}{(1+\alpha)^{j_0}}\le\alpha^3$ whenever α is small enough.

Exercise 17.10 (Skew Shifts) Prove the assertion of Example 17.21.

Exercise 17.11 (Logarithmic Averages) Let $(a_n)_{n\in\mathbb{N}}$ be a bounded sequence in $\mathbb{C}$ and $a\in\mathbb{C}$. Show that the following implication holds:

$$\lim_{N\to\infty}\frac{1}{N}\sum_{n=1}^{N}a_n=a\quad\Longrightarrow\quad\lim_{N\to\infty}\frac{1}{\log N}\sum_{n=1}^{N}\frac{a_n}{n}=a.$$

Hint: Show and use the summation by parts formula

$$\sum_{n=1}^{N}\frac{a_n}{n}=\sum_{M=1}^{N-1}\frac{1}{M(M+1)}\left(\sum_{n=1}^{M}a_n\right)+\frac{1}{N}\sum_{n=1}^{N}a_n.$$

References

1. J. Aaronson, On the ergodic theory of non-integrable functions and infinite measure spaces. Isr. J. Math. **27**(2), 163–173 (1977)
2. J. Aaronson, Rational ergodicity, bounded rational ergodicity and some continuous measures on the circle. Isr. J. Math. **33**(3–4), 181–197 (1979/1980). A collection of invited papers on ergodic theory
3. J. Aaronson, An ergodic theorem with large normalising constants. Isr. J. Math. **38**(3), 182–188 (1981)
4. J. Aaronson, *An Introduction to Infinite Ergodic Theory*. Mathematical Surveys and Monographs, vol. 50 (American Mathematical Society, Providence, RI, 1997)
5. J. Aaronson, M. Hosseini, M. Lemańczyk, IP-rigidity and eigenvalue groups. Ergod. Theory Dyn. Syst. **34**(4), 1057–1076 (2014)
6. J. Aaronson, M. Lin, B. Weiss, Mixing properties of Markov operators and ergodic transformations, and ergodicity of Cartesian products. Isr. J. Math. **33**(3–4), 198–224 (1979/1980). A collection of invited papers on ergodic theory
7. J. Aaronson, M. Lin, B. Weiss, Correction to: "Mixing properties of Markov operators and ergodic transformations, and ergodicity of Cartesian products". Isr. J. Math. **37**(4), 364 (1980)
8. L.M. Abramov, Metric automorphisms with quasi-discrete spectrum. Izv. Akad. Nauk SSSR Ser. Mat. **26**, 513–530 (1962)
9. E. Ackelsberg, V. Bergelson, A. Best,Multiple recurrence and large intersections for abelian group actions. Discrete Anal., pages Paper No. 18, 99 pp. (2021)
10. E.M. Ackelsberg, Rigidity, weak mixing, and recurrence in abelian groups. Discrete Contin. Dyn. Syst. **42**(4), 1669–1705 (2022)
11. T. Adams, Tower multiplexing and slow weak mixing. Colloq. Math. **138**(1), 47–72 (2015)
12. T. Adams, V. Bergelson, W. Sun, Under- and over-independence in measure preserving systems. Isr. J. Math. **235**(1), 349–384 (2020)
13. O.N. Ageev, On the genericity of some nonasymptotic dynamic properties. Uspekhi Mat. Nauk **58**(1(349)), 177–178 (2003)
14. M. Ajtai, I. Havas, J. Komlós, Every group admits a bad topology, in *Studies in Pure Mathematics* (Birkhäuser, Basel, 1983), pp. 21–34
15. M.A. Akcoglu, An ergodic lemma. Proc. Am. Math. Soc. **16**, 388–392 (1965)
16. M.A. Akcoglu, Pointwise ergodic theorems. Trans. Am. Math. Soc. **125**, 296–309 (1966)
17. M.A. Akcoglu, A pointwise ergodic theorem in L_p-spaces. Can. J. Math. **27**(5), 1075–1082 (1975)

T. Eisner, B. Farkas, *A Journey Through Ergodic Theorems*, Birkhäuser Advanced Texts Basler Lehrbücher, https://doi.org/10.1007/978-3-031-96506-7

18. M.A. Akcoglu, A. Bellow, R.L. Jones, V. Losert, K. Reinhold-Larsson, M. Wierdl, The strong sweeping out property for lacunary sequences, Riemann sums, convolution powers, and related matters. Ergod. Theory Dyn. Syst. **16**(2), 207–253 (1996)
19. M.A. Akcoglu, A. Brunel, Contractions on L_1-spaces. Trans. Am. Math. Soc. **155**, 315–325 (1971)
20. M.A. Akcoglu, R.V. Chacon, Ergodic properties of operators in Lebesgue space, in *Advances in Probability and Related Topics*, ed. by P. Ney, vol. 2 (Marcel Dekker, New York, 1970), pp. 1–47
21. M.A. Akcoglu, A. del Junco, Convergence of averages of point transformations. Proc. Am. Math. Soc. **49**, 265–266 (1975)
22. M.A. Akcoglu, M.D. Ha, R.L. Jones, Divergence of ergodic averages, in *Topological Vector Spaces, Algebras and Related Areas (Hamilton, ON, 1994)*. Volume 316 of Pitman Research Notes in Mathematics Series (Longman Scientific & Technical, Harlow, 1994), pp. 175–192
23. M.A. Akcoglu, R.L. Jones, Strong sweeping out for block sequences and related ergodic averages, in *Proceedings of the Conference on Probability, Ergodic Theory, and Analysis (Evanston, IL, 1997)*, vol. 43, pp. 447–456 (1999)
24. M.A. Akcoglu, R.L. Jones, J.M. Rosenblatt, The worst sums in ergodic theory. Mich. Math. J. **47**(2), 265–285 (2000)
25. M.A. Akcoglu, U. Krengel, Ergodic theorems for superadditive processes. J. Reine Angew. Math. **323**, 53–67 (1981)
26. M.A. Akcoglu, L. Sucheston, On operator convergence in Hilbert space and in Lebesgue space. Period. Math. Hungar. **2**, 235–244 (1972)
27. M.A. Akcoglu, L. Sucheston, Remarks on dilations in L_p-spaces. Proc. Am. Math. Soc. **53**(1), 80–82 (1975)
28. M.A. Akcoglu, L. Sucheston, A ratio ergodic theorem for superadditive processes. Z. Wahrsch. Verw. Gebiete **44**(4), 269–278 (1978)
29. M.A. Akcoglu, L. Sucheston, A superadditive version of Brunel's maximal ergodic lemma, in *Measure Theory, Oberwolfach 1981 (Oberwolfach, 1981)*. Volume 945 of Lecture Notes in Mathematics (Springer, Berlin/New York, 1982), pp. 347–351
30. A. Algom, Z. Wang, Arbitrarily slow decay in the Möbius disjointness conjecture. Ergod. Theory Dyn. Syst. **43**(9), 2863–2880 (2023)
31. D. Altman, On Szemerédi's theorem with differences from a random set. Acta Arith. **195**(1), 97–108 (2020)
32. D.V. Anosov, The additive functional homology equation that is connected with an ergodic rotation of the circles. Izv. Akad. Nauk SSSR Ser. Mat. **37**, 1259–1274 (1973)
33. O. Antolín Camarena, B. Szegedy, Nilspaces, nilmanifolds and their morphisms (2010). Preprint, https://arxiv.org/abs/1009.3825
34. R. Arens, Topologies for homeomorphism groups. Am. J. Math. **68**, 593–610 (1946)
35. L. Arnold, *Random Dynamical Systems*. Springer Monographs in Mathematics (Springer, Berlin, 1998)
36. I. Assani, On the punctual and local ergodic theorem for nonpositive power bounded operators in $L^p_{\mathbb{C}}[0, 1]$, $1 < p < +\infty$. Proc. Am. Math. Soc. **96**(2), 306–310 (1986)
37. I. Assani, Sur les opérateurs à puissances bornées et le théorème ergodique ponctuel dans $L^p[0, 1]$, $1 < p < +\infty$. Can. J. Math. **38**(4), 937–946 (1986)
38. I. Assani, Universal weights from dynamical systems to mean-bounded positive operators on L^p, in *Almost Everywhere Convergence, II (Evanston, IL, 1989)* (Academic, Boston, MA, 1991), pp. 9–16
39. I. Assani, The return times and the Wiener-Wintner property for mean-bounded positive operators in L^p. Ergod. Theory Dyn. Syst. **12**(1), 1–12 (1992)
40. I. Assani, Multiple recurrence and almost sure convergence for weakly mixing dynamical systems. Isr. J. Math. **103**, 111–124 (1998)
41. I. Assani, A weighted pointwise ergodic theorem. Ann. Inst. H. Poincaré Probab. Stat. **34**(1), 139–150 (1998)

42. I. Assani, Multiple return times theorems for weakly mixing systems. Ann. Inst. H. Poincaré Probab. Stat. **36**(2), 153–165 (2000)
43. I. Assani, Wiener-Wintner dynamical systems. Ergod. Theory Dyn. Syst. **23**(6), 1637–1654 (2003)
44. I. Assani, *Wiener Wintner Ergodic Theorems* (World Scientific Publishing Co., Inc., River Edge, NJ, 2003)
45. I. Assani, Pointwise convergence of ergodic averages along cubes. J. Anal. Math. **110**, 241–269 (2010)
46. I. Assani, Pointwise double recurrence and nilsequences (2015). Preprint, https://arxiv.org/abs/1504.05732
47. I. Assani, Wiener-Wintner ergodic theorem, in brief. Not. Am. Math. Soc. **69**(2), 198–209 (2022)
48. I. Assani, Z. Buczolich, R.D. Mauldin, An L^1 counting problem in ergodic theory. J. Anal. Math. **95**, 221–241 (2005)
49. I. Assani, D. Duncan, R. Moore. Pointwise characteristic factors for Wiener-Wintner double recurrence theorem. Ergod. Theory Dyn. Syst. **36**(4), 1037–1066 (2016)
50. I. Assani, J. Folks, R. Moore, Higher order Wiener–Wintner systems: Examples and applications (2024). Preprint, https://arxiv.org/abs/2402.17094
51. I. Assani, E. Lesigne, D. Rudolph, Wiener-Wintner return-times ergodic theorem. Isr. J. Math. **92**(1–3), 375–395 (1995)
52. I. Assani, R. Moore, A good universal weight for nonconventional ergodic averages in norm. Ergod. Theory Dyn. Syst. **37**(4), 1009–1025 (2017)
53. I. Assani, R. Moore, Extension of Wiener-Wintner double recurrence theorem to polynomials. J. Anal. Math. **134**(2), 597–613 (2018)
54. I. Assani, K. Nicolaou, Properties of Wiener-Wintner dynamical systems. Bull. Soc. Math. Fr. **129**(3), 361–377 (2001)
55. I. Assani, K. Presser, Pointwise characteristic factors for the multiterm return times theorem. Erg Theory Dyn. Syst. **32**(2), 341–360 (2012)
56. I. Assani, K. Presser, A survey of the return times theorem, in *Ergodic Theory and Dynamical Systems*. De Gruyter Proceedings in Mathematics (De Gruyter, Berlin, 2014), pp. 19–58
57. A. Auffinger, M. Damron, J. Hanson, Rate of convergence of the mean for sub-additive ergodic sequences. Adv. Math. **285**, 138–181 (2015)
58. J. Auslander, On the proximal relation in topological dynamics. Proc. Am. Math. Soc. **11**, 890–895 (1960)
59. L. Auslander, L. Green, F. Hahn, *Flows on Homogeneous Spaces*. With the assistance of L. Markus and W. Massey, and an appendix by L. Greenberg. Annals of Mathematics Studies, vol. 53 (Princeton University Press, Princeton, NJ, 1963)
60. T. Austin, On the norm convergence of non-conventional ergodic averages. Ergod Theory Dyn. Syst. **30**(2), 321–338 (2010)
61. T. Austin, Deducing the density Hales-Jewett theorem from an infinitary removal lemma. J. Theor. Probab. **24**(3), 615–633 (2011)
62. T. Austin, Pleasant extensions retaining algebraic structure, I. J. Anal. Math. **125**, 1–36 (2015)
63. T. Austin, Pleasant extensions retaining algebraic structure, II. J. Anal. Math. **126**, 1–111 (2015)
64. T. Austin, A proof of Walsh's convergence theorem using couplings. Int. Math. Res. Not. IMRN **2015**(15), 6661–6674 (2015)
65. T. Austin, Non-conventional ergodic averages for several commuting actions of an amenable group. J. Anal. Math. **130**, 243–274 (2016)
66. T. Austin, Non-convergence of some non-commuting double ergodic averages. Proc. Am. Math. Soc. **153**(4), 1701–1707 (2025)
67. T. Austin, T. Eisner, T. Tao, Nonconventional ergodic averages and multiple recurrence for von Neumann dynamical systems. Pac. J. Math. **250**(1), 1–60 (2011)

68. J. Avigad, P. Gerhardy, H. Towsner, Local stability of ergodic averages. Trans. Am. Math. Soc. **362**(1), 261–288 (2010)
69. J. Avigad, H. Towsner, Metastability in the Furstenberg-Zimmer tower. Fund. Math. **210**(3), 243–268 (2010)
70. D. Axmann, Struktur- und Ergodentheorie irreduzibler Operatoren auf Banachverbänden. PhD thesis, University of Tübingen, 1980
71. C. Badea, S. Grivaux, Kazhdan constants, continuous probability measures with large Fourier coefficients and rigidity sequences. Comment. Math. Helv. **95**(1), 99–127 (2020)
72. C. Badea, S. Grivaux, E. Matheron, Rigidity sequences, Kazhdan sets and group topologies on the integers. J. Anal. Math. **143**(1), 313–347 (2021)
73. S. Banach, Sur la valeur moyenne des fonctions orthogonales. Bulletin international de L'Académie Polonaise des Sciences et des Lettres, Classe des Sciences Mathématiques et Naturalles. Série A: Sciences Mathématiques, 67–72 (1919). Reprinted in S. Banach, Oeuvres, vol. 1, Warsaw, 1967, pp. 40–46; PDF file at http://matwbn.icm.edu.pl/ksiazki/or/or1/or116.pdf
74. S. Banach, *Théorie des opérations linéaires* (Warsaw, 1932)
75. L. Barreira, *Lyapunov Exponents* (Birkhäuser/Springer, Cham, 2017)
76. L. Barreira, Y. Pesin, *Nonuniform Hyperbolicity*. Volume 115 of Encyclopedia of Mathematics and Its Applications (Cambridge University Press, Cambridge, 2007). Dynamics of systems with nonzero Lyapunov exponents
77. L. Barreira, Y. Pesin, *Introduction to Smooth Ergodic Theory*. Volume 148 of Graduate Studies in Mathematics (American Mathematical Society, Providence, RI, 2013)
78. L. Barreira, Y.B. Pesin, *Lyapunov Exponents and Smooth Ergodic Theory*. Volume 23 of University Lecture Series (American Mathematical Society, Providence, RI, 2002)
79. M. Barthmann, S. Farhangi, Uniform vector-valued pointwise ergodic theorems for operators. Discrete Contin. Dyn. Syst. (To appear)
80. W. Bartoszek, A. Śpiewak, A note on a Wiener-Wintner theorem for mean ergodic Markov amenable semigroups. Proc. Am. Math. Soc. **145**(7), 2997–3003 (2017)
81. A.I. Bashtanov, Generic mixing transformations are rank 1. Math. Notes **93**(1–2), 209–216 (2013). Translation of Mat. Zametki **93**(2), 163–171 (2013)
82. A. Bátkai, M. Kramar Fijavž, A. Rhandi, *Positive Operator Semigroups*. Volume 257 of Operator Theory: Advances and Applications (Birkhäuser/Springer, Cham, 2017). From Finite to Infinite Dimensions, With a foreword by Rainer Nagel and Ulf Schlotterbeck
83. J. Baxter, R. Jones, M. Lin, J. Olsen, SLLN for weighted independent identically distributed random variables. J. Theor. Probab. **17**(1), 165–181 (2004)
84. J.R. Baxter, J.H. Olsen, Weighted and subsequential ergodic theorems. Can. J. Math. **35**(1), 145–166 (1983)
85. F. Bayart, Z. Buczolich, Y. Heurteaux, Fast and slow points of Birkhoff sums. Ergod. Theory Dyn. Syst. **40**(12), 3236–3256 (2020)
86. V. Becher, S. Figueira, An example of a computable absolutely normal number. Theor. Comput. Sci. **270**(1–2), 947–958 (2002)
87. M.E. Becker, A condition equivalent to uniform ergodicity. Studia Math. **167**(3), 215–218 (2005)
88. F. Beleznay, M. Foreman, The complexity of the collection of measure-distal transformations. Ergod. Theory Dyn. Syst. **16**(5), 929–962 (1996)
89. A. Bellow, Sur la structure des suites "mauvaises universelles" en théorie ergodique. C. R. Acad. Sci. Paris Sér. I Math. **294**(1), 55–58 (1982)
90. A. Bellow, On "bad universal" sequences in ergodic theory. II, in *Measure Theory and Its Applications (Sherbrooke, QC, 1982)*. Volume 1033 of Lecture Notes in Mathematics (Springer, Berlin, 1983), pp. 74–78
91. A. Bellow, Perturbation of a sequence. Adv. Math. **78**(2), 131–139 (1989)
92. A. Bellow, H. Furstenberg, An application of number theory to ergodic theory and the construction of uniquely ergodic models. Isr. J. Math. **33**(3–4), 231–240 (1979/1980). A collection of invited papers on ergodic theory

93. A. Bellow, R.L. Jones, A Banach principle for L^∞. Adv. Math. **120**(1), 155–172 (1996)
94. A. Bellow, R.L. Jones, J. Rosenblatt, Almost everywhere convergence of powers, in *Almost Everywhere Convergence (Columbus, OH, 1988)* (Academic, Boston, MA, 1989), pp. 99–120
95. A. Bellow, R.L. Jones, J. Rosenblatt, Convergence for moving averages. Ergod. Theory Dyn. Syst. **10**(1), 43–62 (1990)
96. A. Bellow, V. Losert, On sequences of density zero in ergodic theory, in *Conference in Modern Analysis and Probability (New Haven, CT, 1982)*. Volume 26 of Contemporary Mathematics (American Mathematical Society, Providence, RI, 1984), pp. 49–60
97. A. Bellow, V. Losert, The weighted pointwise ergodic theorem and the individual ergodic theorem along subsequences. Trans. Am. Math. Soc. **288**(1), 307–345 (1985)
98. M.J. Beltrán-Meneu, M.C. Gómez-Collado, E. Jordá, D. Jornet, Mean ergodic composition operators on Banach spaces of holomorphic functions. J. Funct. Anal. **270**(12), 4369–4385 (2016)
99. D. Berend, M. Lin, J. Rosenblatt, A. Tempelman, Modulated and subsequential ergodic theorems in Hilbert and Banach spaces. Ergod. Theory Dyn. Syst. **22**(6), 1653–1665 (2002)
100. V. Bergelson, private communication
101. V. Bergelson, Sets of recurrence of $\mathbf{Z}^m$-actions and properties of sets of differences in $\mathbf{Z}^m$. J. Lond. Math. Soc. (2) **31**(2), 295–304 (1985)
102. V. Bergelson, Ergodic Ramsey theory, in *Logic and Combinatorics (Arcata, CA, 1985)*. Volume 65 of Contemporary Mathematics (American Mathematical Society, Providence, RI, 1987), pp. 63–87
103. V. Bergelson, Weakly mixing PET. Ergod. Theory Dyn. Syst. **7**(3), 337–349 (1987)
104. V. Bergelson, Ergodic Ramsey theory—an update, in *Ergodic Theory of $\mathbf{Z}^d$ Actions (Warwick, 1993–1994)*. Volume 228 of London Mathematical Society Lecture Note Series (Cambridge University Press, Cambridge, 1996), pp. 1–61
105. V. Bergelson, Ergodic theory and Diophantine problems, in *Topics in Symbolic Dynamics and Applications (Temuco, 1997)*. Volume 279 of London Mathematical Society Lecture Note Series (Cambridge University Press, Cambridge, 2000), pp. 167–205
106. V. Bergelson, The multifarious Poincaré recurrence theorem, in *Descriptive Set Theory and Dynamical Systems (Marseille-Luminy, 1996)*. Volume 277 of London Mathematical Society Lecture Note Series (Cambridge University Press, Cambridge, 2000), pp. 31–57
107. V. Bergelson, Minimal idempotents and ergodic Ramsey theory, in *Topics in Dynamics and Ergodic Theory*. Volume 310 of London Mathematical Society Lecture Note Series (Cambridge University Press, Cambridge, 2003), pp. 8–39
108. V. Bergelson, Some historical comments and modern questions around the Ergodic Theorem, in *Dynamics of Complex Systems* (Research Institute for Math. Sciences, Kyoto, 2004), pp. 1–11
109. V. Bergelson, Multiplicatively large sets and ergodic Ramsey theory. Probab. Math. **148**, 23–40 (2005)
110. V. Bergelson, Combinatorial and Diophantine applications of ergodic theory, in *Handbook of Dynamical Systems, Vol. 1B* (Elsevier B. V., Amsterdam, 2006), pp. 745–869. Appendix A by A. Leibman and Appendix B by Anthony Quas and Máté Wierdl
111. V. Bergelson, Ergodic Ramsey theory: A dynamical approach to static theorems, in *International Congress of Mathematicians, Vol. II* (European Mathematical Society, Zürich, 2006), pp. 1655–1678
112. V. Bergelson, Ultrafilters, IP sets, dynamics, and combinatorial number theory, in *Ultrafilters across Mathematics*. Volume 530 of Contemporary Mathematics (American Mathematical Society, Providence, RI, 2010), pp. 23–47
113. V. Bergelson, Equidistribution and related topics (2015). A course at the Ross Mathematics Program, lectures by Vitaly Bergelson, notes by Daniel Shapiro and Sohail Farhangi
114. V. Bergelson, M. Boshernitzan, J. Bourgain, Some results on nonlinear recurrence. J. Anal. Math. **62**, 29–46 (1994)

115. V. Bergelson, A. del Junco, M. Lemańczyk, J. Rosenblatt, Rigidity and non-recurrence along sequences. Ergod. Theory Dyn. Syst. **34**(5), 1464–1502 (2014)
116. V. Bergelson, T. Downarowicz, Large sets of integers and hierarchy of mixing properties of measure preserving systems. Colloq. Math. **110**(1), 117–150 (2008)
117. V. Bergelson, H. Furstenberg, R. McCutcheon, IP-sets and polynomial recurrence. Ergod. Theory Dyn. Syst. **16**(5), 963–974 (1996)
118. V. Bergelson, I.J. Håland, Sets of recurrence and generalized polynomials, in *Convergence in Ergodic Theory and Probability (Columbus, OH, 1993)*. Volume 5 of Ohio State University Mathematical Research Institute Publications (de Gruyter, Berlin, 1996), pp. 91–110
119. V. Bergelson, I.J. Håland Knutson, R. McCutcheon, IP-systems, generalized polynomials and recurrence. Ergod. Theory Dyn. Syst. **26**(4), 999–1019 (2006)
120. V. Bergelson, N. Hindman, Nonmetrizable topological dynamics and Ramsey theory. Trans. Am. Math. Soc. **320**(1), 293–320 (1990)
121. V. Bergelson, N. Hindman, Some topological semicommutative van der Waerden type theorems and their combinatorial consequences. J. Lond. Math. Soc. (2) **45**(3), 385–403 (1992)
122. V. Bergelson, N. Hindman, On IP* sets and central sets. Combinatorica **14**(3), 269–277 (1994)
123. V. Bergelson, B. Host, B. Kra, Multiple recurrence and nilsequences. Invent. Math. **160**(2), 261–303 (2005). With an appendix by Imre Ruzsa
124. V. Bergelson, B. Host, R. McCutcheon, F. Parreau, Aspects of uniformity in recurrence. Colloq. Math. **84/85**, 549–576 (2000). Dedicated to the memory of Anzelm Iwanik
125. V. Bergelson, G. Kolesnik, M. Madritsch, Y. Son, R. Tichy, Uniform distribution of prime powers and sets of recurrence and van der Corput sets in $\mathbb{Z}^k$. Isr. J. Math. **201**(2), 729–760 (2014)
126. V. Bergelson, G. Kolesnik, Y. Son, Uniform distribution of subpolynomial functions along primes and applications. J. Anal. Math. **137**(1), 135–187 (2019)
127. V. Bergelson, J. Kułaga-Przymus, M. Lemańczyk, and F.K. Richter, Rationally almost periodic sequences, polynomial multiple recurrence and symbolic dynamics. Ergod. Theory Dyn. Syst. **39**(9), 2332–2383 (2019)
128. V. Bergelson, J. Kułaga-Przymus, M. Lemańczyk, F.K. Richter, A structure theorem for level sets of multiplicative functions and applications. Int. Math. Res. Not. IMRN **2020**(5), 1300–1345 (2020)
129. V. Bergelson, A. Leibman, Polynomial extensions of van der Waerden's and Szemerédi's theorems. J. Am. Math. Soc. **9**(3), 725–753 (1996)
130. V. Bergelson, A. Leibman, Set-polynomials and polynomial extension of the Hales-Jewett theorem. Ann. Math. (2) **150**(1), 33–75 (1999)
131. V. Bergelson, A. Leibman, A nilpotent Roth theorem. Invent. Math. **147**(2), 429–470 (2002)
132. V. Bergelson, A. Leibman, Distribution of values of bounded generalized polynomials. Acta Math. **198**(2), 155–230 (2007)
133. V. Bergelson, A. Leibman, IP_r^*-recurrence and nilsystems. Adv. Math. **339**, 642–656 (2018)
134. V. Bergelson, A. Leibman, Sets of large values of correlation functions for polynomial cubic configurations. Ergod. Theory Dyn. Syst. **38**(2), 499–522 (2018)
135. V. Bergelson, A. Leibman, E. Lesigne, Intersective polynomials and the polynomial Szemerédi theorem. Adv. Math. **219**(1), 369–388 (2008)
136. V. Bergelson, A. Leibman, T. Ziegler, The shifted primes and the multidimensional Szemerédi and polynomial van der Waerden theorems. C. R. Math. Acad. Sci. Paris **349**(3–4), 123–125 (2011)
137. V. Bergelson, E. Lesigne, Van der Corput sets in $\mathbb{Z}^d$. Colloq. Math. **110**(1), 1–49 (2008)
138. V. Bergelson, R. McCutcheon, Uniformity in the polynomial Szemerédi theorem, in *Ergodic Theory of* $\mathbf{Z}^d$ *Actions (Warwick, 1993–1994)*. Volume 228 of London Mathematical Society Lecture Note Series (Cambridge University Press, Cambridge, 1996), pp. 273–296
139. V. Bergelson, R. McCutcheon, An ergodic IP polynomial Szemerédi theorem. Mem. Am. Math. Soc. **146**(695), viii+106 (2000)

140. V. Bergelson, R. McCutcheon, Central sets and a non-commutative Roth theorem. Am. J. Math. **129**(5), 1251–1275 (2007)
141. V. Bergelson, R. McCutcheon, Q. Zhang, A Roth theorem for amenable groups. Am. J. Math. **119**(6), 1173–1211 (1997)
142. V. Bergelson, J. Moreira, Van der Corput's difference theorem: Some modern developments. Indag. Math. (N.S.) **27**(2), 437–479 (2016)
143. V. Bergelson, J. Moreira, Ergodic theorem involving additive and multiplicative groups of a field and $\{x + y, xy\}$ patterns. Ergod. Theory Dyn. Syst. **37**(3), 673–692 (2017)
144. V. Bergelson, J. Moreira, F.K. Richter, Single and multiple recurrence along non-polynomial sequences. Adv. Math. **368**, 107146, 69 (2020)
145. V. Bergelson, F.K. Richter, Dynamical generalizations of the prime number theorem and disjointness of additive and multiplicative semigroup actions. Duke Math. J. **171**(15), 3133–3200 (2022)
146. V. Bergelson, I. Ruzsa, Squarefree numbers, IP sets and ergodic theory, in *Paul Erdős and His Mathematics, I (Budapest, 1999)*. Volume 11 of Bolyai Society Mathematical Studies (János Bolyai Society Mathematical, Budapest, 2002), pp. 147–160
147. V. Bergelson, T. Tao, T. Ziegler, Multiple recurrence and convergence results associated to $\mathbb{F}_P^{\omega}$-actions. J. Anal. Math. **127**, 329–378 (2015)
148. V. Bergelson, R. Zelada, Strongly mixing systems are almost strongly mixing of all orders. Ergod. Theory Dyn. Syst. **44**(6), 1489–1530 (2024)
149. A. Berman, R. J. Plemmons, *Nonnegative Matrices in the Mathematical Sciences*. Volume 9 of Classics in Applied Mathematics (Society for Industrial and Applied Mathematics (SIAM), Philadelphia, PA, 1994). Revised reprint of the 1979 original
150. T. Bermúdez, A. Bonilla, V. Müller, A. Peris, Cesàro bounded operators in Banach spaces. J. Anal. Math. **140**(1), 187–206 (2020)
151. T. Bewley, Extension of the Birkhoff and von Neumann ergodic theorems to semigroup actions. Ann. Inst. H. Poincaré Sect. B (N.S.) **7**, 283–291 (1971)
152. P. Billingsley, *Ergodic Theory and Information* (Wiley, New York/London/Sydney, 1965)
153. P. Billingsley, *Probability and Measure*. Wiley Series in Probability and Mathematical Statistics, 3rd edn (Wiley, New York, 1995). A Wiley-Interscience Publication
154. G. Birkhoff, A note on topological groups. Compos. Math. **3**, 427–430 (1936)
155. G.D. Birkhoff, Quelques théorèmes sur le mouvement des systèmes dynamiques. S. M. F. Bull. **40**, 305–323 (1912)
156. G.D. Birkhoff, Proof of the ergodic theorem. Proc. Natl. Acad. Sci. USA **17**, 656–660 (1931)
157. G.D. Birkhoff, B.O. Koopman, Recent contributions to the ergodic theory. Proc. Nat. Acad. USA **18**(3), 279–282 (1932)
158. A. Biró, On L_∞ universally bad sequences in ergodic theory. Ill. J. Math. **41**(2), 309–314 (1997)
159. E. Bishop, A constructive ergodic theorem. J. Math. Mech. **17**, 631–639 (1967/1968)
160. J. Blum, B. Eisenberg, Generalized summing sequences and the mean ergodic theorem. Proc. Am. Math. Soc. **42**, 423–429 (1974)
161. J.R. Blum, S.L.M. Christiansen, D. Quiring, Sequence mixing and α-mixing. Ill. J. Math. **18**, 131–135 (1974)
162. J.R. Blum, R. Cogburn, On ergodic sequences of measures. Proc. Am. Math. Soc. **51**, 359–365 (1975)
163. J.R. Blum, D.L. Hanson, On the mean ergodic theorem for subsequences. Bull. Am. Math. Soc. **66**, 308–311 (1960)
164. J.R. Blum, D.L. Hanson, An elementary proof that automorphisms of Kolmogorov are mixing of all orders, in *Ergodic Theory (Proceedings of the International Symposium, Tulane University, New Orleans, LA, 1961)* (Academic, New York/London, 1963), pp. 71–73
165. J.R. Blum, J.I. Reich, Strongly ergodic sequences of integers and the individual ergodic theorem. Proc. Am. Math. Soc. **86**(4), 591–595 (1982)

166. J. Bochi, The basic ergodic theorems, yet again. Cubo **20**(3), 81–95 (2018)
167. S. Bochner, Monotone Funktionen, Stieltjessche Integrale und harmonische Analyse. Math. Ann. **108**, 378–410 (1933)
168. P. Bohl, Über ein in der Theorie der säkularen Störungen vorkommendes Problem. J. Reine Angew. Math. **135**, 189–283 (1909)
169. L. Boltzmann, Ueber die Eigenschaften monocyklischer und anderer damit verwandter Systeme. J. Reine Angew. Math. **98**, 68–94 (1885)
170. R. Bombelli, *L'Algebra Parte Maggiore dell'Arithmetica Divisa in Tre Libri* (Bologna, 1572)
171. A. Bonilla, V. Müller, Kreiss bounded and uniformly Kreiss bounded operators. Rev. Mat. Complut. **34**(2), 469–487 (2021)
172. E. Borel, Les probabilités dénombrables et leurs applications arithmétiques. Rendiconti del Circolo Matematico di Palermo **27**, 247–271 (1909)
173. M. Boshernitzan, Homogeneously distributed sequences and Poincaré sequences of integers of sublacunary growth. Monatsh. Math. **96**(3), 173–181 (1983)
174. M. Boshernitzan, N. Frantzikinakis, M. Wierdl, Under-recurrence in the Khintchine recurrence theorem. Isr. J. Math. **222**(2), 815–840 (2017)
175. M. Boshernitzan, G. Kolesnik, A. Quas, M. Wierdl, Ergodic averaging sequences. J. Anal. Math. **95**, 63–103 (2005)
176. M. Boshernitzan, M. Wierdl, Ergodic theorems along sequences and Hardy fields. Proc. Nat. Acad. Sci. U.S.A. **93**(16), 8205–8207 (1996)
177. M.D. Boshernitzan, Uniform distribution and Hardy fields. J. Anal. Math. **62**, 225–240 (1994)
178. J. Bourgain, Almost sure convergence and bounded entropy. Isr. J. Math. **63**(1), 79–97 (1988)
179. J. Bourgain, An approach to pointwise ergodic theorems, in *Geometric Aspects of Functional Analysis (1986/87)*. Volume 1317 of Lecture Notes in Mathematics (Springer, Berlin, 1988), pp. 204–223
180. J. Bourgain, On the maximal ergodic theorem for certain subsets of the integers. Isr. J. Math. **61**(1), 39–72 (1988)
181. J. Bourgain, On the pointwise ergodic theorem on L^p for arithmetic sets. Isr. J. Math. **61**(1), 73–84 (1988)
182. J. Bourgain, Temps de retour pour les systèmes dynamiques. C. R. Acad. Sci. Paris Sér. I Math. **306**(12), 483–485 (1988)
183. J. Bourgain, Pointwise ergodic theorems for arithmetic sets. Inst. Hautes Études Sci. Publ. Math. (69), 5–45 (1989). With an appendix by the author, Harry Furstenberg, Yitzhak Katznelson and Donald S. Ornstein
184. J. Bourgain, Double recurrence and almost sure convergence. J. Reine Angew. Math. **404**, 140–161 (1990)
185. J. Bourgain, On the correlation of the Moebius function with rank-one systems. J. Anal. Math. **120**, 105–130 (2013)
186. J. Bourgain, H. Furstenberg, Y. Katznelson, D.S. Ornstein, Appendix on return-time sequences. Inst. Hautes Études Sci. Publ. Math. (69), 42–45 (1989)
187. J. Bourgain, M. Mirek, E.M. Stein, J. Wright, On a multi-parameter variant of the Bellow-Furstenberg problem. Forum Math. Pi **11**, Paper No. e23, 64 (2023)
188. J. Bourgain, P. Sarnak, T. Ziegler, Disjointness of Moebius from horocycle flows, in *From Fourier Analysis and Number Theory to Radon Transforms and Geometry*. Volume 28 of Developments in Mathematics (Springer, New York, 2013), pp. 67–83
189. L. Bowen, A. Nevo, Pointwise ergodic theorems beyond amenable groups. Ergod. Theory Dyn. Syst. **33**(3), 777–820 (2013)
190. R. Bowen, Entropy for group endomorphisms and homogeneous spaces. Trans. Am. Math. Soc. **153**, 401–414 (1971)
191. R. Bowen, *Equilibrium States and the Ergodic Theory of Anosov Diffeomorphisms*. Lecture Notes in Mathematics, vol. 470 (Springer, Berlin/New York, 1975)

192. L. Breiman, The individual ergodic theorem of information theory. Ann. Math. Stat. **28**, 809–811 (1957)
193. J. Briët, B. Green, Multiple correlation sequences not approximable by nilsequences. Ergod. Theory Dyn. Syst. **42**(9), 2711–2722 (2022)
194. A. Brogini Bratti, Extension of Kingman's ergodic theorem for superadditive stochastic processes with range in Banach spaces. Rend. Accad. Naz. Sci. XL Mem. Mat. (5) **9**(1), 323–329 (1985)
195. M. Broise, Y. Déniel, Y. Derriennic, Réarrangement, inégalités maximales et théorèmes ergodiques fractionnaires. Ann. Inst. Fourier (Grenoble) **39**(3), 689–714 (1989)
196. F.E. Browder, On the iteration of transformations in noncompact minimal dynamical systems. Proc. Am. Math. Soc. **9**, 773–780 (1958)
197. A. Brunel, Sur un lemme ergodique voisin du lemme de E. Hopf, et sur une de ses applications. C. R. Acad. Sci. Paris **256**, 5481–5484 (1963)
198. A. Brunel, Sur un lemme de la théorie ergodique individuelle. C. R. Acad. Sci. Paris **259**, 4211–4212 (1964)
199. A. Brunel, M. Keane, Ergodic theorems for operator sequences. Z. Wahrscheinlichkeitstheorie und Verw. Gebiete **12**, 231–240 (1969)
200. Z. Buczolich, Arithmetic averages of rotations of measurable functions. Ergod. Theory Dyn. Syst. **16**(6), 1185–1196 (1996)
201. Z. Buczolich, Universally L^1 good sequences with gaps tending to infinity. Acta Math. Hung. **117**(1–2), 91–140 (2007)
202. Z. Buczolich, Almost everywhere convergence of ergodic averages. Real Anal. Exch. **34**(1), 1–15 (2009)
203. Z. Buczolich, Averages along the squares on the torus, in *Ergodic Theory and Dynamical Systems*. De Gruyter Proceedings in Mathematics (De Gruyter, Berlin, 2014), pp. 67–79
204. Z. Buczolich, Ergodic averages with prime divisor weights in L^1. Ergod. Theory Dyn. Syst. **39**(4), 889–897 (2019)
205. Z. Buczolich, T. Eisner, Divergence of weighted square averages in L^1. Adv. Math. **384**, Paper No. 107727, 19 (2021)
206. Z. Buczolich, R.D. Mauldin, Concepts behind divergent ergodic averages along the squares, in *Ergodic Theory and Related Fields*. Volume 430 of Contemporary Mathematics (American Mathematical Society, Providence, RI, 2007), pp. 41–56
207. Z. Buczolich, R.D. Mauldin, Divergent square averages. Ann. Math. (2) **171**(3), 1479–1530 (2010)
208. Y. Bugeaud, *Distribution Modulo One and Diophantine Approximation*. Volume 193 of Cambridge Tracts in Mathematics (Cambridge University Press, Cambridge, 2012)
209. D.L. Burkholder, Semi-Gaussian subspaces. Trans. Am. Math. Soc. **104**, 123–131 (1962)
210. D.L. Burkholder, Y.S. Chow, Iterates of conditional expectation operators. Proc. Am. Math. Soc. **12**, 490–495 (1961)
211. S.V. Butler, J.M. Rosenblatt, Moving averages. Colloq. Math. **113**(2), 251–266 (2008)
212. P.L. Butzer, U. Westphal, The mean ergodic theorem and saturation. Indiana Univ. Math. J. **20**, 1163–1174 (1970/71)
213. L. Cadilhac, S. Wang, Noncommutative maximal ergodic inequalities for amenable groups (2022). Preprint, https://arxiv.org/abs/2206.12228
214. F. Cai, Measure-theoretic equicontinuity and rigidity. Nonlinearity **33**(8), 3739–3762 (2020)
215. A.P. Calderon, A general ergodic theorem. Ann. Math. **58**(2), 182–191 (1953)
216. A.-P. Calderón, Ergodic theory and translation-invariant operators. Proc. Natl. Acad. Sci. U.S.A. **59**, 349–353 (1968)
217. J.T. Campbell, *Variations on the Ergodic Hilbert Transform*. Ph.D. thesis, The University of North Carolina at Chapel Hill, 1984
218. J.T. Campbell, Spectral analysis of the ergodic Hilbert transform. Indiana Univ. Math. J. **35**(2), 379–390 (1986)
219. P. Candela, D. González-Sánchez, B. Szegedy. On nilspace systems and their morphisms. Ergod. Theory Dyn. Syst. **40**(11), 3015–3029 (2020)

220. P. Candela, B. Szegedy, Regularity and inverse theorems for uniformity norms on compact abelian groups and nilmanifolds. J. Reine Angew. Math. **789**, 1–42 (2022)
221. P. Candela, B. Szegedy, Nilspace factors for general uniformity seminorms, cubic exchangeability and limits. Mem. Am. Math. Soc. **287**(1425), v+101 (2023)
222. C. Carathéodory, Über den Variabilitätsbereich der Koeffizienten von Potenzreihen, die gegebene Werte nicht annehmen. Math. Ann. **64**, 95–115 (1907)
223. C. Carathéodory, Über den Variabilitätsbereich der Fourierschen Konstanten von positiven harmonischen Funktionen. Rend. Circ. Mat. Palermo **32**, 193–217 (1911)
224. C. Carathéodory, Über den Wiederkehrsatz von Poincaré. *Berl. Sitzungsber.* 580–584 (1919)
225. T. Carleman, Application de la théorie des équations intégrales linéaires aux équations différentielles de la dynamique. Arkiv för Mat., Astr. och Fys. **22B**(7), 1–5 (1932)
226. T. Carleman, Application de la théorie des équations intégrales linéaires aux systèmes d'équations différentielles non linéaires. Acta. Math. **59**(1), 63–87 (1932)
227. P.A. Cataldi, *Trattato del Modo Brevissimo di Trovare la Radice Quadra delli Numeri*, (Bologna, 1613)
228. D. Çömez, An ergodic theorem for multidimensional superadditive processes. Can. J. Math. **37**(4), 612–634 (1985)
229. D. Çömez, Modern ergodic theory: From a physics hypothesis to a mathematical theory with transformative interdisciplinary impact, in *Handbook of the Mathematics of the Arts and Sciences* (Springer, Cham, 2021), pp. 1969–1992
230. D. Çömez, M. Lin, Multiparameter superadditive ergodic theorems for mean ergodic L_1-contractions. Probab. Math. Stat. **23**(1), 173–187 (2003)
231. D. Çömez, M. Lin, J. Olsen, Weighted ergodic theorems for mean ergodic L_1-contractions. Trans. Am. Math. Soc. **350**(1), 101–117 (1998)
232. D. Çömez, S.N. Litvinov, Ergodic averages with generalized weights. Stud. Math. **173**(2), 103–128 (2006)
233. R.V. Chacon, Identification of the limit of operator averages. J. Math. Mech. **11**, 961–968 (1962)
234. R.V. Chacon, Operator averages. Bull. Am. Math. Soc. **68**, 351–353 (1962)
235. R.V. Chacon, A class of linear transformations. Proc. Am. Math. Soc. **15**, 560–564 (1964)
236. R.V. Chacon, Weakly mixing transformations which are not strongly mixing. Proc. Am. Math. Soc. **22**, 559–562 (1969)
237. R.V. Chacon, J. Olsen, Dominated estimates of positive contractions. Proc. Am. Math. Soc. **20**, 266–271 (1969)
238. R.V. Chacon, D.S. Ornstein, A general ergodic theorem. Ill. J. Math. **4**, 153–160 (1960)
239. J. Chaika, Every ergodic transformation is disjoint from almost every interval exchange transformation. Ann. Math. (2) **175**(1), 237–253 (2012)
240. J. Chaika, A. Eskin, Möbius disjointness for interval exchange transformations on three intervals. J. Mod. Dyn. **14**, 55–86 (2019)
241. J. Chatard, Applications des propriétés de moyenne d'un groupe localement compact à la théorie ergodique. Ann. Inst. H. Poincaré Sect. B (N.S.) **6**, 307–326; erratum, ibid. (N.S.) **7**(1971), 81–82 (1970)
242. C. Chou, Minimal sets and ergodic measures for $\beta N\backslash N$. Ill. J. Math. **13**, 777–788 (1969)
243. S. Chowla, in *The Riemann Hypothesis and Hilbert's Tenth Problem*. Mathematics and Its Applications, vol. 4 (Gordon and Breach Science Publishers, New York/London/Paris, 1965)
244. Q. Chu, Convergence of weighted polynomial multiple ergodic averages. Proc. Am. Math. Soc. **137**(4), 1363–1369 (2009)
245. Q. Chu, Multiple recurrence for two commuting transformations. Ergod. Theory Dyn. Syst. **31**(3), 771–792 (2011)
246. Q. Chu, N. Frantzikinakis, Pointwise convergence for cubic and polynomial multiple ergodic averages of non-commuting transformations. Ergod. Theory Dyn. Syst. **32**(3), 877–897 (2012)

247. Q. Chu, N. Frantzikinakis, B. Host, Ergodic averages of commuting transformations with distinct degree polynomial iterates. Proc. Lond. Math. Soc. (3) **102**(5), 801–842 (2011)
248. Q. Chu, P. Zorin-Kranich, Lower bound in the Roth theorem for amenable groups. Ergod. Theory Dyn. Syst. **35**(6), 1746–1766 (2015)
249. G. Cohen, Doubly power-bounded operators on L^p, $2 \neq p > 1$. J. Math. Anal. Appl. **466**(2), 1327–1336 (2018)
250. G. Cohen, C. Cuny, T. Eisner, M. Lin, Resolvent conditions and growth of powers of operators. J. Math. Anal. Appl. **487**(2), 124035, 24, (2020)
251. G. Cohen, M. Lin, Extensions of the Menchoff-Rademacher theorem with applications to ergodic theory. Isr. J. Math. **148**, 41–86 (2005). Probability in mathematics
252. G. Cohen, M. Lin, Operator ergodic theory, in *Ergodic Theory*, Encyclopedia of Complexity and Systems Science (Springer, New York, 2023), pp. 459–489
253. L.W. Cohen, On the mean ergodic theorem. Ann. Math. **41**(2), 505–509 (1940)
254. J.B. Conway, *A Course in Functional Analysis, Volume 96 of Graduate Texts in Mathematics*, 2nd edn. (Springer, New York, 1990)
255. J.-P. Conze, Convergence des moyennes ergodiques pour des sous-suites, in *Contributions au calcul des probabilités*, pp. 7–15. Bull. Soc. Math. France, Mém. No. 35 (1973)
256. J.-P. Conze, E. Lesigne, Théorèmes ergodiques pour des mesures diagonales. Bull. Soc. Math. Fr. **112**(2), 143–175 (1984)
257. J.-P. Conze, E. Lesigne, Sur un théorème ergodique pour des mesures diagonales. C. R. Acad. Sci. Paris Sér. I Math. **306**(12), 491–493 (1988)
258. J.-P. Conze, E. Lesigne, Sur un théorème ergodique pour des mesures diagonales, in *Probabilités, Volume 1987 of Publications de l'Institut de Recherche Mathématique de Rennes* (University of Rennes I, Rennes, 1988), pp. 1–31
259. B. Cook, A. Magyar, T. Titichetrakun, A multidimensional Szemerédi theorem in the primes via combinatorics. Ann. Comb. **22**(4), 711–768 (2018)
260. I.P. Cornfeld, S.V. Fomin, Y.G. Sinaĭ, *Ergodic Theory*. Volume 245 of Grundlehren der mathematischen Wissenschaften [Fundamental Principles of Mathematical Sciences] (Springer, New York, 1982). Translated from the Russian by A.B. Sosinskiĭ
261. M. Cotlar, A unified theory of Hilbert transforms and ergodic theorems. Rev. Mat. Cuyana **1**, 105–167 (1955)
262. Y. Coudène, *Ergodic Theory and Dynamical Systems*. Universitext (Springer, London; EDP Sciences, [Les Ulis], 2016). Translated from the 2013 French original [MR3184308] by Reinie Erné
263. C. Cuny, Some optimal pointwise ergodic theorems with rate. C. R. Math. Acad. Sci. Paris **347**(15–16), 953–958 (2009)
264. C. Cuny, Pointwise ergodic theorems with rate with applications to limit theorems for stationary processes. Stoch. Dyn. **11**(1), 135–155 (2011)
265. C. Cuny, Y. Derriennic, On the convergence of bilateral ergodic averages. Colloq. Math. **164**(2), 327–339 (2021)
266. C. Cuny, T. Eisner, B. Farkas, Wiener's lemma along primes and other subsequences. Adv. Math. **347**, 340–383 (2019)
267. C. Cuny, M. Lin, Pointwise ergodic theorems with rate and application to the CLT for Markov chains. Ann. Inst. Henri Poincaré Probab. Stat. **45**(3), 710–733 (2009)
268. C. Cuny, F. Parreau, Good sequences with uncountable spectrum and singular asymptotic distribution. Annales de l'Institut Fourier (To appear)
269. C. Cuny, M. Weber, Ergodic theorems with arithmetical weights. Isr. J. Math. **217**(1), 139–180 (2017)
270. K. Dajani, C. Kalle, *A First Course in Ergodic Theory*. (Chapman & Hall/CRC, Boca Raton, FL, 2021)
271. L. Daskalakis, Weak-type (1,1) inequality for discrete maximal functions and pointwise ergodic theorems along thin arithmetic sets. J. Fourier Anal. Appl. **30**(3), Paper No. 37, 50 pp. (2024)

272. H. Davenport, On some infinite series involving arithmetical functions. II. Q. J. Math. Oxf. Ser. **8**, 313–320 (1937)
273. T. de la Rue, 2-fold and 3-fold mixing: Why 3-dot-type counterexamples are impossible in one dimension. Bull. Braz. Math. Soc. (N.S.) **37**(4), 503–521 (2006)
274. T. de la Rue, J. de Sam Lazaro, Une transformation générique peut être insérée dans un flot. Ann. Inst. H. Poincaré Probab. Stat. **39**(1), 121–134 (2003)
275. A. de la Torre, A simple proof of the maximal ergodic theorem. Can. J. Math. **28**(5), 1073–1075 (1976)
276. K. de Leeuw, I. Glicksberg, Almost periodic compactifications. Bull. Am. Math. Soc. **65**, 134–139 (1959)
277. K. de Leeuw, I. Glicksberg, Applications of almost periodic compactifications. Acta Math. **105**, 63–97 (1961)
278. J. de Vries, Two applications of topological dynamics in combinatorial number theory. The shift system and the Stone-Čech compactification of the nonnegative integers. CWI Newslett. (2), 2–17 (1984)
279. A. Deitmar, S. Echterhoff, *Principles of Harmonic Analysis*. Universitext (Springer, New York, 2009)
280. J.G. Deken, Some limit results for longest common subsequences. Discret. Math. **26**(1), 17–31 (1979)
281. A. del Junco, On the decomposition of a subadditive stochastic process. Ann. Probab. **5**(2), 298–302 (1977)
282. A. del Junco, M. Lemańczyk, Generic spectral properties of measure-preserving maps and applications. Proc. Am. Math. Soc. **115**(3), 725–736 (1992)
283. A. del Junco, J. Rosenblatt, Counterexamples in ergodic theory and number theory. Math. Ann. **245**(3), 185–197 (1979)
284. C. Demeter, The best constants associated with some weak maximal inequalities in ergodic theory. Can. J. Math. **56**(3), 449–471 (2004)
285. C. Demeter, Pointwise convergence of the ergodic bilinear Hilbert transform. Ill. J. Math. **51**(4), 1123–1158 (2007)
286. C. Demeter, Improved range in the return times theorem. Can. Math. Bull. **55**(4), 708–722 (2012)
287. C. Demeter, R.L. Jones, Besicovitch weights and the necessity of duality restrictions in the weighted ergodic theorem, in *Chapel Hill Ergodic Theory Workshops*. Volume 356 of Contemporary Mathematics (American Mathematical Society, Providence, RI, 2004), pp. 127–135
288. C. Demeter, M.T. Lacey, T. Tao, C. Thiele, Breaking the duality in the return times theorem. Duke Math. J. **143**(2), 281–355 (2008)
289. M. Denker, On strict ergodicity. Math. Z. **134**, 231–253 (1973)
290. R. Derndinger, R. Nagel, G. Palm, *Ergodic Theory in the Perspective of Functional Analysis* (1987)
291. J.-M. Derrien, E. Lesigne, Un théorème ergodique polynomial ponctuel pour les endomorphismes exacts et les K-systèmes. Ann. Inst. H. Poincaré Probab. Stat. **32**(6), 765–778 (1996)
292. Y. Derriennic, On the integrability of the supremum of ergodic ratios. Ann. Probab. **1**, 338–340 (1973)
293. Y. Derriennic, Sur le théorème ergodique sous-additif. C. R. Acad. Sci. Paris Sér. A-B **281**(22), Aii, A985–A988 (1975)
294. Y. Derriennic, Lois “zéro ou deux” pour les processus de Markov. Applications aux marches aléatoires. Ann. Inst. H. Poincaré Sect. B (N.S.) **12**(2), 111–129 (1976)
295. Y. Derriennic, Quelques applications du théorème ergodique sous-additif, in *Conference on Random Walks (Kleebach, 1979) (French)*. Volume 74 of Astérisque, 4 (Social Mathematic France, Paris, 1980), pp. 183–201
296. Y. Derriennic, Un théorème ergodique presque sous-additif. Ann. Probab. **11**(3), 669–677 (1983)

297. Y. Derriennic, On the mean ergodic theorem for Cesàro bounded operators. Colloq. Math. **84/85**, 443–455 (2000)
298. Y. Derriennic, Some aspects of recent works on limit theorems in ergodic theory with special emphasis on the "central limit theorem". Discrete Contin. Dyn. Syst. **15**(1), 143–158 (2006)
299. Y. Derriennic, Ergodic theorem, reversibility and the filling scheme. Colloq. Math. **118**(2), 599–608 (2010)
300. Y. Derriennic, U. Krengel, Subadditive mean ergodic theorems. Ergod. Theory Dyn. Syst. **1**(1), 33–48 (1981)
301. Y. Derriennic, M. Lin, On invariant measures and ergodic theorems for positive operators. J. Funct. Anal. **13**, 252–267 (1973)
302. Y. Derriennic, M. Lin, Fractional Poisson equations and ergodic theorems for fractional coboundaries. Isr. J. Math. **123**, 93–130 (2001)
303. E.I. Dinaburg, A correlation between topological entropy and metric entropy. Dokl. Akad. Nauk SSSR **190**, 19–22 (1970)
304. J. Ding, A. Zhou, *Statistical Properties of Deterministic Systems*. Tsinghua University Texts (Springer, Berlin; Tsinghua University Press, Beijing, 2009)
305. Y. Do, R. Oberlin, E.A. Palsson, Variation-norm and fluctuation estimates for ergodic bilinear averages. Indiana Univ. Math. J. **66**(1), 55–99 (2017)
306. P. Dodos, V. Kanellopoulos, K. Tyros, A simple proof of the density Hales-Jewett theorem. Int. Math. Res. Not. IMRN (12), 3340–3352 (2014)
307. S. Donoso, A. Koutsogiannis, W. Sun, Seminorms for multiple averages along polynomials and applications to joint ergodicity. J. Anal. Math. **146**(1), 1–64 (2022)
308. S. Donoso, A.N. Le, J. Moreira, W. Sun, Optimal lower bounds for multiple recurrence. Ergod. Theory Dyn. Syst. **41**(2), 379–407 (2021)
309. S. Donoso, A.N. Le, J. Moreira, W. Sun, Averages of completely multiplicative functions over the Gaussian integers—a dynamical approach. Trans. Am. Math. Soc. **377**(10), 7081–7115 (2024)
310. S. Donoso, A. Maass, V. Saavedra-Araya, A pointwise ergodic theorem along return times of rapidly mixing systems (2025). Preprint, https://arxiv.org/abs/2502.17746
311. S. Donoso, W. Sun, Pointwise convergence of some multiple ergodic averages. Adv. Math. **330**, 946–996 (2018)
312. J.L. Doob, Stochastic processes with an integral-valued parameter. Trans. Am. Math. Soc. **44**(1), 87–150 (1938)
313. J.L. Doob, Asymptotic properties of Markoff transition prababilities. Trans. Am. Math. Soc. **63**, 393–421 (1948)
314. J.L. Doob, *Stochastic Processes*. Wiley Classics Library (Wiley, New York, 1990). Reprint of the 1953 original, A Wiley-Interscience Publication
315. A.H. Dooley, V.Y. Golodets, G. Zhang, Sub-additive ergodic theorems for countable amenable groups. J. Funct. Anal. **267**(5), 1291–1320 (2014)
316. Y.N. Dowker, A new proof of the general ergodic theorem. Acta Sci. Math. (Szeged) **12**, 162–166 (1950)
317. T. Downarowicz, S. Kasjan. Odometers and Toeplitz systems revisited in the context of Sarnak's conjecture. Stud. Math. **229**(1), 45–72 (2015)
318. T. Downarowicz, J. Serafin, Almost full entropy subshifts uncorrelated to the Möbius function. Int. Math. Res. Not. IMRN (11), 3459–3472 (2019)
319. T. Downarowicz, J. Serafin, A strictly ergodic, positive entropy subshift uniformly uncorrelated to the Möbius function. Stud. Math. **251**(2), 195–206 (2020)
320. N. Dunford, Spectral theory. I. Convergence to projections. Trans. Am. Math. Soc. **54**, 185–217 (1943)
321. N. Dunford, An individual ergodic theorem for non-commutative transformations. Acta Sci. Math. (Szeged) **14**, 1–4 (1951)
322. N. Dunford, J. Schwartz, Convergence almost everywhere of operator averages. Proc. Nat. Acad. Sci. U.S.A. **41**, 229–231 (1955)

323. N. Dunford, J.T. Schwartz, Convergence almost everywhere of operator averages. J. Ration. Mech. Anal. **5**, 129–178 (1956)
324. N. Dunford, J.T. Schwartz, *Linear Operators. Part I*. Wiley Classics Library (Wiley, New York, 1988). General theory, With the assistance of William G. Bade and Robert G. Bartle, Reprint of the 1958 original, A Wiley-Interscience Publication
325. F. Durand, D. Schneider, Ergodic averages with deterministic weights. Ann. Inst. Fourier (Grenoble) **52**(2), 561–583 (2002)
326. P. Durcik, R. Greenfeld, A. Iseli, A. Jamneshan, J. Madrid, An uncountable ergodic Roth theorem and applications. Discrete Contin. Dyn. Syst. **42**(11), 5509–5540 (2022)
327. P. Durcik, V. Kovač, K.A. Škreb, C. Thiele, Norm variation of ergodic averages with respect to two commuting transformations. Ergod. Theory Dyn. Syst. **39**(3), 658–688 (2019)
328. R. Durrett, *Probability—Theory and Examples*. Volume 49 of Cambridge Series in Statistical and Probabilistic Mathematics, 5th edn. (Cambridge University Press, Cambridge, 2019)
329. W.F. Eberlein, Abstract ergodic theorems and weak almost periodic functions. Trans. Am. Math. Soc. **67**, 217–240 (1949)
330. N. Edeko, On the isomorphism problem for non-minimal transformations with discrete spectrum. Discrete Contin. Dyn. Syst. **39**(10), 6001–6021 (2019)
331. N. Edeko, M. Haase, H. Kreidler, A decomposition theorem for unitary group representations on Kaplansky-Hilbert modules and the Furstenberg–Zimmer structure theorem. Anal. Math. **50**(2), 377–454 (2024)
332. N. Edeko, H. Kreidler, Distal systems in topological dynamics and ergodic theory. Ergod. Theory Dyn. Syst. **43**(8), 2651–2672 (2023)
333. N. Edeko, H. Kreidler, R. Nagel, A dynamical proof of the van der Corput inequality. Dyn. Syst. **37**(4), 648–665 (2022)
334. P. Ehrenfest, T. Ehrenfest. *Begriffliche Grundlagen der Statistischen Auffassung in der Mechanik*. Encykl. d. Math. Wissensch. IV 2 II, Heft 6 (1912)
335. S. Eigen, A. Hajian, Y. Ito, V. Prasad, *Weakly Wandering Sequences in Ergodic Theory*. Springer Monographs in Mathematics (Springer, Tokyo, 2014)
336. M. Einsiedler, T. Ward, *Ergodic Theory with a View Towards Number Theory*. Volume 259 of Graduate Texts in Mathematics (Springer, London, 2011)
337. T. Eisner, *Stability of Operators and Operator Semigroups*. Volume 209 of Operator Theory: Advances and Applications (Birkhäuser Verlag, Basel, 2010)
338. T. Eisner, Linear sequences and weighted ergodic theorems. Abstr. Appl. Anal. Art. ID 815726, 5 pp. (2013)
339. T. Eisner, A polynomial version of Sarnak's conjecture. C. R. Math. Acad. Sci. Paris **353**(7), 569–572 (2015)
340. T. Eisner, Nilsystems and ergodic averages along primes. Ergod. Theory Dyn. Syst. **40**(10), 2769–2777 (2020)
341. T. Eisner, A view on multiple recurrence. Indag. Math. (N.S.) **34**(2), 231–247 (2023)
342. T. Eisner, B. Farkas, M. Haase, R. Nagel, *Operator Theoretic Aspects of Ergodic Theory*. Volume 272 of Graduate Texts in Mathematics (Springer, Cham, 2015)
343. T. Eisner, S. Grivaux, Hilbertian Jamison sequences and rigid dynamical systems. J. Funct. Anal. **261**(7), 2013–2052 (2011)
344. T. Eisner, J. Konieczny, Automatic sequences as good weights for ergodic theorems. Discrete Contin. Dyn. Syst. **38**(8), 4087–4115 (2018)
345. T. Eisner, B. Krause, (Uniform) convergence of twisted ergodic averages. Ergod. Theory Dyn. Syst. **36**(7), 2172–2202 (2016)
346. T. Eisner, M. Lin, On modulated ergodic theorems. J. Nonlin. Var. Anal. **2**, 131–154 (2018). Special issue dedicated to Simeon Reich
347. T. Eisner, V. Müller, Power bounded operators and the mean ergodic theorem for subsequences. J. Math. Anal. Appl. **493**(1), Paper No. 124523, 25 (2021)
348. T. Eisner, T. Tao, Large values of the Gowers-Host-Kra seminorms. J. Anal. Math. **117**, 133–186 (2012)

349. T. Eisner, P. Zorin-Kranich, Uniformity in the Wiener-Wintner theorem for nilsequences. Discrete Contin. Dyn. Syst. **33**(8), 3497–3516 (2013)
350. E.H. El Abdalaoui, Simple proof of Bourgain bilinear ergodic theorem and its extension to polynomials and polynomials in primes. Preprint, https://arxiv.org/abs/1908.02281
351. E.H. El Abdalaoui, Corrigendum to "On the homogeneous ergodic bilinear averages with Möbius and Liouville weights" by E. H. el Abdalaoui. Ill. J. Math. **64**(4), 645 (2020)
352. E.H. El Abdalaoui, On the homogeneous ergodic bilinear averages with Möbius and Liouville weights. Ill. J. Math. **64**(1), 1–19 (2020)
353. E.H. El Abdalaoui, Van der Corput inequality for real line and Wiener-Wintner theorem for amenable groups. Constr. Math. Anal. **4**(4), 420–427 (2021)
354. E.H. El Abdalaoui, T. Adams, Mixing sequences for non-mixing transformations and group actions (2022). Preprint, https://arxiv.org/abs/2206.01177
355. E.H. El Abdalaoui, J. Devianne, A large class of dendrite maps for which Möbius disjointness property of Sarnak is fulfilled. Qual. Theory Dyn. Syst. **23**(2), Paper No. 83, 18 (2024)
356. E.H. El Abdalaoui, S. Kasjan, M. Lemańczyk, 0–1 sequences of the Thue-Morse type and Sarnak's conjecture. Proc. Am. Math. Soc. **144**(1), 161–176 (2016)
357. E.H. El Abdalaoui, J. Kułaga-Przymus, M. Lemańczyk, T. de la Rue, The Chowla and the Sarnak conjectures from ergodic theory point of view. Discrete Contin. Dyn. Syst. **37**(6), 2899–2944 (2017)
358. E.H. El Abdalaoui, J. Kułaga-Przymus, M. Lemańczyk, T. de la Rue, Möbius disjointness for models of an ergodic system and beyond. Isr. J. Math. **228**(2), 707–751 (2018)
359. E.H. El Abdalaoui, M. Lemańczyk, T. de la Rue, On spectral disjointness of powers for rank-one transformations and Möbius orthogonality. J. Funct. Anal. **266**(1), 284–317 (2014)
360. E.H. El Abdalaoui, M. Lemańczyk, T. de la Rue, Automorphisms with quasi-discrete spectrum, multiplicative functions and average orthogonality along short intervals. Int. Math. Res. Not. IMRN (14):4350–4368 (2017)
361. R. Ellis, Distal transformation groups. Pac. J. Math. **8**, 401–405 (1958)
362. R. Ellis, A semigroup associated with a transformation group. Trans. Am. Math. Soc. **94**, 272–281 (1960)
363. R. Ellis, W.H. Gottschalk, Homomorphisms of transformation groups. Trans. Am. Math. Soc. **94**, 258–271 (1960)
364. W.R. Emerson, The pointwise ergodic theorem for amenable groups. Am. J. Math. **96**, 472–487 (1974)
365. R. Émilion. Opérateurs à moyennes bornées et théorèmes ergodiques en moyenne. C. R. Acad. Sci. Paris Sér. I Math. **296**(15), 641–643 (1983)
366. R. Émilion, Mean-bounded operators and mean ergodic theorems. J. Funct. Anal. **61**(1), 1–14 (1985)
367. R. Émilion, On the pointwise ergodic theorems in L_p $(1 < p < \infty)$. Stud. Math. **81**(2), 171–178 (1985)
368. R. Émilion, B. Hachem, A multiparameter strongly superadditive ergodic theorem. Math. Z. **189**(1), 11–17 (1985)
369. K.-J. Engel, R. Nagel, *One-Parameter Semigroups for Linear Evolution Equations*. Volume 194 of Graduate Texts in Mathematics (Springer, New York, 2000). With contributions by S. Brendle, M. Campiti, T. Hahn, G. Metafune, G. Nickel, D. Pallara, C. Perazzoli, A. Rhandi, S. Romanelli and R. Schnaubelt
370. R. Engelking, *General Topology*. Volume 6 of Sigma Series in Pure Mathematics, 2nd edn. (Heldermann Verlag, Berlin, 1989). Translated from the Polish by the author
371. L. Ephremidze, On the uniqueness of the ergodic maximal function. Fund. Math. **174**(3), 217–228 (2002)
372. M. Etedadialiabadi, S. Gao, Sarnak's conjecture for a class of rank-one subshifts. Proc. Am. Math. Soc. Ser. B **9**, 460–471 (2022)
373. A. Fan, Fully oscillating sequences and weighted multiple ergodic limit. C. R. Math. Acad. Sci. Paris **355**(8), 866–870 (2017)

374. A. Fan, D. Schneider, Recurrence properties of sequences of integers. Sci. China Math. **53**(3), 641–656 (2010)
375. A.-H. Fan, Topological Wiener-Wintner ergodic theorem with polynomial weights. Chaos Solitons Fractals **117**, 105–116 (2018)
376. A.-H. Fan, Weighted Birkhoff ergodic theorem with oscillating weights. Ergod. Theory Dyn. Syst. **39**(5), 1275–1289 (2019)
377. A.-H. Fan, Y. Jiang, Oscillating sequences, MMA and MMLS flows and Sarnak's conjecture. Ergod. Theory Dyn. Syst. **38**(5), 1709–1744 (2018)
378. S. Farhangi, A generalization of van der Corput's difference theorem with applications to recurrence and multiple ergodic averages. Dyn. Syst. **39**(1), 5–30 (2024)
379. B. Farkas, H. Kreidler, Relative compactness of orbits and geometry of Banach spaces. J. Math. Anal. Appl. **495**(1), Paper No. 124660, 12 (2021)
380. N.A. Fava, Weak type inequalities for product operators. Stud. Math. **42**, 271–288 (1972)
381. B. Fayad, A. Kanigowski, Rigidity times for a weakly mixing dynamical system which are not rigidity times for any irrational rotation. Ergod. Theory Dyn. Syst. **35**(8), 2529–2534 (2015)
382. B. Fayad, A. Kanigowski, Multiple mixing for a class of conservative surface flows. Invent. Math. **203**(2), 555–614 (2016)
383. B. Fayad, J.-P. Thouvenot, On the convergence to 0 of $m_n\xi$ mod 1. Acta Arith. **165**(4), 327–332 (2014)
384. M. Feder, On power-bounded operators and the pointwise ergodic property. Proc. Am. Math. Soc. **83**(2), 349–353 (1981)
385. S. Ferenczi, C. Holton, L.Q. Zamboni, Joinings of three-interval exchange transformations. Ergod. Theory Dyn. Syst. **25**(2), 483–502 (2005)
386. S. Ferenczi, J. Kułaga-Przymus, M. Lemańczyk, Sarnak's conjecture: what's new, in *Ergodic Theory and Dynamical Systems in their Interactions with Arithmetics and Combinatorics*. Volume 2213 of Lecture Notes in Mathematics (Springer, Cham, 2018), pp. 163–235
387. S. Ferenczi, J. Kułaga-Przymus, M. Lemańczyk, C. Mauduit, Substitutions and Möbius disjointness, in *Ergodic Theory, Dynamical Systems, and the Continuing Influence of John C. Oxtoby*. Volume 678 of Contemporary Mathematics (American Mathematical Society, Providence, RI, 2016), pp. 151–173
388. S. Ferenczi, C. Mauduit, On Sarnak's conjecture and Veech's question for interval exchanges. J. Anal. Math. **134**(2), 545–573 (2018)
389. A. Ferré Moragues, Properties of multicorrelation sequences and large returns under some ergodicity assumptions. Discrete Contin. Dyn. Syst. **41**(6), 2809–2828 (2021)
390. E. Fischer, Über das *Carathéodory*sche Problem, Potenzreihen mit positivem reellen Teil betreffend. Rend. Circ. Mat. Palermo **32**, 240–256 (1911)
391. A. Fish, S. Skinner, An inverse of furstenberg's correspondence principle and applications to nice recurrence (2024). Preprint, https://arxiv.org/abs/2407.19444
392. S.R. Foguel, Powers of a contraction in Hilbert space. Pac. J. Math. **13**, 551–562 (1963)
393. S.R. Foguel, *The Ergodic Theory of Markov Processes*. Van Nostrand Mathematical Studies, vol. 21 (Van Nostrand Reinhold Co., New York/Toronto/London, 1969)
394. C. Foiaş, Sur les mesures spectrales qui interviennent dans la théorie ergodique. J. Math. Mech. **13**, 639–658 (1964)
395. G.B. Folland, *A Course in Abstract Harmonic Analysis*. Studies in Advanced Mathematics (CRC Press, Boca Raton, FL, 1995)
396. V.P. Fonf, M. Lin, P. Wojtaszczyk, Ergodic characterizations of reflexivity of Banach spaces. J. Funct. Anal. **187**(1), 146–162 (2001)
397. V.P. Fonf, M. Lin, P. Wojtaszczyk, A non-reflexive Banach space with all contractions mean ergodic. Isr. J. Math. **179**, 479–491 (2010)
398. V.P. Fonf, M. Lin, P. Wojtaszczyk, Poisson's equation and characterizations of reflexivity of Banach spaces. Colloq. Math. **124**(2), 225–235 (2011)
399. H. Fong, Ratio and stochastic ergodic theorems for superadditive processes. Can. J. Math. **31**(2), 441–447 (1979)

400. H. Fong, M. Lin, On the convergence of ergodic ratios for positive operators. J. Math. Anal. Appl. **55**(3), 667–672 (1976)
401. G. Forni, A. Kanigowski, Multiple mixing and disjointness for time changes of bounded-type Heisenberg nilflows. J. Éc. Polytech. Math. **7**, 63–91 (2020)
402. A.H. Forrest, The construction of a set of recurrence which is not a set of strong recurrence. Isr. J. Math. **76**(1–2), 215–228 (1991)
403. J. Fox, Y. Zhao, A short proof of the multidimensional Szemerédi theorem in the primes. Am. J. Math. **137**(4), 1139–1145 (2015)
404. N. Frantzikinakis, The structure of strongly stationary systems. J. Anal. Math. **93**, 359–388 (2004)
405. N. Frantzikinakis, Uniformity in the polynomial Wiener-Wintner theorem. Ergod. Theory Dyn. Syst. **26**(4), 1061–1071 (2006)
406. N. Frantzikinakis, Multiple ergodic averages for three polynomials and applications. Trans. Am. Math. Soc. **360**(10), 5435–5475 (2008)
407. N. Frantzikinakis, Equidistribution of sparse sequences on nilmanifolds. J. Anal. Math. **109**, 353–395 (2009)
408. N. Frantzikinakis, Multiple recurrence and convergence for Hardy sequences of polynomial growth. J. Anal. Math. **112**, 79–135 (2010)
409. N. Frantzikinakis, A multidimensional Szemerédi theorem for Hardy sequences of different growth. Trans. Am. Math. Soc. **367**(8), 5653–5692 (2015)
410. N. Frantzikinakis, Multiple correlation sequences and nilsequences. Invent. Math. **202**(2), 875–892 (2015)
411. N. Frantzikinakis, Ergodicity of the Liouville system implies the Chowla conjecture. Discrete Anal. **19**, 41 (2017)
412. N. Frantzikinakis, Furstenberg systems of Hardy field sequences and applications. J. Anal. Math. **147**(1), 333–372 (2022)
413. N. Frantzikinakis, Joint ergodicity of fractional powers of primes. Forum Math. Sigma **10**(e30), 30 (2022)
414. N. Frantzikinakis, B. Host, Multiple ergodic theorems for arithmetic sets. Trans. Am. Math. Soc. **369**(10), 7085–7105 (2017)
415. N. Frantzikinakis, B. Host, The logarithmic Sarnak conjecture for ergodic weights. Ann. Math. (2) **187**(3), 869–931 (2018)
416. N. Frantzikinakis, B. Host, Weighted multiple ergodic averages and correlation sequences. Ergod. Theory Dyn. Syst. **38**(1), 81–142 (2018)
417. N. Frantzikinakis, B. Host, Furstenberg systems of bounded multiplicative functions and applications. Int. Math. Res. Not. IMRN **8**, 6077–6107 (2021)
418. N. Frantzikinakis, B. Host, Multiple recurrence and convergence without commutativity. J. Lond. Math. Soc. (2) **107**(5), 1635–1659 (2023)
419. N. Frantzikinakis, B. Host, B. Kra, Multiple recurrence and convergence for sequences related to the prime numbers. J. Reine Angew. Math. **611**, 131–144 (2007)
420. N. Frantzikinakis, B. Host, B. Kra, The polynomial multidimensional Szemerédi theorem along shifted primes. Isr. J. Math. **194**(1), 331–348 (2013)
421. N. Frantzikinakis, M. Johnson, E. Lesigne, M. Wierdl, Powers of sequences and convergence of ergodic averages. Ergod. Theory Dyn. Syst. **30**(5), 1431–1456 (2010)
422. N. Frantzikinakis, B. Kra, Convergence of multiple ergodic averages for some commuting transformations. Ergod. Theory Dyn. Syst. **25**(3), 799–809 (2005)
423. N. Frantzikinakis, B. Kra, Polynomial averages converge to the product of integrals. Isr. J. Math. **148**, 267–276 (2005)
424. N. Frantzikinakis, B. Kra, Ergodic averages for independent polynomials and applications. J. Lond. Math. Soc. (2) **74**(1), 131–142 (2006)
425. N. Frantzikinakis, B. Kuca, Seminorm control for ergodic averages with commuting transformations along pairwise dependent polynomials. Ergod. Theory Dyn. Syst. **43**(12), 4074–4137 (2023)

426. N. Frantzikinakis, B. Kuca, Joint ergodicity for commuting transformations and applications to polynomial sequences. Invent. Math. **239**(2), 621–706 (2025)
427. N. Frantzikinakis, M. Lemańczyk, T. de la Rue, Furstenberg systems of pretentious and mrt multiplicative functions. Ergod. Theory Dyn. Syst. (To appear)
428. N. Frantzikinakis, E. Lesigne, M. Wierdl, Sets of k-recurrence but not $(k+1)$-recurrence. Ann. Inst. Fourier (Grenoble) **56**(4), 839–849 (2006)
429. N. Frantzikinakis, E. Lesigne, M. Wierdl, Powers of sequences and recurrence. Proc. Lond. Math. Soc. (3) **98**(2), 504–530 (2009)
430. N. Frantzikinakis, E. Lesigne, M. Wierdl, Random sequences and pointwise convergence of multiple ergodic averages. Indiana Univ. Math. J. **61**(2), 585–617 (2012)
431. N. Frantzikinakis, E. Lesigne, M. Wierdl, Random differences in Szemerédi's theorem and related results. J. Anal. Math. **130**, 91–133 (2016)
432. N. Frantzikinakis, M. Wierdl, A Hardy field extension of Szemerédi's theorem. Adv. Math. **222**(1), 1–43 (2009)
433. N.A. Friedman, J.L. King, Rank one lightly mixing. Isr. J. Math. **73**(3), 281–288 (1991)
434. N.A. Friedman, D.S. Ornstein, On isomorphism of weak Bernoulli transformations. Adv. Math. **5**, 365–394 (1970)
435. N.A. Friedman, D.S. Ornstein, On partially mixing transformations. Indiana Univ. Math. J. **20**, 767–775 (1970/71)
436. N.A. Friedman, D.S. Ornstein, On mixing and partial mixing. Ill. J. Math. **16**, 61–68 (1972)
437. N.A. Friedman, E.S. Thomas, Higher order sweeping out. Ill. J. Math. **29**(3), 401–417 (1985)
438. S. Fritzsch, A detailed proof of Bourgain's Return Times Theorem. Diploma Thesis, University of Leipzig, https://arxiv.org/abs/1901.04228, 2019
439. K. Frączek, A. Kanigowski, M. Lemańczyk, Prime number theorem for regular Toeplitz subshifts. Ergod. Theory Dyn. Syst. **42**(4), 1446–1473 (2022)
440. K. Frączek, M. Lemańczyk, On mild mixing of special flows over irrational rotations under piecewise smooth functions. Ergod. Theory Dyn. Syst. **26**(3), 719–738 (2006)
441. K. Frączek, M. Lemańczyk, Ratner's property and mild mixing for special flows over two-dimensional rotations. J. Mod. Dyn. **4**(4), 609–635 (2010)
442. K. Frączek, M. Lemańczyk, E. Lesigne, Mild mixing property for special flows under piecewise constant functions. Discrete Contin. Dyn. Syst. **19**(4), 691–710 (2007)
443. H. Furstenberg, Strict ergodicity and transformation of the torus. Am. J. Math. **83**, 573–601 (1961)
444. H. Furstenberg, The structure of distal flows. Am. J. Math. **85**, 477–515 (1963)
445. H. Furstenberg, Ergodic behavior of diagonal measures and a theorem of Szemerédi on arithmetic progressions. J. Anal. Math. **31**, 204–256 (1977)
446. H. Furstenberg, *Recurrence in Ergodic Theory and Combinatorial Number Theory*, M. B. Porter Lectures (Princeton University Press, Princeton, NJ, 1981)
447. H. Furstenberg, IP-systems in ergodic theory, in *Conference in Modern Analysis and Probability (New Haven, Connecticut, 1982)*. Volume 26 of Contemporary Mathematics (American Mathematical Society, Providence, RI, 1984), pp. 131–148
448. H. Furstenberg, Nonconventional ergodic averages, in *The Legacy of John von Neumann (Hempstead, NY, 1988)*. Volume 50 of Proceedings of Symposia in Pure Mathematics (American Mathematical Society, Providence, RI, 1990), pp. 43–56
449. H. Furstenberg, IP-systems and recurrence in ergodic theory: An update. Isr. J. Math. **251**(2), 423–441 (2022)
450. H. Furstenberg, Y. Katznelson, An ergodic Szemerédi theorem for commuting transformations. J. Anal. Math. **34**, 275–291 (1979, 1978)
451. H. Furstenberg, Y. Katznelson, An ergodic Szemerédi theorem for IP-systems and combinatorial theory. J. Anal. Math. **45**, 117–168 (1985)
452. H. Furstenberg, Y. Katznelson, IP_r-sets, Szemerédi's theorem, and Ramsey theory. Bull. Am. Math. Soc. (N.S.) **14**(2), 275–278 (1986)

453. H. Furstenberg, Y. Katznelson, Idempotents in compact semigroups and Ramsey theory. Isr. J. Math. **68**(3), 257–270 (1989)
454. H. Furstenberg, Y. Katznelson, A density version of the Hales-Jewett theorem. J. Anal. Math. **57**, 64–119 (1991)
455. H. Furstenberg, Y. Katznelson, D.S. Ornstein, The ergodic theoretical proof of Szemerédi's theorem. Bull. Am. Math. Soc. (N.S.) **7**(3), 527–552 (1982)
456. H. Furstenberg, B. Weiss, The finite multipliers of infinite ergodic transformations, in *The Structure of Attractors in Dynamical Systems (Proceedings of a Conference, North Dakota State University, Fargo, North Dakota, 1977)*. Volume 668 of Lecture Notes in Mathematics (Springer, Berlin, 1978), pp. 127–132
457. H. Furstenberg, B. Weiss, Topological dynamics and combinatorial number theory. J. Anal. Math. **34**, 61–85 (1979, 1978)
458. H. Furstenberg, B. Weiss, A mean ergodic theorem for $(1/N)\sum_{n=1}^{N} f(T^n x)g(T^{n^2} x)$, in *Convergence in Ergodic Theory and Probability (Columbus, OH, 1993)*. Volume 5 of Ohio State University Mathematical Research Institute Publication (de Gruyter, Berlin, 1996), pp. 193–227
459. G. Gallavotti, Ergodicity, ensembles, irreversibility in Boltzmann and beyond. J. Stat. Phys. **78**, 1571–1589 (1975)
460. V.F. Gaposhkin, The dependence of the rate of convergence in the strong law of large numbers for stationary processes on the rate of diminution of the correlation function. Teor. Veroyatnost. i Primenen. **26**(4), 720–733 (1981)
461. A.M. Garsia, A simple proof of E. Hopf's maximal ergodic theorem. J. Math. Mech. **14**, 381–382 (1965)
462. A.M. Garsia, More about the maximal ergodic lemma of Brunel. Proc. Natl. Acad. Sci. U.S.A. **57**, 21–24 (1967)
463. A.M. Garsia, *Topics in Almost Everywhere Convergence*. Volume 4 of Lectures in Advanced Mathematics (Markham Publishing Co., Chicago, IL, 1970)
464. I. Gelfand, M. Naimark, On the imbedding of normed rings into the ring of operators in Hilbert space. Rec. Math. [Mat. Sbornik] N.S. **12**(54), 197–213 (1943)
465. M. Gerlach, M. Kunze, Mean ergodic theorems on norming dual pairs. Ergod. Theory Dyn. Syst. **34**(4), 1210–1229 (2014)
466. N. Ghoussoub, Subadditive ergodic theory in general function spaces, in *Seminar on Functional Analysis, 1983/1984*. Volume 20 of Publications Mathématiques de l'Université Paris VII (Université Paris VII, Paris, 1984), pp. 157–177
467. N. Ghoussoub, J.M. Steele, Vector-valued subadditive processes and applications in probability. Ann. Probab. **8**(1), 83–95 (1980)
468. E. Glasner, *Ergodic Theory via Joinings*. Volume 101 of Mathematical Surveys and Monographs (American Mathematical Society, Providence, RI, 2003)
469. E. Glasner, Y. Gutman, X. Ye, Higher order regionally proximal equivalence relations for general minimal group actions. Adv. Math. **333**, 1004–1041 (2018)
470. E. Glasner, B. Weiss, Relative weak mixing is generic. Sci. China Math. **62**(1), 69–72 (2019)
471. S. Glasner, D. Maon, Rigidity in topological dynamics. Ergod. Theory Dyn. Syst. **9**(2), 309–320 (1989)
472. R.-N. Gologan, On the subadditive ergodic theorem. Stud. Cerc. Mat. **47**(5–6), 405–408 (1995)
473. A. Gomilko, M. Haase, Y. Tomilov, On rates in mean ergodic theorems. Math. Res. Lett. **18**(2), 201–213 (2011)
474. A. Gomilko, D. Kwietniak, M. Lemańczyk, Sarnak's conjecture implies the Chowla conjecture along a subsequence, in *Ergodic Theory and Dynamical Systems in Their Interactions with Arithmetics and Combinatorics*. Volume 2213 of Lecture Notes in Mathematics (Springer, Cham, 2018), pp. 237–247
475. A. Gomilko, M. Lemańczyk, T. de la Rue, Möbius orthogonality in density for zero entropy dynamical systems. Pure Appl. Funct. Anal. **5**(6), 1357–1376 (2020)

476. A. Gomilko, M. Lemańczyk, T. de la Rue, On Furstenberg systems of aperiodic multiplicative functions of Matomäki, Radziwiłł, and Tao. J. Mod. Dyn. **17**, 529–555 (2021)
477. L.W. Goodwyn, Comparing topological entropy with measure-theoretic entropy. Am. J. Math. **94**, 366–388 (1972)
478. A. Gorodnik, A. Nevo, *The Ergodic Theory of Lattice Subgroups*. Volume 172 of Annals of Mathematics Studies (Princeton University Press, Princeton, NJ, 2010)
479. W.H. Gottschalk, G.A. Hedlund, *Topological Dynamics*. Volume 36 of American Mathematical Society Colloquium Publications (American Mathematical Society, Providence, RI, 1955)
480. S. Gouëzel, A. Karlsson, Subadditive and multiplicative ergodic theorems. J. Eur. Math. Soc. (JEMS) **22**(6), 1893–1915 (2020)
481. W.T. Gowers, A new proof of Szemerédi's theorem. Geom. Funct. Anal. **11**(3), 465–588 (2001)
482. W.T. Gowers, Hypergraph regularity and the multidimensional Szemerédi theorem. Ann. Math. (2) **166**(3), 897–946 (2007)
483. R. Graham, S. Butler, *Rudiments of Ramsey Theory*. Volume 123 of CBMS Regional Conference Series in Mathematics. Published for the Conference Board of the Mathematical Sciences, Washington, DC; by the American Mathematical Society, Providence, RI, second edition, 2015
484. B. Green, T. Tao, An inverse theorem for the Gowers $U^3(G)$ norm. Proc. Edinb. Math. Soc. (2) **51**(1), 73–153 (2008)
485. B. Green, T. Tao, The primes contain arbitrarily long arithmetic progressions. Ann. Math. (2) **167**(2), 481–547 (2008)
486. B. Green, T. Tao, Linear equations in primes. Ann. Math. (2) **171**(3), 1753–1850 (2010)
487. B. Green, T. Tao, Yet another proof of Szemerédi's theorem, in *An Irregular Mind*. Volume 21 of Bolyai Society Mathematical Studies (János Bolyai Mathematical Society, Budapest, 2010), pp. 335–342
488. B. Green, T. Tao, The Möbius function is strongly orthogonal to nilsequences. Ann. Math. (2) **175**(2), 541–566 (2012)
489. B. Green, T. Tao, The quantitative behaviour of polynomial orbits on nilmanifolds. Ann. Math. (2) **175**(2), 465–540 (2012)
490. B. Green, T. Tao, On the quantitative distribution of polynomial nilsequences—erratum. Ann. Math. (2) **179**(3), 1175–1183 (2014)
491. B. Green, T. Tao, T. Ziegler, An inverse theorem for the Gowers U^4-norm. Glasg. Math. J. **53**(1), 1–50 (2011)
492. B. Green, T. Tao, T. Ziegler, An inverse theorem for the Gowers $U^{s+1}[N]$-norm. Ann. Math. (2) **176**(2), 1231–1372 (2012)
493. L.W. Green, Spectra of nilflows. Bull. Am. Math. Soc. **67**, 414–415 (1961)
494. F.P. Greenleaf, Ergodic theorems and the construction of summing sequences in amenable locally compact groups. Commun. Pure Appl. Math. **26**, 29–46 (1973)
495. F.P. Greenleaf, W.R. Emerson, Group structure and the pointwise ergodic theorem for connected amenable groups. Adv. Math **14**, 153–172 (1974)
496. J.T. Griesmer, Separating measurable recurrence and strong recurrence. Discrete Contin. Dyn. Syst. (To appear)
497. J.T. Griesmer, *Ergodic Averages, Correlation Sequences, and Sumsets* (ProQuest LLC, Ann Arbor, MI, 2009). Ph.D. Thesis, The Ohio State University
498. J.T. Griesmer, Recurrence, rigidity, and popular differences. Ergod. Theory Dyn. Syst. **39**(5), 1299–1316 (2019)
499. J.T. Griesmer, Separating topological recurrence from measurable recurrence: Exposition and extension of Kriz's example. Australas. J. Combin. **91**, 1–19 (2025)
500. S. Grivaux, IP-Dirichlet measures and IP-rigid dynamical systems: an approach via generalized Riesz products. Studia Math. **215**(3), 237–259 (2013)
501. Y. Guivarc'h, Sur la loi des grands nombres et le rayon spectral d'une marche aléatoire, in *Conference on Random Walks (Kleebach, 1979) (French)*. Volume 74 of Astérisque (Société Mathématique de France, Paris, 1980), pp. 47–98, 3.

502. Y. Gutman, W. Huang, S. Shao, X.D. Ye, Almost sure convergence of the multiple ergodic average for certain weakly mixing systems. Acta Math. Sin. (Engl. Ser.) **34**(1), 79–90 (2018)
503. Y. Gutman, Z. Lian, Maximal pronilfactors and a topological Wiener-Wintner theorem. Isr. J. Math. **251**, 495–526 (2022)
504. Y. Gutman, Z. Lian, Strictly ergodic distal models and a new approach to the Host-Kra factors. J. Funct. Anal. **284**(4), Paper No. 109779 (2023)
505. Y. Gutman, F. Manners, P.P. Varjú, The structure theory of nilspaces II: Representation as nilmanifolds. Trans. Am. Math. Soc. **371**(7), 4951–4992 (2019)
506. Y. Gutman, F. Manners, P.P. Varjú, The structure theory of nilspaces I. J. Anal. Math. **140**(1), 299–369 (2020)
507. Y. Gutman, F. Manners, P.P. Varjú, The structure theory of nilspaces III: Inverse limit representations and topological dynamics. Adv. Math. **365**, 107059, 53 (2020)
508. M. Haase, *Functional Analysis*. Volume 156 of Graduate Studies in Mathematics (American Mathematical Society, Providence, RI, 2014). An elementary introduction
509. M. Haase, N. Moriakov, On systems with quasi-discrete spectrum. Studia Math. **241**(2), 173–199 (2018)
510. P. Hagelstein, D. Herden, A. Stokolos, A theorem of Besicovitch and a generalization of the Birkhoff ergodic theorem. Proc. Am. Math. Soc. Ser. B **8**, 52–59 (2021)
511. G. Halász, Remarks on the remainder in Birkhoff's ergodic theorem. Acta Math. Acad. Sci. Hung. **28**(3–4), 389–395 (1976)
512. A.W. Hales, R.I. Jewett, Regularity and positional games. Trans. Am. Math. Soc. **106**, 222–229 (1963)
513. P.R. Halmos, In general a measure preserving transformation is mixing. Ann. Math. (2) **45**, 786–792 (1944)
514. P.R. Halmos, *Lectures on Ergodic Theory* (Chelsea Publishing Company, New York, 1960)
515. P.R. Halmos, What does the spectral theorem say? Am. Math. Mon. **70**, 241–247 (1963)
516. P.R. Halmos, *Measure Theory, 2nd printing*. Volume 18 of Graduate Texts in Mathematics (Springer, New York, NY, 1974)
517. P.R. Halmos, J. von Neumann, Operator methods in classical mechanics. II. Ann. Math. (2) **43**, 332–350 (1942)
518. J.M. Hammersley, Postulates for subadditive processes. Ann. Probab. **2**, 652–680 (1974)
519. G. Hansel, J.P. Raoult, Ergodicity, uniformity and unique ergodicity. Indiana Univ. Math. J. **23**, 221–237 (1973/1974)
520. D.M. Hardcastle, The three-dimensional Gauss algorithm is strongly convergent almost everywhere. Exp. Math. **11**(1), 131–141 (2002)
521. D.M. Hardcastle, K. Khanin, The d-dimensional Gauss transformation: strong convergence and Lyapunov exponents. Exp. Math. **11**(1), 119–129 (2002)
522. G.H. Hardy, J.E. Littlewood, A maximal theorem with function-theoretic applications. Acta Math. **54**(1), 81–116 (1930)
523. J. Hawkins, *Ergodic Dynamics—From Basic Theory to Applications*. Volume 289 of Graduate Texts in Mathematics (Springer, Cham, 2021)
524. X. He, K. Wang, A note on Möbius disjointness for skew products on a circle and a nilmanifold. Bull. Aust. Math. Soc. **107**(3), 471–482 (2023)
525. H.A. Helfgott, Expansion, divisibility and parity: An explanation, in *Combinatorial and Additive Number Theory V*. Volume 395 of Springer Proceedings in Mathematics & Statistics (Springer, Cham, 2022), pp. 199–237
526. H.A. Helfgott, M. Radziwiłł, Expansion, divisibility and parity (2021). Preprint, https://arxiv.org/abs/2103.06853
527. G. Helmberg, On the converse of Hopf's ergodic theorem. Z. Wahrscheinlichkeitstheorie und Verw. Gebiete **21**, 77–80 (1972)
528. G. Herglotz, Über Potenzreihen mit positivem, reellem Teil im Einheitskreis. Leipz. Ber. **63**, 501–511 (1911)
529. P. Hermle, H. Kreidler, A Halmos—von Neumann theorem for actions of general groups. Appl. Categ. Struct. **31**(5), 38 (2023)

530. E. Hewitt, K.A. Ross, *Abstract Harmonic Analysis. Vol. I.* Volume 115 of Grundlehren der Mathematischen Wissenschaften, 2nd edn. (Springer, Berlin, 1979)
531. E. Hewitt, K.A. Ross, *Abstract Harmonic Analysis. Vol. II: Structure and Analysis for Compact Groups. Analysis on Locally Compact Abelian Groups. 2nd printing*. Volume 152 of Grundlehren der Mathematischen Wissenschaften (Springer, Berlin, 1994)
532. E. Hewitt, K. Stromberg, *Real and Abstract Analysis. A Modern Treatment of the Theory of Functions of a Real Variable* (Springer, New York, 1969).
533. E. Hille, Remarks on ergodic theorems. Trans. Am. Math. Soc. **57**, 246–269 (1945)
534. E. Hille, R.S. Phillips, *Functional Analysis and Semi-groups*. American Mathematical Society Colloquium Publications, Vol. XXXI (American Mathematical Society, Providence, RI, 1974). Third printing of the revised edition of 1957
535. N. Hindman, Finite sums from sequences within cells of a partition of N. J. Combin. Theory Ser. A **17**, 1–11 (1974)
536. N. Hindman, D. Strauss, *Algebra in the Stone—Čech Compactification*. Volume 27 of de Gruyter Expositions in Mathematics (Walter de Gruyter & Co., Berlin, 1998)
537. M. Hirayama, D.H. Kim, Y. Son, On the multiple recurrence properties for disjoint systems. Isr. J. Math. **247**(1), 405–431 (2022)
538. M. Hochman, Note on one-sided and two-sided ergodic theorems. Einstein Institute Note (2010), available at http://math.huji.ac.il/~mhochman/preprints/two-sided-averages.pdf
539. K.H. Hofmann, S.A. Morris, *The Structure of Compact Groups—a Primer for the Student—a Handbook for the Expert*. Volume 25 of De Gruyter Studies in Mathematics (De Gruyter, Berlin, 2020). Fourth edition [of 1646190]
540. J. Hölz, A note on the uniform ergodicity of dynamical systems (2024). Preprint, https://arxiv.org/abs/2402.08482
541. E. Hopf, On the time average theorem in dynamics. Proc. Natl. Acad. USA **18**, 93–100 (1932)
542. E. Hopf, *Ergodentheorie*. Volume 5 of Ergeb. Math. Grenzgeb (Springer, Berlin, 1937)
543. E. Hopf, The general temporally discrete Markoff process. J. Ration. Mech. Anal. **3**, 13–45 (1954)
544. E. Hopf, On the ergodic theorem for positive linear operators. J. Reine Angew. Math. **205**, 101–106 (1960/1961)
545. S. Horowitz, Strong ergodic theorems for Markov processes. Proc. Am. Math. Soc. **23**, 328–334 (1969)
546. B. Host, Mixing of all orders and pairwise independent joinings of systems with singular spectrum. Isr. J. Math. **76**(3), 289–298 (1991)
547. B. Host, Ergodic seminorms for commuting transformations and applications. Studia Math. **195**(1), 31–49 (2009)
548. B. Host, B. Kra, Convergence of polynomial ergodic averages. Probab. Math. 149, 1–19 (2005)
549. B. Host, B. Kra, Nonconventional ergodic averages and nilmanifolds. Ann. Math. (2) **161**(1), 397–488 (2005)
550. B. Host, B. Kra, Analysis of two step nilsequences. Ann. Inst. Fourier (Grenoble) **58**(5), 1407–1453 (2008)
551. B. Host, B. Kra, Parallelepipeds, nilpotent groups and Gowers norms. Bull. Soc. Math. Fr. **136**(3), 405–437 (2008)
552. B. Host, B. Kra, Uniformity seminorms on ℓ^∞ and applications. J. Anal. Math. **108**, 219–276 (2009)
553. B. Host, B. Kra, Nilpotent Structures in Ergodic Theory. Volume 236 of Mathematical Surveys and Monographs (American Mathematical Society, Providence, RI, 2018)
554. B. Host, B. Kra, A. Maass, Nilsequences and a structure theorem for topological dynamical systems. Adv. Math. **224**(1), 103–129 (2010)
555. B. Host, B. Kra, A. Maass, Complexity of nilsystems and systems lacking nilfactors. J. Anal. Math. **124**, 261–295 (2014)

556. B. Host, A. Maass, Nilsystèmes d'ordre 2 et parallélépipèdes. Bull. Soc. Math. Fr. **135**(3), 367–405 (2007)
557. W. Huang, J. Liu, K. Wang, Möbius disjointness for skew products on a circle and a nilmanifold. Discrete Contin. Dyn. Syst. **41**(8), 3531–3553 (2021)
558. W. Huang, S. Shao, X. Ye, Pointwise convergence of multiple ergodic averages and strictly ergodic models. J. Anal. Math. **139**(1), 265–305 (2019)
559. W. Huang, S. Shao, X. Ye, Polynomial Furstenberg joinings and its applications (2023). Preprint, https://arxiv.org/abs/2301.07881
560. W. Huang, S. Shao, X. Ye, A counterexample on polynomial multiple convergence without commutativity. Bull. Soc. Math. Fr. **152**(1), 149–168 (2024)
561. W. Huang, Z. Wang, X. Ye, Measure complexity and Möbius disjointness. Adv. Math. **347**, 827–858 (2019)
562. W. Huang, Z. Wang, G. Zhang, Möbius disjointness for topological models of ergodic systems with discrete spectrum. J. Mod. Dyn. **14**, 277–290 (2019)
563. W. Huang, L. Xu, X. Ye, Polynomial mean complexity and logarithmic Sarnak conjecture. Ergod. Theory Dyn. Syst. **44**(3), 769–798 (2024)
564. W. Huang, X. Ye, Topological complexity, return times and weak disjointness. Ergod. Theory Dyn. Syst. **24**(3), 825–846 (2004)
565. A.D. Ionescu, E.M. Stein, A. Magyar, S. Wainger, Discrete Radon transforms and applications to ergodic theory. Acta Math. **198**(2), 231–298 (2007)
566. A.D. Ionescu, S. Wainger, L^p boundedness of discrete singular Radon transforms. J. Am. Math. Soc. **19**(2), 357–383 (2006)
567. A. Ionescu Tulcea, Analytic Continuation of Random Serie. J. Math. Mech. **9**(3), 399–410 (1960)
568. A. Ionescu Tulcea, Random series and spectra of measure-preserving transformations, in *Ergodic Theory (Proceedings of an International Symposium, Tulane University, New Orleans, LA, 1961)* (Academic, New York, 1963), pp. 273–292
569. A. Ionescu Tulcea, Ergodic properties of isometries in L^p spaces, $1 < p < \infty$. Bull. Am. Math. Soc. **70**, 366–371 (1964)
570. A. Ionescu Tulcea, On the category of certain classes of transformations in ergodic theory. Trans. Am. Math. Soc. **114**, 261–279 (1965)
571. H. Ishitani, A central limit theorem for the subadditive process and its application to products of random matrices. Publ. Res. Inst. Math. Sci. **12**(3), 565–575 (1976/1977)
572. K. Jacobs, Ergodentheorie und fastperiodische Funktionen auf Halbgruppen. Math. Z. **64**, 298–338 (1956)
573. J. James, T. Koberda, K. Lindsey, C.E. Silva, P. Speh, On ergodic transformations that are both weakly mixing and uniformly rigid. New York J. Math. **15**, 393–403 (2009)
574. G.J.O. Jameson, *The Prime Number Theorem*. Volume 53 of London Mathematical Society Student Texts (Cambridge University Press, Cambridge, 2003)
575. B. Jamison, S. Orey, W. Pruitt, Convergence of weighted averages of independent random variables. Z. Wahrscheinlichkeitstheorie und Verw. Gebiete **4**, 40–44 (1965)
576. A. Jamneshan, An uncountable Furstenberg-Zimmer structure theory. Ergod. Theory Dyn. Syst. **43**(7), 2404–2436 (2023)
577. A. Jamneshan, H. Kreidler, R. Srivastava, Ergodic structure theory and applications (2025). Internet Seminar, https://sites.google.com/view/isem28
578. A. Jamneshan, M. Pan, Uniform syndeticity in multiple recurrence. Ergod. Theory Dyn. Syst. **45**(2), 504–525 (2025)
579. A. Jamneshan, O. Shalom, T. Tao, The structure of arbitrary Conze-Lesigne systems. Commun. Am. Math. Soc. **4**, 182–229 (2024)
580. A. Jamneshan, T. Tao, An uncountable Mackey-Zimmer theorem. Studia Math. **266**(3), 241–289 (2022)
581. R.I. Jewett, The prevalence of uniquely ergodic systems. J. Math. Mech. **19**, 717–729 (1969/1970)

582. M.C.R. Johnson, Convergence of polynomial ergodic averages of several variables for some commuting transformations. Ill. J. Math. **53**(3), 865–882 (2009/2010)
583. L. Jones, V. Kuftinec, A note on the Blum-Hanson theorem. Proc. Am. Math. Soc. **30**, 202–203 (1971)
584. L.K. Jones, A mean ergodic theorem for weakly mixing operators. Adv. Math. **7**, 211–216 (1971)
585. L.K. Jones, M. Lin, Ergodic theorems of weak mixing type. Proc. Am. Math. Soc. **57**(1), 50–52 (1976)
586. L.K. Jones, M. Lin, Unimodular eigenvalues and weak mixing. J. Funct. Anal. **35**(1), 42–48 (1980)
587. R.L. Jones, New proofs for the maximal ergodic theorem and the Hardy-Littlewood maximal theorem. Proc. Am. Math. Soc. **87**(4), 681–684 (1983)
588. R.L. Jones, Ergodic theory and connections with analysis and probability. New York J. Math. **3A**(Proceedings of the New York Journal of Mathematics Conference, June 9–13, 1997), 31–67 (1997/1998)
589. R.L. Jones, On the uniqueness of the ergodic maximal function. Proc. Am. Math. Soc. **132**(4), 1087–1090 (2004)
590. R.L. Jones, R. Kaufman, J.M. Rosenblatt, M. Wierdl, Oscillation in ergodic theory. Ergod. Theory Dyn. Syst. **18**(4), 889–935 (1998)
591. R.L. Jones, M. Lacey, M. Wierdl, Integer sequences with big gaps and the pointwise ergodic theorem. Ergod. Theory Dyn. Syst. **19**(5), 1295–1308 (1999)
592. R.L. Jones, M. Lin, J. Olsen, Weighted ergodic theorems along subsequences of density zero. New York J. Math. **3A**(Proceedings of the New York Journal of Mathematics Conference, June 9–13, 1997), 89–98 (1997/1998)
593. R.L. Jones, J. Olsen, Subsequence pointwise ergodic theorems for operators in L^p. Isr. J. Math. **77**(1–2), 33–54 (1992)
594. R.L. Jones, J. Olsen, M. Wierdl, Subsequence ergodic theorems for L^p contractions. Trans. Am. Math. Soc. **331**(2), 837–850 (1992)
595. R.L. Jones, J.M. Rosenblatt, M. Wierdl, Oscillation inequalities for rectangles. Proc. Am. Math. Soc. **129**(5), 1349–1358 (2001)
596. R.L. Jones, J.M. Rosenblatt, M. Wierdl, Oscillation in ergodic theory: Higher dimensional results. Isr. J. Math. **135**, 1–27 (2003)
597. R.L. Jones, M. Wierdl, Convergence and divergence of ergodic averages. Ergod. Theory Dyn. Syst. **14**(3), 515–535 (1994)
598. R.L. Jones, M. Wierdl, Convergence of ergodic averages, in *Convergence in Ergodic Theory and Probability (Columbus, OH, 1993)*. Volume 5 of Ohio State University Mathematical Research Institute Publications (de Gruyter, Berlin, 1996), pp. 229–247
599. M. Kac, On the notion of recurrence in discrete stochastic processes. Bull. Am. Math. Soc. **53**, 1002–1010 (1947)
600. A.G. Kachurovskiĭ, Rates of convergence in ergodic theorems. Uspekhi Mat. Nauk **51**(4(310)), 73–124 (1996)
601. A.G. Kachurovskiĭ, I.V. Podvigin, Large deviations and the rate of convergence in the Birkhoff ergodic theorem. Math. Notes **94**(3–4), 524–531 (2013). Russian version appears in Mat. Zametki **94**(4), 569–577 (2013)
602. A.G. Kachurovskiĭ, I.V. Podvigin, Estimates of the rate of convergence in the von Neumann and Birkhoff ergodic theorems. Trans. Moscow Math. Soc. 1–53 (2016)
603. A.G. Kachurovskiĭ, I.V. Podvigin, On measuring the rate of convergence in the Birkhoff ergodic theorem. Mat. Zametki **106**(1), 40–52 (2019)
604. A.G. Kachurovskiĭ, V.V. Sedalishchev, On the constants in the estimates for the rate of convergence in von Neumann's ergodic theorem. Mat. Zametki **87**(5), 756–763 (2010)
605. A.G. Kachurovskiĭ, V.V. Sedalishchev, Constants of estimates for the rate of convergence in the von Neumann and Birkhoff ergodic theorems. Mat. Sb. **202**(8), 21–40 (2011)
606. V. Kadets, *A Course in Functional Analysis and Measure Theory*. Universitext (Springer, Cham, 2018). Translated from the 2006 Russian edition by Andrei Iacob

607. J.-P. Kahane, Sur les coefficients de Fourier-Bohr. Studia Math. **21**, 103–106 (1961/1962)
608. J.-P. Kahane, *Some Random Series of Functions*. Volume 5 of Cambridge Studies in Advanced Mathematics, 2nd edn. (Cambridge University Press, Cambridge, 1985)
609. S. Kakutani, Über die Metrisation der topologischen Gruppen. Proc. Imp. Acad. Tokyo **12**(4), 82–84 (1936)
610. S. Kakutani, Iteration of linear operations in complex Banach spaces. Proc. Imp. Acad. Tokyo **14**(8), 295–300 (1938)
611. S. Kakutani, Two fixed-point theorems concerning bicompact convex sets. Proc. Imp. Acad. Tokyo **14**(7), 242–245 (1938)
612. S. Kakutani, Ergodic theorems and the Markoff process with a stable distribution. Proc. Imp. Acad. Tokyo **16**, 49–54 (1940)
613. S. Kakutani, Examples of ergodic measure preserving transformations which are weakly mixing but not strongly mixing, in *Recent Advances in Topological Dynamics (Proceedings of the Conference on Topological Dynamics, Yale University, New Haven, CT, 1972; in honor of Gustav Arnold Hedlund)*. Lecture Notes in Mathematics, vol. 318 (Springer, Berlin, 1973), pp. 143–149
614. S. Kakutani, W. Parry, Infinite measure preserving transformations with "mixing". Bull. Am. Math. Soc. **69**, 752–756 (1963)
615. S. Kakutani, K. Petersen, The speed of convergence in the ergodic theorem. Monatsh. Math. **91**(1), 11–18 (1981)
616. S.A. Kalikow, Twofold mixing implies threefold mixing for rank one transformations. Ergod. Theory Dyn. Syst. **4**(2), 237–259 (1984)
617. S.A. Kalikow, R. McCutcheon, *An Outline of Ergodic Theory*. Volume 122 of Cambridge Studies in Advanced Mathematics (Cambridge University Press, Cambridge, 2010)
618. T. Kamae, A simple proof of the ergodic theorem using nonstandard analysis. Isr. J. Math. **42**(4), 284–290 (1982)
619. T. Kamae, M. Keane, A simple proof of the ratio ergodic theorem. Osaka J. Math. **34**(3), 653–657 (1997)
620. C.H. Kan, Ergodic properties of Lamperti operators. Can. J. Math. **30**(6), 1206–1214 (1978)
621. A. Kanigowski, J. Kułaga-Przymus, Ratner's property and mild mixing for smooth flows on surfaces. Ergod. Theory Dyn. Syst. **36**(8), 2512–2537 (2016)
622. A. Kanigowski, J. Kułaga-Przymus, M. Lemańczyk, T. de la Rue, On arithmetic functions orthogonal to deterministic sequences. Adv. Math. **428**, Paper No. 109138, 68 (2023)
623. A. Kanigowski, M. Lemańczyk, M. Radziwiłł, Rigidity in dynamics and Möbius disjointness. Fund. Math. **255**(3), 309–336 (2021)
624. A. Kanigowski, M. Lemańczyk, M. Radziwiłł, Prime number theorem for analytic skew products. Ann. Math. (2) **199**(2), 591–705 (2024)
625. A. Kanigowski, M. Lemańczyk, C. Ulcigrai, On disjointness properties of some parabolic flows. Invent. Math. **221**(1), 1–111 (2020)
626. D. Karagulyan, On Möbius orthogonality for interval maps of zero entropy and orientation-preserving circle homeomorphisms. Ark. Mat. **53**(2), 317–327 (2015)
627. D. Karagulyan, On Möbius orthogonality for subshifts of finite type with positive topological entropy. Studia Math. **237**(3), 277–282 (2017)
628. D. Karagulyan, Hausdorff dimension of a class of three-interval exchange maps. Discrete Contin. Dyn. Syst. **40**(3), 1257–1281 (2020)
629. S. Karlin, Positive operators. J. Math. Mech. **8**, 907–937 (1959)
630. A. Karlsson, A proof of the subadditive ergodic theorem, in *Groups, Graphs and Random Walks*. Volume 436 of London Mathematical Society Lecture Note Series (Cambridge University Press, Cambridge, 2017), pp. 343–354
631. A. Karlsson, G.A. Margulis, A multiplicative ergodic theorem and nonpositively curved spaces. Commun. Math. Phys. **208**(1), 107–123 (1999)
632. I. Kátai, A remark on a theorem of H. Daboussi. Acta Math. Hung. **47**(1–2), 223–225 (1986)
633. A. Katok, Approximation and generity in abstract ergodic theory. Notes (1985)

634. A. Katok, B. Hasselblatt, *Introduction to the Modern Theory of Dynamical Systems*. Volume 54 of Encyclopedia of Mathematics and Its Applications (Cambridge University Press, Cambridge, 1995). With a supplementary chapter by Katok and Leonardo Mendoza
635. Y. Katznelson, B. Weiss, A simple proof of some ergodic theorems. Isr. J. Math. **42**(4), 291–296 (1982)
636. V. Kazakevičius, Subadditive ergodic theorem for double sequences. J. Theor. Probab. **34**(1), 307–330 (2021)
637. M. Keane, K. Petersen, Easy and nearly simultaneous proofs of the ergodic theorem and maximal ergodic theorem, in *Dynamics & Stochastics*. Volume 48 of IMS Lecture Notes Monograph Series (Institute of Mathematical Statistics, Beachwood, OH, 2006), pp. 248–251
638. A.S. Kechris, *Classical Descriptive Set Theory*. Volume 156 of Graduate Texts in Mathematics (Springer, New York, 1995)
639. W.T. Kelvin, On some test case for the Maxwell–Boltzmann doctrine regarding distribution of energy. Roy. Soc. Proc. **L**, 79–88 (1891). Nature **XLIV**, 355–358
640. D. Kerr, H. Li, *Ergodic Theory*. Springer Monographs in Mathematics (Springer, Cham, 2016). Independence and dichotomies
641. M. Kesseböhmer, S. Munday, B.O. Stratmann, *Infinite Ergodic Theory of Numbers* (De Gruyter, Berlin, 2016)
642. H. Kesten, Sums of stationary sequences cannot grow slower than linearly. Proc. Am. Math. Soc. **49**, 205–211 (1975)
643. A.Y. Khinchin, *Three Pearls of Number Theory* (Graylock Press, Rochester, NY, 1952)
644. A.Y. Khinchin, *Continued Fractions* (Dover Publications, Inc., Mineola, NY, 1997). Translated from the Russian edition. With a preface by B. V. Gnedenko, Reprint of the 1964 translation
645. A. Khintchine, Zu Birkhoffs Lösung des Ergodenproblems. Math. Ann. **107**(1), 485–488 (1933)
646. A. Khintchine, Eine Verschärfung des Poincaréschen "Wiederkehrsatzes". Compositio Math. **1**, 177–179 (1935)
647. Y. Kifer, *Ergodic Theory of Random Transformations*. Volume 10 of Progress in Probability and Statistics (Birkhäuser Boston, Inc., Boston, MA, 1986)
648. J. Kim, Ergodic averages with the Hecke eigenvalue square weights and the Piltz divisor function weights. J. Math. Anal. Appl. **531**(2), Paper No. 127877 (2024)
649. J.F.C. Kingman, The ergodic theory of subadditive stochastic processes. J. Roy. Stat. Soc. Ser. B **30**, 499–510 (1968)
650. J.F.C. Kingman, Subadditive ergodic theory. Ann. Probab. **1**, 883–909 (1973)
651. J.F.C. Kingman, Subadditive processes, in *École d'Été de Probabilités de Saint-Flour, V–1975*. Lecture Notes in Mathematics, vol. 539 (1976), pp. 167–223
652. E.M. Klimko, L. Sucheston, An operator ergodic theorem for sequences of functions. Proc. Am. Math. Soc. **20**, 272–276 (1969)
653. O. Klurman, A.P. Mangerel, J. Teräväinen, On Elliott's conjecture and applications (2024). Preprint, https://arxiv.org/abs/2304.05344
654. A. Kolmogoroff, Ein vereinfachter Beweis des Birkhoff–Khintchinchen Ergodensatzes. Recueil Math. **44**, 366–368 (1937)
655. A.N. Kolmogorov, Sur la loi forte des grands nombres. CRAS Paris **191**, 910–912 (1930)
656. A.N. Kolmogorov, A new metric invariant of transient dynamical systems and automorphisms in Lebesgue spaces. Dokl. Akad. Nauk SSSR (N.S.) **119**, 861–864 (1958)
657. J. Konieczny, Möbius orthogonality for q-semimultiplicative sequences. Monatsh. Math. **192**(4), 853–882 (2020)
658. B.O. Koopman, Hamiltonian systems and transformations in Hilbert space. Proc. Natl. Acad. Sci. U.S.A. **17**, 315–318 (1931)
659. B.O. Koopman, J. von Neumann, Dynamical systems of continuous spectra. Proc. Natl. Acad. Sci. U.S.A. **18**, 255–263 (1932)

660. I. Kornfeld, W. Kosek, Positive L^1 operators associated with nonsingular mappings and an example of E. Hille. Colloq. Math. **98**(1), 63–77 (2003)
661. I.P. Kornfeld, The additive homological equation. Funkcional. Anal. i Priložen. **10**(2), 73–74 (1976)
662. W. Kosek, Example of a mean ergodic L^1 operator with the linear rate of growth. Colloq. Math. **124**(1), 15–22 (2011)
663. Z. Kosloff, S. Sanadhya, Multidimensional local limit theorem in deterministic systems and an application to non-convergence of polynomial multiple averages (2024). Preprint, https://arxiv.org/abs/2409.05087
664. D. Kosz, M. Mirek, S. Peluse, J. Wright, The multilinear circle method and a question of Bergelson (2024). Preprint, https://arxiv.org/abs/2411.09478
665. A. Koutsogiannis, Closest integer polynomial multiple recurrence along shifted primes. Ergod. Theory Dyn. Syst. **38**(2), 666–685 (2018)
666. A. Koutsogiannis, A.N. Le, J. Moreira, F.K. Richter, Structure of multicorrelation sequences with integer part polynomial iterates along primes. Proc. Am. Math. Soc. **149**(1), 209–216 (2021)
667. A. Koutsogiannis, K. Tsinas, Ergodic averages for sparse sequences along primes (2023). Preprint, https://arxiv.org/abs/2309.04939
668. B. Kra, The Green-Tao theorem on arithmetic progressions in the primes: An ergodic point of view. Bull. Am. Math. Soc. (N.S.) **43**(1), 3–23 (2006)
669. B. Kra, Ergodic methods in additive combinatorics, in *Additive Combinatorics*. Volume 43 of CRM Proceedings & Lecture Notes (American Mathematical Society, Providence, RI, 2007), pp. 103–143
670. B. Krause, Polynomial ergodic averages converge rapidly: Variations on a theorem of Bourgain (2014). Preprint, https://arxiv.org/abs/1402.1803
671. B. Krause, *Discrete Analogues in Harmonic Analysis—Bourgain, Stein, and Beyond*. Volume 224 of Graduate Studies in Mathematics (American Mathematical Society, Providence, RI, 2022). With a contribution by Mariusz Mirek
672. B. Krause, Pointwise ergodic theory: Examples and entropy. Astérisque **446**(1199), 87–120 (2023)
673. B. Krause, Multi-frequency oscillation estimates arising in pointwise ergodic theory (2025). Preprint, https://arxiv.org/abs/2502.12887
674. B. Krause, A unified approach to two pointwise ergodic theorems: double recurrence and return times (2025). Preprint, https://arxiv.org/abs/2501.06877
675. B. Krause, M. Mirek, T. Tao, Pointwise ergodic theorems for non-conventional bilinear polynomial averages. Ann. Math. (2) **195**(3), 997–1109 (2022)
676. B. Krause, H. Mousavi, T. Tao, J. Teräväinen, Pointwise convergence of bilinear polynomial averages over the primes (2024). Preprint, https://arxiv.org/abs/2409.10510
677. B. Krause, P. Zorin-Kranich, A random pointwise ergodic theorem with Hardy field weights. Ill. J. Math. **59**(3), 663–674 (2015)
678. B. Krause, P. Zorin-Kranich, A uniform random pointwise ergodic theorem (2017). Preprint, https://arxiv.org/abs/1708.05022
679. U. Krengel, On the global limit behaviour of Markov chains and of general nonsingular Markov processes. Z. Wahrscheinlichkeitstheorie und Verw. Gebiete **6**, 302–316 (1966)
680. U. Krengel, On the individual ergodic theorem for subsequences. Ann. Math. Stat. **42**, 1091–1095 (1971)
681. U. Krengel, Weakly wandering vectors and weakly independent partitions. Trans. Am. Math. Soc. **164**, 199–226 (1972)
682. U. Krengel, On the speed of convergence in the ergodic theorem. Monatsh. Math. **86**(1), 3–6 (1978/1979)
683. U. Krengel, *Ergodic Theorems*. Volume 6 of De Gruyter Studies in Mathematics (Walter de Gruyter & Co., Berlin, 1985). With a supplement by Antoine Brunel

684. U. Krengel, *Einführung in die Wahrscheinlichkeitstheorie und Statistik*. Volume 59 of Vieweg Studium: Aufbaukurs Mathematik [Vieweg Studies: Mathematics Course] (Friedr. Vieweg & Sohn, Braunschweig, 1988)
685. W. Krieger, On unique ergodicity, in *Proceedings of the Sixth Berkeley Symposium on Mathematical Statistics and Probability (University of California, Berkeley, CA, 1970/1971), Vol. II: Probability Theory* (University of California Press, Berkeley, CA, 1972), pp. 327–346
686. L. Kronecker, Quelques remarques sur la détermination des valeurs moyennes. C. R. Acad. Sci. Paris **103**, 980–987 (1887)
687. N. Kryloff, N. Bogoliouboff, La théorie générale de la mesure dans son application à l'étude des systèmes dynamiques de la mécanique non linéaire. Ann. Math. (2) **38**(1), 65–113 (1937)
688. R. Kuang, X. Ye, Mixing via families for measure preserving transformations. Colloq. Math. **110**(1), 151–165 (2008)
689. B. Kuca, On several notions of complexity of polynomial progressions. Ergod. Theory Dyn. Syst. **43**(4), 1269–1323 (2023)
690. L. Kuipers, H. Niederreiter, *Uniform Distribution of Sequences*. Pure and Applied Mathematics (Interscience Publishers [Wiley, Inc.], New York, 1974)
691. J. Kułaga-Przymus, M. Lemańczyk, The Möbius function and continuous extensions of rotations. Monatsh. Math. **178**(4), 553–582 (2015)
692. J. Kułaga-Przymus, M. Lemańczyk, Möbius disjointness along ergodic sequences for uniquely ergodic actions. Ergod. Theory Dyn. Syst. **39**(10), 2793–2826 (2019)
693. J. Kułaga-Przymus, M. Lemańczyk, Sarnak's conjecture from the ergodic theory point of view, in *Ergodic Theory*. Encyclopedia of Complexity and Systems Science (Springer, New York, 2023), pp. 293–311
694. A. Kulik, *Introduction to Ergodic Rates for Markov Chains and Processes*. Volume 2 of Lectures in Pure and Applied Mathematics (Potsdam University Press, Potsdam, 2015). With applications to limit theorems
695. P. Kunde, Uniform rigidity sequences for weak mixing diffeomorphisms on $\mathbb{D}^2$, $\mathbb{A}$ and $\mathbb{T}^2$. J. Math. Anal. Appl. **429**(1), 111–130 (2015)
696. P. Kunde, Uniform rigidity sequences for weakly mixing diffeomorphisms on $\mathbb{D}^m$, $\mathbb{T}^m$ and $\mathbb{S}^1 \times [0, 1]^{m-1}$. J. Math. Anal. Appl. **462**(2), 1398–1424 (2018)
697. D. Kunszenti-Kovács, On the limit of square-Cesàro means of contractions on Hilbert spaces. Arch. Math. (Basel) **94**(5), 459–466 (2010)
698. D. Kunszenti-Kovács, Almost weak polynomial stability of operators. Houst. J. Math. **41**(3), 901–913 (2015)
699. D. Kunszenti-Kovács, R. Nittka, M. Sauter, On the limits of Cesàro means of polynomial powers. Math. Z. **268**(3–4), 771–776 (2011)
700. K.V. Küster, *Topological Dynamics via Structured Koopman Subsystems*. PhD thesis, Eberhard Karls Universität Tübingen (2021)
701. J. Kwiatkowski, Classification of nonergodic dynamical systems with discrete spectra. Comment. Math. Prace Mat. **22**(2), 263–274 (1980/1981)
702. M. Lacey, On an inequality due to Bourgain. Ill. J. Math. **41**(2), 231–236 (1997)
703. J. Lamperti, On the isometries of certain function-spaces. Pac. J. Math. **8**, 459–466 (1958)
704. E. Landau, New proof of the equation $\sum_{k=1}^{\infty} \frac{\mu(k)}{k} = 0$. Inaugural-Dissertation, Berlin, 1899
705. E. Landau, Über einige neuere Grenzwertsätze. Rend. Circ. Mat. Palermo (34), 121–131 (1912). Collected Works, vol. 5 (Thales Verlag, Essen, 1986), pp. 145–155
706. S. Lang, *Algebra*. Volume 211 of Graduate Texts in Mathematics, 3rd edn. (Springer, New York, 2005)
707. P. LaVictoire, An L^1 ergodic theorem for sparse random subsequences. Math. Res. Lett. **16**(5), 849–859 (2009)
708. P. LaVictoire, Universally L^1-bad arithmetic sequences. J. Anal. Math. **113**, 241–263 (2011)

709. A.N. Le, Nilsequences and multiple correlations along subsequences. Ergod. Theory Dyn. Syst. **40**(6), 1634–1654 (2020)
710. A.N. Le, J. Moreira, F.K. Richter, A decomposition of multicorrelation sequences for commuting transformations along primes. Discrete Anal., Paper No. 4, 27 (2021)
711. T.H. Lê, J. Wolf, Polynomial configurations in the primes. Int. Math. Res. Not. IMRN (23), 6448–6473 (2014)
712. J.L. Lebowitz, O. Penrose, Modern ergodic theory. Phys. Today **26**(2), 23–29 (1973)
713. F. Ledrappier, Un champ markovien peut être d'entropie nulle et mélangeant. C. R. Acad. Sci. Paris Sér. A–B **287**(7), A561–A563 (1978)
714. A. Leibman, Multiple recurrence theorem for measure preserving actions of a nilpotent group. Geom. Funct. Anal. **8**(5), 853–931 (1998)
715. A. Leibman, Polynomial sequences in groups. J. Algebra **201**(1), 189–206 (1998)
716. A. Leibman, Lower bounds for ergodic averages. Ergod. Theory Dyn. Syst. **22**(3), 863–872 (2002)
717. A. Leibman, Convergence of multiple ergodic averages along polynomials of several variables. Isr. J. Math. **146**, 303–315 (2005)
718. A. Leibman, Host–Kra and Ziegler factors and convergence of multiple averages, in *Handbook of Dynamical Systems*, ed. by B. Hasselblatt, A. Katok, vol. 1B (Elsevier, 2005), pp. 841–853
719. A. Leibman, Pointwise convergence of ergodic averages for polynomial sequences of translations on a nilmanifold. Ergod. Theory Dyn. Syst. **25**(1), 201–213 (2005)
720. A. Leibman, Multiple polynomial correlation sequences and nilsequences. Ergod. Theory Dyn. Syst. **30**(3), 841–854 (2010)
721. A. Leibman, Orbit of the diagonal in the power of a nilmanifold. Trans. Am. Math. Soc. **362**(3), 1619–1658 (2010)
722. A. Leibman, Nilsequences, null-sequences, and multiple correlation sequences. Ergod. Theory Dyn. Syst. **35**(1), 176–191 (2015)
723. M. Lemańczyk, C. Müllner, Automatic sequences are orthogonal to aperiodic multiplicative functions. Discrete Contin. Dyn. Syst. **40**(12), 6877–6918 (2020)
724. M. Lemańczyk, F. Parreau, Rokhlin extensions and lifting disjointness. Ergod. Theory Dyn. Syst. **23**(5), 1525–1550 (2003)
725. M. Lemańczyk, F. Parreau, Lifting mixing properties by Rokhlin cocycles. Ergod. Theory Dyn. Syst. **32**(2), 763–784 (2012)
726. J. Leng, Efficient equidistribution of nilsequences (2023). Preprint, https://arxiv.org/abs/2312.10772v5
727. J. Leng, Efficient equidistribution of periodic nilsequences and applications (2023). Preprint, https://arxiv.org/abs/2306.13820
728. J. Leng, Structured extensions and multi-correlation sequences (2025). Preprint, https://arxiv.org/abs/2504.07038
729. J. Leng, A. Sah, M. Sawhney, Improved bounds for Szemerédi's theorem (2024). Preprint, https://arxiv.org/abs/2402.17995
730. J. Leng, A. Sah, M. Sawhney, Quasipolynomial bounds on the inverse theorem for the Gowers $U^{s+1}[N]$-norm (2024). Preprint, https://arxiv.org/abs/2402.17994
731. E. Lesigne, Théorèmes ergodiques pour une translation sur un nilvariété. Ergod. Theory Dyn. Syst. **9**(1), 115–126 (1989)
732. E. Lesigne, Un théorème de disjonction de systèmes dynamiques et une généralisation du théorème ergodique de Wiener-Wintner. Ergod. Theory Dyn. Syst. **10**(3), 513–521 (1990)
733. E. Lesigne, Sur une nil-variété, les parties minimales associées à une translation sont uniquement ergodiques. Ergod. Theory Dyn. Syst. **11**(2), 379–391 (1991)
734. E. Lesigne, Spectre quasi-discret et théorème ergodique de Wiener-Wintner pour les polynômes. Ergod. Theory Dyn. Syst. **13**(4), 767–784 (1993)
735. E. Lesigne, C. Mauduit, Propriétés ergodiques des suites q-multiplicatives. Compositio Math. **100**(2), 131–169 (1996)

736. E. Lesigne, C. Mauduit, B. Mossé, Le théorème ergodique le long d'une suite q-multiplicative. Compositio Math. **93**(1), 49–79 (1994)
737. S. Levental, A proof of Liggett's version of the subadditive ergodic theorem. Proc. Am. Math. Soc. **102**(1), 169–173 (1988)
738. Z. Lian, R. Shi, A counter-example for polynomial version of Sarnak's conjecture. Adv. Math. **384**, Paper No. 107765, 14 (2021)
739. P. Liardet, D. Volný, Sums of continuous and differentiable functions in dynamical systems. Isr. J. Math. **98**, 29–60 (1997)
740. T.M. Liggett, An improved subadditive ergodic theorem. Ann. Probab. **13**(4), 1279–1285 (1985)
741. Y. Lima, $\mathbb{Z}^d$-actions with prescribed topological and ergodic properties. Ergod. Theory Dyn. Syst. **32**(1), 191–209 (2012)
742. M. Lin, Problems in the ergodic theory of Markov processes. Ph.D. thesis, Hebrew University, 1971
743. M. Lin, On the uniform ergodic theorem. Proc. Am. Math. Soc. **43**, 337–340 (1974)
744. M. Lin, Quasi-compactness and uniform ergodicity of Markov operators. Ann. Inst. H. Poincaré Sect. B (N.S.) **11**(4), 345–354 (1975/1976)
745. M. Lin, Quasi-compactness and uniform ergodicity of positive operators. Isr. J. Math. **29**(2–3), 309–311 (1978)
746. M. Lin, Dowker's ergodic theorem by the Chacon–Ornstein theorem. Anal. Math. **50**, 625–628 (2024)
747. M. Lin, J. Olsen, A. Tempelman, On modulated ergodic theorems for Dunford-Schwartz operators. Ill. J. Math. **43**(3), 542–567 (1999)
748. M. Lin, D. Shoikhet, L. Suciu, Remarks on uniform ergodic theorems. Acta Sci. Math. (Szeged) **81**(1–2), 251–283 (2015)
749. M. Lin, R. Sine, Ergodic theory and the functional equation $(I - T)x = y$. J. Operator Theory **10**(1), 153–166 (1983)
750. M. Lin, M. Weber, Weighted ergodic theorems and strong laws of large numbers. Ergod. Theory Dyn. Syst. **27**(2), 511–543 (2007)
751. E. Lindenstrauss, Pointwise theorems for amenable groups. Invent. Math. **146**(2), 259–295 (2001)
752. J. Liu, P. Sarnak, The Möbius function and distal flows. Duke Math. J. **164**(7), 1353–1399 (2015)
753. J. Liu, P. Sarnak, The Möbius disjointness conjecture for distal flows, in *Proceedings of the Sixth International Congress of Chinese Mathematicians. Vol. I.* Volume 36 of Advanced Lectures in Mathematics (ALM) (International Press, Somerville, MA, 2017), pp. 327–335
754. P.-D. Liu, M. Qian, *Smooth Ergodic Theory of Random Dynamical Systems*. Volume 1606 of Lecture Notes in Mathematics (Springer, Berlin, 1995)
755. S.P. Lloyd, On the mean ergodic theorem of Sine. Proc. Am. Math. Soc. **56**, 121–126 (1976)
756. S.P. Lloyd, On the uniform ergodic theorem of Lin. Proc. Am. Math. Soc. **83**(4), 710–714 (1981)
757. E.R. Lorch, Means of iterated transformations in reflexive vector spaces. Bull. Am. Math. Soc. **45**, 945–947 (1939)
758. V. Losert, A remark on the strong sweeping out property, in *Convergence in Ergodic Theory and Probability (Columbus, OH, 1993)*. Volume 5 of Ohio State University Mathematical Research Institute Publications (de Gruyter, Berlin, 1996), pp. 291–294
759. H.P. Lotz, Uniform ergodic theorems for Markov operators on $C(X)$. Math. Z. **178**(2), 145–156 (1981)
760. K. Loyd, A dynamical approach to the asymptotic behavior of the sequence $\Omega(n)$. Ergod. Theory Dyn. Syst. **43**(11), 3685–3706 (2023)
761. K. Loyd, S. Mondal, Ergodic averages along sequences of slow growth (2024). Preprint, https://arxiv.org/abs/2410.09769
762. C. Luo, Y. Zhao, A note on measure-theoretic equicontinuity and rigidity. Acta Math. Sci. Ser. B (Engl. Ed.) **42**(2), 769–773 (2022)

763. R. Lyons, Seventy years of Rajchman measures. J. Fourier Anal. Appl., Special Issue, 363–377 (1995). Proceedings of the Conference in Honor of Jean-Pierre Kahane (Orsay, 1993)
764. D. Maharam, Automorphisms of products of measure spaces. Proc. Am. Math. Soc. **9**, 702–707 (1958)
765. P. Major, A counterexample in ergodic theory. Acta Sci. Math. (Szeged) **62**(1–2), 247–258 (1996)
766. P.T. Maker, The ergodic theorem for a sequence of functions. Duke Math. J. **6**, 27–30 (1940)
767. A.I. Mal'cev, On a class of homogeneous spaces. Izvestiya Akad. Nauk. SSSR. Ser. Mat. **13**, 9–32 (1949)
768. R. Mañé, *Ergodic Theory and Differentiable Dynamics*. Volume 8 of Ergebnisse der Mathematik und ihrer Grenzgebiete (3) [Results in Mathematics and Related Areas (3)] (Springer, Berlin, 1987). Translated from the Portuguese by Silvio Levy
769. F. Manners, Quantitative bounds in the inverse theorem for the Gowers U^{s+1}-norms over cyclic groups (2018). Preprint, https://arxiv.org/abs/1811.00718
770. B. Marcus, The horocycle flow is mixing of all degrees. Invent. Math. **46**(3), 201–209 (1978)
771. B. Marcus, K. Petersen, Balancing ergodic averages, in *Ergodic Theory (Proceedings of the Conference on Math. Forschungsinst., Oberwolfach, 1978)*. Volume 729 of Lecture Notes in Mathematics (Springer, Berlin, 1979), pp. 126–143
772. A. Markov, Quelques théoremes sur les ensembles abeliens. C. R. (Dokl.) Acad. Sci. URSS, n. Ser. **1936**(1), 311–313 (1936)
773. F.J. Martín-Reyes, E. de la Rosa, Ergodic theorems for Cesàro bounded operators in L^1. J. Math. Anal. Appl. **509**(2), Paper No. 125973, 26 (2022)
774. F.J. Martín-Reyes, A. De la Torre, The dominated ergodic estimate for mean bounded, invertible, positive operators. Proc. Am. Math. Soc. **104**(1), 69–75 (1988)
775. M. Mathieu, On the origin of the notion 'ergodic theory'. Expo. Math. **6**, 373–377 (1988)
776. K. Matomäki, M. Radziwiłł, Multiplicative functions in short intervals. Ann. Math. (2) **183**(3), 1015–1056 (2016)
777. K. Matomäki, M. Radziwiłł, T. Tao, An averaged form of Chowla's conjecture. Algebra Number Theory **9**(9), 2167–2196 (2015)
778. K. Matomäki, X. Shao, T. Tao, J. Teräväinen, Higher uniformity of arithmetic functions in short intervals I. All intervals. Forum Math. Pi **11**, Paper No. e29, 97 (2023)
779. L. Matthiesen, J. Teräväinen, M. Wang, Quantitative asymptotics for polynomial patterns in the primes (2024). Preprint, https://arxiv.org/abs/2405.12190
780. C. Mauduit, J. Rivat, Prime numbers along Rudin-Shapiro sequences. J. Eur. Math. Soc. (JEMS) **17**(10), 2595–2642 (2015)
781. M. Mbekhta, J. Zemánek, Sur le théorème ergodique uniforme et le spectre. C. R. Acad. Sci. Paris Sér. I Math. **317**(12), 1155–1158 (1993)
782. R. McCutcheon, Three results in recurrence, in *Ergodic Theory and Its Connections with Harmonic Analysis (Alexandria, 1993)*. Volume 205 of London Mathematical Society Lecture Note Series (Cambridge University Press, Cambridge, 1995), pp. 349–358
783. R. McCutcheon, *Elemental Methods in Ergodic Ramsey Theory*. Volume 1722 of Lecture Notes in Mathematics (Springer, Berlin, 1999)
784. R. McCutcheon, FVIP systems and multiple recurrence. Isr. J. Math. **146**, 157–188 (2005)
785. R. McCutcheon, A. Quas, Generalized polynomials and mild mixing. Can. J. Math. **61**(3), 656–673 (2009)
786. R. McNamara, Sarnak's conjecture for sequences of almost quadratic word growth. Ergod. Theory Dyn. Syst. **41**(10), 3060–3115 (2021)
787. N. Mehlhop, Vector-valued maximal inequalities and multiparameter oscillation inequalities for the polynomial ergodic averages along multi-dimensional subsets of primes. J. Fourier Anal. Appl. **30**(6), Paper No. 61, 30 (2024)
788. D. Menchoff, Sur les séries de fonctions orthogonales. I: La convergence. Fundam. Math. **4**, 82–105 (1923)

789. L.D. Mešalkin, A case of isomorphism of Bernoulli schemes. Dokl. Akad. Nauk SSSR **128**, 41–44 (1959)
790. S. Meyn, R.L. Tweedie, *Markov Chains and Stochastic Stability*, 2nd edn. (Cambridge University Press, Cambridge, 2009). With a prologue by Peter W. Glynn
791. M. Mirek, $\ell^p(\mathbb{Z})$-boundedness of discrete maximal functions along thin subsets of primes and pointwise ergodic theorems. Math. Z. **279**(1–2), 27–59 (2015)
792. M. Mirek, Weak type (1, 1) inequalities for discrete rough maximal functions. J. Anal. Math. **127**, 247–281 (2015)
793. M. Mirek, Square function estimates for discrete Radon transforms. Anal. PDE **11**(3), 583–608 (2018)
794. M. Mirek, E.M. Stein, B. Trojan, $\ell^p(\mathbb{Z}^d)$-estimates for discrete operators of Radon type: Variational estimates. Invent. Math. **209**(3), 665–748 (2017)
795. M. Mirek, E.M. Stein, B. Trojan, $\ell^p(\mathbb{Z}^d)$-estimates for discrete operators of Radon type: Maximal functions and vector-valued estimates. J. Funct. Anal. **277**(8), 2471–2521 (2019)
796. M. Mirek, E.M. Stein, P. Zorin-Kranich, Jump inequalities for translation-invariant operators of Radon type on $\mathbb{Z}^d$. Adv. Math. **365**, 107065, 57 (2020)
797. M. Mirek, T.Z. Szarek, J. Wright, Oscillation inequalities in ergodic theory and analysis: One-parameter and multi-parameter perspectives. Rev. Mat. Iberoam. **38**(7), 2249–2284 (2022)
798. M. Mirek, B. Trojan, Discrete maximal functions in higher dimensions and applications to ergodic theory. Am. J. Math. **138**(6), 1495–1532 (2016)
799. M. Mirek, B. Trojan, P. Zorin-Kranich, Variational estimates for averages and truncated singular integrals along the prime numbers. Trans. Am. Math. Soc. **369**(8), 5403–5423 (2017)
800. S. Mondal, Grid method for divergence of averages. Ergod. Theory Dyn. Syst. **44**(2), 646–664 (2024)
801. S. Mondal, J. Rosenblatt, M. Wierdl, Fluctuation of ergodic averages and other stochastic processes (2024). Preprint, https://arxiv.org/abs/2404.11507
802. S. Mondal, M. Roy, M. Wierdl, Sublacunary sequences that are strong sweeping out. New York J. Math. **29**, 1060–1074 (2023)
803. D. Montgomery, L. Zippin, *Topological Transformation Groups* (Robert E. Krieger Publishing Co., Huntington, NY, 1974). Reprint of the 1955 original
804. C.C. Moore, Ergodic theorem, ergodic theory, and statistical mechanics. Proc. Natl. Acad. Sci. USA **112**(7), 1907–1911 (2015)
805. J. Moreira, Monochromatic sums and products in $\mathbb{N}$. Ann. Math. (2) **185**(3), 1069–1090 (2017)
806. J. Moreira, F.K. Richter, A spectral refinement of the Bergelson-Host-Kra decomposition and new multiple ergodic theorems. Ergod. Theory Dyn. Syst. **39**(4), 1042–1070 (2019); Corrigendum in: Ergod. Theory Dyn. Syst. **44**(6), 1724–1728 (2024)
807. J. Moulin Ollagnier, Théorème ergodique presque sous-additif et convergence en moyenne de l'information. Ann. Inst. H. Poincaré Sect. B (N.S.) **19**(3), 257–266 (1983)
808. J. Moulin Ollagnier, *Ergodic Theory and Statistical Mechanics*. Volume 1115 of Lecture Notes in Mathematics (Springer, Berlin, 1985)
809. F. Mukhamedov, A. Al-Rawashdeh, Generalized Dobrushin ergodicity coefficient and uniform ergodicities of Markov operators. Positivity **24**(4), 855–890 (2020)
810. V. Müller, Y. Tomilov, Quasisimilarity of power bounded operators and Blum-Hanson property. J. Funct. Anal. **246**(2), 385–399 (2007)
811. C. Müllner, Automatic sequences fulfill the Sarnak conjecture. Duke Math. J. **166**(17), 3219–3290 (2017)
812. L. Nachbin, *The Haar Integral* (Robert E. Krieger Publishing Co., Huntington, N.Y., 1976). Translated from the Portuguese by Lulu Bechtolsheim, Reprint of the 1965 edition
813. M.G. Nadkarni, *Spectral Theory of Dynamical Systems*. Birkhäuser Advanced Texts: Basler Lehrbücher. [Birkhäuser Advanced Texts: Basel Textbooks] (Birkhäuser Verlag, Basel, 1998)

814. M.G. Nadkarni, *Basic Ergodic Theory*. Volume 6 of Texts and Readings in Mathematics, 3rd edn. (Hindustan Book Agency, New Delhi, 2013)
815. R.J. Nagel, Mittelergodische Halbgruppen linearer Operatoren. Ann. Inst. Fourier (Grenoble) **23**(4), 75–87 (1973)
816. B. Nagle, V. Rödl, M. Schacht, The counting lemma for regular k-uniform hypergraphs. Random Struct. Algoritm. **28**(2), 113–179 (2006)
817. R. Nair, On polynomials in primes and J. Bourgain's circle method approach to ergodic theorems. Ergod. Theory Dyn. Syst. **11**(3), 485–499 (1991)
818. R. Nair, On polynomials in primes and J. Bourgain's circle method approach to ergodic theorems. II. Stud. Math. **105**(3), 207–233 (1993)
819. R. Nair, P. Zaris, On certain sets of integers and intersectivity. Math. Proc. Camb. Philos. Soc. **131**(1), 157–164 (2001)
820. J. Neveu, Sur le théorème ergodique ponctuel. C. R. Acad. Sci. Paris **252**, 1554–1556 (1961)
821. J. Neveu, *Mathematical Foundations of the Calculus of Probability* (Holden-Day, Inc., San Francisco, CA/London/Amsterdam, 1965). Translated by Amiel Feinstein
822. J. Neveu, The filling scheme and the Chacon-Ornstein theorem. Isr. J. Math. **33**(3–4), 368–377 (1980, 1979). A collection of invited papers on ergodic theory
823. J. Neveu, Courte démonstration du théorème ergodique sur-additif. Ann. Inst. H. Poincaré Sect. B (N.S.) **19**(1), 87–90 (1983)
824. X.-X. Nguyen, Ergodic theorems for subadditive spatial processes. Z. Wahrsch. Verw. Gebiete **48**(2), 159–176 (1979)
825. C.P. Niculescu, A. Ströh, A Hilbert approach to Poincaré recurrence theorem, vol. 44, (2000)(1999), pp. 799–805. In honour of Academician Romulus Cristescu on his 70th birthday
826. C.P. Niculescu, A. Ströh, L. Zsidó, Noncommutative extensions of classical and multiple recurrence theorems. J. Oper. Theory **50**(1), 3–52 (2003)
827. J.R. Norris, *Markov Chains*. Volume 2 of Cambridge Series in Statistical and Probabilistic Mathematics (Cambridge University Press, Cambridge, 1998). Reprint of 1997 original
828. G.L. O'Brien, Obtaining prescribed rates of convergence for the ergodic theorem. Can. J. Math. **35**(6), 1129–1146 (1983)
829. J.H. Olsen, Akcoglu's ergodic theorem for uniform sequences. Can. J. Math. **32**(4), 880–884 (1980)
830. J.H. Olsen, The individual ergodic theorem for Lamperti contractions. C. R. Math. Rep. Acad. Sci. Can. **3**(2), 113–118 (1981)
831. J.H. Olsen, The individual weighted ergodic theorem for bounded Besicovitch sequences. Can. Math. Bull. **25**(4), 468–471 (1982)
832. J.H. Olsen, A multiple sequence ergodic theorem. Can. Math. Bull. **26**(4), 493–497 (1983)
833. D. Ornstein, On the pointwise behavior of iterates of a self-adjoint operator. J. Math. Mech. **18**, 473–477 (1968/1969)
834. D.S. Ornstein, Bernoulli shifts with the same entropy are isomorphic. Adv. Math. **4**, 337–352 (1970)
835. D.S. Ornstein, A remark on the Birkhoff ergodic theorem. Ill. J. Math. **15**, 77–79 (1971)
836. D.S. Ornstein, *Ergodic Theory, Randomness, and Dynamical Systems*. Yale Mathematical Monographs, vol. 5 (Yale University Press, New Haven, CT/London, 1974). James K. Whittemore Lectures in Mathematics given at Yale University
837. D.S. Ornstein, B. Weiss, The Shannon-McMillan-Breiman theorem for a class of amenable groups. Isr. J. Math. **44**(1), 53–60 (1983)
838. D.S. Ornstein, B. Weiss, Subsequence ergodic theorems for amenable groups. Isr. J. Math. **79**(1), 113–127 (1992)
839. A. Parrish, Pointwise convergence of ergodic averages in Orlicz spaces. Ill. J. Math. **55**(1), 89–106 (2012, 2011)
840. W. Parry, *Entropy and Generators in Ergodic Theory* (W. A. Benjamin, Inc., New York/Amsterdam, 1969)

841. W. Parry, Ergodic properties of affine transformations and flows on nilmanifolds. Am. J. Math. **91**, 757–771 (1969)
842. W. Parry, Dynamical systems on nilmanifolds. Bull. Lond. Math. Soc. **2**, 37–40 (1970)
843. W. Parry, *Topics in Ergodic Theory*. Volume 75 of Cambridge Tracts in Mathematics (Cambridge University Press, Cambridge/New York, 1981)
844. A. Paszkiewicz, On a unitary operator with pathological ergodic means. Acta Sci. Math. (Szeged) **71**(3–4), 741–749 (2005)
845. A. Patrascioiu, The ergodic hypothesis: A complicated problem in mathematics and physics. Los Alamos Sci. **15**, 263–279 (1987). Stanislaw Ulam 1909–1984
846. R. Pavlov, Some counterexamples in topological dynamics. Ergod. Theory Dyn. Syst. **28**(4), 1291–1322 (2008)
847. S. Peluse, Bounds for sets with no polynomial progressions. Forum Math. Pi **8**(e16), 55 (2020)
848. K. Petersen, *Ergodic Theory*. Volume 2 of Cambridge Studies in Advanced Mathematics (Cambridge University Press, Cambridge, 1989). Corrected reprint of the 1983 original
849. C. Pilatte, Improved bounds for the two-point logarithmic Chowla conjecture (2023). Preprint, https://arxiv.org/abs/2310.19357
850. M. Plancherel, Beweis der Unmöglichkeit ergodischer dynamischer Systeme. Ann. Phys. **42**, 1061–1063 (1913)
851. F. Pogorzelski, Almost-additive ergodic theorems for amenable groups. J. Funct. Anal. **265**(8), 1615–1666 (2013)
852. F. Pogorzelski, Corrigendum and Addendum to "Almost-additive ergodic theorems for amenable groups" [J. Funct. Anal. 265 (2013) 1615–1666] [MR3079231]. J. Funct. Anal. **271**(11), 3393–3401 (2016)
853. H. Poincaré, Sur le problème des trois corps et les équations de la dynamique. Acta Math. **13**, 1–270 (1890)
854. H. Poincaré, Sur une objection à la théorie cinétique des gaz. C. R. Acad. Sci., Paris **116**, 1017–1021 (1893)
855. M. Pollicott, *Lectures on Ergodic Theory and Pesin Theory on Compact Manifolds*. Volume 180 of London Mathematical Society Lecture Note Series (Cambridge University Press, Cambridge, 1993)
856. D.H.J. Polymath, A new proof of the density Hales-Jewett theorem. Ann. Math. (2) **175**(3), 1283–1327 (2012)
857. S. Prendiville, Quantitative bounds in the polynomial Szemerédi theorem: The homogeneous case. Discrete Anal. **5**, 34 (2017)
858. M. Qian, J.-S. Xie, S. Zhu, *Smooth Ergodic Theory for Endomorphisms*. Volume 1978 of Lecture Notes in Mathematics (Springer, Berlin, 2009)
859. A. Quas, M. Wierdl, Rates of divergence of non-conventional ergodic averages. Ergod. Theory Dyn. Syst. **30**(1), 233–262 (2010)
860. M. Queffélec, *Substitution Dynamical Systems—Spectral Analysis*. Volume 1294 of Lecture Notes in Mathematics, 2nd edn. (Springer, Berlin, 2010)
861. H. Rademacher, Einige Sätze über Reihen von allgemeinen Orthogonalfunktionen. Math. Ann. **87**, 112–138 (1922)
862. R.A. Raimi, Minimal sets and ergodic measures in $\beta N - N$. Bull. Am. Math. Soc. **70**, 711–712 (1964)
863. A. Rajchman, Das starke Gesetz der großen Zahlen. Mathesis Polska **6**, 145–161 (1931)
864. O. Ramaré, Chowla's conjecture: From the Liouville function to the Moebius function, in *Ergodic Theory and Dynamical Systems in their Interactions with Arithmetics and Combinatorics*. Volume 2213 of Lecture Notes in Mathematics (Springer, Cham, 2018), pp. 317–323
865. M. Ratner, Raghunathan's topological conjecture and distributions of unipotent flows. Duke Math. J. **63**(1), 235–280 (1991)
866. J.I. Reich, On the individual ergodic theorem for subsequences. Ann. Probab. **5**(6), 1039–1046 (1977)

867. K. Reinhold-Larsson, Discrepancy of behavior of perturbed sequences in L^p spaces. Proc. Am. Math. Soc. **120**(3), 865–874 (1994)
868. K.B. Reinhold-Larsson, *Almost everywhere convergence of weighted averages*, Ph.D. thesis, The Ohio State University (ProQuest LLC, Ann Arbor, MI, 1991)
869. G. Rhin, Sur la répartition modulo 1 des suites $f(p)$. Acta Arith. **23**, 217–248 (1973)
870. F. Riesz, Sur certains systèmes singuliers d'équations intégrales. Ann. Sci. Éc. Norm. Supér. **28**(3), 33–62 (1911)
871. F. Riesz, Some mean ergodic theorems. J. Lond. Math. Soc. **13**(4), 274–278 (1938)
872. F. Riesz, Sur la théorie ergodique. Comment. Math. Helv. **17**, 221–239 (1945)
873. D. Robertson, Mild mixing of certain interval-exchange transformations. Ergod. Theory Dyn. Syst. **39**(1), 248–256 (2019)
874. E.A. Robinson, Jr. On uniform convergence in the Wiener-Wintner theorem. J. Lond. Math. Soc. (2) **49**(3), 493–501 (1994)
875. V. Rödl, J. Skokan, Regularity lemma for k-uniform hypergraphs. Random Struct. Algoritm. **25**(1), 1–42 (2004)
876. V. Rödl, J. Skokan, Applications of the regularity lemma for uniform hypergraphs. Random Struct. Algoritm. **28**(2), 180–194 (2006)
877. V.A. Rokhlin, A "general" measure-preserving transformation is not mixing. Doklady Akad. Nauk SSSR (N.S.) **60**, 349–351 (1948)
878. V.A. Rokhlin, On endomorphisms of compact commutative groups. Izv. Akad. Nauk SSSR. Ser. Mat. **13**, 329–340 (1949)
879. J. Rosenblatt, Universally bad sequences in ergodic theory, in *Almost Everywhere Convergence, II (Evanston, IL, 1989)* (Academic, Boston, 1991), pp. 227–245
880. J. Rosenblatt, Norm convergence in ergodic theory and the behavior of Fourier transforms. Can. J. Math. **46**(1), 184–199 (1994)
881. J.M. Rosenblatt, M. Wierdl, Pointwise ergodic theorems via harmonic analysis, in *Ergodic Theory and Its Connections with Harmonic Analysis (Alexandria, 1993)*. Volume 205 of London Mathematical Society Lecture Note Series (Cambridge University Press, Cambridge, 1995), pp. 3–151
882. A. Rosenthal, Beweis der Unmöglichkeit ergodischer Gassysteme. Ann. Phys. **42**, 796–806 (1913)
883. G.-C. Rota, An "Alternierende Verfahren" for general positive operators. Bull. Am. Math. Soc. **68**, 95–102 (1962)
884. J.-P. Roth, Reformulation et extension de certains théorèmes ergodiques. Ann. Inst. H. Poincaré Probab. Stat. **26**(3), 437–450 (1990)
885. J.-P. Roth, Généralisation des inégalités ergodiques maximales. Potential Anal. **1**(2), 141–154 (1992)
886. K.F. Roth, On certain sets of integers. J. Lond. Math. Soc. **28**, 104–109 (1953)
887. H. Royden, P. Fitzpatrick, *Real Analysis*, 4th edn. (Prentice Hall, 2010)
888. W. Rudin, *Fourier Analysis on Groups*. Interscience Tracts in Pure and Applied Mathematics, vol. 12 (Interscience Publishers (a division of Wiley), New York/London, 1962)
889. W. Rudin, *Real and Complex Analysis* (McGraw-Hill Book Co., New York, International Student Edition, 1970)
890. W. Rudin, *Functional Analysis*. International Series in Pure and Applied Mathematics, 2nd edn. (McGraw-Hill, Inc., New York, 1991)
891. D.J. Rudolph, k-fold mixing lifts to weakly mixing isometric extensions. Ergod. Theory Dyn. Syst. **5**(3), 445–447 (1985)
892. D.J. Rudolph, *Fundamentals of Measurable Dynamics* (Oxford Science Publications. The Clarendon Press, Oxford University Press, New York, 1990). Ergodic theory on Lebesgue spaces
893. D.J. Rudolph, A joinings proof of Bourgain's return time theorem. Ergod. Theory Dyn. Syst. **14**(1), 197–203 (1994)

894. D.J. Rudolph, Fully generic sequences and a multiple-term return-times theorem. Invent. Math. **131**(1), 199–228 (1998)
895. H.H. Rugh, D. Thomine, Ratio ergodic theorems: From Hopf to Birkhoff and Kingman (2018). Preprint, https://arxiv.org/abs/1802.08439
896. C. Ryll-Nardzewski, On fixed points of semigroups of endomorphisms of linear spaces, in *Proceedings of the Fifth Berkeley Symposium on Mathematical Statistics and Probability, (Berkeley, CA, 1965/66), Vol. II: Contributions to Probability Theory, Part 1* (University of California Press, Berkeley, CA, 1967), pp. 55–61
897. C. Ryll-Nardzewski, Topics in ergodic theory, in *Probability—Winter School Probability—Winter School (Proceedings of Fourth Winter School, Karpacz, 1975)*, Vol. 472. Lecture Notes in Mathematics (1975), pp. 131–156
898. V.V. Ryzhikov, Joinings and multiple mixing of the actions of finite rank. Funktsional. Anal. i Prilozhen. **27**(2), 63–78, 96 (1993)
899. V.V. Ryzhikov, Mixing sets for rigid transformations. Mat. Zametki **110**(4), 576–583 (2021)
900. V.V. Ryzhikov, Multiple mixing, 75 years of Rokhlin's problem (2024). Preprint, https://arxiv.org/abs/2411.07234
901. V.V. Ryzhikov, Multiple mixing in ergodic theory (2024). Preprint, https://arxiv.org/abs/2404.06476
902. V.V. Ryzhikov, Non-commuting transformations with non-converging 2-fold ergodic averages (2024). Preprint, https://arxiv.org/abs/2407.13741
903. V.V. Ryzhikov, Non-recurrence and divergent $(p(n), q(n))$-averages for deterministic automorphisms (2024). Preprint, https://arxiv.org/abs/2410.11787
904. L.A.B. San Martin, *Lie Groups*. Latin American Mathematics Series (Springer, Cham, 2021). Translated from the 2016 Portuguese original by José Emílio Maiorino and Carlos Augusto Bassani Varea with revisions by Simon Chiossi
905. S.I. Santos, C. Walkden, Topological Wiener-Wintner ergodic theorems via non-abelian Lie group extensions. Ergod. Theory Dyn. Syst. **27**(5), 1633–1650 (2007)
906. O. Sarig, *Introduction to the Transfer Operator Method* (Winter School on Dynamics, Hausdorff Research Institute for Mathematics, Bonn 2020)
907. A. Sárközy, On difference sets of sequences of integers. I. Acta Math. Acad. Sci. Hungar. **31**(1–2), 125–149 (1978)
908. A. Sárközy, On difference sets of sequences of integers. III. Acta Math. Acad. Sci. Hung. **31**(3–4), 355–386 (1978)
909. P. Sarnak, Three lectures on the Möbius function: Randomness and dynamics (2010). IAS Lecture Notes. http://publications.ias.edu/sites/default/files/MobiusFunctionsLectures(2).pdf
910. R. Sato, An individual ergodic theorem. Proc. Am. Math. Soc. **65**(2), 235–241 (1977)
911. R. Sato, On abstract mean ergodic theorems. Tohoku Math. J. (2) **30**(4), 575–581 (1978)
912. R. Sato, The Hahn-Banach theorem implies Sine's mean ergodic theorem. Proc. Am. Math. Soc. 77(3), 426 (1979)
913. R. Sato, Another proof of Derriennic's reverse maximal inequality for the supremum of ergodic ratios. Comment. Math. Univ. Carol. **47**(1), 155–158 (2006)
914. S. Sawyer, Maximal inequalities of weak type. Ann. Math. (2) **84**, 157–174 (1966)
915. H.H. Schaefer, *Banach Lattices and Positive Operators* (Springer, New York/Heidelberg, 1974). Die Grundlehren der mathematischen Wissenschaften, Band 215
916. J.J. Schäffer, On unitary dilations of contractions. Proc. Am. Math. Soc. **6**, 322 (1955)
917. E. Scheffold, Das Spektrum von Verbandsoperatoren in Banachverbänden. Math. Z. **123**, 177–190 (1971)
918. K. Schmidt, Asymptotic properties of unitary representations and mixing. Proc. Lond. Math. Soc. (3) **48**(3), 445–460 (1984)
919. K. Schmidt, P. Walters, Mildly mixing actions of locally compact groups. Proc. Lond. Math. Soc. (3) **45**(3), 506–518 (1982)
920. K. Schmidt, T. Ward, Mixing automorphisms of compact groups and a theorem of Schlickewei. Invent. Math. **111**(1), 69–76 (1993)

921. D. Schneider, Théorèmes ergodiques perturbés. Isr. J. Math. **101**, 157–178 (1997)
922. M. Schnurr, Generic properties of extensions. Ergod. Theory Dyn. Syst. **39**(11), 3144–3168 (2019)
923. M. Schreiber, Uniform families of ergodic operator nets. Semigroup Forum **86**(2), 321–336 (2013)
924. M. Schreiber, Topological Wiener-Wintner theorems for amenable operator semigroups. Ergod. Theory Dyn. Syst. **34**(5), 1674–1698 (2014)
925. I. Schur, Über die Kongruenz $x^m + y^m = z^m \mod p$. Jahresber. Dtsch. Math.-Ver **25**, 114–117 (1916)
926. K. Schürger, A limit theorem for almost monotone sequences of random variables. Stoch. Process. Appl. **21**(2), 327–338 (1986)
927. K. Schürger, Almost subadditive multiparameter ergodic theorems. Stoch. Process. Appl. **29**(2), 171–193 (1988)
928. K. Schürger, On Derriennic's almost subadditive ergodic theorem. Acta Math. Hung. **52**(1–2), 71–82 (1988)
929. K. Schürger, Almost subadditive extensions of Kingman's ergodic theorem. Ann. Probab. **19**(4), 1575–1586 (1991)
930. M. Schwartz, Polynomially moving ergodic averages. Proc. Am. Math. Soc. **103**(1), 252–254 (1988)
931. N.A. Shah, Invariant measures and orbit closures on homogeneous spaces for actions of subgroups generated by unipotent elements, in *Lie Groups and Ergodic Theory (Mumbai, 1996)*. Volume 14 of Tata Institute of Fundamental Research Studies in Mathematics (Tata Institute of Fundamental Research, Bombay, 1998), pp. 229–271
932. O. Shalom, Multiple ergodic averages in abelian groups and Khintchine type recurrence. Trans. Am. Math. Soc. **375**(4), 2729–2761 (2022)
933. O. Shalom, Host–Kra theory for $\oplus_{p\in P} \mathbb{F}_p$-systems and multiple recurrence. Ergod. Theory Dyn. Syst. **43**(1), 299–360 (2023)
934. O. Shalom, Host-Kra factors for $\bigoplus_{p\in P} \mathbb{Z}/p\mathbb{Z}$ actions and finite-dimensional nilpotent systems. Anal. PDE **17**(7), 2379–2449 (2024)
935. S. Shao, X. Ye, Regionally proximal relation of order d is an equivalence one for minimal systems and a combinatorial consequence. Adv. Math. **231**(3–4), 1786–1817 (2012)
936. P.C. Shields, The ergodic and entropy theorems revisited. IEEE Trans. Inform. Theory **33**(2), 263–266 (1987)
937. P.C. Shields, *The Ergodic Theory of Discrete Sample Paths*. Volume 13 of Graduate Studies in Mathematics (American Mathematical Society, Providence, RI, 1996)
938. A. Shulman, Maximal ergodic theorems on groups. Dep. Lit. NIINTI, No.2184, pp. 1–22 (1988)
939. W. Sierpiński, Sur la valeur asymptotique d'une certaine somme. *Bull. Intl. Acad. Polonaise des Sci. et des Lettres (Cracovie) series A*, 9–11 (1910)
940. W. Sierpiński, Démonstration élémentaire du théorème de M. Borel sur les nombres absolument normaux et détermination effective d'une tel nombre. Bull. Soc. Math. France **45**, 125–132 (1917)
941. R. Sikorski, On the inducing of homomorphisms by mappings. Fundam. Math. **36**(1), 7–22 (1949)
942. C.E. Silva, *Invitation to Ergodic Theory*. Volume 42 of Student Mathematical Library (American Mathematical Society, Providence, RI, 2008)
943. C.E. Silva, A.I. Danilenko (eds.), *Ergodic Theory*. Encyclopedia of Complexity and Systems Science (Springer, New York, 2023)
944. C.E. Silva, P. Thieullen, The subadditive ergodic theorem and recurrence properties of Markovian transformations. J. Math. Anal. Appl. **154**(1), 83–99 (1991)
945. D. Silvestrov, *Perturbed Semi-Markov Type Processes II—Ergodic Theorems for Multi-Alternating Regenerative Processes* (Springer, Cham, 2022)
946. N. Simányi, Conditional proof of the Boltzmann-Sinai ergodic hypothesis. Invent. Math. **177**(2), 381–413 (2009)

947. N. Simányi, D. Szász, Hard ball systems are completely hyperbolic. Ann. Math. (2) **149**(1), 35–96 (1999)
948. J.G. Sinaĭ, Dynamical systems with countable Lebesgue spectrum. I. Izv. Akad. Nauk SSSR Ser. Mat. **25**, 899–924 (1961)
949. Y.G. Sinaĭ, *Topics in Ergodic Theory*. Volume 44 of Princeton Mathematical Series (Princeton University Press, Princeton, NJ, 1994)
950. R. Sine, A mean ergodic theorem. Proc. Am. Math. Soc. **24**, 438–439 (1970)
951. M.D. Smeltzer, Subadditive stochastic processes. Bull. Am. Math. Soc. **83**(5), 1054–1056 (1977)
952. R.T. Smythe, Multiparameter subadditive processes. Ann. Probab. **4**(5), 772–782 (1976)
953. R.T. Smythe, J.C. Wierman, *First-Passage Percolation on the Square Lattice*. Volume 671 of Lecture Notes in Mathematics (Springer, Berlin, 1978)
954. R.T. Smythe, J.C. Wierman, First-passage percolation on the square lattice. III. Adv. Appl. Probab. **10**(1), 155–171 (1978)
955. J. Solymosi, Roth-type theorems in finite groups. Eur. J. Comb. **34**(8), 1454–1458 (2013)
956. A.N. Starkov, Multiple mixing of homogeneous flows. Dokl. Akad. Nauk **333**(4), 442–445 (1993)
957. J.M. Steele, Empirical discrepancies and subadditive processes. Ann. Probab. **6**(1), 118–127 (1978)
958. J.M. Steele, Kingman's subadditive ergodic theorem. Ann. Inst. H. Poincaré Probab. Stat. **25**(1), 93–98 (1989)
959. E.M. Stein, On the maximal ergodic theorem. Proc. Natl. Acad. Sci. U.S.A. **47**, 1894–1897 (1961)
960. E.M. Stein, S. Wainger, Discrete analogues in harmonic analysis. I. l^2 estimates for singular Radon transforms. Am. J. Math. **121**(6), 1291–1336 (1999)
961. L. Sucheston, On one-parameter proofs of almost sure convergence of multiparameter processes. Z. Wahrsch. Verw. Gebiete **63**(1), 43–49 (1983)
962. W. Sun, Multiple recurrence and convergence for certain averages along shifted primes. Ergod. Theory Dyn. Syst. **35**(5), 1592–1609 (2015)
963. R.E. Svetic, A function with locally uncountable rotation set. Acta Math. Hung. **81**(4), 305–314 (1998)
964. B. Sz.-Nagy, Sur les contractions de l'espace de Hilbert. Acta Sci. Math. (Szeged) **15**, 87–92 (1953)
965. B. Sz.-Nagy, On Schäffer's construction of unitary dilations. Ann. Univ. Sci. Budapest. Eötvös Sect. Math. **3/4**, 343–346 (1960/61)
966. B. Sz.-Nagy, C. Foiaş, Sur les contractions de l'espace de Hilbert. IV. Acta Sci. Math. Szeged **21**, 251–259 (1960)
967. D. Szász, Boltzmann's ergodic hypothesis, a conjecture for centuries? Studia Sci. Math. Hung. **31**(1–3), 299–322 (1996)
968. B. Szegedy, On higher order Fourier analysis (2012). Preprint, https://arxiv.org/abs/1203.2260
969. B. Szegedy, From graph limits to higher order Fourier analysis, in *Proceedings of the International Congress of Mathematicians—Rio de Janeiro 2018. Vol. IV. Invited Lectures* (World Scientific Publishing, Hackensack, NJ, 2018), pp. 3213–3233
970. E. Szemerédi, On sets of integers containing no k elements in arithmetic progression. Acta Arith. **27**, 199–245 (1975)
971. T. Tao, The gowers uniformity norm of order 1+. Blog post https://terrytao.wordpress.com/2021/07/25/the-gowers-uniformity-norm-of-order-1/
972. T. Tao, Ratner's theorems. Blog post https://terrytao.wordpress.com/2007/09/29/ratners-theorems/
973. T. Tao, A quantitative ergodic theory proof of Szemerédi's theorem. Electron. J. Combin. **13**(1), Research Paper 99, 49 (2006)
974. T. Tao, A variant of the hypergraph removal lemma. J. Combin. Theory Ser. A **113**(7), 1257–1280 (2006)

975. T. Tao, The dichotomy between structure and randomness, arithmetic progressions, and the primes, in *International Congress of Mathematicians. Vol. I* (European Mathematical Society, Zürich, 2007), pp. 581–608
976. T. Tao, Norm convergence of multiple ergodic averages for commuting transformations. Ergod. Theory Dyn. Syst. **28**(2), 657–688 (2008)
977. T. Tao, *Poincaré's Legacies, Pages from Year Two of a Mathematical Blog. Part II* (American Mathematical Society, Providence, RI, 2009)
978. T. Tao, Yet another proof of Szemerédi's theorem (2010). Blog post, https://terrytao.wordpress.com/2010/02/13/yet-another-proof-of-szemeredis-theorem/
979. T. Tao, The Chowla conjecture and the Sarnak conjecture (2012). Blog post https://terrytao.wordpress.com/2012/10/14/the-chowla-conjecture-and-the-sarnak-conjecture/
980. T. Tao, *Higher Order Fourier Analysis*. Volume 142 of Graduate Studies in Mathematics (American Mathematical Society, Providence, RI, 2012)
981. T. Tao, https://mathoverflow.net/questions/294395/correlation-of-the-mobius-function (2012)
982. T. Tao, The logarithmically averaged Chowla and Elliott conjectures for two-point correlations. Forum Math. Pi **4**(e8), 36 (2016)
983. T. Tao, Equivalence of the logarithmically averaged Chowla and Sarnak conjectures, in *Number Theory—Diophantine Problems, Uniform Distribution and Applications* (Springer, Cham, 2017), pp. 391–421
984. T. Tao, The logarithmically averaged and non-logarithmically averaged Chowla conjectures (2017). Blog post https://terrytao.wordpress.com/2017/10/20/the-logarithmically-averaged-and-non-logarithmically-averaged-chowla-conjectures/
985. T. Tao, Szemerédi's proof of Szemerédi's theorem. Acta Math. Hung. **161**(2), 443–487 (2020)
986. T. Tao, Exploring the toolkit of Jean Bourgain. Bull. Am. Math. Soc. (N.S.) **58**(2), 155–171 (2021)
987. T. Tao, J. Teräväinen, Odd order cases of the logarithmically averaged Chowla conjecture. J. Théor. Nombres Bordeaux **30**(3), 997–1015 (2018)
988. T. Tao, J. Teräväinen, The structure of correlations of multiplicative functions at almost all scales, with applications to the Chowla and Elliott conjectures. Algebra Number Theory **13**(9), 2103–2150 (2019)
989. T. Tao, J. Teräväinen, The structure of logarithmically averaged correlations of multiplicative functions, with applications to the Chowla and Elliott conjectures. Duke Math. J. **168**(11), 1977–2027 (2019)
990. T. Tao, J. Teräväinen, Quantitative bounds for Gowers uniformity of the Möbius and von Mangoldt functions (2021). Preprint, https://arxiv.org/abs/2107.02158
991. T. Tao, T. Ziegler, The primes contain arbitrarily long polynomial progressions. Acta Math. **201**(2), 213–305 (2008)
992. T. Tao, T. Ziegler, Erratum to "The primes contain arbitrarily long polynomial progressions". Acta Math. **210**(2), 403–404 (2013)
993. T. Tao, T. Ziegler, A multi-dimensional Szemerédi theorem for the primes via a correspondence principle. Isr. J. Math. **207**(1), 203–228 (2015)
994. T. Tao, T. Ziegler, Concatenation theorems for anti-Gowers-uniform functions and Host-Kra characteristic factors. Discrete Anal. pages Paper No. 13, 60 (2016)
995. T. Tao, T. Ziegler, Polynomial patterns in the primes. Forum Math. Pi **6**(e1), 60 (2018)
996. A. Tempelman, *Ergodic Theorems for Group Actions*. Volume 78 of Mathematics and Its Applications (Kluwer Academic Publishers Group, Dordrecht, 1992). Informational and thermodynamical aspects, Translated and revised from the 1986 Russian original
997. A.A. Tempelman, Ergodic theorems for general dynamical systems. Trudy Moskov. Mat. Obšč. **26**, 95–132 (1972)
998. A.A. Tempelman, An ergodic theorem for amplitude modulated homogeneous random fields. Litovsk. Mat. Sb. **14**(4), 221–229, 243 (1974)

999. A.F.M. ter Elst, V. Müller, A van der Corput-type lemma for power bounded operators. Math. Z. **285**(1–2), 143–158 (2017)
1000. J. Teräväinen, Pointwise convergence of ergodic averages with Möbius weight (2024). Preprint, https://arxiv.org/abs/2401.03174
1001. J.-P. Thouvenot, La convergence presque sûre des moyennes ergodiques suivant certaines sous-suites d'entiers (d'après Jean Bourgain). Number 189–190, pages Exp. No. 719, 133–153. 1990. Séminaire Bourbaki, vol. 1989/90
1002. E.C. Titchmarsh, *The Theory of the Riemann Zeta-Function*, 2nd edn. (The Clarendon Press, Oxford University Press, New York, 1986). Edited and with a preface by D. R. Heath-Brown
1003. O. Toeplitz, Über die Fouriersche Entwicklung positiver Funktionen. Rend. Circ. Mat. Palermo **32**, 191–192 (1911)
1004. Y. Tomilov, J. Zemánek, A new way of constructing examples in operator ergodic theory. Math. Proc. Camb. Philos. Soc. **137**(1), 209–225 (2004)
1005. H. Totoki, The mixing property of Gaussian flows. Mem. Fac. Sci. Kyushu Univ. Ser. A **18**, 136–139 (1964)
1006. H. Towsner, Convergence of diagonal ergodic averages. Ergod. Theory Dyn. Syst. **29**(4), 1309–1326 (2009)
1007. B. Trojan, Endpoint estimates for the maximal function over prime numbers. J. Fourier Anal. Appl. **25**(6), 3123–3153 (2019)
1008. B. Trojan, Variational estimates for operators over some thin subsets of primes. Math. Res. Lett. **27**(2), 591–628 (2020)
1009. B. Trojan, Weak type (1, 1) estimates for maximal functions along 1-regular sequences of integers. Stud. Math. **261**(1), 103–108 (2021)
1010. R. Urban, J. Zienkiewicz, Weak type (1, 1) estimates for a class of discrete rough maximal functions. Math. Res. Lett. **14**(2), 227–237 (2007)
1011. M. Urbański, M. Roy, S. Munday, *Non-invertible Dynamical Systems. Vol. 1. Ergodic Theory—Finite and Infinite, Thermodynamic Formalism, Symbolic Dynamics and Distance Expanding Maps*. Volume 69 of De Gruyter Expositions in Mathematics (De Gruyter, Berlin, 2022)
1012. J.G. van der Corput, Diophantische Ungleichungen. I. Zur Gleichverteilung Modulo Eins. Acta Math. **56**(1), 373–456 (1931)
1013. B.L. Vander Waerden, Beweis einer Baudetschen Vermutung. Nieuw Archief **15**, 212–216 (1927)
1014. P. Varnavides, On certain sets of positive density. J. Lond. Math. Soc. **34**, 358–360 (1959)
1015. R.C. Vaughan, *The Hardy-Littlewood Method*. Volume 125 of Cambridge Tracts in Mathematics, 2nd edn. (Cambridge University Press, Cambridge, 1997)
1016. W.A. Veech, Well distributed sequences of integers. Trans. Am. Math. Soc. **161**, 63–70 (1971)
1017. W.A. Veech, The metric theory of interval exchange transformations. I. Generic spectral properties. Am. J. Math. **106**(6), 1331–1359 (1984)
1018. W.A. Veech, Möbius Dynamics (2016). Unpublished Lecture Notes
1019. W.A. Veech, Möbius orthogonality for generalized Morse-Kakutani flows. Am. J. Math. **139**(5), 1157–1203 (2017)
1020. M. Viana, *Lectures on Lyapunov Exponents*. Volume 145 of Cambridge Studies in Advanced Mathematics (Cambridge University Press, Cambridge, 2014)
1021. M. Viana, K. Oliveira, *Foundations of Ergodic Theory*. Volume 151 of Cambridge Studies in Advanced Mathematics (Cambridge University Press, Cambridge, 2016)
1022. I.M. Vinogradov, The method of trigonometrical sums in the theory of numbers. Trav. Inst. Math. Stekloff **23**, 109 (1947)
1023. I.M. Vinogradov, On an estimate of trigonometric sums with prime numbers. Izv. Akad. Nauk. SSSR. Ser. Mat. **12**, 225–248 (1948)
1024. I.M. Vinogradov, Trigonometric sums involving values of a polynomial. Izv. Akad. Nauk SSSR. Ser. Mat. **21**, 145–170 (1957)

1025. I.M. Vinogradov, A special case of estimation of trigonometric sums involving prime numbers. Izv. Akad. Nauk SSSR. Ser. Mat. **22**, 3–14 (1958)
1026. J. von Neumann, Einige Sätze über messbare Abbildungen. Ann. Math. (2) **33**(3), 574–586 (1932)
1027. J. von Neumann, Proof of the quasi-ergodic hypothesis. Proc. Natl. Acad. Sci. USA **18**, 70–82 (1932)
1028. J. von Neumann, Zusätze zur Arbeit "zur Operatorenmethode...". Ann. Math. (2) **33**(4), 789–791 (1932)
1029. J. von Plato, Boltzmann's ergodic hypothesis. Arch. Hist. Exact Sci. **42**, 71–89 (1991)
1030. H.-D. Wacker, Über die Verallgemeinerung eines Ergodensatzes von Dunford. Arch. Math. (Basel) **44**(6), 539–546 (1985)
1031. M.N. Walsh, Norm convergence of nilpotent ergodic averages. Ann. Math. (2) **175**(3), 1667–1688 (2012)
1032. P. Walters, Some invariant σ-algebras for measure-preserving transformations. Trans. Am. Math. Soc. **163**, 357–368 (1972)
1033. P. Walters, *An Introduction to Ergodic Theory*. Volume 79 of Graduate Texts in Mathematics (Springer, New York/Berlin, 1982)
1034. P. Walters, Topological Wiener-Wintner ergodic theorems and a random L^2 ergodic theorem. Ergod. Theory Dyn. Syst. **16**(1), 179–206 (1996)
1035. Z. Wang, Möbius disjointness for analytic skew products. Invent. Math. **209**(1), 175–196 (2017)
1036. M. Weber, Uniform bounds under increment conditions. Trans. Am. Math. Soc. **358**(2), 911–936 (2006)
1037. M.J.G. Weber, Criteria of divergence almost everywhere in ergodic theory. *Zap. Nauchn. Sem. S.-Peterburg. Otdel. Mat. Inst. Steklov. (POMI)* **441**(Veroyatnost' i Statistika. 22), 73–118 (2015)
1038. A. Weil, *Œuvres Scientifiques. Collected Papers. Vol. I (1926–1951)* (Berlin: Springer, paperback reprint of the 1979 edition, 2009)
1039. B. Weiss, The isomorphism problem in ergodic theory. Bull. Am. Math. Soc. **78**, 668–684 (1972)
1040. B. Weiss, Actions of amenable groups, in *Topics in Dynamics and Ergodic Theory*. Volume 310 of London Mathematical Society Lecture Note Series (Cambridge University Press, Cambridge, 2003), pp. 226–262
1041. D. Werner, Rajchmans starkes Gesetz der großen Zahlen. Unpublished note, downloaded on May 2022
1042. H. Weyl, Über die Gibbs'sche Erscheinung und verwandte Konvergenzphänomene. Rend. Circ. Mat. Palermo **30**(1), 377–407 (Dec 1910)
1043. H. Weyl, Über die Gleichverteilung von Zahlen mod. Eins. Math. Ann. **77**, 313–352 (1916)
1044. N. Wiener, The homogeneous chaos. Am. J. Math. **60**(4), 897–936 (1938)
1045. N. Wiener, The ergodic theorem. Duke Math. J. **5**(1), 1–18 (1939)
1046. N. Wiener, A. Wintner, Harmonic analysis and ergodic theory. Am. J. Math. **63**, 415–426 (1941)
1047. M. Wierdl, Pointwise ergodic theorem along the prime numbers. Isr. J. Math. **64**(3), 315–336 (1989, 1988)
1048. M. Wierdl, Almost everywhere convergence and recurrence along subsequences in ergodic theory (1989). Ph.D. thesis, Ohio State University
1049. T.D. Wooley, T.D. Ziegler, Multiple recurrence and convergence along the primes. Am. J. Math. **134**(6), 1705–1732 (2012)
1050. J. Woś, Analogues of the Denjoy-Young-Saks and Denjoy-Khintchine theorems in ergodic theory. Bull. Pol. Acad. Sci. Math. **32**(1–2), 23–29 (1984)
1051. J. Woś, Approximate convergence in ergodic theory. Proc. Lond. Math. Soc. (3) **53**(1), 65–84 (1986)
1052. J. Woś, The filling scheme and the ergodic theorems of Kesten and Tanny. Colloq. Math. **52**(2), 263–276 (1987)

1053. W.B. Wu, Strong invariance principles for dependent random variables. Ann. Probab. **35**(6), 2294–2320 (2007)
1054. R. Xiao, Multilinear Wiener-Wintner type ergodic averages and its application. Discrete Contin. Dyn. Syst. **44**(2), 425–446 (2024)
1055. R. Xiao, Polynomial ergodic averages of measure-preserving systems acted by $\mathbb{Z}^d$. J. Differ. Equ. **388**, 403–420 (2024)
1056. K.B. Yancey, On weakly mixing homeomorphisms of the two-torus that are uniformly rigid. J. Math. Anal. Appl. **399**(2), 524–541 (2013)
1057. K.B. Yancey, Topologically weakly mixing homeomorphisms of the Klein bottle that are uniformly rigid. Topol. Appl. **160**(14), 1853–1861 (2013)
1058. T. Yoshimoto, Dirichlet series and uniform ergodic theorems for linear operators in Banach spaces. Stud. Math. **141**(1), 69–83 (2000)
1059. K. Yosida, Mean ergodic theorem in Banach spaces. Proc. Imp. Acad. Tokyo **14**(8), 292–294 (1938)
1060. K. Yosida, S. Kakutani, Birkhoff's ergodic theorem and the maximal ergodic theorem. Proc. Imp. Acad. Tokyo **15**, 165–168 (1939)
1061. Q. Zhang, Sequence entropy and mild mixing. Can. J. Math. **44**(1), 215–224 (1992)
1062. Q. Zhang, On convergence of the averages $(1/N)\sum_{n=1}^{N} f_1(R^n x) f_2(S^n x) f_3(T^n x)$. Monatsh. Math. **122**(3), 275–300 (1996)
1063. O. Zhao, M. Woodroofe, Law of the iterated logarithm for stationary processes. Ann. Probab. **36**(1), 127–142 (2008)
1064. T. Ziegler, A non-conventional ergodic theorem for a nilsystem. Ergod. Theory Dyn. Syst. **25**(4), 1357–1370 (2005)
1065. T. Ziegler, Universal characteristic factors and Furstenberg averages. J. Am. Math. Soc. **20**(1), 53–97 (2007)
1066. R.J. Zimmer, Ergodic actions with generalized discrete spectrum. Ill. J. Math. **20**(4), 555–588 (1976)
1067. P. Zorin-Kranich, Compact extensions are isometric (2013). Unpublished note
1068. P. Zorin-Kranich, Cube spaces and the multiple term return times theorem. Ergod. Theory Dyn. Syst. **34**(5), 1747–1760 (2014)
1069. P. Zorin-Kranich, A nilpotent IP polynomial multiple recurrence theorem. J. Anal. Math. **123**, 183–225 (2014)
1070. P. Zorin-Kranich, Return times theorem for amenable groups. Isr. J. Math. **204**(1), 85–96 (2014)
1071. P. Zorin-Kranich, A uniform nilsequence Wiener–Wintner theorem for bilinear ergodic averages (2015). Preprint, https://arxiv.org/abs/1504.04647
1072. P. Zorin-Kranich, Variation estimates for averages along primes and polynomials. J. Funct. Anal. **268**(1), 210–238 (2015)
1073. P. Zorin-Kranich, Ergodic theory lecture notes (2015/2016). University of Bonn
1074. P. Zorin-Kranich, Norm convergence of multiple ergodic averages on amenable groups. J. Anal. Math. **130**, 219–241 (2016)
1075. P. Zorin-Kranich, A double return times theorem. Isr. J. Math. **229**(1), 255–267 (2019)
1076. J.D. Zund, George David Birkhoff and John von Neumann: A question of priority and the ergodic theorems, 1931–1932. Hist. Math. **29**(2), 138–156 (2002)
1077. R. Zweimüller, Hopf's ratio ergodic theorem by inducing. Colloq. Math. **101**(2), 289–292 (2004)
1078. R. Zweimüller, Uniqueness of the maximal function in the ratio ergodic theorem. Ill. J. Math. **48**(4), 1259–1265 (2004)
1079. A. Zygmund, An individual ergodic theorem for non-commutative transformations. Acta Sci. Math. (Szeged) **14**, 103–110 (1951)
1080. A. Zygmund, *Trigonometric Series*, vols. I, II, 2nd edn. (Cambridge University Press, New York, 1959)

Index

T. Eisner, B. Farkas, *A Journey Through Ergodic Theorems*, Birkhäuser Advanced Texts Basler Lehrbücher, https://doi.org/10.1007/978-3-031-96506-7

U

Symbol Index

Sets of numbers and number theory

Measure theory and Fourier analysis

T. Eisner, B. Farkas, *A Journey Through Ergodic Theorems*, Birkhäuser Advanced Texts Basler Lehrbücher, https://doi.org/10.1007/978-3-031-96506-7

Dynamical systems

Topology and functional analysis

MIX
Papier aus verantwortungsvollen Quellen
Paper from responsible sources
FSC® C105338

If you have any concerns about our products,
you can contact us on
ProductSafety@springernature.com

In case Publisher is established outside the EU,
the EU authorized representative is:
Springer Nature Customer Service Center GmbH
Europaplatz 3, 69115 Heidelberg, Germany

Printed by Libri Plureos GmbH
in Hamburg, Germany